Sound Insulation

Carl Hopkins

 Routledge
Taylor & Francis Group

LONDON AND NEW YORK

First published 2007 by Butterworth-Heinemann

Published 2014 by Routledge
2 Park Square, Milton Park, Abingdon, Oxon, OX14 4RN
711 Third Avenue, New York, NY 10017

Routledge is an imprint of the Taylor & Francis Group, an informa business

Notice
No responsibility is assumed by the publisher for any injury and/or damage to persons or property as a matter of products liability, negligence or otherwise, or from any use or operation of any methods, products, instructions or ideas contained in the material herein.

British Library Cataloguing in Publication Data
A catalogue record for this book is available from the British Library

Library of Congress Cataloging-in-Publication Data
A catalog record for this book is available from the Library of Congress

ISBN: 978-0-7506-6526-1

Typeset by Charon Tec Ltd (A Macmillan Company), Chennai, India
www.charontec.com

Preface

The most effective approach to sound insulation design involves the use of measured data along with statistical and/or analytical models, blended with a combination of empiricism, experience, and pragmatism. Engineering design is predominantly experiential in nature; applying past experience to new problems. This often impedes rapid progress when applying a general knowledge of acoustics to a specific area such as sound insulation. This book is intended for students, engineers, consultants, building designers, researchers and those involved in the manufacture and design of building products. It uses theory and measurements to explain concepts that are important for the application, interpretation and understanding of guidance documents, test reports, product data sheets, published papers, regulations, and Standards. The intention is to enable the reader to tackle many different aspects of sound insulation by providing a textbook and a handbook within a single cover. Readers with a background in acoustics can jump straight to the topic of interest in later chapters and, if needed, return to earlier chapters for fundamental aspects of the theory. This book draws on a wealth of published literature that is relevant to sound insulation, but it does not document a historical review of every incremental step in its development, or cover all possible approaches to the prediction of sound transmission. The references will provide many starting points from which the reader can dive into the vast pool of literature themselves.

All prediction models and measurement methods have their limitations, but with knowledge of their strengths and weaknesses it becomes much easier to make design decisions and to find solutions to sound insulation problems. A model provides more than just a procedure for calculating a numerical result. The inherent assumptions should not simply be viewed as limitations from which use of the model is quickly dismissed; the assumptions may well shed light on a solution to the problem at hand. The fact that we need to deal with a relatively wide frequency range in building acoustics means that there is no single theoretical approach that is suitable for all problems. We rarely know all the variables; but the simple models often identify the ones which are most important. A model provides no more than the word implies; in this sense, every model is correct within the confines of its assumptions. The models and theories described in this book have been chosen because of the insight they give into the sound transmission process. There are many different approaches to the prediction of sound transmission. In general, the more practical theories are included and these will provide the necessary background for the reader to pursue more detailed and complex models when required. Occasionally a more complex theory is introduced but usually with the intention of showing that a simpler method may be adequate. Choosing the most accurate and complex model for every aspect of sound transmission is unnecessary. There are so many transmission paths between two rooms in a building that decisions can often be made whilst accepting relatively high levels of uncertainty in the less important transmission paths. Before doing any calculations it is worth pausing for a second to picture the scene at the point of construction. The uncertainty in describing a building construction tends to differ between walls, floors, or modular housing units that are produced on a factory assembly line compared with a building site that is exposed to the weather and a wide range in the quality of workmanship. Uncertainty is often viewed rather negatively as being at the crux of all design problems, but uncertainty in the form of statistics and probability is part of many theoretical solutions for the transmission of sound and vibration in built-up structures. To avoid sound insulation problems in a completed building, uncertainty needs to be considered at an early stage in the design process. In the words of Francis Bacon (Philosopher, 1561–1626) "If a man will begin with certainties, he shall end in doubts, but if he will be content to begin with doubts, he shall end in certainties".

Sound not only travels via direct transmission across the separating wall or floor, but via the many other walls and floors, as well as via other building elements such as cavities, ceiling voids, and beams; we refer to this indirect transmission as flanking transmission. To predict both direct and flanking transmission, statistical models based upon Statistical Energy Analysis (SEA) are particularly practical because they tend to make gross assumptions about the building elements. This is important because specific details on the material properties and dimensions are not always available in the early (and sometimes late) stages of the design. In fact, during the construction phase a variety of similar building products are often substituted for the one that was originally specified. In addition the quality of workmanship can be highly variable within a single building, let alone between different buildings. SEA or SEA-based models allow an assessment of the different sound transmission paths to determine which paths are likely to be of most importance. These models are also attractive because laboratory sound insulation measurements of complex wall or floor elements can be incorporated into the models. Some construction elements or junction details are not well suited to SEA or SEA-based models. Analytical models and finite element methods therefore have a role to play, although they tend to be more orientated to research work due to the time involved in creating and validating each model. Engineers involved in laboratory measurements become painfully aware that significant changes in the sound insulation can sometimes be produced by small changes to the test element (such as extra screws, different layouts for the framework, or different positions for the porous material in the cavity). Hypersensitivity to certain changes in the construction can often be explained using statistical or deterministic models; but the latter may be needed to help gain an insight into the performance of one specific test element.

Sound insulation tends to be led by regulations, where the required performance is almost always described using a single integer number in decibels. It is common to draw a 'line in the sand' for an acceptable level of sound insulation, such that if the construction fails to achieve this by one decibel, the construction is deemed to have failed. This needs to be considered in the context of a specific pair of rooms in one specific building; we can rarely predict the sound insulation in a specific situation to plus or minus one decibel. However, we can often make reasonable estimates of the average sound insulation for a large number of nominally identical constructions. The design process must therefore consider the sound insulation that can be provided on average, as well as the performance of individual constructions on a particular site.

The fact that we need to design and predict the sound insulation in the field on a statistical basis does not mean that we can accept low precision measurements; quite the opposite. Every decibel is important to the builder having the sound insulation measured to check that the building satisfies the regulations; to the manufacturer who wants to claim an advantage over a competitor's product; to the engineer trying to assess sound transmission mechanisms with laboratory measurements; to the designer trying to choose between two different building products, and to the house builder trying to reduce costs by avoiding over specification in the design. It is important to keep in perspective those situations where the highest level of accuracy is necessary, along with those situations where a rough estimate is more than sufficient. The intention is to provide the reader with a background from which they can decide the appropriate level for the problem at hand. In an engineering context the words 'reasonable' and 'adequate' will quite often be used to describe equations, prediction models, assumptions, and rules of thumb.

Overview of contents

Chapters 1 and 2 deal with theoretical aspects of sound fields in spaces and vibration fields on structures. Sound transmission in buildings is fundamentally concerned with the coupling between these fields. For the reader who is relatively new to acoustics it should be sufficient to start these chapters with a basic background in acoustics terminology, wave theory, and room acoustics. It is assumed that the reader has more experience in room acoustics, or is perhaps more comfortable with these concepts. Sound and vibration are discussed in a similar style so that the reader can see the similarities, and the many differences, between them. The layout of these chapters is intended to simplify its use as a handbook when solving problems that are specific to room acoustics, vibration in buildings, and sound insulation. Sound and vibration fields are described in terms of both waves and modes. It is useful to be able to think in terms of waves and modes interchangeably, taking the most convenient approach to solve the problem at hand.

Chapter 3 looks at sound and vibration measurements relating to sound insulation and material properties. This chapter deals with the underlying theory behind the measurements, and the reasons for adopting different measurement methods. This chapter forms a bridge between the sound and vibration theory in Chapters 1 and 2 and the prediction of sound insulation in Chapters 4 and 5. However, it is not possible to explain all aspects of measurements without referring to some of the theory in Chapters 4 and 5. Some readers may choose to start the book in Chapter 3 and it will sometimes be necessary to refer forward as well as back.

Chapter 4 looks at direct sound transmission across individual building elements. Sound and vibration theory from Chapters 1 and 2 is combined with material property measurements from Chapter 3 to look at prediction models for different sound transmission mechanisms. There is no single theoretical model that can deal with all aspects of sound insulation. Many constructions are so complex that reliance is ultimately placed on measurements. The aim of this chapter is to give insight and understanding into sound transmission for relatively simple constructions. These form a basis from which measurement, prediction, and design decisions can be tackled on more complex constructions. The chapter is based around prediction using SEA augmented by classical theories based on infinite plates. Some aspects of sound transmission are not suitable for SEA models, but the SEA framework can conveniently be used to highlight these areas such that other models can be sought.

Chapter 5 concerns sound insulation *in situ* where there is both direct and flanking transmission. Prediction of vibration transmission across idealized plate junctions is used to illustrate issues that are relevant to measurement and prediction with other types of plates and more complex junction connections. Following on from Chapter 4 the application of SEA is extended to the prediction of direct and flanking transmission. In addition, a simplification of SEA results in an SEA-based model that facilitates the inclusion of laboratory sound insulation measurements.

Generalizations

In building acoustics the main frequency range used to assess sound insulation lays between the 100 and 3150 Hz one-third-octave-bands; an optional extended frequency range is defined between the 50 and 5000 Hz one-third-octave-bands. In this book the range between 50 and 5000 Hz will be referred to as the building acoustics frequency range. For sound and vibration in buildings, it is possible to describe many general trends by defining the low-, mid-, and high-frequency ranges using one-third-octave-band centre frequencies as follows:

- Low-frequency range : 50–200 Hz.
- Mid-frequency range : 250–1000 Hz.
- High-frequency range : 1250–5000 Hz.

The only exact boundaries in these ranges correspond to the 50 and 5000 Hz bands; the intermediate boundaries need to be considered with a degree of flexibility; usually within plus or minus one-third-octave-band.

It is also useful to try and define a range of room volumes that are typically encountered in buildings. For the purpose of making general statements it will be assumed that 'typical rooms' have volumes between 20 and 200 m^3; this covers the majority of practical situations.

Constructions throughout the world are primarily built with concrete, masonry, timber, steel, glass and plasterboard. In a very general sense, the term 'heavyweight' is used for concrete, masonry, and heavy-steel elements and 'lightweight' is used for timber, glass, plasterboard, and light-steel elements. There are also combinations that form a separate lightweight/heavyweight category, such as timber floors with a surface layer of concrete screed. For generic and proprietary materials that are commonly used in lightweight and heavyweight constructions, material properties, and sound insulation values are included in this book to help the reader get a feel for realistic values and to assess general trends.

Contents

Contents

Contents

Contents

Acknowledgements

I would like to thank Les Fothergill (Office of the Deputy Prime Minister, ODPM), Martin Wyatt (Building Research Establishment Ltd, BRE), and John Burdett (BRE Trust) for making the arrangements that allowed me to write this book. I'm very grateful to Les Fothergill (ODPM), Richard Daniels (Department for Education and Skills, DfES), and John Seller (BRE) who provided me with many opportunities to work in research, consultancy, regulations, and standardization over the years. I have also been fortunate enough to have met experts working in the field of building acoustics from all around the world; many of their insights into the subject are referenced within the pages of this book. I extend my thanks to Yiu Wai Lam, Werner Scholl, Ole-Herman Bjor, and Michael Vorländer who made time available to read and comment on various sections.

List of symbols

a	Acceleration (m/s^2), equivalent absorption length (m)
c	Phase velocity (m/s)
c_0	Phase velocity of sound in air (m/s)
c_g	Group velocity (m/s)
$c_{g(B)}$	Group velocity for bending waves (m/s)
$c_{g(L)}$	Group velocity for quasi-longitudinal waves (m/s)
$c_{g(T)}$	Group velocity for torsional waves on beams, or transverse shear waves on plates (m/s)
c_m	Phase velocity for a membrane (m/s)
c_{pm}	Complex phase velocity for sound in a porous material
c_B	Bending phase velocity (m/s)
c_D	Dilatational wave phase velocity (m/s)
c_L	Quasi-longitudinal phase velocity (m/s)
c_T	Torsional phase velocity (m/s), transverse shear phase velocity (m/s)
d	Distance (m), thickness of porous material (m)
d_{mfp}	Mean free path (m)
e	Normalized error
f	Frequency (Hz)
f_c	Critical frequency (Hz)
f_{co}	Cut-off frequency (Hz)
f_d	Dilatational resonance frequency (Hz)
f_{ms}	Mass–spring resonance frequency (Hz)
f_{msm}	Mass–spring–mass resonance frequency (Hz)
f_S	Schroeder cut-off frequency (Hz)
f_l and f_u	Lower and upper limits of a frequency band (Hz)
$f_p, f_{p,q}, f_{p,q,r}$	Mode frequency/eigenfrequency (Hz)
$f_{B(thin)}$	Thin plate limit for bending waves (Hz)
$f_{L(thin)}$	Thin plate limit for quasi-longitudinal waves (Hz)
g	Acceleration due to gravity (m/s^2)
h	Plate thickness (m), height (m)
$h(t)$	impulse response
i	$\sqrt{-1}$
k	Wavenumber (radians/m), spring stiffness (N/m or N/m^3)
k_{pm}	Complex wavenumber for an equivalent gas
k_B	Bending wavenumber (radians/m)
m	Mass (kg), attenuation coefficient in air (Neper/m)
$n(f)$	Modal density (modes per Hz)
$n_{2D}(f)$	Modal density for a cavity – two-dimensional space (modes per Hz)
$n_{3D}(f)$	Modal density for a cavity – three-dimensional space (modes per Hz)
$n(\omega)$	Modal density (modes per radians/s)
p	Sound pressure (Pa)
r	Radius (m), airflow resistivity (Pa.s/m^2)

r_{rd}	Reverberation distance (m)
s	Sample standard deviation (–)
s'	Dynamic stiffness per unit area for an installed resilient material (N/m^3)
s'_a, s'_g	Dynamic stiffness per unit area for enclosed air(a) or other gas(g) (N/m^3)
s'_t	Apparent dynamic stiffness per unit area for a resilient material (N/m^3)
s'_{soil}	Compression stiffness per unit area for soil (N/m^3)
s_{Xmm}	Dynamic stiffness for a wall tie in a cavity of width, Xmm (N/m)
t	Time (s)
u	Sound particle velocity (m/s)
v	Velocity (m/s)
w	Energy density (J/m^3)
A	Absorption area (m^2)
A_T	Total absorption area (m^2)
B	Bandwidth (Hz), filter bandwidth (Hz)
B_b	Bending stiffness for a beam (Nm^2)
B_p	Bending stiffness per unit width for a plate (Nm)
$B_{p,eff}$	Effective bending stiffness per unit width for an orthotropic plate (Nm)
C_W	Waterhouse correction (dB)
D	Sound pressure level difference (dB), damping factor
$D_{I,n}$	Intensity normalized level difference (dB)
$D_{I,n,e}$	Intensity element-normalized level difference (dB)
D_n	Normalized level difference (dB)
$D_{n,e}$	Element-normalized level difference (dB)
$D_{n,f}$	Normalized flanking level difference (dB)
D_{nT}	Standardized level difference (dB)
$D_{v,ij}$	Velocity level difference between source element, i, and receiving element, j (dB)
$\overline{D_{v,ij}}$	Direction-averaged velocity level difference (dB)
E	Energy (J), Young's modulus (N/m^2)
F	Force (N), shear force (N)
F_{pI}	Surface pressure-intensity indicator (dB)
G	Shear modulus (N/m^2)
G_{soil}	Shear stiffness per unit area for soil (N/m^3)
I	Sound intensity (W/m^2), structural intensity (W/m), moment of inertia of the cross-sectional-area about the y- or z-axis (m^4)
I_n	Normal intensity component (W/m^2)
I_x, I_y, I_z	Intensity in the x-, y-, and z-directions
I_θ	Polar moment of inertia about the longitudinal axis of a beam (m^4)
J	Torsional moment of rigidity for a beam (m^4)
K	Stiffness (N/m), contact stiffness (N/m), bulk compression modulus of a gas (Pa)
K_{ij}	Vibration reduction index (dB)
L_d	Dynamic capability index (dB)
L_{ij}	Junction length between elements i and j (m)
L_{In}	Temporal and spatial average sound intensity level over the measurement surface (dB)
L_n	Normalized impact sound pressure level (dB)

$L_{n,f}$	Normalized flanking impact sound pressure level (dB)
L'_n	Normalized impact sound pressure level – field measurement (dB)
L'_{nT}	Standardized impact sound pressure level (dB)
$L_p(t)$	Instantaneous sound pressure level in a space (dB)
L_p	Temporal and spatial average sound pressure level in a space (dB)
$L_{p,s}$	Temporal and spatial average sound pressure level next to a surface (dB)
$L_{p,A}$	A-weighted sound pressure level (dB)
L_x, L_y, L_z	x-, y-, z- dimensions (m)
L_T	Total length of all room edges (m)
L_W	Temporal and spatial average sound power level radiated by a surface (dB)
M	Moment (Nm), moment per unit width (N), modal overlap factor (–), molar mass (kg/mol)
M_{av}	Geometric mean of the modal overlap factors for subsystems i and j (–)
$N(k)$	Number of modes below wavenumber, k (–)
N	Mode count in a frequency band (–), number of reflections, positions, samples, etc.
N_s	Statistical mode count in a frequency band (–)
P	Static pressure (Pa)
P_0	Static pressure for air at atmospheric pressure (Pa)
Q	Shear force per unit width (N)
R	Reflection coefficient (–), sound reduction index (dB), universal gas constant (J/mol.K), airflow resistance (Pa.s/m^3), auto-correlation functions, damping constant (–)
R_0	Normal incidence sound reduction index (dB)
R'	Apparent sound reduction index (dB)
R_I	Intensity sound reduction index (dB)
R'_I	Apparent intensity sound reduction index (dB)
R_{ij}	Flanking sound reduction index (dB)
R_s	Specific airflow resistance (Pa.s/m)
$R'_{tr,s}, R'_{rt,s}, R'_{at,s}$	Apparent sound reduction index for road (tr), railway (rt), and aircraft (at) traffic (dB)
S	Area (m^2)
S_M	Area of the measurement surface (m^2)
S_T	Total area of the room surfaces (m^2)
T	Period (s), averaging time (s), reverberation time (s), temperature (°C), torsional stiffness (Nm^2), tension per unit length around the edge of a membrane (N/m)
T_{int}	Integration time (s)
T_s	Structural reverberation time (s)
T_X	Reverberation time determined from linear regression over a range of X dB (s)
U	Perimeter (m)
V	Volume (m^3)
W	Power (W)
Y_{dp}	Driving-point mobility (m/Ns)
Z_0	Characteristic impedance of air (Pa.s/m)
$Z_{0,pm}$	Characteristic impedance for an equivalent gas (Pa.s/m)
$Z_{a,n}$	Normal acoustic surface impedance (Pa.s/m)

List of symbols

$Z_{a,s}$	Specific acoustic impedance (–)
Z_{dp}	Driving-point impedance (Ns/m)
Z_p	Surface impedance of a plate (Pa.s/m)
α_0	Normal incidence sound absorption coefficient (–)
α_θ	Angle-dependent sound absorption coefficient (–)
α_{st}	Statistical sound absorption coefficient (–)
α_s	Sound absorption coefficient (–)
$\beta_{a,s}$	Specific acoustic admittance (–)
$\delta(t)$	Dirac delta function
δf	Average frequency spacing between modes (Hz)
δ_{pI0}	Pressure-residual intensity index (dB)
ε	Strain (–), normalized standard deviation (–), absolute error
ε^2	Normalized variance (–)
ϕ	Porosity (–)
γ	Shear strain (–), ratio of specific heats (–)
η	Displacement (m), loss factor (–)
η_{int}	Internal loss factor (–)
η_{ii}	Internal loss factor for subsystem i (–)
η_{ij}	Coupling loss factor from subsystem i to subsystem j (–)
η_i	Total loss factor for subsystem i (–)
κ	Gas compressibility (Pa^{-1})
κ_{eff}	Effective gas compressibility for an equivalent gas (Pa^{-1})
λ	Wavelength (m), Lamé constant
μ	Mean (–), Lamé constant
ν	Poisson's ratio (–)
θ	Angular torsional displacement (radians)
ρ	Density (kg/m^3)
ρ_{bulk}	Bulk density (kg/m^3)
ρ_{eff}	Effective gas density for an equivalent gas (kg/m^3)
ρ_0	Density of air (kg/m^3)
ρ_s	Mass per unit area/surface density (kg/m^2)
ρ_l	Mass per unit length (kg/m)
σ	Standard deviation (–), stress (N/m^2), radiation efficiency (–)
τ	Transmission coefficient (–), shear stress (N/m^2), time constant (s)
ω	Angular frequency $\omega = 2\pi f$ (radians/s), angular velocity (radians/s)
ξ	Displacement (m)
ψ	Mode shape/eigenfunction
ζ	Displacement (m)
ζ_{cdr}	Constant damping ratio (–)
Δf	Frequency spacing between eigenfrequencies (Hz), frequency bandwidth (Hz)
Δf_{3dB}	3 dB bandwidth (or half-power bandwidth) (Hz)
ΔL	Decrease in sound pressure level along a corridor (dB), improvement of impact sound insulation (dB)
ΔR	Sound reduction improvement index (dB)
$\Delta R_{Resonant}$	Resonant sound reduction improvement index (dB)
Φ	Potential function (m^2)
Γ	Complex propagation constant

Ψ	Stream function (m^2)
$< >$	Mean value
$< >_t$	Temporal average
$< >_s$	Spatial average
$< >_{t,s}$	Temporal and spatial average
$< >_f$	Frequency average
$\hat{X}$	Denotes peak value of variable X

Sound fields

1.1 Introduction

This opening chapter looks at aspects of sound fields that are particularly relevant to sound insulation; the reader will also find that it has general applications to room acoustics.

The audible frequency range for human hearing is typically 20 to 20 000 Hz, but we generally consider the building acoustics frequency range to be defined by one-third-octave-bands from 50 to 5000 Hz. Airborne sound insulation tends to be lowest in the low-frequency range and highest in the high-frequency range. Hence significant transmission of airborne sound above 5000 Hz is not usually an issue. However, low-frequency airborne sound insulation is of particular importance because domestic audio equipment is often capable of generating high levels below 100 Hz. In addition, there are issues with low-frequency impact sound insulation from footsteps and other impacts on floors. Low frequencies are also relevant to façade sound insulation because road traffic is often the dominant external noise source in the urban environment. Despite the importance of sound insulation in the low-frequency range it is harder to achieve the desired measurement repeatability and reproducibility. In addition, the statistical assumptions used in some measurements and prediction models are no longer valid. There are some situations such as in recording studios or industrial buildings where it is necessary to consider frequencies below 50 Hz and/or above 5000 Hz. In most cases it should be clear from the text what will need to be considered at frequencies outside the building acoustics frequency range.

1.2 Rooms

Sound fields in rooms are of primary importance in the study of sound insulation. This section starts with the basic principles needed to discuss the more detailed aspects of sound fields that are relevant to measurement and prediction. In the laboratory there is some degree of control over the sound field in rooms due to the validation procedures that are used to commission them. Hence for at least part of the building acoustics frequency range, the sound field in laboratories can often be considered as a diffuse sound field; a very useful idealized model. Outside of the laboratory there are a wide variety of rooms with different sound fields. These can usually be interpreted with reference to two idealized models: the modal sound field and the diffuse sound field.

1.2.1 Sound in air

Sound in air can be described as compressional in character due to the compressions and rarefactions that the air undergoes during wave propagation (see Fig. 1.1). Air particles move to and fro in the direction of propagation, hence sound waves are referred to as longitudinal waves. The compressions and rarefactions cause temporal variation of the air density compared to the

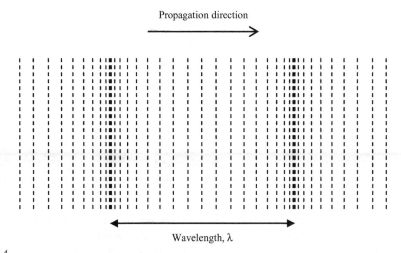

Figure 1.1

Longitudinal wave – compression and rarefaction of air particles.

density at equilibrium. The result is temporal variation of the air pressure compared to the static air pressure. Sound pressure is therefore defined by the difference between the instantaneous pressure and the static pressure.

The phase velocity for sound in air, c_0 (or as it is more commonly referred to, the speed of sound) is dependent upon the temperature, T, in °C and for most practical purposes can be calculated using

$$c_0 = 331 + 0.6T \tag{1.1}$$

for temperatures between 15°C and 30°C and at atmospheric pressure.

The density of air at equilibrium, ρ_0, is also temperature dependent and can be calculated from

$$\rho_0 = \frac{353.2}{273 + T} \tag{1.2}$$

For calculations in buildings it is often assumed that the temperature is 20°C, for which the speed of sound is 343 m/s and the density of air is 1.21 kg/m^3. This will be assumed throughout the book.

The fundamental relationship between the phase velocity, the frequency, f, and the wavelength, λ, is

$$c_0 = f\lambda \tag{1.3}$$

The wavelength is the distance from peak to peak (or trough to trough) of a sinusoidal wave; this equals the distance between identical points of compression (see Fig. 1.1) or rarefaction. For the building acoustics frequency range, the wavelength in air at 20°C is shown in Fig. 1.2. If we consider these wavelengths relative to typical room dimensions, it is clear that we are dealing with a very wide range. For this reason it is useful to describe various aspects of sound fields by referring to low-, mid-, and high-frequency ranges; corresponding to 50 to 200 Hz, 250 to 1000 Hz, and 1250 to 5000 Hz respectively.

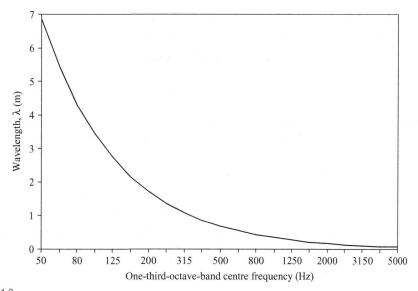

Figure 1.2

Wavelength of sound in air at 20°C.

As we need to describe the spatial variation of sound pressure as well as the temporal variation, it is necessary to use the wavenumber, k, which is defined as

$$k = \frac{\omega}{c_0} = \frac{2\pi}{\lambda} \tag{1.4}$$

where the angular frequency, ω, is

$$\omega = 2\pi f \tag{1.5}$$

and the period, T, of the wave is

$$T = \frac{2\pi}{\omega} = \frac{1}{f} \tag{1.6}$$

The wavenumber is useful for discussing aspects relating to spatial variation in both sound and vibration fields in terms of kd, where d is the distance between two points.

Two types of wave need to be considered both inside and outside of buildings: plane waves and spherical waves. Before reviewing these waves we will briefly review the use of complex notation that simplifies many derivations for sound and structure-borne sound waves.

1.2.1.1 Complex notation

For both sound and vibration, it is useful to look at wave motion or signals at single frequencies; these are defined using harmonic sine and cosine functions, e.g. $p(x, t) = \cos(\omega t - kx)$. It is usually more convenient to describe these simple harmonic waves using complex exponential notation, where

$$\exp(iX) = \cos X + i \sin X \tag{1.7}$$

Equations using complex notation are often easier to manipulate than sines and cosines, and can be written in a more compact form. A brief review of complex notation is given here as this is covered in general acoustic textbooks (e.g. see Fahy, 2001).

The most commonly used complex exponentials are those that describe the temporal and spatial variation of harmonic waves. For the convention used in this book, these are

$$\exp(i\omega t) = \cos(\omega t) + i\sin(\omega t) \tag{1.8}$$

and

$$\exp(-kx) = \cos(kx) - i\sin(kx)$$

Complex notation also simplifies differentiation and integration. For example, differentiation or integration with respect to time becomes equivalent to multiplication or division by $i\omega$ respectively.

Whilst it can be convenient to work with complex notation, the final result that corresponds to a physical quantity (sound pressure, velocity, etc.) must be real, rather than imaginary. In general, it is the real part of the solution that represents the physical quantity.

The time-average of harmonic waves is frequently needed for practical purposes and is denoted by $\langle\ \rangle_t$. The following time-averages often occur in derivations,

$$\lim_{T\to\infty}\frac{1}{T}\int_0^T\cos^2(\omega t)\mathrm{d}t = \lim_{T\to\infty}\frac{1}{T}\int_0^T\sin^2(\omega t)\mathrm{d}t = 0.5 \tag{1.9}$$

and

$$\lim_{T\to\infty}\frac{1}{T}\int_0^T\sin(\omega t)\mathrm{d}t = \lim_{T\to\infty}\frac{1}{T}\int_0^T\cos(\omega t)\mathrm{d}t = \lim_{T\to\infty}\frac{1}{T}\int_0^T\sin(\omega t)\cos(\omega t)\mathrm{d}t = 0 \tag{1.10}$$

where T is the averaging time.

The time-average of the product of two waves, $p_1(t)$ and $p_2(t)$, that are written in complex exponential notation can be calculated using

$$\langle p_1 p_2\rangle_t = \frac{1}{2}\mathrm{Re}\{p_1 p_2^*\} \tag{1.11}$$

where * denotes the complex conjugate.

1.2.1.2 Plane waves

To gain an insight into the sound field in rooms we often assume that it is comprised of plane waves; so called, because in any plane that is perpendicular to the propagation direction, the sound pressure and the particle velocity are uniform with constant phase. These planes are referred to as wavefronts. In practice, plane waves can be realized (approximately) in a long hollow cylinder which has rigid walls. A sound source is placed at one end of the cylinder that generates sound with a wavelength that is larger than the diameter of the cylinder. This results in a plane wave propagating in the direction away from the source. The longitudinal wave shown in Fig. 1.1 can also be seen as representing a plane wave in this cylinder. This one-dimensional scenario may seem somewhat removed from real sound fields in typical rooms. However, the plane wave model can often be used to provide a perfectly adequate description of the complex sound fields that are encountered in practice.

Using a Cartesian coordinate system, the wave equation that governs the propagation of sound through three-dimensional space is

$$\frac{\partial^2 p}{\partial x^2} + \frac{\partial^2 p}{\partial y^2} + \frac{\partial^2 p}{\partial z^2} - \frac{1}{c_0^2}\frac{\partial^2 p}{\partial t^2} = 0 \tag{1.12}$$

where p is the sound pressure.

For a plane wave that is propagating in the positive x, y, and z-direction across this space, the sound pressure is described by an equation of the form

$$p(x, y, z, t) = \hat{p} \exp(-ik_x x) \exp(-ik_y y) \exp(-ik_z z) \exp(i\omega t) \qquad (1.13)$$

where $\hat{p}$ is an arbitrary constant for the peak value, and k_x, k_y, and k_z are constants relating to the wavenumber.

As we are using harmonic time dependence, $\exp(i\omega t)$, the wave equation can now be written in terms of the wavenumber as

$$\frac{\partial^2 p}{\partial x^2} + \frac{\partial^2 p}{\partial y^2} + \frac{\partial^2 p}{\partial z^2} + k^2 p = 0 \qquad (1.14)$$

The relationship between the wavenumber and the constants, k_x, k_y, and k_z is found by inserting Eq. 1.13 into the wave equation, which gives

$$k^2 = k_x^2 + k_y^2 + k_z^2 \qquad (1.15)$$

We will need to make use of this relationship to describe the sound field in rooms. However, the wave equation only governs sound propagation across three-dimensional space. It does not describe the sound field in a room because it does not take account of the waves impinging upon the room surfaces. In other words, this equation does not take account of boundary conditions. Hence the constants, k_x, k_y, and k_z can only be determined once we have defined these boundary conditions.

The particle motion gives rise to sound pressure; hence we can relate the sound particle velocities, u_x, u_y, and u_z (in the x, y, and z directions respectively) to the sound pressure by using the following equations of motion,

$$\frac{\partial p}{\partial x} = -\rho_0 \frac{\partial u_x}{\partial t} \qquad \frac{\partial p}{\partial y} = -\rho_0 \frac{\partial u_y}{\partial t} \qquad \frac{\partial p}{\partial z} = -\rho_0 \frac{\partial u_z}{\partial t} \qquad (1.16)$$

and therefore the particle velocities are

$$u_x = \frac{k_x}{\omega \rho_0} p \qquad u_y = \frac{k_y}{\omega \rho_0} p \qquad u_z = \frac{k_z}{\omega \rho_0} p \qquad (1.17)$$

The ratio of the complex sound pressure to the complex sound particle velocity at a single point is the specific acoustic impedance, Z_a. For a plane wave propagating in a single direction (we will choose the x-direction, so that $k = k_x$ and $u = u_x$), this impedance is referred to as the characteristic impedance of air, Z_0, and is defined as

$$Z_0 = \frac{p}{u} = \rho_0 c_0 = \sqrt{\frac{\rho_0}{\kappa}} \qquad (1.18)$$

where κ is the gas compressibility (adiabatic).

The particle velocity is related to the sound pressure by a real constant that is independent of frequency. Therefore, the sound pressure and the particle velocity always have the same phase on the plane that lies perpendicular to the direction of propagation.

In order to predict or measure sound transmission we will need to quantify the sound intensity, I; the energy that flows through unit surface area in unit time. The sound intensity is the time-averaged value of the product of sound pressure and particle velocity,

$$I = \langle pu \rangle_t = \frac{\langle p^2 \rangle_t}{\rho_0 c_0} \qquad (1.19)$$

where $\langle p^2 \rangle_t$ is the temporal average mean-square sound pressure given by

$$\langle p^2 \rangle_t = \frac{1}{T} \int_0^T p^2 \, dt \tag{1.20}$$

and T is the averaging time. Note that $\sqrt{\langle p^2 \rangle_t}$ is described as the root-mean-square (rms) sound pressure.

1.2.1.3 Spherical waves

For spherical waves the sound pressure and the particle velocity over a spherical surface are uniform with constant phase; these surfaces are referred to as wavefronts (see Fig. 1.3). For a sound source such as a loudspeaker used in sound insulation measurements, a useful idealized model is to treat the loudspeaker as a point source that generates spherical waves. A point source is one for which the physical dimensions are much smaller than the wavelength of the sound, and the sound radiation is omnidirectional.

We now need to make use of a spherical coordinate system defined by a distance, r, from the origin at $r = 0$. For spherically symmetrical waves, the wave equation that governs the propagation of sound through three-dimensional space is

$$\frac{\partial^2 p}{\partial r^2} + \frac{2}{r} \frac{\partial p}{\partial r} - \frac{1}{c_0^2} \frac{\partial^2 p}{\partial t^2} = 0 \tag{1.21}$$

For a spherical wave propagating across this space, the sound pressure can be described by an equation of the form

$$p(r, t) = \frac{\hat{p}}{r} \exp(-ikr) \exp(i\omega t) \tag{1.22}$$

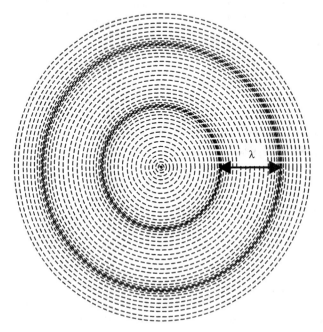

Figure 1.3

Spherical wavefronts produced by a point source.

where $\hat{p}$ is an arbitrary constant for the peak value and r is the distance from a spherical wavefront to the origin.

Substitution of Eq. 1.22 into the following equation of motion,

$$\frac{\partial p}{\partial r} = -\rho_0 \frac{\partial u_r}{\partial t}$$

(1.23)

gives the radial particle velocity, u_r, as

$$u_r = \frac{p}{\rho_0 c_0}\left(1 - \frac{i}{kr}\right) = \frac{p}{\rho_0 c_0}\left(1 - i\frac{c_0}{2\pi f r}\right)$$

(1.24)

This gives the acoustic impedance, Z_a, as

$$Z_a = \frac{p}{u_r} = \frac{\rho_0 c_0}{\left(1 - \frac{i}{kr}\right)} = \frac{\rho_0 c_0 k^2 r^2}{1 + k^2 r^2} + i\frac{\rho_0 c_0 kr}{1 + k^2 r^2}$$

(1.25)

In contrast to plane waves, the particle velocity is related to the sound pressure by a complex variable that is dependent on both the wavenumber and distance. So although the phase of the sound pressure and the phase of the particle velocity are constant over a spherical surface at a specific frequency, they do not have the same phase over this surface.

The time-averaged sound intensity for a harmonic spherical wave is

$$I = \frac{1}{2\rho_0 c_0}\left(\frac{\hat{p}}{r}\right)^2 = \frac{\langle p^2 \rangle_t}{\rho_0 c_0}$$

(1.26)

For spherical waves, the intensity is seen to be proportional to $1/r^2$; this feature is often referred to as spherical divergence. The sound power associated with a point source producing spherical waves can now be calculated from the intensity using

$$W = 4\pi r^2 I = \frac{2\pi \hat{p}^2}{\rho_0 c_0}$$

(1.27)

Rather than use $\hat{p}$ in Eqs 1.26 and 1.27, a point source can be described using a peak volume velocity, $\hat{Q}$, given by

$$\hat{Q} = \frac{4\pi \hat{p}}{\omega \rho_0}$$

(1.28)

When $kr \gg 1$ (i.e. at high frequencies and/or large distances) the imaginary part of Z_a is small, therefore the particle velocity has almost the same phase as the sound pressure and Z_a tends towards Z_0. The time-averaged sound intensity for the harmonic spherical wave then tends towards the value for a plane wave (Eq. 1.19). These links between plane waves and spherical waves indicate why we are able to use the simpler plane wave model in many of the derivations involved in sound insulation. Any errors incurred through the assumption of plane waves are often negligible or insignificant compared to those that are accumulated from other assumptions.

1.2.1.4 Acoustic surface impedance and admittance

As rooms are formed by the surfaces at the boundaries of the space we need to know the acoustic impedance of a room surface as seen by an impinging sound wave. The normal acoustic surface impedance, $Z_{a,n}$, is defined as the ratio of the complex sound pressure at a surface, to the component of the complex sound particle velocity that is normal to this surface,

$$Z_{a,n} = \frac{p}{u_n} \qquad (1.29)$$

Although we are mainly interested in plates (representing walls or floors) that form the room boundaries, the above definition applies to any surface, including sheets of porous materials such as mineral wool or foam.

The specific acoustic impedance, $Z_{a,s}$, is defined using the characteristic impedance of air,

$$Z_{a,s} = \frac{Z_{a,n}}{\rho_0 c_0} \qquad (1.30)$$

In some calculations it is more appropriate or convenient to use the specific acoustic admittance, $\beta_{a,s}$, rather than the specific acoustic impedance, where

$$\beta_{a,s} = \frac{1}{Z_{a,s}} \qquad (1.31)$$

When calculating sound fields in rooms it is often convenient to assume that the room surfaces are rigid. This is reasonable for many hard surfaces in buildings. At a rigid surface, the particle velocity that is normal to this surface is zero; hence $Z_{a,n}$ and $Z_{a,s}$ become infinitely large and $\beta_{a,s}$ is taken to be zero.

1.2.1.5 Decibels and reference quantities

The human ear can detect a wide range of sound intensities. The decibel scale (dB) is commonly used to deal with the wide range in pressure, intensity, power, and energy that are encountered in acoustics. Levels in decibels are defined using the preferred SI reference quantities for acoustics in Table 1.1 (ISO 1683); these reference quantities are used for all figures in the book.

Table 1.1. Sound – definitions of levels in decibels

Level	Definition	Reference quantity
Sound pressure	$L_p = 20 \lg \left(\dfrac{p}{p_0} \right)$ where p is the rms pressure	$p_0 = 20 \times 10^{-6}$ Pa NB only for sound in air
Energy	$L_E = 10 \lg \left(\dfrac{E}{E_0} \right)$	$E_0 = 10^{-12}$ J
Intensity	$L_I = 10 \lg \left(\dfrac{I}{I_0} \right)$	$I_0 = 10^{-12}$ W/m^2
Sound power	$L_W = 10 \lg \left(\dfrac{W}{W_0} \right)$	$W_0 = 10^{-12}$ W
Loss factors (Internal, Coupling, Total)	$L_{ILF} / L_{CLF} / L_{TLF} = 10 \lg \left(\dfrac{\eta}{\eta_0} \right)$	$\eta_0 = 10^{-12}$

1.2.1.6 A-weighting

A-weighting is used to combine sound pressure levels from a range of frequencies into a single value. This is the A-weighted sound pressure level, $L_{p,A}$. It is intended to represent the frequency response of human hearing and is often used to try and make a simple link between the objective and subjective assessment of a sound. A-weighting accounts for the fact that with the same sound pressure level, we do not perceive all frequencies as being equally loud. In terms of the building acoustics frequency range it weights the low-frequency range as being less significant than the mid- and high-frequency range. This does not mean that the low-frequency range is unimportant for sound insulation, usually quite the opposite is true; the A-weighted level depends upon the spectrum of the sound pressure level. Although it is common to measure and predict sound insulation in frequency bands, assessment of the sound pressure level in the receiving room is often made in terms of the A-weighted level.

For N frequency bands, the sound pressure level $L_p(n)$ in each frequency band, n, is combined to give an A-weighted level using

$$L_{p,A} = 10 \lg \left(\sum_{n=1}^{N} 10^{(L_p(n)+A(n))/10} \right) \qquad (1.32)$$

where the A-weighting values, $A(n)$, are shown in Fig. 1.4 for one-third-octave-bands (IEC 61672-1).

For regulatory and practical purposes, the airborne sound insulation is often described using a single-number quantity that corresponds to the difference between the A-weighted level in the source room and the A-weighted level in the receiving room for a specific sound spectrum (e.g. pink noise) in the source room (ISO 717 Part 1). Use of this A-weighted level difference

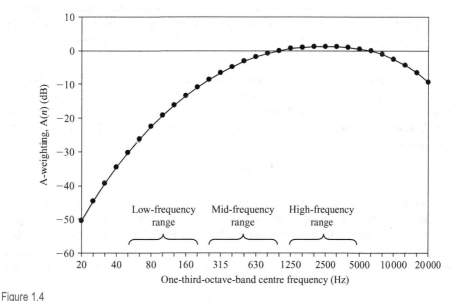

Figure 1.4

A-weighting values over the range of human hearing indicating the low-, mid-, and high-frequency ranges for the building acoustics frequency range.

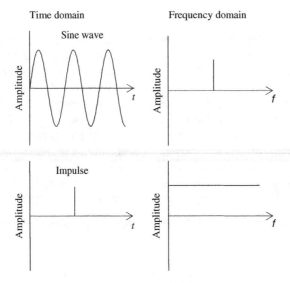

Figure 1.5

Illustration of a sine wave and an impulse in the time and frequency domains.

simplifies calculation of the A-weighted level in the receiving room. It can also be used to make a link to subjective annoyance (Vian *et al.*, 1983).

1.2.2 Impulse response

In sound insulation as well as in room acoustics, we need to make use of the impulse response in both measurement and theory. The frequency spectrum of an impulse is flat. It therefore contains energy at all frequencies, whereas a sine wave only has energy at a single frequency (see Fig. 1.5).

The general principle for an impulse response applies to any acoustic system, whether it is sound pressure in a room or a cavity, or the vibration of a plate or a beam. It is based upon the response of an acoustic system to a Dirac delta function, $\delta(t)$, sometimes called a unit impulse. The delta function is infinite at $t = 0$ and infinitely narrow, such that $\delta(t) = 0$ when $t \neq 0$, and it has the property

$$\int_{-\infty}^{\infty} \delta(t)\mathrm{d}t = 1 \tag{1.33}$$

Excitation of a linear time-invariant (LTI) acoustic system with a delta function results in the impulse response of the system, $h(t)$. The delta function is important because any kind of signal can be described by using a train of impulses that have been appropriately scaled and shifted in time. Hence, the impulse response completely describes the response of an LTI system to any input signal, $x(t)$. The output signal, $y(t)$, can then be found from the convolution integral

$$y(t) = \int_{-\infty}^{\infty} h(u)x(t - u)\mathrm{d}u = \int_{-\infty}^{\infty} h(t - u)x(u)\mathrm{d}u \tag{1.34}$$

where u is a dummy time variable.

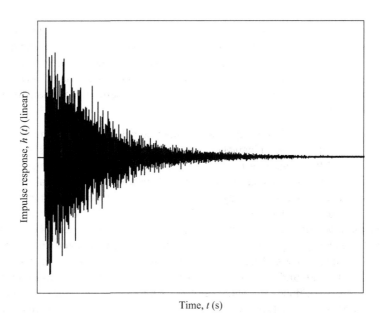

Time, t (s)

Figure 1.6

Example of a measured impulse response in a room.

For brevity, convolution is often written as $y(t) = x(t) * h(t)$. The convolution integral uses the dummy time variable to multiply the time-reversed input signal by the impulse response (or the time-reversed impulse response by the input signal), and integrate over all possible values of t to give the output signal.

An example of a measured impulse response for sound pressure in a room is shown in Fig. 1.6.

1.2.3 Diffuse field

One of the assumptions commonly made in the measurement and prediction of sound insulation is that the sound field in rooms can be considered as being diffuse. A diffuse sound field can be considered as one in which the sound energy density is uniform throughout the space (i.e. the sound field can be considered to be homogeneous), and, if we choose any point in the space, sound waves arriving at this point will have random phase, and there will be equal probability of a sound wave arriving from any direction. The diffuse field is a concept; in practice there must be dissipation of energy, so there cannot be equal energy flow in all directions, there must be net energy flow from a sound source towards part(s) of the space where sound is absorbed.

In diffuse fields it is common to refer to diffuse reflections; this means that the relationship between the angle of incidence and the angle of reflection is random. This is in contrast to specular reflection, where the angle of incidence equals the angle of reflection. Walls and floors commonly found in buildings (excluding spaces specially designed for music performance such as studios or concert halls) tend to be flat and smooth, from which one might assume that specular reflections were the norm, and that diffuse reflections were the exception. However, walls commonly have objects placed near them that partially obscure the wall from the incident sound wave, such as tables, chairs, bookcases, filing cabinets, and cupboards. These can cause the incident wave to be scattered in non-specular directions. Non-specular reflection also occurs when the acoustic impedance varies across the surface; for example a wall where the majority of the surface area is concrete but with areas of glazing, wooden doors, or recessed cupboards, each of which have different impedances. Hence there will usually be a degree of non-specular reflection, such that some of the incident energy is specularly reflected and some is diffusely reflected.

The diffuse field is a very useful concept. It allows many simplifications to be made in the measurement and prediction of sound insulation, as well as in other room acoustics calculations. These make use of the mean free path that will be defined in the following section. In the laboratory we can create close approximations to a diffuse field in the central zone of a room. However, the sound field does not always bare a close resemblance to a diffuse field over the entire building acoustics frequency range. In the low-frequency range this is primarily due to the fact that sound waves arriving at any point come from a limited number of directions. In the mid- and high-frequency ranges, waves arriving at any point tend to come from many different directions. In the central zone of typical rooms it is often reasonable to assume that there is a diffuse field in the mid- and high-frequency ranges. However it is not always appropriate to assume that there is a diffuse field when: (a) there are regular room shapes without diffusing elements, (b) there are non-diffuse reflections from room surfaces, and (c) there is non-uniform distribution of absorption over the room surfaces. For the above reasons, we need to note the limitations in applying diffuse field theory to the real world.

1.2.3.1 Mean free path

The mean free path, d_{mfp}, is the average distance travelled by a sound wave between two successive diffuse reflections from the room surfaces. From the basic relationship, $c_0 = d_{mfp}/t$, we can calculate the time, t, taken to travel this distance. Upon each reflection, a fraction of the sound energy is absorbed; hence the mean free path allows us to calculate the build-up or decay of sound energy in a room over time. It will therefore be needed later on when we derive the reverberation time in diffuse fields as well as when calculating the power incident upon walls or floors that face into a room with a diffuse sound field. The following derivation is taken from Kosten (1960) and starts by deriving the mean free path in a two-dimensional space before extending it to three dimensions. This two-dimensional space has an area, S, and a perimeter length, U. An arbitrary two-dimensional space can be defined by a closed curve as shown in Fig. 1.7; note that although the space is defined by curved lines we assume that all reflections are diffuse. The dashed lines within this curve represent free paths in a single direction, where each free path has a length, l.

Projective geometry is now used to transform points along the perimeter of the space onto a projection plane. Each of the free paths lies perpendicular to a projection plane that defines the apparent length of the surface, L_a. When the space is uniformly filled with free paths, the surface area of the space can be written in terms of the free path lengths using

$$S = \int_{L_a} l dL_a = L_a \bar{l} \tag{1.35}$$

where $\bar{l}$ is the mean free path in one direction.

The number of paths in a single direction is proportional to the apparent length, so using a fixed number of paths per unit of the projection length, and accounting for all N possible directions

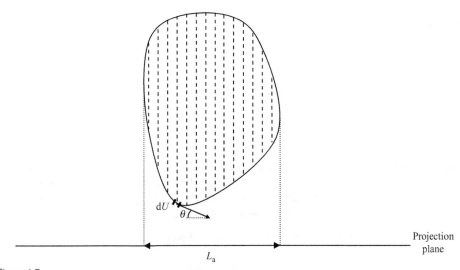

Figure 1.7

Two-dimensional space showing some of the free paths (dashed lines) in a single direction that lie perpendicular to the projection plane. The apparent length, L_a, is calculated by using a line integral to sum the projection of the small perimeter length, dU, onto the projection plane.

gives the mean free path, d_{mfp}, as

$$d_{mfp} = \frac{\lim\limits_{N \to \infty} \sum_{n=1}^{N} L_{a,n} \bar{l}_n}{\lim\limits_{N \to \infty} \sum_{n=1}^{N} L_{a,n}} \qquad (1.36)$$

From Eq. 1.35, $S = L_{a,n} \bar{l}_n$ in each direction, n, so Eq. 1.36 can be rewritten as

$$d_{mfp} = \frac{S}{\bar{L}_a} \qquad (1.37)$$

where $\bar{L}_a$ is the average apparent length.

The next step is to determine this average apparent length, but first we just look at a single direction and calculate the apparent length. This is done by using a line integral for the closed curve. At each point along the closed curve, the vector in the direction of the curve makes an angle, θ, with the projection plane (see Fig. 1.7). The projection of each small perimeter length, dU, onto the projection plane is a positive value, $|\cos\theta|dU$. By integrating around the entire closed curve, the integral is effectively counting each free path twice, so a multiplier of one-half is needed. The apparent length is therefore given by

$$L_a = \frac{1}{2} \oint_C |\cos\theta|dU \qquad (1.38)$$

To find the average apparent length, it is necessary to average over all possible directions. This results in an average cosine term,

$$\overline{|\cos\theta|} = \frac{1}{\pi} \int_0^\pi |\cos\theta|d\theta = \frac{2}{\pi} \qquad (1.39)$$

which gives the average apparent length as

$$\bar{L}_a = \frac{1}{2} \oint_C \overline{|\cos\theta|} dU = \frac{U}{\pi} \tag{1.40}$$

The mean free path for a two-dimensional space can now be found from Eqs 1.37 and 1.40, giving

$$d_{mfp} = \frac{\pi S}{U} \tag{1.41}$$

We now consider a three-dimensional space with a volume, V, and a total surface area, S_T. Moving to three dimensions means that the projection plane becomes a surface (rather than a line) onto which small parts of the surface area, dS_T, are projected (rather than small perimeter lengths). Hence we need to define an apparent surface area, S_a.

The volume of the space can be written in terms of the free path lengths using

$$V = \int_{S_a} l dS_a = S_a \bar{l} \tag{1.42}$$

where $\bar{l}$ is the mean free path in one direction.

The number of paths in a single direction is proportional to the apparent surface area, so using a fixed number of paths per unit area of the projection surface, and accounting for all N possible directions gives the mean free path, d_{mfp}, as

$$d_{mfp} = \frac{\lim\limits_{N\to\infty} \sum_{n=1}^{N} S_{a,n} \bar{l}_n}{\lim\limits_{N\to\infty} \sum_{n=1}^{N} S_{a,n}} \tag{1.43}$$

From Eq. 1.42, $V = S_{a,n}\bar{l}_n$ in each direction, n, so Eq. 1.43 can be rewritten as

$$d_{mfp} = \frac{V}{\bar{S}_a} \tag{1.44}$$

where $\bar{S}_a$ is the average apparent surface area.

The average apparent surface area is found from the surface integral,

$$S_a = \frac{1}{2} \oint_S |\cos\theta| dS_T \tag{1.45}$$

Averaging over all possible directions gives the average cosine term. For any enclosed volume with convex surfaces, the average apparent surface area is given by

$$\bar{S}_a = \frac{S_T}{4} \tag{1.46}$$

The assumption that the volume effectively forms a convex solid does not limit its applicability to real rooms as long as the surface area associated with any concave surfaces within the volume are included in the calculation of S_T (Kosten, 1960). Note once again that it is assumed that all of these curved surfaces result in diffuse reflections.

For any shape of room in which all room surfaces diffusely reflect sound waves, the mean free path for a three-dimensional space is given by Eqs 1.44 and 1.46, hence

$$d_{\text{mfp}} = \frac{4V}{S_T} \qquad (1.47)$$

where S_T is the total area of the room surfaces and V is the room volume.

It is important to note that Eq. 1.47 gives the mean value; as with any random process, there will be a spread of results. The mean free path applies to any shape of room with diffusely reflecting surfaces. However, the statistical distribution of the mean free path in rooms with diffusely reflecting surfaces depends upon the room shape and its dimensions as well as the presence of scattering objects within the room (Kuttruff, 1979).

1.2.4 Image sources

A geometrical approach to room acoustics allows calculation of the room response using image sources. It is briefly described here to introduce the concept of image sources for specular reflections from surfaces. This will be needed in later sections to describe sound fields within rooms, as well as sound incident upon a building façade from outside.

This approach assumes that the wavelength is small compared with the dimensions of the surface that the wave hits. In the study of room acoustics in large rooms and/or at high frequencies this allows sound to be considered in terms of rays rather than waves. Using rays means that diffraction and phase information that causes interference patterns is ignored. In a similar way to the study of optics, a ray can be followed from a point source to the boundary where it undergoes specular reflection, such that the angle of incidence equals the angle of reflection.

Image sources are defined by treating every boundary (e.g. wall, floor, ground) as mirrors in which the actual source can be reflected (see Fig. 1.8). The length of the propagation path from source to receiver is then equal to the distance along the straight line from the image source to the receiver. As we are considering spherical waves from a point source it is necessary to use this distance to take account of spherical divergence when calculating the intensity (Eq. 1.26).

For certain receiver positions in rooms with shapes that are much more complex than a simple box, some of the image sources generated by the reflection process will correspond to paths that cannot physically exist in practice. Hence for rooms other than box-shaped rooms, it is necessary to check the validity of each image source for each receiver position.

1.2.4.1 Temporal density of reflections

For a box-shaped room containing a single point source, the image source approach that was described above can be used to create an infinitely large number of image rooms each containing a single image source. A small portion of this infinite matrix of image rooms in two-dimensional space is shown in Fig. 1.9. Assuming that the point source generates an impulse at $t = 0$, each image source must also generate an identical impulse at $t = 0$. This ensures that all propagation paths have the correct time lag/gain relative to each other. A circle of radius, $c_0 t$, with its origin in the centre of the source room will therefore enclose image sources (i.e. propagation paths with reflections) with propagation times less than t. Moving on to consider

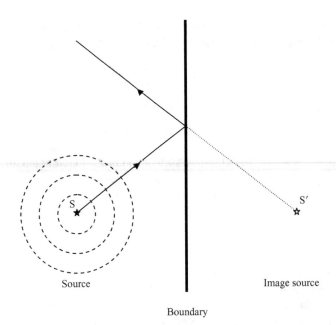

Figure 1.8

Source and image source.

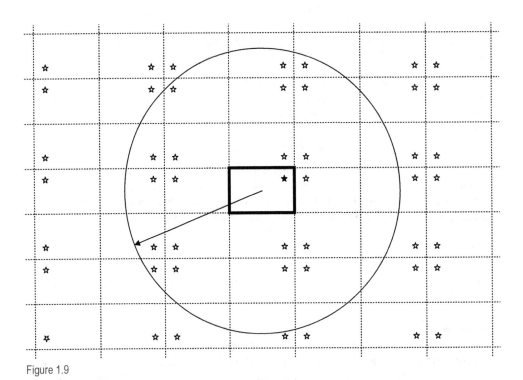

Figure 1.9

Source (★) and image sources (☆) for a box-shaped room (dark solid lines) and some of its image rooms (dotted lines).

three-dimensional space, it follows that the volume of a sphere of radius, $c_0 t$, divided by the volume associated with each image source (i.e. the room volume, V) will equal the number of reflections, N, arriving at a point in the room within time, t. Hence,

$$N = \frac{4\pi(c_0 t)^3}{3V} \qquad (1.48)$$

and the temporal density of reflections, dN/dt (i.e. the number of reflections arriving per second at time, t) is

$$\frac{dN}{dt} = \frac{4\pi c_0^3 t^2}{V} \qquad (1.49)$$

This equation applies to any shape of room with a diffuse field; the derivation simply uses a box-shaped room to simplify use of the image source approach.

1.2.5 Local modes

Having looked at the diffuse field, we will now look at the other idealized model, the modal sound field. We start by defining room modes. To do this we can follow the journey of a plane wave as it travels around a box-shaped room. To simplify matters we assume that all the room surfaces are perfectly reflecting and rigid. Therefore the incident and reflected waves have the same magnitude and the sound pressure is reflected from the surface without any change in phase. A rigid wall or floor is defined as one which is not caused to vibrate when a sound wave impinges upon it; hence the particle velocity normal to the surface is zero. In practice, walls and floors do vibrate because this is the mechanism that is responsible for sound transmission; however, this assumption avoids having to consider the wide range of acoustic surface impedances that are associated with real surfaces.

We now follow the path that is travelled by a sound wave as it travels across a box-shaped room (see Fig. 1.10). At some point in time it will hit one of the room boundaries from which it will be reflected before continuing on its journey to be reflected from other room boundaries.

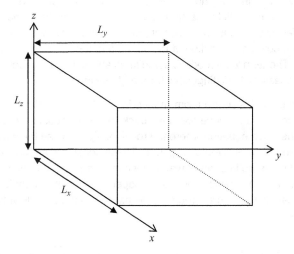

Figure 1.10

Box-shaped room.

Figure 1.11

Room modes. Plan view of a box-shaped room showing one possible journey taken by a plane wave. A room mode occurs when the wave travels through the same starting point (•) travelling in exactly the same direction as when it first left, whilst achieving phase closure.

These reflections are assumed to be specular as would occur with smooth walls and floors that have uniform acoustic surface impedance over their surface. We can also follow the journeys taken by other sound waves travelling in other directions. Some of these waves will return to the starting point travelling in exactly the same direction as when they first left. In some instances the length of their journey, in terms of phase, will correspond to an integer multiple of 2π such that there will be continuity of phase; we will refer to this as phase closure. Each journey that returns to the same starting point travelling in the same direction whilst achieving phase closure defines a mode with a specific frequency (see Fig. 1.11).

The term 'local mode' is used because the modes are 'local' to a space that is defined by its boundaries; in a similar way we will define local modes of vibration for structure-borne sound on plates and beams in Chapter 2. For rooms this definition assumes that there is no interaction between the sound waves in the room and the structure-borne sound waves on the walls and floors that face into that room. The walls and floors are only considered as boundaries that determine the fraction of wave energy that is reflected and the phase change that occurs upon reflection. It is also assumed that there is no sound source exciting these modes; we have simply followed the journey of a plane wave without considering how it was generated. Hence it is important to note that local modes of spaces and structures (e.g. rooms, walls, and floors) are a concept; they do not actually exist in real buildings where the spaces and structures are coupled together. Although the definition of local modes is slightly removed from reality, the concept is very useful in studying certain features of sound or vibration fields, as well as the interaction between these fields using methods such as Statistical Energy Analysis. Local modes are also referred to as natural modes or pure standing waves; they are a property of the space, rather than a combined function of the space and the excitation. The latter is referred to as a resonance. The term local mode is sometimes abbreviated to mode; only using the full name where it is necessary to distinguish it from a global mode.

To calculate the frequencies of the room modes in this box-shaped room it is necessary to calculate the wavenumbers. Hence we refer back to our discussion in Section 1.2.1.2 on plane waves and the wave equation where the relationship between the wavenumber and the constants, k_x, k_y, and k_z, was given by Eq. 1.15. These constants are calculated by using the equation for sound pressure in a plane wave which must satisfy both the wave equation (Eq. 1.14) and the boundary conditions. For a box-shaped room with dimensions L_x, L_y, and L_z, the following boundary conditions are required to ensure that the particle velocity normal to the rigid room surfaces is zero,

$$\frac{\partial p}{\partial x} = 0 \text{ at } x = 0 \text{ and } x = L_x \qquad \frac{\partial p}{\partial y} = 0 \text{ at } y = 0 \text{ and } y = L_y \qquad \frac{\partial p}{\partial z} = 0 \text{ at } z = 0 \text{ and } z = L_z.$$

$$(1.50)$$

By taking the real part of Eq. 1.13 that describes the sound pressure for a plane wave and ignoring time dependence we have the following solution

$$p(x, y, z) = \hat{p} \cos(k_x x) \cos(k_y y) \cos(k_z z) \tag{1.51}$$

This will only satisfy the boundary conditions when $\sin(k_x L_x) = \sin(k_y L_y) = \sin(k_z L_z) = 0$, hence

$$k_x = \frac{p\pi}{L_x} \qquad k_y = \frac{q\pi}{L_y} \qquad k_z = \frac{r\pi}{L_z} \tag{1.52}$$

where the variables p, q, and r can take zero or positive integer values.

Each combination of values for $p, q,$ and r describes a room mode for which the mode wavenumber, $k_{p,q,r}$, (also called an eigenvalue) is found from Eqs 1.15 and 1.52 to be

$$k_{p,q,r} = \pi \sqrt{\left(\frac{p}{L_x}\right)^2 + \left(\frac{q}{L_y}\right)^2 + \left(\frac{r}{L_z}\right)^2} \tag{1.53}$$

Therefore the mode frequency, $f_{p,q,r}$ (also called an eigenfrequency) is

$$f_{p,q,r} = \frac{c_0}{2} \sqrt{\left(\frac{p}{L_x}\right)^2 + \left(\frac{q}{L_y}\right)^2 + \left(\frac{r}{L_z}\right)^2} \tag{1.54}$$

where $p, q,$ and r take zero or positive integer values.

In a box-shaped room there are three different types of room mode: axial, tangential, and oblique modes.

Axial modes describe the situation where wave propagation is parallel to the $x, y,$ or z axis. They have one non-zero value for $p, q,$ or r, and zero values for the other two variables (e.g. $f_{1,0,0}, f_{0,3,0}, f_{0,0,2}$).

Tangential modes can be described by defining a 'pair of surfaces' as two surfaces that lie opposite each other, where each pair of surfaces partially defines the box-shaped room. Hence, tangential modes describe wave propagation at an angle that is oblique to two pairs of surfaces, and is tangential to the other pair of surfaces. They have non-zero values for two of the variables $p, q,$ or r, and a zero value for the other variable (e.g. $f_{1,2,0}, f_{3,0,1}, f_{0,2,2}$).

Oblique modes describe the situation where wave propagation occurs at an angle that is oblique to all surfaces; hence they have non-zero values for $p, q,$ and r (e.g. $f_{2,3,1}$).

We have assumed that all the room surfaces are perfectly reflecting and rigid, in practice there is interaction between the sound pressure in the room and the vibration of the walls and floors facing into that room. However, the assumption of rigid walls and floors is reasonable in many rooms because this interaction results in relatively minor shifts in the eigenfrequencies.

1.2.5.1 Modal density

It is often useful to calculate the first 10 or so modes to gain an insight into their distribution between the frequency bands in the low-frequency range. However, in a room of approximately 50 m³ there are almost one-million modes in the building acoustics frequency range. Fortunately there is no need to calculate all of these modes because we can adopt a statistical viewpoint. A statistical approach also helps us to deal with the fact that very few rooms are

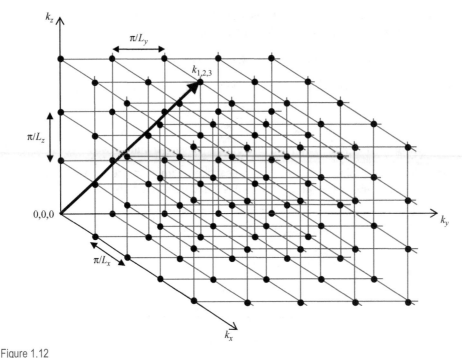

Figure 1.12

Mode lattice for a three-dimensional space. The vector corresponding to eigenvalue, $k_{1,2,3}$, is shown as an example.

perfectly box-shaped with rigid boundaries. For this reason the calculation of any individual mode frequency will rarely be accurate when the wavelength is smaller than any of the room dimensions. We can usually expect to estimate the one-third-octave-band in which a mode frequency will fall to an accuracy of plus or minus one-third-octave-band.

The statistical descriptor for modes is the statistical modal density, $n(f)$, the number of modes per Hertz. To calculate the modal density it is necessary to arrange the eigenvalues in such a way that facilitates counting the modes in a chosen frequency range. It is implicit in the form of Eq. 1.15 that this can be achieved by creating a lattice in Cartesian coordinates where the x, y, and z axes represent k_x, k_y, and k_z (Kuttruff, 1979). This lattice of eigenvalues in k-space is shown in Fig. 1.12, where each intersection in the lattice represents an eigenvalue indicated by the symbol •. The length of the vector from the origin to an eigenvalue equals $k_{p,q,r}$. Eigenvalues that lie along each of the three axes represent axial modes; those that lie on the coordinate planes $k_x k_y, k_x k_z$, and $k_y k_z$ (excluding the eigenvalues on the axes) represent tangential modes; all other eigenvalues (i.e. all eigenvalues excluding those on the axes and the coordinate planes) represent oblique modes. From Eq. 1.52 it is evident that the distance between adjacent eigenvalues in the k_x, k_y, and k_z directions are $\pi/L_x, \pi/L_y$, and π/L_z respectively. Hence the volume associated with each eigenvalue is a cube with a volume of $\pi^3/L_x L_y L_z$, which equals π^3/V.

The number of modes below a specified wavenumber, k, is equal to the number of eigenvalues that are contained within one-eighth of a spherical volume with radius, k, as indicated in Fig. 1.13. If there were only oblique modes this would simply be carried out by dividing $(4\pi k^3/3)/8$ by π^3/V; however, the existence of axial and tangential modes means that this

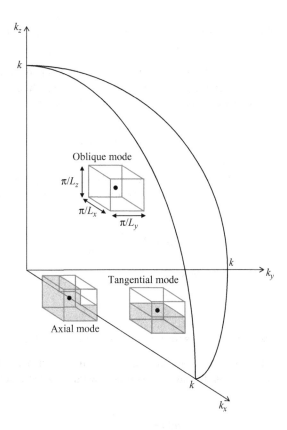

Figure 1.13

Sketch indicating how the volumes associated with the eigenvalues for axial, tangential, and oblique modes fall inside or outside the permissible volume in k-space. The shaded volumes indicate those fractions of the volumes associated with the axial and tangential modes that fall outside the permissible volume in k-space. The octant volume with radius, k, encloses eigenvalues below wavenumber, k.

would be incorrect. This is because part of the cube volume that is associated with these mode types falls outside of the permissible volume in k-space that can only have zero or positive values of k_x, k_y, and k_z. From Fig. 1.13 we also see that for tangential modes on the coordinate planes, one-half of the cube volume falls outside this permissible volume and for axial modes, three-quarters of the cube volume falls outside. Therefore calculating the number of modes is a three-step process. The first step is to divide $(4\pi k^3/3)/8$ by π^3/V to give an estimate for the number of oblique modes that also includes one-quarter of the axial modes and one-half of the tangential modes. The second step is to account for the other halves of the tangential modes that lie in the area on the three coordinate planes; this fraction of the total number of modes is calculated by taking one-half of $(\pi k^2/4)/(\pi^2/(L_x L_y + L_x L_z + L_y L_z))$. The latter step included the axial modes on each of the three coordinate axes as halves. Hence there only remains one-quarter of the axial modes that have not yet been accounted for. The third step determines this remaining fraction of the total number of modes by taking one-quarter of $k/(\pi/(L_x + L_y + L_z))$. The sum of these three components gives the number of modes, $N(k)$, below the wavenumber, k,

$$N(k) = \frac{k^3 V}{6\pi^2} + \frac{k^2 S_T}{16\pi} + \frac{k L_T}{16\pi} \tag{1.55}$$

where the total area of the room surfaces, S_T is $2(L_xL_y + L_xL_z + L_yL_z)$ and the total length of all the room edges, L_T, is $4(L_x + L_y + L_z)$.

As we are working in k-space we calculate the modal density, $n(\omega)$, in modes per radian, and then convert to the modal density, $n(f)$, in modes per Hertz, which is more convenient for practical calculations. The general equation for the modal density, $n(\omega)$, in terms of ω is

$$n(\omega) = \frac{dN(k)}{d\omega} = \frac{dN(k)}{dk}\frac{dk}{d\omega} \tag{1.56}$$

To calculate $n(\omega)$ we now need to find $dk/d\omega$ which is equal to the reciprocal of the group velocity, c_g. The group velocity is the velocity at which wave energy propagates across the space. For sound waves in air, the group velocity is the same as the phase velocity, c_0. Hence the general equation to convert $n(\omega)$ to $n(f)$ is

$$n(f) = 2\pi n(\omega) = \frac{2\pi}{c_g}\frac{dN(k)}{dk} \tag{1.57}$$

which gives the modal density for a box-shaped room as

$$n(f) = \frac{4\pi f^2 V}{c_0^3} + \frac{\pi f S_T}{2c_0^2} + \frac{L_T}{8c_0} \tag{1.58}$$

where the modal density for each frequency band is calculated using the band centre frequency.

For rooms that are not box-shaped, and for typical rooms in the high-frequency range, a reasonable estimate of the modal density can be found by using only the first term in Eq. 1.58, to give

$$n(f) = \frac{4\pi f^2 V}{c_0^3} \tag{1.59}$$

Estimates for the statistical modal density of axial, tangential, and oblique modes can be estimated from (Morse and Ingard, 1968)

$$n_{axial}(f) = \frac{6V^{1/3}}{c_0} \tag{1.60}$$

$$n_{tangential}(f) = \frac{6\pi f V^{2/3}}{c_0^2} - \frac{6V^{1/3}}{c_0} \tag{1.61}$$

$$n_{oblique}(f) = \frac{4\pi f^2 V}{c_0^3} - \frac{3\pi f V^{2/3}}{c_0^2} + \frac{3V^{1/3}}{2c_0} \tag{1.62}$$

for which it is assumed that $L_T \approx 12V^{1/3}$ and $S_T \approx 6V^{2/3}$ (Jacobsen, 1982). This simplifies the calculation for rooms that are almost (but not exactly) box-shaped.

1.2.5.2 Mode count

The mode count, N, in a frequency band with a bandwidth, B, can be determined in two ways. Either by using Eq. 1.54 to calculate the individual mode frequencies and then by counting the number of modes that fall within the band or by using the statistical modal density to determine a statistical mode count, N_s, in that band, where

$$N_s = n(f)B \tag{1.63}$$

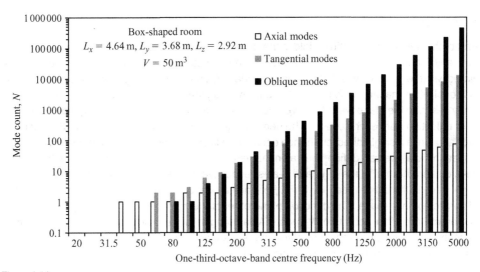

Figure 1.14

Mode count for axial, tangential, and oblique modes in a 50 m³ box-shaped room.

In a box-shaped room, the mode with the lowest frequency will always be an axial mode. As the band centre frequency increases, the number of oblique modes in each band increases at a faster rate than the number of axial modes or the number of tangential modes. As an example we can look at the trends in the mode count for a 50 m³ room. The room dimensions are determined using the ratio $4^{1/3}:2^{1/3}:1$ for $x:y:z$. This ratio is sometimes used in the design of reverberation rooms to avoid dimensions that are integer multiples of each other; this avoids different modes having the same frequency. As we usually work in one-third-octave-bands it is of interest to know the number of modes that fall within each band. The mode counts for the axial, tangential, and oblique modes are shown in Fig. 1.14. For typical rooms we can describe the mode count using three different ranges: A, B, and C. In range A, the frequency bands either contain no modes or a few axial and/or a few tangential modes. In range B, the blend of the three different mode types varies between adjacent frequency bands depending on the room dimensions. In range C, the mode count is always highest for oblique modes and always lowest for axial modes. For this particular example, range A corresponds to one-third-octave-bands below 80 Hz, range B lies between the 80 and 200 Hz bands, and range C corresponds to bands above 200 Hz.

1.2.5.3 Mode spacing

The average frequency spacing between adjacent modes, δf, is calculated from the modal density using

$$\delta f = \frac{1}{n(f)} \tag{1.64}$$

As sound insulation calculations are almost always carried out in one-third-octave or octave-bands it tends to be more informative to calculate the mode counts in these frequency bands rather than use the mode spacing.

1.2.5.4 Equivalent angles

Part of the definition of a diffuse field is that there is equal probability of a sound wave arriving from any direction, i.e. from any angle. Hence it is instructive to look at the range of angles associated with the plane waves that form local modes. We have previously described a local mode in a qualitative manner by following the journey of a plane wave around a room. To quantitatively describe the plane wave field we need to account for the different propagation directions after reflection from each surface. The general equation for a plane wave (Eq. 1.13) describes propagation in a single direction; hence each mode is comprised of more than one plane wave. For each mode we can define equivalent angles, $\theta_x, \theta_y,$ and θ_z; these angles are defined from lines that are normal to the x, y, and z-axis respectively. They are defined in k-space for any eigenvalue in the lattice (see Fig. 1.15). For each mode, one plane wave points in the direction of this vector in k-space. The direction of the other plane waves can be found by reflecting the vector into the other octants of k-space. Axial, tangential, and oblique modes are therefore described by two, four, and eight plane waves respectively. For each mode, the equivalent angles are related to the mode wavenumber, $k_{p,q,r}$, and the constants, $k_x, k_y,$ and k_z by

$$k_x = k_{p,q,r} \sin\theta_x \qquad k_y = k_{p,q,r} \sin\theta_y \qquad k_z = k_{p,q,r} \sin\theta_z \qquad (1.65)$$

hence, from the constant definitions in Eq. 1.52, the equivalent angles for each mode are

$$\theta_x = \mathrm{asin}\left(\frac{pc_0}{2L_x f_{p,q,r}}\right) \qquad \theta_y = \mathrm{asin}\left(\frac{qc_0}{2L_y f_{p,q,r}}\right) \qquad \theta_z = \mathrm{asin}\left(\frac{rc_0}{2L_z f_{p,q,r}}\right) \qquad (1.66)$$

Later on we will need to consider the angles of incidence for the waves that impinge upon a room surface in the calculation of sound transmission. Here we are only assessing the range of equivalent angles for plane waves that propagate across the space to form room modes. Figure 1.16 shows the equivalent angles for the same $50\,\mathrm{m}^3$ room that was used for the mode count, where each point corresponds to a single room mode. Note that we are ignoring the fact that specular reflection would not occur in real rooms at high frequencies. In the low-frequency range, where there are relatively few modes, there is a limited range of angles. As the frequency increases, the number of modes increases (the majority tending to be oblique modes), and the range expands to cover the full range of angles between $0°$ and $90°$. For axial modes, one angle is $90°$ and the other two angles are $0°$ (e.g. $f_{1,0,0}$ has $\theta_x = 90°$, $\theta_y = 0°$, and $\theta_z = 0°$). For tangential modes, one angle is $0°$, and the other two angles are oblique. For oblique modes, all three angles are oblique.

Equivalent angles do not in themselves identify a frequency above which the modal sound field approximates to a diffuse field; we have already noted other important features that define a diffuse field. However, they do illustrate how one aspect of a diffuse field concerning sound arriving from all directions can potentially be satisfied in a modal sound field. In the study of sound transmission it is useful to be able to switch between thinking in terms of modes, and in terms of waves travelling at specific angles.

1.2.5.5 Irregularly shaped rooms and scattering objects

The description of local modes was based on an empty box-shaped room. Rooms in real buildings are not all box-shaped and they usually contain scattering objects such as furniture.

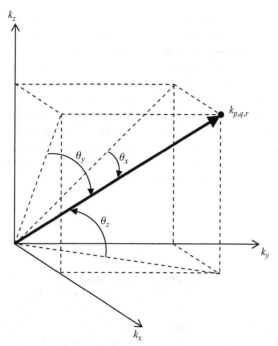

Figure 1.15

Equivalent angles in *k*-space.

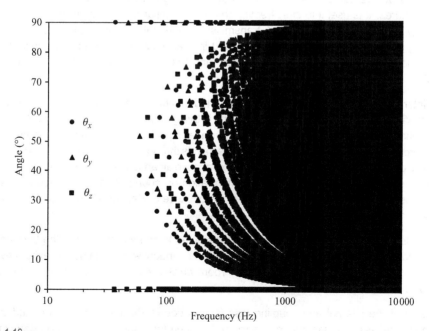

Figure 1.16

Equivalent angles for the modes of a 50 m³ box-shaped room.

In the laboratory, it is common to use non-parallel walls and diffusers to try and create a diffuse field in the central zone of the room. This does not mean that the local mode approach is instantly irrelevant; far from it. Scattering objects can be seen as coupling together the local modes of the empty room, giving rise to hybrid versions of the original mode shapes (Morse and Ingard, 1968). These hybrid versions no longer have the symmetrical sound pressure fields associated with individual local modes in an empty box-shaped room. In the limit, as the room shape becomes increasingly irregular (or sufficient scattering objects are placed inside a box-shaped room) and the room surfaces have a random distribution of acoustic surface impedance, we can effectively consider all the room modes to be some form of oblique mode. As we approach this limit we can leave the local mode model behind us and assume there is a close approximation to a diffuse field in the central zone of the room.

1.2.6 Damping

In our discussion on room modes we assumed that there was a perfect reflection each time the plane wave was reflected from a room boundary. In reality there will always be damping mechanisms that reduce the sound pressure level. When discussing room acoustics we usually refer to absorption and reverberation times, rather than damping. From a room acoustics perspective, the sound source and the listener are located within one space, so from the point-of-view of a listener in the source room, any sound that doesn't return to them has been absorbed. However, with sound insulation our concern is usually for the person that hears the sound in the receiving room, and from their point-of-view the sound has been transmitted. As we are particularly interested in the exchange of sound energy between spaces and structures, it is useful to start treating them in a similar manner by using the same terminology to describe absorption and transmission. Hence it is convenient to relate different damping mechanisms to the loss factors used in Statistical Energy Analysis; these are the internal loss factor, the coupling loss factor, and the total loss factor (Lyon and DeJong, 1995).

With internal losses the sound energy is converted into heat. Hence high internal loss factors are beneficial for the noise control engineer who is trying to reduce sound levels. Internal losses occur when the sound wave hits absorptive surfaces or objects (e.g. sound absorbent ceiling tiles, carpet, porous materials) and as the wave travels through the air due to air absorption. The former is usually more important than the latter because air absorption only becomes significant at high frequencies and in large rooms. Information on sound absorption mechanisms and sound absorbers can be found in a number of textbooks (e.g. Mechel, 1989/1995/1998; Mechel and Vér, 1992; Kuttruff, 1979).

With coupling losses, the sound energy is transmitted to some other part of the building that faces into the room. This could be open door or window where the sound exits, never to return. It could also be a wall or a floor in the room which is caused to vibrate by the impinging sound waves.

The sum of the internal and coupling loss factors equals the total loss factor, and this is related to the reverberation time of the room. We therefore start this section on damping by deriving reflection and absorption coefficients for room surfaces that will lead to a discussion of reverberation times and loss factors for rooms.

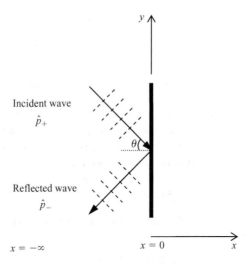

Figure 1.17

Plane wave incident at an angle, θ, upon a surface, and the specularly reflected wave.

1.2.6.1 Reflection and absorption coefficients

A plane wave incident upon a room surface at $x = 0$ can be described using the term, $\hat{p}_+ \exp(-ikx)$ where $\hat{p}_+$ is an arbitrary constant. The term for the reflected wave is $R\hat{p}_+ \exp(ikx)$ where R is defined as the reflection coefficient,

$$R = |R| \exp(i\gamma) \tag{1.67}$$

As seen from Eq. 1.67, the reflection coefficient is complex. It describes the magnitude and phase change that occurs upon reflection. In diffuse fields the waves that are incident upon a surface have random phase, so the information on the phase change is usually ignored.

We now consider a plane wave that is incident upon a surface at an angle, θ; defined such that $\theta = 0°$ when the wave propagates normal to the surface. The aim here is to relate the reflection coefficient to the specific acoustic impedance or admittance. It is assumed that the surface is locally reacting so that the normal component of the particle velocity only depends on the region at the surface where the sound pressure is incident.

The incident wave propagates in the xy plane towards a surface at $x = 0$ as shown in Fig. 1.17. The incident wave is described by

$$p_+(x, y, t) = [\hat{p}_+ \exp(-ik(x \cos \theta + y \sin \theta))] \exp(i\omega t) \tag{1.68}$$

The incident wave is specularly reflected from the surface and the reflection coefficient is used to describe the amplitude of the reflected wave, $\hat{p}_- = R\hat{p}_+$. Hence the reflected wave is

$$p_-(x, y, t) = [R\hat{p}_+ \exp(-ik(-x \cos \theta + y \sin \theta))] \exp(i\omega t) \tag{1.69}$$

The particle velocity in the x-direction (i.e. normal to the surface) is found using Eq. 1.16, which gives

$$u_x = -\frac{1}{i\omega} \frac{1}{\rho_0} \frac{\partial p}{\partial x} \tag{1.70}$$

Therefore the particle velocities for the incident and reflected waves are

$$u_{x+}(x, y, t) = \frac{\cos\theta}{\rho_0 c_0}[\hat{p}_+ \exp(-ik(x\cos\theta + y\sin\theta))]\exp(i\omega t) \qquad (1.71)$$

$$u_{x-}(x, y, t) = -\frac{\cos\theta}{\rho_0 c_0}[R\hat{p}_+ \exp(-ik(-x\cos\theta + y\sin\theta))]\exp(i\omega t) \qquad (1.72)$$

At the surface (i.e. at $x = 0$), the resultant pressure is $p_+ + p_-$, and the resultant particle velocity normal to the surface is $u_{x+} + u_{x-}$. The ratio of the resultant pressure to this resultant particle velocity equals the normal acoustic surface impedance (Eq. 1.29). Hence the specific acoustic impedance of a surface is related to the reflection coefficient by

$$Z_{a,s} = \frac{1}{\cos\theta}\frac{1+R}{1-R} \qquad (1.73)$$

which is re-arranged to give the reflection coefficient in terms of either the specific acoustic impedance or admittance

$$R = \frac{Z_{a,s}\cos\theta - 1}{Z_{a,s}\cos\theta + 1} = \frac{\cos\theta - \beta_{a,s}}{\cos\theta + \beta_{a,s}} \qquad (1.74)$$

In practice it is usually more convenient to work in terms of absorption rather than reflection. The sound absorption coefficient, α, is defined as the ratio of the intensity absorbed by a surface to the intensity incident upon that surface; hence it takes values between 0 and 1. The intensity in a plane wave is proportional to the mean-square pressure (Eq. 1.19), so the absorption coefficient is related to the reflection coefficient by

$$\alpha = 1 - |R|^2 \qquad (1.75)$$

The absorption coefficient can be calculated using Eqs 1.74 and 1.75 in terms of either the specific acoustic impedance or admittance. For a plane wave that is incident upon a locally reacting surface at an angle, θ, the angle-dependent absorption coefficient, α_θ, is

$$\alpha_\theta = \frac{4Z_{Re}\cos\theta}{(Z_{Re}^2 + Z_{Im}^2)\cos^2\theta + 2Z_{Re}\cos\theta + 1} = \frac{4\beta_{Re}\cos\theta}{(\beta_{Re} + \cos\theta)^2 + \beta_{Im}^2} \qquad (1.76)$$

where the real and imaginary parts of the specific acoustic impedance are

$$Z_{a,s} = Z_{Re} + iZ_{Im} \qquad (1.77)$$

and the real and imaginary parts of the specific acoustic admittance are

$$\beta_{a,s} = \frac{1}{Z_{a,s}} = \beta_{Re} - i\beta_{Im} \qquad (1.78)$$

At normal incidence, $\theta = 0°$, hence the normal incidence absorption coefficient, α_0, is

$$\alpha_0 = \frac{4\beta_{Re}}{(\beta_{Re} + 1)^2 + \beta_{Im}^2} \qquad (1.79)$$

There can be significant variation in the absorption coefficient with angle. However, when there is a diffuse sound field incident upon a surface we assume that there is equal probability of sound waves impinging upon the surface from all directions. For diffuse fields we therefore use the statistical sound absorption coefficient, α_{st}, given by

$$\alpha_{st} = \int_0^{\pi/2} \alpha_\theta \sin(2\theta)d\theta \qquad (1.80)$$

The statistical absorption coefficient is calculated from the specific acoustic admittance using (Morse and Ingard, 1968)

$$\alpha_{st} = 8\beta_{Re} \left(1 + \frac{\beta_{Re}^2 - \beta_{Im}^2}{\beta_{Im}} \operatorname{atan}\left(\frac{\beta_{Im}}{\beta_{Re}^2 + \beta_{Im}^2 + \beta_{Re}} \right) - \beta_{Re} \ln\left(\frac{(\beta_{Re} + 1)^2 + \beta_{Im}^2}{\beta_{Re}^2 + \beta_{Im}^2} \right) \right) \quad (1.81)$$

1.2.6.2 Absorption area

Rooms not only have absorbent surfaces, but they also contain absorbent objects (e.g. furniture, people) and there will be air absorption. For practical purposes, the absorption area, A, in m^2 is useful in describing the absorption provided by surfaces, objects, and air. The absorption area is defined as the ratio of the sound power absorbed by a surface or object, to the sound intensity incident upon the surface or object. For a surface the absorption area is the product of the absorption coefficient and the surface area. The absorption area essentially describes all the absorption in the room using a single area; hence an absorption area of $10\,m^2$ corresponds to an area of $10\,m^2$ that is totally absorbing.

For a room with I surfaces, J objects, and air absorption, the total absorption area, A_T is

$$A_T = \sum_{i=1}^{I} S_i \alpha_i + \sum_{j=1}^{J} A_{obj,j} + A_{air} \quad (1.82)$$

Air absorption depends upon frequency, temperature, relative humidity, and static pressure. The absorption area for air is calculated from the attenuation coefficient in air, m, in Neper/m and the volume of air in the space, V, using

$$A_{air} = 4mV \quad (1.83)$$

The attenuation coefficient in dB/m can be calculated according to ISO 9613-1 and converted to Neper/m by dividing by $10\lg(e)$.

Calculated values for A_{air} at $20°C$, 70% RH and $P_0 = 1.013 \times 10^5$ Pa are shown in Fig. 1.18. For rooms with an absorption area of at least $10\,m^2$ due to surfaces and objects, A_{air} will only usually form a significant fraction of A in the high-frequency range. For furnished, habitable rooms (such as those in dwellings, commercial buildings, and schools), air absorption in the building acoustics frequency range can often be ignored in volumes $<150\,m^3$.

1.2.6.3 Reverberation time

When a sound source in a room is stopped abruptly, the sound energy decays away due to the damping mechanisms that are present in the room. This feature is called reverberation and is assessed by plotting a decay curve. This is a plot of the decaying sound pressure level against time, starting from the time at which the sound source is stopped, usually denoted as the time, $t = 0$.

For sound insulation, the reverberation time is needed to relate the sound power radiated into a space to the average sound pressure level in that space and to quantify either the absorption

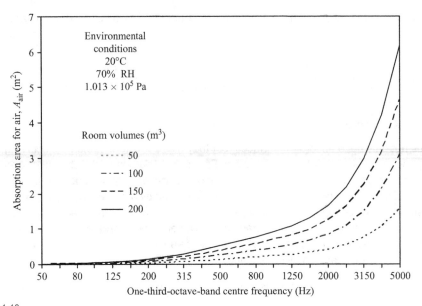

Figure 1.18

Absorption area for air in different room volumes.

in a space or the total loss factor of a space. In room acoustics, several other parameters are used to describe different aspects of the sound field relating to reverberation and the subjective evaluation of sound in spaces (ISO 3382).

The reverberation time, T, is the time in seconds that is taken for the sound pressure level to decay by 60 dB, or in terms of energy, for the sound energy to decay to one-millionth of its initial value. This definition is well-suited to the decay curve that occurs in a diffuse field; a straight line decay as shown in Fig. 1.19. Some decay curves can be approximated by a single straight line over the full 60 dB decay, but there are many that do not follow this simple form. In addition, it is not always possible to measure a 60 dB decay due to the presence of background noise. Therefore we need a definition that can be used when decay curves have more than one slope over the 60 dB decay range. This definition also needs to quantify the time taken for the level to decay by 60 dB by using linear regression over a specified decay range (e.g. 30 dB) so that it is not imperative to use the full 60 dB decay. Hence, the reverberation time is more usefully defined as the time in seconds that would be required for the sound pressure level to decay by 60 dB when using linear regression over a specified part of the decay curve. As we can now use any range for the linear regression, such as 10, 15, 20, or 30 dB, it is necessary to use the notation, T_X, where X identifies the evaluation range used in the linear regression, i.e. T_{10}, T_{15}, T_{20}, T_{30}. With measured decay curves, the starting point for the linear regression is usually 5 dB below the initial level, to the end point at $X + 5$ dB (ISO 354 and ISO 3382). In Section 3.8.3 we will see that 5 dB is used as the starting point primarily because the signal processing distorts the initial part of the decay curve.

We can now look at reverberation times with a diffuse field and a non-diffuse field in a box-shaped room.

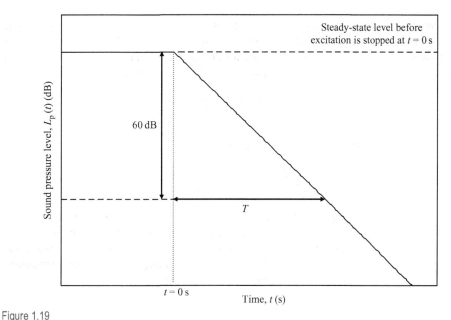

Figure 1.19

Ideal straight line decay curve showing the decrease in the sound pressure level with time after the excitation has stopped.

1.2.6.3.1 Diffuse field

In a diffuse sound field, the mean free path, d_{mfp}, can be used to calculate the average time, Δt, between two successive diffuse reflections from the room boundaries,

$$\Delta t = \frac{d_{mfp}}{c_0} \qquad (1.84)$$

we can then use Δt to calculate the number of diffuse reflections, N, in time, t, using

$$N = \frac{t}{\Delta t} = \frac{c_0 t}{d_{mfp}} \qquad (1.85)$$

The decay process can now be assessed by looking at the probability of waves impinging upon surface areas in the room with different absorption coefficients. This uses the approach taken by Kuttruff (1979). We take a room containing a sound source fed by a stationary signal such as white noise and assume that the resulting sound field is diffuse. When the sound source is stopped at time, $t = 0$, the waves continue to travel across the room volume, and each time they impinge upon a room boundary, a fraction of the energy is reflected, with the remaining fraction being absorbed. The binomial probability distribution is used to assess two possible outcomes when a sound wave impinges upon a room boundary: either the wave is reflected from (and absorbed by) surface area, S_1, or it is reflected from (and absorbed by) the remaining surface area in the room, $S_2 = S_T - S_1$. These outcomes must be statistically independent; hence the probability that a wave is reflected from (and absorbed by) surface area, S_1, is the same every time that the wave impinges upon a room boundary. This is conceivable in a room where each surface area with a different absorption coefficient is uniformly distributed over the total surface area of the room, and there are diffuse reflections from all the room surfaces.

The binomial probability distribution, $P(N_1 \backslash N)$ gives the number of times, N_1, that a wave is reflected from surface area, S_1, out of a total number of reflections, N. The probability of a reflection from surface area, S_1, is S_1/S_T, hence the probability of a reflection from the remaining surface area in the room, S_2, is S_2/S_T or $1 - (S_1/S_T)$. The binomial probability distribution is

$$P(N_1 \backslash N) = \binom{N}{N_1} \left(\frac{S_1}{S_T}\right)^{N_1} \left(1 - \frac{S_1}{S_T}\right)^{N-N_1} \qquad (1.86)$$

where the binomial coefficient is calculated using

$$\binom{N}{N_1} = \frac{N!}{N_1!(N - N_1)!} \qquad (1.87)$$

Each time a wave impinges upon surface area, S_1, the mean-square pressure is reduced by the factor $(1 - \alpha_1)$ where α_1 is the diffuse field sound absorption coefficient for surface area, S_1. The same process occurs with surface area, S_2, with the coefficient, α_2. So, after N reflections, of which N_1 reflections are from surface area, S_1, and $N - N_1$ reflections are from the remaining surface area, S_2, the mean-square pressure as a function of N_1, $p^2(N_1)$, is

$$p^2(N_1) = p^2(0)(1 - \alpha_1)^{N_1}(1 - \alpha_2)^{N-N_1} \qquad (1.88)$$

where $p^2(0)$ is the mean-square pressure at time, $t = 0$.

Taking into account all possible values of N_1 from zero to N, we can calculate the expected value, $E(N_1)$ of random variable, N_1, which has the probability distribution, $P(N_1 \backslash N)$. As we are particularly interested in the decay of the sound pressure we note that $E(N_1)$ equals the population mean (i.e. the average value) of the mean-square pressure,

$$E(N_1) = \sum_{N_1=0}^{N} p^2(N_1)P(N_1 \backslash N) = p^2(0)\left[\left(\frac{S_1}{S_T}\right)(1 - \alpha_1) + \left(1 - \frac{S_1}{S_T}\right)(1 - \alpha_2)\right]^N \qquad (1.89)$$

The term within the square bracket is equal to $(1 - \overline{\alpha})$ where $\overline{\alpha}$ is the average diffuse field sound absorption coefficient

$$\overline{\alpha} = \frac{S_1\alpha_1 + \left(1 - \frac{S_1}{S_T}\right)\alpha_2}{S_T} \qquad (1.90)$$

Hence for rooms with I surface areas that each have a different absorption coefficient (instead of $I = 2$ as has just been assumed), the expression can be generalized to

$$\overline{\alpha} = \frac{\sum_{i=1}^{I} S_i\alpha_i}{\sum_{i=1}^{I} S_i} = \frac{1}{S_T}\sum_{i=1}^{I} S_i\alpha_i \qquad (1.91)$$

The average diffuse field sound absorption coefficient, $\overline{\alpha}$, is a weighted arithmetic average of the absorption coefficients, where each coefficient has been weighted according to its surface area. From Eq. 1.89 the mean-square pressure $p^2(t)$ at time, t, is therefore described by

$$p^2(t) = p^2(0)(1 - \overline{\alpha})^N = p^2(0)\exp(N \ln(1 - \overline{\alpha})) \qquad (1.92)$$

It is important to note from Eq. 1.92 that the mean-square pressure in a diffuse field has an exponential decay. Therefore when plotting the sound pressure level in decibels against time, the decay curve is a straight line. We will soon look at decay curves in non-diffuse fields, where the mean-square pressure does not have an exponential decay, and the decay curve is not a single straight line across the 60 dB decay range.

At time $t = T$, the definition of the reverberation time is such that $p^2(T)$ is one-millionth of $p^2(0)$, hence combining Eqs 1.85 and 1.92 gives

$$\frac{p^2(T)}{p^2(0)} = 10^{-6} = \exp\left(\frac{c_0 T}{d_{mfp}} \ln(1-\overline{\alpha})\right) \tag{1.93}$$

This gives the reverberation time formula that is commonly referred to as Eyring's equation (Eyring, 1930),

$$T = \frac{-d_{mfp} 6 \ln 10}{c_0 \ln(1-\overline{\alpha})} = \frac{-24V \ln 10}{c_0 S_T \ln(1-\overline{\alpha})} \tag{1.94}$$

and by taking air absorption into account, this becomes

$$T = \frac{-24V \ln 10}{c_0(S_T \ln(1-\overline{\alpha}) - 4mV)} \tag{1.95}$$

When considering the effect of the room volume, V, and the total surface area, S_T, on the diffuse field reverberation time, it is useful to think in terms of the mean free path. The reverberation time is proportional to the mean free path for a diffuse field (Eq. 1.47), hence the longer the mean free path, the longer the time between successive reflections from the room surfaces. So if we choose a point in this room to measure the reverberation time, it will have taken longer for the waves to travel around the room before returning to our chosen point. Each time the wave hits a surface, a fraction of the wave energy will be absorbed. Assuming a fixed value for the absorption coefficient of room surfaces, the reverberation time will therefore increase with increasing room volume.

For a diffuse field where the average diffuse field absorption coefficient, $\overline{\alpha}$, is much smaller than unity, we can assume that $\overline{\alpha} \approx -\ln(1-\overline{\alpha})$. Using this approximation in Eq. 1.94 leads to Sabine's equation (Sabine, 1932),

$$T = \frac{d_{mfp} 6 \ln 10}{c_0 \overline{\alpha}} = \frac{24V \ln 10}{c_0 S_T \overline{\alpha}} \tag{1.96}$$

To take account of absorption from objects and the air, as well as from the room surfaces, Eq. 1.96 is more conveniently written in terms of the total absorption area, A_T (Eq. 1.82) as

$$T = \frac{24V \ln 10}{c_0 A_T} \tag{1.97}$$

Assuming that the steady-state sound pressure level in the diffuse field is 60 dB at time $t=0$, the time-varying sound pressure level in decibels that defines the idealized decay curve is

$$L_p(t) = 10 \lg\left(\frac{p^2(t)}{p^2(0)}\right) = 60 - \frac{60t}{T} \tag{1.98}$$

The Sabine equation is based on the assumption that $\overline{\alpha}$ is sufficiently small that $\overline{\alpha} \approx -\ln(1-\overline{\alpha})$, whereas the Eyring equation is applicable to any value of $\overline{\alpha}$. In general, Eyring's equation gives reasonable estimates in rooms where there is uniform surface absorption and diffuse surface reflections, however, it is also appropriate in box-shaped rooms with uniform surface absorption and specular surface reflections (Hodgson, 1993, 1996).

1.2.6.3.2 Non-diffuse field: normal mode theory

In non-diffuse fields the decay curves cannot usually be approximated by a straight line across the entire 60 dB decay range. To understand some of the reasons for this, we return to consider local room modes in a box-shaped room. In each frequency band, the decay curve will be determined by the individual room modes that are decaying within that band, and the interaction between these modes.

We will focus on decay curves in the low-frequency range. When we consider the mode count in one-third-octave-bands for typical rooms, bands in the low-frequency range have relatively few modes compared to those in the mid- and high-frequency ranges (refer back to the 50 m³ room in Fig. 1.14). In reality, we cannot strictly compartmentalize the decaying modes into individual frequency bands. This is due to the damping associated with each mode; a decaying mode may influence the decay in the two bands that are adjacent to the band in which the mode is strictly assumed to fall. However, compartmentalization is used here to provide some insight into the way that axial, tangential, and oblique modes determine the decay curve for a band. The normal mode theory used to calculate the decay curves is taken from Kuttruff (1979) and Bodlund (1980); the latter reference also provides corrections to earlier investigations by Larsen (1978).

Assume that we have a box-shaped room with locally reacting surfaces. This room contains a sound source that is fed by a sinusoidal signal with the same frequency as the mode of interest. In reality, most rooms have modes that are relatively closely spaced. This means that a sinusoidal signal will also excite other room modes unless all the surfaces have very low absorption coefficients and the modes are all well-separated in terms of frequency. However, here we will assume that we are able to excite only a single mode. When the sound source is stopped at time, $t = 0$, the waves continue to travel along the path that is defined for this particular room mode. Upon each reflection from a room surface, a fraction of the energy is reflected, and the remaining fraction is absorbed.

When looking at the reverberant decay of an individual mode, m, the mean-square pressure decays away exponentially according to

$$p^2(t) = p^2(0) \exp(-2\delta_m c_0 t) \tag{1.99}$$

where $\delta_m = \beta_{a,s} \left(\varepsilon_{p,m}/L_x + \varepsilon_{q,m}/L_y + \varepsilon_{r,m}/L_z \right)$ in which $\beta_{a,s}$ is the specific acoustic admittance, and $\varepsilon_{p,m}$, $\varepsilon_{q,m}$ and $\varepsilon_{r,m}$ correspond to mode, $f_{p,q,r}$ (if $p=0$, then $\varepsilon_{p,m} = 1$ else $\varepsilon_{p,m} = 2$; if $q = 0$, then $\varepsilon_{q,m} = 1$ else $\varepsilon_{q,m} = 2$; if $r = 0$ then $\varepsilon_{r,m} = 1$ else $\varepsilon_{r,m} = 2$).

The reverberation time, T_m, for an individual mode can be calculated from Eq. 1.99 at time $t = T_m$ using

$$\frac{p_m^2(T_m)}{p_m^2(0)} = 10^{-6} = \exp\left(-2c_0 T_m \beta_{a,s} \left(\frac{\varepsilon_{p,m}}{L_x} + \frac{\varepsilon_{q,m}}{L_y} + \frac{\varepsilon_{r,m}}{L_z} \right) \right) \tag{1.100}$$

which gives,

$$T_m = \frac{3 \ln 10}{c_0 \beta_{a,s} \left(\frac{\varepsilon_{p,m}}{L_x} + \frac{\varepsilon_{q,m}}{L_y} + \frac{\varepsilon_{r,m}}{L_z} \right)} \tag{1.101}$$

The denominator in Eq. 1.101 is referred to as the damping constant of the mode. From Eq. 1.101 it is possible to identify three trends for the different mode types when the specific acoustic admittance is independent of frequency: (1) the axial modes associated with each room dimension have different reverberation times to each other when $L_x \neq L_y \neq L_z$;

(2) when the tangential modes are considered in three groups defined by $p=0$, $q=0$, and $r=0$, each group will have the same reverberation time; and (3) all oblique modes have the same reverberation time.

For engineering calculations it is convenient to relate the reverberation time directly to the absorption coefficient. Normal mode theory uses the specific acoustic admittance, which is the reciprocal of the specific acoustic impedance; hence it is linked to the absorption coefficient. From Eq. 1.76 we see that the absorption coefficient is dependent upon the angle of incidence and the specific acoustic impedance. Depending on the mode, the waves will be incident upon the room surfaces at different angles. For axial modes the waves always impinge upon two opposite surfaces at an angle of incidence that is normal to these surfaces. For oblique and tangential modes, the angle of incidence varies depending upon the mode and the room boundary upon which the waves are impinging. For simplicity, the angle dependence is ignored in the following examples and a single value for the specific acoustic admittance is used for all of the room surfaces. This still allows us to see the general effect of the modes on the decay curves; we simply acknowledge that the situation is more complex in reality.

Equation 1.101 can now be used to calculate the decay curve for each of the M individual modes within a frequency band. This can be compared with the decay curve for the frequency band itself. It is assumed that the sound source is fed with a white noise signal with an rms volume velocity spectral density, Q_{sd}. It is convenient to set the sound pressure level at $t=0$ to a level of 60 dB for the frequency band, hence we need to establish the level at $t=0$ for each of the M modes in that band. For the mth mode, the spatial average mean-square sound pressure at time $t=0$ is (Bodlund, 1980)

$$\langle p_m^2 \rangle_s = \frac{\rho_0^2 c_0^4 Q_{sd}^2 \pi}{12 V^2 \ln 10} T_m \qquad (1.102)$$

for which it is assumed that the specific acoustic admittance for the room surfaces is a real value, much less than unity, and uniform over all the surfaces.

Using Eq. 1.102 to set the level for each mode at $t=0$, the sound pressure level, $L_{p,m}(t)$, in decibels, for the mth decaying mode in a frequency band is

$$L_{p,m}(t) = 10 \lg \left(\frac{p_m(t)}{p_m(0)} \right)^2 = 60 - \frac{60t}{T_m} + 10 \lg \left(\frac{\langle p_m^2 \rangle_s}{\sum_{m=1}^M \langle p_m^2 \rangle_s} \right) = 60 - \frac{60t}{T_m} + 10 \lg \left(\frac{T_m}{\sum_{m=1}^M T_m} \right)$$

$$(1.103)$$

The sound pressure level, $L_p(t)$, in decibels for the frequency band can then be calculated from the energetic sum of the decay curves for the individual modes in the band,

$$L_p(t) = 10 \lg \left(\sum_{m=1}^M 10^{L_{p,m}(t)/10} \right) \qquad (1.104)$$

We can now look at decays in rooms with the same volume (50 m^3) but different L_x, L_y, and L_z dimensions. The specific acoustic admittance for all room surfaces is assumed to be real, independent of frequency, and independent of the angle of incidence. Although it does not correspond to any particular material commonly used for walls and floors, a value of $\beta_{a,s} = 0.01$ is used to give reverberation times less than 2 s. Note that smooth, heavy concrete walls and floors would usually have much smaller values, which would lead to longer reverberation times. In contrast to measured decay curves, the decay curves from this model can be evaluated from

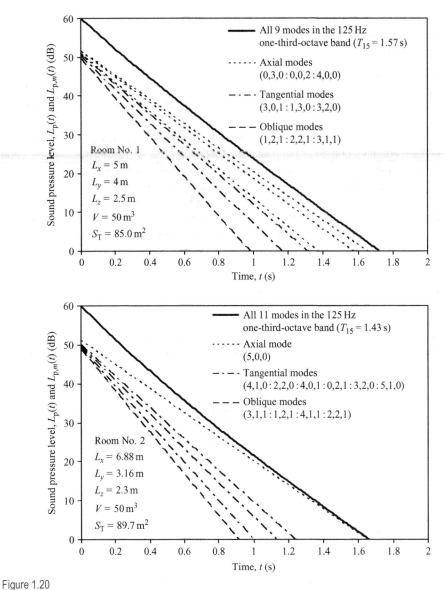

Figure 1.20

Decay curves for box-shaped rooms No. 1 and No. 2. Each room has a volume of 50 m³ but with different L_x, L_y, and L_z dimensions. The curves are shown for individual modes in the 125 Hz one-third-octave-band along with the resulting curve for that band.

$t = 0$ to calculate the reverberation time. This is because the model does not include the effect of direct sound from the sound source, and, unlike a measurement, there is no distortion of the initial part of the decay curve from the signal processing.

As sound fields in practice can rarely be considered as diffuse in the low-frequency range we will initially look at the 125 Hz one-third-octave-band. The decay curves from two different 50 m³ rooms (No. 1 and No. 2) are shown in Fig. 1.20 for the frequency band and the individual modes in this band. For each individual mode the decay curves are straight lines. In contrast, the resulting decay curve for the frequency band is a curve; this is easier to see if a straight edge is placed against it. The decay curves for these two rooms are different. However, in this particular example the initial part of the decay curve has a similar slope in both rooms, hence the reverberation time, T_{15}, is similar too.

The axial modes tend to have longer reverberation times than the tangential modes, which, in turn, tend to be higher than for the oblique modes. For this reason the energy of individual axial modes at $t = 0$ is slightly higher than individual tangential or oblique modes. We recall that the decay curve for the frequency band is calculated from the energetic sum of the decays for the individual modes (Eq. 1.104). Therefore, it is only in the early part of the decay, say within the initial 20 or 30 dB, that the majority of the different room modes play a role in determining the decay curve of the frequency band. In the later part of the decay, the decay curve for the frequency band is primarily determined by the modes with the longest reverberation times. These are always axial modes. This is clearly seen with room No. 2 where there is only a single axial mode, $f_{5,0,0}$ in the frequency band. In the late part of the decay curve, the slope is primarily determined by this one axial mode. Hence for typical rooms in the low-frequency range, the axial modes play an important role in determining the decay curve of the frequency band. This will be more apparent when one room dimension is significantly longer than the other two dimensions. In this situation, the decay curve of the frequency band is predominantly determined by the axial mode(s) with wave propagation along the longest dimension. This allows some insight into which surfaces require low-frequency absorbers to reduce the reverberation time of the frequency band by reducing the reverberation time of specific modes.

In practice, the measured decay curve for the frequency band will fluctuate about the predicted straight line decay due to interaction between the modes causing beating. Also, in using a single value for the specific acoustic admittance we have effectively assumed a single absorption coefficient for all angles of incidence which is not appropriate for many common walls and floors. In the low-frequency range the absorption coefficient can be lower at normal incidence than at oblique incidence. This would make the curvature more distinct due to even longer reverberation times for the axial modes.

The model also allows us to compare the decay curve for a diffuse field with the decay curves for individual frequency bands as the band centre frequency increases. To do this we will choose the 100, 1000, and 5000 Hz one-third-octave-bands for a different 50 m^3 room (room No. 3). It is important to note that the assumption of purely specular reflection for the 1000 and 5000 Hz bands is unrealistic in practice as walls and floors are often slightly irregular with scattering objects near the room surfaces. However, it gives us a useful insight into the effects of different mode counts and the different blends of mode types in each frequency band. Figure 1.21a shows the decay curves for the three frequency bands. The curvature of the decay in the 100 Hz band is in marked contrast to the approximately straight decay of the 5000 Hz band. The reasons for this difference can be seen by grouping together the decay curves for each mode type (axial, tangential, or oblique) as shown in Fig. 1.21b. In the 100 Hz band the initial 20 dB decay is determined by all three mode types. However, the later part of the decay curve is predominantly determined by the axial modes with minimal influence from the tangential and oblique modes. This is in contrast to the 1000 and 5000 Hz bands where the number of axial modes is small compared to the number of tangential or oblique modes; hence, there is only a minor influence from the axial modes on the decay curve for the frequency band. As the frequency increases, there are many more oblique modes than axial or tangential modes. As all oblique modes have the same reverberation time, and the decay curve for each individual mode is a straight line, it follows that the decay curve at high frequencies tends towards a straight line determined by the oblique modes. This is seen in Fig. 1.21b where the decay curve for the 5000 Hz band is very similar to the grouped decay curve for the oblique modes.

For a room with uniform locally reacting surfaces, normal mode theory gives a useful insight into the reasons for curvature of the decay curves. However, it does not fully describe the degree of curvature that occurs in practice. This is partly due to the fact that walls and floors are not purely locally reacting; they can act as surfaces of extended reaction.

Locally reacting surfaces and surfaces of extended reaction: It is very convenient to be able to consider the reverberation time as independent of any interaction between the room modes and the structural modes of the walls and floors. So far it has been assumed that although the sound waves impinging upon a surface are absorbed, this absorption is a 'local matter' between the sound wave and the point on the surface from which it is reflected. From the point-of-view of an impinging sound wave there are two types of room surface that are responsible for absorption: locally reacting surfaces and surfaces of extended reaction (Morse and Ingard, 1968). These two types can be defined by referring to the normal acoustic surface impedance; this is the ratio of the complex sound pressure at the surface to the component of the complex sound particle velocity that is normal to the surface. If a wave impinges upon a point on the surface, a locally reacting surface is one where the particle velocity normal to the surface is only affected by the sound pressure at that point, and is unaffected by the pressure at adjacent points on the surface. In contrast, a surface of extended reaction is one where the particle velocity normal to the surface is affected by the pressure at adjacent points on the surface. Surfaces of extended reaction therefore include plates undergoing bending wave motion, and porous surfaces where the sound propagates inside the porous material in a direction parallel to the surface.

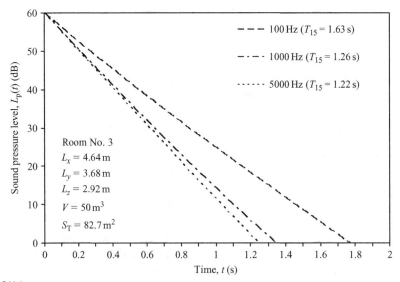

Figure 1.21(a)

Decay curves for box-shaped room No. 3 (100, 1000, and 5000 Hz one-third-octave-bands).

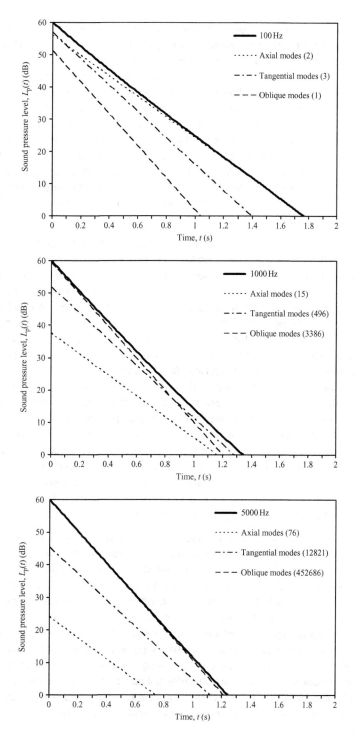

Figure 1.21(b)

Decay curves for the 100, 1000, and 5000 Hz one-third-octave-bands along with the grouped decay curve for each mode type. The number of modes corresponding to each mode type in the one-third-octave-band are shown in brackets.

For practical purposes, the assumption of locally reacting room surfaces is very useful in simplifying the calculation of reverberation time. In many rooms, small areas of rigid frame, porous, absorbent material are fixed to the room surfaces to reduce the reverberation time. Many of these absorbers can be considered as locally reacting. The assumption of locally reacting surfaces is often reasonable due to other factors that introduce uncertainty into the calculation; these are non-uniform distribution of absorption, application of laboratory measurements of absorption coefficients to rooms with non-diffuse fields, specific room geometry, and scattering from objects and surfaces in the room. Bare walls and floors undergoing bending wave motion are surfaces of extended reaction, responsible for transmitting sound to other parts of the building. In this situation the reverberation time is not only determined by the room modes, it is determined by the interaction between the room modes and the structural modes of the walls and floors (Pan and Bies, 1988). This blurs the boundary between the study of room acoustics and structure-borne sound.

For a room with locally reacting surfaces that have a frequency-independent value for the specific acoustic admittance, the normal mode model indicates that the group of oblique modes will have the same reverberation time, and each group of tangential modes with $p = 0$ or $q = 0$ or $r = 0$ have the same reverberation time. However, experimental evidence from a reverberation chamber with 280 mm thick concrete walls and floors indicates that there can be significant variation between the reverberation times of individual modes within these groups (Munro, 1982). This is primarily because the walls and floors are not locally reacting (Pan and Bies, 1988). It has been shown both theoretically and experimentally that the reverberation time in a room can be altered by changing the total loss factor and/or the modal density of its walls and floors (Pan and Bies, 1990). Experiments in the same reverberation chamber demonstrate that by increasing the total loss factor of a bending wave mode on one concrete wall (by wedging wooden blocks between this wall and another wall to increase the structural coupling losses), it is possible to change the reverberation time of an individual room mode (Pan and Bies, 1988). However, this needs to be kept in perspective when predicting reverberation times in rooms. There are other reasons why it is difficult to accurately predict reverberation times; mainly the existence of non-diffuse sound fields and the application of laboratory measurements of absorption or scattering coefficients to a specific situation in the field. The fact that walls and floors are not locally reacting is simply one more reason.

In practice, walls and floors are usually partly or completely covered with a locally reacting absorber, such as carpet on a heavy concrete floor. Therefore it is not always necessary for calculations to consider the effect of modal interaction; reasonable estimates can often be obtained by assigning an absorption coefficient to the areas of wall and floor that are not covered by the locally reacting absorber. This absorption coefficient may be based on measurements or empiricism. In many cases the wall or floor will act as both a surface of extended reaction and a locally reacting surface, although one of these may be more important than the other. For example, some masonry/concrete walls have highly porous surfaces. A reasonable estimate of the reverberation time can often be found by using a measured absorption coefficient and simply treating the wall as a locally reacting surface; it may be unnecessary to consider the fact that it also acts as a surface of extended reaction due to bending wave vibration. This allows calculation of the room reverberation time using absorption coefficients in equations such as Eq. 1.94 or 1.96.

Despite the fact that real walls and floors are not purely locally reacting, normal mode theory shows that the curved decay for a frequency band is due to the different reverberation

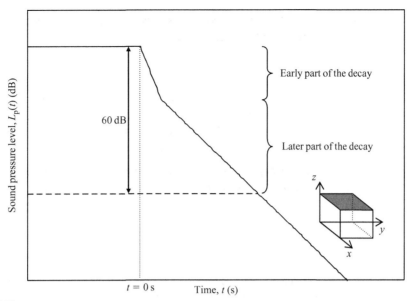

Figure 1.22

Example of a decay curve with a double slope; this can occur in rooms with a highly absorptive ceiling, but where the walls and the floor have relatively low absorption.

times for the modes within that band. Interaction between room modes and structural modes therefore results in a range of reverberation times for individual modes; hence there will still be curved decays in some frequency bands. In fact, measured data suggest that the range of reverberation times for individual modes is much larger than calculated from normal mode theory, resulting in decay curves with a greater degree of curvature (Bodlund, 1980).

1.2.6.3.3 Non-diffuse field: non-uniform distribution of absorption

Non-diffuse fields also occur due to non-uniform distribution of absorption over the room surfaces; one common example occurs when there is a highly absorptive ceiling but the walls and the floor have relatively low absorption. In these situations the decay curve can also show curvature or a distinct double slope as illustrated in Fig. 1.22. Considering the different modes, it is possible to make a basic qualitative assessment of the reasons for this double slope. When the early part of the decay is predominantly determined by the oblique modes (as in the previous example for the 5000 Hz band) we can expect large numbers of these modes to be rapidly attenuated as they impinge upon the highly absorbent ceiling. This gives rise to the fast decay rate in the early part of the decay curve. However, some of the axial and tangential modes will only be reflected from the side walls which have low absorption. Hence we can expect these modes to have relatively long reverberation times and contribute to the late part of the decay, which compared to the early part, will have a much slower rate of decay. The main features of the decay curve can be predicted by dividing the modes into two groups (Nilsson, 2004). The first group contains modes where the waves propagate almost parallel to the ceiling (grazing waves). In the second group the modes propagate at angles that are oblique to the ceiling (non-grazing waves). Using this grouping, the non-grazing waves determine the early part of the decay curve and the grazing waves determine the late part of the curve. Other prediction formulae for rooms with non-uniform distribution of absorption can be found in work by Fitzroy (1959), Arau-Puchades (1988), and Neubauer (2001).

1.2.6.4 Internal loss factor

In later chapters we will look at predicting sound transmission between two rooms using Statistical Energy Analysis (SEA). It will then become useful to denote the different rooms using a subscript. The internal loss factor is usually denoted as η_{int} but here we will start using the notation, η_{ii}, for the internal loss factor of a room subsystem, i, in an SEA model.

Internal losses describe the conversion of sound energy into heat by absorption; if this is the only process that is described by the total absorption area, the internal loss factor is given by

$$\eta_{int} = \eta_{ii} = \frac{c_0 A_T}{8\pi f V} \tag{1.105}$$

1.2.6.5 Coupling loss factor

The coupling loss factor, η_{ij}, describes resonant transmission between a room (subsystem i) and a plate (subsystem j) that faces into the room. This is described in Section 4.3.1.1.

1.2.6.6 Total loss factor

The total loss factor, η_i, of a subsystem, i, is the sum of its internal loss factor and all the coupling loss factors from that subsystem,

$$\eta_i = \eta_{ii} + \sum_{j=1}^{J} \eta_{ij} \quad (i \neq j) \tag{1.106}$$

and is related to the reverberation time by

$$\eta_i = \frac{6 \ln 10}{2\pi f T} = \frac{2.2}{fT} \tag{1.107}$$

For most rooms, the sum of the coupling loss factors is much smaller than the internal loss factor, and the latter provides a reasonable estimate of the total loss factor. For a room where the total absorption area, A_T, is calculated from the measured reverberation time (and hence includes both internal and coupling losses), the total loss factor can be written as

$$\eta_i = \frac{c_0 A_T}{8\pi f V_i} \tag{1.108}$$

1.2.6.7 Modal overlap factor

The modal overlap factor, M, describes the degree of overlap in the modal response. It is defined as the ratio of the 3 dB modal bandwidth, $\Delta f_{3\,dB}$, to the average frequency spacing between mode frequencies, Δf, and is calculated from

$$M = \frac{\Delta f_{3\,dB}}{\Delta f} = f \eta n \tag{1.109}$$

where $\Delta f_{3\,dB}$ (which is also referred to as the half-power bandwidth) is equal to the frequency spacing between the two points on the modal response where the level is 3 dB lower than the peak level, and η is the loss factor.

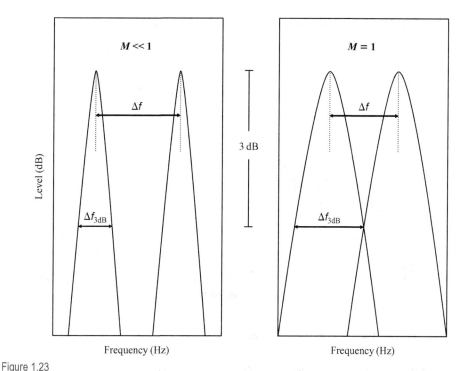

Figure 1.23

Modal response for two adjacent modes with modal overlap factors, $M \ll 1$ and $M = 1$.

An example of the response due to two adjacent modes with a frequency spacing, Δf is shown in Fig. 1.23. This idealized response could represent either the sound pressure level in a room or the velocity level on a wall. When $M \ll 1$ there is no overlap of the 3 dB bandwidths and there can be deep troughs between the two modes. When $M = 1$, the modal responses overlap at the point where the levels are 3 dB below the peak level. As $M \gg 1$ the response becomes increasingly uniform due to the absence of deep troughs.

Figure 1.24 shows the modal overlap factor for different room volumes and reverberation times. The modal overlap factor is often less than unity in the low-frequency range.

A cut-off frequency, f_M, that identifies the lowest frequency associated with a minimum value of the modal overlap factor can be found by substituting Eqs 1.59 and 1.107 in Eq. 1.109 to give

$$f_M = \sqrt{\frac{MTc_0^3}{8.8\pi V}}$$

(1.110)

With a modal overlap factor of three, this cut-off frequency is often referred to as the Schroeder cut-off frequency, f_S, and quoted as (Schroeder, 1962)

$$f_S = 2000\sqrt{\frac{T}{V}}$$

(1.111)

For a room of fixed volume, long reverberation times mean that the damping loss factor is low; hence the modal overlap is also low which results in higher cut-off frequencies. Usually we want to calculate the cut-off frequency from measured reverberation times. When these are approximately constant over the building acoustics frequency range, the average reverberation

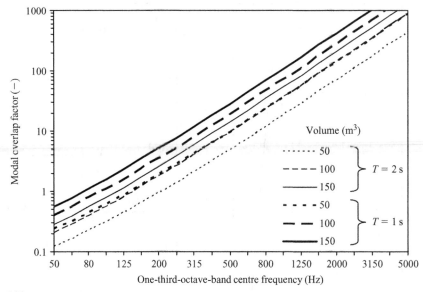

Figure 1.24

Modal overlap factors for different room volumes and reverberation times.

time can be used. Otherwise, an initial estimate for the cut-off frequency can be found from the arithmetic average of the reverberation time over a large part of the frequency range. This will then identify a more relevant part of the frequency range over which the reverberation times can be averaged to refine the estimate.

Figure 1.25 shows the Schroeder cut-off frequency for a range of room volumes and reverberation times. For room volumes less than 60 m³ with reverberation times between 0.5 and 1 s, the lowest cut-off frequency will be in the 200 Hz one-third-octave-band, and often in higher frequency bands.

1.2.7   Spatial variation in sound pressure levels

In the measurement and prediction of sound insulation it is almost always the temporal and spatial average sound pressure level in each room that is of interest rather than the level at a particular point in space at a particular point in time. For this reason, measurement procedures require time-averaged sound pressure levels to be measured at a number of different points in a room and averaged. However, an average value is only useful in the analysis of sound insulation measurements and predictions if we know what it represents. It is therefore necessary to look at the spatial variation of time-averaged sound pressure levels both in theory and in practice.

We start with the theory for the sound field near room boundaries, and then move on to discuss the sound pressure level distribution in a room due to individual modes and in the idealized diffuse sound field. We then consider practical situations where there is a direct sound field near the loudspeaker, and spaces in which the sound pressure level decreases with distance. This allows us to interpret some example measurements of sound fields in frequency bands, and to see the benefit in using statistical descriptions for the spatial variation in the sound pressure level.

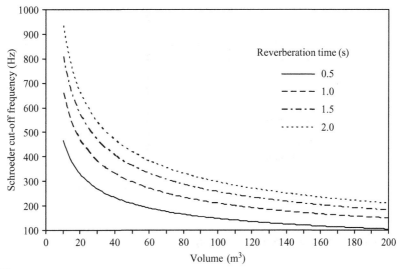

Figure 1.25

Schroeder cut-off frequency for different room volumes and reverberation times.

1.2.7.1 Sound fields near room boundaries

In a diffuse field the phase relationship between all waves passing through a single point in space is random. However, near a room boundary there will be a non-random phase relationship between the incident wave and the reflected wave. When a sound wave is incident upon a room boundary the reflected wave combines with the incident wave to give an interference pattern in the vicinity of this boundary (Waterhouse, 1955).

Initially it is assumed that all the room boundaries are perfectly reflecting and rigid. Under this assumption the sound field close to any surface, edge, or corner in a room can be compared to the sound field far from the surface. This approach is often used to quantify the total sound energy stored in a room where it can be assumed that there is a diffuse field in the central zone of the room. In practice there are a wide range of acoustic surface impedances for walls and floors in buildings and it is necessary to be aware of their effect on the sound field.

1.2.7.1.1 Perfectly reflecting rigid boundaries

To gain an insight into the sound field near a wall or floor in a room, we start with the situation where a harmonic plane wave is incident upon a surface, such as a wall or floor, at an angle that is perpendicular to the surface, i.e. at normal incidence. The surface is positioned at $x = 0$ in the yz plane (see Fig. 1.26). It is assumed that the surface is large compared to the wavelength and that there are no other surfaces that affect the sound field. The incident wave, $\hat{p}_+ \exp(-ikx)$ travels from $-\infty$ towards $x = 0$ where it is reflected from the surface to give the reflected wave, $\hat{p}_- \exp(ikx)$. This conveniently means that the exponential terms equal unity at the surface where $x = 0$. The resulting sound pressure due to the incident and reflected waves is

$$p(x, t) = [\hat{p}_+ \exp(-ikx) + \hat{p}_- \exp(ikx)] \exp(i\omega t) \tag{1.112}$$

where $\hat{p}_+$ is an arbitrary constant for the incident wave. The constant for the reflected wave, $\hat{p}_-$, is related to $\hat{p}_+$ by the reflection coefficient of the surface, R (Eq. 1.67), where $\hat{p}_- = R\hat{p}_+$.

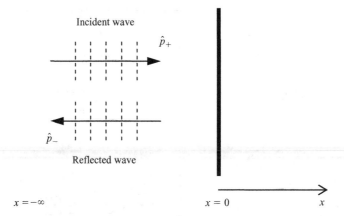

Figure 1.26

Plane waves incident upon, and reflected from a room boundary.

By squaring the real part of Eq. 1.112 and taking the time-average, the mean-square sound pressure at a distance, x, from the surface in the negative x-direction is

$$\langle p^2 \rangle_t = \hat{p}_+^2 (0.5 + 0.5|R|^2 + |R| \cos(2kx + \gamma)) \tag{1.113}$$

For perfect reflection from a rigid surface there is no phase shift (i.e. $\gamma = 0$), so $R = |R| = 1$ and Eq. 1.113 becomes

$$\langle p^2 \rangle_t = \hat{p}_+^2 (1 + \cos(2kx)) \tag{1.114}$$

At $x = 0$, the sound pressure for the incident and reflected waves is in phase, so $\langle p^2 \rangle_t = 2\hat{p}_+^2$. At $x = \lambda/4$ the incident and reflected waves are out of phase with each other and the mean-square pressure is zero.

The particle velocities for the incident and reflected waves are found from the sound pressure terms using Eq. 1.18. This gives $\frac{\hat{p}_+}{\rho_0 c_0} \exp(-ikx)$ for the incident wave and $-\frac{\hat{p}_-}{\rho_0 c_0} \exp(ikx)$ for the reflected wave that travels in the opposite direction. The resulting particle velocity is,

$$u(x,t) = \frac{1}{\rho_0 c_0} [\hat{p}_+ \exp(-ikx) - \hat{p}_- \exp(ikx)] \exp(i\omega t) \tag{1.115}$$

At $x = 0$, the particle velocity is zero when the surface is perfectly reflecting and rigid (i.e. $\hat{p}_- = \hat{p}_+$).

In practice, sound waves in a room are incident from many different directions upon a reflecting surface, so the next step is to consider a single wave that is incident at an angle, θ, to the x-axis. For the reflected wave we will assume specular reflection from the surface. For an oblique angle of incidence, the Cartesian coordinate system is rotated by θ so that x in Eq. 1.112 is replaced by x', where $x' = x \cos\theta + y \sin\theta$. This gives

$$p(x,y,t) = [\hat{p}_+ \exp(-ik(x \cos\theta + y \sin\theta)) + \hat{p}_- \exp(-ik(-x \cos\theta + y \sin\theta))] \exp(i\omega t) \tag{1.116}$$

Hence for oblique incidence, the time-averaged mean-square pressure at a distance, x, from a perfectly reflecting, rigid surface is

$$\langle p^2 \rangle_t = \hat{p}_+^2 (1 + \cos(2kx \cos\theta)) \tag{1.117}$$

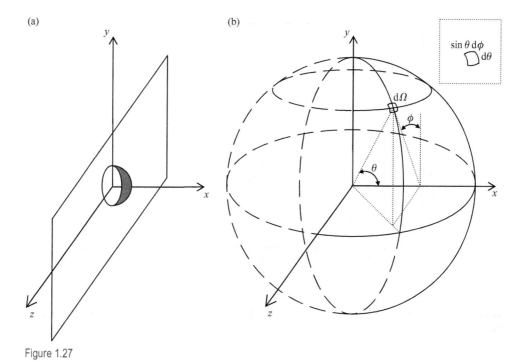

Figure 1.27

(a) Sound waves incident from all possible angles upon a perfectly reflecting surface form a hemisphere around a small area on the surface. (b) Spherical coordinate system with the element of the solid angle used to average the mean-square pressure over the hemisphere.

We can now consider waves that are incident upon the surface from all possible angles. This forms a hemisphere that encloses a small area on the surface as shown in Fig. 1.27. The incident waves are assumed to be incoherent, i.e. to have random phase, therefore the mean-square pressure in Eq. 1.117 can be averaged over all possible angles of incidence using

$$\langle p^2 \rangle_t = \frac{1}{2\pi} \int_0^{2\pi} \int_0^{\pi/2} [\hat{p}_+^2(1 + \cos(2kx\cos\theta))] \sin\theta \, d\theta d\phi = \hat{p}_+^2 \left(1 + \frac{\sin(2kx)}{2kx} \right) \qquad (1.118)$$

where the spherical coordinate system is shown in Fig. 1.27 and the element of the solid angle, $d\Omega$, is $\sin\theta \, d\theta d\phi$.

From Eq. 1.118 the asymptotic value for the mean-square pressure at a distance far from the surface, $\langle p_\infty^2 \rangle_t$, is equal to $\hat{p}_+^2$. At this point it is convenient to change over from using negative x values for the distance and use positive values for the distance, d, along the x-axis. Hence the ratio of the mean-square pressure at a distance, d, from this surface, to the mean-square pressure at a point far away from the surface is (Waterhouse, 1955)

$$\frac{\langle p^2 \rangle_t}{\langle p_\infty^2 \rangle_t} = 1 + \frac{\sin(2kd)}{2kd} \qquad (1.119)$$

This is plotted in Fig. 1.28 as the sound pressure level difference in decibels against $2kd$. The smallest distance at which there is no difference between the level near the surface and the level far away from the surface occurs at $2kd = \pi$, where $d = \lambda/4$. The largest level differences

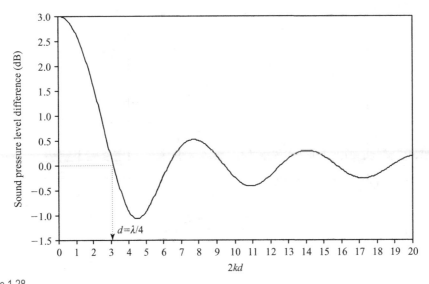

Figure 1.28

Sound pressure level difference between a point at a distance, d, from a perfectly reflecting surface and a point far away from the surface.

(magnitude) occur at distances less than $\lambda/4$ from the surface. At high frequencies and/or large distances from the surface the sound pressure level difference tends towards 0 dB. If we start at a distance $d = \lambda/4$ from the surface and move towards the surface the level difference tends towards 3 dB. Equation 1.119 applies to single frequencies rather than frequency bands, although it also gives reasonable estimates for one-third-octave-bands and octave-bands of white noise at distances up to $\lambda/4$ from the surface (Waterhouse, 1955). To illustrate its practical application, the level difference against frequency at different distances from the reflecting surface is shown in Fig. 1.29.

When measuring the reverberant sound pressure level in a room it is necessary to avoid measuring near the room boundaries in these interference patterns. The requirements for the minimum distance between a microphone and the room boundaries depend on the level of accuracy required, and the practical aspect of finding sufficient measurement positions in small rooms. Measurements are usually carried out simultaneously in all the frequency bands over the building acoustics frequency range. For this reason it is common to quote the minimum distance as a fixed value based upon the lowest frequency of interest; rather than quoting a fraction of a wavelength. In the Standards for field and laboratory sound insulation measurements the minimum measurement distances from the room boundaries are quoted as 0.5, 0.7, and 1.2 m (ISO 140 Parts 3, 4, 5, 6, & 7). Equation 1.119 can be used to estimate the level difference at these distances from walls or floors. At 50 Hz the level difference is 1.4 dB for a distance of 1.2 m. At 100 Hz a level difference of 1.8 dB occurs for a distance of 0.5 m, and 0.8 dB for 0.7 m.

The interference patterns at the edges and the corners also need consideration. The ratio for the mean-square sound pressure at a distance, d, from an edge or a corner due to both the incident and reflected waves, relative to the mean-square sound pressure at a point far away from the edge or corner is given by Waterhouse (1955).

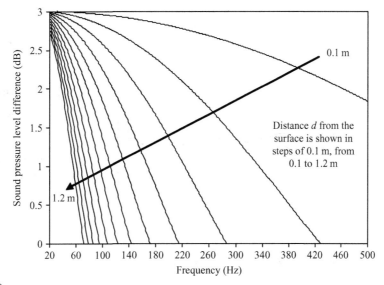

Figure 1.29

Sound pressure level difference between a point at a distance, d, from a perfectly reflecting surface and a point far away. Note that the curves have been truncated at the frequency where $d = \lambda/4$. These single frequency values are also applicable to white noise in one-third-octave or octave-bands.

For an edge that lies along the x-axis, the mean-square sound pressure ratio at a point (y, z) is

$$\frac{\langle p^2 \rangle_t}{\langle p_\infty^2 \rangle_t} = 1 + j_0(2ky) + j_0(2kz) + j_0(2kd_x) \tag{1.120}$$

where $j_0(a) = \sin(a)/a$ and $d_x^2 = y^2 + z^2$.

For a corner positioned at the origin of the Cartesian coordinates the mean-square sound pressure ratio at a point (x, y, z) is

$$\frac{\langle p^2 \rangle_t}{\langle p_\infty^2 \rangle_t} = 1 + j_0(2kx) + j_0(2ky) + j_0(2kz) + j_0(2kd_x) + j_0(2kd_y) + j_0(2kd_z) + j_0(2kd) \tag{1.121}$$

where $d_y^2 = x^2 + z^2$, $d_z^2 = x^2 + y^2$ and $d^2 = x^2 + y^2 + z^2$.

Figure 1.30 allows comparison of the sound pressure level difference for a surface, edge, and a corner. To create this particular example, d is used to represent different distances from the surface, edge, or corner: namely, the distance perpendicular to the surface along the x-axis, the distance from the edge along the line $y = z$, and the distance from the corner along the line $x = y = z$. At the boundary position where $d = 0$, the level differences are 3, 6, and 9 dB for the surface, edge, and corner respectively. Image sources can be used to visualize this finding. Figure 1.31 shows the actual source for a plane wave front near these boundaries along with the image sources. For the surface, edge, and corner there are 1, 3, and 7 reflected waves respectively; this gives the total number of sources as 2, 4, and 8 respectively. The image sources have the same amplitude and phase as the actual source, therefore the sound pressure from the actual source and the image sources is in phase at the boundary position $(d = 0)$. Hence the level differences correspond to 10 times the logarithm (base 10) of the total number of sources.

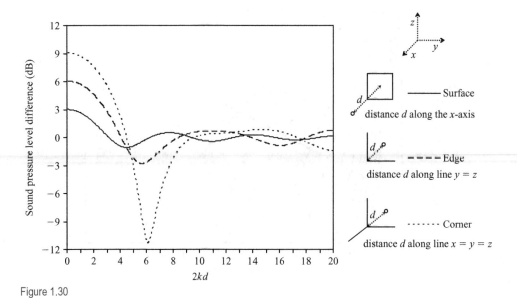

Figure 1.30

Sound pressure level difference between a point at a distance, *d*, from a surface, edge, and corner relative to a point far away.

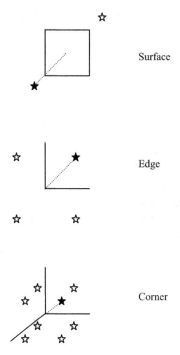

Figure 1.31

Source (★) and image sources (☆) for a plane wave front incident upon a surface, edge, and corner.

The level differences due to interference patterns near edges and corners tend to be larger in magnitude than with a surface and extend to greater distances. However, we will soon look at calculating the total energy stored in a room where it is necessary to account for the energy stored in the interference patterns of surfaces, edges, and corners. When the sound field in the central zone of the room is reasonably diffuse it is found that the energy stored in edge and corner zones is relatively small compared to the energy stored near room surfaces. This is because the surfaces account for large areas in most rooms (Waterhouse, 1955).

1.2.7.1.2 Other boundary conditions

Up till now we have assumed that the room boundaries are perfectly reflecting and rigid, i.e. the normal acoustic surface impedance is infinite. Therefore at the surface, the particle velocity normal to these boundaries is zero. Many walls and floors in buildings have low-absorption coefficients in the low-frequency range where interference patterns in rooms are important; so the assumption that they are perfectly reflecting is reasonable. However, habitable rooms almost always have fairly absorptive surfaces to provide suitable acoustics for the occupants of the building so we need to consider the effect of absorption. In addition we know that real room surfaces are not rigid because they are set into vibration by impinging sound waves. In reality the sound waves that impinge upon the walls and floors cause them to vibrate; hence the particle velocity at, and normal to the room surfaces must be the same as the surface vibration of the wall or floor, and not zero. When we focus on sound transmission, the vibration of these room surfaces becomes particularly important. Room surfaces range from a single sheet of 12.5 mm plasterboard on a timber frame, to a few hundred millimetres of solid concrete; all of these surfaces have finite values for the acoustic surface impedance.

If we restrict our attention to sound waves impinging upon a surface at normal incidence, then the effect of absorptive non-rigid surfaces can be assessed by using Eq. 1.113 to calculate the mean-square pressure. This requires knowledge of the reflection coefficient which can be calculated from the specific acoustic impedance using Eq. 1.74, and can be related to the absorption coefficient using Eq. 1.75.

In order to assess the sound field near a surface it is useful to reference the mean-square pressure from Eq. 1.113 to the mean-square pressure of the free-field incident wave, $\langle p_+^2 \rangle_t$, where

$$\langle p_+^2 \rangle_t = \frac{\hat{p}_+^2}{2} \tag{1.122}$$

The sound field can then be shown using the sound pressure level difference, $10 \lg \left(\langle p^2 \rangle_t / \langle p_+^2 \rangle_t \right)$, where 0 dB corresponds to the level of the free-field incident wave.

Figure 1.32 shows examples of the sound pressure level difference for normal incidence as a function of $2kx$ in front of a surface at $x = 0$. Note that the distance, x, is in the negative x-direction. These examples use a range of values for the specific acoustic impedance to represent different surfaces. For a rigid surface the specific acoustic impedance is infinite. Real values for the specific acoustic impedance are chosen to show a range of absorption coefficients up to unity. The complex value, $1 + 8i$, is used to represent a single sheet of 12.5 mm plasterboard ($10.8 \, \text{kg/m}^2$) at 50 Hz; this plate has a low surface density and is used to provide contrast to the assumption of a rigid surface. The complex value, $1 - 2i$, can occur at a single frequency when a porous material is placed in front of a thick heavy wall to provide absorption in the room.

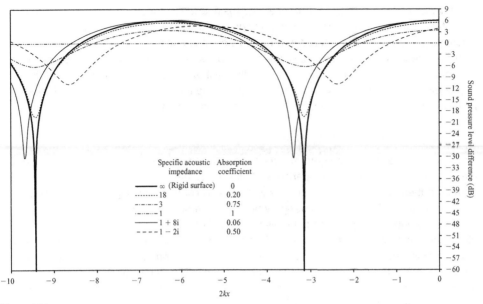

Figure 1.32

Sound field in front of a surface formed by the incident and reflected waves for normal incidence. Surfaces with different specific acoustic impedances are positioned at $x = 0$.

When the specific acoustic impedance is unity then the absorption coefficient is also unity and the incident wave is completely absorbed at the surface. Therefore the sound field in front of the surface only comprises the incident wave, and the sound pressure level difference is 0 dB for all values of $2kx$. In contrast, at the rigid surface there is a perfect reflection so that there is pressure doubling, a level difference of 6 dB. As we move away from the surface, there are sharp interference minima (troughs) where the incident and reflected waves are exactly out of phase with each other. There are also interference maxima with peak values of 6 dB where the waves are in phase with each other. The minima occur when $2kx = -(2n-1)\pi$ where $n = 1, 2, 3$, etc.

When the specific acoustic impedance has real or complex values, resulting in absorption coefficients between 0 and 1, the depths of the minima and the height of the maxima are significantly reduced in comparison to the rigid surface. Also, in comparison to surfaces with real or infinite impedance, complex impedances can significantly change the value of $2kx$ at which the minima and maxima occur. In the next section we will look at the modal sound field by assuming perfectly reflecting and rigid boundaries; hence the features we have seen here for real boundaries will be of relevance again.

1.2.7.2 Sound field associated with a single mode

Before we look at sound fields where there are many modes, it is instructive to look at the sound field associated with an individual mode. We will use the box-shaped room with perfectly reflecting and rigid boundaries and send a sinusoidal signal to a loudspeaker positioned in one of the corners.

For a sound source positioned at x_s, y_s, z_s, the mean-square sound pressure level at a receiver point x, y, z that is associated with mode, $f_{p,q,r}$, is calculated using normal mode theory (Morse and Ingard, 1968)

$$\langle p_{p,q,r}^2(x,y,z,x_s,y_s,z_s)\rangle_t = \left| \frac{\omega \rho_0 c_0^2 Q_{rms} \psi_{p,q,r}(x,y,z)\psi_{p,q,r}(x_s,y_s,z_s)}{V \Lambda_{p,q,r} \sqrt{4\omega_{p,q,r}^2 \zeta_{p,q,r}^2 + (\omega^2 - \omega_{p,q,r}^2)^2}} \right|^2 \qquad (1.123)$$

where ω is the frequency of the sinusoidal signal, Q_{rms} is the rms volume velocity of the source, $\zeta_{p,q,r}$ is the damping constant, and $\Lambda_{p,q,r} = 1/(\varepsilon_p \varepsilon_q \varepsilon_r)$ for which ε_p, ε_q, and ε_r have already been defined next to Eq. 1.99.

The damping constant, $\zeta_{p,q,r}$, can be linked back to δ_m that was previously used to describe the reverberant decay of an individual mode in Eq. 1.99. In terms of the damping constant, the decay of the mean-square sound pressure for mode, $f_{p,q,r}$, is

$$p^2(t) = p^2(0)\exp(-2\zeta_{p,q,r}t) \qquad (1.124)$$

We can now describe the damping constant in terms of the reverberation time or loss factor. From Eq. 1.101 the damping constant is related to the reverberation time, $T_{p,q,r}$, for an individual mode by

$$\zeta_{p,q,r} = \frac{3\ln 10}{T_{p,q,r}} \qquad (1.125)$$

which is related to the loss factor for an individual mode, $\eta_{p,q,r}$, using

$$\eta_{p,q,r} = \frac{6\ln 10}{2\pi f T_{p,q,r}} = \frac{\zeta_{p,q,r}}{\pi f} \qquad (1.126)$$

From Eqs 1.51 and 1.52 the local mode shape (also called an eigenfunction), $\psi_{p,q,r}$, that describes the sound pressure distribution in space for the receiver position is

$$\psi_{p,q,r}(x,y,z) = \cos\left(\frac{p\pi x}{L_x}\right)\cos\left(\frac{q\pi y}{L_y}\right)\cos\left(\frac{r\pi z}{L_z}\right) \qquad (1.127)$$

and for the source position is

$$\psi_{p,q,r}(x_s,y_s,z_s) = \cos\left(\frac{p\pi x_s}{L_x}\right)\cos\left(\frac{q\pi y_s}{L_y}\right)\cos\left(\frac{r\pi z_s}{L_z}\right) \qquad (1.128)$$

We will soon look at how different source positions affect the excitation of individual modes. For the moment it is only necessary to ensure that we can excite any mode; hence the source needs to be positioned at any one of the corners, for example at 0,0,0. When the source is at a corner, $|\psi_{p,q,r}(x_s,y_s,z_s)| = 1$ for any mode. This allows us to focus on the way that $\psi_{p,q,r}(x,y,z)$ affects the spatial distribution of the mean-square sound pressure.

The maximum value that $|\psi_{p,q,r}(x,y,z)|$ can take is 1; hence for an individual mode, the maximum mean-square sound pressure occurs at receiver positions where $|\psi_{p,q,r}(x,y,z)| = 1$. These maximum values occur at positions referred to as anti-nodes and their position in the room depends upon the individual mode, $f_{p,q,r}$. However, when the receiver is positioned at any of the eight corners of the box-shaped room, then $|\psi_{p,q,r}(x,y,z)| = 1$ for all modes. Hence, the corner of a room is an ideal point to detect which modes have been excited.

For any individual mode in a box-shaped room, the sound field on any of the three orthogonal planes forming the room is symmetrical about the lines perpendicular to the axes.

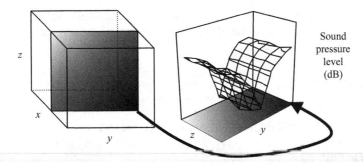

Figure 1.33

Illustration showing how the sound pressure levels from each plane in the box-shaped room is displayed on each three-dimensional surface plot. In this example it is the yz plane midway along the x dimension, L_x. Each plane is shown with a graduated shaded area so that corners with the darkest and the lightest shading can be used as reference points to identify sound pressure levels at different positions on the plane. If more than one plane is shown, then all these planes have the same sound pressure level distribution.

When at least one of the cosine terms in $\psi_{p,q,r}(x,y,z)$ is 0, the sound pressure is also 0, so there will be planes of zero pressure perpendicular to the x, y, or z-axes. In a box-shaped room these are referred to as nodal planes; these exist where $x = nL_x/2p$, $y = nL_y/2q$ and $z = nL_z/2r$ for $n = 1, 3, 5$, etc. For any mode, $f_{p,q,r}$, there will be p nodal planes perpendicular to the x-axis, q nodal planes perpendicular to the y-axis, and r nodal planes perpendicular to the z-axis.

For rooms with volumes less than 30 m³, there will only be one mode or a few modes in individual one-third-octave-bands between 50 and 100 Hz. These modes are usually the axial modes $f_{1,0,0}$, $f_{0,1,0}$, $f_{0,0,1}$, $f_{2,0,0}$, $f_{0,2,0}$, the tangential modes $f_{1,1,0}$, $f_{0,1,1}$, $f_{1,0,1}$ and the oblique mode $f_{1,1,1}$.

The graphs in this section can be interpreted with reference to Fig. 1.33. Examples of the sound pressure level distribution are shown for axial modes $f_{1,0,0}$ and $f_{0,0,1}$ (Fig. 1.34), the tangential mode, $f_{1,1,0}$ (Fig. 1.35), and the oblique mode $f_{1,1,1}$ (Fig. 1.36). To allow a practical interpretation of the sound field, the decibel scale has been used. However the use of decibels is not ideal because the mean-square pressure is zero on the nodal planes. In reality there will not be zero mean-square sound pressure for two reasons. Firstly, there will always be some background noise in the measurement, and secondly, not all real surfaces are perfectly reflecting and rigid. For the latter reason the surfaces will absorb some of the incident sound and there will not be perfect cancellation along nodal planes, i.e. there will not be pure standing waves. This was previously seen in Section 1.2.7.1.2 when we looked at the sound field near room boundaries where the specific acoustic impedance of the surfaces had complex or finite real values. The maximum level for each mode has therefore been normalized to 0 dB, and we will assume that the background noise level is 60 dB below this maximum level; so levels of −60 dB on these graphs represent the nodal planes with zero mean-square pressure.

The main features of the modal sound field are the large spatial variations in the sound pressure level. The highest levels occur at the room boundaries, with the nodal planes sited away from these boundaries.

This model of the sound field gives us a basic insight into the sound field in a room. However by assuming that the room boundaries are perfectly rigid we have not considered the interaction between the sound in the room and the vibration of the walls and floors. In addition we have not yet considered the situation where frequency bands 'contain' zero, one, or more modes.

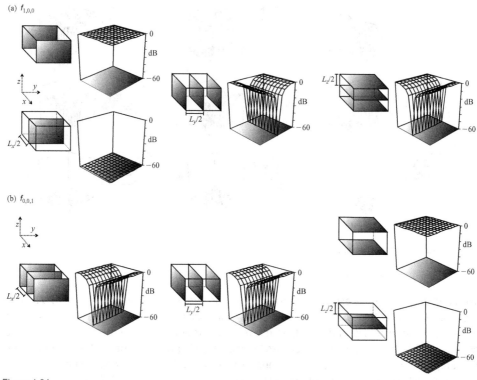

Figure 1.34

Sound pressure level distribution for axial modes: (a) $f_{1,0,0}$ and (b) $f_{0,0,1}$.

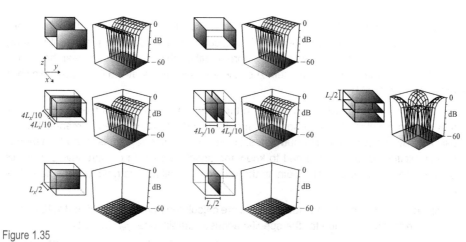

Figure 1.35

Sound pressure level distribution for the tangential mode, $f_{1,1,0}$.

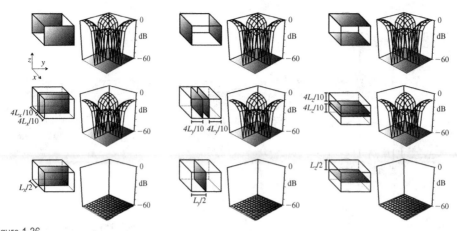

Figure 1.36

Sound pressure level distribution for the oblique mode $f_{1,1,1}$.

1.2.7.3 Excitation of room modes

We need to consider the excitation of room modes in two different situations: one in a source room where the airborne sound insulation is being measured by using a loudspeaker to generate the sound field, and the other in a receiving room where the room surfaces radiate sound into the receiving room, which excites the receiving room modes. The latter situation primarily concerns the coupling between energy stored in walls, floors, or other spaces, and the energy stored in the receiving room modes and is discussed in Chapter 4. At this point, we just consider the effect of using a single loudspeaker at different positions in the source room, and how those different positions affect the excitation of room modes in the low-frequency range. For individual frequencies the mean-square sound pressure due to each room mode can be calculated using Eq. 1.123 and summed to give the overall mean-square sound pressure. This is similar to carrying out a swept-sine measurement with a constant amplitude signal from the loudspeaker. In practice the airborne sound insulation is usually measured using broad-band noise, but single frequencies gives a clearer understanding of the effect of different source positions.

Three different source positions are assessed: one near a corner, another at the mid-point along one wall, and another that is exactly in the centre of the room. In practice it is rarely possible to put the loudspeaker exactly at the corner, although we can place it near the corner; so we will assume that the acoustic centre of the loudspeaker is at 0.5,0.5,0.25 m which is nearest to the corner at 0,0,0.

The receiver position is at L_x,L_y,L_z, which is in the corner opposite the loudspeaker. A microphone would not be placed in a corner for standard sound insulation measurements; it would be positioned away from the room boundaries. However, we only want to assess which modes have been excited so we do not need to know the absolute sound pressure levels. For this reason the receiver position is at a point in the room where all modes have an anti-node (i.e. in a corner).

The damping constants for the individual modes are calculated from Eqs 1.125 and 1.101 using a frequency-independent value for the specific acoustic admittance ($\beta_{a,s} = 0.01$).

Predicted curves for the sound pressure level are shown in Fig. 1.37 for the same 50 m³ box-shaped room that was used to look at the mode count in one-third-octave-bands. The peaks in the sound pressure level are due to the room modes that have been excited. On each curve, the axial, tangential, and oblique modes have been plotted at their respective eigenfrequencies. When there are relatively few modes and the modes are all well-separated we see that if a mode frequency coincides with a peak in the sound pressure level curve, this indicates that this particular mode has been excited; if it coincides with a trough in the curve, then it has not been excited. If a mode frequency occurs on the curve between a trough and a peak, then either the mode has not been excited, or it has been excited but it has such a low sound pressure level that there is no discernible peak. This numerical experiment indicates the difficulty in using physical experiments to find the mode count where a microphone and a sound source are placed in a single position. The problem is that counting the peaks in the response is unlikely to identify all of the modes; this equally applies to vibration measurements used to identify structural modes of vibration.

When the source is near a corner there are many more peaks in the sound pressure level curve than when the source is at the centre of the room; hence many more modes are excited by a corner position than the central point. When the source is near a corner, all the modes below 100 Hz have been excited and are clustered at or near the peaks in the sound pressure level curve. In contrast, when the source is at the centre of the room, the first five modes have not been excited at all, and there are clusters of modes that have not been excited near the troughs of the sound pressure level curve. The modes that have not been excited have one or more nodal planes that cut through the source position at the centre of the room (e.g. see $f_{1,1,1}$ in Fig. 1.37). Similarly, when the source is mid-way along one wall, the modes that have not been excited also have one or more nodal planes that cut through the source position (e.g. see $f_{1,0,0}$ and $f_{1,1,1}$ in Fig. 1.37).

For field airborne sound insulation measurements in non-diffuse sound fields it is necessary to excite the majority of the modes in the source room. For this reason, loudspeaker positions near the corners are used in box-shaped rooms as well as in other shapes of room. In addition it is necessary to take average measurements from more than one source position. However, it must also be ensured that the direct sound from the loudspeaker does not cause significant excitation of the walls or floors compared to excitation by the reverberant sound field.

1.2.7.4 Diffuse and reverberant fields

The Schroeder cut-off frequency is sometimes used to estimate the lowest frequency above which the sound field can be considered to be diffuse. This is the frequency at which the modal overlap factor equals three; hence it identifies the lowest frequency above which the sound energy is relatively uniform in the central zone of a room.

In practice, a diffuse field cannot be realized throughout the entire room volume due to the fact that rooms are defined by the walls and floors that form the room boundaries. These boundaries give rise to interference patterns close to the surfaces, edges, and corners; hence the energy density is not uniform throughout the space. In addition these boundaries absorb sound (to varying degrees) so that there must be a net power flow from the sound source to the boundaries, whereas in a diffuse field the net power flow is zero. Under laboratory conditions it is possible to achieve a suitable degree of diffusivity through careful room design, by using diffusing elements in the room, and by defining measurement positions away from the room boundaries. Hence close approximations to diffuse fields in the building acoustics frequency range only tend to exist in the central zones of large reverberant chambers. Such chambers are carefully designed and validated for the laboratory measurement of absorption or sound

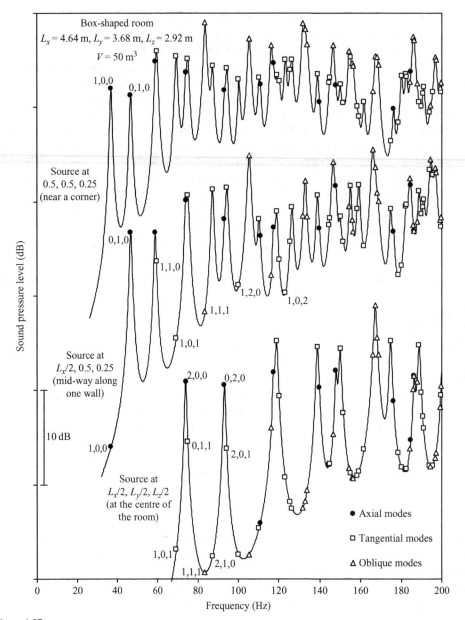

Figure 1.37

Excitation of room modes with three different source positions. Curves for the sound pressure level in the corner position (L_x, L_y, L_z) are shown along with the axial, tangential and oblique mode frequencies to assess which modes are, and which modes are not excited by the source position. Note that the curves have been offset from each other; this allows the relative levels along each individual curve to be assessed, but not the relative levels between different curves.

power levels. In laboratory measurements of airborne and impact sound insulation we also come across close approximations to diffuse sound fields in the source and receiving rooms.

In general it is better to avoid referring to a diffuse sound field in a room because we must add several caveats to any such statement. At frequencies above the lowest room mode it is simpler if we just refer to a reverberant field. This acknowledges the fact that the room response varies over the building acoustics frequency range due to the existence of one, a few, several, or many modes in the different frequency bands. It also serves as a reminder that interference patterns exist at the room boundaries, and that it is only in the central zone of reverberant rooms that sound fields resemble (to varying degrees) a diffuse field.

1.2.7.5 Energy density

The energy density, w, equals the energy per unit volume. For a sound field comprised of plane waves, Eq. 1.19 describes the sound intensity, i.e. the energy that flows through a unit surface area in unit time. Therefore, as a plane wave will travel a distance equal to c_0 in unit time, the energy density in any reverberant sound field comprised of plane waves (which includes diffuse or modal sound fields) is

$$w_r = \frac{I}{c_0} = \frac{\langle p^2 \rangle_{t,s}}{\rho_0 c_0^2} \qquad (1.129)$$

Hence the energy density is directly proportional to the temporal and spatial average mean-square sound pressure, $\langle p^2 \rangle_{t,s}$.

1.2.7.5.1 Diffuse field

As any kind of signal can be described by using many impulses, the steady-state energy density in a diffuse field can be derived by considering the impulse response of a room (Barron, 1973; Kuttruff, 1979). We assume that a room contains a point source that generates an impulse at time, $t = 0$. This point source generates spherical waves for which the intensity is inversely proportional to the square of the distance travelled (Eq. 1.26). At an arbitrary receiver position in the room, the sound intensity is the sum of the intensity from the direct path and the many indirect paths involving at least one reflection from a room boundary. For direct propagation from the source to the receiver, we define the intensity at the receiver to be I_0, after it has travelled the source–receiver distance, r_0. For each propagation path that involves at least one reflection, we can consider an image source that generates an identical impulse to the actual source at $t = 0$. At time, t, impulses from the actual source or an image source will have travelled a distance, $c_0 t$. For the image sources, a fraction of the sound intensity is absorbed upon each reflection from the room boundaries; hence from Eq. 1.92 the intensity is attenuated by the factor

$$\exp\left(\frac{c_0 t}{d_{\text{mfp}}} \ln(1 - \bar{\alpha}) \right) \qquad (1.130)$$

At an arbitrary receiver point, the intensity from an image source at time, t, is therefore given by

$$\frac{I_0 r_0^2}{(c_0 t)^2} \exp\left(\frac{c_0 t}{d_{\text{mfp}}} \ln(1 - \bar{\alpha}) \right) \qquad (1.131)$$

At time, t, we now need to determine how many impulses from image sources will arrive at the receiver in a small time interval, δt. This is equivalent to finding the number of reflections arriving during this time interval, and can be calculated from the temporal density of reflections (Eq. 1.49), using

$$\frac{4\pi c_0^3 t^2}{V} \delta t \qquad (1.132)$$

Therefore the total intensity arriving at the receiver after a specific time, t_1, can be found by integrating the intensity from all reflections according to

$$I = \int_{t_1}^{\infty} \frac{I_0 r_0^2}{(c_0 t)^2} \exp\left(\frac{c_0 t}{d_{mfp}} \ln(1 - \overline{\alpha}) \right) \frac{4\pi c_0^3 t^2}{V} \, dt \qquad (1.133)$$

and the energy density can be found using

$$w_r = \frac{I}{c_0} = \int_{t_1}^{\infty} \frac{W}{V} \exp\left(\frac{c_0 t}{d_{mfp}} \ln(1 - \overline{\alpha}) \right) dt \qquad (1.134)$$

where the sound power, W, for the spherical wave source is

$$W = 4\pi r_0^2 I_0 \qquad (1.135)$$

The choice of time, t_1, for the lower limit of the integral needs to consider the fact that the direct sound does not arrive at the receiver until $t = r_0/c_0$, and that the exponential decay in a diffuse field can only start after the first reflections arrive at the receiver (i.e. when $t > r_0/c_0$). Therefore if the lower limit of the integral is taken to be $t_1 = 0$, the energy density will include the time interval before the direct sound has arrived and the exponential decay has begun. Using $t_1 = 0$ to estimate the energy density in a diffuse field, w_r, gives the classical equation

$$w_r = \frac{4W}{c_0 A} \qquad (1.136)$$

where $A = -S_T \ln(1 - \overline{\alpha})$.

The time interval during which the direct sound travels to the receiver can be excluded by using $t_1 = r_0/c_0$ to give an estimate of the energy density that is dependent upon r_0 (Barron, 1973). In large spaces such as concert halls, this dependence on source–receiver distance is often important. However, for sound insulation in typical rooms it is not necessary (or practical) to relate the energy density to specific source–receiver distances. On the basis that the exponential decay in a diffuse field can only start after the first reflections arrive at the receiver position; it is necessary to find a lower limit for the integration that represents the average time taken to travel from the source to the receiver when there is one reflection. This can be determined by using the mean free path (Kuttruff, 1979; Vorländer, 1995). Although the mean free path is the average distance travelled after leaving one boundary and striking the next boundary, it is reasonable to assume that the average distance from either the source or the receiver to any room boundary is approximately equal to half the mean free path. Therefore the mean free path represents the average path length between the source and receiver when there is one reflection. Taking the lower limit of the integral to be $t_1 = d_{mfp}/c_0$ gives the energy density of the diffuse sound field, w_r, as

$$w_r = \frac{4W}{c_0 A}(1 - \overline{\alpha}) \qquad (1.137)$$

where $A = -S_T \ln(1 - \overline{\alpha})$

If there is significant air absorption then for each image source, the intensity is attenuated by the factor

$$\exp\left(\frac{c_0 t}{d_{mfp}} \ln(1 - \overline{\alpha}) \right) \exp(-mc_0 t) \qquad (1.138)$$

and the resulting energy density is (Vorländer, 1995)

$$w_r = \frac{4W}{c_0 A} \exp\left(-\frac{A}{S_T}\right) \qquad (1.139)$$

where $A = -S_T \ln(1 - \bar{\alpha}) + 4mV$.

Equating Eq. 1.129 to any of the above equations for the diffuse field energy density (Eqs 1.136, 1.137, or 1.139) gives the basic relationship between reverberant sound pressure and absorption area for a fixed sound power input into a room; namely, that the sound pressure in a room is reduced by increasing the absorption area, and vice versa. In diffuse fields with an exponential decay, the decay curve is a straight line over the full 60 dB range; hence there is a simple and unambiguous relationship between the reverberation time and the absorption area.

Having seen that there is more than one equation for the energy density in a diffuse field, we need to discuss the equations that are used in practice. Although Eq. 1.139 is the more accurate equation, it is not always necessary to consider air absorption. In practice, Eq. 1.137 is usually adequate. In most rooms with volumes less than $200\,m^3$ and reverberation times less than 2 s at 20°C and 50% RH the error in neglecting air absorption in the calculation of the diffuse field energy density is only greater than 1 dB in one-third-octave-bands above 3150 Hz. In the measurement Standards for sound insulation, the classical equation (Eq. 1.136) is used to derive equations that link the sound power to the reverberant sound pressure level in a room (Section 3.5.1). This is reasonable in transmission suites (where reverberation times are at least 1 s) because negligible errors are incurred when using Eq. 1.136. However, for field sound insulation measurements (where reverberation times in furnished rooms are approximately 0.5 s), consideration could be given to use of Eq. 1.137 to determine the apparent sound reduction index (Vorländer, 1995).

1.2.7.5.2 Reverberant sound fields with non-exponential decays

In comparison with diffuse fields, it is more awkward to make a link between the sound power radiated into a room and the reverberant sound pressure level for a non-diffuse field. For modal sound fields, some insight can be gained by using normal mode theory and the specific acoustic admittance of the room boundaries; however, this does not give a completely general solution that corresponds to the practical situation (Jacobsen, 1982; Bodlund, 1980). For non-diffuse fields, it is the reverberation time that provides a practical link between the sound power and the reverberant sound pressure level. The difficulty lies in evaluating decay curves from non-diffuse fields because they are not straight lines across the entire 60 dB decay range. For this reason it is necessary to identify which part of the decay curve should be used to calculate the reverberation time.

From normal mode theory in Section 1.2.6.3.2 we have seen that it is only within the initial 20 or 30 dB of the decay curve that the majority of room modes play a role in determining the decay curve of the frequency band. The late part of the decay is determined by a relatively small number of modes with longer decay times. In the steady-state situation, energy is stored in all modes; hence, it is appropriate to determine the reverberation time using an evaluation range in which the majority of the room modes play a role in forming the decay curve. Furthermore, as any signal can be represented by a train of impulses, we can describe the steady-state sound pressure by the energetic sum of a train of impulse responses. For each impulse response, only the initial 20 dB drop of its decay curve will determine the steady-state sound pressure level to within 0.1 dB. Therefore T_{10}, T_{15}, or T_{20} should be used to determine the absorption area, rather than T_{30} or T_{60}. The energy density in reverberant sound fields with non-exponential decays can then be estimated by using the equations for the diffuse field energy density (Eqs 1.136, 1.137, and 1.139).

1.2.7.6   Direct sound field

So far we have considered the sound field in the central zone of a room without considering the sound field close to the sound source, i.e. the direct field. Most sound insulation measurements use an omnidirectional loudspeaker and we need to consider the direct field near the sound source in order to assess how far the microphone should be from the loudspeaker when we want to measure the reverberant sound pressure level in the central zone of the room without any strong influence from the direct field.

For an omnidirectional source that emits a sound power, W, and is positioned away from the room boundaries, the energy density of the direct field, w_d, at a distance, d, from the source is

$$w_d = \frac{W}{4\pi c_0 d^2} \tag{1.140}$$

Figure 1.38 shows the energy density due to the direct and the reverberant fields in a 50 m³ room at distances up to 1 m from the sound source. The energy density due to the direct field decreases by 6 dB every time the distance is doubled. The distance from the sound source at which the energy density in the direct field (Eq. 1.140) equals the energy density in the diffuse field (Eq. 1.137) is described as the reverberation distance, r_{rd}. When $\bar{\alpha} \approx -\ln(1 - \bar{\alpha})$ the reverberation distance can be estimated using

$$r_{rd} \approx \sqrt{\frac{S_T \bar{\alpha}}{16\pi}} \tag{1.141}$$

In rooms with volumes less than 150 m³ and reverberation times greater than 0.5 s, the reverberation distance is usually less than 1 m. To measure the reverberant sound pressure level the preferred option is to position the microphone at distances slightly greater than the reverberation distance. However in most rooms a practical choice for the minimum distance between the microphone and most loudspeakers is 1 m; this is commonly used for airborne sound insulation measurements (ISO 140 Parts 3 & 4).

1.2.7.7   Decrease in sound pressure level with distance

There are two main types of room in which there can be a significant decrease in the sound pressure level with distance from the source: (1) large rooms (often with volumes greater than 200 m³) with absorbent surfaces and/or large scattering objects and (2) corridors or passageways, usually with highly absorbent ceilings. Computer models (usually based around a geometrical ray approach) can be used to calculate the sound pressure level distribution, but for corridors it is still possible to gain some insight using simpler models.

Long corridors, such as those in flats, offices, and schools, are usually broken up into smaller lengths of corridor by fire doors. This typically results in sections of corridor that are less than 30 m in length, and where the dimensions of the cross-section are between 1.5 and 5 m. Noise control measures in the corridor often require absorptive ceilings; hence these elongated spaces can show a significant decrease in the sound pressure level with distance. As the sound propagates down the corridor, there are two types of internal damping that reduce the sound

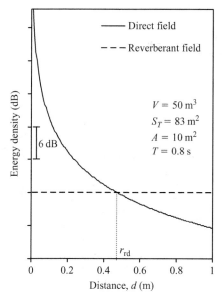

Figure 1.38

Energy density due to the direct and reverberant fields in a room at distances up to 1 m from the sound source.

pressure level: air absorption (which will be considered as negligible) and absorption by the corridor surfaces.

For a thorough overview of models for long enclosures the reader is referred to the book by Kang (2002). A simple model can be based on a corridor of infinite length that is divided into a number of very thin box-shaped sections of depth, dL (Redmore and Flockton, 1977). Effectively we are considering a large number of two-dimensional sound fields that are coupled together along the length of the corridor (see Fig. 1.39). The following derivation and the resulting equation (Eq. 1.145) are different from that in Redmore and Flockton (1977). However, it gives the same equation that was later determined empirically in scale model experiments of corridors by Redmore (1982).

It is assumed that the energy density in each section of volume, dV, is uniform and that the corridor surfaces have an average absorption coefficient, α. We arbitrarily choose a section at $x = 0$ in this infinite corridor, and follow sound propagating in the positive x-direction. In each two-dimensional sound field, sound energy impinges upon the corridor surfaces, c_0/d_{mfp} times every second; the mean free path for a two-dimensional field is given later in Eq. 1.185. The power absorbed by the corridor surfaces is

$$W_{abs} = wdV \frac{c_0}{d_{mfp}} \frac{A}{UdL} = \frac{Ec_0 U\alpha}{\pi L_y L_z} \qquad (1.142)$$

where $A = UdL\alpha$, and the perimeter of the corridor section, $U = 2L_y + 2L_z$.

The absorbed power is related to the loss factor by

$$W_{abs} = \omega \eta E \qquad (1.143)$$

63

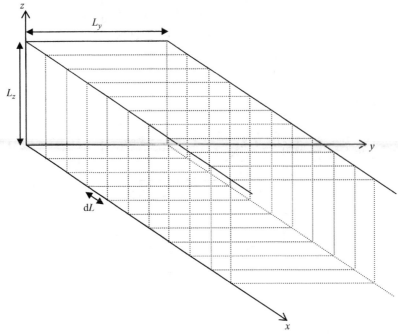

Figure 1.39

Corridor divided into sections.

hence equating Eqs 1.142 and 1.143 gives the loss factor as

$$\eta = \frac{c_0 U \alpha}{2\pi^2 f L_y L_z} \tag{1.144}$$

After travelling a distance, d, down the infinitely long corridor, the sound energy is reduced by the factor $\exp(-k\eta d)$, which gives the decrease in the sound pressure level in decibels as

$$\Delta L_{inf} = \frac{10}{\ln 10} \frac{1}{\pi} \frac{U \alpha d}{L_y L_z} \tag{1.145}$$

Equation 1.145 applies to an infinitely long corridor (i.e. without ends). In practice, sound will be partially reflected and partially absorbed from the end(s) of the corridor. We now consider a corridor that extends to infinity in the negative x-direction, but has a termination at $x = D$ (e.g. at the fire doors). It will be assumed that there are no interference effects between the incident and reflected sound at the receiver. In addition, we will not consider how the sound is injected into the corridor; this avoids consideration of the direct field from the sound source. To do this it is assumed that the thin corridor section at $x = 0$ starts with uniform energy density, and that sound propagates in the positive x-direction. The decrease in sound pressure level after the sound has travelled a distance, d, down the corridor is (Redmore and Flockton, 1977)

$$\Delta L = -10 \lg \left(10^{-\Delta L_{inf}/10} + R 10^{-\Delta L_{inf}((2D-d)/d)/10} \right) \tag{1.146}$$

where the surface at the end of the corridor has a reflection coefficient, R.

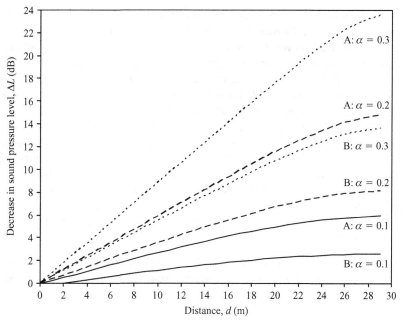

Figure 1.40

Decrease in sound pressure level along a corridor. Two different cross-sections ($L_y \times L_z$) are shown: A (1.5 × 2.5 m) and B (3 × 3 m). The corridor surfaces have average absorption coefficients of 0.1, 0.2, or 0.3. The surface that forms the end of the corridor has a reflection coefficient, $R = 0.95$.

Figure 1.40 shows the calculated decrease in level along two different corridors with different average absorption coefficients. The partial reflection at the end of the corridor causes the initial decrease in sound pressure level per unit distance to be larger than it is towards the end of the corridor.

1.2.7.8 *Sound fields in frequency bands*

So far we have focused on the sound field at single frequencies in box-shaped rooms that have perfectly reflecting rigid boundaries. We usually measure sound insulation in one-third-octave or octave-bands in rooms with a wide variety of boundaries. These bands contain many modes with different decay times and, when considering interaction with the room boundaries, it becomes more complex to predict this sound field. To look at the superposition of modes that occurs in real rooms we now look at some example measurements of sound pressure levels in one-third-octave-bands.

1.2.7.8.1 *Below the lowest mode frequency*

Below the lowest mode frequency we expect the sound field to be homogeneous and uniform throughout the space. This assumption is reasonable when there is no significant overlap from the response of the lowest mode into frequency bands below the lowest mode frequency.

To gain an impression of the sound field we can look at sound pressure level measurements in a 29 m³ source room, and an 18 m³ receiving room, both with timber-frame walls and floors

(Hopkins and Turner, 2005). The lowest mode frequency is calculated to be 39 Hz for the source room and 59 Hz for the receiving room. Therefore we will look at the 20 Hz one-third-octave-band because this is well-below the lowest mode in both rooms. A broad-band noise source was used with measurement positions in a three-dimensional grid (including positions at the room boundaries). To gain a visual impression of the sound field, sound pressure levels from three different measurement planes are shown in Figs 1.41a and 1.41b for the source and receiving rooms respectively. In the source room there is a peak in the sound pressure level immediately next to the loudspeaker. Further away from the loudspeaker in the source room, and throughout the receiving room, the spatial variation within a single measurement plane can be as low as 2 dB or as high as 12 dB. Although the sound field is relatively uniform in some measurement planes it will not always be homogeneous throughout the room volume. The non-uniform sound field in the receiving room may be attributed to structural modes and resonances of the walls and floors that occur below the lowest room mode. The mass–spring–mass resonance frequency of the timber-frame separating wall is estimated to fall in the 31.5 Hz one-third-octave-band.

For field measurements it is worth noting that both source and receiving rooms can have quite large spatial variations in the sound pressure level below the lowest calculated mode frequency.

1.2.7.8.2 Reverberant field: below the Schroeder cut-off frequency

In rooms with volumes less than 30 m³ the sound field in one-third-octave-bands below 100 Hz is sometimes dominated by the response of a single mode. However, in the low-frequency range the situation is usually more complex due to the influence of one, two, or three modes that fall within a frequency band.

In most dwellings the height (L_z) of a room is less than the width (L_x) and the depth (L_y). Therefore the lowest mode frequency will be $f_{1,0,0}$ or $f_{0,1,0}$, and when $L_x \neq L_y$, one of these modes will usually be the only mode that falls exactly within the lower and upper limits of the associated one-third-octave-band. Depending on the amount of damping and the bandwidth, the response from one or more modes in adjacent bands can overlap into this band. As an example we can look at the measured sound pressure level distribution in a 34 m³ receiving room with masonry/concrete walls and floors for the 50 Hz one-third-octave-band; this has a measured reverberation time of 1.2 s. For this room the Schroeder cut-off frequency is in the 500 Hz band. Broad-band excitation was applied in the source room, so it is representative of the situation that is encountered in field sound insulation tests. The sound field is shown in Fig. 1.41c and can generally be described as symmetrical in each plane. The first three modes are $f_{1,0,0}$, $f_{0,1,0}$, and $f_{1,1,0}$ for which the calculated mode frequencies (assuming rigid walls) are 43, 47, and 64 Hz respectively. Although $f_{0,1,0}$ is calculated to be the only mode that falls exactly within the limits of the 50 Hz band, there is evidence of overlapping response from one or both of the $f_{1,0,0}$ and $f_{1,1,0}$ modes. This is evident from the nodal planes along both the x and y-axes where the sound pressure levels are lowest in the middle of both these axes, rather than just the y-axis as would occur if the response was only due to $f_{0,1,0}$. We previously noted that real walls and floors are not rigid and will dissipate energy; hence the sound pressure in nodal planes will not be zero as implied by normal mode theory for sinusoidal excitation of a single mode in a room with rigid boundaries. However, these nodal planes still cause a high degree of spatial variation in the sound field. For the 50 Hz band in this example, there is a difference of 28 dB between the lowest level in the central zone of the room and the highest

(a) Source room (29 m³), 20 Hz one-third-octave-band. The loudspeaker position is indicated by ☆

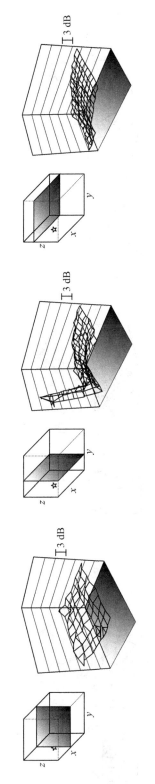

(b) Receiving room (18 m³), 20 Hz one-third-octave-band

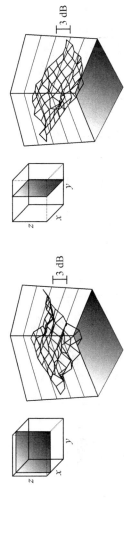

Figure 1.41

Measured sound pressure level distribution in rooms. For each figure, the same scale is used in each of the three plots to allow an assessment of the differences between the sound pressure levels in the different planes; however, different scales are used for different rooms and different frequency bands.

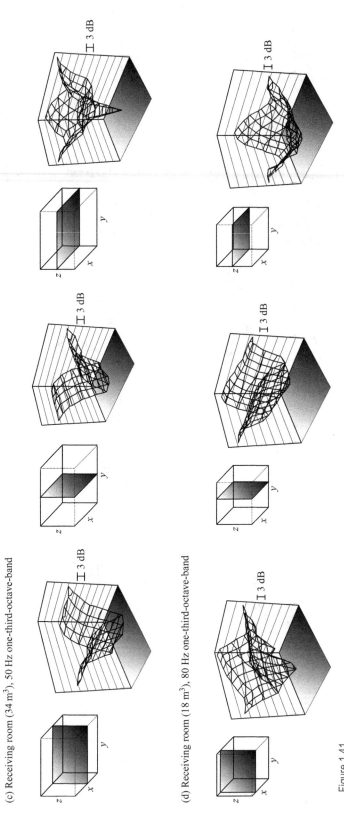

(c) Receiving room (34 m³), 50 Hz one-third-octave-band

(d) Receiving room (18 m³), 80 Hz one-third-octave-band

Figure 1.41

(Continued)

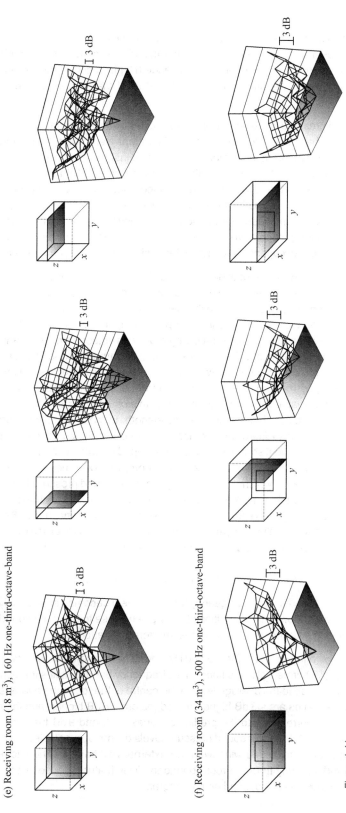

(e) Receiving room (18 m³), 160 Hz one-third-octave-band

(f) Receiving room (34 m³), 500 Hz one-third-octave-band

Figure 1.41

(Continued)

level that is $\approx$0.5 m from the room boundaries. For sound insulation measurements with broad-band noise sources, measured data suggests that this difference will usually be between 17 and 28 dB for typical rooms in the low-frequency range (Hopkins and Turner, 2005; Simmons, 1996); background noise will always limit the lowest level that is measurable in the nodal planes.

The highest sound pressure levels exist in corners and near wall/floor surfaces, with low levels near the centre of the room. This highlights an important issue for spatial average sound pressure level measurements in small rooms. Field sound insulation measurement procedures usually require that the microphone is positioned at a minimum distance of 0.5 m from the boundaries with guidance that this minimum distance should be increased to 1.2 m below 100 Hz (ISO 140 Part 4). In small rooms this means that the microphone is positioned in the central zone of the room where the sound pressure level is lowest. Therefore the spatial average levels are not representative of either the room average sound pressure level or the level perceived by room occupants who often sit and sleep near the room boundaries.

For a room height, L_z, which is between 2.1 and 2.4 m, the $f_{0,0,1}$ mode will fall within the 63 Hz or 80 Hz one-third-octave-band. This mode gives rise to low sound pressure levels on the $z = L_z/2$ measurement plane (the plane that lies in the middle of the z-axis) compared to the levels on the $z = 0$ and the $z = L_z$ measurement planes. Figure 1.41d shows the data for the 80 Hz one-third-octave-band in an 18 m^3 receiving room with timber frame walls and floors with a measured reverberation time of 0.5 s. For this room the Schroeder cut-off frequency is in the 315 Hz band. We might expect the time-averaged sound pressure level in each measurement plane for a box-shaped room to be symmetrical in frequency bands containing the first few modes. However, the spatial variation is asymmetric; this usually occurs when one or more of the room dimensions are equal to at least one wavelength. In most rooms there are recessed windows or lobby areas associated with the door, i.e. smaller volumes connected to the main rectangular space. As the room shape and the surface impedance of the room surfaces become increasingly irregular with scattering objects in the room, the local mode shapes effectively become hybrid mode shapes which do not have symmetrical sound pressure fields. Above the first few modes there is usually a marked degree of asymmetry and complexity in the sound field. An example is shown in Fig. 1.41e for the 160 Hz one-third-octave-band in the 18 m^3 receiving room of timber-frame construction. Asymmetry not only occurs in the receiving room, but also in the source room, although this is partly due to higher levels in the direct field of the loudspeaker.

1.2.7.8.3   Reverberant field: at and above the Schroeder cut-off frequency

At and above the Schroeder cut-off frequency the sound field becomes increasingly uniform in the centre of the room. However, at positions that are very close to the room boundaries there are still higher sound pressure levels due to the interference patterns.

An example is shown in Fig. 1.41f for the 500 Hz one-third-octave-band in a 34 m^3 receiving room with masonry/concrete walls and floors. This frequency band contains the Schroeder cut-off frequency. Compared to the average level in the centre of the room, the average measured levels at the wall surfaces are $\approx$3 dB higher, the edges are $\approx$6 dB higher, and the corners are $\approx$8 dB higher; these correspond to the predicted values 3, 6, and 9 dB that were discussed earlier in Section 1.2.7.1.1. Higher sound pressure levels did not occur at a few grid points near the room surface with a recessed window in the external wall because the microphone was then $\approx$200 mm further away from the room boundary. This feature can be seen in Fig. 1.41f where the window position is indicated on the diagram.

Above the Schroeder cut-off frequency the spatial variation usually decreases significantly so it is more useful to look at the standard deviation rather than plots of the spatial distribution of the sound pressure level.

1.2.7.9 Statistical description of the spatial variation

When measuring and predicting the sound pressure level in rooms, it is almost always the spatial average value that is required. Hence we need to be able to quantify the spatial variation of the sound pressure level in terms of the normalized variance of the mean-square sound pressure, and the standard deviation of the sound pressure level in decibels. The normalized standard deviation, ε, is the ratio of the standard deviation to the mean, which is squared to give the normalized variance, ε^2.

To determine the standard deviation and confidence intervals, it is necessary to know the probability distribution (probability density function) for the mean-square pressure, or identify one that gives a reasonable representation of the actual distribution. The standard deviation depends upon the type of excitation. For sound insulation we almost always use broad-band noise and measure in frequency bands; although sound insulation against pure tones is occasionally of interest with environmental noise sources. For frequency band measurements in rooms containing a single omnidirectional sound source emitting broad-band noise, estimates of the standard deviation can be found in the same way as for sound power measurements in a reverberant chamber (Schroeder, 1969; Lubman, 1974). For airborne sound insulation measurements we carry out spatial sampling of the sound pressure in the room that contains the loudspeaker, the source room. We will assume that the sound pressure is sampled at stationary microphone positions located at random points in the room; these positions are away from the room boundaries and at positions where the direct field from the source is insignificant. In this situation, the spatial variation of the mean-square pressure is represented by a gamma probability distribution for either modal or diffuse sound fields (Bodlund, 1976; Lubman, 1968; Schroeder, 1969; Waterhouse, 1968). This gamma distribution is asymmetric, right-skewed and is bounded at the lower end of the distribution by the minimum possible value for the mean-square pressure.

In practice it is necessary to consider temporal as well as spatial averaging. Here it is assumed that the uncertainty due to time-averaging of the random noise signal at each position is negligible; for further discussion of temporal averaging in measurements see Section 3.3.3.

The valid frequency ranges for the variance and standard deviation formulae in this section are defined in terms of the Schroeder cut-off frequency. At frequencies between $0.2f_S$ and $0.5f_S$ the normalized variance of the mean-square sound pressure can be estimated from (Lubman, 1974)

$$\varepsilon^2(p^2) = \left(1 + \frac{N}{\pi}\right)^{-1} \tag{1.147}$$

where N is the mode count in the frequency band (Section 1.2.5.2).

The corresponding standard deviation of the sound pressure level in decibels, σ_{dB}, can be estimated from Eq. 1.147 by taking account of the gamma distribution using (Craik, 1990)

$$\sigma_{dB} \approx \frac{4.34}{-0.22 + \sqrt{1 + \frac{N}{\pi}}} \tag{1.148}$$

Using the statistical modal density to calculate the mode count for use in Eqs 1.147 and 1.148 gives a non-integer mode count compared to the integer value determined from the individual mode frequencies by counting the number of modes that fall within each frequency band. The difference between the two methods is rarely large but the statistical approach is more robust. It accounts for the fact that most rooms are not perfect box-shaped rooms with rigid boundaries; in practice, a modal response will often occur in a frequency band that is adjacent to the band in which it was predicted to lie. It also avoids arbitrary decisions when an individual mode is calculated to lie very close to the boundary between two adjacent frequency bands. By using a statistical approach at frequencies with such low modal overlap we must expect the actual standard deviation in decibels to fluctuate about the smooth curve predicted by Eq. 1.148.

If the lower limit of $0.2f_S$ does not include the lowest frequency band of interest, then a reasonable estimate can still be calculated with Eq. 1.148 when the limit is lowered to that of the frequency band containing the lowest mode frequency (Hopkins and Turner, 2005). As noted by Lubman, this equation takes no account of other important factors: modal damping (reverberation time); combinations of small numbers of different room modes (axial, tangential, or oblique); and the degree to which the modes are excited. Despite these omissions, it generally gives estimates within ± 1 dB of measured values.

In a diffuse field at frequencies above f_S, the normalized variance can be calculated from the bandwidth, B, and the room reverberation time, T, according to (Schroeder, 1969)

$$\varepsilon^2(p^2) = \frac{1}{1 + 0.145BT} \tag{1.149}$$

for which the corresponding standard deviation in decibels is (Schroeder, 1969)

$$\sigma_{dB} = \frac{5.57}{\sqrt{1 + 0.238BT}} \tag{1.150}$$

In most rooms, Eqs 1.149 and 1.150 will give reasonable estimates at and above $0.5f_S$ (rather than above f_S); this allows continuity from Eqs 1.147 and 1.148 across the entire building acoustics frequency range. Whilst these equations are normally sufficient, it is sometimes necessary to take account of the direct sound field from the omnidirectional sound source. In this situation, the normalized variance above $0.5f_S$ can be calculated from (Michelsen, 1982)

$$\varepsilon^2(p^2) = \frac{1}{1 + 0.145BT} + \left[\frac{\sqrt{\frac{A}{16\pi}}}{160^2 d_{min}} \left(\frac{S_T\sqrt{A}}{V} \right)^3 \right] \tag{1.151}$$

where d_{min} is the minimum distance between the microphone and the sound source.

For airborne sound insulation measurements in both the laboratory and the field, $d_{min} = 1$ m (ISO 140 Parts 3 & 4), and the difference between Eqs 1.149 and 1.151 can become significant in the high-frequency range for low reverberation times and/or large rooms.

For Eq. 1.151, the corresponding standard deviation in decibels is (Michelsen, 1982)

$$\sigma_{dB} \approx 4.34\sqrt{\varepsilon^2(p^2)} \tag{1.152}$$

Calculated standard deviations over the building acoustics frequency range (using Eqs 1.148, 1.151, and 1.152) are shown in Fig. 1.42 for a 50 m³ room with different reverberation times and $d_{min} = 1$ m. Above 100 Hz the standard deviation increases as the reverberation time decreases. Note that these standard deviations are for one-third-octave-bands and that octave-bands will have lower values; hence octave-bands are sometimes used to reduce the required number of microphone positions in a room. As an aside it is worth noting that the standard

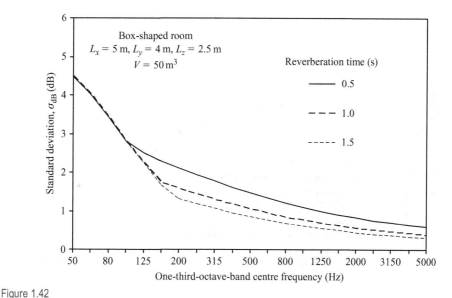

Figure 1.42

Predicted standard deviation for the spatial variation of the sound pressure level for a 50 m³ source room with different reverberation times.

deviation for a pure tone in a diffuse sound field is 5.6 dB (Schroeder, 1969); this is much higher than the standard deviations that typically occur with broad-band noise measured in one-third-octave or octave-bands over the building acoustics frequency range.

Describing the spatial variation in the receiving room is of equal importance to the source room. However, the situation is more complex for three main reasons.

Firstly, the sound transmitted into the receiving room is not always broad-band in nature. It may contain peaks in the sound pressure level at single frequencies, for example at the critical frequencies of walls/floors/windows, or the mass–spring resonances of wall linings. In reality this may only occur with a few types of homogeneous isotropic building elements because the majority of building elements are constructed from small components (e.g. bricks forming a brick wall) such that there will be spatial variation in the dynamic properties of the element due to both workmanship and the material properties. This makes it less likely that there will be a well-defined pure tone for a critical frequency or mass–spring resonance.

Secondly, we are no longer dealing with a single point source; one or more room surfaces are acting as the sound sources. In the laboratory we assume that all the sound is radiated into the receiving room by the test element, whereas in the field, it is radiated by both separating and flanking elements. Michelsen (1982) and Olesen (1992) have investigated the standard deviation of sound pressure levels in the source and receiving rooms for sound insulation measurements in both the laboratory and the field. Radiating surfaces in the receiving room can be represented as an equivalent number of uncorrelated point sources, hence the larger the surface, the larger the number of point sources. The implication for the standard deviation in a receiving room is that it should be lower than the source room because of the increased number of uncorrelated point sources. Measured data does not confirm that lower values always occur in practice (Michelsen, 1982; Olesen, 1992; Hopkins and Turner, 2005).

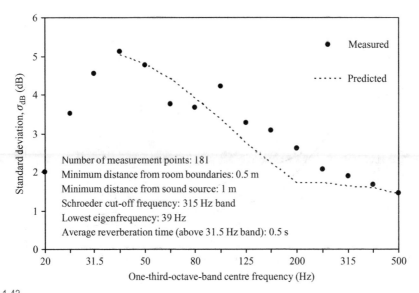

Figure 1.43

Comparison of measured and predicted standard deviations for the spatial variation of the sound pressure level in a 29 m³ source room. Measured data are reproduced with permission from Hopkins and Turner (2005).

Thirdly, the gamma probability distribution may not be a reasonable representation of the actual distribution for mean-square sound pressure in a receiving room. If we consider the interaction between all the radiating surfaces and the space it is clear that the mean-square sound pressure at any point in the room is determined by a large number of variables. By assuming that these are independent random variables, the sound pressure level will be the sum of a large number of random quantities. The central limit theorem can therefore be used to infer that the spatial variation of the mean-square sound pressure will have a log-normal probability distribution, and the sound pressure level in decibels will have a normal (Gaussian) probability distribution (Lyon and DeJong, 1995).

Despite these three complexities, empirical evidence suggests that reasonable estimates for receiving rooms can be found by using the same equations as for source rooms.

Figures 1.43 and 1.44 show measured and predicted standard deviations for a 29 m³ source room and a 34 m³ receiving room (Hopkins and Turner, 2005). The microphone positions are at least 0.5 m from the room boundaries and at least 1 m from the sound source. Generally, there is good agreement between the measurements and the calculated values. The largest discrepancies tend to occur below $0.5 f_S$; in practice, measured standard deviations will rarely be greater than 6 dB for typical rooms within this frequency range. The equations discussed above are only valid for sound fields above the lowest mode frequency. However, the measurements in the source room allow us to see the trend for the standard deviation below the lowest mode frequency in the 20 to 31.5 Hz frequency bands. In this range we assume that the sound field is uniform. Below the lowest mode frequency the standard deviation rapidly decreases due to the fading influence of the modes on the sound field. In practice, it is unlikely that the sound field in both source and receiving rooms can be considered as homogeneous and uniform in the first few frequency bands below the band that contains the lowest mode frequency.

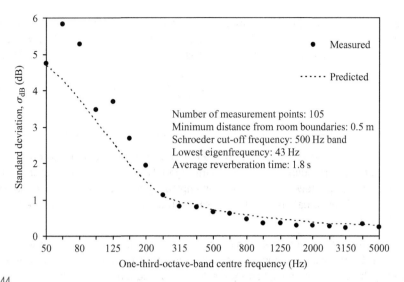

Figure 1.44

Comparison of measured and predicted standard deviations for the spatial variation of the sound pressure level in a 34 m^3 receiving room. Measured data are reproduced with permission from Hopkins and Turner (2005).

1.2.8 Energy

Although we are ultimately interested in the temporal and spatial average sound pressure level in rooms, sound transmission involves energy flow between spaces and structures. This makes it convenient to work with a single variable, energy; so we need to know the relationship between the temporal and spatial average values of sound pressure and energy in a room.

The energy in a room can be derived with two different approaches, one using sound pressure, and the other using sound particle velocity. The latter approach is used to describe the energy stored in structures such as plates and beams; hence it is included here to highlight the link between the way we deal with sound in spaces and the vibration of structures.

In both diffuse and non-diffuse fields, we can assume that the sound field is comprised of plane waves and calculate the sound energy using the plane wave intensity described by Eq. 1.19. The plane wave intensity quantifies the energy travelling through an imaginary surface of unit area in 1 s, where this surface lies perpendicular to the direction of wave propagation. The group velocity, c_g, is defined as the velocity at which wave energy propagates, which for longitudinal waves in air is the same as the phase velocity, c_0. Therefore the energy density in a reverberant field, w_r, that describes the energy in a unit volume is

$$w_r = \frac{I}{c_0} \tag{1.153}$$

Hence the energy stored in volume, V, is

$$E = w_r V = \frac{\langle p^2 \rangle_{t,s} V}{\rho_0 c_0^2} \tag{1.154}$$

where $\langle p^2 \rangle_{t,s}$ is the temporal and spatial average mean-square sound pressure.

An alternative way of deriving the room energy is from the product of the mass of air within the room and the temporal and spatial average mean-square sound particle velocity, $\langle u^2 \rangle_{t,s}$, where the latter can be found from the characteristic impedance of air (Eq. 1.18). This gives

$$E = m\langle u^2 \rangle_{t,s} = \rho_0 V \frac{\langle p^2 \rangle_{t,s}}{\rho_0^2 c_0^2} = \frac{\langle p^2 \rangle_{t,s} V}{\rho_0 c_0^2} \qquad (1.155)$$

1.2.8.1   Energy density near room boundaries: Waterhouse correction

When calculating the sound energy stored in a reverberant room we need to consider the fact that energy density is not uniformly distributed throughout the space. Near the room boundaries the phase relationships between sound waves impinging upon a point are no longer random. This causes interference patterns and an increase in energy density close to the boundaries (Section 1.2.7.1).

To determine the total sound energy stored in a reverberant room from the energy calculated with Eq. 1.154 (using the spatial average sound pressure measured in the central zone of the room) the energy is multiplied by the following frequency-dependent correction term (Waterhouse, 1955)

$$1 + \frac{S_T \lambda}{8V} \qquad (1.156)$$

where S_T is the total surface area of the room.

This term is widely referred to as the Waterhouse correction which is usually more convenient to use in decibels,

$$C_W = 10 \lg \left(1 + \frac{S_T \lambda}{8V} \right) \qquad (1.157)$$

In the derivation of the correction term it is assumed that the room surfaces are perfectly reflecting, which is often a reasonable assumption in the low-frequency range where the term is most important. It is also assumed that sound waves are incident from all directions upon a reflecting surface which has dimensions that are large compared to the wavelength. This will not be a valid assumption where the sound field in the central zone of the room cannot be classified as reasonably diffuse, and the room dimensions are small. Another important assumption is that the energy stored in edge and corner zones is relatively small compared to the energy stored near room surfaces. This is a reasonable assumption in many shapes and sizes of room when the walls and floors have large surface areas. However, it is not necessarily appropriate for modal sound fields in the low-frequency range where there are relatively few oblique modes compared to axial and tangential modes. In this situation, numerical calculations indicate that the Waterhouse correction sometimes appears to give accurate results because it overestimates the energy stored near room surfaces (Agerkvist and Jacobsen, 1993). This compensates for the fact that the energy stored in edge and corner zones is not included in the correction term.

Example values for the Waterhouse correction are shown in Fig. 1.45 for box-shaped rooms with volumes in the range 50 to 200 m^3. For these rooms the correction term is greater than 1 dB in the low-frequency range which is often below the Schroeder cut-off frequency. This means that significant values for the correction term tend to occur at frequencies where the sound field in the central zone of the room is not a close approximation to a diffuse field. However, as a rule-of-thumb for the building acoustics frequency range, the Waterhouse correction term tends

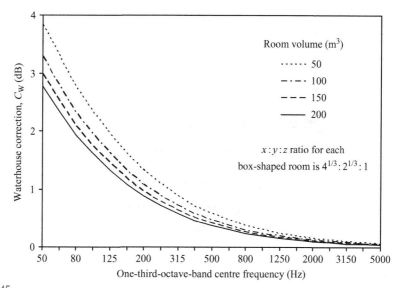

Figure 1.45

Waterhouse correction, C_W, for different room volumes.

to give a reasonable estimate for most empty box-shaped rooms with a minimum volume of $50\,\mathrm{m}^3$. This assumes that the sound pressure level in the central zone is adequately sampled.

When calculating the sound power radiated into a room from sound pressure measurements made in the central zone of a reverberant room, the Waterhouse correction in decibels should be added to the sound pressure level in decibels. However, we do not need to account for interference patterns at the room boundaries when we calculate the diffuse field intensity that is incident upon a surface from the diffuse field sound pressure level. Hence there are some situations where we need to use the Waterhouse correction and some where we don't. Specific applications of the Waterhouse correction that apply to the measurement of sound insulation are noted in Chapter 3.

1.3 Cavities

Cavities exist in many different parts of a building, for example: ceiling voids, roof voids, between the joists in timber floors, in thermal glazing units, within cavity walls, and behind wall linings. They can play an important role in sound transmission because vibration is not only transmitted via structural connections between the plates that form a cavity, but also by the sound field in the cavity itself.

As with rooms we will retain use of the convenient box-shaped space but for cavities we will use L_z as the smallest dimension, the cavity depth (see Fig. 1.46).

1.3.1 Sound in gases

Almost all cavities in buildings are filled with air, so Eq. 1.1 for the speed of sound in rooms is also applicable to cavities. However, cavities such as those in insulating glass units are

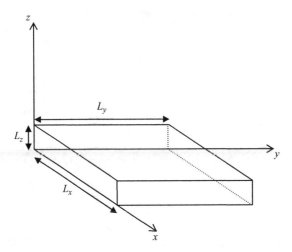

Figure 1.46

Box-shaped cavity.

sometimes filled with other gases. Therefore a more general approach to calculate the phase velocity, c, for any ideal gas is given by

$$c = \sqrt{\frac{1}{\kappa\rho}} = \sqrt{\frac{V}{\kappa n M}} = \sqrt{\frac{\gamma P V}{n M}} = \sqrt{\frac{\gamma R(T + 273.15)}{M}} \quad (1.158)$$

where κ is the gas compressibility (adiabatic), ρ is the gas density, V is the volume occupied by n moles of a gas, M is the molar mass of the gas (kg/mol), γ is the ratio of specific heats at constant pressure and constant volume which is 1.67 for monatomic gases such as helium, 1.41 for diatomic gases such as air, and 1.33 for polyatomic gases, P is the static pressure which is 1.013×10^5 Pa for air at atmospheric pressure, R is the universal gas constant which is 8.314 J/mol.K, and T is the temperature in °C.

Properties of gases that are components of air or gases that are sometimes used in insulating glass units are listed in the Appendix, Table A1.

1.3.2 Sound in porous materials

Cavities in walls and floors are sometimes partly filled or fully filled with porous materials to absorb sound energy and provide other benefits such as thermal insulation. Porous materials are also used around the perimeter of cavities to absorb sound and/or to control the spread of fire. Some examples are shown in Fig. 1.47.

Porous materials essentially consist of a skeletal frame (which could be formed from fibres, granules, or a polymer, etc.) that is surrounded by air. A wide range of porous materials are used in buildings, with a range of frames (e.g. mineral wool, polystyrene balls, open-cell foam, masonry blocks). Sound transmission through porous materials takes place due to airborne propagation through the pores and structure-borne propagation via the frame. However, there are varying degrees of coupling between these types of propagation, and they cannot simply be assumed to occur independently of each other. For this reason, sound propagation through

Lightweight wall (fully filled)

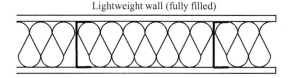

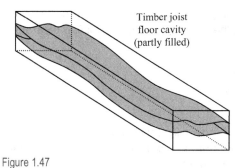

Timber joist
floor cavity
(partly filled)

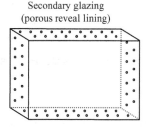

Secondary glazing
(porous reveal lining)

Figure 1.47

Examples of porous materials used in cavities.

porous materials is considerably more complex than in air; the subject is only touched upon here to introduce basic concepts and parameters that are needed in other chapters. For a thorough review of different models used to describe sound in porous materials, the reader is referred to Allard (1993).

1.3.2.1 Characterizing porous materials

Two simple parameters that can be used to describe the properties of porous materials are the porosity and the airflow resistance. For a more complete description of the material, other parameters such as the structure factor, shape factor, and tortuosity can be used to describe aspects relating to the propagation path through the pores. However, these parameters are rarely available, more awkward to measure, and are used in more complex models than will be looked at here.

1.3.2.1.1 Porosity

For porous materials, the porosity, ϕ, is defined as

$$\phi = \frac{V_{\text{air}}}{V_{\text{bulk}}} \tag{1.159}$$

where V_{air} is the volume of air within the material and V_{bulk} is the bulk volume (i.e. total volume) of the material.

For porous materials used in buildings, the porosity is usually in the range, $0.90 < \phi < 0.99$. For mineral wool it is typically $0.95 < \phi < 0.99$. Mineral wool (i.e. glass or rock fibre) is usually made of solid fibres, hence if the material that binds these fibres together has negligible mass, the porosity can be estimated using

$$\phi = 1 - \frac{\rho_{bulk}}{\rho_{fibre}} \tag{1.160}$$

where ρ_{bulk} is the bulk density of the material and ρ_{fibre} is the density of the fibre.

1.3.2.1.2 Airflow resistance

Sound absorption by, and sound transmission through porous materials is partly described by their ability to resist airflow. This is quantified by the following parameters: airflow resistance, specific airflow resistance, and airflow resistivity.

The airflow resistance, R (Pa.s/m^3) is defined as

$$R = \frac{\Delta p}{q_v} \tag{1.161}$$

where Δp is the air pressure difference (referred to as differential pressure) across a layer of porous material with respect to the atmosphere (Pa), and q_v is the volumetric airflow rate passing through the layer (m^3/s). The volumetric airflow rate is

$$q_v = uS \tag{1.162}$$

where u is the linear airflow velocity (m/s) and S is the cross-sectional area of the porous material perpendicular to the direction of airflow (m^2).

The specific airflow resistance, R_s (Pa.s/m) applies to a specific thickness of a porous material; hence it is an appropriate specification parameter for both homogeneous and non-homogeneous materials as well as materials with a porous surface coating or perforated surface layer.

$$R_s = RS \tag{1.163}$$

The airflow resistivity, r (Pa.s/m^2) is the specific airflow resistance per unit thickness, and is only appropriate as a specification parameter for homogeneous materials.

$$r = \frac{S\Delta p}{dq_v} = \frac{RS}{d} = \frac{R_s}{d} \tag{1.164}$$

where d is the thickness of the layer of porous material in the direction of airflow (m).

NB: Specific airflow resistance and airflow resistivity are sometimes quoted in Rayls and Rayls/m respectively. The Rayl is used as a unit for the ratio of sound pressure to particle velocity and is equivalent to Pa.s/m.

For fibrous materials the airflow resistance depends upon the direction of airflow through the material. These materials are usually supplied and used in rectangular sheets, either cut from slabs or from a roll, hence the airflow can be measured in two directions as shown in Fig. 1.48: (1) in the plane of the sheet, the lateral airflow and (2) perpendicular to the plane of the sheet, the longitudinal airflow. In the literature it is usually measurements of the longitudinal airflow that are quoted (e.g. Bies and Hansen, 1980). In rooms or cavities where sheets of material are used to cover a surface it is the longitudinal direction that is needed to calculate the sound absorption coefficient for the surface. However, narrow cavities are sometimes separated by sheets of

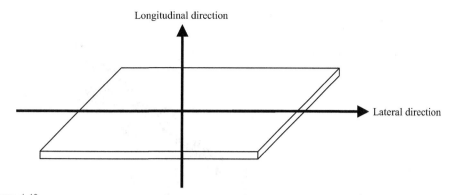

Longitudinal direction

Lateral direction

Figure 1.48

Airflow resistivity of a sheet of porous material – definition of lateral and longitudinal directions.

fibrous materials that form a junction between the different cavities. Depending on the orientation of these sheets it is either the lateral or longitudinal direction that is needed to calculate the absorption coefficient for the cavity boundary or sound transmission between cavities.

1.3.2.1.3 Fibrous materials

Fibrous materials are commonly used in cavities of walls and floors. The airflow resistance of fibrous materials is due to friction between the fibres and the air particles moving between the fibres, hence it can depend upon: size of fibres, shape/type of fibres (e.g. crimped, hollow), density of fibres, number of fibres per unit volume, and fibre orientation/distribution (e.g. random, stratified/layered, stratified with higher fibre density near the surface of the sheet).

Mineral wool is anisotropic as the fibres tend to lie in planes that are parallel to the plane of the sheet; the orientation of the fibres within each plane being random. Therefore the airflow resistivity in the lateral direction is significantly lower than in the longitudinal direction.

For mineral wool (i.e. glass or rock wool) empirical relationships can be found between airflow resistance and bulk density according to (Bies, 1988; Nichols, 1947)

$$r = \frac{k_1 \rho_{\text{bulk}}^{1+k_2}}{d_{\text{fibre}}^2} \tag{1.165}$$

where k_1 is a constant for a material that is manufactured in a particular way, k_2 is a constant that depends upon fibre orientation, and d_{fibre} is the fibre diameter (microns).

For one type of mineral wool with a known average fibre diameter, the constants k_1 and k_2 can be found from measured airflow resistivity data for a range of bulk densities. By plotting $\lg(r)$ against $\lg(\rho_{\text{bulk}})$, the data points should cluster along straight lines, and linear regression can be used to determine k_1 and k_2. An example is shown in Fig. 1.49 for the lateral and longitudinal airflow resistivity of rock wool (random fibre orientation, average $d_{\text{fibre}} = 4.75\,\mu\text{m}$, average $\rho_{\text{fibre}} = 2600\,\text{kg/m}^3$, porosity range was 0.94 (highest bulk density) $\leq \phi \leq 0.99$ (lowest bulk density), two different UK manufacturers). For the lateral airflow resistivity, $k_1 = 353$, $k_2 = 0.63$ over the bulk density range, $31 \leq \rho_{\text{bulk}} \leq 155\,\text{kg/m}^3$. For the longitudinal airflow resistivity, $k_1 = 780$, $k_2 = 0.59$ over the bulk density range, $38 \leq \rho_{\text{bulk}} \leq 162\,\text{kg/m}^3$.

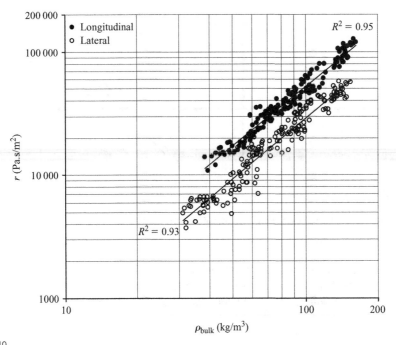

Figure 1.49

Measured airflow resistivity (lateral and longitudinal directions) for rock wool. Individual measurements are shown along with regression lines. Measured data from Hopkins are reproduced with permission from ODPM and BRE.

Measured airflow resistivities and empirical relationships for other porous materials can be found from Bies and Hansen (1980), Mechel and Vér (1992), and Mechel (1995). To cover the full density range for a material it may be necessary to have more than one empirical relationship, this can occur with fibrous materials that can be produced in a wide range of fibre diameters. For materials such as glass wool, the combination of different manufacturing processes and different fibre diameters can lead to empirical relationships that are specific to one manufacturer and/or density range (Bies, 1988).

To determine empirical relationships for materials other than mineral wool, the form of Eq. 1.165 may not be appropriate. For example, with polyester fibre materials it has been shown that better correlation can be found between the airflow resistivity and the number of fibres per unit volume (Narang, 1995).

1.3.2.2 Propagation theory for an equivalent gas

General theory for sound propagation in a fluid-saturated porous elastic material requires consideration of two longitudinal waves and one shear wave (Biot, 1956). Modelling these three waves requires knowledge of the fluid density, frame density, porosity, airflow resistivity, tortuosity, complex shear modulus, and Poisson's ratio. In buildings we are usually interested in air-saturated porous materials, rather than liquid-saturated. This simplifies matters because with gases it can often be assumed that the skeletal frame is not elastic, and is sufficiently rigid that it does not move. This allows use of simpler sound propagation models.

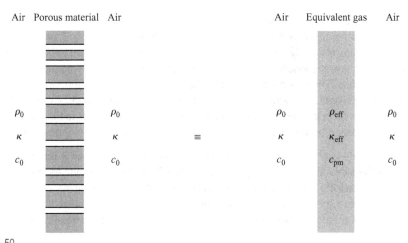

Air Porous material Air   Air Equivalent gas Air

ρ_0 ρ_0 ρ_0 ρ_{eff} ρ_0

κ κ $\equiv$ κ κ_{eff} κ

c_0 c_0 c_0 c_{pm} c_0

Figure 1.50

Equivalent gas model used for a porous material in air.

For porous materials with a rigid skeletal frame and porosities close to unity, sound propagation can be modelled with a single longitudinal wave by using the concept of an equivalent gas to represent the porous material and the gas (usually air) contained within it (Morse and Ingard, 1968). Within a porous material the compressibility of the gas is altered, and its effective mass is increased because the flow of the gas is impeded by the porous structure. Hence, the equivalent gas is described by using an effective gas compressibility, κ_{eff}, and an effective gas density, ρ_{eff}.

The gas compressibility, κ, equals the reciprocal of the bulk compression modulus of a gas, K, such that

$$\kappa = \frac{1}{K} = \frac{1}{\rho}\frac{\partial \rho}{\partial P} \qquad (1.166)$$

where ρ is the gas density.

From this point onwards we will assume that the gas in the porous material is always air. The equivalent gas model is shown in Fig. 1.50. For an infinite medium without internal losses, K takes real values; $K = P_0$ for an isothermal process and $K = 1.4P_0$ for an adiabatic process, where P_0 is the static pressure for air (usually taken as 1.013×10^5 Pa at atmospheric pressure). However, in a porous material it is necessary to use complex values to include the effect of internal damping. For sound propagation in typical rooms, the distances are only usually large enough to require consideration of air absorption in the high-frequency range; this is an internal loss due to the conversion of sound into heat energy. These internal losses occur due to both thermal conduction and viscosity, and result from the molecular constitution of the gas; in an infinite medium the thermal conduction and viscosity contribute almost equally to the internal damping (Morse and Ingard, 1968). In a porous material, sound propagates close to the boundaries of the skeletal frame and the losses due to thermal conduction and viscosity are much larger. Therefore we need to account for these internal losses by using complex values for both the effective gas compressibility and the effective gas density.

The effective gas compressibility varies over the building acoustics frequency range, and depends upon heat transfer between the air and the frame. At 'low' frequencies, the rate of compression and rarefaction for the longitudinal sound wave in a porous material is sufficiently slow that heat is transferred back and forth between the air and the frame. This means that the temperature remains relatively constant and the process can be assumed to be isothermal. At 'high' frequencies there is insufficient time for this heat transfer to take place, so it becomes an adiabatic process. There is no general definition of 'low' and 'high' frequencies. As a rule-of-thumb for fibrous materials over the building acoustics frequency range, it can be assumed that 'low' corresponds to the low-frequency range, and 'high' corresponds to the high-frequency range, with a transition between isothermal and adiabatic in the mid-frequency range.

The effective gas density also varies with frequency. This can be described in terms of the mass impedance of the skeletal frame, $i\omega m_{frame}$ (Beranek, 1947). At 'low' frequencies where the mass impedance is small, the compressions and rarefactions of the air particles cause the frame to move too; hence the effective gas density needs to take account of the mass of the frame. At 'high' frequencies where the mass impedance is large, the frame effectively remains motionless.

The concept of an equivalent gas allows sound propagation in porous materials to be described using two parameters, both of which are complex: the complex wavenumber, k_{pm}, and the characteristic impedance, $Z_{0,pm}$. Assuming harmonic time dependence for a wave using the term $\exp(i\omega t)$, the wave equation for sound propagation in the porous material has the same form as the wave equation for an infinite medium (Eq. 1.14); the difference being that the wavenumber, k, is replaced by k_{pm}.

The complex wavenumber for sound in a porous material, k_{pm}, is

$$k_{pm} = \text{Re}\{k_{pm}\} + i\text{Im}\{k_{pm}\} = \frac{\omega}{c_{pm}} \qquad (1.167)$$

where the phase velocity for sound in the porous material, c_{pm}, is also complex, and equals

$$c_{pm} = \sqrt{\frac{1}{\phi \rho_{eff} \kappa_{eff}}} \qquad (1.168)$$

The complex wavenumber is used here to clarify the link between propagation of sound in air and propagation in a porous material via the wave equation. Note that some texts prefer to use the propagation constant, Γ, which is related to the complex wavenumber by

$$\Gamma = ik_{pm} \qquad (1.169)$$

The characteristic impedance for air in a porous material, $Z_{0,pm}$, is determined in the same way as the characteristic impedance for air in an infinite medium (Eq. 1.18), which gives

$$Z_{0,pm} = \frac{p}{u} = \rho_{eff} c_{pm} = \sqrt{\frac{\rho_{eff}}{\phi \kappa_{eff}}} \qquad (1.170)$$

The complex wavenumber (Eq. 1.167) and the characteristic impedance (Eq. 1.170) are both calculated from the effective density and the effective gas compressibility. The latter two parameters can be calculated if the structure of the porous material can be represented using idealized geometry. For example, representing all pores by cylindrical tubes at a specified angle to the surface of a sheet of porous material, or representing all the fibres in a sheet of fibrous material by long cylindrical tubes that lie in planes parallel to the surface of the sheet. Microstructural models that assume idealized geometry can be quite complicated. However, they can give an effective density and gas compressibility that adequately represents real porous materials as well as giving an insight into which parameters are important for sound propagation (e.g. see Allard, 1993). For many porous materials the geometry is not simple and requires a statistical description. However, an alternative, simpler approach can be taken that avoids direct calculation of the effective density and the effective gas compressibility whilst retaining use of the equivalent gas model. This makes use of empirical relationships to determine the complex wavenumber and the characteristic impedance.

The most widely used empirical equations are those of Delany and Bazley (1969, 1970). These form a benchmark against which many other theories are tested, and other empirical equations are compared. They were derived from a large number of measurements on different fibrous materials. The resulting empirical equations for $Z_{0,pm}$ and k_{pm} only require knowledge of the airflow resistivity which can be measured or determined from other empirical relationships. Although these empirical equations were based upon fibrous materials they can be used to estimate values for porous foams with $r < 10\,000\,\text{Pa.s/m}^2$ (Allard, 1993).

The assumption of a rigid skeletal frame allows empirical laws to be used to calculate sound propagation in isotropic, homogeneous, porous materials. Fibrous materials such as mineral wool can be considered as relatively homogeneous, although they are formed from layers so they are anisotropic. However, by considering propagation through the material in only a single direction, they can be treated as isotropic, homogeneous materials.

Empirical equations are not absolute laws; there are many different materials and there is often more than one way to group or plot the data to carry out regression analysis. Other empirical equations to determine the characteristic impedance and the propagation constant for fibrous materials can be found in the literature (e.g. Mechel and Vér, 1992). A theoretical model for rigid frame fibrous materials from Allard and Champoux (1992) gives similar values to the Delany and Bazley equations in the range of validity but improves the low-frequency trends.

The empirical equations of Delany and Bazley (1969, 1970) are

$$Z_{0,pm} = \rho_0 c_0 (1 + 0.0571 X^{-0.754} - i0.087 X^{-0.732}) \tag{1.171}$$

and

$$k_{pm} = \text{Re}\{k_{pm}\} + i\text{Im}\{k_{pm}\} = \frac{2\pi f}{c_0}(1 + 0.0978 X^{-0.700} - i0.189 X^{-0.595}) \tag{1.172}$$

where the variable, X, is

$$X = \frac{\rho_0 f}{r} \tag{1.173}$$

The range of validity for Eqs 1.171 and 1.172 is (Delany and Bazley, 1969)

$$0.01 < X < 1.0 \tag{1.174}$$

For the equivalent gas model, the wavelength of sound within the porous material, λ_{pm}, is calculated using

$$\lambda_{pm} = \frac{2\pi}{\text{Re}\{k_{pm}\}} \tag{1.175}$$

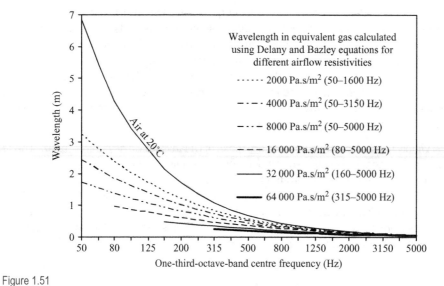

Figure 1.51

Comparison of the wavelength in porous materials using the equivalent gas model with the wavelength in air. The range of validity for the Delany and Bazley equations is shown in brackets in terms of frequency.

In Fig. 1.51 the wavelength in air can be compared with the wavelength for porous materials in air that is calculated using the Delany and Bazley equations. The calculations use a range of airflow resistivities (2000 to 64 000 Pa.s/m^2) that represents porous materials commonly used in buildings. The range of validity for the Delany and Bazley equations (Eq. 1.174) usually allows use of the equivalent gas model for a large part, but not all, of the building acoustics frequency range. The wavelength in the equivalent gas is significantly shorter than air in the low- and mid-frequency ranges, but tends towards the wavelength in air within the high-frequency range.

The sound pressure for a plane wave propagating through a porous material in the positive x-direction is described by

$$p(x, t) = \hat{p} \exp(-ik_{pm}x) \exp(i\omega t) = \hat{p} \exp(-i\text{Re}\{k_{pm}\}x) \exp(\text{Im}\{k_{pm}\}x) \exp(i\omega t) \qquad (1.176)$$

The definition of a complex wavenumber implies attenuation with distance, hence $\text{Im}\{k_{pm}\}$ is negative; this can be seen in the empirical equation for fibrous materials (Eq. 1.172). Therefore the amplitude of the plane wave decreases with distance according to the decaying exponential term, $\exp(\text{Im}\{k_{pm}\}x)$. This gives the decrease in sound pressure level in decibels, ΔL_P, after propagating a distance, x, through the porous material,

$$\Delta L_P = \frac{20}{\ln 10} |\text{Im}\{k_{pm}\}| x \qquad (1.177)$$

The installation of porous materials in air spaces means that it is often necessary to account for the reflection that occurs when sound enters the material from air, and when it exits the material into air. This is described in Section 4.3.9 in the calculation of the normal incidence sound reduction index for porous materials.

In some cases there is no air space between the porous material and the plate that forms part of a wall or floor, such as a cavity wall where a porous material fills the cavity. For a plate undergoing bending wave vibration that is immediately next to a porous material, sound transmission from the plate into and through the porous material may need to use Biot theory for the porous material to take account of the shear wave and two longitudinal waves that can propagate within it. In this case, the simplified assumption of an equivalent gas may no longer be appropriate.

1.3.3 Local modes

From Eq. 1.54 the mode frequencies of closed cavities are calculated using

$$f_{p,q,r} = \frac{c}{2}\sqrt{\left(\frac{p}{L_x}\right)^2 + \left(\frac{q}{L_y}\right)^2 + \left(\frac{r}{L_z}\right)^2} \qquad (1.178)$$

1.3.3.1 Modal density

To calculate the cavity modal density across the building acoustics frequency range we not only need to consider three-dimensional sound fields like in rooms, but also one-dimensional ($p \neq 0$ and $q = r = 0$), and two-dimensional ($p \neq 0$, $q \neq 0$, and $r = 0$) sound fields. Hence we can represent cavities as a one-dimensional space of length, L_x, a two-dimensional space of surface area, $S = L_x L_y$, and a three-dimensional space of volume, $V = L_x L_y L_z$.

One-, two-, and three-dimensional sound fields can occur in lightweight walls and floors where cavities are formed by a framework of studs or joists. In these cases, the one or two-dimensional modal density can be determined by using Eq. 1.178 to calculate the mode frequencies; the number of modes that fall within each band are then divided by the bandwidth. As with rooms it is simpler to use the following statistical approaches.

For a long (L_x), narrow (L_y), and thin (L_z) cavity at low frequencies there is a one-dimensional sound field consisting purely of axial modes. The modal density is calculated in the same way as for structural waves on beams (Section 2.5.1.4), hence

$$n_{1D}(f) = \frac{2L_x}{c} \qquad (1.179)$$

At frequencies at and above $f_{0,1,0}$, but below $f_{0,0,1}$, the cavity acts as a two-dimensional space that supports axial and tangential modes. To count the number of modes the eigenvalues are arranged in a two-dimensional lattice as shown in Fig. 1.52 (Price and Crocker, 1970). Eigenvalues that lie along the x and y-axes represent axial modes; those that lie on the coordinate plane $k_x k_y$ (excluding the eigenvalues on the axes) represent tangential modes. The area associated with each eigenvalue is a rectangle with an area of $\pi^2/L_x L_y$ (which equals π^2/S). The number of modes below a specified wavenumber, k, is equal to the number of eigenvalues that are contained within one-quarter of the area of a circle with radius, k. However, one-half of the area associated with each axial mode falls outside the permissible area in k-space that can only have zero or positive values of k_x and k_y. Therefore calculating the number of modes is a two-step process. The first step is to divide $\pi k^2/4$ by π^2/S to give an estimate for the number of tangential modes that also includes one-half of the axial modes. The second step is to account for the other halves of the axial modes that lie on the x and y-axes by taking one-half

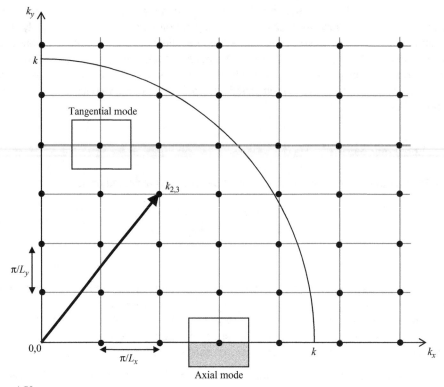

Figure 1.52

Mode lattice for a two-dimensional space. The vector corresponding to eigenvalue, $k_{2,3}$, is shown as an example. The shaded area indicates the fraction of the area associated with axial modes that falls outside the permissible area in k-space. The area enclosed by a circle with radius, k, encloses eigenvalues below wavenumber, k.

of $k/(\pi/(L_x + L_y))$. The sum of these two components gives the number of modes, $N(k)$, below the wavenumber, k, where

$$N(k) = \frac{k^2 S}{4\pi} + \frac{k(L_x + L_y)}{2\pi} \qquad (1.180)$$

Hence, from Eq. 1.57 the modal density is

$$n_{2D}(f) = \frac{2\pi f S}{c^2} + \frac{L_x + L_y}{c} \qquad (1.181)$$

For cavities that are not box-shaped, and for cavities where there is ambiguity about whether it is reasonable to assume rigid boundaries for one or two of the four boundaries (i.e. those that lie along the planes where $x = 0$, $x = L_x$, $y = 0$, and $y = L_y$), the modal density can be calculated by using only the first term in Eq. 1.181,

$$n_{2D}(f) = \frac{2\pi f S}{c^2} \qquad (1.182)$$

The crossover point from a two-dimensional to a three-dimensional sound field occurs at the frequency where there is a half wavelength across the smallest dimension, L_z, which is usually the cavity depth. This corresponds to the axial mode $f_{0,0,1}$, the first cross-cavity mode, where

$$f_{0,0,1} = \frac{c}{2L_z} \tag{1.183}$$

At and above $f_{0,0,1}$ there are axial, tangential, and oblique modes, hence the cavity acts as a three-dimensional space for which the modal density is

$$n_{3D}(f) = \frac{4\pi f^2 V}{c^3} + \frac{\pi f S_T}{2c^2} + \frac{L_T}{8c} \tag{1.184}$$

where S_T is $2(L_x L_y + L_x L_z + L_y L_z)$ and L_T is $4(L_x + L_y + L_z)$.

For cavities that are not box-shaped, and for box-shaped cavities in the high-frequency range, a reasonable estimate of the modal density is found by using only the first term in Eq. 1.184.

The statistical mode count in a frequency band is calculated from the modal density using Eq. 1.63. Mode counts are now used to gain an insight into the distribution of modes for two common cavities, a timber joist floor cavity and a wall cavity (see Fig. 1.53). The timber joist floor cavity is long, narrow, and thin; in the low-frequency range this results in only axial modes along the longest dimension, L_x. Above the first cross-cavity mode in the 800 Hz band there is then a rapid increase in the number of modes with increasing frequency. In contrast, the wall cavity has a two-dimensional sound field over the majority of the building acoustics frequency range with the first cross-cavity mode in the 2500 Hz band.

As with rooms, the distribution of the different mode types is useful in determining which internal cavity surfaces should be lined with absorbent material to reduce the sound level in the cavity. We can take the timber joist floor cavity as an example. To absorb sound in the low-frequency range where there are only axial modes along L_x, absorbent material could be positioned over the surfaces perpendicular to the x-axis at the ends of the cavity where $L_x = 0$ m and $L_x = 4$ m. In practice, floor cavities are often partially or fully filled with absorbent material along their entire length to absorb sound energy stored in axial, tangential, and oblique modes.

1.3.3.2 Equivalent angles

Equivalent angles for local modes in rooms were introduced in Section 1.2.5.4. Figure 1.54 shows equivalent angles for the timber joist floor cavity and wall cavity described in Fig. 1.53. These can be compared with the equivalent angles for a 50 m^3 room (refer back to Fig. 1.16). Below the first cross-cavity mode, $\theta_z = 0°$, because the sound field is two-dimensional and there is a limited range of angles. Above the first cross-cavity mode the range of angles tends to cover the full range from 0° to 90°; however, the elongated shape of the timber joist floor cavity means that the distribution of angles between θ_x, θ_y, and θ_z is uneven when compared with the 50 m^3 room.

Compared with rooms, the small volumes and elongated shapes associated with typical cavities means that in the building acoustics range there is often a limited range of angles from which the sound waves will arrive at any point in the space.

(a) Timber joist floor cavity

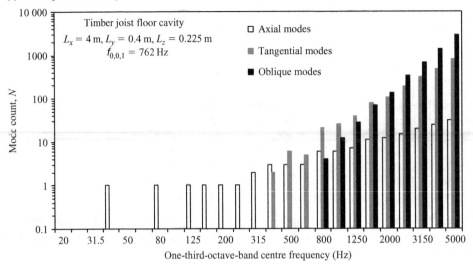

(b) Wall cavity

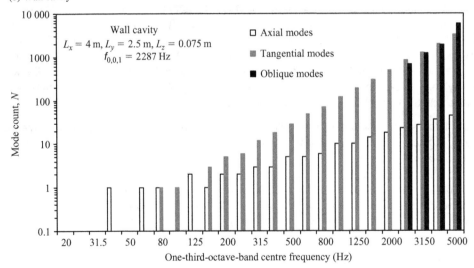

Figure 1.53

Mode count for a timber joist floor cavity and a wall cavity.

1.3.4 Diffuse field

A diffuse field in a cavity is defined in the same way as for rooms. However, when there is a two-dimensional sound field we need to account for the fact that waves can only arrive from directions within one plane rather than from all possible directions in three-dimensional space.

Compared to rooms, cavities have much smaller volumes and the sound field can only usually be considered as diffuse over a narrow part of the building acoustics frequency range.

(a) Timber joist floor cavity

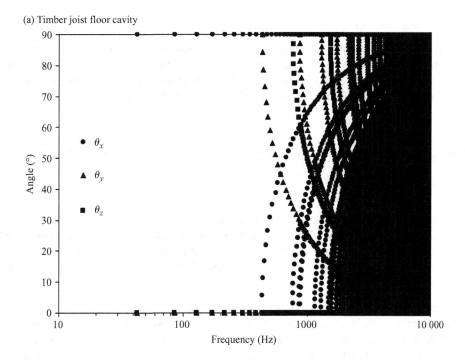

(b) Wall cavity

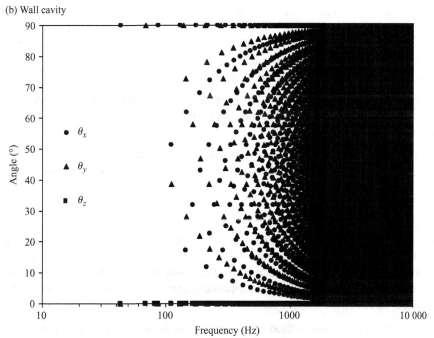

Figure 1.54

Equivalent angles for the modes of a timber joist floor cavity and a wall cavity.

1.3.4.1 Mean free path

As with rooms, the mean free path is only defined for the situation where all reflections from the boundaries are diffuse. When a cavity acts as a three-dimensional space, the mean free path is the same as for rooms and is defined in Eq. 1.47. The mean free path for a two-dimensional space has already been derived in Section 1.2.3.1 and is given by (Kosten, 1960)

$$d_{mfp} = \frac{\pi S}{U} \tag{1.185}$$

where U is the perimeter of the cavity ($U = 2L_x + 2L_y$ for a rectangular cavity with a depth, L_z).

1.3.5 Damping

In rooms, absorptive material is often distributed in one of two ways: either it is distributed over all the surfaces, or one or two of the room surfaces provide the majority of the absorption area (e.g. highly absorbent tiles that cover the ceiling). In cavities there is more scope to vary the distribution of absorbent material; it can be placed within the cavity volume as well as over the surfaces.

The implications of one, two, and three-dimensional sound fields in cavities becomes apparent when we consider the position of the absorption within the cavity. Below the first cross-cavity mode there are only axial and tangential modes in the cavity, hence sound waves are only incident upon the perimeter of the cavity. To absorb sound in this frequency range, absorptive material needs to be placed around the perimeter of the cavity. In fact, this is sometimes the only practical place to position the absorption. An example of this is high performance windows in music studios, where two or more glazing units are separated by wide cavities. To increase the absorption of sound at and above the first cross-cavity mode the two main surfaces that face into the cavity also need to be absorptive.

Cavities within plasterboard and masonry walls are often filled or partially filled with absorbent porous material. This introduces additional internal losses as sound waves propagate through the porous material.

Cavities tend to have relatively small volumes which often contain additional absorbent material so it is not usually necessary to consider air absorption for the building acoustics frequency range.

1.3.5.1 Reverberation time

Sound fields in cavities rarely approximate a diffuse field in either two or three dimensions, hence the decay curves tend to show various degrees of curvature. The reasons for this are similar to those previously discussed for non-diffuse fields in rooms; normal mode theory indicates that the degree of curvature varies depending upon the combination of axial, tangential, and oblique modes in a frequency band.

Reverberation times in cavities tend to be shorter than those in rooms; examples are shown in Fig. 1.55 which were measured using T_{10}, T_{15}, or T_{20}.

1.3.5.2 Internal losses

Below the frequency of the first cross-cavity mode, the internal loss factor is determined by the absorption of the surface at the cavity perimeter. For locally reacting surfaces, Eq. 1.76

(a) Timber joist floor cavity

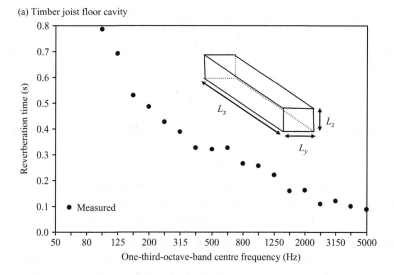

(b) Masonry wall cavity

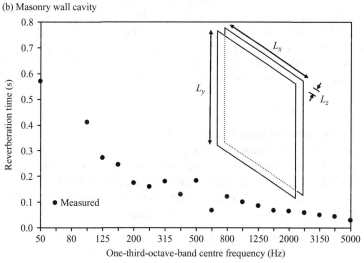

Figure 1.55

Examples of measured reverberation times in cavities. (a) Timber joist floor cavity. $L_x = 4.2\,\text{m}$, $L_y = 0.4\,\text{m}$, and $L_z = 0.225\,\text{m}$. Areas L_yL_z are fair-faced masonry. Areas L_xL_z are timber joists. Area L_xL_y (upper) is chipboard. Area L_xL_y (lower) is plasterboard. (b) Masonry wall cavity. $L_x = 3.6\,\text{m}$, $L_y = 5.0\,\text{m}$, and $L_z = 0.075\,\text{m}$. Areas L_xL_y, L_xL_z and L_yL_z (left side) are fair-faced masonry. Area L_yL_z (right side) is 455 mm, 28 kg/m³ mineral wool (cavity stop). Measured data from Hopkins are reproduced with permission from ODPM and BRE.

gives the angle-dependent sound absorption coefficient, however, to simplify the calculation it is assumed that $\alpha_\theta = \alpha_0 \cos(\theta)$, which gives (Price and Crocker, 1970)

$$\eta_{ii} = \frac{S_P \bar{\alpha}_P c_0}{2\pi^2 f V} \tag{1.186}$$

where S_P is the surface area of the cavity perimeter and $\bar{\alpha}_P$ is the average statistical sound absorption coefficient for the cavity perimeter. For box-shaped cavities, $S_P = 2(L_xL_z + L_yL_z)$

and $\bar{\alpha}_P = \sum_{k=1}^{4} S_k \alpha_k / S_P$ where S_k and α_k correspond to the area and statistical absorption coefficient for each side of the cavity perimeter. If the statistical absorption coefficients are not available and the perimeter surface is locally reacting, then $\bar{\alpha}_P$ can be estimated from the normal incidence absorption coefficient, α_0, using (Price and Crocker, 1970)

$$\bar{\alpha}_P = \frac{\pi}{4}\alpha_0 \tag{1.187}$$

At and above the frequency of the first cross-cavity mode, the internal loss factor is

$$\eta_{ii} = \frac{S_i \bar{\alpha} c_0}{8\pi fV} \tag{1.188}$$

where S_T is the total area of all the cavity surfaces and $\bar{\alpha}$ is the average statistical sound absorption coefficient for all the cavity surfaces. For box-shaped cavities, $S_T = 2(L_x L_y + L_x L_z + L_y L_z)$ and $\bar{\alpha} = \sum_{k=1}^{6} S_k \alpha_k / S_T$ where S_k and α_k correspond to each surface of the cavity.

Near the frequency of the first cross-cavity mode an issue arises in using Eqs 1.186 and 1.53 to calculate the internal loss factor. This is because there is often a significant difference between the values for two-dimensional and three-dimensional sound fields. Over the building acoustics frequency range, this causes a sharp transition in the predicted internal loss factor and the predicted reverberation time. In practice, damping measurements inside real cavities indicate a more gradual transition. In the prediction of sound transmission, this is not usually a problem as a sharp transition will not normally occur in the predicted sound insulation because of the existence of many other sound transmission paths.

1.3.5.2.1 Sound absorption coefficient: Locally reacting porous materials

Calculation of the internal loss factor requires the normal incidence or statistical sound absorption coefficient for the cavity boundaries. For porous materials the absorption coefficient can be calculated by treating the material as an equivalent gas and using wave theory to calculate the specific acoustic impedance or admittance (e.g. see Allard, 1993). This can make use of equations such as those of Delany and Bazley (Section 1.3.2.2) or Allard and Champoux (1992) to determine $Z_{0,pm}$ and k_{pm} for the equivalent gas.

It is assumed that the porous material is locally reacting with a thickness, h, and is positioned a distance, d, from a rigid non-porous surface that has an infinite impedance (see Fig. 1.56). For this calculation, most masonry/concrete walls and floors can be assumed to be rigid. The calculations in this section are equally applicable to rooms where locally reacting porous materials are placed near masonry/concrete walls or floors.

The normal incidence and statistical absorption coefficients can be calculated using Eqs 1.79 and 1.81 respectively where the specific acoustic admittance is calculated using

$$\beta_{a,s} = \beta_{Re} - i\beta_{Im} = \frac{1}{Z_{a,s}} = \left(\frac{iZ_{0,pm}}{\rho_0 c_0} \frac{\tan(k_{pm}h)\tan\left(\frac{2\pi fd}{c_0}\right) - \frac{\rho_0 c_0}{Z_{0,pm}}}{\tan\left(\frac{2\pi fd}{c_0}\right) + \frac{\rho_0 c_0}{Z_{0,pm}}\tan(k_{pm}h)} \right)^{-1} \tag{1.189}$$

When $d = 0$, the porous layer is next to the rigid surface (often referred to as rigid backing). Equation 1.189 then reduces to

$$\beta_{a,s} = \beta_{Re} - i\beta_{Im} = \frac{1}{Z_{a,s}} = \left(\frac{-iZ_{0,pm}}{\rho_0 c_0} \frac{1}{\tan(k_{pm}h)} \right)^{-1} \tag{1.190}$$

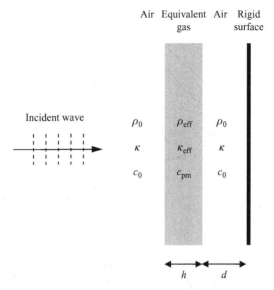

Air Equivalent Air Rigid
gas surface

Incident wave ρ_0 ρ_{eff} ρ_0

κ κ_{eff} κ

c_0 c_{pm} c_0

h d

Figure 1.56

Absorber: porous material – air gap – rigid surface. Equivalent gas model used to represent the porous material.

When $d = n\lambda/2$ for $n = 1, 2, 3$, etc., the specific acoustic admittance calculated from Eq. 1.189 is the same as Eq. 1.82, and the porous material can be considered as rigidly backed.

Examples for the statistical absorption coefficient are shown in Fig. 1.57 for a range of airflow resistivities from 2000 to 64 000 Pa.s/m². Two thicknesses of porous material are considered, $h = 0.025$ m and $h = 0.1$ m, each of which have air gaps of $d = 0$ m and $d = 0.1$ m. For rigid backing, increasing the thickness of the material from 25 to 100 mm significantly increases the absorption coefficient in the low- and mid-frequency ranges. However, by using a 100 mm air gap with the 25 mm material it is possible to achieve similarly high values to the 100 mm material with rigid backing in the low- and mid-frequency ranges; this is at the expense of lower absorption coefficients in the high-frequency range. With an air gap, the curve for the absorption coefficient has a ripple with troughs that tend to become less pronounced with increasing airflow resistivity. In practice, this ripple is less pronounced due to the use of frequency bands, variation in material properties, and variation in d due to workmanship.

The airflow resistance of porous materials tends to increase with increasing bulk density, but there is no simple rule that porous materials with low or high airflow resistivity will always give the highest absorption coefficients over the building acoustics frequency range. To determine suitable values of r, h, and d, it is necessary to identify which part of the frequency range requires the highest absorption coefficients. There are a large number of permutations for these three variables, and measured absorption coefficients for a specific combination are not always available. In order to assess their effect it is usually sufficient to calculate the absorption coefficient as described in this section. For fibrous materials, a wide range of densities are available (typically 10 to 200 kg/m³) which gives a wide range of airflow resistivities from which to choose a specific material. However, commonly available materials come in a limited range of thicknesses, which, in combination with the cavity dimensions will limit the choice of h and d.

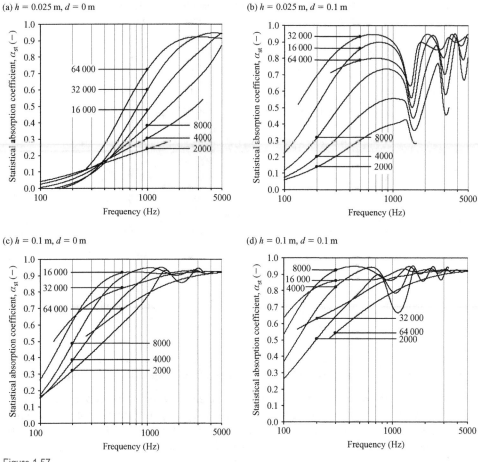

Figure 1.57

Statistical absorption coefficients of porous materials for a range of airflow resistivities in Pa.s/m^2.

1.3.5.3 Coupling losses

Calculation of the coupling loss factors involving the cavity are discussed in Section 4.3.5.3.

1.3.5.4 Total loss factor

The total loss factor equals the sum of the internal and coupling loss factors. For most cavities in walls and floors that have absorptive surfaces, the coupling loss factors are much smaller than the internal loss factor, and the latter provides a reasonable estimate of the total loss factor. As with rooms, Eq. 1.107 can be used to calculate the total loss factor from the reverberation time and vice versa.

1.3.5.5 Modal overlap factor

The modal overlap factor for cavities is calculated using Eq. 1.109.

1.3.6 Energy

Calculation of the sound energy stored in a cavity is calculated using Eq. 1.154 in the same way as for rooms.

1.4 External sound fields near building façades

To assess the airborne sound insulation of the building façade from external sound sources it is necessary to measure the sound pressure levels both inside and outside the building. Having looked at the internal sound field, we will now look at the external sound field near a façade.

The sound pressure level near the façade depends upon: the position of the microphone in relation to the façade and the ground, diffraction effects from the edges of the façade, diffraction effects from protruding or recessed elements on the building façade (e.g. balconies), sound propagation from the source (including the effects of ground impedance, façade impedance, and meteorological conditions), the orientation of the sound source, and the type of sound source outside the building (e.g. point source, line source).

Microphone positions relative to the façade and the ground often differ depending upon whether the primary aim is to measure the façade sound insulation, or measure/predict the environmental noise near the façade. In the latter case, the measurements/predictions are often used at a later point in time to estimate sound transmission into the building via the façade; it is clearly advantageous if the microphone positions are the same or the levels can be accurately converted. For field measurements of façade sound insulation, the microphone is usually attached to the surface of the façade at variable heights that depend upon the building element that is being measured, or positioned 2 m in front of the façade at a height of 1.5 m above the floor of the receiving room (ISO 140 Part 5). Environmental noise measurements are taken at a variety of different positions; often at a height of 1.2, 1.5, or 4 m above floor level, and at distances between 1 and 2 m in front of the façade (ISO 1996 Part 1).

In practice we often need to convert sound pressure levels near the building façade to free-field levels in the absence of the façade and vice versa. This section therefore looks at the difference between the external sound pressure level with the façade to the level without the façade (i.e. the change in level due to the presence of the façade).

1.4.1 Point sources and semi-infinite façades

For façade sound insulation measurements made with a loudspeaker and some environmental noise sources it is appropriate to consider a point source. We therefore start by looking at the sound field generated by a point source in the vicinity of a façade. By creating the image sources for this situation as shown in Fig. 1.58, we see that the sound travels from the source (S) to the receiver (R) via four different paths: the first path is the direct path from the source to the receiver, the second path involves a single reflection from the ground, the third path involves a single reflection from the ground and a single reflection from the façade, and the fourth path involves a single reflection from the façade. The path lengths in terms of the distance, d, from the source, or image source, to the receiver are also indicated in this diagram.

We will assume that: (a) the source emits spherical waves, (b) the ground and façade are perfectly reflecting with no phase change upon reflection (c) all reflections are specular

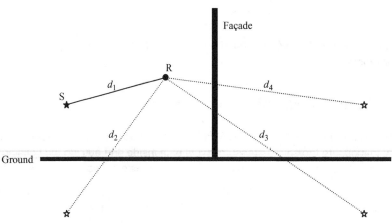

Figure 1.58

External façade sound pressure level measurements. Source (S) and receiver (R) orientation with image sources (☆) for the different propagation paths.

(d) the façade has dimensions that are very large compared to the wavelength (i.e. a semi-infinite plate), and (e) there are no other façades nearby that significantly affect the sound field. Therefore we will not concern ourselves with diffraction from the edges of the wall or with different impedances for the ground and the façade. The assumption of specular reflection is reasonable for this situation, particularly below 1000 Hz; it can generally be assumed that real façades have small scattering coefficients (Ismail and Oldham, 2005).

We are interested in the difference between the sound pressure level in front of the façade and the free-field level without the façade. This requires the ratio of the total mean-square sound pressure, p^2, to the mean-square sound pressure, $(p_1 + p_2)^2$; the latter term corresponds to the combination of the direct path between source and receiver (path length d_1), and the path in which the sound is reflected directly from the ground to the receiver (path length d_2). The sound pressure for spherical waves at single frequencies is taken from Eq. 1.22, hence the required ratio is

$$\frac{\langle p^2 \rangle_t}{\langle (p_1 + p_2)^2 \rangle_t} = \frac{\left| \dfrac{\exp(-ikd_1)}{d_1} + \dfrac{\exp(-ikd_2)}{d_2} + \dfrac{\exp(-ikd_3)}{d_3} + \dfrac{\exp(-ikd_4)}{d_4} \right|^2}{\left| \dfrac{\exp(-ikd_1)}{d_1} + \dfrac{\exp(-ikd_2)}{d_2} \right|^2} \qquad (1.191)$$

Now we can calculate the change in level due to the presence of the façade for different distances of the receiver from the façade. For façade sound insulation measurements, the external microphone is usually at a distance of 2 m from the façade or on the surface of the façade (ISO 140 Part 5); hence we will use these to define the minimum and maximum distances for the range of interest. To illustrate the effect of intermediate distances we will look at 300 mm and 1 m.

For measurements on the surface of the façade there are usually physical limitations that determine how close the microphone can be positioned to the surface. For a half-inch microphone (12.7 mm diameter) attached to the façade with the axis of the microphone parallel to the plane of the façade, we can assume that the façade–receiver distance is 6.35 mm (i.e. the distance from the façade surface to the centre of the microphone diaphragm). Figure 1.59 shows the

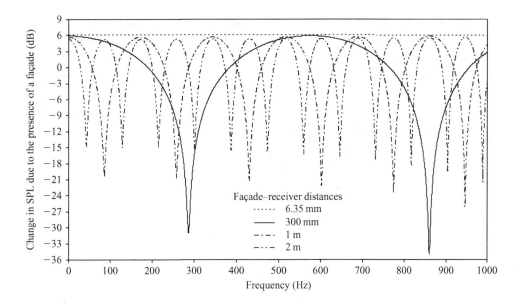

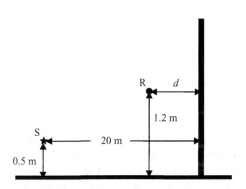

Figure 1.59

Change in the sound pressure level due to the presence of the façade for receiver positions at four different distances, d, from the façade (single frequencies from 0.25 to 1000 Hz). Source-receiver-façade geometry is indicated in the sketch.

calculated level difference for four different façade-receiver distances at frequencies up to 1000 Hz. For a half-inch microphone attached to the surface of a façade there is a constant level difference of 6 dB, this is referred to as pressure doubling. As the microphone is moved further away from the façade we see that there are interference minima in the spectrum due to destructive interference between the different propagation paths. These occur due to the different distances travelled by the sound waves along each of the different paths. For the various combinations of paths, the path difference in metres corresponds to a phase difference in radians. Destructive interference occurs where the path length difference, Δd_{pq}, between paths p and q, corresponds to a phase difference of an odd number of π radians,

$$2\pi \frac{\Delta d_{pq}}{\lambda} = (2n + 1)\pi \tag{1.192}$$

where $n = 0, 1, 2, 3$, etc.

The upper frequency shown in this example has been limited to 1000 Hz because at higher frequencies, turbulent air in the outdoor environment tends to reduce the coherence between the waves that travel along the different propagation paths (Attenborough, 1988; Quirt, 1985). As a result, this simple model is no longer appropriate, and sharp minima in the spectrum due to destructive interference are less likely to occur above 1000 Hz.

In practice we usually deal with frequency bands rather than single frequencies. For frequency bands the same ratio can be calculated from the band centre frequency using

$$\frac{\langle p^2 \rangle_t}{\langle (p_1 + p_2)^2 \rangle_t} = \frac{1 + \left(\frac{d_1}{d_2}\right)^2 + \left(\frac{d_1}{d_3}\right)^2 + \left(\frac{d_1}{d_4}\right)^2 + \frac{2d_1}{d_2}R(\Delta d_{12}) + \frac{2d_1}{d_3}R(\Delta d_{13}) + \frac{2d_1}{d_4}R(\Delta d_{14}) + \frac{2d_1^2}{d_2 d_3}R(\Delta d_{23}) + \frac{2d_1^2}{d_2 d_4}R(\Delta d_{24}) + \frac{2d_1^2}{d_3 d_4}R(\Delta d_{34})}{1 + \left(\frac{d_1}{d_2}\right)^2 + \frac{2d_1}{d_2}R(\Delta d_{12})} \tag{1.193}$$

where the autocorrelation function, $R(\Delta d_{pq})$ for each path length difference (magnitude), Δd_{pq}, is (Delany et al., 1974)

$$R(\Delta d_{pq}) = \frac{\lambda}{2\pi B_L \Delta d_{pq}} \cos\left(\frac{2\pi \Delta d_{pq}}{\lambda}\right) \sin\left(\frac{2\pi B_L \Delta d_{pq}}{\lambda}\right) \tag{1.194}$$

for which λ is the wavelength corresponding to the band centre frequency, and B_L is calculated from the lower and upper frequency limits of the band, f_l and f_u, using $B_L = (f_u - f_l)/(f_u + f_l)$. $B_L = 0.115$ for one-third-octave-bands.

For one-third-octave-bands between 50 and 1000 Hz the change in level due to the presence of the façade is shown in Fig. 1.60 (source and receiver positions are the same as in Fig. 1.59). The change in level is 6 dB for a half-inch microphone attached to the surface of the façade. One advantage of using surface measurements is that if the microphone is positioned very close to the surface, we can avoid interference minima in the building acoustics frequency range, although there will be small departures from pressure doubling in the high-frequency range. This allows us to make the convenient assumption of pressure doubling. In contrast, the façade–receiver distance of 300 mm provides an example of the variation that can be introduced when measurements are not made on the surface of the façade. With this particular combination of source–receiver–façade geometry there is an interference dip around the 315 Hz band. If, for example, we were to change the façade-receiver distance from 300 to 200 mm, we would shift the interference dip into a different frequency band. This dependence on the specific geometry of each situation illustrates the importance of well-defined measurement positions for comparative measurements of the sound field near façades. For a façade-receiver distance of 1 or 2 m there are dominant interference minima in the low-frequency range. However the one-third-octave-bands get wider as the frequency increases and the interference effects begin to average out. For façade-receiver distances of 1 or 2 m in the mid-frequency range, the change in level tends towards 3 dB; this is referred to as energy doubling.

For a point sound source, such as a loudspeaker in the low-frequency range, the above discussion indicates that the sound pressure level will also vary over the surface of a façade due to the different interference patterns that occur with different source–receiver–façade geometries. In the mid- and high-frequency ranges, loudspeakers tend to become increasingly directional

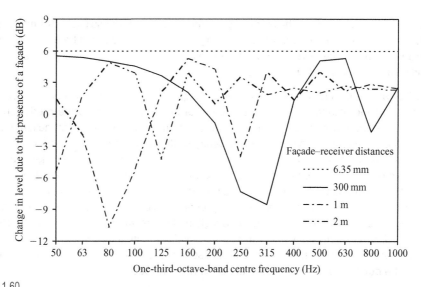

Figure 1.60

Change in the sound pressure level due to the presence of the façade for receiver positions at different distances from the façade.

and no longer act as point sources, hence the variation in sound pressure level over the façade is also affected by the directionality of the loudspeaker.

Although several assumptions have been made in this basic model, it adequately illustrates the general trends. In practice there are other factors that affect the depth and frequency of the interference minima. The finite impedance of the ground causes a phase change upon reflection from the ground, so to improve the model it is necessary to incorporate measurements of the ground impedance (e.g. see Ogren and Jonasson, 1998). Compared to the ground, relatively little information is available on the impedance of façades. However, façade surfaces are rarely highly porous and tend to have low-absorption coefficients (typically less than 0.1 in the low- and mid-frequency ranges). For this reason, the assumption of a perfectly reflecting surface is often reasonable.

1.4.1.1 Effect of finite reflector size on sound pressure levels near the façade

As real façades are of finite size, we need to look at the effects of diffraction from the edges of a façade. The sound field in front of finite size reflectors can be considered as the combination of the four geometrical wave paths (as previously considered for the semi-infinite reflector), combined with edge or boundary diffraction waves. To assess diffraction we will look at indoor scale-model measurements because it is awkward to control all the relevant parameters with outdoor measurements near real buildings. Results are taken from scale model experiments in a semi-anechoic chamber with a concrete floor, and a 30 mm thick square reflector (varnished board) to represent the façade. Good agreement between these measurements and predictions using Integral Equation Methods (IEM) allow conclusions to be drawn purely by using measured data (Hopkins and Lam, 2008). A 1:5 scale model was used for the measurements, but all the results shown and discussed in this section are scaled-up to the situation for real façades (i.e. full-size). The source was a small loudspeaker positioned in the vertical plane perpendicular to the center line of the reflector. The receiver position was offset from this plane by one-twelfth of the receiver dimension to avoid perfect symmetry in the set-up that might be unrepresentative of the situation in practice. Five square reflectors were tested that represented full-size façades with side dimensions of 2, 3, 4, 5, and 6 m.

Figure 1.61 shows the change in level due to two square reflectors (6×6 m and 2×2 m) with a façade–receiver distance of 2 m. Measured data is shown alongside the prediction for a semi-infinite reflector (Eq. 1.191). Compared to the semi-infinite reflector, diffraction from the edges of the finite reflector affects the frequency of the peaks and troughs as well as their values. As one would expect, this is more pronounced for the smaller reflector. For small reflectors, the receiver will be relatively close to the edges and the edge diffracted pressure can significantly change the interference pattern in comparison to the semi-infinite reflector. The 6×6 m reflector can be taken as being representative of the façade of a detached house, and diffraction can be considered to have negligible effect on measured levels above 100 Hz. For the 2×2 m reflector, diffraction can have a significant effect below 1000 Hz; in practice, most façades are much larger than this, but it is used here to represent small square protruding sections of a building (e.g. bay window, entrance hall, enclosed balcony).

For practical purposes we need to assess the difference between finite size reflectors and a semi-infinite reflector in one-third-octave-bands; this is done using the difference between the measured and the predicted (Eq. 1.193) change in level due to the presence of the façade. Figure 1.62 shows this level difference for a façade–receiver distance of 2 m. For square reflectors with side dimensions between 3 and 6 m, the level differences are generally less than 3 dB in the low-frequency range. The differences are larger with the 2×2 m reflector, particularly at 63 Hz, but they are generally less than 3 dB across the low- and mid-frequency ranges. Environmental noise measurements are often taken using a façade–receiver distance between 1 and 2 m. Figure 1.63 shows the level difference for 11 different façade–receiver distances in 0.1 m steps from 1 to 2 m for each of four different square reflectors (side dimensions between 3 and 6 m). In the low-frequency range there are significant differences between the semi-infinite and the finite reflectors due to diffraction. In the mid-frequency range these differences are negligible and these finite reflectors can be treated as semi-infinite. The level differences for the 2×2 m reflector are shown separately in Fig. 1.63; these indicate that it is not appropriate to treat this small reflector as semi-infinite in both the low- and mid-frequency ranges.

In practice there are so many permutations of source–receiver–façade geometry that it is difficult to make a definitive statement about the conditions in which diffraction effects will be negligible. As a rule-of-thumb for a point source near the ground, and façade–receiver distances between 1 and 2 m, diffraction effects are only likely to be significant in the low-frequency range for façades with dimensions <5 m.

1.4.1.2 Spatial variation of the surface sound pressure level

Façade elements such as windows or doors often have lower airborne sound insulation than the wall around them. Hence, field measurement of the apparent sound reduction index is often needed for these elements. This requires measurement of the average surface sound pressure level over the element. A spatial average is needed for all elements regardless of the source–receiver–façade geometry; usually between three and ten microphone positions on the surface of the element (ISO 140 Part 5).

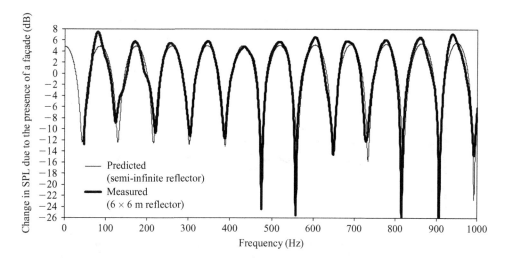

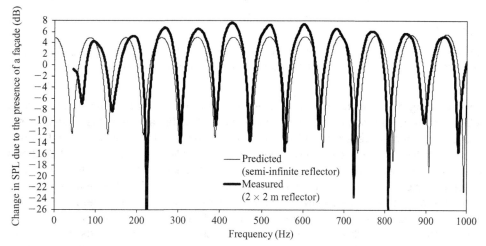

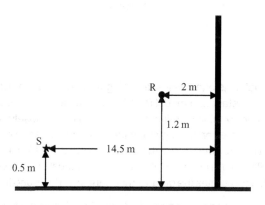

Figure 1.61

Comparison of measured and predicted data for the change in the sound pressure level due to the presence of the façade. Source–receiver–façade geometry is indicated in the sketch. Measured data reproduced with permission from Hopkins and Lam (2008).

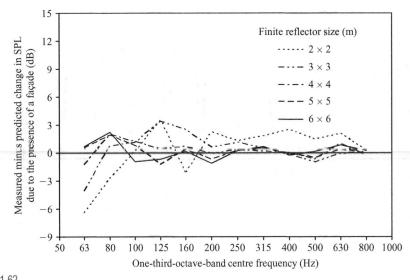

Figure 1.62

Difference between the measured (finite reflector) and predicted (semi-infinite reflector) change in level due to the presence of the façade. Façade–receiver distance of 2 m. Measured data are reproduced with permission from Hopkins and Lam (2008).

For protruding or recessed building elements, the spatial variation over the surface can be affected by a combination of diffraction, shielding, and, within a recess, the existence of a sound field that partly resembles a two-dimensional reverberant field (sometimes referred to as a niche effect). It is quite common for windows to be installed in a recess. Figure 1.64 shows the effect of measuring the surface sound pressure level within a 200 mm deep frame (1 × 1 m) attached to the surface of a masonry façade (Quirt, 1985). A single measurement within the frame is seen to be unrepresentative of the average from eight positions. Measurements on a 1.2 × 1.2 m window with recess depths of 120 and 320 mm indicate that the spatial variation over the surface of a window is larger with a deeper recess (Jonasson and Carlsson, 1986). To get a more accurate estimate of the average surface sound pressure level, more microphone positions may be needed with deep recesses ($\approx$300 mm), than with shallower ones ($\approx$100 mm).

1.4.2 Line sources

Façade sound insulation is often assessed using road traffic noise, which can be represented by a line source. The details of a model for a line source are not discussed here, but the basic principle involves approximating a line source by a line of closely spaced incoherent point sources. For a line source comprising many incoherent point sources, air absorption starts to become significant towards the ends of the line source and therefore needs to be included in the model (ISO 9613 Part 1). An overview of a suitable spherical wave propagation model for each point source which incorporates the ground impedance can be found from Attenborough (1988).

To gain a practical insight into the sound field near façades with a line source it is more useful to look at the statistics of measured data. Hall et al. (1984) took measurements at houses on 33 different sites with road traffic as the sound source to assess the level measured at a distance of 2 m from the façade using a microphone on the façade surface (one position

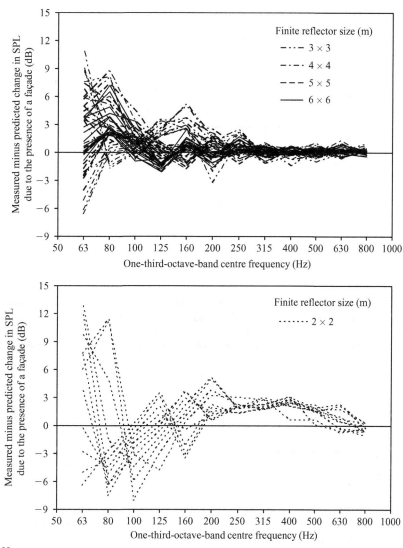

Figure 1.63

Difference between the measured (finite reflector) and predicted (semi-infinite reflector) change in level due to the presence of the façade. For each reflector size the 11 curves correspond to façade–receiver distance in 0.1 m steps from 1 to 2 m inclusive. Measured data are reproduced with permission from Hopkins and Lam (2008).

only). The microphone height above ground level was unspecified although it was the same for both the surface and the 2 m measurement. The results are shown in Fig. 1.65. By assuming pressure doubling (6 dB) for the façade microphone, the assumption of energy doubling (3 dB) for the microphone that is 2 m from the façade can be assessed by comparing the difference between these two microphone positions with a value of 3 dB. In the low-frequency range the assumption of energy doubling is invalid due to large fluctuations caused by the interference pattern. In the mid- and high-frequency ranges the assumption of energy doubling is reasonable when we consider the mean of many measurements; however, from the minimum and maximum values in Fig. 1.65 we see that this assumption is not always valid for an individual

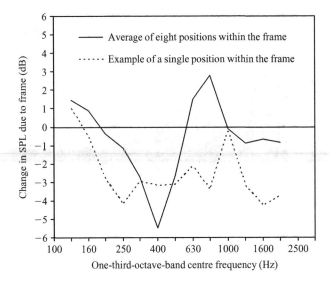

Figure 1.64

Change in the surface sound pressure level on a wall due to the addition of a 200 mm deep frame (1 × 1 m). The loudspeaker was placed on the ground at a distance of 25 m from the mid-point of the frame, with sound incident upon the surface at an angle of 60°. NB: The angle prescribed for façade insulation measurements with a loudspeaker in ISO 140 Part 5 is 45° ± 5° rather than 60°. Measured data are reproduced with permission from Quirt (1985) and the National Research Council of Canada.

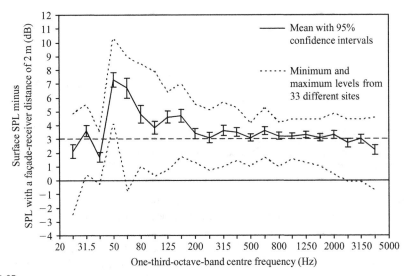

Figure 1.65

Sound pressure levels measured at houses on 33 different sites to assess the level that is measured 2 m from the façade with road traffic as the sound source. Measurements were made with a microphone on the façade surface, and at a distance of 2 m from the façade. Measured data are reproduced with permission from Hall et al. (1984).

measurement. This presents a problem if we need to accurately convert individual measurements in frequency bands from the 2 m microphone position to a different microphone position near the façade. This will rarely be possible due to the uncertainty in the many factors that affect the sound propagation paths. Usually we can only make reasonable estimates when we want to convert the mean value of many measurements for either frequency bands, or an A-weighted level.

References

Allard, J.F. (1993). *Propagation of sound in porous media: modelling sound absorbing materials*, Elsevier Science Publishers Ltd, London and New York. ISBN: 185166887X.

Allard, J.F. and Champoux, Y. (1992). New empirical equations for sound propagation in rigid frame fibrous materials, *Journal of the Acoustical Society of America*, **91** (6), 3346–3353.

Arau-Puchades, H. (1988). An improved reverberation formula, *Acustica*, **65**, 163–180.

Attenborough, K. (1988). Review of ground effects on outdoor sound propagation from continuous broadband sources, *Applied Acoustics*, **24**, 289–319.

Barron, M. (1973). Growth and decay of sound intensity in rooms according to some formulae of geometric acoustics theory, *Journal of Sound and Vibration*, **27** (2), 183–196.

Beranek, L.L. (1947). Acoustical properties of homogeneous, isotropic rigid tiles and flexible blankets, *Journal of the Acoustical Society of America*, **19** (4), 556–568.

Bies, D.A. (1988). Acoustical properties of porous materials. In Beranek, L.L (ed.), *Noise and vibration control*, Washington, DC. Institute of Noise Control Engineering, 245–269. ISBN: 0962207209.

Bies, D.A. and Hansen, C.H. (1980). Flow resistance information for acoustical design, *Applied Acoustics*, **13**, 357–391.

Biot, M.A. (1956). Theory of propagation of elastic waves in a fluid-saturated porous solid. I. Low-frequency range. II. Higher frequency range, *Journal of the Acoustical Society of America*, **28** (2), 168–191.

Bodlund, K. (1976). Statistical characteristics of some standard reverberant sound field measurements, *Journal of Sound and Vibration*, **45** (4), 539–557.

Bodlund, K. (1980). Monotonic curvature of low frequency decay records in reverberation chambers, *Journal of Sound and Vibration*, **73** (1), 19–29.

Craik, R.J.M. (1990). On the accuracy of sound pressure level measurements in rooms, *Applied Acoustics*, **29**, 25–33.

Delany, M.E. and Bazley, E.N. (1969). Acoustical characteristics of fibrous absorbent materials. *NPL AERO Report Ac37 March 1969*, National Physical Laboratory, UK.

Delany, M.E. and Bazley, E.N. (1970). Acoustical properties of fibrous absorbent materials, *Applied Acoustics*, **3**, 105–116.

Delany, M.E., Rennie, A.J. and Collins, K.M. (1974). Model evaluation of the noise shielding of aircraft ground-running pens. *NPL Report Ac67 April 1974*, National Physical Laboratory, UK.

Eyring, C.F. (1930). Reverberation time in dead rooms, *Journal of the Acoustical Society of America*, **1**, 217–241.

Fahy, F. (2001). *Foundations of engineering acoustics*, Elsevier Ltd, Oxford. ISBN: 0122476654.

Fitzroy, D. (1959). Reverberation formulae which seems to be more accurate with non-uniform distribution of absorption, *Journal of the Acoustical Society of America*, **31**, 893–897.

Hall, F.L., Papakyriakou, M.J. and Quirt, J.D. (1984). Comparison of outdoor microphone locations for measuring sound insulation of building façades, *Journal of Sound and Vibration*, **92** (4), 559–567.

Hodgson, M. (1993). Experimental evaluation of the accuracy of the Sabine and Eyring theories in the case of non-low surface absorption, *Journal of the Acoustical Society of America*, **94** (2), 835–840.

Hodgson, M. (1996). When is diffuse-field theory applicable?, *Applied Acoustics*, **49** (3), 197–207.

Hopkins, C. and Turner, P. (2005). Field measurement of airborne sound insulation between rooms with non-diffuse sound fields at low frequencies, *Applied Acoustics*, **66** (12), 1339–1382.

Hopkins, C. and Lam, Y. (2008). Sound fields near building façades – comparision of finite and semi-infinite reflectors on a rigid ground plane, *Applied Acoustics*. (Submitted)

Ismail, M.R. and Oldham, D.J. (2005). A scale model investigation of sound reflection from building façades, *Applied Acoustics*, **66**, 123–147.

Jacobsen, F. (1982). Decay rates and wall absorption at low frequencies, *Journal of Sound and Vibration*, **81** (3), 405–412.

Jonasson, H. and Carlsson, C. (1986). Measurement of sound insulation of windows in the field. *Nordtest Project 556–85, Technical Report SP-RAPP 1986:37*, Swedish National Testing Institute SP.

Kang, J. (2002). *Acoustics of long spaces: Theory and design practice*, Thomas Telford, London. ISBN: 0727730134.

Kosten, C.W. (1960). The mean free path in room acoustics, *Acustica*, **10**, 245–250.

Kuttruff, H. (1979). *Room acoustics*, Second Edition, Applied Science Publishers Ltd, Barking, England. ISBN: 0853348138.

Larsen, H. (1978). *Technical review No.4: Reverberation at low frequencies*, Bruel & Kjaer. (Refer to the paper by Bodlund (1980) for corrections.)

Lubman, D. (1968). Fluctuations of sound with position in a reverberant room, *Journal of the Acoustical Society of America*, **44**, 1491–1502.

Lubman, D. (1974). Precision of reverberant sound power measurements, *Journal of the Acoustical Society of America*, **56** (2), 523–533.

Lyon, R.H. and DeJong, R.G. (1995). *Theory and application of statistical energy analysis*, Butterworth-Heinemann, MA, USA. ISBN: 0750691115.

Mechel, F.P. (1989/1995/1998). *Schallabsorber Band I, II and III*, S. Hirzel Verlag, Stuttgart. ISBN: 3777604259/ISBN: 3777605727/ISBN: 3777608092.

Mechel, F.P. and Vér, I.L. (1992). Sound-absorbing materials and sound absorbers. In Beranek, L.L. and Vér, I.L. (eds.), *Noise and vibration control engineering: principles and applications*, John Wiley & Sons, 203–243. ISBN: 0471617512.

Mechel, F.P. (1995). Absorberparameter, *Schallabsorber Band II: Innere Schallfelder, Strukturen, Kennwerte*, S. Hirzel, Verlag, Stuttgart, 103–123. ISBN: 3777605727.

Michelsen, N. (1982). Repeatability of sound insulation measurements. *Technical Report No. 36*, Danish Acoustical Laboratory (now DELTA Acoustics), Denmark.

Morse, P.M. and Ingard, K.U. (1968). *Theoretical acoustics*, McGraw-Hill, New York. ISBN: 0691084254.

Munro, T.J. (1982). Alternative models of the sound field in a reverberation room, M.Eng thesis, Department of Mechanical Engineering, University of Adelaide, Adelaide, South Australia.

Narang, P.P. (1995). Material parameter selection in polyester fibre insulation for sound transmission and absorption, *Applied Acoustics*, **45**, 335–358.

Neubauer, R.O. (2001). Estimation of reverberation time in rectangular rooms with non-uniformly distributed absorption using a modified Fitzroy equation, *Building Acoustics*, **8** (2), 115–137.

Nichols, R.H. (1947). Flow-resistance characteristics of fibrous acoustical materials, *Journal of the Acoustical Society of America*, **19** (5), 866–871.

Nilsson, E. (2004). Decay processes in rooms with non-diffuse sound fields. Part I: Ceiling treatment with absorbing material, *Building Acoustics*, **11** (1), 39–60.

Ogren, M. and Jonasson, H. (1998). Measurement of the acoustic impedance of ground. *SP Report 1998:28, Swedish National Testing and Research Institute SP*.

Olesen, H.S. (1992). Measurements of the acoustical properties of buildings – additional guidelines. Nordtest Technical Report No. 203 (www.nordtest.org). ISSN: 0283-7234.

Pan, J. and Bies, D.A. (1988). An experimental investigation into the interaction between a sound field and its boundaries, *Journal of the Acoustical Society of America*, **83** (4), 1436–1444.

Pan, J. and Bies, D.A. (1990). The effect of fluid-structural coupling on sound waves in an enclosure – Theoretical part, *Journal of the Acoustical Society of America*, **87** (2), 691–707.

Pan, J. and Bies, D.A. (1990). The effect of fluid-structural coupling on sound waves in an enclosure – Experimental part, *Journal of the Acoustical Society of America*, **87** (2), 708–717.

Price, A.J. and Crocker, M.J. (1970). Sound transmission through double panels using statistical energy analysis, *Journal of the Acoustical Society of America*, **47** (3), 683–693.

Quirt, J.D. (1985). Sound fields near exterior building surfaces, *Journal of the Acoustical Society of America*, **77** (2), 557–566.

Redmore, T.L. (1982). A method to predict the transmission of sound through corridors, *Applied Acoustics*, **15**, 133–146.

Redmore, T.L. and Flockton, S.J. (1977). A design formula for predicting the attenuation of sound along a corridor, *Acoustic Letters*, **1**, 21–24.

Sabine, C. (1932) *Collected papers on acoustics*, (1964 Edition) Dover Publications, New York.

Schroeder, M.R. (1962). Frequency-correlation functions of frequency responses in rooms, *Journal of the Acoustical Society of America*, **34** (12), 1819–1823.

Schroeder, M.R. (1969). Effect of frequency and space averaging on the transmission responses of multimode media, *Journal of the Acoustical Society of America*, **46** (1), (2), 277–283.

Simmons, C. (1996). Measurement of low frequency sound in rooms. *SP Report 1996:10*, Swedish National Testing and Research Institute SP.

Vian, J.P., Danner, W.F. and Bauer, J.W. (1983). Assessment of significant acoustical parameters for rating sound insulation of party walls, *Journal of the Acoustical Society of America*, **73** (4), 1236–1243.

Vorländer, M. (1995). Revised relation between the sound power and the average sound pressure level in rooms and consequences for acoustic measurements, *Acustica*, **81**, 332–343.

Waterhouse, R.V. (1955). Interference patterns in reverberant sound fields, *Journal of the Acoustical Society of America*, **27** (2), 247–258.

Waterhouse, R.V. (1968). Statistical properties of reverberant sound fields, *Journal of the Acoustical Society of America*, **43**, 1436–1444.

Vibration fields

2.1 Introduction

When airborne sound from the human voice is transmitted from one room to another, the resulting vibration on the walls and floors in the receiving room is at a sufficiently low level that it can rarely be perceived with our fingertips. Ears are rather more sensitive, so if the background noise level in the receiving room is low, and the airborne sound insulation is quite low too, we can detect the sound pressure radiated by these vibrating walls and floors. This chapter looks at the theory describing vibration fields and relates it to plates and beams that are used to build walls and floors in buildings. An understanding of vibration fields, power input into structures, and sound radiation from vibrating structures is essential to the study of sound insulation. Compared to sound fields the degree of complexity increases due to the existence of different wave types as well as a wide range of wavelengths over the building acoustics frequency range.

The main structural building components for which we need to describe the vibration fields are beams and plates (see Fig. 2.1). Rooms are bounded by plates in the form of walls and floors; hence plates play the key role in both sound radiation and structure-borne sound transmission. Many walls and floors, or their constituent parts, can be represented as solid plates hence these form a fundamental building element. In practice there are many plates with more complicated forms, such as walls built from masonry blocks with large internal voids, profiled concrete floors, and studwork walls; however, the solid plate forms an important benchmark against which other plates can be assessed. Beams, such as columns and joists, play a minor role in terms of sound radiation but play an important role in the transmission of structure-borne sound between plates. Buildings typically contain a variety of solid beams (e.g. concrete columns, timber joists) along with more complex profiles (e.g. resilient metal channels in lightweight walls and floors).

2.2 Vibration

To describe the vibration of structures we make use of three vector quantities: (i) displacement (η), (ii) velocity (v), and (iii) acceleration (a). Their time signals are related to each other by differentiation or integration using,

$$\eta(t) = \int v(t)\,dt \qquad v(t) = \frac{d\eta(t)}{dt} \qquad v(t) = \int a(t)\,dt \qquad a(t) = \frac{dv(t)}{dt} \qquad (2.1)$$

For a sinusoid or narrow bandwidths with an angular centre frequency, ω, the amplitudes of these quantities ($\hat{\eta}, \hat{v}, \hat{a}$) are related to each other by

$$\hat{\eta} = \frac{\hat{v}}{\omega} = \frac{\hat{a}}{\omega^2} \qquad (2.2)$$

(a) Rectangular and circular cross-section beams

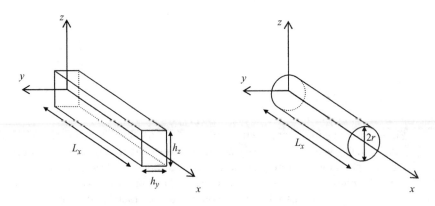

(b) Plates

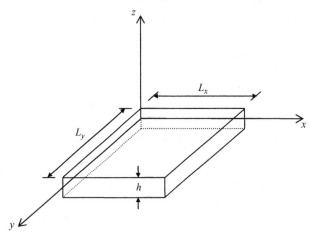

Figure 2.1

General coordinate conventions and dimensions used for beams and plates.

2.2.1   Decibels and reference quantities

The decibel scale (dB) is commonly used to deal with the wide range of values encountered with vibration. Table 2.1 gives the reference quantities that are used for all figures in the book.

2.3   Wave types

For sound in air and other gases there is only one wave type that needs to be considered, namely longitudinal waves. In contrast, there are three wave types for vibrations on homogeneous beams and plates in buildings.

For beams the three important wave types are bending, quasi-longitudinal, and torsional waves. For plates they are bending, quasi-longitudinal, and transverse shear waves. We are primarily

Table 2.1. Vibration – definitions of levels in decibels

Level	Definition	Reference quantity
Displacement	$L_d = 20 \lg \left(\dfrac{\eta}{\eta_0} \right)$ where η is the rms displacement	$\eta_0 = 10^{-12}$ m
Velocity[#]	$L_v = 20 \lg \left(\dfrac{v}{v_0} \right)$ where v is the rms velocity	$v_0 = 10^{-9}$ m/s
Acceleration[#]	$L_a = 20 \lg \left(\dfrac{a}{a_0} \right)$ where a is the rms acceleration	$a_0 = 10^{-6}$ m/s^2
Energy	$L_E = 10 \lg \left(\dfrac{E}{E_0} \right)$	$E_0 = 10^{-12}$ J
Structure-borne sound power	$L_W = 10 \lg \left(\dfrac{W}{W_0} \right)$	$W_0 = 10^{-12}$ W
Loss factors (Internal, Coupling, Total)	$L_{\eta_{ii}} / L_{\eta_{ij}} / L_{\eta_j} = 10 \lg \left(\dfrac{\eta}{\eta_0} \right)$	$\eta_0 = 10^{-12}$
Impedance*	$L_Z = 20 \lg \left(\dfrac{Z}{Z_0} \right)$	$Z_0 = 1$ Ns/m
Mobility*	$L_Y = 20 \lg \left(\dfrac{Y}{Y_0} \right)$	$Y_0 = 1$ m/(Ns)

[#] Indicates use of preferred SI reference quantities for acoustics (ISO 1683).
*Applies to real part, imaginary part, or magnitude.

interested in sound radiation from plates, for which bending waves are the most important; although all three types play a role in structure-borne sound transmission. For brevity, quasi-longitudinal and transverse shear waves are sometimes referred to as in-plane waves. It is not usually necessary to consider in-plane waves for direct sound transmission across a wall, floor, door, or window. However when flanking transmission involves several connected walls and floors in the mid- and high-frequency range, in-plane waves start to play an important role.

Compared to sound in gases it is clearly more complicated to have three different wave types to consider, as well as some differences between the wave motion on beams and plates. There are two practical options for modelling walls and floors as simple plates over the building acoustics frequency range; either bending waves are considered to be the only important wave type (thin and/or thick plate theory), which is the simpler and often perfectly adequate option, or all three wave types are considered together because quasi-longitudinal and transverse shear waves always co-exist. However, it is better to avoid the temptation to jump straight to the section on bending waves. There is a logical flow in the sequential derivation of quasi-longitudinal waves, followed by transverse shear waves, and finally, bending waves. In addition it is easier to grasp the concepts of the different wave types on beams before moving on to look at plates. It is important to note that the range of wavelengths for structural waves is very different to sound in air; some examples for different plates are shown in Fig. 2.2.

For a thorough review of structure-borne sound the reader is referred to the book by Cremer et al. (1973).

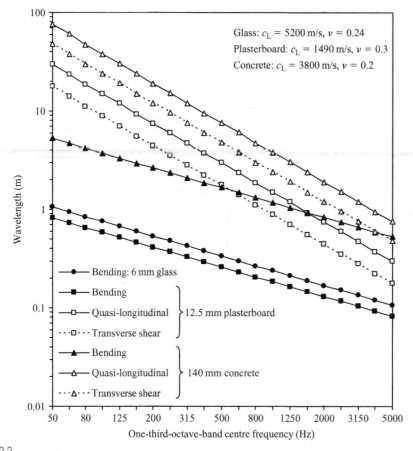

Figure 2.2

Wavelengths for bending, quasi-longitudinal, and transverse shear waves on different plates (thin plate theory).

2.3.1 Quasi-longitudinal waves

For longitudinal waves, propagation is in the same direction as the particle motion. In Chapter 1 we looked at longitudinal waves for sound in air (or other gases) in rooms and cavities. This pure form of longitudinal wave motion can also occur in structures too, but only when all the dimensions of the structure are much larger than the wavelength of the longitudinal wave. Over the building acoustics frequency range, most beams and plates in buildings are 'thin' in terms of their thickness when compared to the wavelength. For this reason the majority of beams and plates support a different type of longitudinal wave, a quasi-longitudinal wave. As with pure longitudinal waves propagating in the x-direction, quasi-longitudinal waves have in-plane displacements in the x-direction, ξ, which give rise to longitudinal strains. However these displacements not only cause longitudinal strains, but also lateral strains (see Fig. 2.3). The out-of-plane displacements that are associated with these lateral strains are ζ and η for the y- and z-directions respectively.

In contrast to bending waves, the lateral strains associated with quasi-longitudinal waves produce small lateral displacements. For this reason, sound radiation from quasi-longitudinal

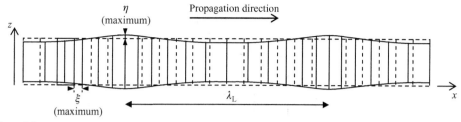

Figure 2.3

Quasi-longitudinal wave on a beam or plate.

waves is usually insignificant compared to bending waves, as is airborne excitation of these quasi-longitudinal waves from a reverberant sound field (Heckl, 1981). However, quasi-longitudinal waves do play an important role in the transmission of structure-borne sound between connected plates where they can be generated by bending waves that are incident upon a junction with other plates.

For a quasi-longitudinal wave propagating in the x-direction, the lateral strains, ε_y and ε_z, resulting from the longitudinal strain, ε_x, are determined from the Poisson's ratio of the material, v. The general set of equations relating Young's modulus and Poisson's ratio to stress and strain in Cartesian coordinates is

$$E = \frac{\sigma_x - v(\sigma_y + \sigma_z)}{\varepsilon_x}$$

$$E = \frac{\sigma_y - v(\sigma_x + \sigma_z)}{\varepsilon_y} \qquad (2.3)$$

$$E = \frac{\sigma_z - v(\sigma_x + \sigma_y)}{\varepsilon_z}$$

For a beam with a quasi-longitudinal wave propagating in the x-direction there are no constraints on the sides of the beam; hence substituting the conditions $\sigma_y = 0$ and $\sigma_z = 0$ into Eq. 2.3 gives

$$E = \frac{\sigma_x}{\varepsilon_x} \qquad (2.4)$$

and

$$\varepsilon_y = \varepsilon_z = -v\varepsilon_x \qquad (2.5)$$

The in-plane displacements and stresses on a beam are shown in Fig. 2.4 for a small rectangular element with a width, dx, for which the longitudinal strain is written in terms of the displacement, ξ, as

$$\varepsilon_x = \frac{\partial \xi}{\partial x} \qquad (2.6)$$

The stress is related to the displacement using the equation of motion,

$$\left(\sigma_x + \frac{\partial \sigma_x}{\partial x} dx\right) - \sigma_x = \rho \, dx \frac{\partial^2 \xi}{\partial t^2} \qquad (2.7)$$

hence,

$$\frac{\partial \sigma_x}{\partial x} = \rho \frac{\partial^2 \xi}{\partial t^2} \qquad (2.8)$$

115

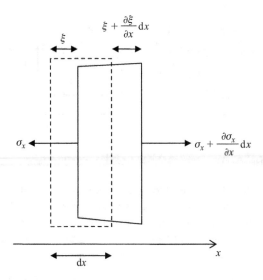

Figure 2.4

Small rectangular element of a beam undergoing quasi-longitudinal wave motion in the x-direction. The position at rest is shown in dashed lines along with the altered shape and position in solid lines due to the wave motion.

where ρ is the density.

For a beam, the longitudinal force, F_x, on the cross-sectional area of the element, S, is defined such that a compressive stress, $-\sigma_x$, gives a positive force, and a tensile stress, σ_x, gives a negative force:

$$F_x = -S\sigma_x \qquad (2.9)$$

It is now possible to determine the wave equation; from Eqs 2.8 and 2.9 we find that

$$\frac{\partial F_x}{\partial x} = -\rho S \frac{\partial^2 \xi}{\partial t^2} \qquad (2.10)$$

and from Eqs 2.4, 2.6 and 2.9 we have

$$\frac{\partial F_x}{\partial t} = -ES \frac{\partial^2 \xi}{\partial x \partial t} \qquad (2.11)$$

Equations 2.10 and 2.11 therefore define the wave equation for quasi-longitudinal waves on a beam as

$$E\frac{\partial^2 \xi}{\partial x^2} - \rho \frac{\partial^2 \xi}{\partial t^2} = 0 \qquad (2.12)$$

Having established the basic equations for a beam, they can now be adapted to a plate. For a quasi-longitudinal wave propagating along a plate in the x-direction the cross-section is constrained by the plate material in the y-direction, but unconstrained in the z-direction, hence substituting the conditions $\varepsilon_y = 0$ and $\sigma_z = 0$ into Eq. 2.3 gives

$$\frac{E}{(1 - \nu^2)} = \frac{\sigma_x}{\varepsilon_x} \qquad (2.13)$$

where the strain is written in terms of the displacement, ξ, as

$$\varepsilon_x = \frac{\partial \xi}{\partial x} \qquad (2.14)$$

For a plate it is necessary to reclassify F_x as the longitudinal force per unit width, to give

$$F_x = -h\sigma_x \tag{2.15}$$

where h is the plate thickness.

Equations 2.8 and 2.13–2.15 therefore give

$$\frac{\partial F_x}{\partial x} = -\rho h \frac{\partial^2 \xi}{\partial t^2} \tag{2.16}$$

and

$$\frac{\partial F_x}{\partial t} = \frac{-Eh}{(1-\nu^2)} \frac{\partial^2 \xi}{\partial x \, \partial t} \tag{2.17}$$

From Eqs 2.16 and 2.17 the wave equation for quasi-longitudinal waves on a plate is

$$\frac{E}{(1-\nu^2)} \frac{\partial^2 \xi}{\partial x^2} - \rho \frac{\partial^2 \xi}{\partial t^2} = 0 \tag{2.18}$$

For a quasi-longitudinal wave propagating along a beam or plate in the x-direction, the displacement in the x-direction, ξ, can be described by

$$\xi(x, t) = \hat{\xi} \exp(-ik_x x) \exp(i\omega t) \tag{2.19}$$

where $\hat{\xi}$ is an arbitrary constant, and k_x equals the quasi-longitudinal wavenumber, $k_L = \omega / c_L$.

The phase velocity, c_L, for quasi-longitudinal waves is determined by inserting Eq. 2.19 into the appropriate wave equation for beams or plates. Hence the phase velocity is

$$c_{L,b} = \sqrt{\frac{E}{\rho}} \quad \text{for beams} \tag{2.20}$$

and

$$c_{L,p} = \sqrt{\frac{E}{\rho(1-\nu^2)}} \quad \text{for plates} \tag{2.21}$$

Note that the subscript b for beams and p for plates is used in this chapter to make a distinction between them when it is considered necessary or helpful; later chapters predominantly discuss plates so the subscript is dropped and reference is simply made to c_L.

For most homogeneous materials used to build walls and floors in buildings, quasi-longitudinal phase velocities are >1400 m/s (see the Appendix, Table A2 for examples of material properties). We will soon see that the phase velocities for other wave types can be calculated from the quasi-longitudinal phase velocity; this makes it a very useful property.

The group velocity, c_g, is the velocity at which the wave energy propagates across the beam or plate. For quasi-longitudinal waves where the wavelength is much greater than the beam or plate thickness, the group velocity $c_{g(L)}$, is the same as the phase velocity, c_L.

2.3.1.1 Thick plate theory

The phase velocity can be assumed to be independent of frequency, i.e. non-dispersive, when the wavelength is much greater than the plate thickness. However, the dispersive nature of quasi-longitudinal waves can no longer be ignored in the high-frequency range with thick masonry/concrete plates that are sometimes used in buildings. An error of $X\%$ in the phase

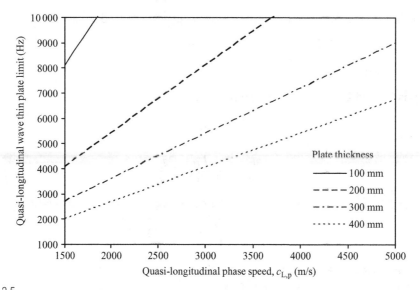

Figure 2.5

Thin plate limits for quasi-longitudinal waves.

velocity can be used to define a limiting frequency, $f_{L(thin)}$; this frequency can be considered as a thin plate limit above which the plate can no longer be considered as a thin plate for quasi-longitudinal waves (Cremer *et al.*, 1973).

$$f_{L(thin)} = \frac{c_{L,p}}{h} \sqrt{\frac{\frac{X\%}{100}}{\frac{\pi^2}{6}\left(\frac{\nu}{1-\nu}\right)^2}} \tag{2.22}$$

For an error of 3% and a Poisson's ratio of 0.3, this corresponds to the frequency at which $\lambda_L \approx 3h$.

Figure 2.5 shows the thin plate limit for a common range of phase velocities, assuming an error of 3% and a Poisson's ratio of 0.2. The thin plate limits tend to be in or above the high-frequency range. Due to other aspects which limit accurate prediction of structure-borne sound transmission on building elements (particularly at high frequencies where many plates cannot be considered as homogeneous) it is usually sufficient to use thin plate theory over the entire building acoustics frequency range.

2.3.2   Transverse waves

Solids in the form of beams and plates can support transverse waves due to their ability to support shear stresses. On beams these waves are referred to as torsional waves, whereas those on plates are referred to as transverse shear waves.

2.3.2.1   *Beams: torsional waves*

Torsional waves are generated in beams where a time-varying moment is applied via an axis that passes through the axis of the beam. As a torsional wave propagates along a beam in the x-direction, the cross-sections rotate about the axis of the beam by an angle, θ. It is therefore

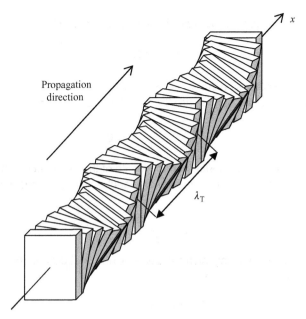

Figure 2.6

Torsional wave on a beam.

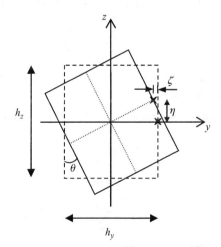

Figure 2.7

Cross-section of a rectangular beam in the yz plane undergoing torsional wave motion. The position at rest is shown in dashed lines along with the altered shape and position in solid lines due to the wave motion. Example displacements (exaggerated) are indicated by ✗ for one point on the surface of the beam.

apparent that for beams with circular cross-sections there will be no displacement outside of the cross-section. These waves are easier to visualize for a beam with a rectangular cross-section as shown in Fig. 2.6. For rectangular cross-sections the displacements in the y- and z-directions are ζ and η respectively (see Fig. 2.7), and vary depending on the distance from the axis,

$$\zeta = -\theta z \qquad \eta = \theta y \qquad (2.23)$$

The moment about the x-axis, M_x, for any shape of cross-section is (Cremer $et\ al.$, 1973)

$$M_x = T\frac{\partial \theta}{\partial x} \qquad (2.24)$$

This moment is used to determine the wave equation that governs the propagation of torsional waves in the x-direction and is defined as (Cremer $et\ al.$, 1973)

$$\frac{\partial^2 \theta}{\partial x^2} - \frac{\rho I_\theta}{GJ}\frac{\partial^2 \theta}{\partial t^2} = 0 \qquad (2.25)$$

where I_θ is the polar moment of inertia (per unit length of the beam) about the x-axis, J is the torsional moment of rigidity, and G is the shear modulus of the material which is related to Young's modulus by

$$G = \frac{E}{2(1+\nu)} \qquad (2.26)$$

For a torsional wave propagating along a beam in the x-direction, the angular torsional displacement, θ, can be described by

$$\theta(x,t) = \hat{\theta}\exp(-ik_x x)\exp(i\omega t) \qquad (2.27)$$

where $\hat{\theta}$ is an arbitrary constant, and k_x equals the torsional wavenumber, $k_T = \omega/c_{T,b}$.

Substitution of Eq. 2.27 into the wave equation gives the phase velocity as

$$c_{T,b} = \sqrt{\frac{GJ}{\rho I_\theta}} = \sqrt{\frac{T}{\rho I_\theta}} \qquad (2.28)$$

The product GJ equals the torsional stiffness, T. The product ρI_θ is sometimes referred to as the mass moment of inertia about the x-axis.

For a solid beam with a circular cross-section of radius, r, the torsional stiffness is

$$T = G\frac{\pi r^4}{2} \qquad (2.29)$$

and the polar moment of inertia about the x-axis is

$$I_\theta = \frac{\pi r^4}{2} \qquad (2.30)$$

For a solid beam of rectangular cross-section, the torsional stiffness can be calculated using (Timoshenko and Goodier, 1970)

$$T = G\frac{h_z h_y^3}{3}\left[1 - \frac{192h_y}{\pi^5 h_z}\tanh\left(\frac{\pi h_z}{2h_y}\right)\right] \qquad (2.31)$$

where $h_z \geq h_y$, and the polar moment of inertia about the x-axis is

$$I_\theta = \frac{(h_y h_z^3 + h_z h_y^3)}{12} \qquad (2.32)$$

Torsional waves are non-dispersive, so the group velocity, $c_{g(T)}$, is the same as the phase velocity, $c_{T,b}$.

2.3.2.2 Plates: transverse shear waves

For transverse shear waves propagating in the x-direction, the in-plane displacement, ζ, takes place in the y-direction, i.e. perpendicular to the direction of propagation (see Fig. 2.8). Because the only motion of the plate surface is tangential to the adjacent air (or other gas) these waves are not able to radiate sound, or to be excited by airborne sound that is incident upon the plate. So, as with quasi-longitudinal waves, their primary role is in structure-borne sound transmission.

Transverse motion is assessed using a small rectangular element on the plate which lies in the xy plane (see Fig. 2.9) with an area of $dxdy$ (Cremer *et al.*, 1973). For a wave propagating in the x-direction there will be different displacements on the two sides of the element that are parallel to the y-axis, namely ζ and $\zeta + (\partial\zeta/\partial x)dx$. These displacements cause the element to change shape from its rectangular form to that of a parallelogram. Therefore the sides of the element that were parallel to the x-axis at rest have moved through the small angle, γ_{xy}, (where $\tan \gamma_{xy} \approx \gamma_{xy}$) although the volume of the plate element has remained the same. This angle is

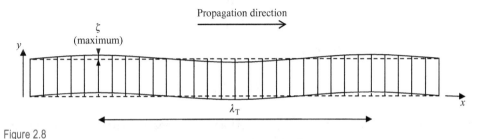

Figure 2.8

Transverse shear wave on a plate.

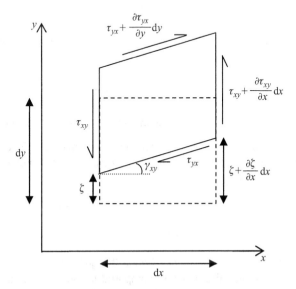

Figure 2.9

Small rectangular element in the xy plane of a plate undergoing transverse shear wave motion. The position at rest is shown in dashed lines along with the altered shape and position in solid lines due to the wave motion.

referred to as the shear strain and is related to the displacement by

$$\gamma_{xy} = \frac{\partial \zeta}{\partial x} \tag{2.33}$$

The displacements result in shear stresses: τ_{xy} (a stress acting in the y-direction on the plane that lies perpendicular to the x-axis, i.e. the yz plane) and τ_{yx} (a stress acting in the x-direction on the plane that lies perpendicular to the y-axis, i.e. the xz plane). These shear stresses are of equal magnitude when the displaced element is at equilibrium in terms of the moments that act upon the yz and xz planes; they are also proportional to the shear strain, hence

$$\tau_{xy} = \tau_{yx} = G\gamma_{xy} \tag{2.34}$$

where G is the shear modulus, the ratio of the shear stress to the shear strain. The shear modulus is related to Young's modulus by

$$G = \frac{E}{2(1 + v)} \tag{2.35}$$

The equation of motion for the element is

$$\frac{\partial \tau_{xy}}{\partial x} = \rho \frac{\partial^2 \zeta}{\partial t^2} \tag{2.36}$$

The wave equation governing the propagation of transverse shear waves in the x-direction can now be found from Eqs 2.33, 2.34, and 2.36 as

$$G\frac{\partial^2 \zeta}{\partial x^2} - \rho \frac{\partial^2 \zeta}{\partial t^2} = 0 \tag{2.37}$$

For a transverse shear wave propagating along a plate in the x-direction, the displacement in the y-direction, ζ, can be described by an equation of the form

$$\zeta(x, t) = \hat{\zeta} \exp(-ik_x x) \exp(i\omega t) \tag{2.38}$$

where $\hat{\zeta}$ is an arbitrary constant, and k_x equals the transverse shear wavenumber, $k_T = \omega/c_{T,p}$.

The phase velocity, $c_{T,p}$ for transverse shear waves is determined by inserting Eq. 2.38 into the wave equation; hence it is related to the shear modulus, which can be calculated from the quasi-longitudinal phase velocity,

$$c_{T,p} = \sqrt{\frac{G}{\rho}} = \sqrt{\frac{E}{2\rho(1 + v)}} = c_{L,p}\sqrt{\frac{1 - v}{2}} \tag{2.39}$$

Transverse shear waves are non-dispersive hence the group velocity $c_{g(T)}$, is the same as the phase velocity, $c_{T,p}$. Plates that are thin compared to the wavelength do not affect the transverse shear waves which propagate in the plane of the plate. So, unlike bending and quasi-longitudinal waves there is no limiting frequency for the phase velocity relating to the assumption of a thin plate where the wavelength is large compared to the plate thickness.

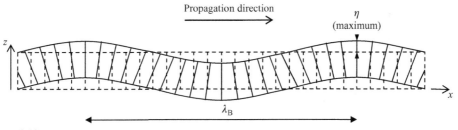

Figure 2.10

Bending wave on a beam or plate.

2.3.3 Bending waves

Pure bending waves occur where the bending wavelength is large compared to the beam or plate thickness; hence these waves only occur on 'thin beams' and 'thin plates'. A propagating bending wave causes both rotation and lateral displacement of the beam or plate elements (see Fig. 2.10). When compared with in-plane waves, bending waves have large lateral displacements; hence they play the main role in the radiation of sound.

The wave equation for a beam is determined by considering a bending wave propagating in the x-direction (Cremer et al., 1973). For the purpose of this derivation the beam is positioned so that the x-axis runs along the centroid of the cross-section. We will choose the z-direction as the direction in which the lateral displacement occurs but note that a beam of rectangular cross-section can support a bending wave in both the y- and z-directions. Figure 2.11 shows that a small rectangular element on the beam will undergo lateral displacement, η, in the z-direction as well as rotation through the small angle, β (where $\tan \beta \approx \beta$). Hence we need to consider the lateral velocity, $v_z = \partial \eta / \partial t$, in the z-direction, as well as the angular velocity, ω_y, about the y-axis. The former is defined as positive in the positive z-direction, and the latter is positive for anti-clockwise rotation. The angle of rotation is related to the displacement by

$$\beta = \frac{\partial \eta}{\partial x} \tag{2.40}$$

from which the angular velocity is related to the lateral velocity by

$$\omega_y = \frac{\partial \beta}{\partial t} = \frac{\partial^2 \eta}{\partial t\, \partial x} = \frac{\partial v_z}{\partial x} \tag{2.41}$$

In comparing the element before and after bending deformation we see that the line on the upper surface is shortened due to compression (i.e. AD becomes A′D′), and that the line on the lower surface is elongated due to tension (i.e. BC becomes B′C′). Between the upper and lower surfaces there is a neutral axis along which there is neither compression nor tension; therefore the element length, dx, along this axis is unchanged. The magnitude of the stresses and strains are zero on the neutral axis and linearly increase towards the largest values on the upper and lower surfaces of the element (see Fig. 2.12).

For a homogeneous beam of rectangular or circular cross-section, the neutral axis lies at the mid-height of the beam. The neutral axis on the deformed element lies on an arc of radius, R_0, hence

$$dx = R_0 \theta \tag{2.42}$$

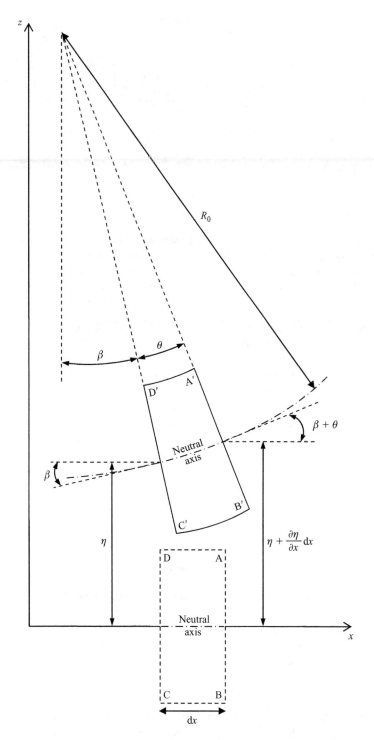

Figure 2.11

Small rectangular element of a beam undergoing bending wave motion with displacement in the z-direction. The position at rest is shown in dashed lines along with the altered shape and position in solid lines due to the wave motion.

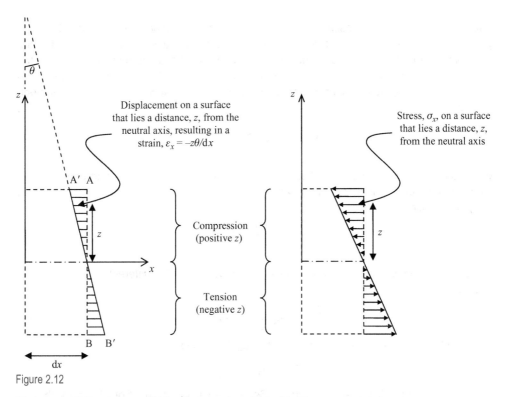

Figure 2.12

Displacements (left) and stresses (right) on the beam element using line AB as an example.

It is assumed that plane surfaces in the beam remain plane when they are rotated by the small angle, β. This important assumption means that lines such as AB or CD on the element at rest, which are perpendicular to the neutral axis, will remain normal to the neutral axis on the element during bending wave motion. Therefore lines AB and CD rotate in opposite directions such that the resulting angle, θ, between A′B′ and C′D′ on the deformed element is

$$\theta = \frac{\partial \beta}{\partial x}dx = \frac{\partial^2 \eta}{\partial x^2}dx \tag{2.43}$$

and from Eq. 2.42, the radius of the neutral axis is related to the lateral displacement by

$$\frac{1}{R_0} = \frac{\theta}{dx} = \frac{\partial^2 \eta}{\partial x^2} \tag{2.44}$$

Hence we can describe the strain in the x-direction on any surface that lies a distance, z, from the neutral axis (see Fig. 2.12) using

$$\varepsilon_x = \frac{[(R_0 - z)\theta] - R_0\theta}{R_0\theta} = -\frac{z}{R_0} = -z\frac{\partial^2 \eta}{\partial x^2} \tag{2.45}$$

The accompanying stress for the unconstrained cross-section of a beam is calculated using the conditions $\sigma_y = 0$ and $\sigma_z = 0$ in Eq. 2.3, and using Eq. 2.45 for the strain to give

$$\sigma_x = E\varepsilon_x = -Ez\frac{\partial^2 \eta}{\partial x^2} \tag{2.46}$$

The stresses from compression and tension result in a bending moment, M_y about the y-axis. This moment rotates in the same direction as the angular velocity, ω_y. The individual moments that result in the bending moment are the product of the surface stress and the distance of the surface, z, from the neutral axis (see Fig. 2.12). This bending moment is calculated by summing up the individual moments over the cross-sectional area of the element, S, using

$$M_y = \int_S \sigma_x z \, dS = -El_b \frac{\partial^2 \eta}{\partial x^2} \tag{2.47}$$

where the moment of inertia for a beam, l_b, is a measure of its resistance to bending. The moment of inertia is defined about an axis. In this case we need the moment of inertia about the y-axis which is

$$l_b = \int_S z^2 \, dS \tag{2.48}$$

To relate the bending moment to the angular velocity, Eq. 2.47 is differentiated with respect to time,

$$\frac{\partial M_y}{\partial t} = -El_b \frac{\partial^3 \eta}{\partial t \, \partial x^2} = -El_b \frac{\partial \omega_y}{\partial x} \tag{2.49}$$

for which the bending stiffness of a solid beam, B_b, is defined as

$$B_b = El_b = \rho c_{L,b}^2 l_b \tag{2.50}$$

For solid circular section beams, $l_b = \pi(2r)^4/64$, where r is the radius. Solid beams with rectangular sections can support bending waves in both the y- and z-directions. Therefore it is necessary to identify the direction of the lateral displacement because the moment of inertia depends upon the orientation of the beam. For lateral displacement in the z-direction as we have used in this derivation, the moment of inertia of the cross-sectional area about the y-axis is $l_b = h_y h_z^3/12$; whereas for lateral displacement in the y-direction the moment of inertia about the z-axis would be $l_b = h_z h_y^3/12$.

In addition to the stresses within the beam element due to compression and tension, there will also be shear stresses due to a shear force, F_z, acting in the z-direction (see Fig. 2.13). The shear force is found in terms of the lateral displacement using the equation of motion

$$F_z - \left(F_z + \frac{\partial F_z}{\partial x} dx \right) = \rho_l \, dx \frac{\partial^2 \eta}{\partial t^2} \tag{2.51}$$

which gives

$$\frac{\partial F_z}{\partial x} = -\rho_l \frac{\partial^2 \eta}{\partial t^2} \tag{2.52}$$

where ρ_l is the mass per unit length.

Balancing the shear forces and moments on the element at equilibrium (see Fig. 2.13) yields,

$$F_z dx = M_y - \left(M_y + \frac{\partial M_y}{\partial x} dx \right) \tag{2.53}$$

hence the shear force is related to the bending moment by

$$F_z = -\frac{\partial M_y}{\partial x} \tag{2.54}$$

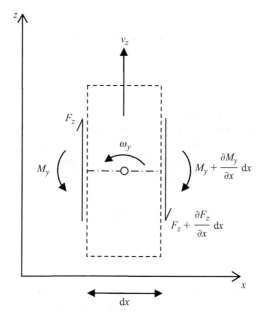

Figure 2.13

Moments, shear forces, angular and transverse velocities for the beam element undergoing bending wave motion.

Note that this equilibrium equation has ignored the rotatory inertia of the element. This is a valid assumption for pure bending wave motion on thin beams and thin plates (Cremer *et al.*, 1973). For thick beams the rotatory inertia is accounted for by replacing F_z with $F_z + \rho I_b(\partial \omega_y / \partial t)$ in Eq. 2.54.

The one-dimensional wave equation for bending waves on a thin homogeneous beam is determined from Eqs 2.49, 2.52, and 2.54,

$$B_b \frac{\partial^4 \eta}{\partial x^4} + \rho_l \frac{\partial^2 \eta}{\partial t^2} = 0 \tag{2.55}$$

The bending wave equation is different to the other wave equations in that it contains a fourth order (rather than second order) derivative term for the displacement. This results in a more complex solution to the wave equation. For a sinusoidal bending wave propagating along a beam in the *x*-direction we initially describe the displacement in the *z*-direction by an equation of the form

$$\eta(x, t) = \hat{\eta} \exp(-ik_x x) \exp(i\omega t) \tag{2.56}$$

where $\hat{\eta}$ is an arbitrary constant and k_x is a constant relating to the wavenumber.

Substituting Eq. 2.56 into the wave equation for the beam gives the bending wavenumber, k_B, as

$$k_x^4 = k_B^4 = \frac{\omega^2 \rho_l}{B_b} \tag{2.57}$$

which has four roots: $+k_B$, $-k_B$, $+ik_B$, and $-ik_B$. Hence for bending wave motion in the x-direction the general solution for the displacement is written in the form

$$\eta(x, t) = [\hat{\eta}_+ \exp(-ik_B x) + \hat{\eta}_- \exp(ik_B x) + \hat{\eta}_{n-} \exp(-k_B x) + \hat{\eta}_{n+} \exp(k_B x)] \exp(i\omega t) \quad (2.58)$$

where $\hat{\eta}_+$, $\hat{\eta}_-$, $\hat{\eta}_{n-}$, and $\hat{\eta}_{n+}$ are constants.

The first and second terms within the square brackets represent a bending wave propagating in the positive and the negative x-directions respectively. The third and fourth terms represent vibration fields called nearfields that decay away exponentially in the positive and negative x-directions respectively. An exponential decay means that the nearfield contribution to the overall displacement is usually only significant at positions close to the point at which the nearfields are generated.

From Eq. 2.57, the phase velocity for the propagating bending waves on a solid beam is

$$c_{B,b} = \sqrt[4]{\frac{\omega^2 B_b}{\rho_l}} = \sqrt[4]{\frac{4\pi^2 f^2 c_{L,b}^2 I_b}{S}} \quad (2.59)$$

In contrast to the phase velocity for quasi-longitudinal waves on thin plates, transverse shear waves on plates, and longitudinal sound waves in air, the phase velocity for bending waves is frequency-dependent. Bending waves are therefore described as dispersive. The dispersive nature of bending waves means that the group velocity, $c_{g(B),b}$, at which bending wave energy travels is not the same as the phase velocity and is calculated from

$$c_{g(B),b} = \frac{d\omega}{dk} = 2c_{B,b} \quad (2.60)$$

The one-dimensional wave equation for a thin beam can be adapted to apply to a bending wave propagating in the x-direction on a thin homogeneous isotropic plate. As we have already seen with quasi-longitudinal waves on plates, E must be replaced by $E/(1 - \nu^2)$ because, unlike the beam, the plate element cross-section is constrained by the material in the y-direction, hence Eq. 2.46 describing the internal stresses becomes

$$\sigma_x = \frac{E\varepsilon_x}{1 - \nu^2} = \frac{-Ez}{1 - \nu^2} \frac{\partial^2 \eta}{\partial x^2} \quad (2.61)$$

and the bending stiffness for a thin homogeneous isotropic plate, B_p, is defined as

$$B_p = \frac{EI_p}{1 - \nu^2} = \frac{Eh^3}{12(1 - \nu^2)} = \frac{\rho c_{L,p}^2 h^3}{12} \quad (2.62)$$

where I_p is the moment of inertia per unit width, $h^3/12$.

Equation 2.52 is adapted by temporarily reclassifying F_z as the shear force per unit width of the plate, and by replacing the mass per unit length with the mass per unit area, ρ_s, to give

$$\frac{\partial F_z}{\partial x} = -\rho_s \frac{\partial v_z}{\partial t} = -\rho_s \frac{\partial^2 \eta}{\partial t^2} \quad (2.63)$$

Hence, the one-dimensional wave equation for bending waves propagating in the x-direction on a thin homogeneous plate is

$$B_p \frac{\partial^4 \eta}{\partial x^4} + \rho_s \frac{\partial^2 \eta}{\partial t^2} = 0 \quad (2.64)$$

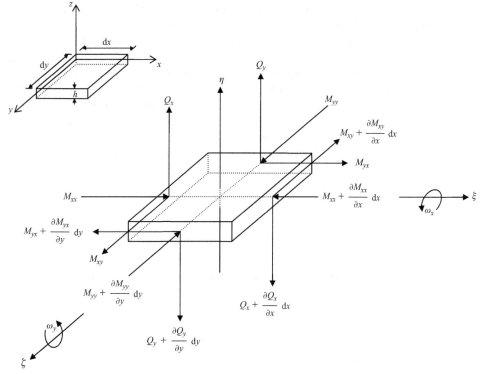

Figure 2.14

Bending and twisting moments per unit width, shear forces per unit width, displacements and angular velocities for an element on a plate undergoing bending wave motion.

The phase velocity on a thin plate is determined using the same approach as for beams, where the wavenumber is

$$k_B^4 = \frac{\omega^2 \rho_s}{B_p}$$ (2.65)

which gives the phase velocity as

$$c_{B,p} = \sqrt[4]{\frac{\omega^2 B_p}{\rho_s}} = \sqrt[4]{\frac{4\pi^2 f^2 h^2 E}{12\rho(1-\nu^2)}} = \sqrt{\frac{2\pi f h c_{L,p}}{\sqrt{12}}}$$ (2.66)

The group velocity, $c_{g(B),p}$, which describes the transport of bending wave energy is determined from Eq. 2.65, where

$$c_{g(B),p} = \frac{d\omega}{dk_B} = 2c_{B,p}$$ (2.67)

In practice there will be bending waves on a plate propagating in the x- and y-directions simultaneously. Therefore we need to derive the two-dimensional wave equation for a thin plate that resists bending deformation about the x- and y-axes and has a twisting moment. We start by positioning the plate so that the x-axis is aligned at mid-height of the cross-section. A rectangular element in the xy plane of the plate with the dimensions, dx by dy, undergoes displacements, ξ, ζ, and η in the x-, y- and z-directions respectively. For the element in equilibrium the moments and shear forces per unit width are shown in Fig. 2.14.

The shear forces per unit width are now defined as Q_x and Q_y. The bending moments per unit width are M_{xy} and M_{yx} where the moment acts perpendicular to the axis denoted by the first subscript letter, in the direction denoted by the second subscript letter. The twisting (or torsional) moments per unit width are M_{xx} and M_{yy}. The direction of rotation for the moments is found by taking a viewpoint looking directly into the arrowhead, along the line of the arrow; the rotation from this viewpoint is in an anti-clockwise direction. There are now two angular velocities: ω_x, about the x-axis, and ω_y, about the y-axis, defined as positive in the anti-clockwise direction as shown in Fig. 2.14. We have already formulated ω_y in Eq. 2.41 for an angle of rotation, β, in the anti-clockwise direction about the y-axis. To determine ω_x the plate element rotates anti-clockwise about the x-axis by an angle, α, such that

$$\alpha = -\frac{\partial \eta}{\partial y} \tag{2.68}$$

and the angular velocity is

$$\omega_x = \frac{\partial \alpha}{\partial t} = -\frac{\partial^2 \eta}{\partial t\, \partial y} = -\frac{\partial v_z}{\partial y} \tag{2.69}$$

To find the bending moments it is necessary to determine the stresses. The stress–strain relationships are found by substituting the condition $\sigma_z = 0$ into Eq. 2.3 to give

$$E = \frac{\sigma_x - v\sigma_y}{\varepsilon_x} \quad \text{and} \quad E = \frac{\sigma_y - v\sigma_x}{\varepsilon_y} \tag{2.70}$$

from which we obtain each stress in terms of the strains

$$\sigma_x = \frac{E}{1 - v^2}(\varepsilon_x + v\varepsilon_y) \quad \text{and} \quad \sigma_y = \frac{E}{1 - v^2}(\varepsilon_y + v\varepsilon_x) \tag{2.71}$$

These strains can now be written in terms of the lateral displacement, η, using the same approach as for a beam element (see Eq. 2.45) as

$$\varepsilon_x = -z\frac{\partial^2 \eta}{\partial x^2} \quad \text{and} \quad \varepsilon_y = -z\frac{\partial^2 \eta}{\partial y^2} \tag{2.72}$$

to give the stresses

$$\sigma_x = -\frac{Ez}{1 - v^2}\left(\frac{\partial^2 \eta}{\partial x^2} + v\frac{\partial^2 \eta}{\partial y^2}\right) \quad \text{and} \quad \sigma_y = -\frac{Ez}{1 - v^2}\left(\frac{\partial^2 \eta}{\partial y^2} + v\frac{\partial^2 \eta}{\partial x^2}\right) \tag{2.73}$$

To find the twisting moments we need to determine the shear stress, τ_{xy}, in terms of the lateral displacement. The shear stress is determined from the shear strain, γ_{xy}, which is a result of the in-plane displacements as shown in Fig. 2.15. The sides of the element at rest that were parallel to the x- and y-axes have moved through the small angles, $\partial \zeta/\partial x$ and $\partial \xi/\partial y$ respectively. The sum of these angles equals the shear strain, therefore the shear stress is

$$\tau_{xy} = G\gamma_{xy} = G\left(\frac{\partial \xi}{\partial y} + \frac{\partial \zeta}{\partial x}\right) \tag{2.74}$$

From Eq. 2.72 we already have the strains in terms of the lateral displacement; hence these can be equated to the following strains in terms of the in-plane displacement,

$$\varepsilon_x = \frac{\partial \xi}{\partial x} \quad \text{and} \quad \varepsilon_y = \frac{\partial \zeta}{\partial y} \tag{2.75}$$

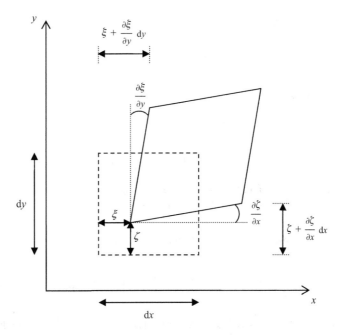

Figure 2.15

Small rectangular element in the *xy* plane of a plate undergoing shear deformation as part of bending wave motion. The position at rest is shown in dashed lines along with the altered shape and position in solid lines due to shear deformation.

for which the appropriate differentiation and integration gives

$$\frac{\partial \xi}{\partial y} = \frac{\partial \zeta}{\partial x} = -z \frac{\partial^2 \eta}{\partial x \, \partial y} \tag{2.76}$$

The shear stress can now be written in terms of E and v by using Eq. 2.35 for the shear modulus to give

$$\tau_{xy} = -\frac{Ez}{1+v} \frac{\partial^2 \eta}{\partial x \, \partial y} = -\frac{Ez(1-v)}{1-v^2} \frac{\partial^2 \eta}{\partial x \, \partial y} \tag{2.77}$$

We now assume that the shear strains, γ_{xz} and γ_{yz} are so small as to be negligible. This is consistent with the assumption that any point on the element which is normal to the neutral axis when at rest, remains normal to the neutral axis after bending deformation.

The bending moments per unit width are calculated by summing up the individual moments over the plate thickness,

$$M_{xy} = \int\limits_{-h/2}^{h/2} \sigma_x z \, dz \quad \text{and} \quad M_{yx} = -\int\limits_{-h/2}^{h/2} \sigma_y z \, dz \tag{2.78}$$

The rotation of M_{yx} about the *x*-axis is defined as being in the same direction as the rotation of the vertical edge of the plate element that lies in the *xz* plane due to compression and tension. This is opposite to the convention used for M_{xy}, and therefore a negative sign is included in the equation for M_{yx} above.

The moment of inertia for a plate is,

$$I_p = \int\limits_{-h/2}^{h/2} z^2 dz = \frac{h^3}{12} \tag{2.79}$$

hence these moments are

$$M_{xy} = -B_p \left(\frac{\partial^2 \eta}{\partial x^2} + v \frac{\partial^2 \eta}{\partial y^2} \right) \tag{2.80}$$

and

$$M_{yx} = B_p \left(\frac{\partial^2 \eta}{\partial y^2} + v \frac{\partial^2 \eta}{\partial x^2} \right) \tag{2.81}$$

A similar approach is used to find the twisting moments from the shear stress, where $\tau_{xy} = \tau_{yx}$. Taking account of the different moment directions these are

$$M_{xx} = -M_{yy} = -\int\limits_{-h/2}^{h/2} \tau_{xy} z \, dz = B_p(1 - v) \frac{\partial^2 \eta}{\partial x \, \partial y} \tag{2.82}$$

The two equilibrium equations for the moments of all forces acting on the element about the y-axis and the x-axis are

$$\left[Q_x x - \left(Q_x + \frac{\partial Q_x}{\partial x} dx \right)(x + dx) \right] dy + \left[M_{xy} - \left(M_{xy} + \frac{\partial M_{xy}}{\partial x} dx \right) \right] dy$$

$$+ \left[M_{yy} - \left(M_{yy} + \frac{\partial M_{yy}}{\partial y} dy \right) \right] dx = 0 \tag{2.83}$$

$$\left[\left(Q_y + \frac{\partial Q_y}{\partial y} dy \right)(y + dy) - Q_y y \right] dx + \left[M_{yx} - \left(M_{yx} + \frac{\partial M_{yx}}{\partial y} dy \right) \right] dx$$

$$+ \left[M_{xx} - \left(M_{xx} + \frac{\partial M_{xx}}{\partial x} dx \right) \right] dy = 0 \tag{2.84}$$

For the shear force terms, the plate is orientated in the coordinate system so that $x = y = 0$, and the term involving $\partial Q_x / \partial x$ can be assumed to be negligible. Hence, substituting the moments (Eqs 2.80, 2.81, and 2.82) into the above equations results in

$$Q_x = -\frac{\partial M_{xy}}{\partial x} - \frac{\partial M_{yy}}{\partial y} = B_p \frac{\partial}{\partial x} \left(\frac{\partial^2 \eta}{\partial x^2} + \frac{\partial^2 \eta}{\partial y^2} \right) \tag{2.85}$$

$$Q_y = \frac{\partial M_{yx}}{\partial y} + \frac{\partial M_{xx}}{\partial x} = B_p \frac{\partial}{\partial y} \left(\frac{\partial^2 \eta}{\partial x^2} + \frac{\partial^2 \eta}{\partial y^2} \right) \tag{2.86}$$

For thin plates, as with thin beams, it is appropriate to ignore the rotatory inertia of the element in these equilibrium equations. For thick plates, rotatory inertia is accounted for by replacing Q_x with $Q_x + \rho I_p (\partial \omega_y / \partial t)$ in Eq. 2.85 and replacing Q_y with $Q_y + \rho I_p (\partial \omega_x / \partial t)$ in Eq. 2.86.

The shear force is now described in terms of the lateral displacement using the equation of motion,

$$-\frac{\partial Q_x}{\partial x} dx dy - \frac{\partial Q_y}{\partial y} dy dx = \rho_s dx \, dy \frac{\partial^2 \eta}{\partial t^2} \tag{2.87}$$

The two-dimensional wave equation for bending waves on a thin homogeneous isotropic plate is now given by substituting Eqs 2.85 and 2.86 into Eq. 2.87 to yield

$$B_p \left(\frac{\partial^4 \eta}{\partial x^4} + 2 \frac{\partial^4 \eta}{\partial x^2 \partial y^2} + \frac{\partial^4 \eta}{\partial y^4} \right) + \rho_s \frac{\partial^2 \eta}{\partial t^2} = 0 \tag{2.88}$$

In Eq. 2.58 a general solution was derived for bending wave motion in the x-direction which involved both bending waves and nearfields. In a similar way to the representation of sound fields using plane waves, we can assume a plane wave field for bending waves on a plate without nearfields. The displacement can therefore be described by an equation of the form

$$\eta(x, y, t) = \hat{\eta} \exp(-ik_x x) \exp(-ik_y y) \exp(i\omega t) \tag{2.89}$$

where $\hat{\eta}$ is an arbitrary constant, and k_x and k_y are constants relating to the wavenumber.

The relationship between the wavenumber and the constants, k_x and k_y is determined by inserting Eq. 2.89 into the wave equation, which gives

$$k_B^2 = k_x^2 + k_y^2 \tag{2.90}$$

In a similar way to sound fields in rooms and cavities, Eq. 2.90 can be used to determine the plate modes.

2.3.3.1 Thick beam/plate theory

At frequencies where the bending wavelength is not large compared to the beam or plate thickness, pure bending waves no longer occur and account needs to be taken of the shear deformation and rotatory inertia that occurs with thick plates (Mindlin, 1951). For thin plates it is assumed that plane surfaces in the plate remain plane when they are rotated through the small angles, α and β. For thin beams this only applies to the angle, β. This assumption becomes invalid as the bending wavelength decreases with increasing frequency due to shear deformation of the cross-section of the element. Referring back to the beam element in Fig. 2.11 this shear deformation means that lines A′B′ and C′D′ in the deformed element are no longer straight lines. For homogeneous beams and plates the effect of shear deformation tends to be more significant than rotatory inertia (Cremer et al., 1973).

The theory for thick beams and plates is more complex than that for thin beams or plates. In addition, it does not allow simple calculation of the mode frequencies, although estimates can be made for the modal density of thick beams and plates (Lyon and DeJong, 1995). This complicates matters for masonry/concrete structures because we will soon see that the crossover frequency from thin to thick beam/plate theory usually occurs in the building acoustics frequency range. Hence it is not always possible to justify using pure bending wave theory across the entire frequency range. This transition from thin to thick can be observed in practice when measuring the modal response of a single beam or plate when it is uncoupled from other structural elements. However a building only consists of coupled beams and plates, and the point at which it is essential to switch from thin to thick beam/plate theory in the prediction of sound transmission is not clear-cut. For some masonry/concrete buildings, thin plate theory can be used across the building acoustics frequency range without incurring errors that are larger than those from other assumptions. The simplest approach is to use thin beam/plate theory and always calculate a crossover frequency to indicate the frequency above which there is increased uncertainty in the use of thin plate theory.

For beams and plates, a general rule for the validity of pure bending wave motion is that the bending wavelength must be much larger than the thickness. We will focus on thick plates (rather than thick beams) as these tend to be more important in determining the sound insulation in many buildings. A specific crossover frequency for plates can be calculated by considering the percentage difference in the phase velocity between pure bending waves on thin plates,

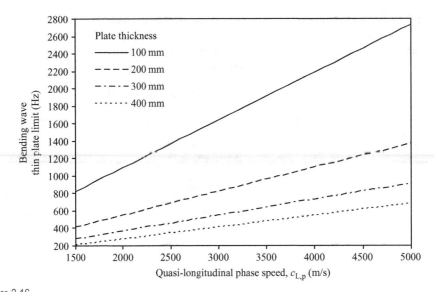

Figure 2.16

Thin plate limits for bending waves.

and bending waves on thick plates (Cremer *et al.*, 1973). This frequency is described as the thin plate limit for pure bending waves, $f_{B(thin)}$, and can be calculated for an $X\%$ difference in phase velocity using

$$f_{B(thin)} = \frac{X\%}{100} \frac{21.6c_{L,p}}{\pi^2 h} \left(1 + \frac{2.4(1+\nu)}{1-\nu^2}\right)^{-1} \qquad (2.91)$$

Cremer *et al.* (1973) propose that the bending wave correction terms for shear deformation and rotatory inertia can be important when the difference in the phase velocity is $>10\%$. This corresponds to the frequency at which $\lambda_B = 6h$. Most building materials have a Poisson's ratio between 0.2 and 0.3, hence the thin plate limit can be estimated using

$$f_{B(thin)} \approx \frac{0.05c_{L,p}}{h} \qquad (2.92)$$

The frequency at which $\lambda_B = 6h$ can be used to define a thin beam limit, hence Eq. 2.92 also applies to rectangular section beams where h corresponds to the direction of lateral displacement, h_y or h_z, and $c_{L,p}$ is replaced with $c_{L,b}$.

The thin plate limit for a single sheet of 12.5 mm plasterboard is above the building acoustics frequency range. However, for masonry/concrete walls and floors it is quite common for the thin plate limit to occur in the mid- or high-frequency range; the thin plate limit for a common range of quasi-longitudinal phase velocities is shown in Fig. 2.16 for $X = 10\%$ and a Poisson's ratio of 0.2.

The bending wave equation can be modified for thick plates to include terms that account for shear deformation and rotatory inertia (Mindlin, 1951). The bending wave effectively becomes a combination of pure bending waves and transverse shear waves. In practice we need to calculate the phase velocity and group velocity across a frequency range in which the plate is either described as thin or thick depending upon the frequency. To cover such a frequency range with a single smooth curve, a single phase velocity, $c_{B,thick,p}$, is given by (Rindel, 1994)

$$c_{B,thick,p} = \left(\frac{1}{c_{B,p}^3} + \frac{1}{\gamma^3 c_{T,p}^3} \right)^{-\frac{1}{3}}$$

(2.93)

where γ is defined for different values of the Poisson's ratio as $\gamma = 0.689$ for $\nu = 0.2$, and $\gamma = 0.841$ for $\nu = 0.3$ (Craik, 1996a).

The corresponding group velocity is (Craik, 1996)

$$c_{g(B,thick),p} = \left(\frac{c_{B,thick,p}^2}{c_{B,p}^2 \, c_{g(B),p}} + \frac{c_{B,thick,p}^2}{\gamma^3 c_{T,p}^3} \right)^{-1}$$

(2.94)

2.3.3.2 Orthotropic plates

For bending wave motion on a homogeneous plate it is necessary to establish whether the plate should be classified as isotropic (uniform stiffness), orthotropic (different stiffness in the x- and y-directions – assuming that the axes of stiffness align with the plate axes), or anisotropic (different stiffness in multiple directions). The majority of homogeneous plates in buildings can be considered as isotropic or orthotropic. The transmission of sound and vibration involving orthotropic plates is more complex than with isotropic plates, but if the properties in the two orthogonal directions are not very different, it is often possible to treat them as isotropic plates.

The wave equation for bending waves on a thin homogeneous orthotropic plate is (Cremer et al., 1973)

$$B_{p,x} \frac{\partial^4 \eta}{\partial x^4} + 2B_{p,xy} \frac{\partial^4 \eta}{\partial x^2 \partial y^2} + B_{p,y} \frac{\partial^4 \eta}{\partial y^4} + \rho_s \frac{\partial^2 \eta}{\partial t^2} = 0$$

(2.95)

where $B_{p,x}$ and $B_{p,y}$ are the bending stiffness in the x- and y-directions respectively, and

$$B_{p,xy} = \frac{\nu_{xy} E_{xy} h^3}{12} + \frac{2G_{xy} h^3}{12}$$

(2.96)

The elastic properties for orthotropic materials needed to calculate $B_{p,xy}$ (Eq. 2.96) are not always known. However, $B_{p,xy}$ is approximately equal to the geometric mean of the bending stiffness in the two orthogonal directions. This is referred to as the effective bending stiffness, $B_{p,eff}$, and is given by (Cremer et al., 1973)

$$B_{p,xy} \approx B_{p,eff} = \sqrt{B_{p,x} B_{p,y}}$$

(2.97)

Use of the effective bending stiffness often allows an orthotropic plate to be approximated by an isotropic plate. Equation 2.62 can then be used to calculate an effective quasi-longitudinal phase velocity, $c_{L,eff}$, if required for other calculations.

Masonry blocks or bricks are usually rectangular in cross-section; hence a wall will have significantly different numbers of joints per metre in the horizontal and vertical directions. For solid block/brick walls with both the horizontal and vertical joints mortared, the bending stiffness is often up to 20% higher in the horizontal direction than the vertical direction. However, in most calculations of airborne and structure-borne sound transmission it is reasonable to treat these plates as isotropic.

Beam component   Beam components forming a profiled plate

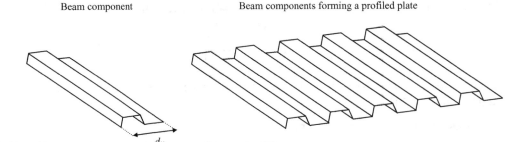

Figure 2.17

Profiled plate and one of its constituent beams, where the repetition distance is d_R.

For flat plates (isotropic or orthotropic) it is appropriate to refer to either the bending stiffness or the quasi-longitudinal phase velocity, but for profiled, corrugated or ribbed plates it is only appropriate to refer to the bending stiffness in the two orthogonal directions. The bending stiffness in the two orthogonal directions for profiled, corrugated and ribbed plates is calculated in the following sections on the assumption that the bending wavelength is much larger than the profile, corrugation, or rib spacing.

2.3.3.2.1   Profiled plates

Many orthotropic plates have profiles or corrugations running in one direction. When viewed in cross-section, there is usually a single profile that repeats at regular intervals with a repetition distance, d_R. Hence a plate can effectively be built-up from beams that have this profile, where the beams are connected along their lengths and each beam has a width, d_R (see Fig. 2.17). To determine the bending stiffness of such a plate we need to determine the moment of inertia for one of these beams. We will continue to use the coordinate conventions in Fig. 2.1, hence the cross-section of each beam lies in the yz plane with its length running along the x-axis. This means we need to determine the moment of inertia about the y-axis. For many profiled plates, such as steel cladding, the cross-section of each beam can be represented by six solid homogenous rectangular elements (Bies and Hansen, 1988). A generalized cross-section for a beam formed by N solid homogenous rectangular elements (for which $N \geq 2$) is shown in Fig. 2.18. Although rectangular elements are most commonly used to form the cross-section (see Fig. 2.19), other shapes of element could also be used in the following analysis.

For any shape of element, the moments of inertia about the y- and z-axes (I_y and I_z, respectively) are

$$I_y = \int_S z^2 \, dS \quad \text{and} \quad I_z = \int_S y^2 \, dS \qquad (2.98)$$

These are centroidal moments of inertia and are calculated using the axes convention shown in Fig. 2.20. Hence for a rectangular element (base length, b, and height, h), I_y can be calculated from Eq. 2.98 yielding the well-known equation,

$$I_y = \int_{-h/2}^{h/2} \int_{-b/2}^{b/2} z^2 \, dy \, dz = \frac{bh^3}{12} \qquad (2.99)$$

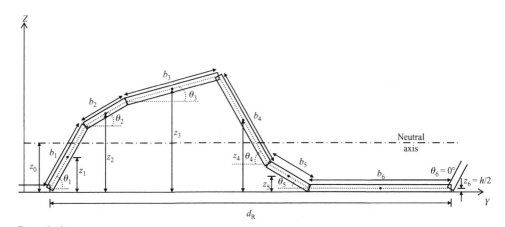

Figure 2.18

Generalized profiled beam section formed from rectangular elements ($N = 6$) of thickness, h, with a repetition distance, d_R.

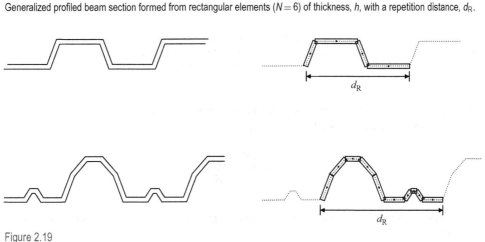

Figure 2.19

Examples of profiled plates with a repeating pattern that can be built up from a beam consisting of rectangular elements.

For other shapes, the centroidal moments of inertia can be calculated in a similar way, or found in standard textbooks. To create a plate profile, some of its constituent elements need to be rotated. For elements that are inclined at an angle, θ, the moment of inertia needs to be transformed to axes (y', z') that are inclined at an anti-clockwise angle, θ, from the (y, z) axes (see Fig. 2.20). This is done by replacing y with y', where $y' = y \cos\theta + z \sin\theta$, and replacing z with z', where $z' = z \cos\theta - y \sin\theta$. Hence the transformed moment of inertia about the y'-axis is

$$I_{y'} = I_y \cos^2\theta + I_z \sin^2\theta - I_{yz}\sin 2\theta = \frac{I_y + I_z}{2} + \frac{I_y - I_z}{2}\cos 2\theta - I_{yz}\sin 2\theta \qquad (2.100)$$

where I_{yz} is the product of inertia given by

$$I_{yz} = \int_S yz \, dS \qquad (2.101)$$

Note that when either the y- or z-axis is an axis of symmetry of the element, then $I_{yz} = 0$.

137

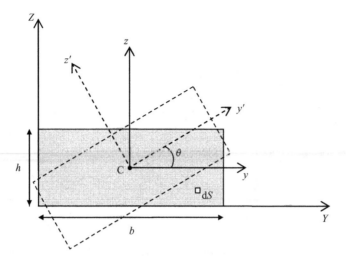

Figure 2.20

Axes used for calculating the moment of inertia of an element; this example shows a rectangular element. (Y, Z) are the global rectangular axes. (y, z) are rectangular axes with their origin at the centroid (C) of the element. (y′, z′) correspond to rectangular axes (y, z) rotated anti-clockwise by an angle, θ, whilst keeping the origin at the centroid.

Returning to the generalized cross-section for the beam (Fig. 2.18), the neutral axis of the beam cross-section is at a height, z_0, which is calculated using

$$z_0 = \frac{\sum_{n=1}^{N} S_n z_n}{\sum_{n=1}^{N} S_n} \tag{2.102}$$

where S_n is the area of each element (hb_n), and z_n is the z-coordinate of the centroid for each element.

The moment of inertia, I_y, for each of the N individual elements now needs to be transferred from the y-axis to the neutral axis for the beam cross-section; this gives I_{y0}. As these two axes are parallel, this is done using the parallel axis theorem,

$$I_{y0} = I_y + Sd_n^2 \tag{2.103}$$

where d_n is the distance of the centroid of each element from the neutral axis and is given by

$$d_n = |z_0 - z_n| \tag{2.104}$$

For N rectangular elements, the moment of inertia for the beam cross-section can now be calculated using Eq. 2.103 to give

$$I_b = \sum_{n=1}^{N} I_y + Sd_n^2 = h \sum_{n=1}^{N} d_n^2 + \frac{h^2 + b_n^2}{24} + \frac{h^2 - b_n^2}{24} \cos 2\theta_n \tag{2.105}$$

As the profiled plate is formed from beams running in the x-direction, the moment of inertia, $I_{p,x}$, for the profiled plate is $I_{p,x} = I_b/d_R$. We can now calculate the bending stiffness for the profiled

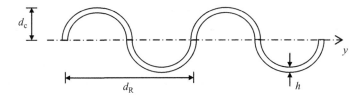

Figure 2.21

Cross-section of corrugated plate.

plate under the assumption that $\lambda_B \gg d_R$. So the plate bending stiffness in the x-direction (along the profiles) is

$$B_{p,x} = \frac{E}{(1 - \nu^2)} \frac{I_b}{d_R} \tag{2.106}$$

and the plate bending stiffness in the y-direction (perpendicular to the profiles) is the same as an isotropic plate but takes account of the effective increase in plate width due to the profiles,

$$B_{p,y} = \frac{Eh^3}{12(1 - \nu^2)} \frac{d_R}{\sum_{n=1}^{N} b_n} \tag{2.107}$$

2.3.3.2.2 Corrugated plates

For corrugated plates of thickness, h, (see Fig. 2.21) with the profiles running in the x-direction, the bending stiffness in the x- and y-directions is given by (Cremer *et al.*, 1973)

$$B_{p,x} = \frac{Ehd_c^2}{2} \left[1 - \frac{0.81}{1 + 2.5 \left(\frac{d_c}{2d_R} \right)^2} \right]$$

$$\text{for } \lambda_B \gg d_R \tag{2.108}$$

$$B_{p,y} = \frac{Eh^3}{12(1 - \nu^2)} \frac{1}{1 + \left(\frac{\pi d_c}{2d_R} \right)^2}$$

2.3.3.2.3 Ribbed plates

For a plate of thickness, h, formed from a single material with rectangular ribs running in one direction (see Fig. 2.22) the bending stiffness in the x and y directions is given by (Cremer *et al.*, 1973)

$$B_{p,x} = E \left[\frac{d_R}{3} [C_1^2 - (C_1 - h)^2] + \frac{d_y}{3} [(C_1 - h)^2 + (d_z - C_1)^2] \right]$$

$$\text{for } \lambda_B \gg d_R \tag{2.109}$$

$$B_{p,y} = \frac{Eh^3}{12} \frac{d_R}{d_R - d_y \left(1 - \frac{h^3}{d_z^3} \right)}$$

where $C_1 = \dfrac{1}{2} \dfrac{d_R d_z^2 + (d_R - d_y)h^2}{d_R d_z + (d_R - d_y)h}$

In Section 2.8.3.3, calculation of the driving-point mobility on ribbed plates requires the moment of inertia about the y-axis for the repeating T-shape beam; this is given by

$$I_y = \frac{1}{3} \left[d_y z_C^3 + d_R (d_z - z_C)^3 - (d_R - d_y)(d_z - z_C - h)^3 \right] \tag{2.110}$$

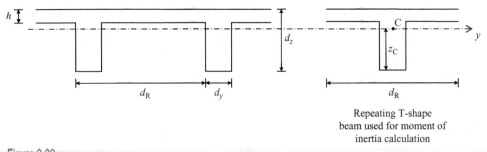

Repeating T-shape
beam used for moment of
inertia calculation

Figure 2.22

Cross-section of ribbed plate with ribs running in the x-direction.

where

$$z_C = d_z - \frac{d_y d_z^2 + h^2(d_R - d_y)}{2(hd_R + d_y d_z - hd_y)}$$

2.4 Diffuse field

A diffuse field for vibration can be defined in a similar way to that for sound. However, it is necessary to account for the fact that with some structure-borne sound waves, such as bending waves, the intensity is not constant throughout the cross-section of the structure. For this reason it is useful to refer to imaginary lines on the surface that run through the thickness dimension of the structure. A diffuse field therefore has uniform intensity per unit angle for all possible angles when waves are incident upon any such line within the structure. Waves that are incident upon this line must have random phase with equal probability of a wave arriving from any of the possible directions.

On a beam, a wave can only arrive from two possible directions, and as the boundary conditions do not change with time, the phase is not random. For this reason there is no need to define a diffuse field for beams; the modal overlap determines whether the wave field on the beam can be considered as reverberant. On a sufficiently large plate, waves can potentially arrive with uniform intensity from many directions in the plane within which it lies. Hence there is the potential for a diffuse field when there are diffuse reflections from the plate boundaries. Most walls and floors in buildings are rectangular with plate boundaries that are assumed to be uniform, so the likelihood of diffuse reflections from these boundaries could be considered as rather low. In practice the impedance along these boundaries will vary due to different material properties and workmanship, and in walls there will be additional boundaries formed by the perimeter of windows and doors. It can therefore be assumed that there will be a degree of non-specular reflection. However, the assumption of diffuse vibration fields on walls and floors is often harder to justify than with diffuse sound fields in rooms.

2.4.1 Mean free path

For beams or plates, the mean free path, d_{mfp}, is the average distance travelled by a structure-borne sound wave between two successive reflections from the boundaries. It is important because it is used to determine the power that is incident upon the junction of connected beams and plates.

For a beam, the mean free path is equal to its length,

$$d_{mfp} = L_x \qquad (2.111)$$

For a plate, the mean free path is the same as for a two-dimensional sound field in a cavity. Hence assuming that there are diffuse reflections from the plate boundaries

$$d_{mfp} = \frac{\pi S}{U} \qquad (2.112)$$

where U is the perimeter of the plate ($U = 2L_x + 2L_y$ for a rectangular plate).

2.5 Local modes

In a similar manner to sound waves in enclosed spaces, local modes occur when a structure-borne sound wave travels around a beam or plate and, after reflection from various boundaries, returns to the starting point travelling in exactly the same direction as when it first left, whilst achieving phase closure. When calculating local modes for sound waves in rooms and cavities it is usually sufficient to consider only one idealized boundary condition, namely rigid surfaces. For structure-borne sound waves there are a variety of idealized boundary conditions: free (i.e. no constraint), simply supported (also called pinned), clamped (also called fixed, or a rigid constraint), and guided boundaries. The magnitude of the reflected wave from these idealized boundaries is the same as the incident wave (i.e. a perfect reflection) but the boundary can introduce phase shifts in the reflected wave. The actual boundary condition of a beam or plate in a building is often unknown. This makes it difficult to gain accurate estimates for their mode frequencies; but a rough estimate is often sufficient. Mode frequencies may only be needed to indicate the frequency range in which it is reasonable to adopt a statistical approach to the prediction of sound transmission. Fortunately at frequencies above the tenth local mode (approximately), estimates of the statistical modal density are not significantly affected by different boundary conditions.

We will only calculate the mode frequencies for solid homogeneous, isotropic beams and plates with a limited set of relevant boundary conditions. Formulae to calculate the mode frequencies for other combinations of boundary conditions can be found from Blevins (1979) and Leissa (1973). For beams and plates with complex shapes, such as lightweight steel components with cut-outs, the mode frequencies can be predicted using Finite Element Methods (FEM) (Zienkiewicz, 1977) or measured using modal analysis. For laminated plates, details on mode shapes and mode frequencies can be found from Qatu (2004).

2.5.1 Beams

To determine the conditions for phase closure that define the local modes on a beam it is convenient to follow the journey of a propagating wave as it is reflected from each end of the beam. For bending waves this also provides some insight into the vibration field near the boundaries which is more easily obtained for wave propagation in one-dimension (beams) than in two-dimensions (plates). The same approach will be useful when we look at vibration fields near plate boundaries.

Simply supported ends   **Clamped ends**

(a) Bending wave leaves an arbitrary starting point (•) on the beam and travels towards $x = 0$

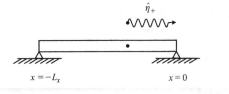
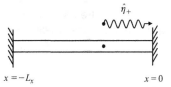

(b) Bending wave is reflected at $x = 0$

(b) Bending wave is reflected at $x = 0$, generating a nearfield

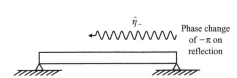
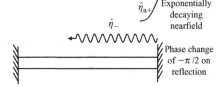

(c) Bending wave is reflected at $x = -L_x$

(c) Bending wave is reflected at $x = -L_x$ generating a nearfield

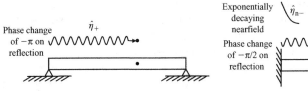

Figure 2.23

Bending wave modes on beams. A mode occurs when the bending wave returns to the starting point travelling in exactly the same direction as when it first left, whilst achieving phase closure.

2.5.1.1 Bending waves

The effect of the boundaries can be seen by positioning the beam so that one end is at $x = -L_x$, and the other is at $x = 0$. We will assume that lateral displacement is in the z-direction. At some arbitrary starting point near the middle of the beam we consider a bending wave, for which the lateral displacement is described by $\hat{\eta}_+ \exp(-ik_B x)$, propagating in the positive x-direction towards the boundary at $x = 0$ (see Fig. 2.23). We start by assuming that the ends of the beam are simply supported; for this boundary condition the lateral velocity, $\partial \eta / \partial t$, and the bending moment, M_y (Eq. 2.47) at these ends must be zero. Hence the boundary conditions are

$$\frac{\partial \eta}{\partial t} = 0 \quad \text{and} \quad \frac{\partial^2 \eta}{\partial x^2} = 0 \quad \text{at } x = 0 \text{ and } x = -L_x \tag{2.113}$$

Using the boundary conditions at $x = 0$ in the general solution for bending wave motion on a beam (Eq. 2.58) gives

$$\hat{\eta}_+ + \hat{\eta}_- + \hat{\eta}_{n+} = 0 \tag{2.114}$$

$$-\hat{\eta}_+ - \hat{\eta}_- + \hat{\eta}_{n+} = 0 \tag{2.115}$$

where $\hat{\eta}_-$ corresponds to the reflected bending wave travelling in the negative x-direction, and $\hat{\eta}_{n+}$ corresponds to the nearfield that decays in the negative x-direction.

Combining Eqs 2.114 and 2.115 yields two pieces of information about what happens at the boundary. For the propagating bending waves, $\hat{\eta}_- = -\hat{\eta}_+$, which means that there is a phase shift of $-\pi$ when the bending wave is reflected at the end. As $\hat{\eta}_{n+} = 0$, there is no nearfield accompanying the reflected wave that travels in the negative x-direction. The fact that no nearfield is generated at a simply supported boundary is very convenient as it simplifies the analysis, although real boundary conditions for beams and plates often generate nearfields. We now know that when this reflected wave reaches $x = -L_x$, it will undergo another phase shift of $-\pi$ and that there will not be a nearfield accompanying the wave that returns to the starting point. Hence modes occur at frequencies where phase closure is achieved for the propagating bending wave after it has travelled a distance of $2L_x$, during which there will have been two phase shifts of $-\pi$. Recalling the analysis of room modes, we see that we have a one-dimensional version of the three-dimensional situation that was used for rooms with perfectly reflecting boundaries. As the phase shifts due to reflection amount to -2π, phase closure occurs when

$$k_x (2L_x) = p(2\pi) \tag{2.116}$$

where p takes positive integer values 1, 2, 3, etc., and the relationship to the wavenumber is

$$k_x = k_B = \frac{2\pi}{\lambda_B} = \frac{p\pi}{L_x} \tag{2.117}$$

Substituting k_B from Eq. 2.57 into Eq. 2.117 gives the bending mode frequency, $f_{p(B)}$, for thin beams with simply supported ends as

$$f_{p(B)} = \frac{\pi}{2} \sqrt{\frac{B_b}{\rho_l}} \left(\frac{p}{L_x}\right)^2 \tag{2.118}$$

For beams connected to other building elements at both ends, such as structural (load-bearing) columns, it can be useful to calculate the mode frequency for both simply supported and clamped ends; in practice, the modal response often occurs somewhere between these two calculated frequencies. At clamped ends the lateral velocity, $\partial\eta/\partial t$, and the angular velocity, ω_y (Eq. 2.41) must be zero. Hence the boundary conditions are

$$\frac{\partial\eta}{\partial t} = 0 \quad \text{and} \quad \frac{\partial^2\eta}{\partial t\partial x} = 0 \quad \text{at } x = 0 \text{ and } x = -L_x \tag{2.119}$$

As before we take a starting point somewhere in the middle of the beam and consider a bending wave propagating in the positive x-direction towards the boundary at $x = 0$ (see Fig. 2.23). Substitution of the boundary conditions into the general solution at $x = 0$ gives

$$\hat{\eta}_+ + \hat{\eta}_- + \hat{\eta}_{n+} = 0 \tag{2.120}$$

$$-i\hat{\eta}_+ + i\hat{\eta}_- + \hat{\eta}_{n+} = 0 \tag{2.121}$$

Equations 2.120 and 2.121 can be combined to provide information on the reflected bending wave and the nearfield that return from this clamped end. For the bending waves, $\hat{\eta}_- = -i\hat{\eta}_+$, which means that there has been a phase shift of $-\pi/2$ on reflection. At $x = 0$ the magnitude of the nearfield relative to the incident bending wave is given by $\hat{\eta}_{n+} = (i - 1)\hat{\eta}_+$. So the reflected bending wave that travels in the negative x-direction is accompanied by a nearfield which is exponentially decaying from $x = 0$ towards $x = -L_x$. The existence of the nearfield means that defining the modes through phase closure will be more awkward than with simply supported boundaries. We now have to follow a reflected bending wave and a nearfield towards the boundary at $x = -L_x$ from which we will again have bending waves and accompanying nearfields to follow on their way back to the starting point. However, due to the exponential decay of the nearfield it is reasonable to assume that when the beam is sufficiently long, the nearfield will be negligible by the time it reaches $x = -L_x$. A 'sufficiently long' beam is a frequency-dependent description because the distance over which the nearfield decays to a negligible level will decrease with increasing frequency. However, this assumption allows us to ignore the nearfield and find the mode frequencies where phase closure is achieved after the bending wave has travelled a distance of $2L_x$, during which it will have undergone two phase shifts of $-\pi/2$. As the phase shifts due to reflection amount to $-\pi$, phase closure occurs when

$$k_x(2L_x) = (2p - 1)\pi \tag{2.122}$$

where p can only take positive integer values 2, 3, 4, etc.

The bending mode frequency, $f_{p(B)}$, for thin beams with clamped ends can be estimated from

$$f_{p(B)} \approx \frac{\pi}{8} \sqrt{\frac{B_b}{\rho_l}} \left(\frac{2p - 1}{L_x} \right)^2 \tag{2.123}$$

Hence a beam with clamped ends will have higher mode frequencies than the same beam with simply supported ends.

Another idealized boundary condition is the free boundary where the bending moment, M_y (Eq. 2.47) and the shear force, F_z (Eq. 2.54) must be zero. However, this condition is less relevant to beams in buildings because most beams which have a free end will be supported at various points along their length; hence it will be the combination of these supports along with the free ends that defines the modes. It is therefore sufficient to be aware that reflection of a bending wave from a free boundary also gives rise to a nearfield. In passing, it is worth noting that a thin beam with free ends has the same mode frequencies as a thin beam with clamped ends and can be estimated using Eq. 2.123.

2.5.1.2 Torsional waves

For torsional waves propagating along a beam in both directions, the angular torsional displacement is described by

$$\theta(x, t) = [\hat{\theta}_+ \exp(-ik_x x) + \hat{\theta}_- \exp(ik_x x)] \exp(i\omega t) \tag{2.124}$$

For torsional waves the idealized boundary conditions that are most common are free or clamped ends. At free ends the moment, M_x (Eq. 2.224), must be zero. At clamped ends the angle of rotation, θ, must be zero. Clamped ends are more relevant to beams in buildings, for which the boundary condition at $x = 0$ yields the relation, $\hat{\theta}_- = -\hat{\theta}_+$. This means there will be a phase shift of $-\pi$ upon reflection from each end. Hence we have exactly the same situation as with bending waves on a beam with simply supported ends, and phase closure is defined by Eq. 2.116, for which the torsional wavenumber is

$$k_x = k_{T,b} = \frac{2\pi f}{c_{T,b}} = \frac{p\pi}{L_x} \tag{2.125}$$

Therefore the torsional mode frequency, $f_{p(T)}$, for beams with both ends clamped is

$$f_{p(T)} = \frac{pc_{T,b}}{2L_x} \qquad (2.126)$$

where p takes positive integer values 1, 2, 3, etc.

2.5.1.3 Quasi-longitudinal waves

For quasi-longitudinal waves propagating along a beam in both directions, the displacement is described by

$$\xi(x,t) = [\hat{\xi}_+ \exp(-ik_x x) + \hat{\xi}_- \exp(ik_x x)]\exp(i\omega t) \qquad (2.127)$$

As with torsional waves there are two idealized boundary conditions that are commonly considered, free or clamped ends. For free ends the longitudinal force, F_x (Eq. 2.9) must be zero. For clamped ends the in-plane displacement, ξ, must be zero. Both conditions give rise to the same mode frequencies, but with different mode shapes. We will look at the clamped condition, which is more relevant to structural beams and columns in buildings. Substitution of the boundary conditions into the solution at $x = 0$ therefore gives $\hat{\xi}_- = -\hat{\xi}_+$, which means a phase shift of $-\pi$ upon reflection from each end. Phase closure is therefore defined by Eq. 2.116. Hence the quasi-longitudinal mode frequency, $f_{p(L)}$, for beams with both ends clamped (or free) is

$$f_{p(L)} = \frac{pc_{L,b}}{2L_x} \qquad (2.128)$$

where p takes positive integer values 1, 2, 3, etc.

2.5.1.4 Modal density

The modal density of a beam is calculated by arranging the eigenvalues along a line. This is a one-dimensional version of the mode lattice used for rooms. The length associated with each eigenvalue is a line of length π/L_x. An example mode line is shown in Fig. 2.24 for bending modes on a beam with simply supported ends.

Below a specified wavenumber, k, the number of modes is equal to the number of eigenvalues that lie upon a line of length, k. Therefore the number of modes, $N(k)$, below the wavenumber, k, is

$$N(k) = \frac{kL_x}{\pi} \qquad (2.129)$$

The modal density can then be calculated from $N(k)$ and the group velocity using Eq. 1.57. Strictly speaking, $N(k)$ needs a correction factor for the various different boundary conditions (Lyon and DeJong, 1995). However, Eq. 2.129 can be used to give a reasonable estimate

Figure 2.24

Mode line for bending modes on a beam with simply supported ends. A line of length, k, encloses eigenvalues below wavenumber, k.

in most cases as there is always a degree of uncertainty in the actual boundary conditions. Hence the modal densities can be calculated using

$$n_{B,b}(f) = \frac{L_x}{c_{B,b}} \quad \text{for bending waves (thin beam)} \tag{2.130}$$

$$n_{T,b}(f) = \frac{2L_x}{c_{T,b}} \quad \text{for torsional waves} \tag{2.131}$$

$$n_{L,b}(f) = \frac{2L_x}{c_{L,b}} \quad \text{for quasi-longitudinal waves} \tag{2.132}$$

The statistical mode count, N_s, in a frequency band is calculated from the modal density using Eq. 1.63. It is instructive to look at differences between the statistical mode count, and the mode count that is determined by counting the number of modes that fall within each band. To do this we will look at the modes for a common beam found in buildings, a timber stud from a timber frame wall.

This idealized timber stud will be assumed to have simply supported ends for the bending modes, and clamped ends for the torsional and quasi-longitudinal modes. It is also assumed that any fixings along the length of the stud that would occur in a real wall have a negligible effect on the mode frequencies. The mode count and the statistical mode count in one-third-octave-bands for this beam are shown in Fig. 2.25. The statistical mode count for each wave type is only plotted for frequency bands at and above the band that contains the estimated fundamental mode. The frequencies for the thin beam limit are both close to the 5000 Hz one-third-octave-band so it is reasonable to calculate the bending modes using thin beam theory (Eq. 2.118) across the entire building acoustics frequency range.

Figure 2.25 serves as a reminder that there can be bending wave modes with motion in both the y- and z-direction, although the type of excitation and the direction in which it is applied will determine which, if any, are excited. For studs in walls, or joists in floors, airborne or impact sources usually excite the bending wave modes that have motion perpendicular to the surface of the wall or floor. Both of the fundamental bending modes, $f_{1(B)}$, are below the 50 Hz band. However, it is clear from the mode count in Fig. 2.25a that this hardly results in an abundance of bending modes in and above the 50 Hz band. For a beam with simply supported ends the bending mode frequencies, $f_{p(B)}$, (see Eq. 2.118) above the fundamental mode (i.e. where $p > 1$) are multiples of $f_{1(B)}$ with the factor p^2. Hence there is a relatively wide frequency interval between adjacent modes which results in low mode counts. In each frequency band there are zero, one, or two modes for each of the four different types: bending (y-direction), bending (z-direction), torsional, and quasi-longitudinal. These low mode counts are in stark contrast to the large numbers of modes in rooms. The fundamental bending modes occur at much lower frequencies than the fundamental torsional and quasi-longitudinal modes. For this timber stud there are no torsional and quasi-longitudinal modes in the low-frequency range, so only bending modes need to be considered in the modal response of the beam. In the mid- and high-frequency ranges there can be torsional and quasi-longitudinal modes, which, if excited, also need consideration.

We can now use Fig. 2.25 to compare the different mode counts. In the low- and mid-frequency ranges the statistical mode count now takes fractional values <1. To find some physical meaning in a fractional mode count we should first recall that the mode count in Fig. 2.25a was based on mode frequencies for idealized boundary conditions; we rarely know the actual boundary conditions. We then assumed that we had precise knowledge of the dimensions and the

(a) Mode count, N

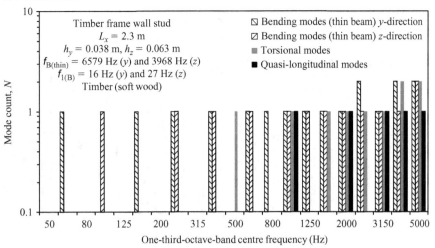

(b) Statistical mode count, N_s

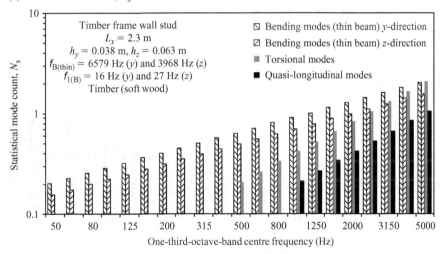

Figure 2.25

Beams: mode counts for a timber frame wall stud.

material properties of a single timber stud. In practice there will be uncertainty in all of these parameters, whether they are estimated or measured. In fact, as a lightweight wall is built from several timber studs we expect there to be natural variation in the properties of different studs; this also gives rise to uncertainty in the mode frequencies. The mode count also ignores the fact that each mode has a finite bandwidth due to damping; hence the 3 dB bandwidth can over-lap two or more bands. For these reasons the mode count is only useful in indicating potential issues with a lack of modes; we remain uncertain as to which band we can attribute each mode in the actual construction. Therefore the statistical modal density tends to be of more practical use. To try and ensure that the statistical modal density is only used in frequency bands above the fundamental mode, this mode can be estimated by assuming idealized boundary conditions.

(a) Timber floor joist

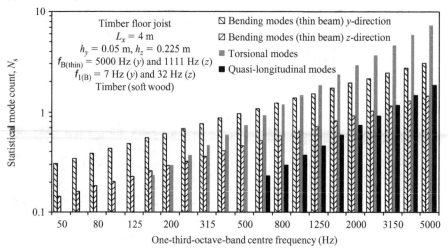

(b) Structural concrete column

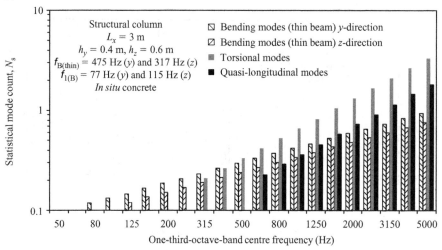

Figure 2.26

Beams: statistical mode counts for a timber floor joist and a structural concrete column.

The statistical mode count for the timber stud can be compared with two other common beams found in buildings, a timber floor joist and a structural column as shown in Fig. 2.26. These beams are much thicker than the timber stud and the thin beam limit occurs within the building acoustics frequency range. Note that the structural column has a very low statistical mode count.

2.5.2 Plates

To simplify the calculation of specific mode frequencies and to derive the modal density in rooms and cavities we assumed box-shaped spaces. In a similar way we will now use rectangular plates to calculate the specific mode frequencies. This is convenient because most plates in buildings are rectangular, but it will also allow us to determine statistical modal densities for any shape of plate.

For beams we looked at the specific phase change associated with different boundary conditions and the generation of nearfields that can occur when bending waves impinge upon certain boundary conditions. There are buildings where each side of a plate that forms a wall or floor has significantly different boundary conditions; however there will always be uncertainty in describing each of these conditions. From our previous analysis of modes on beams, a practical solution for plates in buildings is to assume that all boundaries are simply supported for bending waves so that there is no need to consider nearfields, and all boundaries are clamped for in-plane waves. This assumption is usually reasonable when the boundaries of a plate or beam are rigidly connected to other beams or plates. One exception is a plate that forms a floating floor (e.g. a floating floor screed on a resilient material) for which the plate boundaries can be assumed to be free; in this case the fundamental bending mode will be lower than for simply supported boundaries and can be calculated from tabulated data if required (see Blevins, 1979; Leissa, 1973).

2.5.2.1 Bending waves

On a thin plate with simply supported boundaries there will be a plane bending wave field without nearfields. With this plane wave field we have already used the wave equation to find the relationship between the wavenumber and the constants, k_x and k_y in Eq. 2.90. To determine these constants we can compare a rectangular plate with simply supported boundaries to two beams with simply supported ends. We will orientate these beams with lengths, L_x and L_y, along the x- and y-axis respectively. For the beam aligned along the x-axis, we already have the equation that describes k_x (Eq. 2.116) for phase closure. For the beam aligned along the y-axis it therefore follows that phase closure is achieved when $k_y(2L_y) = q(2\pi)$ where q takes positive integer values 1, 2, 3, etc. Hence from Eq. 2.90, phase closure on the plate must be satisfied when

$$k_B^2 = \left(\frac{p\pi}{L_x}\right)^2 + \left(\frac{q\pi}{L_y}\right)^2 \tag{2.133}$$

The bending mode frequencies for a simply supported, isotropic, rectangular plate can now be calculated from Eqs 2.133, 2.65, and 2.66, to give

$$f_{p,q(B)} = \frac{c_{B,p}}{2}\sqrt{\left(\frac{p}{L_x}\right)^2 + \left(\frac{q}{L_y}\right)^2} = \frac{\pi h c_{L,p}}{4\sqrt{3}}\left[\left(\frac{p}{L_x}\right)^2 + \left(\frac{q}{L_y}\right)^2\right] \tag{2.134}$$

where p and q take positive integer values 1, 2, 3, etc.

Similarly, for a simply supported, orthotropic, rectangular plate,

$$f_{p,q(B)} = \frac{\pi}{2\sqrt{\rho_s}}\left[\sqrt{B_{p,x}}\left(\frac{p}{L_x}\right)^2 + \sqrt{B_{p,y}}\left(\frac{q}{L_y}\right)^2\right] = \frac{\pi h}{2\sqrt{12}}\left[c_{L,p,x}\left(\frac{p}{L_x}\right)^2 + c_{L,p,y}\left(\frac{q}{L_y}\right)^2\right] \tag{2.135}$$

2.5.2.2 Transverse shear waves

For transverse shear waves on a rectangular plate with clamped boundaries, we have the boundary condition that the in-plane displacement, ζ, must be zero at the boundaries.

Hence we use the wavenumber relationship $k_T^2 = k_x^2 + k_y^2$ where $k_x(2L_x) = p(2\pi)$ and $k_y(2L_y) = q(2\pi)$. This gives the transverse shear mode frequencies as

$$f_{p,q(T)} = \frac{c_{T,p}}{2} \sqrt{\left(\frac{p}{L_x}\right)^2 + \left(\frac{q}{L_y}\right)^2} = \frac{c_{L,p}}{2} \sqrt{\frac{1-\nu}{2}} \sqrt{\left(\frac{p}{L_x}\right)^2 + \left(\frac{q}{L_y}\right)^2} \qquad (2.136)$$

where p and q take positive integer values 1, 2, 3, etc.

2.5.2.3 Quasi-longitudinal waves

For quasi-longitudinal waves on a beam with clamped ends, we have already seen that the phase shift upon reflection is the same as for bending waves on a beam with simply supported ends. Hence we use the same expressions for k_x and k_y as with bending waves. For quasi-longitudinal waves on a rectangular plate with clamped boundaries, the wavenumber relationship $k_L^2 = k_x^2 + k_y^2$ gives the quasi-longitudinal mode frequencies as

$$f_{p,q(L)} = \frac{c_{L,p}}{2} \sqrt{\left(\frac{p}{L_x}\right)^2 + \left(\frac{q}{L_y}\right)^2} \qquad (2.137)$$

where p and q take positive integer values 1, 2, 3, etc.

2.5.2.4 Modal density

The modal density is calculated in a similar way to cavities where the eigenvalues are arranged in a two-dimensional lattice as shown in Fig. 2.27. It is assumed that the plate boundaries are simply supported (bending waves) or clamped (in-plane waves) such that p and q only take positive integer values; hence there are no eigenvalues on the x- and y-axes. The area associated with each eigenvalue is a rectangle with an area of π^2/L_xL_y, which is equal to π^2/S. The number of modes below a specified wavenumber, k, is equal to the number of eigenvalues that are contained within one-quarter of the area of a circle with radius, k (see Fig. 2.27). Hence the number of modes, $N(k)$, below the wavenumber, k, is

$$N(k) = \frac{k^2 S}{4\pi} \qquad (2.138)$$

where k is k_B, k_T or k_L for bending, transverse shear, or quasi-longitudinal waves, respectively.

For each of the three wave types the modal density is calculated using Eq. 1.57 where the group velocity, c_g is equal to $2c_B$ for bending waves, c_L for quasi-longitudinal waves, and c_T for transverse shear waves. Hence the modal densities are

$$n_{B,p}(f) = \frac{S\sqrt{3}}{hc_{L,p}} \quad \text{for bending waves (thin plates)} \qquad (2.139)$$

$$n_{T,p}(f) = \frac{2\pi fS}{c_{T,p}^2} = \frac{4\pi fS}{c_{L,p}^2(1-\nu)} \quad \text{for transverse shear waves} \qquad (2.140)$$

$$n_{L,p}(f) = \frac{2\pi fS}{c_{L,p}^2} \quad \text{for quasi-longitudinal waves (thin plates)} \qquad (2.141)$$

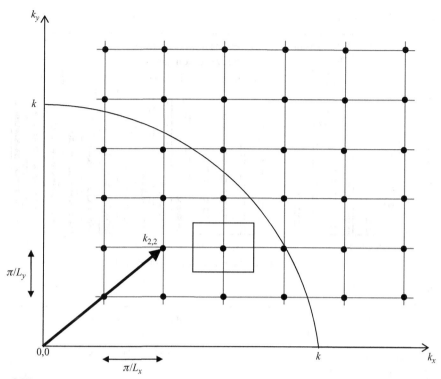

Figure 2.27

Mode lattice for a two-dimensional plate with simply supported boundaries. The vector corresponding to eigenvalue, $k_{2,2}$, is shown as an example. The quadrant with radius, k, encloses eigenvalues below wavenumber, k.

Note that for bending waves on orthotropic plates, the effective bending stiffness can be used to calculate an effective quasi-longitudinal phase velocity for use in Eq. 2.139.

The statistical mode count, N_s, is calculated from the modal density using Eq. 1.63 and is shown for four different plates in Fig. 2.28: a sheet of plasterboard, a concrete floor, and two different masonry walls. It is assumed that the plate boundaries are simply supported for bending waves, and clamped for transverse shear and quasi-longitudinal modes. For each wave type the statistical mode count is only plotted for frequency bands at and above the band that contains the estimated fundamental mode.

As with beams, the fundamental bending modes on plates are at a lower frequency than the fundamental in-plane modes. For these plates this means that the in-plane modes only need to be considered in the mid- and high-frequency ranges.

In the low-frequency range the concrete floor and masonry walls have <5 bending modes in each band. In contrast, a sheet of plasterboard has many more bending modes than masonry/concrete walls and floors. However, sheets of plasterboard that form part of a wall or floor can be fixed in many different ways. Some fixings will significantly change the bending wave modal density from this idealized situation of simply supported boundaries. For example, the change due to a few screw fixings onto a light steel frame will be less significant than large dabs of adhesive that are closely spaced over the entire surface of the sheet.

(a) Sheet of plasterboard

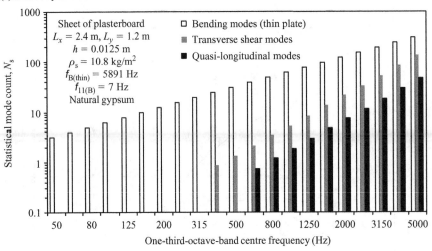

(b) Concrete floor

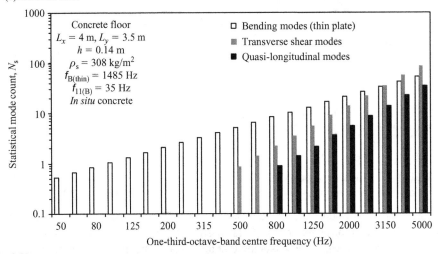

Figure 2.28

Plates: statistical mode counts for a sheet of plasterboard, a concrete floor, and two different masonry walls. Note that values are only shown at and above the frequency band that contains the estimated fundamental mode for each wave type.

2.5.3 Equivalent angles

For modes on beams, the waves propagate along the axis of the beam so there is no need to consider equivalent angles. For modes on plates, the two-dimensional lattice in k-space defines the equivalent angles for the different wave types. They are defined in the same way as for rooms and cavities (Section 1.2.5.4), therefore the equivalent angles are related to the mode wavenumber, $k_{p,q}$, and the constants, k_x and k_y by

$$k_x = k_{p,q} \sin \theta_x \qquad k_y = k_{p,q} \sin \theta_y \qquad (2.142)$$

(c) Masonry wall

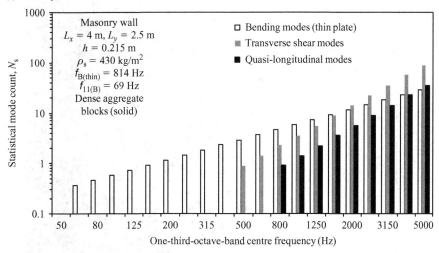

(d) Masonry wall

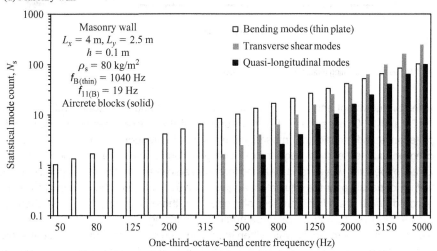

Figure 2.28

(*Continued*)

and the equivalent angles for each plate mode are

$$\theta_x = \text{asin}\left(\frac{pc}{2L_x f_{p,q}}\right) \qquad \theta_y = \text{asin}\left(\frac{qc}{2L_y f_{p,q}}\right) \tag{2.143}$$

where c is $c_{B,p}$, $c_{L,p}$, or $c_{T,p}$ for bending, quasi-longitudinal, or transverse shear waves respectively.

The equivalent angles θ_x and θ_y are defined from lines normal to the x- and y-axis respectively; hence, θ_x and θ_y also represent the equivalent angles of incidence upon the plate boundaries, L_x and L_y respectively. The ratio, $c/f_{p,q}$, in Eq. 2.143 is the same for all three wave types, hence for modes with the same values of p and q, the angles θ_x and θ_y are identical.

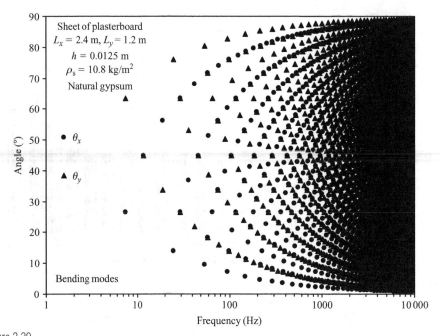

Figure 2.29

Equivalent angles for the bending modes of a sheet of plasterboard.

Examples of equivalent angles for a sheet of plasterboard and a masonry wall are shown in Figs. 2.29 and 2.30 respectively. For the purpose of these examples we can overlook the thin plate limits and the fact that masonry walls are not homogeneous plates with specularly reflecting boundaries at high frequencies. The plate boundaries are assumed to be simply supported for bending waves, and clamped for transverse shear and quasi-longitudinal modes; hence it is not possible for the equivalent angles to be 0° or 90°. For the bending modes of the plasterboard, there are a large number of modes and there is symmetry in the pattern about the 45° line because $L_x = 2L_y$. The large number of bending modes for the plasterboard is in marked contrast to the masonry wall. When the angles are considered as equivalent angles of incidence upon each boundary, it is clear that there is a limited range of angles incident upon the plate boundaries in the low- and mid-frequency ranges. This is relevant to the calculation of structure-borne sound transmission between coupled plates using the wave approach where it is assumed that waves are incident at all possible angles.

2.6   Damping

Damping is described here for vibration in a similar style to sound in Chapter 1 with emphasis on the internal, coupling, and total loss factors that are defined for use with Statistical Energy Analysis (SEA).

2.6.1   Structural reverberation time

The structural reverberation time, T_s, is defined in the same way as for reverberation times in rooms. In practice we only need to measure and use the structural reverberation time for

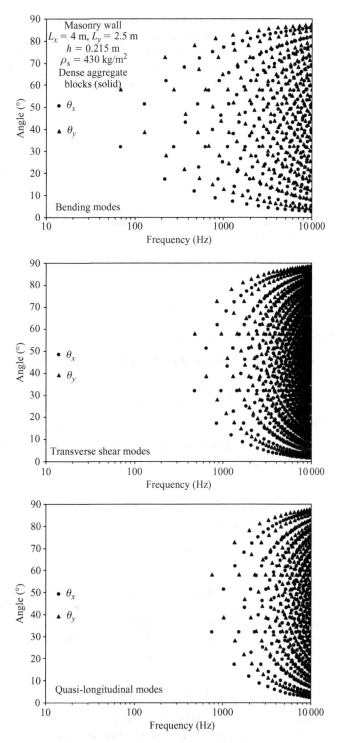

Figure 2.30

Equivalent angles for bending, transverse shear, and quasi-longitudinal modes of a 215 mm dense aggregate wall.

bending wave fields although we can separately calculate structural reverberation times for bending, quasi-longitudinal, and transverse shear waves. In comparison with reverberation times in rooms, structural reverberation times of building elements can be quite short. As with rooms, non-diffuse fields on plates also cause curvature of the decay curve when the decaying modes have very different decay times. For the same reasons as discussed for rooms (Section 1.2.7.5.2), T_{10}, T_{15}, or T_{20} are used to relate the decay time to reverberant vibration levels rather than T_{30} or T_{60}.

2.6.2 Absorption length

For rooms, the absorption area is a convenient way to describe all the absorption in a room. For bending waves on a plate it is possible to use a similar concept with an absorption length, a, in metres. This can be used to describe all the absorption at the plate boundaries by using a single length which is totally absorbing.

The absorption length is defined as the ratio of the power absorbed by the plate boundaries, to the intensity incident upon them. The power absorbed is

$$W_{abs} = \omega \eta E = \omega \frac{2.2}{f T_s} E \qquad (2.144)$$

where η is the total loss factor, and E is the plate energy.

Assuming a diffuse vibration field, the intensity that is incident upon the boundaries (in W/m) is

$$I_{inc} = \frac{c_{g(B),p}}{d_{mfp}} \frac{E}{U} = \frac{2c_{B,p}}{d_{mfp}} \frac{E}{U} \qquad (2.145)$$

where d_{mfp} is given by Eq. 2.112.

The bending phase velocity can be written in terms of the critical frequency of the plate, f_c, using Eqs 2.66 and 2.201 to give

$$c_{B,p} = c_0 \sqrt{\frac{f}{f_c}} \qquad (2.146)$$

Hence, the ratio of Eq. 2.144 to Eq. 2.145 defines the absorption length as

$$\frac{2.2\pi^2 S}{c_0 T_s} \sqrt{\frac{f_c}{f}} \qquad (2.147)$$

Equation 2.147 is only suitable for homogeneous plates where the critical frequency is known. This is slightly awkward because a single critical frequency is not well-defined for non-homogeneous and/or orthotropic plates. Therefore to generalize the definition for a plate, i, an equivalent absorption length, a_i, is defined by using a reference frequency rather than the critical frequency (Gerretsen, 1996; also see EN 12354 Parts 1 & 2)

$$a_i = \frac{2.2\pi^2 S_i}{c_0 T_{s,i}} \sqrt{\frac{f_{ref}}{f}} = \frac{\pi^2 S_i \eta_i}{c_0} \sqrt{f_{ref} f} \qquad (2.148)$$

where f_{ref} is a reference frequency of 1000 Hz. Note that the equivalent absorption length only corresponds to a totally absorbing length for a plate which actually has a critical frequency of 1000 Hz, and that measured structural reverberation times or total loss factors will include both radiation and internal losses.

2.6.3 Internal loss factor

The internal losses describe the inherent material damping. When a beam or plate deforms whilst undergoing wave motion, the internal losses convert vibration energy into heat; therefore high internal losses are desirable for sound insulation purposes. In comparison with other parameters that describe the material properties, such as density or longitudinal phase velocity, the internal loss factor (denoted as η_{int}, or, η_{ii} for SEA subsystem i), is more awkward to quantify. Internal losses can vary depending upon the type of wave motion, frequency, temperature, amplitude of vibration, and manufacturing process. This is rarely a problem when predicting vibration transmission between adjacent rooms in buildings. Uncertainty in the internal loss factor can often be tolerated when the sum of the coupling loss factors is much greater than the internal loss factor; this occurs with many plates and beams in buildings.

Measurements used to determine internal loss factors for beams and plates are discussed in Section 3.11.3.4. Measured internal loss factors only tend to be available for bending waves; example values for common building materials are listed in the Appendix, Table A2. Due to a lack of data it is often assumed that the internal loss factors for in-plane waves are the same as those for bending waves; measurements indicate that this is a reasonable assumption for materials such as concrete or bricks (Kuhl and Kaiser, 1952).

For solid homogeneous plates formed from concrete or masonry it is reasonable to assume a frequency-independent internal loss factor (bending or in-plane waves) over the building acoustics frequency range. Frequency-dependence has been measured for solid concrete covered with a granular material (e.g. sand) and hollow masonry/concrete units filled with sand; with increasing frequency, the granular material tends to increase the internal losses (Kuhl and Kaiser, 1952). The internal loss factor for laminated glass is also frequency-dependent and tends to increase significantly with frequency (see Section 3.11.3.4). The effect of temperature on the internal loss factor varies between materials; it is of minor importance for concrete, but very important for laminated glass.

If the internal loss factor changes with the amplitude of vibration, the internal damping is described as non-linear. This has been observed in the measurement of internal damping with both bricks and hollow blocks that were mortared together; increasing the vibration amplitude resulted in higher internal loss factors (Kuhl and Kaiser, 1952; Watters, 1959). It can be tentatively assumed that it applies to most mortared masonry blocks. However, this non-linearity needs to be kept in perspective. The relatively high vibration amplitudes that can be used in the laboratory to measure the internal loss factor do not usually correspond to the vibration amplitudes that occur on building elements excited by typical airborne and structure-borne sound sources. For structure-borne sound transmission in buildings it is reasonable to assume that Hooke's law is obeyed in the majority of situations (Cremer *et al.*, 1973). Internal loss factors should therefore be determined from measurements at low levels of excitation where the internal damping can be considered as linear.

Another form of internal damping occurs due to losses at the edges of a plate caused by bending wave motion. This is sometimes referred to as edge damping; although it should not be confused with coupling losses to the surrounding structure along the plate edges. Edge damping is usually specific to the edge fixing for a specific material. For panes of glass mounted in rubber/neoprene gaskets, dissipation can occur due to frictional losses at the edges caused by the gasket mounting (Utley and Fletcher, 1973). For metal beams riveted to metal plates, energy can be dissipated due to viscous losses of air pumping in and out of the small spaces between the beam and the plate (Maidanik, 1966).

For two plates closely connected together over their surface, bending wave motion also causes viscous losses in the thin layer of air between them which can result in a higher internal loss factor that is also frequency-dependent (Trochidis, 1982).

2.6.4   Coupling loss factor

Coupling losses describe the energy losses from a plate or beam via structural connections to other plates or beams and via sound radiation to the surrounding air (or other gas). For example, the coupling losses for one leaf of a cavity masonry wall are due to: connections around the perimeter to other masonry walls, connections such as wall ties, and sound radiation into a room on one side and a cavity on the other. For masonry/concrete walls and floors, the radiation losses are often insignificant compared to the structural coupling losses.

The role of the coupling loss factor, η_{ij}, in quantifying vibration transmission is discussed in Chapters 4 and 5. At this stage we simply focus on vibration transmission from one plate or beam (referred to with subscript i) to another plate or beam (referred to with subscript j) to which it is coupled. We will assume that these plates or beams have reverberant vibration fields.

For any type of wave on plate or beam, i, that is incident upon the junction connecting i to j, the transmission coefficient, τ_{ij}, is

$$\tau_{ij} = \frac{W_{ij}}{W_{\text{inc},i}} = \frac{\omega \eta_{ij} E_i}{W_{\text{inc},i}} \qquad (2.149)$$

where W_{ij} is the power transmitted from i to j, and $W_{\text{inc},i}$ is the incident power on i.

The coupling loss factor can now be written as

$$\eta_{ij} = \frac{\tau_{ij} W_{\text{inc},i}}{\omega E_i} \qquad (2.150)$$

In this section we are only concerned with quantifying $W_{\text{inc},i}$ on beams and plates; the transmission coefficient for different types of junction will be discussed in Section 5.2. The incident power upon a boundary is determined by using the mean free path to quantify the number of times that vibrational energy is reflected from the boundaries of a beam or a plate every second.

For a beam i that is connected at both of its ends to either a plate or a beam, the power incident upon one of its two ends is

$$W_{\text{inc},i} = \frac{c_{g,i} E_i}{d_{\text{mfp},i}} \frac{1}{2} \qquad (2.151)$$

where d_{mfp} is the mean free path given by Eq. 2.111 and from Eq. 2.150 the coupling loss factor is

$$\eta_{ij} = \frac{c_{g,i} \tau_{ij}}{4\pi f L_i} \qquad (2.152)$$

where L_i is the length of beam i. For a plate i with a junction length, L_{ij}, along its perimeter, U_i, that connects plate i to plate j, the power incident upon the junction is

$$W_{\text{inc},i} = \frac{c_{g,i} E_i}{d_{\text{mfp},i}} \frac{L_{ij}}{U_i} \qquad (2.153)$$

where d_{mfp} is the mean free path given by Eq. 2.112 and from Eq. 2.150 the coupling loss factor is

$$\eta_{ij} = \frac{c_{g,i} L_{ij} \tau_{ij}}{2\pi^2 f S_i} \qquad (2.154)$$

2.6.5 Total loss factor

The total loss factor for subsystem i (denoted as η_i) is the sum of the internal loss factor for subsystem i and all the coupling loss factors from subsystem i to other subsystems (Eq. 1.106). The coupling loss factors are due to sound radiation (Section 4.3.1.1) and/or structural coupling to other plates/beams (Section 5.2).

Total loss factors can be calculated by predicting or measuring the internal and coupling loss factors. If accurate prediction of the total loss factor is not possible, it is usually necessary to measure the structural reverberation time (Section 3.11.3.3) which is simply related to the total loss factor by Eq. 1.107. For some calculations, an estimate for the total loss factor is adequate. For bending wave motion on masonry/concrete plates that are rigidly connected on all sides, estimates for the total loss factor usually take the form (Craik, 1981)

$$\eta_i = \eta_{ii} + \frac{X}{\sqrt{f}} \qquad (2.155)$$

The second term in Eq. 2.155 represents the sum of the coupling loss factors for structural coupling; note that for these connected plates the coupling losses due to sound radiation are negligible. When the plate boundaries are rigidly connected to other parts of the structure, the following estimates can be used for X. Field measurements indicate that $X = 1$ is a reasonable estimate for masonry/concrete walls and floors (Craik, 1981). The structural coupling losses in the laboratory are not usually as high as in the field. To try and minimize variation between measurements in different laboratories, the requirement in the measurement standard is based on $X > 0.3$ (ISO 140 Part 1). In the field and laboratory, $0.3 \leq X \leq 1$ gives a reasonable indication of the range for the total loss factor of masonry/concrete walls and floors (see Fig. 2.31).

For bending wave motion on single sheets/boards (e.g. plasterboard, chipboard) that form a lightweight wall or floor, the measured total loss factor can be highly variable depending on the type of frame and whether there is an absorbent material in the cavity. Rough estimates for sheets/boards can be calculated using Eq. 2.155 with $X = 0.4$ for timber frames with empty cavities, and $X = 0.8$ for timber or steel frames with mineral wool very close or touching the sheet/board (100–5000 Hz). If a wall and floor has a significant decrease in vibration across it, then the total loss factor is of questionable use.

2.6.6 Modal overlap factor

The modal overlap factor is calculated using Eq. 1.109.

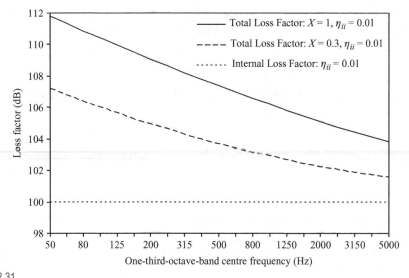

Figure 2.31

Total loss factor estimates for masonry/concrete plates that are rigidly connected on all sides along with the internal loss factor used in this example.

2.7 Spatial variation in vibration level: bending waves on plates

Measurement and prediction of bending wave vibration on walls and floors requires an awareness of the spatial variation. The focus here is mainly on reverberant bending wave fields although it should be noted that there will also be a net flow of bending wave energy across the plate for which structural intensity can give an insight into the energy flow (Section 3.12.3.1.4).

2.7.1 Vibration field associated with a single mode

For a rectangular plate with simply supported boundaries the local mode shape, $\psi_{p,q}$, describes the plate displacement for bending mode, $f_{p,q(B)}$. By satisfying the boundary conditions and the wave equation (Eq. 2.88) this gives

$$\psi_{p,q} = \sin\left(\frac{p\pi x}{L_x}\right)\sin\left(\frac{q\pi y}{L_y}\right) \tag{2.156}$$

When at least one of the sine terms is zero, the displacement is also zero, so there will be nodal lines of zero displacement perpendicular to the x- or y-axis. For a solid homogenous wall or floor with simply supported boundaries we can use Eq. 2.156 to visualize the bending wave vibration field as shown in Fig. 2.32. Real walls and floors are not all homogenous, isotropic and rectangular plates with simply supported boundary conditions; they also have openings (e.g. windows, doors) for which there can be a variety of boundary conditions. Uncertainty in some or all of the boundary conditions for a wall or floor (and any openings) means that mode shapes can rarely be accurately predicted.

The mode shapes for many real structures are too complex to calculate with analytical models but they can be calculated using FEM (Zienkiewicz, 1977). This approach is used to calculate the first eight local mode shapes for a wall with and without a window opening as shown in

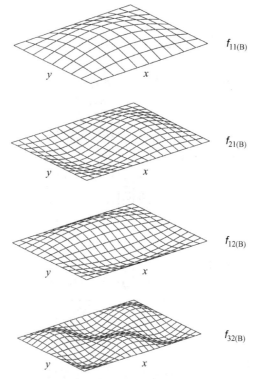

Figure 2.32

Mode shapes for bending modes on a plate with simply supported boundaries.

Fig. 2.33 (Hopkins, 2003b). Without an opening, this wall can be seen as representing one leaf of a cavity separating wall between dwellings. With and without an opening this wall can be seen as representing the inner leaf of the external flanking wall. The figure shows that introducing an opening can significantly alter the mode shape depending on the boundary conditions of the opening. Mode shapes for plates without openings and the same boundary conditions on all sides have simple symmetry; this no longer occurs when there is an opening. The thin strips of wall between the wall boundaries and the boundaries of the opening tend to have very low vibration levels when both boundaries are simply supported and the distance between them is $<\lambda_B/2$.

Flanking walls contain openings such as doors and windows in many different positions with a wide range of boundary conditions. In conjunction with the uncertainty in the wall dimensions and material properties this also implies that a statistical approach to plate vibration is more practical than deterministic calculations of the modal response.

2.7.2 Nearfields near the plate boundaries

When measuring bending wave vibration it is necessary to be aware of nearfields near the boundaries of a beam or plate. In most situations we are only interested in quantifying the reverberant vibration level due to propagating bending waves because, unlike the nearfield, they transport energy. It is therefore necessary to exclude measurement positions close to the

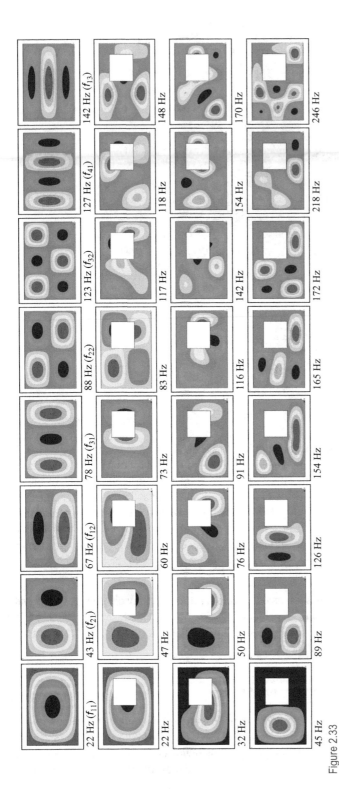

Figure 2.33

Mode shapes for a wall with simply supported boundaries, with and without a window opening.
Row 1: Wall without an opening.
Row 2: Wall with an opening that has free boundaries.
Row 3: Wall with an opening where the top boundary is simply supported and other boundaries are free.
Row 4: Wall with an opening that has simply supported boundaries.
In each row the eight lowest mode frequencies are shown. For all modes except those in the first column, red and dark blue indicate maximum displacement r opposite directions.
Wall properties: $x = 3.5\,\text{m}$, $y = 2.4\,\text{m}$, $h = 0.1\,\text{m}$, $\rho = 600\,\text{kg/m}^3$, $c_L = 1900\,\text{m/s}$, $\nu = 0.2$.
Reproduced with permission from Hopkins (2003).

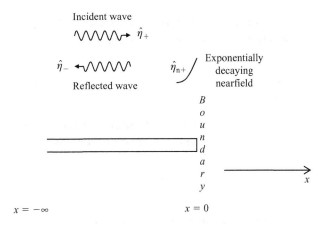

Figure 2.34

Bending waves incident upon, and reflected from the beam boundary.

boundaries that are affected by nearfield vibration. Hence it is useful to estimate the minimum distance from the boundaries that should be used when taking measurements so that the nearfield can be assumed to be negligible. This also needs to take into consideration the fact that the exact boundary conditions are rarely known.

Bending waves on a plate impinge upon the boundaries from a variety of angles, but to simplify the analysis only normal incidence is considered here. This conveniently allows us to follow a bending wave propagating along a beam from $-\infty$ towards the boundary at $x = 0$ (see Fig. 2.34). This analysis will apply to beams if the bending stiffness for a beam is used to determine the bending wavenumber, but we are looking at plates so the plate bending stiffness is used. As we are restricting our attention to the end of the beam near $x = 0$ the relevant solution for bending wave motion on a beam is taken from Eq. 2.58 as

$$\eta(x,t) = [\hat{\eta}_+ \exp(-ik_Bx) + \hat{\eta}_- \exp(ik_Bx) + \hat{\eta}_{n+}\exp(k_Bx)]\exp(i\omega t) \tag{2.157}$$

where the complex wave coefficients are

$$\hat{\eta}_+ = |\hat{\eta}_+| \exp(i\gamma_+) \qquad \hat{\eta}_- = |\hat{\eta}_-| \exp(i\gamma_-) \qquad \hat{\eta}_{n+} = |\hat{\eta}_{n+}| \exp(i\gamma_{n+}) \tag{2.158}$$

Differentiating the real part of Eq. 2.157 with respect to time gives the velocity, which is squared and time-averaged to give the mean-square velocity

$$\langle v_z^2 \rangle_t = \left\langle \left(\frac{\partial \eta}{\partial t}\right)^2 \right\rangle_t = \frac{\omega^2}{2}(|\hat{\eta}_+|^2 + |\hat{\eta}_-|^2 + |\hat{\eta}_{n+}|^2\exp(2k_Bx))$$

$$+ \omega^2|\hat{\eta}_+||\hat{\eta}_-|\cos(2k_Bx - \gamma_+ + \gamma_-)$$

$$+ \omega^2|\hat{\eta}_+||\hat{\eta}_{n+}|\exp(k_Bx)\cos(k_Bx - \gamma_+ + \gamma_{n+})$$

$$+ \omega^2|\hat{\eta}_-||\hat{\eta}_{n+}|\exp(k_Bx)\cos(k_Bx + \gamma_- - \gamma_{n+}) \tag{2.159}$$

From Eq. 2.159 we see that the mean-square velocity is described by the interaction between the incident wave, the reflected wave and the nearfield. At this point we recall that the nearfield must be considered in the solution to ensure that the wave equation and the boundary condition are satisfied. However, to assess the effect of the nearfield on the vibration field we will ignore the nearfield components of Eq. 2.159 to give a mean-square velocity that is only valid at distances from the boundary where the nearfield is negligible

$$\langle v_{z^*}^2 \rangle_t = \left\langle \left(\frac{\partial \eta}{\partial t} \right)^2 \right\rangle_t = \frac{\omega^2}{2} (|\hat{\eta}_+|^2 + |\hat{\eta}_-|^2) + \omega^2 |\hat{\eta}_+||\hat{\eta}_-| \cos(2k_{\mathrm{B}}x - \gamma_+ + \gamma_-) \qquad (2.160)$$

To compare the mean-square velocities from Eqs 2.159 and 2.160 it is useful to reference each of them to the mean-square velocity of the free-field incident bending wave,

$$\langle v_+^2 \rangle_t = \frac{\omega^2}{2} |\hat{\eta}_+|^2 \qquad (2.161)$$

The vibration field can now be shown using the velocity level differences,

$$10 \lg \left(\frac{\langle v_z^2 \rangle_t}{\langle v_+^2 \rangle_t} \right) \quad \text{and} \quad 10 \lg \left(\frac{\langle v_{z^*}^2 \rangle_t}{\langle v_+^2 \rangle_t} \right)$$

where 0 dB corresponds to the level of the free-field incident wave.

To calculate Eqs 2.159 and 2.160, the wave coefficients can either be determined by assuming perfect reflection at idealized boundary conditions or by using wave analysis to determine the amplitudes of the reflected wave, transmitted wave and nearfields at a specific type of junction (Craik, 1996; Cremer et al., 1973).

The wave coefficients for bending waves that are incident upon idealized boundaries can be found in terms of the arbitrary constant, $\hat{\eta}_+$, where (Mead, 1982a)

$$\hat{\eta}_- = -\hat{\eta}_+ \text{ and } \hat{\eta}_{n+} = 0 \quad \text{for a simply supported boundary} \qquad (2.162)$$

$$\hat{\eta}_- = -i\hat{\eta}_+ \text{ and } \hat{\eta}_{n+} = (i - 1)\hat{\eta}_+ \quad \text{for a clamped boundary} \qquad (2.163)$$

$$\hat{\eta}_- = -i\hat{\eta}_+ \text{ and } \hat{\eta}_{n+} = (1 - i)\hat{\eta}_+ \quad \text{for a free boundary} \qquad (2.164)$$

There is a nearfield generated at clamped and free boundaries, but not at simply supported boundaries. Out of these three idealized boundaries it is only the free (i.e. unconnected) boundary that can be visually identified in a building. Free plate boundaries can occur along the top or side edge of a wall, near door or window openings in a wall, and along the edges of a floating floor. For any plate boundary that forms a junction with another beam or plate we have seen that when calculating mode frequencies or mode shapes it is convenient to represent it as either simply supported or clamped. In practice the boundary condition will rarely (if ever) be either of these and will be more complex. This is partly due to the amplitude of the reflected wave and nearfield varying with the angle at which the incident wave impinges upon the junction.

As a more realistic boundary condition we take the situation where the plate of interest (plate 1) is connected to another plate (plate 2). For this example we will choose a right-angled corner junction, referred to as an L-junction. The complex reflection coefficients for the bending wave and the nearfield are R and R_n respectively, where

$$\hat{\eta}_- = R\hat{\eta}_+ \quad \text{and} \quad \hat{\eta}_{n+} = R_n\hat{\eta}_+ \qquad (2.165)$$

For a bending wave incident upon an L-junction the reflection coefficients are taken from wave theory as (Cremer *et al.*, 1973)

$$R = \frac{\psi(1 - 2\beta_2 - \beta_1\beta_2) + \chi(1 + 2\beta_1 - \beta_1\beta_2) + i[\psi(1 + \beta_1 - \beta_1\beta_2) + \chi(-1 + \beta_2 + \beta_1\beta_2)]}{[\psi(-1 - \beta_1 - 2\beta_2 - \beta_1\beta_2) + \chi(-1 - 2\beta_1 - \beta_2 - \beta_1\beta_2)] + i[(\psi + \chi)(1 - \beta_1\beta_2)]}$$

(2.166)

$$R_n = \frac{-1 + \beta_2 - R(1 + \beta_2)}{1 + i\beta_2}$$

(2.167)

where

$$\chi = \frac{k_{B2}}{k_{B1}} \quad \psi = \frac{B_{p,2}k_{B2}^2}{B_{p,1}k_{B1}^2} \quad \beta_1 = \frac{c_{B,p2}\rho_{s2}}{c_{L,p1}\rho_{s1}} \quad \text{and} \quad \beta_2 = \frac{c_{B,p1}\rho_{s1}}{c_{L,p2}\rho_{s2}} \text{ for plates.}$$

The minimum distance from the boundary at which the nearfield is negligible is frequency-dependent. For free or clamped boundaries the magnitude of the difference between velocity levels with and without the nearfield is $<1\,dB$ at distances $>\lambda_B/3\,m$ from the boundary, and $<0.1\,dB$ at distances $>3\lambda_B/4\,m$ from the boundary. Measurements are usually carried out in all frequency bands simultaneously, and the minimum distance will be largest at the lowest frequency; a practical solution is to calculate a minimum distance based on 100 Hz.

Figure 2.35 shows the velocity level difference near a free boundary, a clamped boundary, and an L-junction relative to the incident bending wave. These are shown for three different plates that commonly form walls or floors. For bending waves that are reflected from free and clamped boundaries the reflection coefficients are identical in magnitude and phase. Therefore Eq. 2.160, which assumes that the nearfield is negligible, gives the same curve for both of these idealized boundaries.

Close to the boundary at $x = 0$ the interference pattern is due to the interaction between the incident wave, reflected wave and the nearfield. In this region, the large difference between velocities calculated with and without the nearfield indicates the significant effect of the nearfield. However, the nearfield has an exponential decay. Therefore it soon becomes negligible with increasing distance from the boundary until there is no difference between the mean-square velocity calculated using Eq. 2.159 or 2.160. For the three plates used in this example at 100 Hz, the magnitude of this difference is $<1\,dB$ at distances between 0.2 and 0.8 m from the boundary. In practice, 0.25 m is chosen as a suitable distance for the majority of constructions over the building acoustics frequency range (ISO 10848 Part 1). This takes account of the fact that some walls and floors have relatively small dimensions, such that larger distances would limit measurements to a very small area in the middle of the wall or floor.

At distances where the nearfield is negligible, there are sharp minima in the interference pattern with the idealized boundary conditions. These are due to destructive interference between the incident and reflected waves that have identical amplitudes. This is in contrast to the L-junction where a fraction of the power that is incident upon the boundary is transmitted to the other plate. Therefore the reflected wave has a lower amplitude than the incident wave which results in minima which are shallower, and maxima that are reduced in height compared to the idealized boundaries which are perfectly reflecting. This is a better representation of the real situation in buildings where walls and floors are coupled together.

The focus here has been on the plate boundaries. However over the central area of many lightweight walls and floors there are sheets of plasterboard or timber boards that are supported

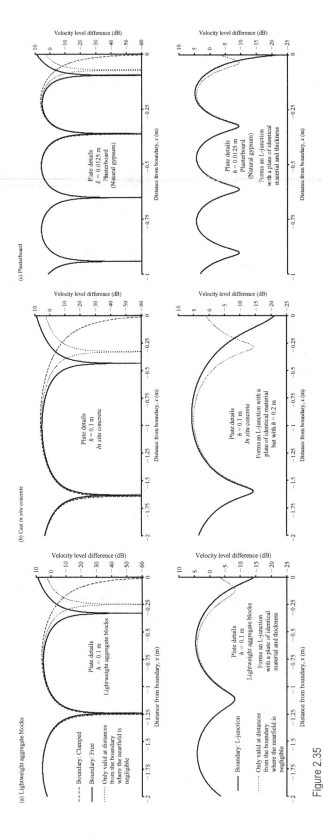

Figure 2.35

Velocity levels at different distances from various boundaries at $x = 0$ (free boundary, clamped boundary, and an L-junction); these levels are referenced to the incident bending wave (normal incidence) to give a velocity level difference. The plates are formed from different materials (lightweight aggregate blockwork, cast *in situ* concrete, and plasterboard).

by a frame of studs or joists. This frame can represent a discontinuity to the propagating wave on the plate and also generate nearfields.

2.7.3 Diffuse and reverberant fields

As with diffuse sound fields in rooms, diffuse vibration fields on plates represent the ideal situation rather than the reality. In practice we often need to assume that there are diffuse fields in order to simplify the calculations for sound transmission. It is difficult to make general statements about the likelihood of finding close approximations to diffuse vibration fields on walls and floors without appearing to be overly pessimistic. This is partly because of the wide variety of materials, plate dimensions, and plate geometries. It is also because the vibration field can rarely be described as diffuse across the entire building acoustics frequency range. At frequencies above the lowest mode it is usually more appropriate to refer to the vibration field as reverberant rather than diffuse.

2.7.4 Reverberant field

For practical purposes, measurements are usually taken at random positions over the plate surface to calculate the spatial average vibration level in frequency bands. An example is used here to illustrate various aspects that give rise to the spatial variation.

Figure 2.36 shows vibration contour plots from measured velocity levels on a masonry wall. The masonry wall is 100 mm thick with a 13 mm plaster finish on one side. The side with the plaster finish faces into a reverberant room and is directly excited by the sound field in this room where the sound source is broad-band noise from a loudspeaker. The left and right hand side boundaries of the wall form corner junctions with other masonry walls. The lower boundary forms an in-line junction with a similar masonry wall below but there are built-in timber floor joists along this junction line. Vibration measurements were taken with different boundary conditions for the top boundary of the wall; firstly with a free (unconnected) boundary, and secondly after the masonry wall was extended upwards to form an in-line junction.

In the 50 Hz one-third-octave-band (Fig. 2.36a) there is evidence of the f_{21} mode shape when there are junctions at all four boundaries. The mode frequency for the f_{21} mode is calculated to be 37 Hz which is just outside this band but we should note that it is not unusual to predict to an accuracy of plus or minus one-third-octave-band. The vibration level at 50 Hz varies by up to 17 dB over the surface. This large variation needs to be borne in mind when determining the number of measurement positions. At 1000 Hz (Fig. 2.36b) where $M \geq 1$ and $N \geq 5$ for bending modes we see that regardless of the top boundary condition, the vibration levels are relatively uniform in the central area of the wall. The vibration levels vary by < 8 dB. In the 3150 Hz band (Fig. 2.36c) we might expect to find greater uniformity due to the large number of modes and high modal overlap. Whilst this is generally true, the existence of very minor damage to the plaster finish, and positions where the plaster did not bond strongly to the masonry result in high vibration levels at a few points. In addition there is variation in fixing the accelerometers with bees wax in the high-frequency range. Hence despite the increased uniformity of the vibration field at high frequencies it is still necessary to average measurements from a number of different positions to ensure that the spatial average value is representative.

(a) 50 Hz one-third-octave-band ($M \approx 0.5$, $N \approx 1$)

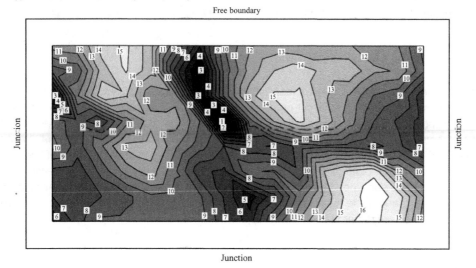

Figure 2.36

Vibration contour plot produced from a grid of measurements over the surface of a 100 mm masonry wall ($x = 5.45$ m, $y = 2.45$ m, $\rho_s = 166$ kg/m², $c_L = 2690$ m/s, built from solid blocks mortared together on all sides, 13 mm plaster finish). The lower, left, and right wall boundaries are rigidly connected to other masonry walls. Two different boundary conditions are shown for the top wall boundary: free boundary (upper plot) and a rigidly connected boundary forming an in-line junction to an identical masonry wall (lower plot). The contours indicate the velocity levels in decibels relative to the lowest level that was measured on the wall in that one-third-octave-band. The estimated modal overlap factor (M) and mode count (N) for bending modes are given for each one-third-octave-band.

Measured data from Hopkins are reproduced with permission from ODPM and BRE.

2.7.5 Direct vibration field

For point excitation of bending waves on a plate there will be a direct vibration field near the excitation point. In a similar way to rooms we can calculate the distance, r_{rd}, at which the

(b) 1000 Hz one-third-octave-band ($M \approx 2.5$, $N \approx 20$)

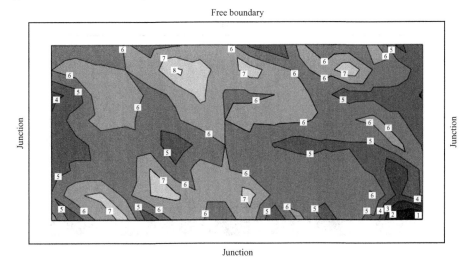

Figure 2.36

(*Continued*)

energy density in the direct field equals that in the reverberant field (Lyon and DeJong, 1995). A cylindrical wave radiates out from the excitation position, and attenuation with distance occurs due to both geometrical spreading of the wavefront and internal damping (see Fig. 2.37).

Internal damping, as described by the internal loss factor, η_{int}, causes an exponential decay of vibration with distance. The decrease in bending wave amplitude due to internal damping after travelling a distance, d, is determined by the factor (Cremer *et al.*, 1973)

$$\exp\left(\frac{-\omega\eta_{\text{int}}d}{c_{\text{g(B),p}}}\right) \tag{2.168}$$

(c) 3150 Hz one-third-octave-band ($M \approx 5$, $N \approx 62$)

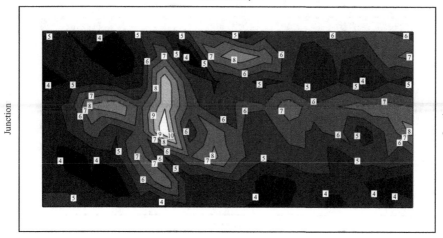

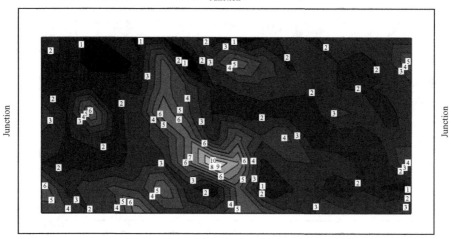

Figure 2.36

(*Continued*)

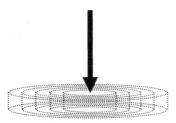

Figure 2.37

Cylindrical wavefronts from point excitation of bending waves on a thin plate.

The power input due to point excitation is W. At a distance, d, from the excitation point, the energy density of the direct field due to both geometrical spreading and internal damping is

$$w_d = \rho_s \langle v_d^2 \rangle_t = \frac{W}{2\pi d c_{g(B),p}} \exp\left(\frac{-\omega\eta_{int}d}{c_{g(B),p}} \right) \tag{2.169}$$

On a finite plate, the direct wave is reflected from the plate boundaries. Ideally we need to assume that there are irregularly shaped boundaries so that there is no coherence between the reflected wave field and the direct wave field. In practice the boundaries of walls and floors are usually regular but we can assume that that there are a sufficiently large number of reflections to ensure that any coherence is lost. The mean free path is defined as the average distance travelled by a wave after leaving one boundary and striking the next boundary (Eq. 2.112). To determine the reverberant power input, we can assume that the average distance from the excitation position to a plate boundary is approximately equal to half the mean free path.

The reverberant power is calculated by assuming that the direct field is attenuated with distance across the plate and then undergoes partial reflection at a plate boundary which has an absorption coefficient, α. As we are currently interested in the vibration field on the plate we refer to an absorption coefficient, whereas in Section 5.2 we will look at it from the perspective of vibration transmission to other connected plates, the same value will then be referred to as the transmission coefficient. The energy density of the reverberant field is therefore dependent on the total loss factor, η, and is calculated from

$$w_r = \rho_s \langle v_r^2 \rangle_t = \frac{W(1-\alpha)}{\omega\eta S} \exp\left(\frac{-\omega\eta_{int}d_{mfp}}{2c_{g(B),p}} \right) \tag{2.170}$$

As an example, Fig. 2.38 shows the energy density at 1000 Hz due to the direct and the reverberant fields for a 10 m^2 masonry wall at distances up to 1 m from the excitation position where the total loss factor is representative of that in a transmission suite. The velocity level due to

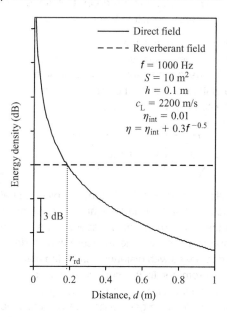

Figure 2.38

Energy density due to the direct and reverberant fields on a plate at distances up to 1 m from the excitation point.

the direct field decreases by 3 dB every time the distance is doubled. The distance from the excitation point at which the energy density in the direct field (Eq. 2.169) equals that in the reverberant field (Eq. 2.170) is the reverberation distance, r_{rd}. Due to the dispersive nature of bending waves, r_{rd} varies with frequency. The reverberation distance can either be calculated by using a numerical solution to find the distance at which $w_d = w_r$, or, by using thin plate theory and assuming that there is perfect reflection at the plate boundaries to give the following approximation

$$r_{rd} \approx \frac{\omega \eta S}{4\pi c_{B,p}} \qquad (2.171)$$

Over the building acoustics frequency range, walls and floors with surface areas $<20\,\mathrm{m}^2$ that act as homogenous, isotropic plates typically have reverberation distances $<0.75\,\mathrm{m}$. When measuring the reverberant velocity level a practical choice for the minimum distance between an accelerometer and the excitation position is usually taken as 1 m (ISO 10848 Part 1).

2.7.6   Statistical description of the spatial variation

In a similar way to a room excited by a point source, the spatial variation of the mean-square velocity on a plate excited by a point force can also be described by a gamma probability distribution (Lyon, 1969). However, we are also interested in wave fields on plates that have been excited by sound fields, and/or by structural waves that impinge upon the plate junctions from connected plates or beams. This means that the mean-square velocity at any point on the plate is determined by a large number of variables. If these are independent random variables (or not too strongly associated), then, by invoking the central limit theorem, the probability distribution of a sum or average of many small random quantities can be approximated by a normal (Gaussian) distribution. It can therefore be assumed that the spatial variation of the mean-square velocity will have a log-normal probability distribution, and the velocity level in decibels will have a normal probability distribution (Lyon and DeJong, 1995).

Compared with sound pressure in rooms, it is more difficult to calculate reasonable estimates for the standard deviation for the spatial variation of vibration on walls and floors. We will assume that the uncertainty due to time-averaging is negligible. In frequency bands above the fundamental mode of the plate, the standard deviation of the velocity level in decibels, σ_{dB}, is dependent upon the mode count in the frequency band, N, and the modal overlap factor, M, (Lyon and DeJong, 1995)

$$\sigma_{dB} \approx \sqrt{43 \lg \left(1 + \left[\left(1 + \frac{3}{\pi M}\right)\left(1 + \frac{N}{\pi M}\right)^{-1}\right]\right)} \qquad (2.172)$$

For masonry/concrete walls or floors, Eq. 2.172 tends to overestimate the standard deviation at low- and mid-frequencies, and underestimate it at high frequencies. For masonry walls with or without a plaster finish the standard deviation usually starts to increase above 2000 Hz due to three main factors: imperfections over the measurement surface, a non-reverberant field where the vibration level decreases with distance across the plate, and difficulties in obtaining a strong fixing of the accelerometer to the surface using beeswax. An example is shown in Fig. 2.39 for airborne excitation of a 100 mm masonry wall with a 13 mm plaster finish. For masonry/concrete walls or floors a generalized curve is included on the figure because the curve shape can usually be described using just four points. This generalized curve for masonry/concrete elements only indicates the general shape, not the typical values for these

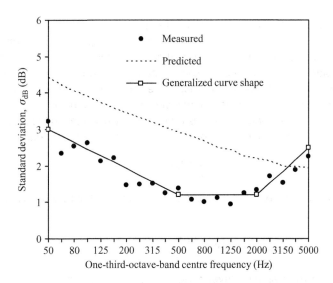

Figure 2.39

Comparison of measured and predicted standard deviations for the velocity level (airborne excitation) on a 100 mm masonry wall with a 13 mm plaster finish.
Measured data from Hopkins are reproduced with permission from ODPM and BRE.

four points; these depend on the specific type of masonry (e.g. solid bricks, hollow blocks), surface finish, material properties, wall dimensions and type of excitation. For measurement purposes it is useful to be aware of the increase in the standard deviation in the high-frequency range. The increase in the mode count and modal overlap at these high frequencies does not necessarily lead to a reduction in the standard deviation.

2.7.7 Decrease in vibration level with distance

With structural excitation there can be a significant decrease in vibration across a wall or floor. This has implications for modeling vibration transmission between structures as well as when measuring impact sound insulation and flanking transmission. Bending waves are of particular importance so we will only consider the decrease that occurs for bending wave motion perpendicular to the surface of the plate.

Any decrease in bending wave vibration with distance due to internal damping only tends to become apparent in the high-frequency range because most materials used for walls and floors have relatively low internal loss factors. For a plane bending wave propagating on a plate, Eq. 2.168 is used to give the decrease in the vibration level, ΔL_{int}, in decibels; this is due to internal damping after travelling a distance, d, where

$$\Delta L_{\text{int}} = \frac{10}{\ln 10} \frac{\omega \eta_{\text{int}} d}{c_{g(B),p}} \qquad (2.173)$$

As a plane bending wave travels across a plate it will also be attenuated each time it impinges upon a plate boundary. This will be referred to as excess attenuation, i.e. the attenuation that occurs in addition to the internal damping. We will take the situation where bending waves on a plate are excited along one of its boundaries due to wave transmission from another plate. For bending wave transmission across the plate in the direction away from the junction, the plate can be considered as a series of one-dimensional (beam-like) elements that are orientated parallel to the junction line; each element is of length, L. Absorption occurs at both ends of these elements as the waves are partially absorbed by the plate boundaries due to wave transmission to other connected plates. The lower and upper boundaries of the plate have absorption coefficients, α_L and α_U respectively, hence the average value is $(\alpha_L + \alpha_U)/2$. For these one-dimensional elements, energy impinges upon the boundaries, $c_{g(B),p}/d_{mfp}$ times every second, where the mean free path is equal to L. By considering the power absorbed by the boundaries, W_{abs}, a loss factor can be determined for the excess attenuation, η_{excess} using

$$W_{abs} = E \frac{c_{g(B),p}}{L} \frac{(\alpha_L + \alpha_U)}{2} = \omega \eta_{excess} E \qquad (2.174)$$

The decrease in the vibration level in decibels due to excess attenuation, ΔL_{excess}, after travelling a distance, d, is

$$\Delta L_{excess} = 10 \lg \left[\exp\left(\frac{\omega \eta_{excess} d}{c_{g(B),p}} \right) \right] = \frac{10}{\ln 10} \frac{(\alpha_L + \alpha_U)}{2L} d \qquad (2.175)$$

The total decrease in the vibration level in decibels, ΔL_{total}, due to internal damping and excess attenuation is then given by

$$\Delta L_{total} = \Delta L_{int} + \Delta L_{excess} \qquad (2.176)$$

Significant decreases usually only occur at frequencies above the thin plate limit for bending waves. In these situations the group velocity for bending waves on thick plates, $c_{g,B(thick)}$, needs to be used instead of the thin plate group velocity, $c_{g,B}$. An example is shown in Fig. 2.40 for a 100 mm masonry wall in the 3150, 4000, and 5000 Hz one-third-octave-bands that are all above the thin plate limit (Hopkins, 2000). For the 4000 and 5000 Hz bands, the attenuation due to internal damping does not account for all the decrease in vibration; hence it is necessary to include the excess attenuation. The excess attenuation is important here because the upper plate boundary is a straight junction to an identical plate for which α_U is estimated to be unity. However, for thick plates it is difficult to obtain reasonable estimates for the power absorbed at the plate boundaries. Figure 2.41 shows the vibration contour plot for the 5000 Hz one-third-octave-band to illustrate the decrease in vibration level across the surface of the same wall with and without a window (Hopkins, 2000).

Although ΔL_{int} and ΔL_{excess} can account for the decrease in vibration level across homogeneous walls and floors, the modular and periodic nature of many building elements causes other loss mechanisms. Apart from relatively homogeneous building elements such as those built from cast *in situ* concrete, the majority of heavyweight walls and floors are built from components such as bricks, blocks, beams, or slabs which are fixed together in a variety of ways. The connections between these individual components can be highly variable, and/or very weak. A common example of a non-homogeneous floor is a beam and block floor as shown in Fig. 2.42. For this particular floor the blocks are not bonded to the beams or to adjacent blocks so the decrease in vibration level is shown using measurements on the beams. In practice, the vibration level on the beams does not keep decreasing with distance because flanking paths to the more distant beams become more important with increasing distance. Large floors ($\approx 200\,m^2$) built from individual concrete slabs with a screed finish can also show significant attenuation with distance (Steel *et al.*, 1994).

A decrease in vibration with distance can also occur with spatially periodic plates. An example is a ribbed plate where the cross-section repeats with a certain repetition distance. In buildings,

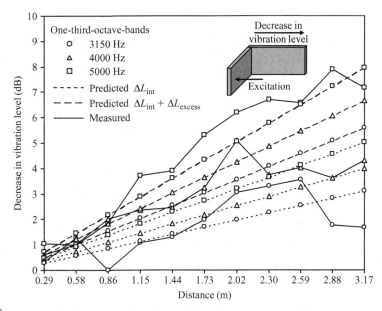

Figure 2.40

Decrease in vibration level with distance across a 100 mm masonry wall ($L_x = 4.04$ m, $L_y = L = 2.38$ m, $\rho_s = 70$ kg/m², $c_L = 2370$ m/s, $\eta_{int} = 0.0125$, solid aircrete blocks with mortar on each side, 13 mm plaster finish). Each measured value corresponds to the average of seven measurement points along a line parallel to the junction. To calculate a decrease in level these values are referenced to the line that is 0.58 m from the junction. The wall was excited via vibration transmission across a corner junction (junction line: L_y) using hammer excitation of the other masonry wall that formed the junction. Note that both walls were rigidly connected to other walls/floors on all sides.

Measured data are reproduced with permission from Hopkins (2000).

ribbed plates produced from a single material can be found in the form of profiled cast *in situ* concrete floors. Lightweight walls and floors also form periodic plates where sheet materials are connected to a framework of beams; examples include timber joist floors or studwork walls. These may be periodic in one or two directions. In addition, the beams and the plates usually have different material properties.

Here we will briefly consider plates that are periodic in a single direction and look at the propagation of bending waves in the direction perpendicular to the ribs/beams. This is the direction that usually has the highest attenuation with distance. When the bending wavelength is much larger than the repetition distance, a periodic plate can be modelled as an orthotropic plate (Section 2.3.3.2). When the wavelength is similar, or smaller than the repetition distance, propagation of bending waves becomes increasingly complex. At these frequencies, periodic structures can generally be considered as one of two types: a precise periodic structure where there is exact periodicity with exactly the same material properties in each repeating section, or a non-precise periodic structure where there is irregularity in the periodicity, dimensions, boundary conditions, and/or material properties. For precise periodic structures, structure-borne sound propagation is characterized by 'pass bands' and 'stop bands' at different frequencies (Cremer et al., 1973; Mead, 1982b). As implied by their names, waves either propagate freely across the ribs/beams, or they are highly attenuated. However, high levels of attenuation can only occur across a relatively small number of ribs/beams. This is due to flanking transmission when the plate is connected to other walls and floors and/or conversion between wave

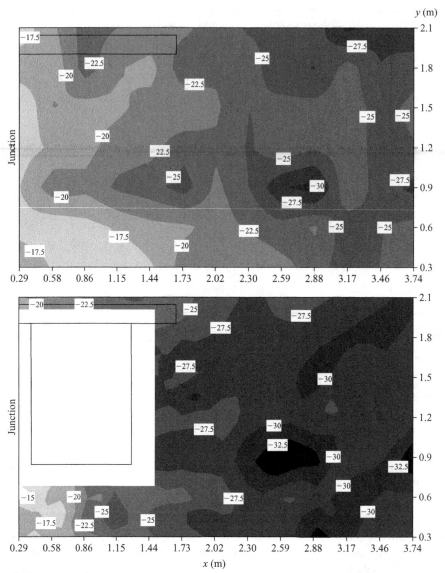

Figure 2.41

Vibration contour plot for the 5000 Hz one-third-octave-band produced from a grid of measurements over the surface of a 100 mm masonry wall. The wall is connected on all sides but only excited along one junction line (indicated on the diagram) by mechanical excitation of the connected wall. Results are shown for the wall with and without a window opening. The decrease in vibration level with distance is shown by plotting $-D_{v,ij}$ (where $D_{v,ij}$ is the velocity level difference), i.e. the -15 dB contour line corresponds to vibration levels on this wall that are 15 dB lower than the spatial average vibration level on the connected wall that is being excited. Refer to Fig. 2.40 for wall properties.

Measured data are reproduced with permission from Hopkins (2000).

types at each junction of the plate and the beams. We are usually interested in the situation where bending waves are excited on a plate. These are converted to in-plane waves at the junction with each beam or rib. Point excitation of bending waves gives a rapid decrease in the bending wave vibration away from the excitation point. This initially steep gradient

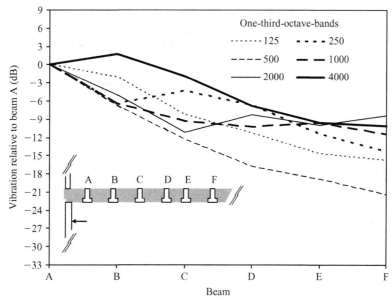

Figure 2.42

Decrease in vibration level with distance across the beams of a beam and block floor. The propagation direction in the measurements is perpendicular to the 150 mm thick solid concrete beams ($\approx 85 \times 50$ mm, $\rho_l = 31$ kg/m). The 150 mm deep blocks are solid masonry ($\rho = 2000$ kg/m^3) with a width of 440 mm – except between beams D and E where the width is 215 mm. Each measured value corresponds to the average of four points on a beam. To show the decrease in level the vibration on each beam is referenced to beam A which is closest to the junction. The floor was excited via vibration transmission across the junction using hammer excitation on one of the masonry walls that formed the junction.

Measured data from Hopkins are reproduced with permission from ODPM and BRE.

becomes much shallower with increasing distance due to conversion from bending to in-plane waves at the beam/rib junctions; these are converted back to bending waves at more distant beam/rib junctions (Heckl, 1964; Mead and Markus, 1983). For non-precise periodic structures, there is potential for a localization effect which confines the highest vibration levels to the vicinity of the excitation (Hodges, 1982). For the above reasons the change in vibration level across a periodic plate is rarely a simple function of frequency; an example is shown in Fig. 2.43 for a timber raft floating floor that represents a non-precise periodic plate.

Figure 2.44 shows measurements on a timber joist floor from Nightingale and Bosmans (1999). There is a rapid decrease in the vibration level with distance up to the butt joint between adjacent boards of OSB. After this the vibration level becomes more uniform with increasing distance. These features are indicative of a periodic plate as discussed above. The fact that the plate surface is formed from individual boards means that there will almost always be a joint or discontinuity between boards at a short distance from any excitation point. Different types of connection between sheets or boards will alter the transmission of bending wave across these joints as well as the conversion to other wave types. In the direction parallel to the joists the attenuation with distance is much lower; the only significant attenuation occurs in the high-frequency range across tongue and grooved joints between adjacent boards (Nightingale and Bosmans, 1999).

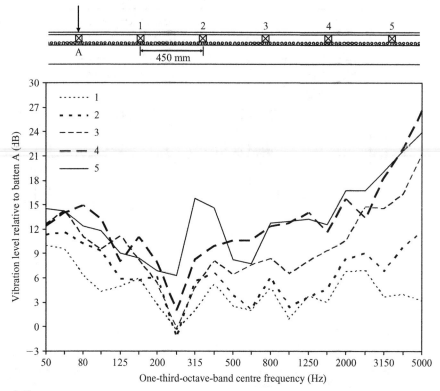

Figure 2.43

Decrease in vibration level with distance across a timber raft floating floor. Measurements used impulse excitation with a plastic headed hammer at two different positions above batten A. Two random positions were used for the accelerometers on the chipboard above each batten.

Timber raft details: 18 mm tongue and grooved chipboard ($\rho_s = 10.6\,\text{kg/m}^2$, $c_L = 2410\,\text{m/s}$) screwed at 300 mm centres to $45 \times 45\,\text{mm}$ timber battens ($\rho_l = 0.9\,\text{kg/m}$, $c_L = 5490\,\text{m/s}$), floating on mineral wool on 150 mm concrete slab.

Measured data from Hopkins are reproduced with permission from ODPM and BRE.

2.8 Driving-point impedance and mobility

To quantify the power input into a plate or beam from a mechanical source it is necessary to know its impedance or mobility. The driving-point impedance or mobility that applies to point force excitation of bending waves is of particular importance. Point force excitation occurs with many structure-borne sound sources in buildings including the hammers of the ISO tapping machine. For impact sound insulation we need to consider random excitation positions in frequency bands and it is the spatial average value over the floor surface that is of particular interest. For other applications, such as the positioning of machinery or other equipment it is sometimes necessary to look at the driving-point impedance or mobility at specific positions and at single frequencies.

A force can be considered as being applied at a single point when the excitation area is much smaller than the structural wavelength. For excitation by a point force, the driving-point impedance, Z_{dp}, is the ratio of the complex force to the complex velocity at the excitation point,

$$Z_{dp} = \frac{F}{v} \tag{2.177}$$

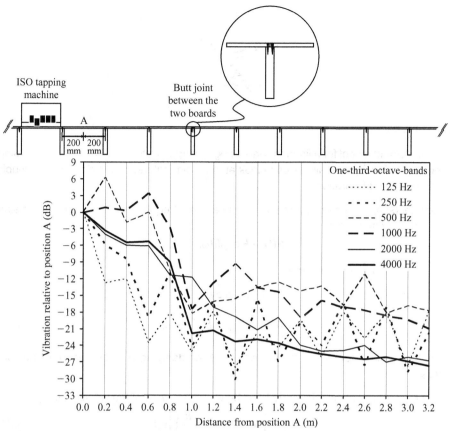

Figure 2.44

Decrease in vibration level with distance across the walking surface of a timber joist floor. The propagation direction for the measurements is perpendicular to the solid timber joists (38 × 235 mm, joist spacing: 400 mm centres) and includes a butt joint between two adjacent boards of 16 mm OSB (screwed to joists at 300 mm centres). Each gridline corresponds to a measurement point on the OSB that is either above or midway between the joists. Excitation using the ISO tapping machine. Measured data are reproduced with permission from Nightingale and Bosmans (1999) and the National Research Council of Canada.

and the driving-point mobility, Y_{dp}, is defined as

$$Y_{dp} = \frac{1}{Z_{dp}} = \frac{v}{F} \qquad (2.178)$$

For a point force, the structural power input can be written in terms of either the impedance or the mobility,

$$W_{in} = \langle Fv \rangle_t = \frac{1}{2} \text{Re}\{Fv^*\} = v_{rms}^2 \text{Re}\{Z_{dp}\} = F_{rms}^2 \text{Re}\left\{\frac{1}{Z_{dp}}\right\} = F_{rms}^2 \text{Re}\{Y_{dp}\} \qquad (2.179)$$

where * denotes the complex conjugate. This is only applicable to a source which has negligible impedance compared to the structure itself; discussion of this point with regards to the hammers of the ISO tapping machine is in Section 3.6.3.

For finite reverberant structures, the driving-point mobility is determined by the damped modes of the structure. Before looking at a multi-modal structure such as a plate it is useful to consider

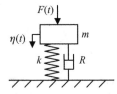

Figure 2.45

Mass–spring–dashpot system.

a lumped element model for a linear mass–spring–dashpot system (see Fig. 2.45). This single degree-of-freedom system allows us to look at the main features of the driving-point mobility for a single mode of vibration.

A sinusoidal (simple harmonic) force, $F(t)$, with angular frequency, ω, is applied to the single degree-of-freedom system,

$$F(t) = \hat{F}\exp(i(\omega t + \phi)) \tag{2.180}$$

where ϕ is an arbitrary phase term.

The equation of motion for this system is given by

$$F(t) = m\frac{\partial^2 \eta}{\partial t^2} + R\frac{\partial \eta}{\partial t} + k\eta \tag{2.181}$$

where η is the displacement, m is the mass, R is the damping constant, and k is the spring stiffness.

For simple harmonic motion, Eq. 2.181 can be written in terms of the velocity as

$$F(t) = \left[R + i\left(\omega m - \frac{k}{\omega}\right)\right]v(t) \tag{2.182}$$

If the system has no damping (i.e. $R = 0$), Eq. 2.182 indicates that the system response is a maximum at the undamped resonance frequency, f_0, where

$$f_0 = \frac{\omega_0}{2\pi} = \frac{1}{2\pi}\sqrt{\frac{k}{m}} \tag{2.183}$$

At this point, η is no longer needed to represent the displacement. Therefore we can use it to describe the damping in terms of the loss factor. This is more practical than using the damping constant as we almost always discuss the modal response of plates and beams with reference to the various loss factors. The damping constant is related to the loss factor, η, by

$$R = \eta\sqrt{km} = \eta\omega_0 m \tag{2.184}$$

The driving-point mobility is now given by Eqs 2.182 and 2.184 as

$$Y_{dp} = \frac{v}{F} = \frac{1}{\eta\omega_0 m + i\left(\omega m - \frac{k}{\omega}\right)} \tag{2.185}$$

From Eqs 2.179 and 2.185, the power input into the system is given by

$$W_{in} = F_{rms}^2 \text{Re}\{Y_{dp}\} = F_{rms}^2 \eta\omega_0 m|Y_{dp}|^2 \tag{2.186}$$

Hence the real part and the magnitude of the driving-point mobility are needed for practical purposes (see Fig. 2.46). The response at resonance is characterized by a peak in the

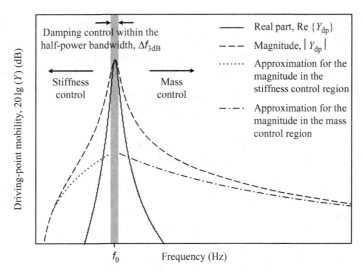

Figure 2.46

Driving-point mobility for a mass–spring–dashpot system.

driving-point mobility. When the damping is sufficiently low ($\eta < 0.3$), the peak in either the real part or the magnitude of the driving-point mobility occurs at the undamped resonance frequency given by Eq. 2.183. From Eq. 2.186 we see that by calculating the magnitude in decibels using $20\lg|Y_{dp}|$, the 3 dB down points from the peak will give the half-power bandwidth, Δf_{3dB}. Note that the curve for $|Y_{dp}|$ is much broader than for Re$\{Y_{dp}\}$. Within the half-power bandwidth, $|Y_{dp}|$ is under damping control. At lower frequencies $|Y_{dp}|$ tends towards pure stiffness control and at higher frequencies towards pure mass control. Approximations for the driving-point mobility under stiffness or mass control are found from Eq. 2.185, giving

$$Y_{dp} \approx i\frac{\omega}{k} \quad \text{under stiffness control} \tag{2.187}$$

$$Y_{dp} \approx \frac{1}{i\omega m} \quad \text{under mass control} \tag{2.188}$$

Characteristics shown by this single degree-of-freedom system form a useful background for the analysis of multi degree-of-freedom systems such as beams and plates.

In this section the focus has been on the driving-point mobility due to its relevance to the power input. If the velocity is measured at a point on the structure that is different to the excitation point, it is referred to as the transfer mobility, Y_{tr}.

2.8.1 Finite plates (uncoupled): Excitation of local modes

Earlier sections discussed local modes on plates with idealized boundary conditions that were uncoupled and isolated from any other structure. A plate has an infinite number of modes; hence it has an infinite number of degrees-of-freedom. However, the single degree-of-freedom system that was discussed in the previous section can be used to represent each mode of vibration. Assuming linearity, the response of the uncoupled plate can be found by superposing the response of all the local modes.

Of particular interest for impact sound insulation is the excitation of bending modes on a plate by a point force acting perpendicular to the surface. The principle of superposition can be used to give an analytic solution for bending vibration of a thin rectangular plate. This can then be used to give the driving-point mobility at a position (x, y) on a rectangular plate (Cremer et al., 1973)

$$Y_{dp} = \frac{i4\omega}{\rho_s S} \sum_{p=1}^{\infty} \sum_{q=1}^{\infty} \frac{\psi_{p,q}^2(x, y)}{[\omega_{p,q}^2(1 + i\eta)] - \omega^2} \tag{2.189}$$

where the local bending mode shape, $\psi_{p,q}(x, y)$ for angular mode frequency, $\omega_{p,q}$, is given by Eq. 2.156 for a rectangular plate with simply supported boundaries.

Figure 2.47 shows the real part of the driving-point mobility for a concrete slab. Equation 2.189 is initially used to calculate the mobility at two chosen positions, A and B. Position A in the centre of the plate excites f_{11} and f_{22} as indicated by the two peaks in the mobility spectrum; it does not excite f_{21} and f_{12} because position A lies on the nodal line of these modes. In contrast, position B excites all of the first four modes as seen by the peaks in the mobility spectrum. The fact that the peaks correspond to excited modes is important; for a given mean-square force input, peaks in the driving-point mobility will correspond to peaks in the power input (Eq. 2.179). Equation 2.189 is then used to calculate a spatial average value for comparison with the measured mobility; this shows that the local mode assumption gives a reasonable estimate of the mobility spectrum. In general, the fluctuations significantly decrease with frequency and the mobility tends towards a single value. This value corresponds to the driving-point mobility for an infinite plate; it is shown in Fig. 2.47 for comparison with the measured values because it will soon be introduced in Section 2.8.3.

2.8.2 Finite plates (coupled): Excitation of global modes

Local modes are extremely useful in the analysis of sound and vibration, but the local mode viewpoint deliberately takes a blinkered view by ignoring the interaction with other parts of the structure to which the plate or beam is coupled. For example, if we take a rectangular separating wall in a building, its four boundaries will be connected to flanking walls and floors. The local modes can be calculated for each of the uncoupled and isolated walls and floors. However, the system of coupled walls and floors also has its own natural modes of vibration; these can be referred to as global modes. For complex built-up structures, the global mode frequencies and mode shapes can be determined using analytic solutions or FEM. Taking the global mode approach to extremes will result in a model of the entire building. Whilst this may be appropriate for very low-frequency vibration in buildings (below 20 Hz), it is not necessary for the prediction of sound insulation where we deal with smaller wavelengths in the building acoustics frequency range. Local modes remain a very convenient way of modelling vibration; in Chapters 4 and 5 we will look at the prediction of sound insulation based on SEA, which relies on the local mode assumption. However, it is worth noting that natural modes of vibration can be modelled from both a local and a global viewpoint.

Figure 2.48 illustrates the effect of global modes on the driving-point mobility (real part) of five connected walls that form an H-block (Hopkins, 2003a). Below 100 Hz, the separating wall only has two local modes (assuming three simply supported boundaries and one free boundary); hence the spatial average mobility for the uncoupled separating wall only has two peaks. In contrast to this there are 22 global modes of the H-block below 100 Hz. These global modes result in the mobility curve for the coupled separating wall in the H-block having many more peaks

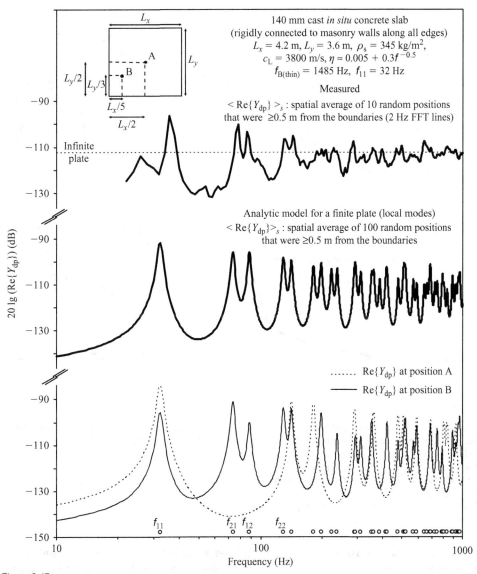

Figure 2.47

Real part of the driving-point mobility for a finite plate. The symbol o next to the x-axis indicates the local mode frequencies for the uncoupled plate with simply supported boundaries.
Measured data from Hopkins are reproduced with permission from ODPM and BRE.

than the uncoupled wall. The existence of many peaks is also seen in the measured curve, although only the general trend is followed, not the precise details. When the local mode density is low, the predicted mobility has deep, wide troughs between the modes that are not under damping control. The abundance of global modes tends to reduce the depth and width of these troughs. Similar features are seen for the flanking wall. In this case the fundamental local mode (assuming two simply supported boundaries and two free boundaries) is just below the lowest global mode. The global modes cause the first peak in the mobility to be shifted upwards in frequency; this is in agreement with the measured data. If there is uncertainty in classifying the

(a) Separating wall

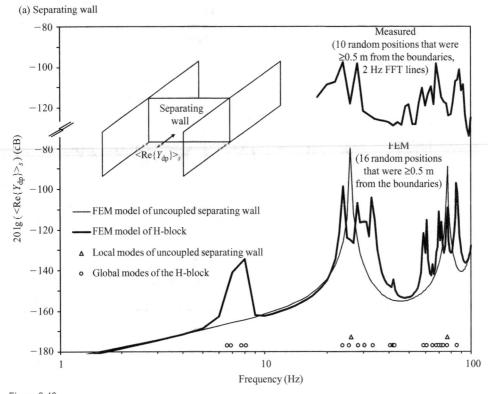

Figure 2.48

Effect of global modes on the spatial average driving-point mobility for coupled plates. This example uses an H-block of five rigidly connected masonry walls which are free-standing on a wide 300 mm thick concrete floor. The 215 mm separating wall ($L_x = 4.5$ m, $L_y = 2.5$ m) and the four 100 mm flanking walls ($L_x = 3.6$–4.1 m, $L_y = 2.5$ m) were all built from solid dense aggregate blocks ($\rho = 2000$ kg/m^3). The walls were slightly orthotropic; this was modelled in the FEM model using the measured longitudinal wavespeed for each wall in the two orthogonal directions. Local and global mode frequencies were also calculated using the FEM model.
Reproduced with permission from Hopkins (2003).

boundary conditions of a plate it would be useful if the peaks in the mobility could be used to esti- mate local mode frequencies for comparison with different idealized boundary conditions. This example indicates that this approach is prone to error; in some cases a rigidly connected bound- ary might be classified as clamped rather than simply supported due to the effect of the global modes. When masonry/concrete plates are rigidly connected on all sides, the total loss factor of each connected wall will be quite high and the global modes will not usually be prominent in the measured driving-point mobility. An example of this was seen with the concrete slab in Fig. 2.47.

The above example for coupled masonry walls indicates that even under laboratory conditions, with measurement of the quasi-longitudinal phase velocity in both directions, and plates with some indisputable boundary conditions (i.e. free), it is still not possible to predict the fine structure of the driving-point mobility using deterministic models. This provides an incentive to try and use statistical models (wherever possible) to describe the vibration of building structures. Such models often make use of the fact that for the purpose of calculating the power input many finite plates can be treated as infinite plates.

(b) Flanking wall

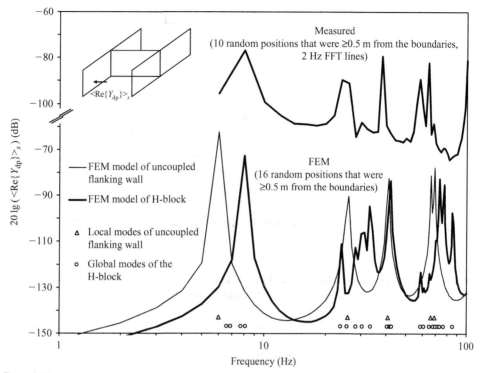

Figure 2.48

(*Continued*)

2.8.3 Infinite beams and plates

In many situations it can be assumed that the structure is of infinite or semi-infinite extent, such that waves travelling out from the excitation point do not return to that point. A simple link can therefore be made to excitation of a finite plate that has completely absorbing boundaries so that no waves return to the excitation point. In practice, waves emanating from the excitation point are only partly absorbed at the boundaries and the reflected waves can return to the excitation point. However, the infinite plate assumption is still appropriate when these reflected waves are incoherent with the excitation signal and at a significantly lower level than at the excitation point.

Excitation of bending waves is now considered for the two situations shown in Fig. 2.49: (a) in the central part of an infinite plate or beam to represent positions on finite structures that are far away from the boundaries and (b) at a position on the edge of a semi-infinite plate or beam.

2.8.3.1 Excitation in the central part

For a thin plate of infinite extent, the driving-point impedance for excitation of bending waves in the central part is real and given by (Cremer *et al.*, 1973)

$$Z_{dp} = 8\sqrt{B_p \rho h} = 2.3\rho c_{L,p} h^2 \tag{2.190}$$

(*Note*: For an orthotropic plate, the effective bending stiffness (Eq. 2.97) can be used.)

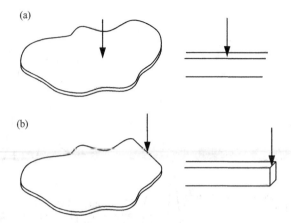

Figure 2.49

Excitation of bending waves: (a) in the central part of an infinite plate or beam and (b) at the edge of a semi-infinite plate or beam.

For a thin beam of infinite extent, the driving-point impedance for excitation of bending waves in the central part is complex and given by (Cremer et al., 1973)

$$Z_{dp} = (1 + i)2\rho S c_{B,b} = (1 + i)2.67\rho S\sqrt{c_{L,b}hf} \tag{2.191}$$

where S is the cross-sectional area of the beam.

When predicting the impact sound insulation of floors it is usually the low and mid-frequency range that are of most importance. At these frequencies most floors or floating floors act as thin plates. The driving-point mobility for a thick plate is complex rather than real (Cremer et al., 1973). However, it is the real part that determines the power input. As the high-frequency range is usually of lesser importance, any errors incurred using thin plate theory above the thin plate limit can usually be neglected.

We have already seen that peaks in the driving-point mobility occur due to excited modes. As these peaks are under damping control, the infinite plate assumption becomes more appropriate as the plate damping increases and the mode spacing decreases.

Figure 2.50 compares infinite plate theory (Eq. 2.190) with the measured real part of the mobility on different finite plates in buildings that can be considered as homogeneous. The measured data is shown in one-third-octave-bands as these are relevant to impact sound insulation measurements. In general, the assumption of a thin infinite plate becomes more appropriate with increasing frequency (assuming $f < f_{B(thin)}$). The largest differences between finite and infinite plates tend to occur in the low-frequency range where there are relatively few bending modes. These trends can be seen with the 140 mm concrete slab; this is the same slab as previously seen in Fig. 2.47. The fundamental local mode is below 50 Hz, and there are only two local modes between 50 and 100 Hz (f_{21} and f_{12}). The resulting deep trough and high peaks between 50 and 100 Hz causes the measured mobility to be significantly different to the infinite plate.

Measurements on the two floating sand–cement screed floors illustrate a number of points. Firstly we note that these plates are square shaped. Square isotropic plates have pairs of

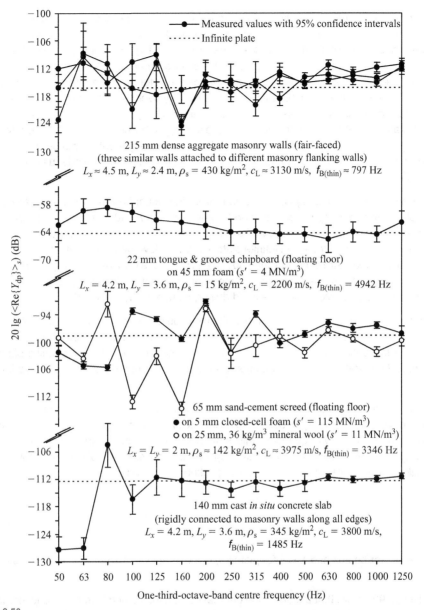

Figure 2.50

Spatial average, one-third-octave-band values for the real part of the driving-point mobility. Comparison of measurements with infinite plate theory for different plates that form walls and floors.
Measured data from Hopkins are reproduced with permission from ODPM and BRE.

modes with identical mode frequencies and identical mode shapes. When different modes have identical mode frequencies the plates are described as having degenerate modes. These can cause pronounced peaks and troughs in the plate response when the mode spacing is wide and damping is low. Secondly, the resilient material that is under the screed affects the plate mobility. The dynamic stiffness of the closed-cell foam is much higher than the mineral wool; hence the structural coupling losses from the screed to the base floor are much higher with the foam than the mineral wool. This means that the damping in terms of the total loss factor is higher for the screed on the closed-cell foam; this higher damping reduces the modal fluctuations. Thirdly, there are variations in the screed properties due to workmanship and materials. Despite a nominally identical sand–cement mix for both screeds, the screed on the mineral wool had a more variable thickness and density over its cross-section. This is due to manual laying of screeds on resilient materials that have significantly different compression under a static load. All of these factors cause the mobility to be significantly different in the low-frequency range for what might be considered to be nominally identical plates. However, as the mode count increases in the mid-frequency range, both screeds are adequately modelled by an infinite plate.

The 22 mm chipboard floating floor is formed from individual boards. If this floor acted as a single plate, all frequency bands between 50 and 1250 Hz would have mode counts greater than five. With distributed damping over its surface due to the resilient material this floor it could be expected to be modelled as an infinite plate in both the low- and mid-frequency ranges. The only significant departure occurs around 80 Hz; this coincides with the mass–spring resonance frequency. However, mass–spring resonance frequencies of floating floors (timber or screed) are not usually detectable in the measured driving-point mobility (real part).

Three similar 215 mm dense aggregate walls were built under laboratory conditions by the same builder using the same batch of blocks. The measurements indicate that it is unlikely that modal fluctuations in the mobility can be accurately predicted for a specific plate and subsequently validated by measurements when the uncertainty due to the spatial variation is so large.

These examples illustrate that for many calculations it is adequate to estimate the spatial average driving-point mobility by assuming an infinite plate. As it is rarely possible to predict the modal fluctuations at low frequencies, this is often assumed over the entire building acoustics frequency range. However, for individual positions it is possible to estimate the envelope of the peaks in the mobility spectrum, $\hat{Y}_{dp}$, using (Skudrzyk, 1981)

$$\hat{Y}_{dp} = \frac{X}{\omega \eta m} \tag{2.192}$$

where $X = 2$ for beams and $X = 4$ for plates, η is the total loss factor and m is the mass of the beam or plate.

Equation 2.192 shows that the envelope decreases with increasing loss factor and increasing frequency; eventually tending towards the infinite plate equation. Figure 2.51 shows the envelope of peaks for the 140 mm concrete slab for two individual excitation positions. When the local mode assumption is reasonable, these estimates work well for individual positions, but for masonry/concrete plates they tend to overestimate the envelope for the spatial average value (Moorhouse and Gibbs, 1995).

For concrete floors, static loads (e.g. furniture) do not significantly affect the driving-point mobility. On lightweight floating floors, static loads of 20–25 kg/m² typically change the driving-point mobility by up to ± 3 dB in an individual one-third-octave-band. These fluctuations tend to average out across the frequency range so the effect of the static load can usually be ignored when the infinite plate model is appropriate for the unloaded floor.

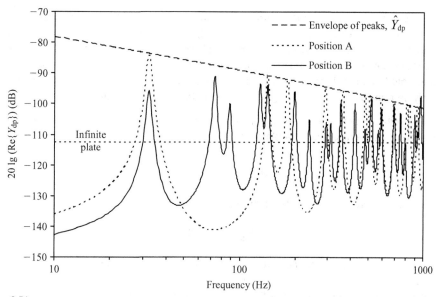

Figure 2.51

Envelope of the peaks in the driving-point mobility at individual positions on the 140 mm cast *in situ* concrete slab described in Fig. 2.47 (positions A and B).

2.8.3.2 Excitation at the edge

Structure-borne sound sources can sometimes be placed at the edge of a wall or floor although it is more common for them to be in the central part.

For a thin plate of semi-infinite extent, the driving-point impedance for excitation of bending waves at an edge is real and given by (Cremer *et al.*, 1973)

$$Z_{dp} = 3.5\sqrt{B_p \rho h} = \rho c_{L,p} h^2 \qquad (2.193)$$

For a thin beam of semi-infinite extent, the driving-point impedance for excitation of bending waves at an edge is complex and given by (Cremer *et al.*, 1973)

$$Z_{dp} = (1 + i)\frac{1}{2}\rho S c_{B,b} = (1 + i)0.67\rho S\sqrt{c_{L,b}hf} \qquad (2.194)$$

where S is the cross-sectional area of the beam.

2.8.3.3 Finite beams and plates with more complex cross-sections

Many beams and plates have more complex cross-sections than a solid, homogeneous rectangle. For one-dimensional beams with solid, non-rectangular cross-sections, the moment of inertia can usually be found in standard tables. This allows calculation of the phase velocity for bending waves (Eq. 2.59) and use of the infinite beam equation to calculate the driving-point impedance (Eq. 2.191). For point excitation in the central area of any two- or three-dimensional isotropic wave-bearing structure (i.e. away from the boundaries), a general rule-of-thumb is that the driving-point impedance equals the product of the angular frequency and the mass of the structure contained within a sphere (centred at the excitation point) with a radius of one-third of the structural wavelength (Heckl, 1988). For orthotropic structures, it is the mass contained within an ellipsoid (centred at the excitation point) where the semi-axes correspond to the wavelengths in the different directions. For thin plates, estimates can be found by considering the mass contained below a circle (isotropic plate) or an ellipse (orthotropic plate) on the surface. Note that for orthotropic non-homogeneous plates with complicated cross-sections, it may only be possible to make a rough estimate for the bending wavelength.

For point excitation of bending waves, this approach can be used to estimate the driving-point impedance in the central area of a plate (Heckl, 1988)

$$Z_{dp} \approx \omega \rho_s \pi \left(\frac{\lambda_{B,p}}{3} \right)^2 \qquad (2.195)$$

or in the central part of a beam (Heckl, 1988)

$$Z_{dp} \approx \omega \rho_l \frac{\lambda_{B,b}}{3} \qquad (2.196)$$

Many walls and floors are formed from combinations of plates and beams. The resulting driving-point mobility on the surface of the plate is often frequency-dependent over the building acoustics frequency range. Despite this, reasonable estimates for many plates with complicated cross-sections can often be obtained by using the infinite plate equations. As an example, Fig. 2.52 indicates that solid concrete stairs can be adequately represented by an infinite isotropic plate with an average plate thickness, h_{av}.

Some cast *in situ* concrete floors are profiled and can be modelled using the general form of a ribbed plate, often with a solid trapezoid forming the rib cross-section instead of a solid rectangle. Ribbed plates (as defined in Section 2.3.3.2.3) are often highly orthotropic. The effective bending stiffness can be calculated from the stiffness in the two orthogonal directions (Eq. 2.109). At low frequencies where $\lambda_{B,eff} \gg d_R$, the driving-point mobility can be calculated using the effective bending stiffness in the infinite plate equation. It can also be assumed to be the same at any point on the plate surface (i.e. on top of a rib, or in-between ribs). With increasing frequency the bending wavelength decreases and when $d_R \approx \lambda_{B,eff}/2$, the plate section between the ribs can be considered as vibrating independently of the entire plate. The driving-point mobility will then be significantly different in the area between the ribs compared to the area along the top of the ribs.

In-between the ribs the driving-point mobility tends towards that of an infinite plate where the bending stiffness is determined only by the properties of the plate in-between the ribs. The transition from the infinite orthotropic plate model (using the effective bending stiffness) to an infinite isotropic plate representing the area in-between the ribs can be assumed to start when $d_R \approx \lambda_{B,eff}/4$, and to be completed when $d_R \approx \lambda_{B,eff}/2$ (Gerretsen, 1986). A straight line between one-third-octave-band centre frequencies is usually adequate for this transition range.

At excitation points along the top of the ribs, the driving-point mobility can be modelled using the moment of inertia for a T-shape beam (Gerretsen, 1986). This is the repeating T-shape cross-section that has a repetition distance, d_R (refer back to Fig. 2.22). For the rectangular T-shape beam the relevant moment of inertia for lateral displacement in the z-direction is given by Eq. 2.110. This can be used to calculate the phase velocity for bending waves (Eq. 2.59) followed by the driving-point impedance for a thin infinite beam (Eq. 2.191). This is a general approach that can be used with other shapes for the repeating beam section. The model for an infinite orthotropic plate can switch to the infinite beam model when $d_R \approx \lambda_{B,eff}/4$.

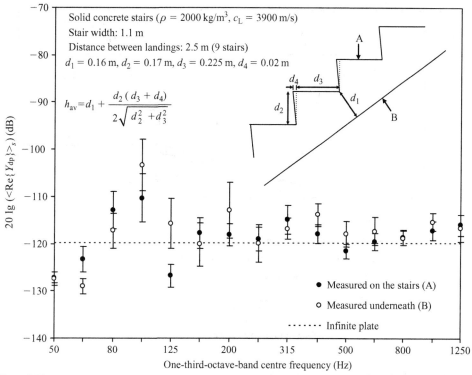

Figure 2.52

Driving-point mobility (real part) on solid concrete stairs. Measured data are shown with 95% confidence intervals. Measured data from Hopkins are reproduced with permission from ODPM and BRE.

When predicting the impact sound insulation of floors, a weighted average driving-point mobility is usually needed at frequencies above $d_R \approx \lambda_{B,eff}/4$ (Gerretsen, 1986). This is because most impact sources (e.g. footsteps, ISO tapping machine) excite the floor at random positions both above and in-between the ribs. This weighted average can be calculated using

$$\langle \text{Re}\{Y_{dp}\}\rangle_w = \frac{d_y}{d_R}\text{Re}\{Y_{dp}\}_{\substack{\text{above}\\\text{ribs}}} + \frac{d_R - d_y}{d_R}\text{Re}\{Y_{dp}\}_{\substack{\text{in-between}\\\text{ribs}}} \qquad (2.197)$$

An example is shown in Fig. 2.53 for a 100 mm thick ribbed floor made of solid concrete; 100 and 200 mm thick solid concrete floors are also shown for comparison. Although it appears that the 100 mm ribbed plate has a significantly lower driving-point mobility than the 100 mm solid plate, this needs to be considered along with the modal peaks that will occur on different finite size plates (Section 2.8.1).

Some floors incorporate beams within the cross-section; this usually results in an orthotropic plate. A common example is a beam and block floor built from concrete beams and rows of loose-laid masonry blocks. If the masonry blocks are not rigidly bonded together and there are narrow gaps between some of them, the bending stiffness in the direction parallel to the concrete beams will mainly be determined by the beams themselves. Perpendicular to

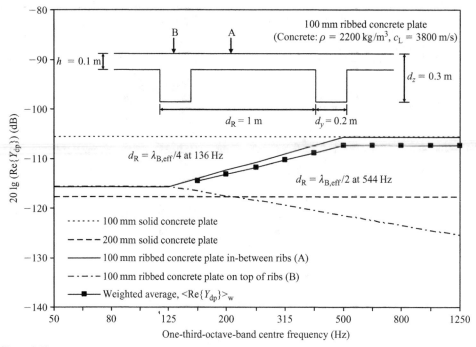

Figure 2.53

Driving-point mobility (real part) on ribbed concrete plates.

the beams, the bending stiffness is usually lower, but highly variable because the plates are non-homogeneous. For floors built with solid concrete beams and solid masonry blocks the bending stiffness can be 20–50% higher in the direction parallel to the beams compared to perpendicular to the beams (Hopkins, 2004). Figure 2.54 shows measurements on four similar beam and block floors with different solid blocks. Bending modes on individual solid blocks only occur in the high-frequency range. In the low- and mid-frequency ranges the real part of the driving-point mobility on the blocks can be highly variable and simple models cannot predict the differences between different blocks. In contrast, the beams have a low spatial variation for the driving-point mobility and the infinite thin beam model is reasonable over most of the low- and mid-frequency range. A thin surface finish that bonds the beams and the blocks together may allow the floor to be modelled as an infinite plate for the driving-point mobility either on or in-between the beams. The example shown in Fig. 2.54 indicates that the driving-point mobility on a floor with a thin surface finish is significantly lower than on the beams or the blocks of the same floor when fair-faced. Much thicker surface finishes, such as a bonded screed which is typically >30 mm, do not always cause the floor to behave in the same way. For thicker surface finishes, the driving-point mobility can be estimated using Eq. 2.195 but it is awkward to determine the bending wavelength without measurements. If the blocks have a very low mass per unit area (e.g. hollow clay pots or polystyrene) compared to the bonded screed then it is sometimes possible to ignore them and use the ribbed plate model for the beams and the screed (Gerretsen, 1986). There are a wide variety of beam and block floors of varying complexity. Measurements of the driving-point mobility are often required to identify simple models or to develop empirical equations for a particular type of floor.

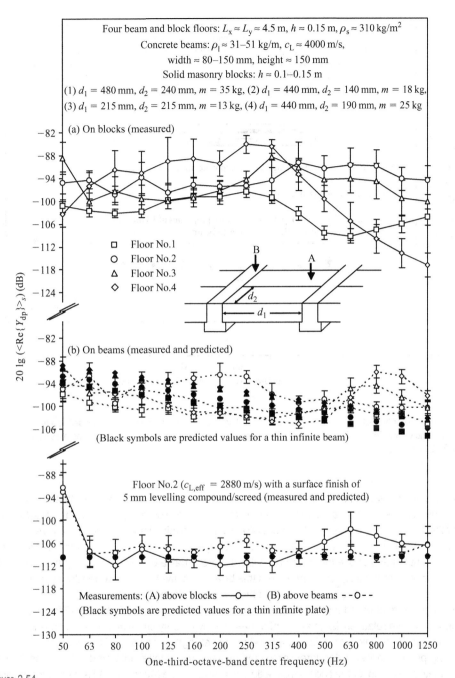

Figure 2.54

Driving-point mobility (real part) on beam and block floors. Measured data are shown with 95% confidence intervals. Measured data from Hopkins are reproduced with permission from ODPM and BRE.

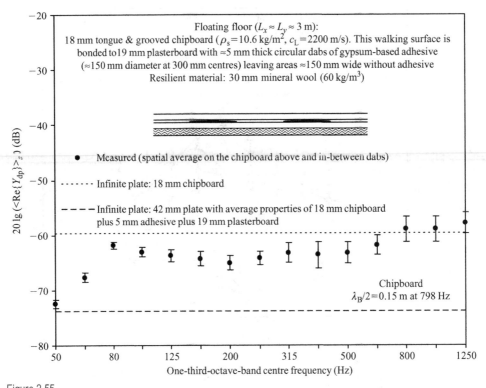

Figure 2.55

Driving-point mobility (real part) on a timber platform floating floor comprised of two boards bonded together. Measured data are shown with 95% confidence intervals. Measured data from Hopkins are reproduced with permission from ODPM and BRE.

Lightweight walls and floors are often constructed with two or three layers of board (e.g. plasterboard, chipboard) that are fixed or bonded together using screws, nails, or dabs of adhesive. For these closely connected plates there is only usually a thin air gap between the layers. When point excitation is applied to one of the boards it begins to support modal vibration between the fixing points at and above the frequency where there is half a bending wavelength between the connections. As the frequency increases this board effectively acts independently of the other board(s). This allows the driving-point mobility to be modelled by assuming an infinite plate for the single layer of board that is being excited. An example is shown in Fig. 2.55 for two boards connected with relatively large areas of adhesive leaving only small unconnected areas. Above the frequency at which there is half a bending wavelength over each of these small areas, the driving-point mobility is adequately predicted by using the infinite plate model. In fact as the infinite plate prediction for the two boards acting as a single plate (42 mm thick) is clearly inappropriate; the infinite plate assumption for the single board can be used to give a reasonable estimate across the entire low- and mid-frequency range. Any differences need to be considered alongside the fact that apart from floors built in the laboratory or under factory conditions there will be so much variation due to workmanship that the infinite plate assumption will be quite adequate.

Lightweight walls and floors are usually built from boards that are nailed or screwed to a framework of beams. We previously looked at the driving-point mobility on ribbed plates formed

(a) Spatial average values

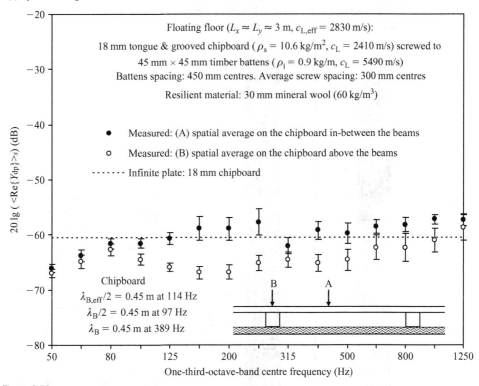

Figure 2.56

Driving-point mobility (real part) on a timber raft floating floor. Measured data are shown with 95% confidence intervals. Measured data from Hopkins are reproduced with permission from ODPM and BRE.

from a single material, such as concrete. This model is less useful for lightweight walls and floors when the beams and plates have significantly different material properties and there are only point connections between them. In addition, although some frames (e.g. timber) can significantly change the bending stiffness of the board, others (e.g. light steel) have a lesser effect. To illustrate some of the main features, measurements on a floating floor comprising single boards screwed to solid timber battens are shown in Fig. 2.56. The battens run in a single direction giving a bending stiffness for the floor that is ≈350% higher in the direction parallel to the battens than perpendicular to them. Below the frequency at which there is approximately $\lambda_{B,eff}/2$ between the battens, the spatial average mobility above or in-between the battens is approximately the same. The relatively uniform response above and in-between the battens can also be seen on the contour plots (region A). However, at these low frequencies the floating floor does not simply act as an infinite orthotropic plate. The measured mobility in region A is affected by the constraint to motion and the damping that is provided by the resilient material and the base floor. These no longer have a significant effect when there is approximately $\lambda_B/2$ between the battens and modes occur on the sub-panels that are formed by the chipboard in-between the battens (region B). In region B there are several modes with half a bending wavelength between the battens; the chipboard above the battens then lies along nodal lines and the mobility is lower than in-between the battens. At frequencies where

(b) Spatial variation along two measurement lines between battens
    (line 1 is approximately mid-way between screws and line 2 is close to the screws)

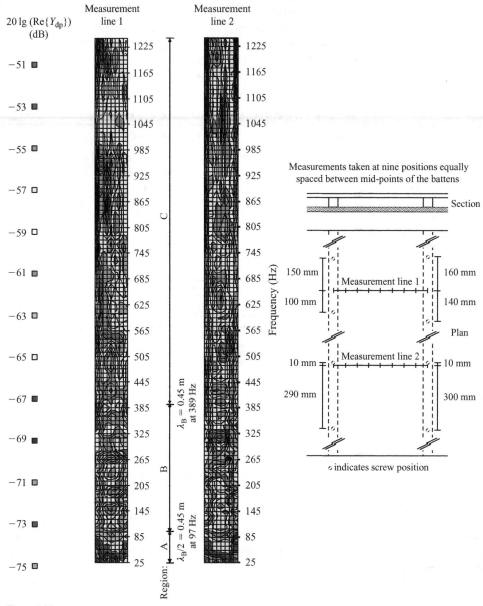

Figure 2.56

(*Continued*)

there is at least one bending wavelength between the battens, the distinction between the mobility above and in-between the battens starts to break down (region C). The infinite plate model can then be used to estimate the driving-point mobility. For design purposes, different screw spacings, different board sizes, different connections between boards (straight edge or tongue and grooved) as well as variation in the material properties imply that there is little to be gained from a single deterministic model.

2.9 Sound radiation from bending waves on plates

Sound radiation concerns the coupling of structure-borne sound waves to sound waves in an adjacent fluid. Our main interest is in bending waves radiating into air. Radiation is usually described using the radiation efficiency, σ. This is defined as the ratio of the radiated power to the power radiated by a large baffled piston ($ka \gg 1$ where a is the piston radius) with a uniform mean-square velocity equal to the temporal and spatial average mean-square velocity of the plate,

$$\sigma = \frac{W}{S\rho_0 c_0 \langle v^2 \rangle_{t,s}}$$ (2.198)

where W is the radiated sound power, and S is the surface area of the plate.

For $ka \gg 1$, the piston dimensions are larger than a wavelength in air, so the air particles cannot escape compression by moving sideways. The particle velocity, u, therefore equals the velocity of the piston surface, v_{piston}. The power radiated by a piston is determined using Eq. 1.18 to give the sound pressure,

$$p = \rho_0 c_0 v_{piston}$$ (2.199)

which is substituted in Eq. 1.19 to give the sound power radiated by the piston,

$$W = SI = S\rho_0 c_0 \langle v^2_{piston} \rangle_t$$ (2.200)

When the radiation efficiency is unity, a plate radiates the same power as a baffled piston (assuming the same $\rho_0 c_0$) which has the same area and mean-square velocity. We will shortly see that plates can be much less or much more efficient in radiating sound than this baffled piston.

This section starts by looking at sound radiation from bending waves on an infinite plate. This provides a general insight into sound radiation and sets the scene for sound radiation from individual bending modes on finite reverberant plates. We then look at the frequency-average radiation efficiency as it is usually necessary to adopt a statistical approach to quantify sound radiation from large numbers of bending modes.

2.9.1 Critical frequency

Sound radiation involves two different types of wave with different phase velocities. Bending waves are dispersive, so for isotropic plates there will be a single frequency at which $c_{B,p} = c_0$. Example phase velocities for various plates are compared with c_0 in Fig. 2.57. The frequency at which $c_{B,p} = c_0$ is called the critical frequency, f_c, and can be calculated from Eq. 2.66, giving

$$f_c = \frac{c_0^2}{2\pi}\sqrt{\frac{\rho_s}{B_p}} = \frac{c_0^2}{\pi}\sqrt{\frac{3\rho_s(1-\nu^2)}{Eh^3}} = \frac{c_0^2\sqrt{3}}{\pi h c_L}$$ (2.201)

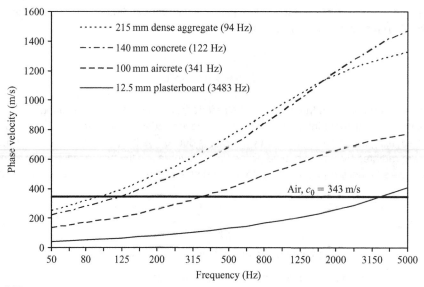

Figure 2.57

Comparison of the phase velocity for longitudinal waves in air with bending waves (thin or thick plate) on various plates. The critical frequency is given in brackets for each plate in the legend. Note that only the 12.5 mm plasterboard acts as a thin plate over the entire building acoustics frequency range, the others act as thin or thick plates depending upon the frequency.

For orthotropic plates, there are two critical frequencies, $f_{c,x}$ and $f_{c,y}$, in the x- and y-directions respectively. To simplify some calculations an effective critical frequency, $f_{c,eff}$, can be calculated from the effective bending stiffness (Eq. 2.97), this is equivalent to

$$f_{c,eff} = \sqrt{f_{c,x}f_{c,y}} \tag{2.202}$$

Note that the critical frequency depends on the phase velocities of sound in air and bending waves on the plate; hence it is dependent upon temperature. All examples relating to sound radiation will assume $c_0 = 343$ m/s and $\rho_0 = 1.21$ kg/m³ unless stated otherwise.

2.9.2   Infinite plate theory

Before looking at finite plates, it is useful to consider sound radiation from a plane bending wave propagating on an infinite plate without damping.

At the critical frequency, equal phase velocities corresponds to $\lambda_B = \lambda$ as illustrated in Fig. 2.58. The critical frequency is sometimes referred to as the lowest coincidence frequency. Coincidence, or trace-matching, occurs when the projection of the wavelength in air onto the surface of the plate equals the bending wavelength (Cremer, 1942). The critical frequency describes the lowest frequency at which coincidence occurs (i.e. at grazing incidence where the angle of incidence $\theta = 90°$). At any frequency above the critical frequency there will always be an angle, θ, that satisfies the coincidence condition (see Fig. 2.59) for which the following relationship applies,

$$\sin \theta = \frac{\lambda}{\lambda_B} \tag{2.203}$$

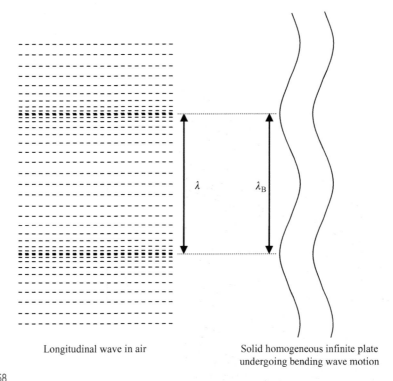

Longitudinal wave in air

Solid homogeneous infinite plate
undergoing bending wave motion

Figure 2.58

Bending wave on an infinite plate and the radiated longitudinal wave in air at the critical frequency.

We will assume that the infinite plate lies in the xy plane and that the plate radiates into air in the positive z-direction (see Fig. 2.60). The negative z-direction is assumed to be *in vacuo*. To simplify matters, the problem can be restricted to one-dimension of the plate by only considering a bending wave propagating in the positive x-direction. This bending wave has a lateral velocity described by

$$v(x, t) = \hat{v} \exp(-ik_{B}x) \exp(i\omega t) \tag{2.204}$$

The radiated sound pressure must have the same dependence on x as the bending wave (i.e. $k_x = k_B$), and propagate in the positive z-direction, hence

$$p(x, z, t) = \hat{p} \exp(-ik_{B}x) \exp(-ik_{z}z) \exp(i\omega t) \tag{2.205}$$

which must satisfy the wave equation for longitudinal waves in air, so inserting Eq. 2.205 into Eq. 1.14 gives

$$k^2 = k_{B}^2 + k_{z}^2 \tag{2.206}$$

Equation 2.206 then gives k_z as

$$k_z = \pm\sqrt{k^2 - k_{B}^2} \tag{2.207}$$

At $z = 0$ the lateral plate velocity must equal the component of the particle velocity that is perpendicular to the plate surface (i.e. the z-component), so

$$v(x, t) = u_{z}(x, t) \quad \text{at } z = 0 \tag{2.208}$$

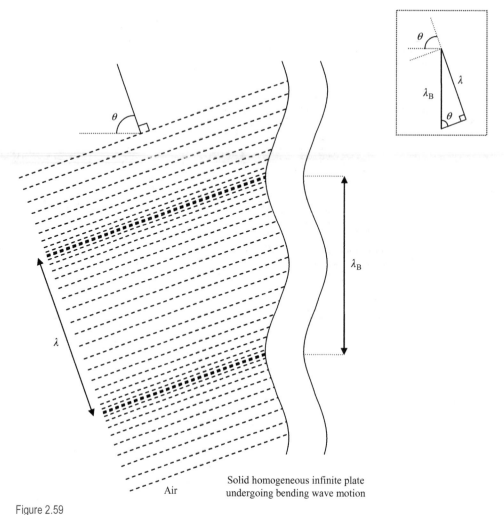

Air — Solid homogeneous infinite plate undergoing bending wave motion

Figure 2.59

Plane wave in air radiated by a bending wave on an infinite plate above the critical frequency.

The particle velocity is taken from Eq. 1.17, which gives

$$\hat{v} = \frac{k_z}{\omega \rho_0} \hat{p}$$

(2.209)

Substituting Eqs 2.207 and 2.209 into Eq. 2.205 yields the radiated sound pressure,

$$p(x, z, t) = \frac{\omega \rho_0 \hat{v}}{\pm\sqrt{k^2 - k_B^2}} \exp(-ik_B x) \exp[-i(\pm\sqrt{k^2 - k_B^2})z] \exp(i\omega t)$$

(2.210)

where the choice of positive or negative sign depends on the ratio of k to k_B and the physics of the situation as discussed below.

Above the critical frequency, where $k > k_B$ (as well as $c_{B,p} > c_0$ and $\lambda_B > \lambda$), the value of k_z is real. So for sound to propagate away from the plate in the positive z-direction, the positive sign

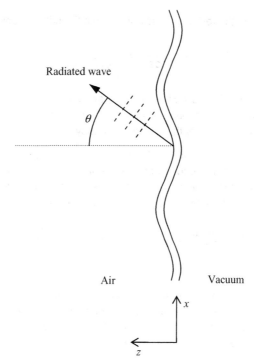

Radiated wave

θ

Air   Vacuum

x

z

Figure 2.60

Plane wave radiated by an infinite plate.

must be chosen in Eq. 2.210. As k_z is real, the pressure at the plate surface is in phase with the plate velocity. From our initial discussion on sound radiation above the critical frequency, we already have a relationship between the wavelength in air and the bending wavelength for the coincidence condition (Eq. 2.203). This can be rewritten in terms of the wavenumbers,

$$\sin\theta = \frac{k_B}{k} \qquad (2.211)$$

Equation 2.210 can now be rewritten in terms of the propagation angle, θ, for the radiated sound wave,

$$p(x, z, t) = \frac{\rho_0 c_0 \hat{v}}{\cos\theta} \exp(-ik_x \sin\theta) \exp(-ik_z \cos\theta) \exp(i\omega t) \qquad (2.212)$$

To determine the radiation efficiency, the plate is temporarily assumed to be very large, rather than infinite, so that S has a finite value and the power radiated from one side of the plate is given by

$$W = \frac{1}{2} \int_S \text{Re}\{pv^*\}dS = \frac{S\rho_0 c_0 \hat{v}^2}{2\cos\theta} \qquad (2.213)$$

The temporal and spatial average mean-square velocity for the plate is

$$\langle v^2 \rangle_{t,s} = \frac{\hat{v}^2}{2} \qquad (2.214)$$

201

Therefore from Eq. 2.198, the radiation efficiency above the critical frequency is

$$\sigma = \frac{1}{\cos\theta} = \frac{k}{\sqrt{k^2 - k_B^2}} \qquad (2.215)$$

At the critical frequency, $k = k_B$ (as well as $c_{B,p} = c_0$ and $\lambda_B = \lambda$). The sound wave radiated by the infinite plate must therefore propagate parallel to the plate surface ($\theta = 90°$). Hence from Eq. 2.215, the radiation efficiency at the critical frequency is infinite.

Below the critical frequency, where $k < k_B$ (as well as $c_{B,p} < c_0$ and $\lambda_B < \lambda$), the value of k_z is imaginary. Equation 2.210 must result in an exponentially decaying nearfield in the positive z-direction, hence the negative sign needs to be chosen. The sound field represents a surface wavefield (Cremer et al., 1973) where the sound pressure is given by

$$p(x, z, t) = \frac{i\omega\rho_0\hat{v}}{\sqrt{k_B^2 - k^2}} \exp(-ik_B x) \exp\left(-z\sqrt{k_B^2 - k^2}\right) \exp(i\omega t) \qquad (2.216)$$

As k_z is imaginary, the pressure at the plate surface is 90° out of phase with the plate velocity; therefore the radiated power and the radiation efficiency are both zero. The nearfield decays away rapidly with distance from the plate surface. The two components of the particle velocity in this nearfield are given by Eqs 1.17 and 2.216,

$$u_x(x, z, t) = \frac{ik_B\hat{v}}{\sqrt{k_B^2 - k^2}} \exp(-ik_B x) \exp\left(-z\sqrt{k_B^2 - k^2}\right) \exp(i\omega t) \qquad (2.217)$$

$$u_z(x, z, t) = \hat{v} \exp(-ik_B x) \exp\left(-z\sqrt{k_B^2 - k^2}\right) \exp(i\omega t) \qquad (2.218)$$

In this nearfield the particle velocity follows closed elliptical paths (Cremer et al., 1973). The sound field is reactive as there is no net energy transport away from the plate surface. Conceptually it is simpler to imagine that the particles move parallel to the plate surface in a back and forth motion. In this way the particles constantly manage to avoid any compression due to the plate motion, and without compression, there can be no radiation of sound.

The above analysis of an infinite plate shows that we can consider sound radiation in three distinct regions: below, at, and above the critical frequency. To illustrate this, the radiation efficiency for an infinite plate is shown in Fig. 2.61. We will find that these three regions are also important for finite plates, although the values for the radiation efficiency will differ. The feature that finite and infinite plates share in common is that above the critical frequency, the radiation efficiency tends towards unity. Below the critical frequency the radiation efficiency for finite plates will often be low, but not zero; at the critical frequency it can be higher than unity, but not infinite.

2.9.3   Finite plate theory: Radiation from individual bending modes

Sound radiation from individual bending modes can be calculated for a homogeneous, isotropic, rectangular plate with simply supported boundaries. This plate lies within the plane of an infinite rigid baffle to avoid interaction between the sound fields on opposite sides of the plate

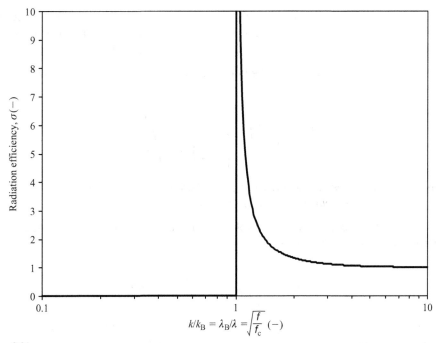

Figure 2.61

Radiation efficiency for an infinite plate.

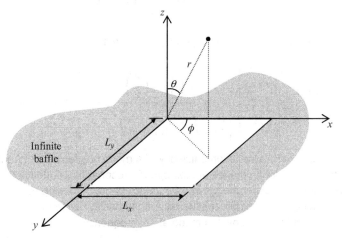

Figure 2.62

Coordinate system for sound radiation from a plate in an infinite baffle to a receiver point in the farfield.

(see Fig. 2.62). At any angular frequency, ω, the sound pressure at a point in the farfield due to radiation from plate mode $f_{p,q}$ is (Wallace, 1972)

$$p_{p,q}(r,\theta,\phi,t) = \frac{i\omega\rho_0}{2\pi r}\exp(-ikr)\exp(i\omega t)\int_0^{L_y}\int_0^{L_x}\hat{v}_{p,q}\psi_{p,q}\exp\left[i\left(\frac{\alpha x}{L_x}\right)+i\left(\frac{\beta y}{L_y}\right)\right]dx\,dy \quad (2.219)$$

203

where $\alpha = kL_x \sin\theta\cos\phi$, $\beta = kL_y \sin\theta\sin\phi$, $\hat{v}_{p,q}$ is a constant relating to the velocity in the z-direction, and $\psi_{p,q}$ is the local mode shape given by Eq. 2.156. The farfield condition applies when $(r - x\sin\theta\cos\phi - z\sin\theta\sin\phi) \gg L_x$ and L_y (Fahy, 1985).

The sound intensity from mode $f_{p,q}$ in the farfield is given by Eqs 1.19 and 2.219 as

$$I_{p,q}(r,\theta,\phi) = 2\rho_0 c_0 \left(\frac{\hat{v}_{p,q}kL_xL_y}{r\pi^3 pq}\right)^2 \left\{\frac{\Gamma\left(\frac{\alpha}{2}\right)\Lambda\left(\frac{\beta}{2}\right)}{\left[\left(\frac{\alpha}{p\pi}\right)^2 - 1\right]\left[\left(\frac{\beta}{q\pi}\right)^2 - 1\right]}\right\}^2 \qquad (2.220)$$

where Γ is cos when p is an odd integer and sin when p is an even integer, Λ is cos when q is an odd integer and sin when q is an even integer.

The power radiated from one side of the plate is calculated from the surface integral over a hemisphere encompassing the plate,

$$W_{p,q} = \int_0^{2\pi} \int_0^{\pi/2} I_{p,q}(r,\theta,\phi)r^2 \sin\theta \, d\theta \, d\phi \qquad (2.221)$$

To calculate the radiation efficiency we also need the temporal and spatial average mean-square velocity, which for mode, $f_{p,q}$, on a plate with simply supported boundaries is

$$\langle v_{p,q}^2 \rangle_{t,s} = \frac{1}{L_xL_y}\int_0^{L_y}\int_0^{L_x} \frac{(\hat{v}_{p,q}\Psi_{p,q})^2}{2} dx \, dy = \frac{\hat{v}_{p,q}^2}{8} \qquad (2.222)$$

Hence, the radiation efficiency, $\sigma_{p,q}$, for mode $f_{p,q}$ is (Eq. 2.198) is

$$\sigma_{p,q} = \frac{64k^2 L_x L_y}{\pi^6 p^2 q^2}\int_0^{\pi/2}\int_0^{\pi/2}\left\{\frac{\Gamma\left(\frac{\alpha}{2}\right)\Lambda\left(\frac{\beta}{2}\right)}{\left[\left(\frac{\alpha}{p\pi}\right)^2 - 1\right]\left[\left(\frac{\beta}{q\pi}\right)^2 - 1\right]}\right\}^2 \sin\theta \, d\theta \, d\phi \qquad (2.223)$$

It is important to note that the radiation efficiency of a mode is a function of frequency. So far we have only been considering modes at their specific resonance frequencies, such as when counting the number of resonant modes in a frequency band, or looking at mode shapes. For sound radiation it is important to note that whilst a mode has its own resonance frequency, we can calculate the sound radiated by that mode at any frequency.

Equation 2.223 can be evaluated numerically, or the following approximations can be used when $kL_x \ll 1$ and $kL_y \ll 1$ (Wallace, 1972)

$$\sigma_{p,q} \approx \frac{32k^2 L_x L_y}{\pi^5 p^2 q^2}\left\{1 - \frac{k^2 L_x L_y}{12}\left[\left(1 - \frac{8}{(p\pi)^2}\right)\frac{L_x}{L_y} + \left(1 - \frac{8}{(q\pi)^2}\right)\frac{L_y}{L_x}\right]\right\} \qquad (2.224)$$

when p and q are both odd integers.

$$\sigma_{p,q} \approx \frac{2k^6 L_x^3 L_y^3}{15\pi^5 p^2 q^2}\left\{1 - \frac{5k^2 L_x L_y}{64}\left[\left(1 - \frac{24}{(p\pi)^2}\right)\frac{L_x}{L_y} + \left(1 - \frac{24}{(q\pi)^2}\right)\frac{L_y}{L_x}\right]\right\} \qquad (2.225)$$

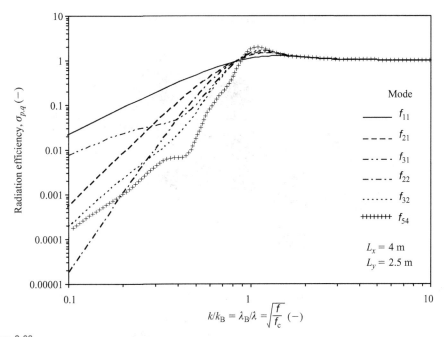

Figure 2.63

Radiation efficiency of individual modes.

when p and q are both even integers.

$$\sigma_{p,q} \approx \frac{8k^4 L_x L_y^3}{3\pi^5 p^2 q^2}\left\{1 - \frac{k^2 L_x L_y}{20}\left[\left(1 - \frac{8}{(p\pi)^2}\right)\frac{L_x}{L_y} + \left(1 - \frac{24}{(q\pi)^2}\right)\frac{L_y}{L_x}\right]\right\} \tag{2.226}$$

when p is an odd integer and q is an even integer (for the opposite combination, swap over p and q, and, L_x and L_y).

Example radiation efficiencies are shown in Fig. 2.63 for a number of modes on a rectangular plate. These have been calculated using numerical integration of Eq. 2.223. They are plotted as a function of k/k_B, hence $k/k_B = 1$ corresponds to radiation at the critical frequency. The range $0.1 \leq k/k_B \leq 10$ covers the building acoustics frequency range for the majority of plates typically used in buildings. Below the critical frequency, the fundamental mode, f_{11}, has the highest radiation efficiency; although in the vicinity of the critical frequency, other modes have slightly higher values. Just above the critical frequency, the radiation efficiency for all modes reaches a peak. At this peak the radiation efficiency is slightly greater than unity. As the frequency increases further, the radiation efficiency asymptotes towards unity. There is significant variation below the critical frequency. In this range, the general tendency is for the fundamental mode to have the highest radiation efficiency, and modes with p and q as odd integers to have high radiation efficiencies when compared against modes with p and q as even integers.

Below the critical frequency, the variation in radiation efficiency between individual modes is due to the sinusoidal nature of the mode shapes. A physical interpretation can be found by treating the plate as an array of point sources (Maidanik, 1962). Each point source is positioned at the centre of an area defined by the nodal lines of the mode. The adjacent point sources are π radians out-of-phase with each other. Figure 2.64 shows an example using the mode f_{54}.

(a) Vibration field of the f_{54} bending mode on a rectangular plate in the plane of an infinite baffle with simply supported boundaries.

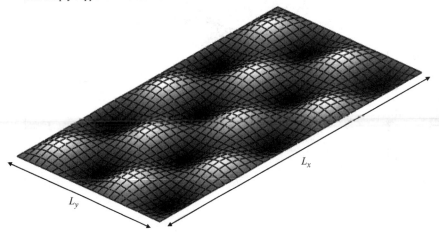

(b) Representation of the f_{54} mode assuming that the areas between the nodal lines act as point sources (rectangular piston radiators). The opposite phase of the point sources is indicated using black and white.

Figure 2.64

Radiation from bending modes on a rectangular plate.

We have already used the sound pressure in the farfield to quantify the radiated sound power over the surface of a hemisphere that encloses the plate. If we look at the path length between a position on this surface and two adjacent point sources of opposite phase on the plate we will find positions where the phase difference due to the different path lengths is negligible compared to the phase difference of π radians between the sources. At such positions there will be almost complete cancellation of the sound pressure from the two point sources. The path lengths depend upon the distances between adjacent point sources (L_x/p and L_y/q). So depending upon frequency, mode numbers and plate dimensions there will be varying degrees of cancellation over the surface of the hemisphere. To consider all the point sources that represent the mode we now need to re-define the array of point sources. Hence we take

(c) Corner radiator below the critical frequency.
Dashed lines within the transparent grey area indicate the areas containing dipole and quadrupole sources.

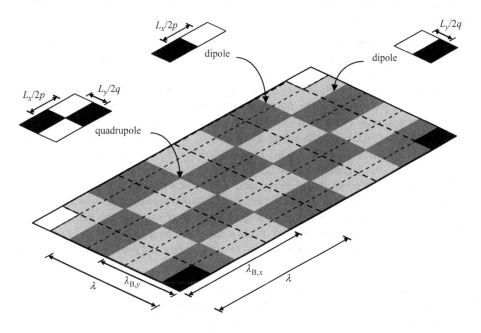

(d) Edge radiator below the critical frequency.

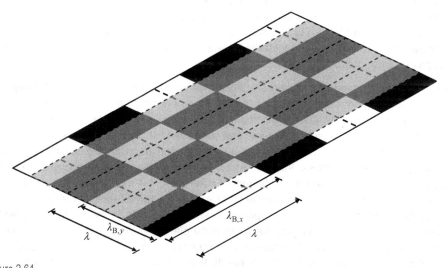

Figure 2.64

(*Continued*)

each area that is demarcated by the nodal lines and split it into four zones of equal area that all have the same phase; each of these represents a point source. When the distance between adjacent point sources is much smaller than a wavelength, the point sources are considered in combination: a dipole is formed from two point sources (one of each phase), and a quadrupole is formed from four point sources (two of each phase). Below the critical frequency, these dipoles and quadrupoles radiate significantly less power than a point source (Morse and Ingard, 1968; Vér and Holmer, 1988).

Below the critical frequency, the wavelength in air is larger than the bending wavelength along one or both of the plate dimensions. We start by considering the frequency at which $\lambda > \lambda_{B,x}$ and $\lambda > \lambda_{B,y}$ for which the point sources combine to form dipoles or quadrupoles as shown in Fig. 2.64c. The quadrupoles lie in the central region of the plate, with dipoles along the plate edges. This leaves point sources in the corners that are responsible for most of the radiated sound power; however, this will be affected by any interaction between the corner sources when the plate dimensions are less than a wavelength. For the same mode we can look at the frequency where $\lambda < \lambda_{B,x}$ and $\lambda > \lambda_{B,y}$ (see Fig. 2.64d). Now there are only two strips left along the edges in the x-dimension that do not form dipoles or quadrupoles. As $\lambda < \lambda_{B,x}$ there are no dipoles formed along these edges, and these strips are responsible for most of the radiated sound power. Below the critical frequency, plate modes can be classified as corner radiators or edge radiators; the former usually radiate less power than the latter. This classification is frequency-dependent. The relatively weak radiation from corner modes and edge modes results in radiation efficiencies that are less than unity, but not zero as with an infinite plate.

As the fundamental mode is represented by a single point source, there are no interactions that form dipoles or quadrupoles. It therefore tends to have the highest radiation efficiency below the critical frequency. Compared to other modes it acts more like a simple piston. However, when the plate dimensions are less than a wavelength in air, the air particles can partly escape compression by moving sideways. This results in radiation efficiencies lower than unity below the critical frequency. For modes with p and q as odd integers acting as corner radiators, the corner sources are in phase. For this reason they tend to have higher radiation efficiencies than modes with p and q as even integers for which there is a degree of cancellation due to the opposite phase of the corner sources.

Below the critical frequency, radiation efficiencies for individual modes vary significantly with different boundary conditions (Gomperts, 1977). When predicting the radiation efficiency for plates in buildings, there is always uncertainty in describing the actual boundary conditions. In practice it is often reasonable to assume that connected wall or floor boundaries are simply supported. However we also need to consider the fact that many walls are built from components such as bricks and blocks; this introduces variation in the material properties over the plate. Combined with variation in the boundary conditions, this results in mode shapes that do not have a distinct, precise, sinusoidal pattern, and we cannot exactly predict the mode frequencies. In addition, the mode shapes of flanking walls with doors or windows may be considerably different to a homogeneous rectangular plate with simply supported boundaries (refer back to Fig. 2.33). The classification of corner and edge radiators therefore becomes rather blurred in practice. For frequency bands that are influenced by a large number of modes, this is less of an issue because we can use a frequency-average radiation efficiency. However, when there are low mode counts in bands below the critical frequency there may be significant errors in using either the individual mode or the frequency-average radiation efficiency.

Above the critical frequency, the entire surface of the plate radiates sound; at these frequencies we can classify the mode as a surface radiator. In Section 4.3.1.5 we will see that this has important implications for modelling the sound insulation of finite plates using an infinite plate model.

2.9.4 Finite plate theory: Frequency-average radiation efficiency

To determine a frequency-average radiation efficiency, it is assumed that the plate is homogeneous and rectangular with sinusoidal mode shapes. It is also assumed that the modal density is sufficiently high that the radiation efficiency can be considered as a continuous function of frequency. This leads to individual expressions for the radiation efficiency below, at, and above the critical frequency. The radiation efficiency is usually required in one-third-octave or octave-bands; for the following equations, the frequency, f, can be taken as the band centre frequency.

Below the critical frequency (Leppington, 1996; Leppington $et\ al.$, 1982, 1984, 1987),

$$\sigma = \frac{U}{2\pi\mu kS\sqrt{\mu^2-1}}\left[\ln\left(\frac{\mu+1}{\mu-1}\right) + \frac{2\mu}{\mu^2-1}\right]\left[C_{BC}C_{OB} - \mu^{-8}\left(C_{BC}C_{OB} - 1\right)\right] \qquad (2.227)$$

$$\text{for } f < f_c$$

where U is the plate perimeter, S is the plate area, C_{BC} is a constant for the plate boundary conditions ($C_{BC} = 1$ for simply supported boundaries, $C_{BC} = 2$ for clamped boundaries), C_{OB} is a constant for the orientation of the baffle that surrounds the edges of the plate ($C_{OB} = 1$ when the plate lies within the plane of an infinite rigid baffle, $C_{OB} = 2$ when the rigid baffles along the plate perimeter are perpendicular to the plate surface), and

$$\mu = \sqrt{\frac{f_c}{f}} \qquad (2.228)$$

Above the critical frequency (Leppington $et\ al.$, 1982; Maidanik, 1962),

$$\sigma = \frac{1}{\sqrt{1-\mu^2}} \quad \text{for } f > f_c \qquad (2.229)$$

At the critical frequency (Leppington $et\ al.$, 1982),

$$\sigma \approx \left(0.5 - \frac{0.15L_1}{L_2}\right)\sqrt{k}\sqrt{L_1} \quad \text{for } f = f_c \qquad (2.230)$$

where L_1 is the smaller and L_2 is the larger of the rectangular plate dimensions, L_x and L_y (for square plates, $L_1 = L_2 = L_x = L_y$). The radiation efficiency for the frequency band that contains the critical frequency can be calculated from Eq. 2.230 using $k = 2\pi f_c/c_0$.

For orthotropic plates modelled as isotropic plates using the effective bending stiffness, the effective critical frequency is used.

2.9.4.1 Method No. 1

As we will now introduce alternative calculations for the radiation efficiency, we will refer to the calculations using Eqs 2.227, 2.229, and 2.230 as method no. 1.

Examples are shown in Fig. 2.65a; the infinite baffle chosen for this example would be appropriate for plates used in the middle of a large wall that are flush with the adjacent wall surface.

(a) Plates set in the plane of an infinite baffle (method No.1)

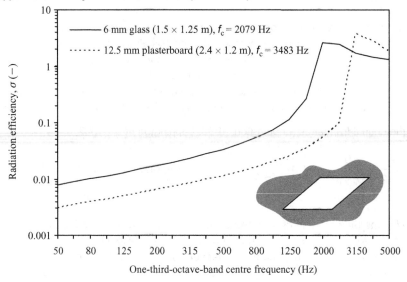

(b) Plate set in the plane of an infinite baffle (method No.2)

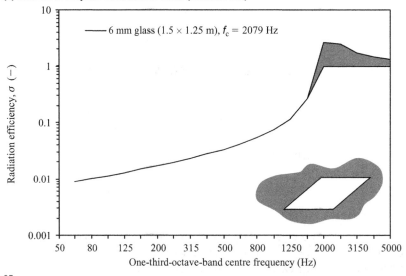

Figure 2.65

Radiation efficiencies for finite plates with simply supported boundaries.

2.9.4.2   *Method No. 2*

Accurate prediction of the radiation efficiency at and near the critical frequency is difficult to achieve. For thin plates such as glass, plasterboard, and metal, it is quite common for the calculated radiation efficiency to overestimate the actual value in the vicinity of the critical frequency. At and near the critical frequency, the radiation efficiency calculated using Eqs 2.227, 2.229, and 2.230 can be considered as an upper limit. A corresponding lower limit can be determined by using the calculated values and setting $\sigma = 1$ in the lowest frequency band for which $\sigma > 1$,

(c) Masonry/concrete plate with baffles perpendicular to all plate edges (method No.3)

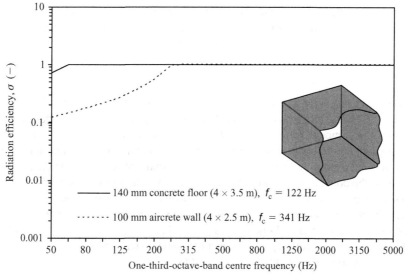

(d) Masonry/concrete plate set in the plane of an infinite baffle. Comparison of method No. 3 with No. 4.

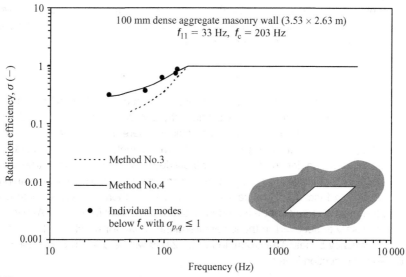

Figure 2.65

(*Continued*)

and then setting $\sigma = 1$ in all the higher frequency bands. In most cases the radiation efficiency will lie between the lower and upper limits; for the example in Fig. 2.65b, this is shown by the shaded area.

For plates commonly used in lightweight walls and floors (e.g. plasterboard, chipboard) that are rigidly connected to a timber frame, the lower limit can be a better estimate than the upper limit.

For laminated plates the radiation at critical frequency can be highly dependent on the number and type of ply groups (Matsikoudi-Iliopoulou and Trochidis, 1992). For some laminates such as plywood the radiation efficiency may not peak above unity at the critical frequency and the lower limit can give a better estimate than the upper limit.

2.9.4.3   Method No. 3 (masonry/concrete plates)

For masonry/concrete walls and floors, there are some important issues relating to methods no. 1 and no. 2. Firstly, these plates usually have a low modal density which goes against the assumption of high modal density. The second is that whilst common thicknesses of glass, metal, and plasterboard have a distinct peak in the radiation efficiency near the critical frequency, this is not usually the case with masonry/concrete plates that are rigidly connected to other plates at the boundaries. An exception is when the plate boundaries are free, or isolated from the surrounding structure by a resilient material and the coupling losses from the plate are negligible. The third is that Eqs 2.227, 2.229, and 2.230 sometimes give irregular, rather than smooth curves near the critical frequency (e.g. double peaks). For masonry/concrete plates (which typically have $f_c < 500\,\mathrm{Hz}$), semi-empirical adjustments that will be now be described by methods no. 3 and no. 4 usually give reasonable estimates.

Method no. 3 requires the full calculation using Eqs 2.227, 2.229, and 2.230, then setting $\sigma = 1$ in the lowest frequency band for which $\sigma > 1$, and then setting $\sigma = 1$ in all higher frequency bands. Examples are shown in Fig. 2.65c; these particular examples represent complete wall or floor surfaces in a box-shaped room, so the chosen orientation for the baffle is perpendicular to the plate surface along the perimeter of the plate.

2.9.4.4   Method No. 4 (masonry/concrete plates)

For homogeneous masonry/concrete plates with $N_s < 3$ in frequency bands below the critical frequency, the radiation efficiency can be significantly underestimated by Eq. 2.227. Better estimates than method no. 3 can be found in situations where the plate can be modelled as being within the plane of an infinite rigid baffle and having simply supported boundaries. This allows calculation of the radiation efficiency for individual modes below the critical frequency using Eq. 2.223. For modes with $\sigma_{p,q} \leq 1$, regression analysis is used to give a smooth radiation efficiency curve as a function of frequency. This curve can then be used to determine the radiation efficiency at the band centre frequencies below the critical frequency. As with method no. 3, it is necessary to set $\sigma = 1$ in the lowest frequency band for which $\sigma > 1$, and set $\sigma = 1$ in all higher frequency bands. The example in Fig. 2.65d will be used in Section 4.3.1.3.4 to predict the sound reduction index.

There are advantages in using regression rather than calculating a band-average radiation efficiency from the modes that lie within each band. Assigning individual modes to particular bands takes no account of uncertainty in calculating the mode frequencies. It also results in bands that are not assigned any modes, even though there will be overlapping response from modes in adjacent bands. These problems are avoided by using regression to give a smooth curve. The result is generally easier to interpret, and easier to compare with individual measurements which will usually have more irregular curves.

This method is well-suited to plates in the laboratory situation where the boundary conditions around the perimeter are uniform, there is minimal interaction between the test element and the laboratory structure, and where the quasi-longitudinal phase velocity or bending stiffness of the plate is known, or can be measured. It is less suitable for predictions in real buildings because the boundary conditions are more variable, and the baffles are usually perpendicular to the wall or floor. It is not suited to orthotropic plates with widely separated critical frequencies because of large differences between the radiation efficiencies of the individual modes; in this situation the band average is more appropriate (Craik, 1996).

2.9.4.5 *Plates connected to a frame*

Thin sheets or boards that are used to form lightweight walls and floors (e.g. plasterboard, chipboard) are usually connected to a framework of studs, battens, or joists. With so many different types of frames and connections, there is no single model for the effect of the frame on sound radiation; in fact more than one model may be needed to cover a range of frequencies.

A light, flexible frame may have negligible effect on the radiation efficiency of the plate. If the frame causes the plate to become slightly orthotropic (i.e. the two critical frequencies are not wide apart) then the effective bending stiffness (Eq. 2.97) can be used to estimate a single critical frequency. This can then be used to calculate the radiation efficiency for the plate.

Stiff, heavy frames can cause the plate to act as a number of sub-panels between the framing members. When these sub-panels have similar spatial average vibration levels (but incoherent vibration fields), then the radiation efficiency of the plate is increased below the critical frequency. The plate can be modelled as a single plate using method no. 1 or no. 2, but Eq. 2.227 needs to be modified by replacing the plate perimeter, U, with $U + L_{frame}$, where L_{frame} is twice the total length of the internal frame (Maidanik, 1962). Another possibility is that sub-panels connected to different parts of the framework will have different boundary conditions and different baffle orientations. The plate can then be modelled by a number of sub-panels (assuming that each sub-panel acts as a reverberant plate). This approach has been used for lightweight cavity walls (e.g. plasterboard walls on a timber frame) so that the boundary conditions and baffle orientation can be dealt with individually at each timber stud (Craik and Smith, 2000a).

Sheets or boards used to form lightweight walls and floors are usually screwed or nailed to the framework. The boundary condition of the sub-panel is affected by whether the junction between the plate and the frame acts as a line connection, or a number of point connections; this affects the radiation efficiency. As a rule-of-thumb the transition between screws/nails acting as a line connection (low frequency model) to individual point connections (high frequency model) starts when the screw/nail spacing is approximately equal to half a bending wavelength (Craik and Smith, 2000b). In practice there will be differences between screw/nail fixings on different boards and different frames. This is partly due to the contact area between the board and the frame being larger than the cross-section of the screw or nail (Bosmans and Nightingale, 1999).

2.9.5 Radiation into a porous material

For a plate undergoing bending wave motion and radiating into air, the radiation efficiency can be used to predict the coupling loss factor between a plate, and a room or an empty cavity. Many cavities in walls and floors are partly or fully filled with porous materials, and walls are sometimes covered with a porous material. If the porous material is very close to the plate, but does not actually touch it, then the porous material can be treated as an equivalent fluid to calculate a radiation efficiency (Tomlinson *et al.*, 2004). This may be important in calculating the total loss factor of the plate because the radiation efficiency into a porous material is usually higher than into air. For plates with low internal loss factors and low coupling losses, this becomes an important loss mechanism. Low internal losses are often associated with thin metal plates hence this is relevant to metal roofing and metal cladding systems. For a thin metal plate, the radiation damping can differ depending on whether the porous material is glued or loose-laid on the surface (Trochidis, 1985). When it is not touching, the damping significantly increases as the porous material is moved closer to the surface of the plate (Cummings *et al.*, 1999). Plasterboard has higher internal losses than most metal plates and so the effect of radiation damping is less significant. However, sheets of plasterboard that form lightweight walls and floors do tend to have higher total loss factors when there is a porous material immediately next to them within the cavity.

To predict the direct sound transmission when a porous material is touching, or is fixed to the plate, the individual aspects of damping, radiation, sound propagation through the porous material, and vibration transmission into the frame of the porous material cannot be considered in isolation from each other. For this kind of sandwich plate (i.e. plate–porous material–plate), Biot theory can be used to model wave propagation in the porous material and predict the sound reduction index (Lauriks *et al.*, 1992).

2.9.6 Radiation into the soil

Concrete ground floor slabs undergoing bending wave motion are often highly damped due to high radiation losses into the soil. For bending wave motion on a plate coupled on one side to a semi-infinite elastic homogeneous medium representing the soil, both compressional (longitudinal) and shear waves are radiated into the soil (Heckl, 1987). The radiation efficiency depends on how the slab is excited. For excitation by a line force on a 600 mm concrete slab with one side coupled to the soil, the coupling loss factor due to radiation into the soil is predicted to be ≈10 dB higher in the low-frequency range than the total loss factor for the same slab that is rigidly connected to other masonry/concrete plates along all edges and only radiates into air (Villot and Chanut, 2000). Measurements on a 125 mm concrete slab using point excitation in Fig. 2.66 indicate similarly high losses in the low- and mid-frequency range. Radiation losses into the ground cause much higher values for the total loss factor than can be attributed to the sum of the internal loss factor and other coupling losses. Note that these high radiation losses do not always cause a significant decrease in vibration with distance across the floor.

Ground floor slabs form part of many flanking paths in a building and it is necessary to account for the high damping of such slabs in a prediction model. This is not easily predicted and can be quite variable which sometimes makes measurement necessary. In some masonry/concrete buildings, errors in estimating the total loss factor of the floor slab will have negligible effect on the prediction of airborne sound insulation between adjacent rooms when all other flanking paths are included (Craik, 1996b).

2.9.7 Nearfield radiation from point excitation

When a mechanical form of point excitation excites bending waves (such as the ISO tapping machine, electrodynamic shaker, force hammer) there is not only sound radiation from the plate bending modes, but also from the nearfield generated in the vicinity of the excitation point.

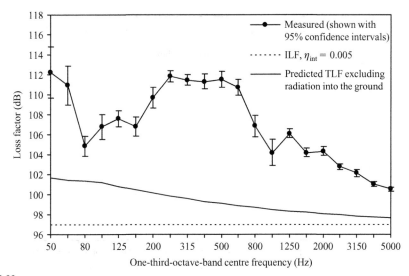

Figure 2.66

Measured total loss factor of a 125 mm cast *in situ* concrete slab ($\rho = 2400\,\text{kg/m}^3$, $c_L = 4200\,\text{m/s}$) on well-consolidated hardcore blinded with sand. (NB Structural coupling losses to connected walls were relatively low for this slab because there were only two connected brick walls, each forming an L-junction with the slab.)
Measured data from Hopkins are reproduced with permission from ODPM and BRE.

This can be important below the critical frequency where the radiation efficiency for bending modes is well-below unity, although its contribution is often negligible.

The sound power radiated by the nearfield, W_n, can be calculated for point excitation on an infinite homogeneous plate and is given by (Cremer *et al.*, 1973)

$$W_n = \frac{\rho_0 c_0 k^2 F_{rms}^2}{2\pi\omega^2 \rho_s^2} \quad \text{for } f \ll f_c \tag{2.231}$$

where F_{rms}^2 is the mean-square force.

Note that this nearfield power is independent of the plate stiffness and damping, and only depends upon the mass per unit area.

For comparison, the power radiated by the bending wave field, W_b, due to point force excitation on a finite homogeneous plate is

$$W_b = \frac{\rho_0 c_0 \sigma}{\omega\eta\rho_s} F_{rms}^2 \text{Re}\left\{\frac{1}{Z_{dp}}\right\} \tag{2.232}$$

Hence W_b can be compared directly with W_n using the ratio,

$$\frac{W_b}{W_n} = \frac{2\pi\omega\rho_s\sigma}{\eta k^2}\text{Re}\left\{\frac{1}{Z_{dp}}\right\} \quad \text{for } f \ll f_c \tag{2.233}$$

and by substituting the driving-point impedance for an infinite homogeneous plate (Eq. 2.190) this can be simplified to (Fahy, 1985)

$$\frac{W_b}{W_n} = \frac{\pi f_c \sigma}{4f\eta} \quad \text{for } f \ll f_c \tag{2.234}$$

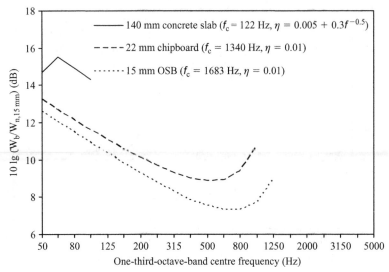

Figure 2.67

Ratio of the sound power radiated by plate bending modes to the sound power radiated by the nearfield due to point excitation by a tapping machine hammer with radius, 15 mm. Plate dimensions are $L_x = 4$ m, $L_y = 3.5$ m and it is assumed that there are baffles along the plate perimeter that are perpendicular to the plate surface. Curves valid for $f \ll f_c$.

Equation 2.234 is sufficient for most estimates such as for point excitation from a narrow rod connected to a shaker. However, to take account of the point force acting over a finite circular contact area with radius, r, the radiated power from the nearfield can be represented by radiation from a piston with a radius, $r + \lambda_B/4$, where (Cremer et al., 1973)

$$W_{n,r} = \frac{1}{2} \rho_0 c_0 k^2 \pi \left(r + \frac{\lambda_B}{4} \right)^4 \frac{F_{rms}^2}{Z_{dp}^2} \quad \text{for } f \ll f_c \tag{2.235}$$

and using the infinite plate impedance, the ratio of W_b to $W_{n,r}$, is

$$\frac{W_b}{W_{n,r}} = \frac{c_0^4 \sigma}{\pi^5 \eta f_c f^3 \left(r + \frac{\lambda_B}{4} \right)^4} \quad \text{for } f \ll f_c \tag{2.236}$$

For point excitation from the ISO tapping machine it can be assumed that the hammer is flat and circular with a radius of 15 mm. Figure 2.67 shows the ratio of W_b to $W_{n,15\,mm}$ for three different plates calculated using Eq. 2.236. The loss factors used for these plates represent minimum total loss factors; for the chipboard and OSB these are based on the internal loss factor. Therefore the ratio represents the maximum that could occur in practice. The curves are only valid for $f \ll f_c$, but indicate that the power radiated by the nearfield on a concrete floor can usually be ignored because the plate radiation from the bending waves will dominate; this may not be the case with highly damped plates that form lightweight floors.

One way of reducing sound radiation from a plate excited by a mechanical point source is to increase the internal damping of the plate by applying a damping compound, or strips of damping material. This would be relevant to impact sound from thin timber plates that form the walking surface of a floor (sound is radiated into the room with the walker as well as into the cavity of a timber floor), or thin metal sheets used for roofing with rain excitation. In such cases it is useful to estimate the nearfield power because there is a limit as to how much damping material is worth applying; eventually the nearfield radiation will become more important than radiation from the highly damped bending waves.

Similar expressions can be found for line excitation of a plate boundary (see Fahy, 1985).

2.10 Energy

For homogeneous beams or plates, the energy associated with each wave type is given by the product of the mass of the beam or plate and the temporal and spatial average mean-square velocity associated with that wave motion using

$$E = m\langle v^2 \rangle_{t,s} \tag{2.237}$$

For bending waves on plates, this calculation uses the wave motion perpendicular to the plane of the plate, i.e. in the z-direction. However, beams of rectangular cross-section can support a bending wave in both the y- and z-directions; hence energy can either be assigned to each direction separately or the two energies can be combined.

References

Bies, D.A. and Hansen, C.H. (1988). *Engineering noise control: theory and practice*, Unwin Hyman, London. ISBN: 0046200223.

Blevins, R.D. (1979). *Formulas for natural frequency and mode shape*, Van Nostrand Reinhold, New York. ISBN: 0442207107.

Bosmans, I. and Nightingale, T.R.T. (1999). Structure-borne sound transmission in rib-stiffened plate structures typical of wood frame buildings, *Building Acoustics*, **6** (3/4), 289–308.

Craik, R.J.M. (1981). Damping of building structures, *Applied Acoustics*, **14**, 347–359.

Craik, R.J.M. (1996a). *Sound transmission through buildings using statistical energy analysis*. Gower, Aldershot. ISBN: 0566075725.

Craik, R.J.M. (1996b). The effect of design changes on sound transmission through a building, *Building Acoustics*, **3** (3), 145–185.

Craik, R.J.M. and Smith, R.S. (2000a). Sound transmission through double leaf lightweight partitions. Part I: Airborne sound, *Applied Acoustics*, **61**, 223–245.

Craik, R.J.M. and Smith, R.S. (2000b). Sound transmission through lightweight parallel plates. Part II: Structure-borne sound, *Applied Acoustics*, **61**, 247–269.

Cremer, L. (1942). Theorie der Luftschalldämmung dünner Wände bei schrägem Einfall, *Akustische Zeitschrift*, **7**, 81–104.

Cremer, L., Heckl, M. and Ungar, E.E. (1973). *Structure-borne sound*, Springer-Verlag. ISBN: 0387182411.

Cummings, A., Rice, H.J. and Wilson, R. (1999). Radiation damping in plates, induced by porous media, *Journal of Sound and Vibration*, **221** (1), 143–167.

Fahy, F.J. (1985). *Sound and structural vibration. Radiation, transmission and response*, Academic Press, London. ISBN: 0122476700.

Gerretsen, E. (1986). Calculation of airborne and impact sound insulation between dwellings, *Applied Acoustics*, **19**, 245–264.

Gerretsen, E. (1996). Vibration reduction index, K_{ij}, a new quantity for sound transmission at junctions of building elements, *Proceedings of Internoise 96*, Liverpool, UK, 1475–1480.

Gomperts, M.C. (1977). Sound radiation from baffled, thin, rectangular plates, *Acustica*, **37**, 93–102.

Heckl, M. (1981). The tenth Sir Richard Fairey Memorial lecture: Sound transmission in buildings, *Journal of Sound and Vibration*, **77** (2), 165–189.

Heckl, M. (1987). Schallabstrahlung in Medien mit Kompressibilität und Schubsteife bei Anregung durch ebene Strahler, *Acustica*, **64** (5), 229–247.

Heckl, M. (1988). Excitation of sound in structures, *Proceedings of Internoise 88*, Avignon, France, 497–502.

Heckl, M.A. (1964). Investigations on the vibrations of grillages and other simple beam structures, *Journal of the Acoustical Society of America*, **36** (7), 1335–1343.

Hodges, C.H. (1982). Confinement of vibration by structural irregularity, *Journal of Sound and Vibration*, **82** (3), 411–424.

Hopkins, C. (2000). Structure-borne sound transmission between coupled plates, *PhD thesis, Department of Building Engineering and Surveying*, Heriot-Watt University, Edinburgh, Scotland.

Hopkins, C. (2003a). Vibration transmission between coupled plates using finite element methods and statistical energy analysis. Part 1: Comparison of measured and predicted data for masonry walls with and without apertures, *Applied Acoustics*, **64**, 955–973.

Hopkins, C. (2003b). Vibration transmission between coupled plates using finite element methods and statistical energy analysis. Part 2: The effect of window apertures in masonry flanking walls, *Applied Acoustics*, **64**, 975–997.

Hopkins, C. (2004). Airborne sound insulation of beam and block floors: Direct and flanking transmission, *Building Acoustics*, **11** (1), 1–25.

Kuhl, W. and Kaiser, H. (1952). Absorption of structure-borne sound in building materials without and with sand-filled cavities, *Acustica*, **2** (1), 179–188.

Lauriks, W., Mees, P. and Allard, J.F. (1992). The acoustic transmission through layered systems, *Journal of Sound and Vibration*, **155** (1), 125–132.

Leissa, A.W. (1973). The free vibration of rectangular plates, *Journal of Sound and Vibration*, **31** (3), 257–293.

Leppington, F.G. (1996). Acoustic radiation from plates into a wedge-shaped fluid region: Application to the free plate problem, *Proceedings of the Royal Society*, London, **452**, 1745–1764.

Leppington, F.G., Broadbent, E.G. and Heron, K.H. (1982a). The acoustic radiation efficiency of rectangular panels, *Proceedings of the Royal Society*, London. **A382**, 245–271.

Leppington, F.G., Broadbent, E.G. and Heron, K.H. (1982b). Acoustic radiation from rectangular panels with constrained edges, *Proceedings of the Royal Society*, London, **A393**, 67–84.

Leppington, F.G., Heron, K.H., Broadbent, E.G. and Mead, S.M. (1987). Resonant and non-resonant acoustic properties of elastic panels. II. The transmission problem, *Proceedings of the Royal Society*, London, **A412**, 309–337.

Lyon, R.H. (1969). Statistical analysis of power injection and response in structures and rooms, *Journal of the Acoustical Society of America*, **45**, 545–565.

Maidanik, G. (1962). Response of ribbed panels to reverberant acoustic fields, *Journal of the Acoustical Society of America*, **34** (6), 809–826. Corrections to the radiation efficiency formulae published in this paper are given in Crocker, M.J. and Price, A.J. (1969). Sound transmission using statistical energy analysis, *Journal of Sound and Vibration*, **9** (3), 469–486.

Maidanik, G. (1966). Energy dissipation associated with gas-pumping in structural joints, *Journal of the Acoustical Society of America*, **40** (5), 1064–1072.

Matsikoudi-Iliopoulou, M. and Trochidis, A. (1992). Sound transmission through composite laminates, *Acustica*, **76**, 38–44.

Mead, D.J. (1982a). Structural wave motion. In White, R.G. and Walker, J.G. (eds.), *Noise and vibration*, Ellis Horwood, 207–226. ISBN: 0853125023.

Mead, D.J. (1982b). Response of periodic structures to noise fields. In White, R.G. and Walker, J.G. (eds.), *Noise and vibration*, Ellis Horwood, 285–306. ISBN: 0853125023.

Mead, D.J. and Markus, S. (1983). Coupled flexural-longitudinal wave motion in a periodic beam, *Journal of Sound and Vibration*, **90** (1), 1–24.

Mindlin, R.D. (1951). Influence of rotatory inertia and shear on flexural motion of isotropic, elastic plates, *Journal of Applied Mechanics*, **18**, 31–38.

Moorhouse, A.T. and Gibbs, B.M. (1995). Calculation of the mean and maximum mobility for concrete floors, *Applied Acoustics*, **45**, 227–245.

Morse, P.M. and Ingard, K.U. (1968). *Theoretical acoustics*, McGraw-Hill, New York. ISBN: 0691084254.

Nightingale, T.R.T. and Bosmans, I. (1999). Vibration response of lightweight wood frame building elements, *Building Acoustics*, **6** (3/4), 269–288.

Qatu, M.S. (2004). *Vibration of laminated shells and plates*. Elsevier Ltd., Oxford. ISBN: 0080442714.

Rindel, J.H. (1994). Dispersion and absorption of structure-borne sound in acoustically thick plates, *Applied Acoustics*, **41**, 97–111.

Skudrzyk, E. (1981). The mean value method of predicting the dynamic response of complex vibrators, *Journal of the Acoustical Society of America*, **67**, 347–359.

Steel, J.A., Craik, R.J.M. and Wilson, R. (1994). A study of vibration transmission in a framed building, *Building Acoustics*, **1** (1), 49–64.

Timoshenko, S.P. and Goodier, J.N. (1970). *Theory of elasticity*, McGraw-Hill, New York. ISBN: 0070858055.

Tomlinson, D., Craik, R.J.M. and Wilson, R. (2004). Acoustic radiation from a plate into a porous medium, *Journal of Sound and Vibration*, **273**, 33–49.

Trochidis, A. (1982). Körperschalldämpfung mittels Gas-oder Flüssigkeitsschichten, *Acustica*, **51** (4), 201–212.

Trochidis, A. (1985). Panel vibration damping due to sound-absorbing liners, *Proceedings of Internoise 85*, Munich, Germany. 733–736.

Utley, W.A. and Fletcher, B.L. (1973). The effect of edge conditions on the sound insulation of double windows, *Applied Acoustics*, **26** (1), 63–72.

Vér, I.L. and Holmer, C.I. (1988). Interaction of sound waves with solid structures. In Beranek, L.L (ed.), *Noise and vibration control*, Institute of Noise Control Engineering, Washington, DC. 270–361. ISBN: 0962207209.

Villot, M. and Chanut, J. (2000). Vibrational energy analysis of ground/structure interaction in terms of wave type, *Journal of Sound and Vibration*, **231** (3), 711–719.

Wallace, C.E. (1972). Radiation resistance of a rectangular panel, *Journal of the Acoustical Society of America*, **51** (3), 946–952.

Watters, B.G. (1959). Transmission loss of some masonry walls, *Journal of the Acoustical Society of America*, **31** (7), 898–911.

Zienkiewicz, O.C. (1977). *The finite element method*, McGraw-Hill Company, London.

Measurement

3.1 Introduction

Measurements generally fall into three categories: laboratory measurements that provide information at the design stage, field measurements that demonstrate whether the required sound insulation has been achieved in a building, and field measurements that help an engineer solve sound insulation problems in existing buildings.

For many buildings the acoustic requirements are described in building regulations; hence repeatability, reproducibility, and relevance (i.e. the link between the measured sound insulation and the satisfaction of the building occupants) are particularly important for airborne and impact sound insulation. Laboratory measurements of the acoustic properties of materials and building elements (e.g. walls, floors, windows, doors) are primarily used for comparing products and calculating the sound insulation *in situ*. Measurements of material properties are particularly useful in assessing whether one material in the construction could be substituted for a different one, and for use in prediction models.

This chapter gives an overview of measurements that are relevant to sound insulation, outlining the basic principles alongside the underlying assumptions or limitations. For all procedural aspects of measurements, requirements on laboratory facilities and test elements, and calculation of single-number quantities the reader is referred to the relevant National or International Standards. In this chapter the latter are generally referred to as the 'relevant Standard' with a reference to the Standard that was current at the time of writing.

3.2 Transducers

This section gives a brief overview of aspects relating to the transducers commonly used to measure sound pressure and vibration; these are microphones and accelerometers.

3.2.1 Microphones

To measure sound pressure levels or reverberation times in reverberant fields we require an omnidirectional microphone so that the response is independent of the direction of incident sound. Over the building acoustics frequency range we would ideally like to measure sound pressure without altering the sound field due to the presence of the microphone. However, the finite dimensions of a microphone and its acoustic impedance mean that this is only achieved when the wavelengths are large. For small wavelengths, diffraction effects from the microphone become significant.

Microphones are usually designed to give a flat frequency response in a particular sound field; typically a pressure field, free-field, or diffuse (random-incidence) field (Anon, 1996).

In a pressure field, the sound pressure has the same magnitude and phase at all points, when very close to (or flush with) a reflective surface, and within a very small closed cavity such as in a sound level calibrator. In a free-field, waves propagate without any influence from reflecting objects or surfaces, such as in an anechoic chamber. In a diffuse field, there is equal probability of a wave arriving at the microphone from any direction. Although microphones are intended or optimized for one specific field they can sometimes be used with negligible error in other fields.

A diffuse field microphone is specifically designed to give a flat response in a random-incidence sound field. In practice, only close approximations to diffuse fields are encountered. These microphones are suitable for sound pressure measurements in reverberant rooms with varying degrees of diffusivity. However, pressure field and free-field microphones may also be considered for use in reverberant sound fields. Free-field microphones are designed to give a flat frequency response at normal incidence (i.e. perpendicular to the microphone diaphragm). The design aim is usually to achieve this over the majority of audio frequencies, which encompasses the building acoustics frequency range. However, the presence of the microphone in a free-field alters the measured sound pressure depending upon the angle of incidence. For angles of incidence other than normal incidence, the response is no longer flat at frequencies in the high-frequency range. For a half-inch microphone, the deviation at specific angles in the high-frequency range is typically within ± 2 dB of the response at normal incidence. However, when the response of a free-field microphone is averaged over all angles of incidence to calculate its response to random-incidence, the resulting response is often relatively flat over the building acoustics frequency range. If it is not sufficiently flat, a special correcting device can usually be attached to a free-field microphone to achieve a flat response. Similarly to free-field microphones, some half-inch pressure microphones (or smaller) also have a relatively flat response in random-incidence sound fields over the building acoustics frequency range.

To measure façade sound insulation the external microphone is either fixed to the surface of the test element (e.g. window) or positioned 2 m from the façade (ISO 140 Part 5). The incident sound field may be from a loudspeaker at 45°, or from an environmental noise source at a variety of different angles. For the latter it is often reasonable to assume random-incidence; hence either pressure field, free-field or diffuse field microphones with a relatively flat response in random-incidence sound fields can be used. Surface measurements require a half-inch microphone or smaller to avoid interference minima in the building acoustics frequency range; this is because of the distance between the centre of the microphone diaphragm and the test element (refer back to Section 1.4.1). When the microphone diaphragm is almost parallel to the direction of the direct sound, pressure field microphones may be used; otherwise free-field microphones are appropriate if their directivity in the high-frequency range has negligible effect.

For measurements of façade sound insulation, a windshield should be used for outdoor measurements of the sound pressure level. Windshields are not usually necessary for indoor measurements unless there is significant airflow, but they are a useful way of protecting the microphone whilst carrying it around a building.

3.2.2 Accelerometers

It is important to ensure that the accelerometer is capable of accurately quantifying the actual vibration. In particular it is important to consider the fixing of the accelerometer, and how the mass of the accelerometer can alter the vibration of lightweight structures. Details relating to accelerometer mounting are contained in the relevant Standard (ISO 5348).

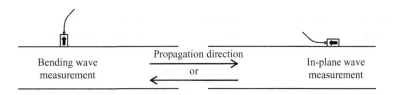

Figure 3.1

Accelerometer alignment for measuring bending or in-plane wave vibration on a beam or plate. The arrow on the accelerometer indicates the main axis of sensitivity.

The type of motion that we want to measure on beams and plates is usually bending or in-plane wave vibration. An accelerometer is designed to be sensitive to motion along one main axis, hence the accelerometer is aligned as shown in Fig. 3.1. Although there is only one main axis of sensitivity, the accelerometer will also exhibit minor sensitivity to transverse motion. When measuring bending wave motion the effect of this transverse sensitivity is usually negligible when the accelerometer is mounted flush to a flat surface. However, it needs to be considered when trying to measure in-plane wave motion in the presence of bending waves.

3.2.2.1 Mounting

Bees wax or petroleum wax is a very convenient method for mounting accelerometers on a building element. However it is important to pay careful attention to the measured levels in the high-frequency range because weak fixing can cause measurement errors. Strong fixing can be obtained by smearing a thin layer of wax over the contact surface of the accelerometer. To fix it to the surface, a fair degree of pressure needs to be applied to the top of the accelerometer (usually with the thumb), and it sometimes helps to use a slight twisting motion. If too much wax is applied, or the surface is fragile, crumbly, dusty, or powdery it can be difficult to achieve a strong fixing. This usually results in low vibration levels being measured in the high-frequency range. With simultaneous measurement of acceleration on the same surface using two or more accelerometers, one accelerometer with weak fixing can usually be identified by significantly lower levels at high frequencies, often with a sharp drop-off. If the surface texture prevents a strong fixing with wax, then small metal washers can be glued/cemented to the surface to allow wax fixing or fixing via a stud on the washer (see ISO 5348).

3.2.2.2 Mass loading

When an accelerometer is fixed to the surface of a structure, a mass has effectively been added to the structure that can reduce the vibration level at the measurement point. This is referred to as mass loading. Assuming a lump impedance model with a force source, the relationship between the measured and the actual velocity is

$$v_{measured} = v_{actual} \frac{Z_{dp}}{Z_{dp} + i\omega m_{acc}} \qquad (3.1)$$

where m_{acc} is the mass of the accelerometer and Z_{dp} is the driving-point impedance of the structure.

State-of-the-art accelerometers that are used to cover the building acoustics frequency range often have a low mass. Hence because plates that form building elements tend to have relatively high surface densities, mass loading is rarely a problem. The effects of mass loading can be avoided when the accelerometer impedance is much less than the plate impedance,

$$\omega m_{acc} \ll Z_{dp} \tag{3.2}$$

For a thin homogeneous isotropic plate this requirement can be calculated using the infinite plate impedance (Eq. 2.190), which gives

$$m_{acc} \ll \frac{0.37 \rho c_L h^2}{f} \tag{3.3}$$

Mass loading is rarely an issue with masonry/concrete elements but it needs to be checked in the high-frequency range with building elements such as glass or plasterboard.

3.3 Signal processing

In any measurement, the signal processing has the potential to affect the measured response. This section contains a brief overview of signal processing for aspects that are relevant to the majority of sound and vibration measurements described in this chapter. It is restricted to filter analysis and does not cover measurement techniques using the Fast Fourier Transform for which many other texts are available (e.g. see Randall, 1987).

From the viewpoint of the analyser, the analogue input signal is a time-varying voltage; therefore the processing applies to both sound and vibration signals. The analogue input signal from sound or vibration transducers is continuous in time, but in the analogue-to-digital conversion it is sampled so that the subsequent analysis is carried out on discrete time data. For filter analysis, the digitized input signal passes through the filters and then the detector as shown in Fig. 3.2.

The effect of signal processing on the measurement of reverberation time is a specific issue that is discussed in Section 3.8.3.

3.3.1 Signals

Signals can be considered in two distinct groups: stationary and non-stationary.

'Stationary' is used to indicate no change over time, or no change during the measurement period. Stationary signals are commonly used in sound insulation measurements and fall into two categories: random and deterministic signals. For stationary random signals, the probability

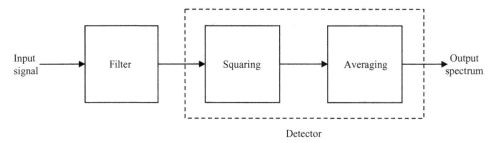

Figure 3.2

Basic components of signal processing using filtering to determine the frequency spectrum.

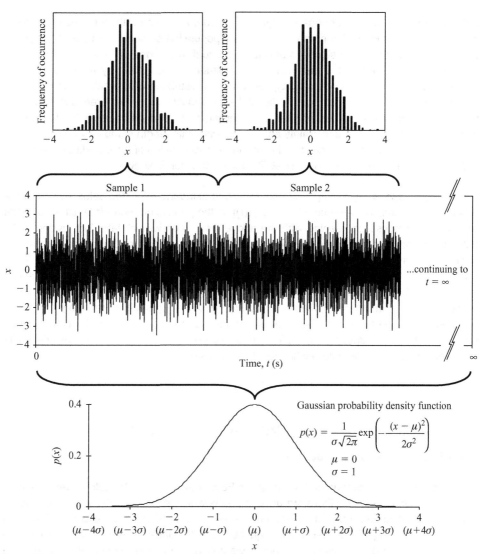

Figure 3.3

Gaussian white noise – finite and infinite length samples.

density function for the signal does not change with time. An example of a stationary random signal is Gaussian white noise. Stationary deterministic signals can be predicted at any point in time, such as a sinusoidal signal or a signal composed of a number of sinusoids. A specific type of deterministic signal is a Maximum Length Sequence; this will be discussed in Section 3.9.

Non-stationary signals used in measurements are typically transients. For example, an impulse generated by a gunshot to measure the reverberation time in a room, or an impulse generated by a hammer hit to measure the velocity level difference between two walls.

White noise is a stationary random signal with constant power spectral density. Gaussian white noise is a particularly useful type of white noise because its statistics are described by the normal (Gaussian) probability distribution. Figure 3.3 shows a record of Gaussian white noise in terms of a positive or negative signal amplitude, x, varying with time. The white noise shown here has a population mean of zero and a population standard deviation of unity. By using a mean of zero, the standard deviation is equal to the rms value of the signal. For an infinite length sample, the distribution of amplitudes is described by the smooth curve of the probability density function, $p(x)$, for a normal distribution. In practice, we use finite length samples of white noise; Fig. 3.3 shows the distributions for two different samples using histograms. The difference between these two histograms serves as a reminder that we will soon need to consider the uncertainty associated with finite length samples when averaging white noise over time in frequency bands. Fortunately, calculation of the statistics for temporal averaging are simplified when the probability density function is known.

A continuous random signal, $x(t)$, can be described using the expected value and the auto-correlation function. The expected value, $E[x(t)]$, is the population mean from all realizations of a random process; in keeping with the above discussion we will take this as zero. The auto-correlation function equals the expected value of the product of the signal with itself at another point in time, i.e. $E[x(t-t_1)x(t-t_2)]$ where t_1 and t_2, are two different points in time. For stationary white noise there will be no statistical correlation between the signal at t_1 and t_2 when $t_1 \neq t_2$; this is due to the random nature of the signal. However, there will be correlation at the same point in time, i.e. when $t_1 = t_2$. $E[x(t-t_1)x(t-t_2)]$ is therefore zero when $t_1 \neq t_2$ and non-zero when $t_1 = t_2$. The auto-correlation function for white noise is therefore defined using the Dirac delta function, $\delta(t_2 - t_1)$. A continuous random signal is defined as white if its expected value and its auto-correlation function, $R_{xx}(t_2 - t_1)$, satisfy

$$E[x(t)] = 0 \tag{3.4}$$

$$R_{xx}(t_2 - t_1) = E[x(t - t_1)x(t - t_2)] = N\delta(t_2 - t_1) \tag{3.5}$$

where N is the power spectral density.

The spectral density is determined by taking the Fourier transform of the auto-correlation function. The Fourier transform therefore gives equal power spectral density at all frequencies. However, constant power per Hertz over an infinitely wide band of frequencies would result in the total power being infinite. Hence although the theory for Gaussian white noise proves to be useful in a number of derivations, in practice we only measure with band-limited white noise signals that cover a broad-band of frequencies. For white noise, the one-third-octave-band or octave-band level increases by 3 dB per doubling of the band centre frequency.

Pink noise is also used for airborne sound insulation measurements. The power spectral density of pink noise is proportional to $1/f$; therefore the one-third-octave-band or octave-band level decreases by 3 dB per doubling of the band centre frequency. Pink noise can be generated from a white noise signal using cascaded filters to give a -3 dB/octave rolloff.

Measurements of airborne sound insulation generally use band-limited white or pink noise signals; although quite often reference is simply made to broad-band noise. Whatever the input signal, the spectrum of the measured sound pressure level in the source room will be affected by the response of the room. For this reason the signal is often shaped using a graphic equalizer as there is a requirement for the source room that the difference between the sound pressure levels in adjacent one-third-octave-bands is no greater than 6 dB (ISO 140 Parts 3 & 4).

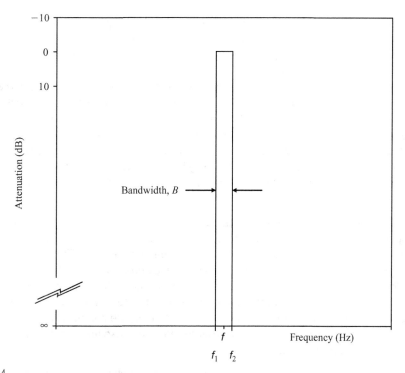

Figure 3.4

Ideal filter.

3.3.2 Filters

Filtering of the signal is almost always required for analysis and other calculations. It is important to ensure that the filtering process does not significantly affect the measurement of a level or reverberation time.

3.3.2.1 Bandwidth

In building acoustics, it is common to use filters that have a constant percentage bandwidth; either one-third-octave or octave-bands. For calculations used to predict sound insulation, the filter bandwidth, B, can be taken as a constant percentage of the band centre frequency, f, where $B = 0.23f$, for one-third-octave-bands, and $B = 0.707f$, for octave-bands. This corresponds to the concept of an ideal filter that has 0 dB attenuation at all frequencies within the passband and infinite attenuation outside the passband (see Fig. 3.4). The bandedge frequencies that define the lower and upper frequencies of this ideal passband are f_1 and f_2 respectively. The centre (or midband) frequency, f, of the filter is the geometric mean of f_1 and f_2,

$$f = \sqrt{f_1 f_2} \qquad (3.6)$$

and the bandwidth is

$$B = f_2 - f_1 \qquad (3.7)$$

Real-time filters that are used in measurement equipment cannot exactly recreate the properties of this ideal filter. There will not be uniform attenuation within the passband with a vertical rolloff slope at the edges of the passband. The performance requirements for filters and tolerances on the attenuation limits inside and outside the passband are given in the relevant Standard (IEC 61260). In this chapter, a 6th order Infinite Impulse Response (IIR) Butterworth filter (six poles) is used to create examples that are indicative of filters that can be used in practice. Butterworth filters don't have a particularly steep rolloff slope, but they can give relatively uniform attenuation in the passband. The attenuation for 1000 Hz one-third-octave and octave-band filters is shown in Fig. 3.5 along with the minimum and maximum attenuation limits for a Class 1 filter (IEC 61260). Both of the 6th order Butterworth filters are within the attenuation limits of the Standard. Although the flanks of the filters cover the building acoustics frequency range, the attenuation is sufficiently high outside the passband that for relatively smooth spectrum shapes (i.e. without discrete tones and steep slopes) this will not significantly affect measurement accuracy.

Measured levels in one-third-octave-bands are sometimes combined to give the octave-band level. Figure 3.6 shows the three one-third-octave-band filters that would be combined to form the 1000 Hz octave-band alongside the 1000 Hz octave-band filter itself. This shows a marked difference between the rolloff slopes for one-third-octave and octave-band filters. Hence there can be differences between levels measured with octave-band filters, and octave-band levels calculated from measurements with one-third-octave-band filters. The extent of these differences depends on the shape of the spectrum. To avoid dispute it is possible to require that measurements be taken in one-third-octave-bands, and that these measured values are used to calculate the octave-band values.

To quantify the effect of the rolloff slope and non-uniform attenuation in the passband it is necessary to introduce other descriptors for the filter bandwidth. These are the effective bandwidth and the statistical bandwidth. For a stationary random signal, the effective bandwidth applies to the power passed by the filter, whereas the statistical bandwidth applies to the statistics of the power passed by the filter, i.e. the variance.

The definition of the effective bandwidth (or noise bandwidth) assumes that a stationary random signal with constant power spectral density is sent as an input signal to the filter. For this particular signal the effective bandwidth equals the equivalent bandwidth of an ideal filter that would pass the same power (mean-square value) as the real filter. It is calculated by numerical evaluation of the following integral (IEC 61260)

$$B_e = \int_0^\infty 10^{-A(f)/10} df \tag{3.8}$$

where $A(f)$ is the filter attenuation in decibels as a function of frequency.

The statistical bandwidth is relevant to the variance of random noise power that is passed by a filter (Bendat and Piersol, 2000; Davy and Dunn, 1987). It is defined for a stationary random signal that is band-limited. The statistical bandwidth equals the bandwidth of this signal when the variance of its power (mean-square value) is equal to the variance of the power passed by the filter. For an nth order Butterworth filter, the statistical bandwidth can be calculated from the effective bandwidth using (Davy and Dunn, 1987)

$$B_s = \frac{2n}{2n-1} B_e \tag{3.9}$$

A 6th order Butterworth filter is the lowest order that would commonly be used in practice, for which Eq. 3.9 gives $B_s = 1.091 B_e$. In the absence of information on specific types of filters it is often necessary (and usually reasonable) to assume that $B_s \approx B_e \approx B$.

(a) 1000 Hz one-third-octave-band filter

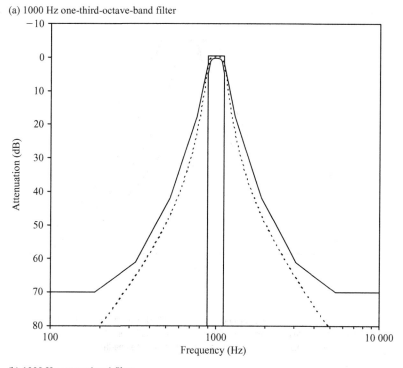

(b) 1000 Hz octave-band filter

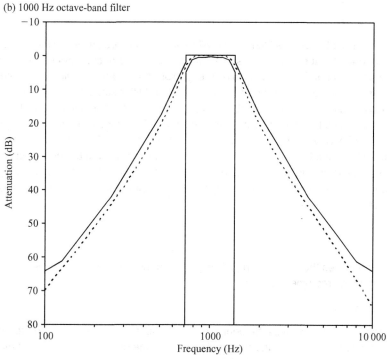

Figure 3.5

Filter attenuation for 6th order Butterworth filters (dotted lines) with a centre frequency of 1000 Hz along with the minimum and maximum attenuation limits for a Class 1 filter according to IEC 61260 (solid lines).

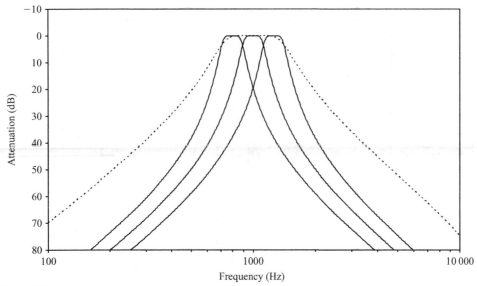

Figure 3.6

Filter attenuation for 6th order Butterworth filters. The 800, 1000, and 1250 Hz one-third-octave-bands are shown in solid lines with the 1000 Hz octave-band in dotted lines. Note that 800, 1000, and 1250 Hz are the preferred centre frequencies (ISO 266) whereas the exact base-ten frequencies (IEC 61260) are used to define the one-third-octave-band filters.

3.3.2.2 *Response time*

When a signal is sent through a filter, there will be a short time delay before the filter responds and gives an output that exactly corresponds to the amplitude of the input signal. This mainly depends on the filter bandwidth, but also on the type of filter. As an input signal we take a single sinusoid, sent through a one-third-octave or octave-band filter starting at $t = 0$. The frequency of the sinusoid must be within the passband of the filter. The input signal and the filtered output signal are shown as a function of normalized time in Fig. 3.7. Normalized time is the product of the filter bandwidth, B, and time, t. It is dimensionless and is useful for looking at general trends. A rough estimate of the time taken to respond to the input signal and to output a sinusoid with exactly the same amplitude is usually taken as $t = 1/B$ s, i.e. normalized time of unity (Randall, 1987). Figure 3.7 indicates that this is a slight underestimate for these filters, but it is usually adequate for practical purposes.

3.3.3 Detector

The detector receives the time-varying AC signal from the filter, squares the signal, and then carries out the required type of temporal averaging.

3.3.3.1 *Temporal averaging*

Linear averaging over time gives the equivalent continuous level, L_{eq}. This corresponds to the level of a continuous steady sound that contains the same energy as the actual time-varying signal during the integration (averaging) time, T_{int}, hence

$$L_{eq} = 10 \lg \left(\frac{1}{T_{int}} \int_0^{T_{int}} \frac{X^2(t)}{X_0^2} dt \right) \tag{3.10}$$

(a) One-third-octave-band filter

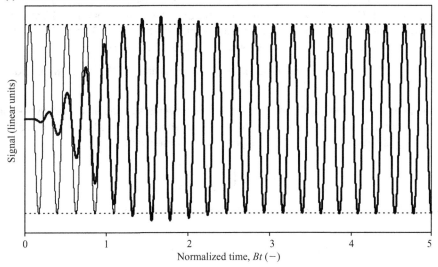

(b) Octave-band filter

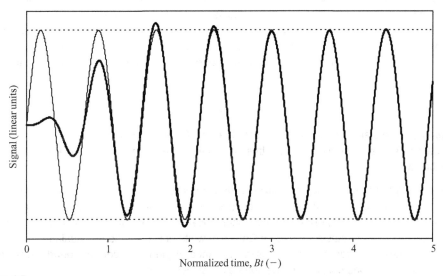

Figure 3.7

Time delay due to filtering of a sinusoidal input signal (thin line) applied at $t = 0$, and the filtered output signal (thick line). Dotted lines indicate the minimum and maximum amplitudes of the input signal.

where X is p, d, v, or a (sound pressure, displacement, velocity, or acceleration) and X_0 is the corresponding reference quantity.

The variation in level with time is assessed using the time-weighted level, $L(t)$; this makes use of an exponential weighting term, and is given by

$$L(t) = 10\lg\left(\frac{1}{\tau}\int_{-\infty}^{t}\frac{X^2(u)}{X_0^2}\exp\left(-\frac{t-u}{\tau}\right)du\right) \tag{3.11}$$

where u is a dummy time variable and τ is the time constant for Fast (F) or Slow (S) time-weighting (IEC 61672 Part 1). The Fast and Slow time constants are 125 ms and 1 s respectively. There is also an Impulse (I) time-weighting where $\tau = 35$ ms, although it is no longer recommended for rating impulsive sounds (IEC 61672 Part 1).

The maximum time-weighted level, L_{Fmax} or L_{Smax}, is the maximum value of $L(t)$ within a defined time interval (IEC 61672 Part 1).

The sound exposure level, L_E, is used for single noise events such as an aircraft flyover or an impulsive sound. It gives the steady sound level that when maintained over 1 s contains the same energy as the actual time-varying signal for the noise event, hence

$$L_E = 10 \lg \left(\frac{1}{t_0} \int_{t_1}^{t_2} \frac{p^2(t)}{p_0^2} dt \right) \tag{3.12}$$

where $t_0 = 1$ s, with the noise event occurring between times t_1 and t_2.

3.3.3.2 *Statistical description of the temporal variation*

For practical calculations relating to the temporal variation of sound pressure levels it is convenient to assume a Gaussian white noise signal. As previously seen in Fig. 3.3, there will be variation between each short sample of noise used in the measurements. This uncertainty can be described using the normalized standard deviation, ε (the ratio of the standard deviation to the mean), which is squared to give the normalized variance, ε^2. For Gaussian white noise passed through a filter, the normalized variance of the signal due to temporal variation is (Bendat and Piersol, 2000)

$$\varepsilon^2 = \frac{1}{B_s T_{int}} \tag{3.13}$$

where B_s is the statistical bandwidth of the filter and T_{int} is the integration time. The statistical bandwidth B_s is not usually known; if this is the case, it is simplest to assume an ideal filter with bandwidth, B.

Equation 3.13 shows that the normalized variance decreases with increasing averaging time and increasing filter bandwidths. However, this equation only applies to the Gaussian white noise itself. It doesn't apply to the sound pressure measured at a stationary microphone position in a reverberant room that is excited by Gaussian white noise. For sound insulation measurements, we need to quantify the normalized variance of the mean-square sound pressure at stationary microphone positions in the source or receiving room.

For a source room excited by Gaussian white noise, the normalized variance of the mean-square sound pressure at a fixed point is twice the normalized variance of the noise itself; this is due to the response of the room (Andres, 1965/1966).

$$\varepsilon^2(p_S^2) = \frac{2}{B_s T_{int}} \tag{3.14}$$

where $B_s T_{int} \gg 1 \gg B_s/f$ (f is the band centre frequency). This assumes that the spatial distribution of the complex Frequency Response Function in the room has a Gaussian distribution, and that the room absorption is independent of frequency over the frequency band.

For a receiving room, Michelsen (1982) has shown experimentally that a reasonable estimate of the normalized variance of the mean-square sound pressure at a fixed point is four times the normalized variance of the Gaussian white noise signal; i.e. the response of two rooms has affected the signal,

$$\varepsilon^2(p_R^2) = \frac{4}{B_s T_{int}} \tag{3.15}$$

The normalized standard deviation from Eq. 3.13, 3.14, or 3.15 can be used to give the standard deviation in decibels using

$$\sigma_{dB} \approx 4.34\sqrt{\varepsilon^2(p^2)} \tag{3.16}$$

For a sound pressure measurement at a stationary microphone position, the minimum integration time is typically $T_{int} = 6\,s$ (ISO 140) for which the standard deviation in decibels is shown in Fig. 3.8. Ideally the minimum integration time would be assessed on a case-by-case basis as it needs to be considered in the context of the uncertainty for the spatial variation of sound pressure in a room (Section 1.2.7.10). Referring back to Fig. 1.42, the standard deviation is shown for a 50 m³ room with reverberation times between 0.5 and 1.5 s. As a general observation from Fig. 1.42 and Fig. 3.8 we see that the standard deviation in decibels for the spatial variation is significantly larger than for the temporal variation. In addition we note that the total variance for the spatial and temporal variation is the sum of the two variances. Therefore the minimum integration time that is needed for measurements can be determined by satisfying the following inequality between the normalized variances for the temporal and spatial variation (Lubman, 1971)

$$\varepsilon^2_{temporal}(p^2) \ll \varepsilon^2_{spatial}(p^2) \tag{3.17}$$

For frequencies between $0.2f_S$ and $0.5f_S$, the normalized variance for the spatial variation can be estimated from the number of modes in the frequency band (Eq. 1.147). A rule-of-thumb for appropriate integration times can now be found for one-third-octave-bands in this frequency range with typical room volumes (20–200 m³). Because the Schroeder frequency

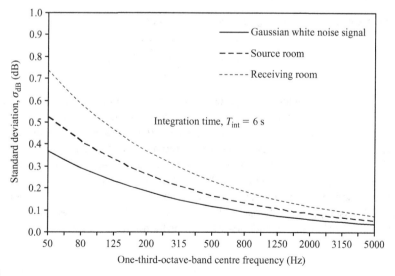

Figure 3.8

Standard deviation for the temporal variation of Gaussian white noise and the temporal variation of the sound pressure level in reverberant source and receiving rooms when the source room is excited with Gaussian white noise.

is calculated from the room volume and the reverberation time, this rule-of-thumb is linked to the reverberation times in the source and receiving rooms, which are denoted as T_S and T_R respectively. If the inequality is interpreted as implying a factor of at least 10, then it will be satisfied when $T_{int} = 6\,s$ in source rooms for $T_S < 2\,s$, and when $T_{int} = 12\,s$ in receiving rooms for $T_R < 2\,s$.

For frequencies at and above $0.5f_S$, the normalized variance for the spatial variation is dependent upon the filter bandwidth and the room reverberation time (Eq. 1.149); this allows rules-of-thumb to be developed by substituting Eq. 1.149 along with Eq. 3.14 or 3.15 into Eq. 3.17. Thus, the integration time for frequency bands above $0.5f_S$ should be chosen such that $T_{int} > 3T_S$ for the source room and $T_{int} > 6T_R$ for the receiving room.

3.4   Spatial averaging

Due to the spatial variation of sound pressure in a space, or vibration over a surface, we almost always need to measure temporal and spatial average levels.

The temporal and spatial average sound pressure level in a room is usually referred to as the average sound pressure level. Temporal and spatial average values for sound pressure and velocity levels don't tend to use different symbols; they simply use L_p and L_v respectively, and rely on accompanying text to make it clear.

For N individual measurement positions, where $L_{eq,n}$ is measured at each position, the temporal and spatial average level is

$$L = 10\lg\left(\frac{1}{N}\sum_{n=1}^{N}10^{L_{eq,n}/10}\right) \tag{3.18}$$

For sound pressure, spatial averaging can also be carried out with a continuously moving microphone. In this case, L_p is equal to the measured L_{eq}.

The sample standard deviation, s_{dB}, can be calculated using

$$s_{dB} = \sqrt{\frac{\sum_{n=1}^{N}L_{eq,n}^2 - \frac{1}{N}\left(\sum_{n=1}^{N}L_{eq,n}\right)^2}{N-1}} \tag{3.19}$$

Calculation of confidence intervals requires knowledge of the probability distribution for the variable of interest; this variable is usually the sound pressure level or the velocity level. Exact probability distributions will vary and are rarely known (Lyon, 1969). For mean-square pressure, a gamma distribution applies to the source room and a log-normal distribution can be assumed in the receiving room (Section 1.2.7.10). For mean-square velocity, it is also reasonable to assume a log-normal distribution (Section 2.7.6). Therefore, it can be assumed that the spatial variation of sound pressure levels and vibration levels in decibels is described by the normal (Gaussian) probability distribution (Craik, 1990; Lyon and DeJong, 1995; Weise, 2003). An estimate of the 95% confidence interval can therefore be calculated from the standard statistical formula,

$$t\frac{s_{dB}}{\sqrt{N}} \tag{3.20}$$

where t is the t-value of the Student's t-distribution for $N-1$ degrees of freedom and a probability of 0.05.

3.4.1 Spatial sampling of sound fields

In a reverberant field, the temporal and spatial average sound pressure level is determined by sampling the sound field at different positions in the central zone of the room; note that these positions must be a certain distance away from the room boundaries and the sound source. Spatial sampling is carried out by using a number of stationary microphone positions or a continuously moving microphone (e.g. a microphone on a rotating boom). The aim is to use sufficient samples to determine the average level with an acceptably low level of uncertainty.

3.4.1.1 Stationary microphone positions

Spatial averaging using a limited number of stationary microphone positions is only effective if a sufficiently large number of positions are used, and these positions provide samples of the mean-square pressure that are uncorrelated in space. Correlation between the sound pressure at two microphone positions separated by a distance, d, in a three-dimensional diffuse sound field can be assessed by using a spatial correlation coefficient; this coefficient will be unity when the positions are the same, i.e. when $d = 0$.

A diffuse sound field can be considered as being comprised of plane waves of equal amplitude and random phase. In this field there is equal probability of a plane wave arriving at the two positions from any angle, θ. By initially considering a single plane propagating wave as shown in Fig. 3.9, the sound pressures at the two positions can be defined as

$$p_1(t) = \hat{p}\exp(i\phi_R)\exp(i\omega t) \tag{3.21}$$

and

$$p_2(t) = \hat{p}\exp(-ikd\cos\theta)\exp(i\phi_R)\exp(i\omega t) \tag{3.22}$$

where $\hat{p}$ is an arbitrary constant for the peak amplitude and ϕ_R is a random phase for the incident plane wave.

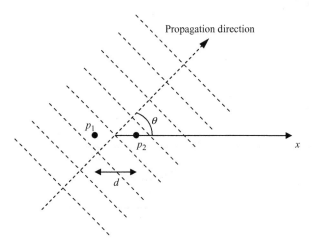

Figure 3.9

Measurement of sound pressure at two points in a propagating plane wave.

The spatial correlation coefficient, $R_{12}(kd)$, is

$$R_{12}(kd) = \left\langle \frac{\langle p_1 p_2 \rangle_t}{\sqrt{\langle p_1^2 \rangle_t \langle p_2^2 \rangle_t}} \right\rangle_{\theta,\phi} \qquad (3.23)$$

where $\langle\ \rangle_{\theta,\phi}$ indicates the average over all possible angles in three-dimensional space.

The mean-square sound pressures are calculated by squaring the real parts of Eqs 3.21 and 3.22, then taking the time-average to give

$$\langle p_1^2 \rangle_t = \langle p_2^2 \rangle_t = 0.5\,\hat{p}^2 \qquad (3.24)$$

The time-average of the product of the real parts of p_1 and p_2 gives

$$\langle p_1 p_2 \rangle_t = 0.5 \hat{p}^2 \cos(kd \cos\theta) \qquad (3.25)$$

Hence the spatial correlation coefficient is (Cook et al., 1955)

$$R_{12}(kd) = \langle \cos(kd \cos\theta) \rangle_{\theta,\phi} = \frac{1}{4\pi} \int_0^{2\pi} \int_0^{\pi} \cos(kd \cos\theta) \sin\theta \, d\theta \, d\phi = \frac{\sin(kd)}{kd} \qquad (3.26)$$

The plot of $R_{12}(kd)$ in Fig. 3.10 shows that there is only weak correlation between two positions when $kd \geq \pi$, i.e. when $d \geq \lambda/2$. The correlation coefficient tends to zero when $kd \gg \pi$. This analysis only applies to single frequencies; however, it can be considered as a reasonable estimate for broad-band signals measured in one-third-octave-bands. Another important caveat is that it only applies to a three-dimensional diffuse field; for typical rooms this will often limit its application to the mid- and high-frequency ranges.

Uncorrelated samples can be taken by choosing microphone positions in the central zone of a reverberant room that are all separated from each other by $d \geq \lambda/2$. In typical rooms the diffuse field assumption is only appropriate in the mid- and high-frequency ranges, i.e. at and

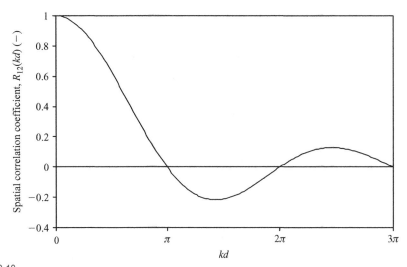

Figure 3.10

Spatial correlation coefficient in a three-dimensional diffuse field.

above 250 Hz. As it is common to use parallel filter measurements over the building acoustics frequency range, the requirement for uncorrelated samples can be based on 250 Hz, which means that $d \geq 0.7$ m. This requirement is feasible in typical rooms, and is the minimum distance quoted in sound insulation measurement Standards (ISO 140).

The advantage in using stationary positions is that the standard deviation (Eq. 3.19) and confidence intervals can be calculated as the measurement progresses. This allows the measurer to make quick decisions as to when the uncertainty in the average sound pressure level is sufficiently low.

Sometimes it is not possible to find sufficient positions that will give uncorrelated samples and it is necessary to use $d < \lambda/2$. In this situation some positions will be correlated and others will be uncorrelated. The equivalent number of uncorrelated samples, $N_{eq,s}$, in the total number of samples, N, can be calculated from (Lubman, 1971, 1974),

$$N_{eq,s} = \frac{N}{1 + \frac{1}{N}\sum_{i=1}^{N}\sum_{j=1}^{N}[R_{ij}(kd_{ij})]^2} \quad \text{for } i \neq j \qquad (3.27)$$

where d_{ij} is the distance between positions i and j.

Inclusion of correlated samples where $d < \lambda/2$, means that $N_{eq,s} < N$. In practice, Eq. 3.27 is of limited use as it is only applicable to a three-dimensional diffuse field. If we use the Schroeder cut-off frequency as an indicator of diffuse fields, we find that in room volumes less than 60 m^3, it will rarely be possible to use Eq. 3.27 to estimate $N_{eq,s}$ in the low-frequency range (refer back to Fig. 1.25).

The aim of spatial averaging is to take sufficient samples to determine the mean value with a low level of uncertainty. In a sound field that can be considered as diffuse, it is better to avoid using correlated samples to increase the size of a data set because this risks introducing a greater level of uncertainty in the mean value (Lubman et al., 1973).

3.4.1.2 Continuously moving microphones

One way of generating large numbers of samples is to use a continuously moving microphone. The most common type of continuously moving microphone uses a rotating boom to trace out a circular path of radius, r, at uniform speed. The boom is tilted at an angle to avoid averaging in a plane that is parallel to a room surface. The effectiveness of this method in a diffuse field depends upon the path radius, the speed, and the total averaging time. An approximation for the equivalent number of uncorrelated samples along a circular path, $N_{eq,c}$, is given by (Lubman et al., 1973)

$$N_{eq,c} = \begin{cases} 1 & \text{for } \frac{2U}{\lambda} < 1 \\ \frac{2U}{\lambda} & \text{for } \frac{2U}{\lambda} \geq 1 \end{cases} \qquad (3.28)$$

where U is the perimeter length of the circular path ($2\pi r$).

Examples for different radii are shown in Fig. 3.11, including the minimum values referred to in the Standards, namely $r = 0.7$ m (ISO 140 Part 4) and $r = 1.0$ m (ISO 140 Part 3). Note that Eq. 3.28 is only applicable to three-dimensional diffuse fields, i.e. above the Schroeder cut-off frequency. We will assume that a rotating boom with a minimum radius of 0.7 m is typically used as a substitute for five stationary microphone positions. From Fig. 3.11 it is clear that the rotating boom becomes a very efficient way of generating large numbers of uncorrelated samples in the mid- and high-frequency ranges. However, in the low-frequency range, the required radius tends to becomes impractically long to achieve a minimum of five uncorrelated positions (2.75 m at 50 Hz).

An alternative path for a continuously moving microphone is along a straight line. The equivalent number of uncorrelated samples, $N_{eq,l}$, along a line of length, L, falls within the range (Lubman et al., 1973)

$$\frac{2L}{\lambda} < N_{eq,l} < 1 + \frac{2L}{\lambda} \tag{3.29}$$

When commissioning a laboratory, there is time available to validate stationary and/or continuously moving microphones. In contrast, time is very limited in the field. The Schroeder frequency can be estimated so the ideal approach might be considered as using stationary positions below the Schroeder frequency to allow calculation of the standard deviation, and a rotating boom above the Schroeder frequency. In practice, using both methods in the field is far too time-consuming. For this reason, one method or the other is chosen; apart from practical considerations the choice may depend upon which part of the frequency range is deemed to be more important or whether calculation of the standard deviation would be useful when interpreting the results.

3.4.2 Measurement uncertainty

The sound insulation descriptors in this chapter are calculated from different temporal and spatial average values. Hence we need to be able to calculate the uncertainty from combinations of measured temporal and spatial average values. Combining the variances is made

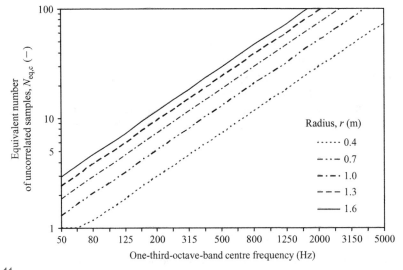

Figure 3.11

Equivalent number of uncorrelated samples for a continuously moving microphone on a circular path of radius, r.

complicated by the different probability distributions (Bodlund, 1976), hence for practical purposes it is necessary to make simplifying assumptions.

The majority of equations take the form,

$$Y(\text{dB}) = X_1(\text{dB}) \pm X_2(\text{dB}) \pm X_3(\text{dB}) \pm \ldots \tag{3.30}$$

For each individual component in Eq. 3.30, the sample standard deviation in decibels, s_{dB}, can be calculated using Eq. 3.19. If the variances for each individual component are uncorrelated, then the combined variance can be estimated from the sum of the individual variances using

$$s_{\text{dB}}^2(Y) \approx s_{\text{dB}}^2(X_1) + s_{\text{dB}}^2(X_2) + s_{\text{dB}}^2(X_3) + \ldots \tag{3.31}$$

To estimate the uncertainty in the single-number quantities that are used to quantify sound insulation (e.g. $D_{nT,w}$, R_w, $L_{n,w}$) it is possible to use Monte-Carlo methods (Goydke et al., 2003).

3.5 Airborne sound insulation

Laboratory airborne sound insulation measurements are primarily used to compare the sound insulation provided by different test elements and to calculate the sound insulation in situ. The role of field measurements is usually to check that a certain level of sound insulation has been achieved.

3.5.1 Laboratory measurements

A transmission suite for airborne sound insulation measurements comprises two rooms; a source room and a receiving room separated by a test element (see Fig. 3.12). At this stage we simply assume that all sound is transmitted via the test element, and that the structure

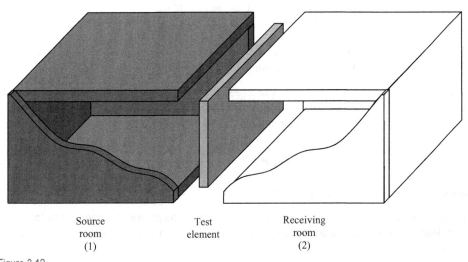

Figure 3.12

Outline sketch of a transmission suite for airborne sound insulation measurements.

of the transmission suite itself plays no role other than defining the space for the source and receiving rooms.

The transmission coefficient, τ, is defined as the ratio of the sound power transmitted by the test element, W_2, to the sound power incident on the test element, W_1,

$$\tau = \frac{W_2}{W_1} \qquad (3.32)$$

The sound reduction index, or transmission loss, R, in decibels is defined as

$$R = 10\lg\left(\frac{1}{\tau}\right) = 10\lg\left(\frac{W_1}{W_2}\right) \qquad (3.33)$$

By assuming that the sound fields in the source and receiving room are diffuse we can calculate the incident and the transmitted sound power from sound pressure level measurements in each room. To find these sound powers, the first step is to calculate the sound intensity incident upon any surface in a diffuse field in terms of the average sound pressure level in the room. This makes use of the mean free path, d_{mfp} (Section 1.2.3.1). For any shape of room with walls and floors that diffusely reflect an incident sound wave, the mean free path is given by Eq. 1.47. Whilst this assumes diffuse reflection from the room surfaces, most walls and floors in a transmission suite are flat and smooth with relatively uniform impedance. For this reason one might assume that specular reflection is more likely to occur than diffuse reflection. In practice, the reflections are neither specular nor diffuse. The procedures used to commission transmission suites aim to provide a sound field in the central zone of the source or receiving room that can be considered as reasonably diffuse. This is often achieved by hanging diffusers in the room, hence, the reflections can be considered to be partially diffuse. On this basis we move forward by assuming diffuse reflections. We can now use the mean free path to infer that sound energy is reflected from the room surfaces c_0/d_{mfp} times every second. The sound energy in the room is equal to the product of the energy density, w, and the room volume, V. Hence the intensity that is incident upon any surface in the room is

$$I = wV\frac{c_0}{d_{mfp}}\frac{1}{S_T} = \frac{wc_0}{4} = \frac{\langle p^2\rangle_{t,s}}{4\rho_0 c_0} \qquad (3.34)$$

where S_T is the total surface area of the room and $\langle p^2\rangle_{t,s}$ is the temporal and spatial average mean-square sound pressure in the diffuse field.

To calculate the diffuse field intensity that is incident upon a surface, it is not necessary to take account of interference patterns at the room boundaries (Vorländer, 1995). Therefore the power that is incident upon the test element in the source room is

$$W_1 = I_1 S = \frac{\langle p_1^2\rangle_{t,s}}{4\rho_0 c_0} S \qquad (3.35)$$

where S is the area of the test element.

The power transmitted into the receiving room must equal the power absorbed in the receiving room. Hence, for an absorption area, A, in the receiving room, Eq. 3.34 gives the transmitted power as:

$$W_2 = \frac{w_2 c_0}{4} A = \frac{\langle p_2^2\rangle_{t,s}}{4\rho_0 c_0} A \qquad (3.36)$$

We now need to consider the interference patterns at the room boundaries that result in non-uniform distribution of energy density in a reverberant room (Section 1.2.7.1). If we use the average sound pressure in the central zone of the receiving room to calculate the transmitted sound power, we will underestimate this sound power and overestimate the sound reduction index. In theory, we could take account of the higher energy density near the boundaries by including the Waterhouse correction in the calculation of the transmitted sound power using

$$W_2 = \frac{\langle p_2^2 \rangle_{t,s} \left(1 + \dfrac{S_T \lambda}{8V}\right)}{4\rho_0 c_0} A \tag{3.37}$$

where S_T is the total surface area of the receiving room and V is the volume of the receiving room.

The Waterhouse correction is more important in the low-frequency range. For room volumes between 50 and 150 m³ it is between 3 and 4 dB at 50 Hz, and less than 1 dB above 200 Hz.

The definition of the sound reduction index does not incorporate the Waterhouse correction. This is reasonable because the correction term is not sufficiently accurate to be used with receiving rooms in transmission suites that have relatively small volumes; the minimum receiving room volume is 50 m³ (ISO 140 Part 1). In addition, errors may arise from using the correction term if the presence of large diffusers significantly changes the distribution of energy density. Hence because no better estimate is available, and the Waterhouse correction is not sufficiently accurate, a correction term is not incorporated that could cause incorrect ranking of test results from different laboratories in the low-frequency range. However we will need to look at this issue again when we want to compare measurements of the sound reduction index made with sound intensity, to those made with sound pressure.

Returning to the derivation, Eqs 3.32, 3.35, and 3.36 give

$$\tau \frac{\langle p_1^2 \rangle_{t,s}}{4\rho_0 c_0} S = \frac{\langle p_2^2 \rangle_{t,s}}{4\rho_0 c_0} A \tag{3.38}$$

hence,

$$\frac{1}{\tau} = \frac{\langle p_1^2 \rangle_{t,s}}{\langle p_2^2 \rangle_{t,s}} \frac{S}{A} \tag{3.39}$$

which is converted to decibels to give the sound reduction index,

$$R = 10 \lg \left(\frac{\langle p_1^2 \rangle_{t,s}}{\langle p_2^2 \rangle_{t,s}} \right) + 10 \lg \left(\frac{S}{A} \right) = L_{p1} - L_{p2} + 10 \lg \left(\frac{S}{A} \right) \tag{3.40}$$

where L_{p1} and L_{p2} are the temporal and spatial average sound pressure levels in the source and receiving rooms respectively.

Ideally, the test element would have the same dimensions as the element being assessed for installation in a building. For practical reasons this is not always possible. Whether it is absolutely necessary can only be established by measuring different size elements and/or with a prediction model for the sound insulation. The relevant Standard (ISO 140 Part 3) requires the following test element areas: ≈10 m² for walls, and between 10 and 20 m² for floors where the smaller dimension is at least 2.3 m for both walls and floors. Hence these wall and floor areas are representative of those in real buildings. The sound reduction index is an appropriate descriptor where the test element area is well-defined and the test result is representative of identical elements with different dimensions. The latter is a reasonable assumption for many walls and floors, and, to a lesser degree, for windows and doors. However, it is not appropriate for many elements that are smaller than $1\,m^2$ such as ventilation devices or cable ducts. For these small building elements or devices the sound reduction index will rarely be representative of one with different dimensions. It is therefore misleading to use this descriptor because it gives the impression that we could use the test result for different sizes of the same element or device. An alternative descriptor to the sound reduction index is therefore required which does not involve the test element area, and which is attributable to a single element or device. It is not appropriate to derive this descriptor using the sound power that is incident upon the element or device because this uses the test element area; so the alternative descriptor is based on the sound pressure level difference, D, between the source and receiving rooms where

$$D = L_{p1} - L_{p2} \tag{3.41}$$

For a given sound power input into the receiving room, the mean-square sound pressure in the room is inversely proportional to the absorption area, A (see Eq. 3.36). Hence the receiving room sound pressure level needs to be 'normalized' to a reference absorption area using the measured absorption area, A, in that room. For small building elements or devices, this gives the element-normalized level difference, $D_{n,e}$,

$$D_{n,e} = L_{p1} - \left(L_{p2} + 10\lg\left(\frac{A}{A_0}\right) \right) = D - 10\lg\left(\frac{A}{A_0}\right) \tag{3.42}$$

where the reference absorption area, A_0, is $10\,m^2$ (ISO 140 Part 10).

3.5.1.1   Sound intensity

The sound reduction index of a test element can also be measured using sound intensity. The sound power incident upon the test element is calculated from sound pressure level measurements in the source room using Eq. 3.35. The transmitted power is calculated from the average normal intensity, I_n, over the measurement surface, S_M, that has been measured with an intensity probe,

$$W_2 = I_n S_M \tag{3.43}$$

An intensity probe measures net intensity. The importance of this becomes apparent if we consider measuring a test element with an absorbent surface because the intensity measurement gives an estimate of the net power, i.e. the transmitted power minus the absorbed power. To avoid underestimating the transmitted power it is necessary to satisfy requirements on F_{pI} and $\delta_{pI0} - F_{pI}$. These requirements are discussed in Section 3.10 and can be satisfied by introducing sufficient absorption into the room where the intensity measurements are carried out.

For intensity measurements the sound reduction index in decibels is calculated by making use of the relationship between $\rho_0 c_0$ and the reference quantities for sound pressure (20×10^{-6} Pa) and sound intensity (10^{-12} W/m^2), where $10\lg(\rho_0 c_0) = 10\lg((20 \times 10^{-6})^2/10^{-12})$. Hence, the intensity sound reduction index, R_I, is given by

$$R_I = L_{p1} - L_{In} - 10\lg\left(\frac{S_M}{S}\right) - 6\,\text{dB} \tag{3.44}$$

where L_{p1} is the average sound pressure level in the source room, L_{In} is the temporal and spatial average normal sound intensity level over the measurement surface, and S is the area of the test element.

For small building elements, the intensity element-normalized level difference, $D_{I,n,e}$, is calculated using

$$D_{I,n,e} = L_{p1} - L_{In} - 10\lg\left(\frac{S_M}{A_0}\right) - 6\,dB \qquad (3.45)$$

where the reference absorption area, A_0, is $10\,m^2$ (ISO 15186 Part 1).

In comparison with the sound pressure level method (ISO 140 Part 3), the sound intensity method has advantages when there is flanking transmission from the laboratory structure, or from the filler wall used with small test elements. This is because the intensity probe can be used to measure only the sound intensity that is radiated by the test element. Therefore in the presence of flanking transmission, R_I tends to be higher than R. However, the definition of R does not take account of the higher energy density near the receiving room boundaries when calculating the transmitted sound power. This means that R_I tends to be lower than R in the low-frequency range when there is no significant flanking transmission.

3.5.1.1.1 Low-frequency range

In Sections 1.2.7.9 and 1.2.7.10 we saw that there can be significant spatial variation of the sound pressure levels in the low-frequency range. In addition we know that when the sound reduction index is determined using sound pressure level measurements in the receiving room, no account is taken of the higher energy density near the room boundaries when calculating the transmitted sound power. Hence, in the low-frequency range the transmitted sound power that is determined from sound intensity measurements in the receiving room should be more accurate than measuring sound pressure levels in the central zone of the receiving room. To measure an even better estimate of the sound reduction index in the low-frequency range it is necessary to make changes to the measurement method in the source room as well as to the sound field in the receiving room. This is the measurement method proposed by Pedersen et al. (2000) that is implemented in ISO 15186 Part 3. Inter-laboratory comparisons (Olesen, 2002; Pedersen et al., 2000) indicate that this approach has better reproducibility than the traditional method (ISO 140 Part 3) in one-third-octave-bands between 50 and 160 Hz. However, it is primarily intended for one-third-octave-bands between 50 and 80 Hz (ISO 15186 Part 3).

The measurement accuracy and reproducibility in the low-frequency range are significantly improved by effectively converting the receiving room into an anechoically terminated duct (Pedersen et al., 2000; Roland, 1995). This is achieved by placing highly absorbent material on the back wall of the receiving room to try and simulate an anechoic termination (see Fig. 3.13). In practice an anechoic termination is not possible, but high levels of absorption can be achieved by using a very thick layer of mineral wool, often up to 1 m in thickness (ISO 15186 Part 3). The test element therefore radiates sound into a duct-like receiving room. We can compare common ducts used for HVAC purposes with this duct-like receiving room. For HVAC ducts, the wavelengths in the low-frequency range are usually large compared to the dimensions of the duct cross-section, and at these frequencies only plane waves will propagate along the duct. At higher frequencies, the duct cross-section can support modes that propagate along the duct, these are referred to as propagating cross-modes. A receiving room in a transmission suite has much larger cross-sectional dimensions than a typical HVAC duct. Hence, for typical receiving rooms there will be propagating cross-modes in each one-third-octave-band from 50 Hz upwards. The measurement environment in this duct-like room is well-suited to determining the transmitted sound power using sound intensity

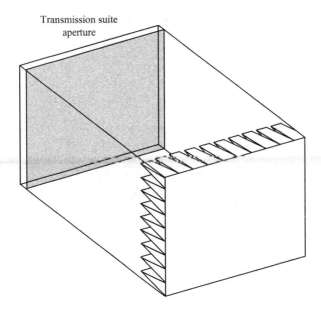

Transmission suite
aperture

Figure 3.13

Sketch of a receiving room with highly absorbent back wall to simulate an anechoically terminated duct. In practice, thick layers of mineral wool are used rather than long anechoic wedges. The shaded area indicates the measurement surface for intensity measurements with a test element that fills the aperture.

measurements; this is in contrast to normal rooms where it can be difficult to achieve acceptable F_{pI} values, particularly in one-third-octave-bands below 100 Hz.

In the source room a different measurement method is used to avoid the errors incurred by calculating the incident sound power from sound pressure levels in the central zone of the source room. This method uses sound pressure measurements near the test element surface. For sound fields near room boundaries, we saw in Section 1.2.7.1 that very close to a perfectly reflecting surface in a reverberant room the sound pressure level is 3 dB higher than the diffuse field level. Hence it can be assumed that the diffuse field sound pressure level will be 3 dB less than the spatial average sound pressure level over the surface of the test element, $L_{p1,s}$, when measured at a distance less than 50 mm from the surface (ISO 15186 Part 3). To account for the different measurement method in the source room, the intensity sound reduction index, $R_{I(LF)}$, is calculated using

$$R_{I(LF)} = L_{p1,s} - L_{In} - 10 \lg \left(\frac{S_M}{S} \right) - 9\,\text{dB} \tag{3.46}$$

and the intensity element-normalized level difference, $D_{I(LF),n,e}$, is calculated using

$$D_{I(LF),n,e} = L_{p1,s} - L_{In} - 10 \lg \left(\frac{S_M}{A_0} \right) - 9\,\text{dB} \tag{3.47}$$

Values of $R_{I(LF)}$ (or $D_{I(LF),n,e}$) in the low-frequency range can be combined with values of R and/or R_I (or $D_{n,e}$ and/or $D_{I,n,e}$) in the mid- and high-frequency ranges to produce a continuous spectrum between 50 and 5000 Hz.

3.5.1.2 *Improvement of airborne sound insulation due to wall linings, floor coverings, and ceilings*

3.5.1.2.1 *Airborne excitation*

For linings used on separating walls or floors, the improvement due to the lining for direct sound transmission is determined by measuring the sound reduction index of a base wall or floor with and without the lining. The sound reduction improvement index, ΔR, is then given by

$$\Delta R = R_{\text{with lining}} - R_{\text{without lining}} \tag{3.48}$$

Wall linings often improve the sound insulation in the high-frequency range at the expense of reducing it in the low-frequency range. Hence whilst it is referred to as an improvement index, it can take negative or positive values.

ΔR depends upon the type of base wall or floor to which the lining is fixed; hence, the relevant Standard (ISO 140 Part 16) describes different base walls or floors to allow a fairer comparison between products.

3.5.1.2.2 *Mechanical excitation*

Linings used on flanking walls or floors are not always directly excited by a sound field but also by structure-borne sound transmitted across one or more junctions. A wall or floor may also be subject to impacts, or machinery/equipment that only acts as a structure-borne sound source (i.e. without airborne excitation) may be fixed to one side. In these situations there is no non-resonant (mass law) transmission. Below the critical frequency, ΔR includes both resonant and non-resonant transmission; therefore, it is sometimes useful to measure the improvement due to only resonant transmission. This is the resonant sound reduction improvement index, $\Delta R_{\text{resonant}}$ and can be measured with mechanical rather than airborne excitation. (Note that at the time of writing there was no published measurement Standard.)

In the laboratory it is convenient to apply mechanical excitation via a shaker to the side of the wall without the lining. The test procedure requires measurement of the vibration level of the base wall or floor (on the side of the wall without the lining) along with the resulting sound pressure level in the receiving room. These measurements are carried out with and without the lining. To avoid problems in measuring low-sound pressure levels it is convenient to take dual-channel measurements using an MLS signal (Section 3.9) rather than broad-band noise. Vibration and sound pressure can then be measured simultaneously; one in each channel.

M excitation positions are used with N different accelerometer positions for each excitation position. The sound pressure is measured in the receiving room at R microphone positions for each accelerometer position; stationary positions are required due to the use of MLS. In a transmission suite, $M = 3$, $N = 4$, and $R = 2$ is usually sufficient. See Section 3.12.3.2.3 for excitation and accelerometer positions.

For walls, the shaker is usually pushed up against the wall without any transducers to measure the power input. This means that the same power input may not be applied at all excitation positions. Therefore the vibration and sound pressure measurements need to be arithmetically averaged to calculate a standardized pressure–vibration level, $L_{\text{pv},T}$, using

$$L_{\text{pv},T} = \left\{ \frac{1}{M} \sum_{m=1}^{M} \left[\left(10 \lg \left(\frac{1}{NR} \sum_{r=1}^{NR} 10^{L_{\text{p},r}/10} \right) \right)_m - \left(10 \lg \left(\frac{1}{N} \sum_{n=1}^{N} 10^{L_{\text{v},n}/10} \right) \right)_m \right] \right\} - 10 \lg \left(\frac{T}{T_0} \right)$$

$$\tag{3.49}$$

where T is the reverberation time in the room where the sound pressure level is measured, and $T_0 = 0.5\,\mathrm{s}$. $L_{pv,T}$ usually takes negative values. Note that for ceilings on floors it is possible to use the ISO tapping machine to provide a constant power input.

The resonant sound reduction improvement index is then given by,

$$\Delta R_{resonant} = L_{pv,T\text{(without lining)}} - L_{pv,T\text{(with lining)}} \qquad (3.50)$$

In practice this measurement is more relevant to point excitation of a wall or floor (e.g. impacts, attached machinery/equipment) than excitation due to structure-borne sound that occurs with flanking transmission. The reason for this is that mechanical excitation at a point tends to overemphasize the adverse effect of any mass–spring–mass resonance (see Section 4.3.8.2). In addition, measurements above the thin plate limit are of limited applicability due to thick plate effects and because they are specific to the contact area used for point excitation.

3.5.1.3 Transmission suites

The ideal test result from a transmission suite is one that is repeatable within the laboratory, reproducible in a different laboratory, allows a fair comparison with different test elements, and is in a form that can easily be used to estimate the sound insulation in situ. It is also important to be able to validate transmission suite measurements against a theoretical model for relatively simple test elements (e.g. non-porous, solid, homogeneous, isotropic plates).

There are many different sizes and types of test elements, so certain aspects of a transmission suite design are often tailored to suit a particular element. For example, test elements such as masonry walls need to dry out before they are tested, so it is convenient if they can be built in a frame outside the transmission suite where they can be left to dry. When they are ready to be tested they can then be 'inserted' between the source and receiving rooms. As sound transmission across this type of element is affected by its total loss factor, the connections that are made between the frame and the laboratory will affect the measured sound reduction index. The relevance of total loss factor measurements will shortly be discussed in more detail.

Laboratory tests on separating walls or floors are often used to calculate the sound insulation in situ. A separating wall or floor in a building usually forms one complete surface in a room, so the connected flanking walls and/or floors effectively form baffles that are orientated perpendicular to its surface. Below the critical frequency of any plate that faces into a room, the radiation efficiency is affected by the baffle orientation (Section 2.9.4). For this reason there is logic in having laboratory walls and floors around the perimeter of the test element that are perpendicular to the test element surface as well as having a test aperture that has dimensions similar to walls or floors in situ (Kihlman and Nilsson, 1972; ISO 140 Part 1). However there are competing objectives in the low-frequency range; for measurement accuracy it can be beneficial to create relatively large source and receiving rooms so that each room has a high mode count in each one-third-octave-band. For a test aperture area that is $\approx 10\,\mathrm{m}^2$, the use of laboratory walls and floors that form perpendicular baffles means that room volumes are limited to $\approx 50\,\mathrm{m}^3$; this is because an elongated room with the longest dimension perpendicular to the test element tends not to provide an incident sound field that can be considered as diffuse. Another factor that requires consideration is that some test elements (such as curtain walling used for facades) are much larger than $\approx 10\,\mathrm{m}^2$ and a laboratory structure that forms perpendicular baffles may be less representative of in situ than baffles that lie in the same plane as the test element.

3.5.1.3.1 Suppressed flanking transmission

Transmission suites are referred to as having "suppressed flanking transmission" (ISO 140 Part 1). This acknowledges the fact that flanking is omnipresent but that it can be suppressed to such a level that its effect can be considered as negligible; thus allowing useful measurements. It is difficult to make general statements about the suppression of flanking because there are many different designs of transmission suite. In addition, flanking transmission is not a fixed property of the transmission suite itself; it depends on the type of test element, and how and where the test element is connected to the laboratory structure that defines the source and receiving rooms.

This is a useful point to introduce the commonly used classifications for transmission paths; these are Dd, Ff, Fd, and Df, as shown in Fig. 3.14. Direct transmission via the test element is denoted as Dd and the flanking paths are denoted as Ff, Fd, and Df. These classifications can be used to consider the different permutations for the installation of test elements in transmission suites. For example, the test element could be structurally isolated from both rooms, connected to certain walls or floors in one room, or connected around all its boundaries to the structure of both rooms. These classifications simplify analysis and discussion of flanking paths, although in Chapter 4 we will introduce SEA transmission paths to allow a more flexible and detailed analysis of different transmission mechanisms.

The different flanking transmission paths can be described as follows:

(1) Transmission paths involving direct sound transmission between the source and receiving rooms other than via the test element.
This transmission can occur via a filler wall that is used to temporarily change the size of the test aperture. In some laboratories the test aperture does not cover one complete room surface; hence, it can also occur via laboratory walls/floors that are adjacent to, and in the same plane as the test aperture.
(2) Transmission paths via the laboratory structure (including any filler wall) that do not enter or involve the test element (path type: Ff).
(3) Transmission paths via the laboratory structure (including any filler wall) that involve energy flow into the test element (path type: Fd).
(4) Transmission paths via the laboratory structure (including any filler wall) that involve energy flow out of the test element (path type: Df).

Consideration of these flanking paths should not lead to the conclusion that a test element needs to be completely isolated from all parts of the laboratory by using resilient connections all around its border. In Section 3.5.1.3.2 we will discuss the effect of changing the connection around the perimeter of heavyweight test elements because this can significantly change the measured sound insulation.

Transmission suites may be heavyweight or lightweight structures, or a combination of the two. For masonry/concrete test elements that are rigidly connected to a heavyweight laboratory structure, the coupling between the laboratory structure and the test element depends on both their thicknesses and material properties. Laboratories with heavyweight structures are usually built from masonry or concrete, this leads to a rule-of-thumb that the mass per unit area of a solid masonry/concrete test element should be less than half the mass per unit area of each connected laboratory wall or floor (Gerretsen, 1990). SEA predictions in a variety of different laboratory designs also indicate that this is reasonable (Craik, 1992). However, energy flow

Source room   Test element   Receiving room

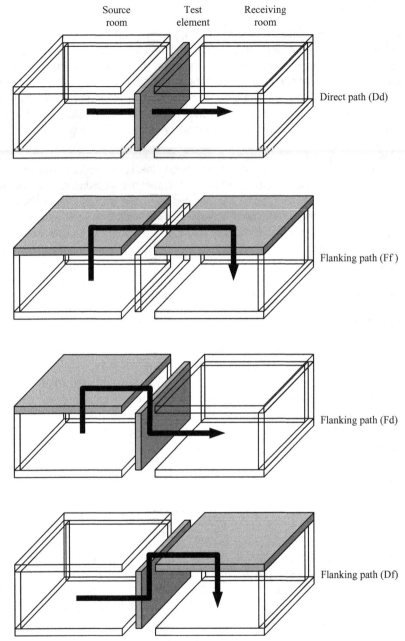

Direct path (Dd)

Flanking path (Ff)

Flanking path (Fd)

Flanking path (Df)

Figure 3.14

Transmission paths between two adjacent rooms. Direct transmission via the test element surface is indicated by D in the source room, and d in the receiving room. Flanking surfaces are indicated by F in the source room and f in the receiving room. In this illustration the ceiling has been chosen as the flanking element; the principle also applies to the side walls or the floor in each room. Each flanking path Ff, Fd, or Df is only defined for vibration transmission across one junction (i.e. a physical connection between the two grey plates); in practice there will also be flanking paths involving more than one junction.

between a test element and a laboratory is sufficiently complex that it is not possible to define a flanking limit for a laboratory in terms of one maximum achievable sound reduction index; this varies depending on the test element (Craik, 1992). The relevant Standard (ISO 140 Part 1) adopts the pragmatic solution of measuring the maximum achievable sound reduction index for a limited number of test elements that are representative of those that are most commonly measured.

3.5.1.3.2 *Total loss factor*

To assess the role of the total loss factor (Section 2.6.5) in the measurement of the sound reduction index we now consider a non-porous, solid, homogeneous plate as the test element. We start by assuming that this plate is only excited by the sound field in the source room. Vibrational energy then leaves the test element via: (a) structural coupling into the laboratory to which it is connected, (b) radiation losses into the rooms, and (c) internal losses into heat. As the laboratory structure will be excited to some degree, we need to assume that there is no significant flow of vibrational energy from the laboratory structure into the test element, and that the test element is the dominant radiating element in the receiving room.

Sound transmission across the plate is determined by both non-resonant and resonant transmission; these terms are discussed in Section 4.3.1. At frequencies where the sound reduction index of the plate is partly or wholly determined by resonant transmission, it is very useful to measure the structural reverberation time of the plate in order to calculate the total loss factor (Section 3.11.3.3). The reason for this is that resonant transmission is directly affected by the plate damping, so it is possible to change the measured sound reduction index of a plate by changing its total loss factor. This is particularly important for masonry/concrete plates where resonant transmission is the dominant transmission mechanism over the majority of the building acoustics frequency range (see examples in Section 4.3.1.3). The simplest way to achieve a change in the total loss factor is to change the connections at the plate boundaries from rigid to resilient connections; for example from rigid mortar to a resilient material such as mineral wool (Gösele, 1961; Kihlman and Nilsson, 1972). The total loss factor for a rigidly connected plate will generally be higher than for a resiliently connected plate. If there is no significant flow of vibrational energy from the laboratory structure into the test element and the resonant transmission dominates, the sound reduction index will be higher for the rigidly connected plate. In practice there is an added complication because significant changes to the plate boundaries (which usually fall into an idealized category of being simply supported, clamped, or free) also change the plate modes and the radiation efficiency below the critical frequency. For a wall, the lower plate boundary usually needs a rigid connection in order to provide structural stability. This introduces an additional complication with orthotropic walls if there is a mixture of rigid and resilient boundary conditions (see Section 4.3.2.2).

The structure of the test aperture varies from one transmission suite to another; usually it is formed from dense concrete, blocks or bricks of various thicknesses. Therefore when one masonry/concrete wall or floor is tested in different transmission suites, the coupling losses to the laboratory structure can vary significantly (Craik, 1992; Kihlman, 1970). The resulting variation in the total loss factor means that it is possible to measure different sound reduction indices for a single test element in different laboratories (Gerretsen, 1990; Kihlman and Nilsson, 1972; Meier *et al.*, 1999). One possibility would be to build identical transmission suites all over the world; opinions on the optimum design would no doubt vary, and such uniformity would not always be helpful for testing non-standard elements.

Above the critical frequency of a non-porous, solid, homogeneous plate, there is a direct relationship between the sound reduction index and the total loss factor of the plate; note that if this plate is orthotropic, then it is the higher of its two critical frequencies. This fundamental relationship can be seen from the equation for an infinite plate with mass, stiffness, and damping (Eq. 4.61), or a three-subsystem SEA model for resonant transmission across a finite plate that separates two rooms (Eq. 4.23). We consider the situation where this plate has been measured in laboratory A and has a sound reduction index, R_A, with a total loss factor, η_A. The same plate is then measured in laboratory B where the total loss factor is η_B. Assuming that all measurement errors are negligible, we can convert R_A to the sound reduction index, R_B, that would be measured in laboratory B using

$$R_B = R_A + 10 \lg \left(\frac{\eta_B}{\eta_A} \right) \tag{3.51}$$

An example is shown in Fig. 3.15 for a masonry wall with a high mass per unit area and a low critical frequency. In the mid- and high-frequency ranges, the conversion adequately accounts for the difference between the measured sound reduction indices. It is not appropriate in the

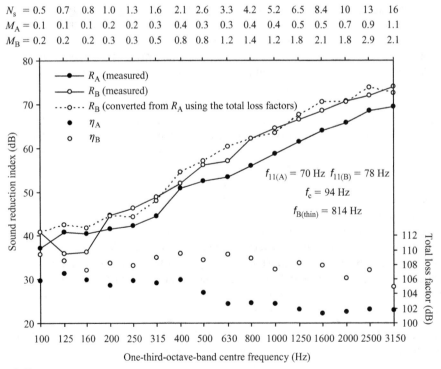

N_s	= 0.5	0.7	0.8	1.0	1.3	1.6	2.1	2.6	3.3	4.2	5.2	6.5	8.4	10	13	16
M_A	= 0.1	0.1	0.1	0.2	0.2	0.3	0.4	0.3	0.3	0.4	0.4	0.5	0.5	0.7	0.9	1.1
M_B	= 0.2	0.2	0.2	0.3	0.3	0.5	0.8	0.8	1.2	1.4	1.2	1.8	2.1	1.8	2.9	2.1

$f_{11(A)} = 70 \text{ Hz}$ $f_{11(B)} = 78 \text{ Hz}$

$f_c = 94 \text{ Hz}$

$f_{B(thin)} = 814 \text{ Hz}$

Figure 3.15

Converting the sound reduction index measured in laboratory A to laboratory B, where the measured total loss factor of the test element is different in each laboratory. Test element: 215 mm masonry wall (solid blocks) with a 13 mm lightweight plaster finish (each side). Wall areas are 9.3 and 8.6 m² in laboratories A and B respectively. Upper x-axis labels show the predicted statistical mode count (average value of both walls is shown) and the modal overlap factor for each wall A and B in each frequency band. Sound insulation measurements according to ISO 15186 Part 1. Measured data from Hopkins are reproduced with permission from ODPM and BRE.

low-frequency range where the plate has very low modal overlap and only two or three bending modes in the entire low-frequency range. These bending modes vary between the two different sized walls; hence, the conversion is not able to account for the measured difference in the sound reduction index.

Measurements on solid concrete block walls with a much lower mass per unit area and a relatively high critical frequency ($\rho_s = 41\,\text{kg/m}^2$, $f_c = 570\,\text{Hz}$) have been tested in five different laboratories (Kihlman and Nilsson, 1972). The results indicated that above the critical frequency the conversion could adequately account for the variation between these laboratories (a range of $\approx 3\,\text{dB}$). An intercomparison study between 12 laboratories on an orthotropic solid block wall ($\rho_s = 440\,\text{kg/m}^2$, $f_{c,x} = 180\,\text{Hz}$, $f_{c,y} = 108\,\text{Hz}$) showed that the conversion was satisfactory above the highest critical frequency (Meier et al., 1999). In these two studies there were a few laboratories that were outliers in the converted data set. It is reasonable to assume that the assumptions in the opening paragraph of this section do not apply to some laboratory designs. Ignoring the few laboratories that are outliers, these studies indicate that the possible range for the sound reduction index in one-third-octave-bands above the critical frequency is $\approx 5\,\text{dB}$ for rigidly connected masonry/concrete plates in different laboratories. These studies indicate that this variation can be reduced to less than $2\,\text{dB}$ in individual frequency bands if R_A from each laboratory is converted to R_B using a reference loss factor, η_B. This reference loss factor may be arbitrarily chosen, or related to a typical value for a certain type of wall or floor. Note that Eq. 3.51 is a conversion, not a correction. We can measure a range of values for the same wall or floor in different laboratories but all of these measured values are 'correct' under the assumptions described in the opening paragraph of this section.

The conversion is valid above the critical frequency for solid homogeneous plates that are non-porous. A bonded surface finish (e.g. plaster) usually removes the non-resonant transmission path across porous blocks and/or mortar joints thus allowing the wall or floor to be treated as a non-porous plate. For fair-faced solid masonry walls the porosity is highly variable and the conversion will not be appropriate in all cases. Another exception occurs at frequencies where sound transmission is primarily determined by thickness resonances; this occurs with some thick hollow brick/block walls. For such plates it will be the internal loss factor relating to the plate thickness that primarily determines the sound reduction index (see Section 4.3.1.4). Note that structural reverberation time measurements will quantify the total loss factor and this may be dominated by coupling losses that have negligible effect on sound transmission for this type of plate.

Below the critical frequency, the theory for airborne sound insulation of an infinite plate with mass, stiffness, and damping implies that there will only be non-resonant (mass law) transmission, and that the total loss factor plays no role in airborne sound transmission. For many solid masonry/concrete plates this is not the case; resonant transmission may dominate below, at, and above the critical frequency (see the examples in Section 4.3.1.3). In order to carry out the conversion below the critical frequency using Eq. 3.51, it is necessary to estimate the non-resonant transmission (Section 4.3.1.2). If this is negligible compared to the resonant transmission the conversion can be made in the same way. If not, the non-resonant component of the sound reduction index needs to be removed so that the conversion can be carried out on the resonant component. The non-resonant component can then be re-introduced after the conversion.

There are other factors that need to be considered in applying the conversion below the critical frequency. This concerns the low-frequency range where repeatability and reproducibility values are highest, i.e. least favourable (see ISO 140 Part 2). With some laboratories it may only be possible to make reliable conversions when intensity measurements are used to determine the sound reduction index (ISO 15186 Parts 1 & 3). In addition, laboratories often have different arrangements of side walls/floors. These walls/floors act as baffles to the test element; hence, they can change the radiation efficiency of the plate below the critical frequency. Another important factor in the low-frequency range is that the bending mode counts are usually very low in one-third-octave or octave-bands with $\approx 10 \, m^2$ masonry/concrete plates. For this reason the uncertainty in the measured structural reverberation time needs to be considered in any conversion.

For solid walls and floors it is clear that the total loss factor is useful in interpreting the measured sound reduction index, and applying the result to the field situation. However, there are a number of issues that limit its widespread application. Originally the problem concerned the difficulty in measuring very short reverberation times (Utley and Pope, 1973). These difficulties have been overcome through signal processing techniques (Section 3.8.3.2.2), although there is still the issue of increased measurement uncertainty with non-diffuse vibration fields. We now need to consider the different types of masonry/concrete plates that are measured in practice. To simplify design decisions it would be ideal if all masonry/concrete plates were non-porous, solid, homogenous, isotropic, and reverberant with high modal densities. All measured values for the sound reduction index could then be supplied with the measured total loss factor; ideally along with the measured internal loss factor. In fact, if all plates were like this we would probably decide not to measure at all; prediction would be quite adequate. In practice, the walls and floors that we need to measure do not satisfy these ideal criteria. Take an example where we need to know whether a fair-faced, solid blockwork wall (slightly porous) is likely to achieve the same sound insulation as a hollow block wall with a plaster finish. A robust assessment can only be made by measuring the sound reduction index, total loss factor, internal loss factor, longitudinal wavespeed or bending stiffness, and airflow resistivity. Even if we do not have prediction models for these specific walls, we can still use models of non-porous, solid, homogeneous plates (Section 4.3.1) to gain some insight into the transmission mechanisms. This will allow us to make a decision as to whether these two walls are likely to provide similar sound insulation *in situ*. On this basis the total loss factor often proves useful in making design decisions on many different types of solid masonry/concrete plate.

Measurement of the sound reduction improvement index is also affected by the total loss factor of the base wall or floor. However with linings there is the added complexity of another non-resonant mechanism which gives mass–spring–mass resonances that sometimes affect the transmission over several one-third-octave-bands. In addition, any change in the total loss factor of the base wall due to addition of the lining is often within the limits of measurement uncertainty. Hence there is no simple conversion that is valid across the entire building acoustics frequency range for all types of base walls and floors and all types of lining.

For cavity walls with masonry/concrete leaves, the variation in the total loss factor of each leaf also accounts for some of the variation in test results between different laboratories. However, there is no simple conversion. The reasons for this will become apparent in Section 4.3.5.2.2 which looks at different sound transmission paths across cavity walls.

From the above discussion it is clear that the main issue concerning total loss factors is that it is not possible to use the conversion with all types of walls and floors. For those types where it can theoretically be used at all frequencies, it may not be possible to accurately measure the structural reverberation time at all frequencies. For these reasons it is not particularly useful to convert a limited number of measurements to a reference total loss factor. However, any assessment of a laboratory measurement for use *in situ* always requires consideration of the *in situ* total loss factor and flanking transmission. Therefore the most helpful approach is to simply provide the measured total loss factor alongside the measured sound reduction index in the laboratory test report. At the very least this allows engineering judgement to be used in converting laboratory measurements to the *in situ* situation; even if it is only possible over part of the building acoustics frequency range.

For the impact sound insulation of solid homogeneous plates, there is only resonant transmission and the normalized impact sound pressure level, $L_{n(A)}$ can be converted to $L_{n(B)}$, using

$$L_{n(B)} = L_{n(A)} - 10 \lg \left(\frac{\eta_B}{\eta_A} \right) \tag{3.52}$$

3.5.1.3.3 Niche effect

The transmission suite aperture, or the aperture of a filler wall which contains the test element, is usually much thicker than the test element itself. This is usually necessary to suppress flanking transmission via the laboratory structure or direct transmission across the filler wall. The aperture therefore forms a niche on one or both sides of the test element, depending on where the element is positioned within the aperture. Unless a specific niche arrangement is to be tested, the Standards (ISO 140 Parts 1 & 3) describe an arrangement based on different niche depths on each side of the element with a ratio of 2:1, and niche boundaries with an absorption coefficient <0.1 over the building acoustics frequency range.

The reason for carefully specifying and reporting the niche arrangement used in the laboratory is that the position of the element within the aperture can significantly change the sound reduction index. This is often referred to as the 'niche effect'. Although this name implies a single effect that is linked purely to the niche geometry, it depends upon both the test element and the niche on both sides; this results in a frequency-dependent effect that varies between test elements. The following discussion is based around the niche effect in laboratory measurements, but the effect is equally important *in situ*. We will assume that the niche has rigid boundaries and that the test element is installed such that the plates or beams forming the niche do not introduce a flanking path or radiate sound themselves.

As there are a number of variables that affect the measured sound reduction index of a test element within a niche, it is quite possible to come across contradictory measurements concerning the niche effect. For this reason we will consider individual factors that contribute to the niche effect whilst acknowledging that all factors need to be considered together. It is difficult to gain a practical insight into the niche effect by considering only one theoretical model. This is to our benefit as the following discussion will provide a useful backdrop to the application of finite and infinite plate models that will be discussed in Section 4.3.1. To assess the niche effect it is simplest to assume that the test element is a solid plate; we can therefore consider sound transmission in terms of non-resonant and resonant transmission paths. The examples in Section 4.3.1.3 can be used to gain an impression of the importance of non-resonant transmission relative to resonant transmission below the critical frequency of a plate.

The factors contributing to the niche effect can be considered as follows:

(1) Shielding of the test element surface from sound waves that impinge upon the element at near-grazing angles of incidence.

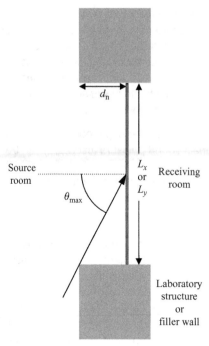

Figure 3.16

Test element mounted in a niche (cross-section).

Tests in different laboratories generally show that the effect of shielding is most significant at and below the critical frequency of the plate facing into the niche (Guy and Sauer, 1984; Kihlman and Nilsson, 1972).

The angle of incidence for an incident plane wave is measured from a line perpendicular to the surface of the test element. When the test element is mounted in a niche as shown in Fig. 3.16, the maximum angle of incidence, θ_{max}, that can impinge upon the centre point of the element is given by $\theta_{max} = \arctan(L/2d_n)$, where d_n is the niche depth and L is either L_x or L_y. If the test element is a full-size wall with dimensions, 4×2.5 m and a niche depth of 25 mm facing into the source room, the average θ_{max} is 89° and the shielding effect is negligible. However, for a sheet of glass with dimensions, 1.5×1.25 m and a niche depth of 150 mm, the average θ_{max} is 78°. At this point we can make a link to the prediction of non-resonant transmission based on an infinite plate acting as a limp mass. This is often calculated by assuming that all angles of incidence are equally probable between 0° and 78°, and is referred to as the field incidence mass law. For an infinite plate, non-resonant transmission is highly dependent upon the angle of incidence with near-grazing angles being very efficient at transmitting sound (refer forward to Fig. 4.6). The infinite plate model indicates that shielding of near-grazing angles by the niche will significantly change the sound reduction index due to non-resonant transmission below the critical frequency.

This approach is too simplistic to fully explain the shielding effect. We have ignored the two-dimensional reverberant sound field within each niche, as well as the effects of diffraction and scattering at the niche edges. Despite this, it is still useful

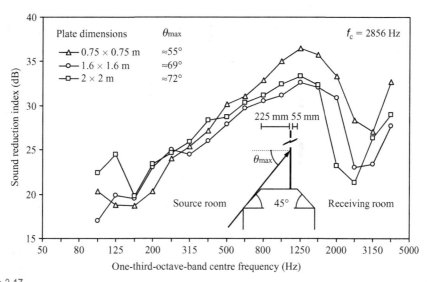

Figure 3.17

Measured airborne sound insulation for different size sheets of 6.25 mm glass within a niche that gives different degrees of shielding. Measurements according to ASTM E90-75. Measured data are reproduced with permission from Guy and Sauer (1984).

because it makes a link to infinite plate models that are often used to predict non-resonant (mass law) transmission; it also shows that there is little basis for assuming that all angles of incidence are between 0° and 78° for any size of plate with any depth of niche. For finite plates, non-resonant transmission is dependent upon the plate dimensions. The non-resonant sound reduction index tends to increase with decreasing plate size (Section 4.3.1.2.2). Hence for relatively small elements within a relatively deep niche, such as glazing or windows, it is the finite size of the plate as well as the shielding of near-grazing angles that affects non-resonant transmission below the critical frequency.

In the vicinity of the critical frequency, the absence of near-grazing angles in the incident sound field due to shielding would be expected to have an effect on the initial part of the coincidence dip. Using the infinite plate model for a plate with mass, stiffness, and damping shows that the coincidence dip for field incidence is slightly higher than for diffuse incidence (refer forward to Fig. 4.17). Measurements on different sizes of glass in Fig. 3.17 also show that the coincidence dip tends to shift upwards in frequency as θ_{max} decreases (Guy and Sauer, 1984; Guy et al., 1985). Shielding may partly account for the fact that using frequency-average radiation efficiency formulae tends to predict deeper coincidence dips in the sound reduction index than are measured in the laboratory (Leppington et al., 1987).

(2) Orientation of the baffle that surrounds the plate.

Plates that form building elements usually have a standard baffle orientation that can be described as a baffle in the same plane as the plate, or a baffle around the plate perimeter that is orientated perpendicular to the plate surface. Laboratory tests on elements such as glazing, windows, doors, skylights, or infill-panels are usually carried out by fixing the element in a filler wall with a niche on one or both sides. The baffle formed by the niche and the filler wall may or may not

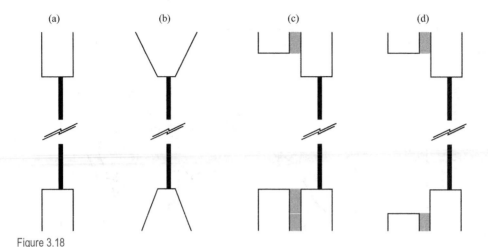

(a)　　　　　　　(b)　　　　　　　(c)　　　　　　　(d)

Figure 3.18

Examples of different types of niche used to mount test elements. (a) box-shaped niches in a solid wall, (b) tapered niches in a solid wall, (c) niche with one step or stagger in a cavity wall, and (d) niche with more than one step or stagger in a cavity wall.

be representative of the element when installed in a building. When flush with the surface of the filler wall, these elements effectively lie within the plane of a baffle formed by the filler wall. When positioned within a niche, the baffle orientation is no longer simple to describe; it depends upon the niche geometry as well as sound radiation from individual bending modes. The niche geometry may be more complex than a simple box-shaped cavity; it can have sloped, stepped, or staggered boundaries (see Fig. 3.18).

Below the critical frequency, the frequency-average radiation efficiency for bending modes on a finite plate depends upon the orientation of the baffle around the plate perimeter (Section 2.9.4). Compared to a plate that lies within the plane of an infinite rigid baffle, the radiation efficiency below the critical frequency doubles when rigid baffles are placed perpendicular to the plate perimeter (Leppington, 1996). Most niche geometries could be described by one of these two idealized baffle orientations, although part of the frequency range will inevitably fall somewhere between the two. We now recall the discussion in Section 2.9.3 in which individual bending modes on a simply supported plate are classified as either corner or edge radiators below the critical frequency; sound radiation from the edge radiators tends to dominate. A crossover frequency between the two idealized baffle orientations can be estimated by describing the niche depth in terms of a fraction of a wavelength. Hence we could assume that niche depths $<\lambda/2$ can be modelled as if the plate lies in the plane of an infinite rigid baffle, and that with niche depths $\geq\lambda/2$, the plate can be modelled as if there were rigid baffles perpendicular to the plate perimeter.

In the laboratory, the depth of a box-shaped niche is typically in the range, $50 \leq d_n \leq 200$ mm. A 50 mm niche depth corresponds to $\lambda/2$ at 3430 Hz. Most plates used in buildings (except thin metal sheets) have critical frequencies below this frequency; hence resonant transmission below the critical frequency will be similar whether the plate is flush with the filler wall or within a 50 mm deep niche.

A 200 mm niche depth corresponds to λ/2 at 858 Hz; hence for plates such as glass or plasterboard (which typically have critical frequencies above 2000 Hz) the niche is likely to increase the radiation efficiency of modes just below the critical frequency compared to when the plate is flush with the filler wall. This may or may not be detectable in the sound reduction index because it depends upon the relative strengths of non-resonant transmission and resonant transmission at these frequencies. This will be specific to the test element and its fixing at the boundaries (e.g. simply supported, clamped) as well as the effect that the fixing has on the coupling losses from the element.

(3) Two-dimensional modal sound field within the niche.

As the niche depth increases, the incident sound field on the test element is distinctly different to when it is flush with a room surface. Although a niche opens out directly into the three-dimensional sound field of the room, we can find an explanation for some aspects of the niche effect by considering the existence of a two-dimensional modal sound field within the niche. When the niche on each side of the test element has identical dimensions, then the niche modes and the modes of the test element are likely to be strongly coupled. This links to the observation that airborne sound transmission in the low-frequency range tends to be lower between rooms with identical dimensions, than between rooms with different dimensions (Heckl and Seifert, 1958). Deep niches with identical dimensions on each side tend to reduce the measured sound reduction index at frequencies above, at and below the critical frequency of the test element.

The effect of equal niche depths is most commonly observed below the critical frequency (Cops et al., 1987; Guy and Sauer, 1984). An example is shown in Fig. 3.19a where the sound reduction index decreases, as the niche depth increases (Yoshimura, 2006). The effects of strong coupling between each niche and the test element can be reduced or avoided by using different niche dimensions on both sides. This can be done by introducing a step and/or stagger into the test aperture, or by using small diffusing elements around the niche boundaries to alter the niche modes in one or both of the niches.

The adverse effects of a deep niche on only one side of the test element (source or receiving room) are less pronounced than with equal niche depths. However, they can still affect important features of the sound reduction index such as the depth of the mass–spring–mass resonance dip as shown in Fig. 3.19b (Yoshimura, 2006).

(4) Absorption of sound energy by the niche boundaries.

Whether we consider transmitted waves at specific angles, or a two-dimensional modal sound field in each niche, it is clear that absorptive surfaces within the niche should increase the sound reduction index. Depending on the niche geometry and the absorber characteristics, this can occur below, at and above the critical frequency (Guy and Mulholland, 1979).

(5) Changes to the structural coupling losses from the test element to the laboratory structure (or filler wall) that forms the test aperture.

This affects resonant transmission below, at and above the critical frequency and can be significant when it is the structural coupling loss factors that primarily determine the total loss factor of the test element. Different positions of the test element within the niche may change the structural coupling losses. Any change can be assessed (and sometimes accounted for) by measuring the total loss factor of the test element.

(a) Equal niche depths

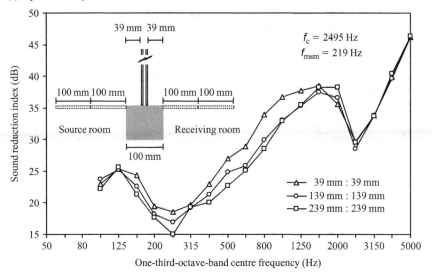

(b) Equal and different niche depths

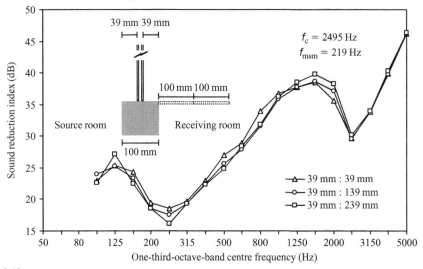

Figure 3.19

Effect of a box-shaped niche on both sides of a 5-12-5 insulating glass unit (1.89 × 1.54 m) with equal and different niche depths. The test aperture is formed by a 100 mm concrete plate, with deeper niches formed using sheets of 12 mm plasterboard. Measurements according to ISO 140 Part 3. Measured data are reproduced with permission from Yoshimura (2006).

3.5.2 Field measurements within buildings

For airborne sound insulation in the field, sound is not only transmitted by the separating wall or floor, but also by flanking transmission. To calculate a sound reduction index for field measurements it is therefore necessary to define a transmission coefficient, τ', that includes the sound power, W_3, transmitted by flanking elements into the receiving room,

$$\tau' = \frac{W_2 + W_3}{W_1} \qquad (3.53)$$

The apparent sound reduction index, R', is therefore defined as

$$R' = 10\lg\left(\frac{1}{\tau'}\right) = 10\lg\left(\frac{W_1}{W_2 + W_3}\right) = L_{p1} - L_{p2} + 10\lg\left(\frac{S}{A}\right) \qquad (3.54)$$

where S is the area of the separating element. In the field there may be a step or a stagger between the rooms, therefore S is the area that is common to both the source and receiving rooms.

Note that whilst R' is primarily intended for field measurements, it is sometimes necessary to quote laboratory test results using R' rather than R when laboratory measurements are known to be significantly affected by flanking transmission.

For lightweight separating elements, the presence of poor workmanship and/or flanking transmission may be indicated by values of R' from field tests that are lower than R for the separating element. However, this difference may also be due to structural coupling around the perimeter of the wall or floor that is different *in situ* compared to the laboratory (see Section 4.3.5.4.4). For heavyweight constructions, any comparison of R' and R that is used to try and identify poor workmanship needs to take account of the total loss factor of the separating element (Section 3.5.1.3.2) as well as flanking transmission (Chapter 5).

For field measurements the airborne sound insulation can be described in terms of the sound pressure level difference, D, between the source and receiving rooms. This can cause problems when setting sound insulation requirements for regulatory purposes, because adding or removing sound absorptive material from the receiving room will change the measured sound pressure level, and hence change the level difference. In some situations the reverberation time in the receiving room may be fixed by other requirements and it may be appropriate just to use the level difference. Otherwise it is necessary to measure the reverberation time in the receiving room and to 'standardize' or 'normalize' the level difference. This provides a fairer basis on which to set performance standards for sound insulation. The level difference, D, is 'standardized' using a reference value for the reverberation time and the level difference is 'normalized' using a reference value for the absorption area.

For a given sound power that is transmitted into the receiving room, the mean-square sound pressure in this room is inversely proportional to the absorption area, A, of that room (Eq. 3.36). On this basis the normalized level difference, D_n, is defined by using a reference absorption area, A_0, of $10\,\text{m}^2$ for the receiving room (ISO 140 Part 4)

$$D_n = L_{p1} - \left(L_{p2} + 10\lg\left(\frac{A}{A_0}\right)\right) = D - 10\lg\left(\frac{A}{A_0}\right) \qquad (3.55)$$

A similar approach is used to give the standardized level difference. For a given sound power transmitted into the receiving room, the mean-square sound pressure in this room is proportional to the reverberation time, T, of that room. The standardized level difference, D_{nT}, is defined by using a reference reverberation time, T_0, for the receiving room, which for dwellings is $0.5\,\text{s}$ (ISO 140 Part 4)

$$D_{nT} = L_{p1} - \left(L_{p2} - 10\lg\left(\frac{T}{T_0}\right)\right) = D + 10\lg\left(\frac{T}{T_0}\right) \qquad (3.56)$$

Note that from Sabine's equation (Eq. 1.97) the relationship between T and A only involves the receiving room volume; hence D_n and D_{nT} will be the same when the volume is 31 m^3.

Regulatory requirements for dwellings are often set using single-number quantities that are calculated with the rating method in ISO 717 Part 1; other descriptors that are specific to individual countries are not discussed here. The choice of single-number quantity depends on what the regulation is aiming to achieve and how the regulation is implemented and enforced. For this reason different countries may need to use different single-number quantities, although the requirements are usually based around D_{nT} or R'. D_n uses a reference absorption area which is not usually appropriate because of the wide range of room dimensions, for which there will be a wide range of absorption areas. R' uses the common area of the separating element; this can be awkward if there is no common area between the rooms (e.g. diagonally offset rooms). D_{nT} is based on a reference reverberation time; however, typical reverberation times in dwellings tend to vary between countries. They vary depending on the preferred type of surface finish (e.g. carpet, ceramic tiles) and whether the room is furnished. For furnished rooms (excluding kitchens and bathrooms) with volumes between 15 and 60 m^3, the reverberation time over the building acoustics frequency range typically lies between 0.4 and 0.6 s (European data from Vorländer, 1995). It is therefore reasonable to use $T_0 = 0.5$ s for furnished dwellings where the reverberation time tends to be independent of the room volume. This is because the absorption area tends to be proportional to the volume; i.e. larger rooms usually have more absorbent furnishings (van den Eijk, 1972). When testing between rooms of unequal volume this means that the measured D_{nT} will depend on the choice of source and receiving room. If the receiving room is the larger room, L_{p2} will be lower than if the smaller room had been chosen; this is due to the higher absorption area in the larger room. Therefore when carrying out testing in a single direction to assess compliance with regulations it is common to require that the smaller room is used as the receiving room so that the lowest D_{nT} values are measured.

When all sound is transmitted between two rooms via a separating element with a sound reduction index, R, the relationship between D_{nT} and R is

$$D_{nT} = R + 10 \lg \left(\frac{V}{S} \right) - 5 \, \text{dB} \tag{3.57}$$

which for box-shaped receiving rooms simplifies to

$$D_{nT} = R + 10 \lg d - 5 \, \text{dB} \tag{3.58}$$

where V is the volume of the receiving room and d is the receiving room dimension that is perpendicular to the separating element. Note that $D_{nT} = R$ when $d = 3.2$ m.

Whilst D_{nT} has some advantages in setting regulatory requirements, Eq. 3.58 indicates that some caution is needed when a data set of field sound insulation tests is used to calculate pass and failure statistics for a certain type of construction. The data set may be biased if all the receiving rooms have very low or very high values of d.

3.5.2.1 Reverberation time

For measurements in diffuse fields it is appropriate to use T_{30} for the reverberation time. In typical rooms the building acoustics frequency range often covers sound fields that are non-diffuse as well as those which can be considered to be diffuse. Previous discussions on reverberation time (Section 1.2.6.3) show that non-diffuse fields can lead to curvature of the decay curve or distinct double slopes. To make the link between sound power and reverberant sound pressure level in non-diffuse fields it is therefore necessary to use T_{10}, T_{15}, or T_{20} (Section 1.2.7.5.2). In practice, a balance can usually be struck by using T_{20} for all of the building acoustics frequency range.

3.5.2.2 Sound intensity

Sound intensity can also be used in the field to measure the sound insulation of a building element in the presence of flanking transmission (ISO 15186 Part 2). Examples of relevant building elements inside buildings include separating walls, flanking walls, doors, or small building elements such as cable ducts.

For the sound power radiated by a separating element, the apparent intensity sound reduction index, R'_I, is defined as

$$R'_I = L_{p1} - L_{In} - 10 \lg \left(\frac{S_M}{S} \right) - 6 \, \text{dB} \tag{3.59}$$

where L_{p1} is the average sound pressure level in the source room, L_{In} is the average sound intensity level over the measurement surface for the separating element, S_M is the total area of the measurement surface, and S is the area of the separating element. When there is a step or a stagger between the source and receiving rooms, S is the area that is common to both rooms.

In the measurement of R'_I, the intensity probe is used to measure only the sound power radiated by the separating element. Hence there is an important difference between R'_I and R'. With R' (ISO 140 Part 4) the sound power radiated by flanking elements is included, but with R'_I (ISO 15186 Part 2) it is not included.

For the sound power radiated by flanking element j, the intensity sound reduction index, $R'_{IF,j}$, is defined as

$$R'_{IF,j} = L_{p1} - L_{In,j} - 10 \lg \left(\frac{S_{M,j}}{S} \right) - 6 \, \text{dB} \tag{3.60}$$

where L_{In} is the average sound intensity level over the measurement surface for flanking element j, S_M is the total area of the measurement surface for flanking element j, and S is the area of the separating element. When there is a step or a stagger between the source and receiving rooms, S is the area that is common to both rooms.

Tho intensity normalized level difference, $D_{I,n}$, is intended for use when there is no common surface area in the source and receiving rooms. This can occur with rooms that are sited diagonally across from each other, or with rooms that are separated from each other by another room. In this case one specific element is chosen for the intensity measurement.

$$D_{I,n} = L_{p1} - L_{In} - 10 \lg \left(\frac{S_M}{A_0} \right) - 6 \, \text{dB} \tag{3.61}$$

where L_{In} is the average sound intensity level over the measurement surface for a chosen element, and S_M is the total area of the measurement surface.

For small building elements or devices, the intensity element-normalized level difference, $D_{I,n,e}$, can also be used in the field situation and is calculated in the same way as in the laboratory.

$$D_{I,n,e} = L_{p1} - L_{In} - 10 \lg \left(\frac{S_M}{A_0} \right) - 6 \, \text{dB} \tag{3.62}$$

where L_{In} is the average sound intensity level over the measurement surface for the small building element or device and S_M is the total area of the measurement surface.

3.5.3   Field measurements of building façades

Airborne sound insulation measurement of building façades are categorized according to the relevant Standard (ISO 140 Part 5). The measurement is either used to determine the sound insulation of a single building element (e.g. window, door) or the entire façade.

3.5.3.1   *Sound insulation of building elements*

Measurements on a building element can either be carried out using a loudspeaker, or the existing environmental noise source such as road, railway, or aircraft traffic.

3.5.3.1.1   *Loudspeaker method*

The apparent sound reduction index of an individual building element, such as a window, can be measured from outside to inside using sound pressure level measurements with a loudspeaker facing towards the façade. Note that we can only measure the apparent sound reduction index because there will inevitably be some sound transmitted by the rest of the façade, which we regard as flanking transmission. The transmission coefficient is defined by Eq. 3.53. However the incident sound power, W_1, is calculated differently; this is because the incident sound field on the element is different to that from within a room.

From Section 1.4.1 we recall that with a point source and sound pressure measurements on a surface, pressure doubling can be assumed compared to the free-field value, and that interference dips can be avoided in the building acoustics frequency range. Hence, it is convenient to measure the average surface sound pressure level, $L_{p1,s}$, on the building element under test, and the average sound pressure level in the receiving room, L_{p2}.

By placing the loudspeaker at a sufficient distance from the test element we can assume that the incident sound field is comprised of plane waves. For measurements according to the relevant Standard (ISO 140 Part 5) the loudspeaker is positioned at an angle, θ, of 45°; θ is the angle between the line normal to the centre of the test element, and the axis of the loudspeaker that points towards the centre of the test element (see Fig. 3.20). The component of the plane wave intensity that is incident upon the test element is $I \cos \theta$. The incident sound power, W_1, can be calculated from the average free-field mean-square sound pressure, $\langle p_1^2 \rangle_{t,s}$, that would exist at the test element position if the test element and the rest of the façade were not there by using

$$W_1 = S I_1 \cos 45° = \frac{\langle p_1^2 \rangle_{t,s}}{\rho_0 c_0} S \cos 45° \qquad (3.63)$$

where the spatial average corresponds to the area that would be occupied by the test element if it were present.

By assuming pressure doubling at the surface we can calculate W_1 from the measurement of the average surface mean-square sound pressure, $\langle p_{1,s}^2 \rangle_{t,s}$, using

$$W_1 = \frac{1}{4} \frac{\langle p_{1,s}^2 \rangle_{t,s}}{\rho_0 c_0} S \cos 45° \qquad (3.64)$$

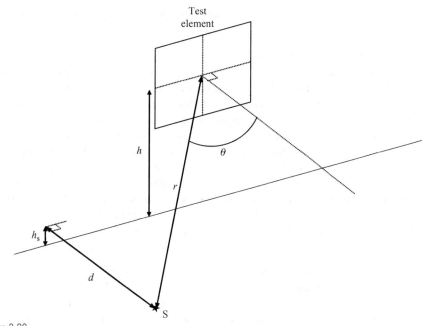

Figure 3.20

Loudspeaker position (S) relative to the centre of the test element.

where the spatial average corresponds to measurements over the surface area of the test element.

The power transmitted into the receiving room is calculated from Eq. 3.36, hence the apparent sound reduction index, $R'_{45°}$, is

$$R'_{45°} = 10\lg\left(\frac{\langle p^2_{1,s}\rangle_{t,s}}{\langle p^2_2\rangle_{t,s}}\right) + 10\lg\left(\frac{S}{A}\right) + 10\lg(\cos 45°) = L_{p1,s} - L_{p2} + 10\lg\left(\frac{S}{A}\right) - 1.5\,dB \quad (3.65)$$

where $L_{p1,s}$ is the average sound pressure level measured on the surface of the façade and L_{p2} is the average sound pressure level in the receiving room.

Angle of incident sound: Airborne sound insulation is dependent upon the angle of the incident sound (Section 4.3.1.5); hence differences are to be expected between the sound reduction index measured with a diffuse incidence sound field and with sound incident at a single angle. There is no general rule for conversion between the two types of incident sound; this is obvious if one considers a single sheet of glass, a side-hung window opened at a narrow angle, and a through-wall ventilator. However, for closed windows, $\theta = 45°$ often gives a reasonable estimate of the sound reduction index with a diffuse incidence sound field. For closed windows, there is some evidence that the apparent sound reduction index measured using $\theta = 60°$ instead of 45°, gives closer agreement with R measured in the laboratory (Jonasson and Carlsson, 1986). However no single angle will give exact agreement over the building acoustics frequency range and it is necessary to standardize a single angle. It is also difficult to justify a change to 60° when for practical purposes it is simpler to position a loudspeaker at an angle of 45°, the sound insulation of different façade elements will have different angle dependence, and there are other variables such as loudspeaker height that also affect the measurement. The relevant measurement Standard therefore uses $\theta = 45°$ (ISO 140 Part 5).

Uniformity of incident sound field: Derivation of the incident sound power (Eq. 3.63) is based around the free-field mean-square sound pressure at the test element position. Ideally there would be a perfectly uniform incident sound field over the entire surface of the test element. In practice this is not possible to achieve. This is only partly due to the directivity of the loudspeaker because this aspect is controlled by the loudspeaker specifications in the Standard. The spatial variation over the surface of the element is primarily determined by the combination of ground impedance, element dimensions, and the source-receiver-façade geometry. The latter is constrained by the requirements of the Standard and the practical restrictions on site that limit the available positions for a loudspeaker. The Standard requires that the loudspeaker height, h_s, and distance from the façade, d, are chosen so that the spatial variation over the surface of the element is minimized. To do this it requires that the loudspeaker should preferably be placed on the ground or as high above the ground as is practical. For a loudspeaker placed on the ground, the acoustic centre of the loudspeaker will typically be at a height, $h_s \approx 0.5\,\text{m}$.

The effect of different loudspeaker heights has been investigated experimentally by Jonasson and Carlsson (1986). Their measurements looked at the spatial variation of the free-field sound pressure level over a $2 \times 2\,\text{m}$ area using nine microphones. Two different loudspeaker heights were used, one on the ground and the other raised by 2.9 m. The distance from the loudspeaker to the centre of the $2 \times 2\,\text{m}$ area was 5 m; this is the minimum distance required by the Standard when measuring a façade element. For windows, typical mid-height values above the ground are approximately 1.5 m for the ground floor, and 4 m for the first floor; in these measurements the mid-point of the $2 \times 2\,\text{m}$ area was at a height of 1.6 or 4.2 m. Figure 3.21 shows the maximum variation in the sound pressure level over the $2 \times 2\,\text{m}$ area. For a ground floor window, the spatial variation in the low-frequency range is significantly lower with the loudspeaker on the ground than when raised off the ground. The sound pressure at any point in the free field is the combination of the direct path and the path involving a single reflection from the ground. For the ground floor window, the first interference dip occurs in the mid-frequency range when the loudspeaker is on the ground, but in the low-frequency range when the loudspeaker is raised off the ground. The latter interference dip is deeper than the former, and significantly increases the spatial variation over the $2 \times 2\,\text{m}$ area. In these experiments the ground cover was short grass; however, the frequency and depth of the interference dips will vary depending on the ground impedance.

3.5.3.1.2 Sound intensity

The apparent sound reduction index of an individual building element in the façade, such as a window or door can also be measured using sound intensity. In this situation the measurements are taken with the loudspeaker inside the room and the test element is scanned with the intensity probe from outside the room to give the apparent intensity sound reduction index, R'_I,

$$R'_I = L_{p1} - L_{In} + 10\lg\left(\frac{S}{S_M}\right) - 6\,\text{dB} \tag{3.66}$$

Note that we refer to this sound reduction index as 'apparent' to indicate that it is a field measurement, although the intensity probe is used to suppress measurement of sound that is radiated by the flanking elements.

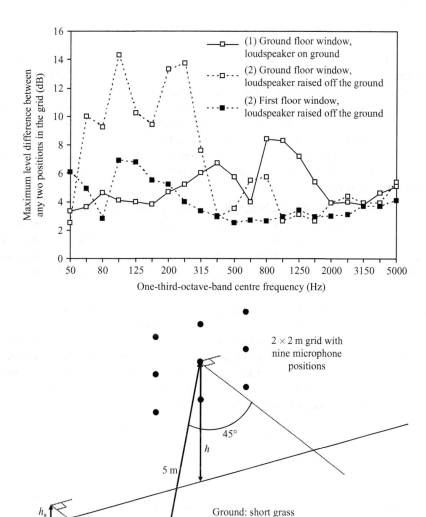

Figure 3.21

Maximum spatial variation of the free-field sound pressure level measured over a 2×2 m area. The key describes the situation to which the measurements correspond, the heights are: (1) $h_s = 0.3$ m, $h = 1.6$ m (2) $h_s = 2.9$ m, $h = 1.6$ m and (3) $h_s = 2.9$ m, $h = 4.2$ m. Measured data are reproduced with permission from Jonasson and Carlsson (1985).

For small building elements or devices, such as a ventilator in the external wall, the intensity element-normalized level difference, $D_{I,n,e}$, is

$$D_{I,n,e} = L_{p1} - L_{In} - 10\lg\left(\frac{S_M}{A_0}\right) - 6\,dB \tag{3.67}$$

where L_{In} is the average sound intensity level over the measurement surface for the small building element or device, and S_M is the total area of the measurement surface.

3.5.3.1.3 Road traffic method

The apparent sound reduction index of an individual building element can also be measured from outside to inside using road traffic noise as the sound source. Requirements on the geometry of the road traffic source in relation to the façade and the number of passing vehicles during the measurement period are described in the Standard. The noise level and spectrum will depend on the geometry of the situation as well as traffic flow, traffic speed, traffic composition in terms of the type of vehicles, road gradient, and road surface. Compared to loudspeaker measurements it is difficult to estimate the background noise level in the receiving room from measurements unless the traffic noise can be stopped. This means that background noise corrections are rarely possible.

For road traffic along streets that are lined with buildings on both sides, the external sound field begins to resemble a reverberant field. However, sound is not usually incident from above the building element unless there are features, such as balconies, which reflect or scatter sound down onto the element from above. In general, wherever a road runs approximately parallel to the façade it is reasonable to assume that sound will be incident upon the building element from a wide range of angles.

As with loudspeaker measurements, the external sound pressure level is measured on the surface of the element. In Section 1.2.7.1 we looked at sound fields near room boundaries and the interference patterns that exist near surfaces in rooms. For road traffic, the theory for sound incident upon a surface from all directions is applied, hence sound pressure levels at the surface are assumed to be 3 dB higher than at a distance far away from the surface. It is therefore necessary to subtract 3 dB from the surface sound pressure level, $L_{p1,s}$, to modify Eq. 3.54; this gives the apparent sound reduction index for a road traffic noise source, $R'_{tr,s}$, as:

$$R'_{tr,s} = L_{p1,s} - L_{p2} + 10\lg\left(\frac{S}{A}\right) - 3\,dB \qquad (3.68)$$

3.5.3.1.4 Aircraft and railway noise

The apparent sound reduction index can also be measured using aircraft or railway traffic as the sound source, giving $R'_{at,s}$ or $R'_{rt,s}$ respectively. In the same way as with loudspeaker measurements, the external sound pressure level is measured on the surface of the element.

In the context of sound insulation, aircraft noise events are particularly complex. During an aircraft flyover, the sound insulation provided by the test element varies with time. This is due to the time-varying incident sound field, as well as the sound insulation of the element varying with the angle of incidence. As an aircraft travels along a certain flight path, the source–receiver–façade geometry changes, which therefore changes the incident sound field and the angle of incidence. Sound insulation is often measured in buildings quite close to airports, and there is usually more than one possible flight path for take-off and landing. In addition, the sound emitted from an aircraft can be highly directional and its level and spectrum will vary depending on the type of aircraft as well as whether it is taking-off, cruising, or landing. A further complication is the fact that we can only measure the apparent sound reduction index. It is difficult to confirm that the dominant sound transmission path throughout the duration of the noise event occurs via the element under test, such as a window. As the aircraft changes its orientation to the test element and the rest of the building, flanking paths such as those involving the roof may become dominant for some or all of the duration of the noise event. For the above reasons, comparison of R from the laboratory with a single measurement of $R'_{at,s}$ in the field is rarely meaningful.

In contrast to loudspeaker and road traffic noise measurements, both aircraft and railway noise consist of individual noise events. Therefore, the sound pressure inside and outside the building is measured in terms of the sound exposure level, L_E. During the noise event it is assumed that the sound will be incident upon the test element from the majority of directions. Therefore the apparent sound reduction index for railway or aircraft noise is calculated using the same equation as for road traffic noise (Eq. 3.68), but by using L_E for the sound pressure levels. For the mean of many measurements, the assumption of energy doubling for the surface sound pressure measurement is a reasonable estimate in the mid-frequency range, but for aircraft there can be significant differences in the low- and high-frequency ranges (Bradley and Chu, 2002).

3.5.3.2 Sound insulation of façades

We now look at measuring the sound insulation of an entire façade, rather than a single building element. The apparent sound reduction index of a façade can be measured from outside to inside, either by using a loudspeaker, or by using the existing environmental noise source. The preferred method is to use the existing environmental noise source because the sound insulation is usually being measured to assess the ability of the façade to reduce the environmental noise level inside the building. In addition, the external sound field associated with environmental noise may be too complex to adequately reproduce with a loudspeaker. For example, the sound incident upon one part of the façade can be significantly different to another part, such as with aircraft flyovers where the sound that is incident on a window in a side wall can be significantly different to that which is incident on the roof. However, if the environmental noise source is not sufficiently loud, or the sound insulation is particularly high, then it may not be possible to measure levels in the receiving room that are above the background noise level, and the appropriate solution is to use a loudspeaker.

For the entire façade it is not practical to measure the spatial average surface sound pressure level over every single façade element. In addition, the spatial average becomes less meaningful when there are large variations in level across the façade, particularly when façades contain elements with significantly different levels of sound insulation (e.g. a thick masonry wall and a single pane of glass). The approach that is adopted in the relevant measurement Standard (ISO 140 Part 5) is to measure the external sound pressure level at the middle of the façade at a height of 1.5 m above the floor of the receiving room, and at a distance of 2 m perpendicular to the plane of the façade. For this scenario we have seen in Section 1.4.1 that there will be interference between the different sound propagation paths from the source to the receiver in the low-frequency range. The interference patterns are affected by the orientation of the source, receiver, and the façade, as well as by the ground and façade impedances. These patterns will therefore vary from site to site. In addition we have seen that in the mid- and high-frequency ranges, the assumption of energy doubling at a distance of 2 m from the façade is reasonable for the mean of many measurements; but not for any individual measurement. For these reasons the incident sound power cannot be accurately calculated from sound pressure level measurements made at a distance of 2 m the façade. The external sound pressure level can therefore be treated purely as a reference level that is needed to calculate the level difference.

For loudspeaker or road traffic noise measurements, the level difference, D_{2m}, is

$$D_{2m} = L_{p1,2m} - L_{p2} \qquad (3.69)$$

where $L_{p1,2m}$ is the sound pressure level measured 2 m in front of the façade.

For railway or aircraft traffic measurements, the level difference, D_{E2m}, is

$$D_{E2m} = L_{E1,2m} - L_{E2} \tag{3.70}$$

where $L_{E1,2m}$ is the sound exposure level measured 2 m in front of the façade.

As with other field sound insulation measurements the reverberation time in the receiving room is measured in order to replace the actual room damping by a reference value so that the level difference can be 'normalized' or 'standardized'.

For loudspeaker or road traffic noise measurements the normalized level difference, $D_{2m,n}$, is defined by using a reference absorption area, A_0, of $10\,m^2$ for the receiving room (ISO 140 Part 5).

$$D_{2m,n} = D_{2m} - 10\lg\left(\frac{A}{A_0}\right) \tag{3.71}$$

For railway or aircraft traffic measurements, the normalized sound exposure level difference, $D_{E2m,n}$, is

$$D_{E2m,n} = D_{E2m} - 10\lg\left(\frac{A}{A_0}\right) \tag{3.72}$$

For loudspeaker or road traffic noise measurements the standardized level difference, $D_{2m,nT}$, is defined by using a reference reverberation time, T_0, for the receiving room, which for dwellings is 0.5 s (ISO 140 Part 5).

$$D_{2m,nT} = D_{2m} + 10\lg\left(\frac{T}{T_0}\right) \tag{3.73}$$

For railway or aircraft traffic measurements, the standardized sound exposure level difference, $D_{E2m,nT}$, is

$$D_{E2m,nT} = D_{E2m} + 10\lg\left(\frac{T}{T_0}\right) \tag{3.74}$$

When measurements of $D_{2m,n}$ or $D_{2m,nT}$ are taken according to the relevant Standard (ISO 140 Part 5) there will not usually be significant differences between results taken using road traffic noise or a loudspeaker as the sound source (EN 12354 Part 3).

3.5.4 Other measurement issues

In this section we look at the effect of background noise on sound pressure measurements and conversion of measurements from one-third-octave to octave-bands. As it is possible to determine the sound reduction index using either sound pressure or sound intensity measurements we see how the two results can be compared. In addition a few practical issues are discussed such as the drying out of test elements and identifying sound leaks.

3.5.4.1 Background noise correction

When measuring constructions that have high levels of airborne or impact sound insulation, the sound pressure levels in the receiving room may not be high above the background noise level. This quite often occurs in the field because occupied buildings or semi-completed buildings

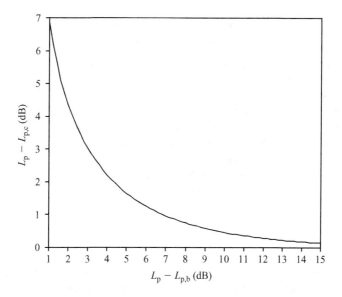

Figure 3.22

Background noise correction. This shows the correction, in terms of $L_p - L_{p,c}$, that is made to the measured sound pressure level, L_p to remove the effect of background noise level, $L_{p,b}$.

with ongoing building work have higher background noise levels than a laboratory. The effect of measuring sound pressure levels that are close to the background noise is to give a higher level than would be measured in the absence of the background noise. The sound pressure level measured in the receiving room can be corrected to remove the effect of steady background noise. The corrected sound pressure level, $L_{p,c}$, is given by:

$$L_{p,c} = 10 \lg (10^{L_p/10} - 10^{L_{p,b}/10}) \tag{3.75}$$

where L_p is the measured sound pressure level in the presence of background noise and $L_{p,b}$ is the background noise level.

The effect of background noise on the measured sound pressure level is shown in Fig. 3.22 using Eq. 3.75.

For airborne and impact sound insulation in the relevant Standards, this correction is only pre-scribed for specific ranges of $L_p - L_{p,b}$. For laboratory measurements (ISO 140) this correction is applied when $6\,\text{dB} < L_p - L_{p,b} < 15\,\text{dB}$. For field measurements (ISO 140) this correction is applied when $6\,\text{dB} < L_p - L_{p,b} < 10\,\text{dB}$.

Under controlled conditions in a laboratory the background noise level is usually steady and can be quantified with greater accuracy than in the field. When $L_p - L_{p,b} \geq 15\,\text{dB}$, we find that $L_p - L_{p,c} \leq 0.1\,\text{dB}$. As we measure sound pressure levels to the nearest 0.1 dB the correction is not used when $L_p - L_{p,b} \geq 15\,\text{dB}$. In the field there is less control over the measurement environment, and estimates of the background noise level during the measurement are more prone to error.

When $L_p - L_{p,b} \leq 6\,\text{dB}$ there is insufficient signal to ensure that the large corrections (see Fig. 3.22) are appropriate. The approach taken in the relevant Standards is to limit the correction to 1.3 dB rather than to continue to use Eq. 3.75 (ISO 140). This value corresponds to the correction that is calculated when $L_p - L_{p,b} = 6$ dB.

The background noise in the receiving room is often measured once before, and once after the measurement with the loudspeaker, to check for any significant change. We therefore have to assume that these background noise levels are representative of the background noise during the measurement. This is reasonable for background noise that can be considered as a stationary signal; one example could be environmental noise from a steady flow of road traffic that is transmitted through the building façade. However, with field measurements in occupied buildings or semi-completed buildings with ongoing building work, there is an increased likelihood of unsteady background noise. Hence, it is useful to monitor the maximum sound pressure level during the measurement period to check for transient noises, such as doors slamming in the building.

3.5.4.2 Converting to octave-bands

Airborne sound insulation descriptors in one-third-octave-bands can be converted to octave-bands using the three one-third-octave-bands from which the octave-band is formed,

$$X_{OB} = -10 \lg \left(\frac{1}{3} \sum_{n=1}^{3} 10^{-X_{TOB,n}/10} \right) \tag{3.76}$$

where X represents R, $R'_{45°}$, D_n, D_{nT}, etc.

3.5.4.3 Comparing the airborne sound insulation measured using sound pressure and sound intensity

When choosing suitable constructions at the design stage it may be necessary to compare the sound reduction index, R, for one product, with the intensity sound reduction index, R_I, for another product. To compare the performance of the two products we need to account for the different methods by which R and R_I are calculated (Jonasson, 1993). In the derivation of R, we noted that the transmitted sound power is underestimated in the low-frequency range because no account is taken of the higher energy density near the room boundaries. The reason that no correction term is used in the definition of the sound reduction index is partly because the Waterhouse correction is not considered sufficiently accurate for the smaller room volumes that can be used in transmission suites (i.e. 50 m³). However, if we want to compare R with R_I then the Waterhouse correction can still be used as an estimate. The modified intensity sound reduction index, $R_{I,M}$, is introduced to represent R_I after it has been modified with the Waterhouse correction (ISO 15186 Part 1),

$$R_{I,M} = R_I + C_W \tag{3.77}$$

where C_W is the Waterhouse correction in decibels (Eq. 1.157) for the sound field in the receiving room. Note that calculation of C_W requires knowledge of the volume and total surface area of the receiving room.

$R_{I,M}$ can now be compared directly with R. Note that R_I in the low-frequency range is the better estimate of the 'true' sound reduction index as defined by Eq. 3.33; it is only for the purpose of this comparison that it is necessary to introduce some additional uncertainty.

3.5.4.4 Variation in the sound insulation of an element due to moisture content and drying time

For some elements, such as masonry walls or screed floating floors, the sound insulation will vary depending on the degree to which they have dried out. These changes are relevant in the field as well as in the laboratory.

In the laboratory, individual components (such as bricks, blocks or sheet material) can be stored in dry conditions before building the test element. However, the moisture content usually increases once they are bonded together with mortar or a water-based adhesive. Surface finishes, such as wet plaster also increase the moisture content of the wall or floor.

Potential changes to the sound insulation can be considered in terms of the mass, stiffness, coupling losses, and internal losses of the element. In most cases we only have information on the increase in the mass per unit area due to moisture when compared to standard conditions. At frequencies where the sound reduction index is only determined by non-resonant (mass law) transmission, estimates for the increase in sound insulation can be calculated (see Section 4.3.1.2). When there is only resonant transmission the effect of moisture on the plate stiffness and internal damping needs to be considered, although information is rarely available to quantify the effect. This is partly due to other confounding factors. As the plate dries it tends to shrink away from the test aperture, which can reduce the plate coupling losses, and hence the total loss factor. This potentially introduces air paths at cracks or gaps around the plate perimeter (Schmitz et al., 1999). Resonant transmission dominates over most of the frequency range for masonry/concrete elements and it is possible for the effect of moisture to increase or decrease the sound reduction index at different frequencies.

3.5.4.5 Identifying sound leaks and airpaths

With most constructions care needs to be taken to avoid gaps or cracks around the perimeter of the element. However, when the airborne sound insulation curve of a wall or floor decreases or forms a plateau in the high-frequency range it does not necessarily mean that there is a sound leak or air path. If problems due to high background noise have been ruled out, this feature of the curve could be due to a number of factors including: a plateau due to the thick plate effect, a porous plate material, a plate with a high critical frequency that forms the surface of the wall or floor, or flanking transmission. These aspects are discussed in Chapters 4 and 5.

High-frequency leaks can often be detected with ones ears, but they can also be detected using a sound intensity probe (Section 3.10.1). The axis of a p–p sound intensity probe can be slowly scanned parallel to the surface that is suspected of containing the leak. This is different to how the probe is normally used to determine the sound power, for which the probe axis is perpendicular to the measurement surface. As the probe moves past a leak, the intensity reading will flip from positive to negative or vice versa. Leaks tend to transmit significant levels of sound in the mid- or high-frequency range, so a 12 mm spacer is usually needed.

With windows and doors there are many possible airpaths around the frame or locking devices. An assessment of their effect on the airborne sound insulation is most simply made by testing before and after sealing all the joints with tape. Airtightness testing may also be used to detect potential problems with window units before they are installed.

271

3.6   Impact sound insulation (floors and stairs)

As an introduction to impact sound insulation it is useful to briefly re-consider airborne sound insulation. To measure airborne sound insulation we use sound pressure level measurements in the source and receiving rooms to relate the transmitted sound to the incident sound. More specifically, for the sound reduction index we determine the ratio of the sound power that is incident on the test element to the sound power transmitted by that same element. Hence when power is 'injected' into the source room via a loudspeaker there is no need to know its sound power output. As long as the sound transmission process from one room to the other is a linear process (which it normally is), we can take measurements using loudspeakers with different sound power outputs and still calculate the same value for the airborne sound insulation. This assumes that there is a relatively flat spectrum in the source room and that we can measure a signal in the receiving room that is well-above the background level.

To measure the impact sound insulation we need to 'inject' power into a floor using a structure-borne sound source. Taking the approach used for airborne sound insulation, the starting point is to consider how we could determine the structure-borne sound power input into the floor. Although there are methods for measuring this power input, it would be difficult and very time-consuming to take accurate measurements on all types of floors. So to avoid measuring the power input, the logical step is to standardize the excitation source. This is the approach that is used for impact sound insulation measurements in the Standards. However, we will soon see that this does not fully take into account the complexities involved with power input into a structure.

There are many different structure-borne sound sources on floors including footsteps, impacts from children playing, dropped objects, chairs or other furniture being dragged across the floor, and vibrations from machinery such as washing machines or mechanical services. Some of these impact sources generate impulses due to a falling mass, such as a dropped object or footsteps; these only apply a force and are often described as force sources. Other sources may apply forces and/or moments to the structure and generate an impulse or a continuous signal. No single artificial source can accurately represent all of these real sources. Even if we restricted ourselves to representing impacts from footsteps, there is such a wide range of body weights, walking styles, footwear, and impact velocities that we would need to identify which type of footstep the structure-borne sound source should represent. This is illustrated by the wide range of different spectra for different walkers in Fig. 3.23 (Vian and Drouin, 1977). The inherent difficulty in defining a measurement and rating system for impacts on floors has occupied researchers ever since the first experimental work in the late 1920s. For a thorough historical review the reader is referred to papers by Schultz (1981) and Cremer (1976/1977); the following papers can then be used as starting points to continue beyond 1980: Bodlund (1985), Rindel and Rasmussen (1996), Warnock (1998), Tachibana et al. (1998), and Scholl (2001).

For the purpose of standardization it is clear that a well-defined impact source is necessary to establish minimum regulatory requirements and to allow the ranking of impact sound insulation provided by different floors. To achieve this it is not only necessary to standardize the source, but also the rating procedure that is used to produce a single-number quantity. The primary source is the ISO tapping machine which is described in International Standards (ISO 140 Parts 6 & 7). The test method (ISO 140 Parts 6 & 7) is defined along with a rating procedure (ISO 717 Part 2) to allow comparison of the impact sound insulation of different floors.

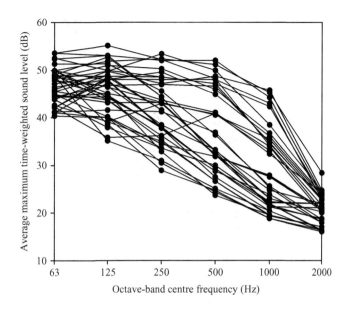

Figure 3.23

Sound pressure level spectra (40 curves) in a receiving room below a 140 mm concrete floor slab from 40 different walkers (male and female) in a variety of footwear walking in a natural manner. For each of the 40 walkers the average value for the maximum time-weighted sound pressure level (time constant, $\tau = 35$ ms) was calculated from the loudest 25% (i.e. upper quartile) of levels measured during a time period containing approximately 50 footsteps. Measured data are reproduced with permission from Vian and Drouin (1977).

3.6.1 Laboratory measurements

Laboratory measurement of impact sound insulation requires measurement of the temporal and spatial average sound pressure level in a room, L_p, when the floor above is excited by the ISO tapping machine (see Fig. 3.24). For a given sound power transmitted into the receiving room, the mean-square sound pressure in this room is inversely proportional to the absorption area, A, of that room (Eq. 3.36). The normalized impact sound pressure level, L_n, is therefore defined by using a reference absorption area, A_0, of 10 m^2 for the receiving room (ISO 140 Part 6).

$$L_n = L_p + 10 \lg \left(\frac{A}{A_0} \right) \tag{3.78}$$

Note that there is no need to normalize the sound pressure level to the surface area of the test element as with the sound reduction index. This can be seen from the derivation of L_n for a solid homogeneous isotropic plate in Section 4.4.1.

3.6.1.1 *Improvement of impact sound insulation due to floor coverings*

There are many different types of floor covering ranging from soft floor coverings (e.g. carpet, vinyl) to rigid walking surfaces (e.g. floating floors, parquet, ceramic tiles). The improvement of impact sound insulation due to a floor covering depends upon the base floor. For this reason the base floor constructions are prescribed in the laboratory measurement Standards; these are categorized as either heavyweight (ISO 140 Part 8) or lightweight (ISO 140 Part 11) base floors.

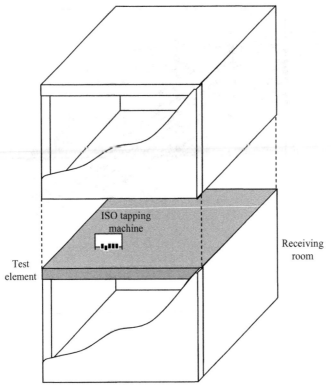

Figure 3.24

Outline sketch of a transmission suite for impact sound insulation measurements (upper room is displaced upwards).

3.6.1.1.1   Heavyweight base floor (ISO)

The heavyweight base floor can generally be referred to as a 140 mm reinforced concrete slab; the exact details are given in the relevant Standard (ISO 140 Part 8). The improvement of impact sound insulation, ΔL, is measured using the ISO tapping machine and is defined as:

$$\Delta L = L_{n0} - L_n \tag{3.79}$$

where L_{n0} is the normalized impact sound pressure level for the ISO heavyweight base floor without the floor covering and L_n is the normalized impact sound pressure level with the floor covering.

3.6.1.1.2   Lightweight base floors (ISO)

Compared with heavyweight base floors, it is more awkward to choose a single lightweight base floor that adequately represents the range of lightweight floors that are built around the world. For this reason, three different timber floor bases are described in the relevant Standard (ISO 140 Part 11); this makes it important to state which base floor has been used. For lightweight base floors, the improvement of impact sound insulation is usually measured with the ISO tapping machine, but to assess heavy impacts, measurements can also be made using the ISO rubber ball (ISO 140 Part 11).

ISO tapping machine: The improvement of impact sound insulation, ΔL_t, measured with the ISO tapping machine is

$$\Delta L_t = L_{n,t,0} - L_{n,t}$$ (3.80)

where $L_{n,t,0}$ is the normalized impact sound pressure level for the ISO lightweight base floor without the floor covering and $L_{n,t}$ is the normalized impact sound pressure level with the floor covering.

Equation 3.80 also applies to the modified ISO tapping machine that is described in the relevant Standard (ISO 140 Part 11); this modification is discussed in Section 3.6.3.4.

ISO rubber ball: For each impact from the rubber ball, the maximum time-weighted (Fast) sound level, L_{Fmax}, is measured and averaged for different excitation and microphone positions to give the impact sound pressure level, $L_{i,Fmax}$. The improvement of impact sound insulation is then calculated using:

$$\Delta L_r = L_{i,Fmax,0} - L_{i,Fmax}$$ (3.81)

where $L_{i,Fmax,0}$ is the impact sound pressure level for the ISO lightweight base floor without the floor covering and $L_{i,Fmax}$ is the impact sound pressure level with the floor covering.

3.6.2 Field measurements

Calculation of the normalized impact sound pressure level for the field situation, L'_n, is identical to that in the laboratory (ISO 140 Part 7) where:

$$L'_n = L_p + 10 \lg \left(\frac{A}{A_0} \right)$$ (3.82)

To find the standardized impact sound pressure level, L'_{nT}, we know that for a given sound power transmitted into the receiving room, the average mean-square sound pressure in this room is proportional to the reverberation time, T, of that room. Hence L'_{nT} is defined using a reference reverberation time, T_0, for the receiving room, which for dwellings is 0.5 s (ISO 140 Part 7).

$$L'_{nT} = L_p - 10 \lg \left(\frac{T}{T_0} \right)$$ (3.83)

3.6.3 ISO tapping machine

The ISO tapping machine (see Fig. 3.25) has a line of five equally spaced hammers that are driven in such a way that there are 10 impacts upon the floor every second, with 100 ms between successive impacts (ISO 140 Parts 6 & 7). The reason for using a train of impacts instead of a single impact was to allow accurate measurements at a time when instrumentation was better suited to measuring continuous signals; nowadays it is possible to accurately measure a single impulse. The requirement for each tapping machine hammer is that the momentum of each hammer impact should represent a free-falling mass of 0.5 kg with a drop-height of 0.04 m. In practice the mass will not be free-falling, and there will be some friction losses in the guidance device used to minimize lateral motion of the hammer. The relevant ISO Standard gives tolerances on the mass of the hammer, and the velocity at impact, hence,

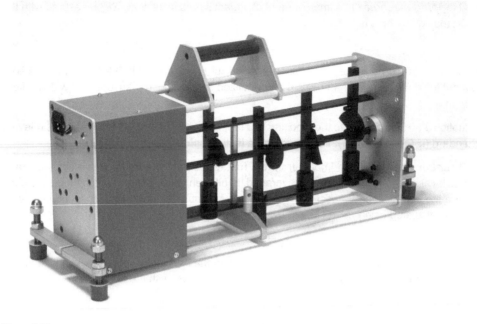

Figure 3.25

ISO tapping machine: an example of a commercially available machine. Photo provided by Norsonic.

where necessary, slightly greater drop-heights than 0.04 m can be used to compensate for friction losses.

The ISO Standards contain a thorough specification for the ISO tapping machine and its hammers because relatively small changes in the design have been found to affect its power input (e.g. see Bodlund and Jonasson, 1983; Gösele, 1956; Goydke and Fischer, 1983). Critical aspects of the design include: the distance between the hammers, the distance between the hammers and the tapping machine supports, vibration isolation of the supports, hammer mass, hammer dimensions, curvature and material of the hammer impact surface, hammer velocity at impact, time between impacts, and the time between impact and the lifting of the hammer.

3.6.3.1 Force

The ISO tapping machine applies a force due to the hammers that impact upon the surface of the floor at regular time intervals. For a free-falling mass, m, which falls from a height, h, under acceleration due to gravity, g, the kinetic energy equals the potential energy, therefore:

$$\frac{mv^2}{2} = mgh \tag{3.84}$$

from which the velocity, v_0, of the mass at impact is

$$v_0 = \sqrt{2gh} \tag{3.85}$$

and the hammer velocity at impact for the ISO tapping machine will be 0.886 m/s.

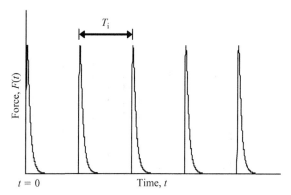

Figure 3.26

Time history of impact forces from the ISO tapping machine.

From Vér (1971) and Cremer *et al.* (1973) the force can now be estimated for impacts upon a surface that occur with a frequency, f_i, and a time period, T_i, between impacts, where $T_i = 1/f_i$. For the tapping machine, f_i is 10 Hz and T_i is 0.1 s. The impact forces are as indicated in Fig. 3.26 where the first impact is positioned at time $t = 0$. In the time domain the force, $F(t)$, is represented by the Fourier series:

$$F(t) = \sum_{n=1}^{\infty} F_n \cos(2\pi f_i n t) \tag{3.86}$$

where $n = 1, 2, 3$, etc. for the sequence of Fourier frequency components, F_n. These components are calculated using:

$$F_n = \frac{2}{T_i} \int_0^{T_i} F(t) \cos(2\pi f_i n t) dt \tag{3.87}$$

For short duration impacts it can be assumed that $\cos(2\pi f_i n t) \approx 1$, hence,

$$F_n \approx \frac{2}{T_i} \int_0^{T_i} F(t) dt \tag{3.88}$$

Over the building acoustics frequency range it is reasonable to assume that short duration impacts occur on homogeneous bare concrete floors with a thickness of at least 100 mm. We will soon look at this assumption in more detail. Returning to Eq. 3.88, the integral is solved with Newton's second law of motion, $F = ma$. This can be written as $F = m\frac{dv}{dt}$ and re-arranged to give $Fdt = mdv$, from which the integral over the time period, T_i, is equal to the change in momentum, such that,

$$\int_0^{T_i} F(t) dt = mv_{T_i} - mv_0 \tag{3.89}$$

At time, T_i, after a hammer impact, the velocity of that hammer should be zero. Therefore the magnitude of the peak force from the tapping machine, $|F_n|$, can be calculated from:

$$|F_n| \approx \frac{2}{T_i} mv_0 = 2f_i m\sqrt{2gh} \tag{3.90}$$

The force spectrum is a line spectrum as shown in Fig. 3.27 where the frequency lines occur at 10 Hz intervals. For the practical purposes of measurement and prediction, we need to use

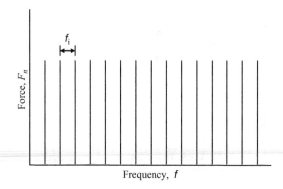

Figure 3.27

Fourier frequency components from the ISO tapping machine.

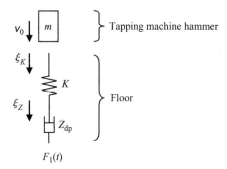

Figure 3.28

Lumped element model representing the hammer impact upon the floor using a mass–spring–dashpot system.

this line spectrum to calculate the mean-square force spectrum in one-third-octave or octave-bands. For a frequency band with a bandwidth, B, there will be B/f_i frequency lines in the band, so the mean-square force in a frequency band is

$$F^2_{rms} = \frac{|F_n|^2 B}{2f_i} \qquad (3.91)$$

where $B = 0.23f$ for one-third-octave-bands, $B = 0.707f$ for octave-bands, and f is the band centre frequency.

Hence, when it can be assumed that the impacts are of short duration, the mean-square force from the tapping machine is

$$F^2_{rms} = 3.9B \qquad (3.92)$$

We will now look at the assumption of short duration impacts. The time-history of a single force pulse from the ISO tapping machine hammer can be calculated using a lumped element model (Brunskog and Hammer, 2003; Lindblad, 1968). This model accounts for the effect of both the hammer impedance and floor impedance on the force pulse using a mass–spring–dashpot system (see Fig. 3.28). The hammer is represented as a lump mass. The floor is represented by a spring for the contact stiffness, K, in series with a dashpot damper for the floor impedance, Z_{dp}. It is assumed that both K and Z_{dp} are frequency-independent. When we assumed a short duration impact, the force was only dependent upon the hammer mass. Now we can use this

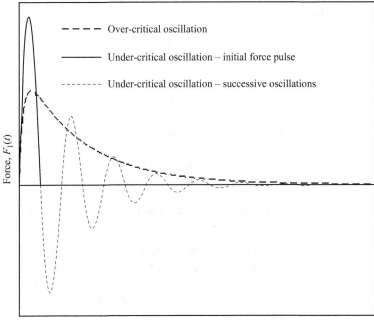

Figure 3.29

Force pulse from a single tapping machine hammer – examples of over-critical and under-critical oscillations.

more complete model to assess the effect of the contact stiffness and the floor impedance on the force applied by the hammer.

The equations of motion for the mass–spring–dashpot system in terms of the displacement, ξ, are:

$$m\frac{d^2\xi_K}{dt^2} = K(\xi_K - \xi_Z)$$ (3.93)

$$K(\xi_Z - \xi_K) = Z_{dp}\frac{d\xi_Z}{dt}$$ (3.94)

From Lindblad (1968), these equations can be solved to give the force pulse for a single hammer impact, $F_1(t)$. The resulting equations for the force pulse depend on whether the mass–spring–dashpot system gives rise to an over-critical oscillation ($Km \geq 4Z_{dp}^2$), or an under-critical oscillation ($Km < 4Z_{dp}^2$). Examples of the different force pulses due to over-critical and under-critical oscillations are shown in Fig. 3.29. For over-critical oscillations, the force pulse decays to zero and takes only positive values. After the initial force pulse of an under-critical oscillation, $F_1(t)$ oscillates with alternate positive and negative values about the zero force line.

When $Km \geq 4Z_{dp}^2$ (over-critical oscillation),

$$F_1(t) = \frac{v_0 K \exp\left(\dfrac{-Kt}{2Z_{dp}}\right) \sinh\left(t\sqrt{\left(\dfrac{K}{2Z_{dp}}\right)^2 - \dfrac{K}{m}}\right)}{\sqrt{\left(\dfrac{K}{2Z_{dp}}\right)^2 - \dfrac{K}{m}}}$$ (3.95)

When $Km < 4Z_{dp}^2$ (under-critical oscillation),

$$F_1(t) = \frac{v_0 K \exp\left(\dfrac{-Kt}{2Z_{dp}}\right) \sin\left(t\sqrt{\dfrac{K}{m} - \left(\dfrac{K}{2Z_{dp}}\right)^2}\right)}{\sqrt{\dfrac{K}{m} - \left(\dfrac{K}{2Z_{dp}}\right)^2}} \tag{3.96}$$

The magnitude of the peak force from the tapping machine, $|F_n|$, can now be calculated by taking the Fourier transform of the force pulse, $F_1(t)$, for this single hammer impact, and accounting for the impact repetition rate from the tapping machine, f_i. Equation 3.91 can then be used to calculate the mean-square force from the tapping machine in a frequency band. For the under-critical oscillations only the initial force pulse that has zero or positive force values is used to determine the force spectrum, with all subsequent values of $F_1(t)$ due to the oscillations set to zero before taking the Fourier transform.

There are two types of contact stiffness that can be used in the calculation of the force pulse. One type is for the contact stiffness of a plate material, such as concrete or chipboard, in the contact area of the hammer. The other type is for a soft floor covering, such as carpet on a heavyweight floor, where it can be assumed that the contact stiffness is only determined by the covering and is not affected by the contact stiffness of the plate underneath the soft covering. Effectively we are treating the soft floor covering or the plate material that deforms in the contact area as a linear spring.

The contact stiffness of the plate material, K, can be estimated using (Timoshenko and Goodier, 1970):

$$K = \frac{2rE}{1 - \nu^2} \tag{3.97}$$

where r is the radius of the circular contact area of the hammer with the plate, E is the Young's modulus of the plate, and ν is the Poisson's ratio of the plate.

The contact stiffness for a soft covering, K, is

$$K = \frac{E\pi r^2}{d} \tag{3.98}$$

where d is the thickness of the soft covering and E is the Young's modulus of the soft covering.

Calculation of the contact stiffness assumes a circular contact area of radius, r, for the tapping machine hammer. The hammer radius, 15 mm, can be used as an estimate because the actual contact area is not well-defined due to the spherical impact surface of the hammer, and deformation of the plate or soft floor covering upon impact.

For the floor we will assume an infinite homogeneous plate, for which the driving-point impedance, Z_{dp}, is frequency-independent and given by Eq. 2.190. It is now possible to look at the force pulse on a range of flooring materials using material properties that are based on concrete, sand-cement screed, chipboard, and oriented strand board (OSB). For a 140 mm concrete slab or a 65 mm sand-cement screed, the plate impedance and the contact stiffness are significantly higher than with 22 mm chipboard or 15 mm OSB. For the concrete slab and screed, the hammer rebounds from the plate with an under-critical oscillation and the force pulse has a short duration (see Fig. 3.30). However, for the chipboard and OSB plates there is no distinct rebound as with a concrete slab, and the pulse has a much longer duration due

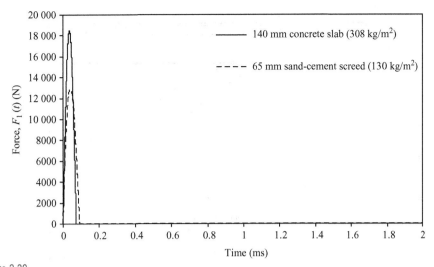

Figure 3.30

Force pulse from the tapping machine hammer on concrete and sand-cement screed plates. (NB: Only the initial part of the under-critical force pulse is shown which has zero or positive force values.)

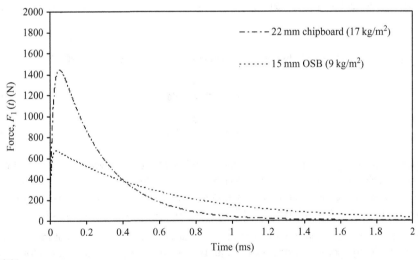

Figure 3.31

Force pulse from the tapping machine hammer on chipboard and OSB plates.

to the over-critical oscillation (see Fig. 3.31). So for common flooring materials the range of values for both the impedance and the contact stiffness has a significant effect on the shape of the force pulse, which, in turn, determines the force spectrum.

Figure 3.32 shows the force spectrum in terms of the magnitude of the peak force, $|F_n|$, calculated from the Fourier transform of the force pulses for the four different plates. We can assess the trends from the viewpoint of both increasing and decreasing frequency. With decreasing

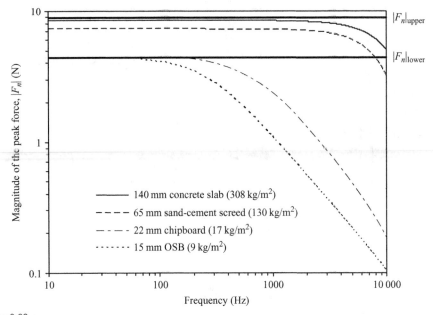

Figure 3.32

Force spectrum for the tapping machine on four different plates calculated using force pulses from the mass–spring–dashpot model.

frequency, $|F_n|$, asymptotes between two limits (Brunskog and Hammer, 2003):

$$|F_n|_{lower} = \frac{1}{T_i}mv_0 \qquad (3.99)$$

$$|F_n|_{upper} = \frac{2}{T_i}mv_0 \qquad (3.100)$$

The upper limit, $|F_n|_{upper}$, corresponds to the situation where the hammer rebounds from the floor with the velocity, v_0, and is the same as Eq. 3.90. The lower limit, $|F_n|_{lower}$, occurs when the hammer momentum is dissipated during the impact such that the hammer does not rebound. In terms of the mean-square force, the difference between the lower and upper limit is 6 dB.

The short duration pulse on the 140 mm concrete slab gives a relatively flat force spectrum across the building acoustics frequency range. Below 4000 Hz the mean-square force values are within 1 dB of the upper limit, $|F_n|_{upper}$. For concrete floor slabs of at least 100 mm thickness it is reasonable to estimate the mean-square force using Eq. 3.92 which assumes short duration impacts.

For the 22 mm chipboard and 15 mm OSB plates, the longer duration pulse means that the force spectrum is not flat and decreases above 100 Hz. In addition, the force is significantly lower than with the concrete slab, with $|F_n|$ tending towards the lower limit, $|F_n|_{lower}$, at frequencies below 100 Hz. This implies that use of Eq. 3.92 will overestimate the mean-square force for the 22 mm chipboard and 15 mm OSB plates.

With increasing frequency, there is a cut-off frequency, f_{co}, above which there is no longer a relatively flat force spectrum and the force is significantly reduced. This cut-off frequency can be calculated from (Brunskog and Hammer, 2003):

$$f_{co} = \frac{1}{2\pi}\left[\frac{K}{2Z_{dp}} - \sqrt{\left(\frac{K}{2Z_{dp}}\right)^2 - \frac{K}{m}}\right] \tag{3.101}$$

when $Km \geq 4Z_{dp}^2$ (over-critical oscillation), and

$$f_{co} = \frac{1}{2\pi}\sqrt{\frac{K}{m}} \tag{3.102}$$

when $Km < 4Z_{dp}^2$ (under-critical oscillation).

The lumped element model gives an insight into the differences between the force spectra from the tapping machine with different plate materials. This assumes that the walking surface behaves as an infinite plate with frequency-independent driving-point impedance and that the estimate of the contact stiffness is reasonable. Examples of measured driving-point impedances can be found in Section 2.8 although they are presented in terms of the driving-point mobility; these indicate when the assumption of an infinite plate is reasonable. Lightweight floors do not always have impedances that are independent of frequency because they often need joists or battens to support the plate; the impedance can vary over the surface depending on whether the hammer makes contact with the plate on top of the joists or battens, or in-between them. For lightweight floors with frequency-dependent impedance, numerical methods can be used to calculate the force spectrum (Brunskog and Hammer, 2003).

3.6.3.2 Power input

In the previous section the mean-square force was calculated using two different approaches. The first approach was purely based around the hammer momentum. This assumed a short duration impact and made no further consideration of the interaction between the hammer and the floor. This resulted in the simple expression in Eq. 3.92. The second approach took account of the interaction between the hammer and the floor by using a mass–spring–dashpot model to calculate the force pulse from a single hammer impact.

When F_{rms}^2 is calculated using the mass–spring–dashpot model the power input is given by:

$$W_{in} = F_{rms}^2 \text{Re}\left\{\frac{1}{Z_{dp}}\right\} \tag{3.103}$$

where Z_{dp} is the driving-point impedance of the floor.

When F_{rms}^2 is calculated from Eq. 3.92, the power input needs to take account of the hammer impedance, Z_h, which is in series with the driving-point impedance of the floor during the impact (Cremer et al., 1973). Hence, the power input is

$$W_{in} = F_{rms}^2 \text{Re}\left\{\frac{1}{Z_{dp} + Z_h}\right\} \tag{3.104}$$

where the impedance of the hammer is that of a lump mass, m, hence $Z_h = i\omega m$, where $m = 0.5\,\text{kg}$.

Using the infinite plate impedance, the power input is therefore:

$$W_{in} = F_{rms}^2 \frac{2.3\rho c_L h^2}{(2.3\rho c_L h^2)^2 + (\omega m)^2} \tag{3.105}$$

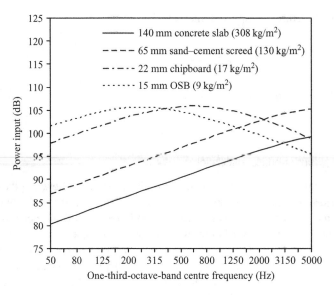

Figure 3.33

Power input for the tapping machine on four different plates calculated using force pulses from the mass–spring–dashpot model.

Figure 3.33 shows the power input calculated using force pulses from the more accurate mass–spring–dashpot model for the four different plates. The power inputs are quite different for the four plates. However, this single fact does not imply a favourable or unfavourable bias towards one type of floor by using the tapping machine; any falling mass that impacts upon a floor will have a different power input on different floors, whether it is the hammer of a tapping machine or a child jumping. When the magnitude of the hammer impedance is negligible in comparison with the driving-point impedance of the floor, the hammer impedance does not affect the power input. Hence, the power input in one-third-octave or octave-bands increases by 3 dB per doubling of the band centre frequency. This occurs for the 140 mm concrete slab and the 65 mm sand-cement screed below 1000 Hz. Above 1000 Hz, the hammer impedance starts to become significant and slightly reduces the power input in the high-frequency range. For the 22 mm chipboard and 15 mm OSB, the magnitude of the hammer impedance is significant in comparison with the plate impedance across the majority of the building acoustics frequency range. Hence, the curves do not increase by 3 dB per doubling of the band centre frequency. Above a limiting frequency, f_{limit}, the power input starts to decrease with increasing frequency. This is defined by the frequency at which the plate impedance equals the magnitude of the hammer impedance (Cremer et al., 1973),

$$f_{\text{limit}} = \frac{1}{2\pi} \frac{Z_{\text{dp}}}{m} \qquad (3.106)$$

3.6.3.3 Issues arising from the effect of the ISO tapping machine hammers

The effect of the hammer impedance on the power input has implications for the comparison of impact sound insulation on different floors. For a floor with a driving-point impedance, Z_{dp}, we can use Eq. 3.104 to calculate the ratio of the power input from the ISO tapping machine hammer, to the power input from any real impact due to a falling mass (Schultz, 1975). If the ratio of the power inputs for any type of floor is constant for each frequency band, then we can simply use the ISO tapping machine to assess the impact sound insulation against the real impact by comparing the normalized (or standardized) impact sound pressure levels for different floors. To calculate this ratio we will assume the same impact repetition rate, f_i, for the real impact as for the ISO tapping machine. This gives the ratio of the power inputs as:

$$\frac{W_{in(h)}}{W_{in(r)}} = \frac{F^2_{rms(h)}}{F^2_{rms(r)}} \frac{Re\left\{\dfrac{1}{Z_{dp} + Z_h}\right\}}{Re\left\{\dfrac{1}{Z_{dp} + Z_r}\right\}} \qquad (3.107)$$

where the subscript, h, is used for the ISO tapping machine hammer and the subscript, r, is used for the real impact.

It is reasonable to assume that footsteps are the most common source of impacts on floors; hence, we will use this source as our real impact. However, without evidence we cannot simply assume that footsteps are the most common source of annoyance when people give a subjective evaluation of the impact sound insulation. The ratio of the power inputs is the product of two components, the force ratio, $F^2_{rms(h)}/F^2_{rms(r)}$, and the impedance ratio, $Re\{1/(Z_{dp} + Z_h)\}/Re\{1/(Z_{dp} + Z_r)\}$. To appreciate the issues involved with footsteps and other real impacts it is useful if we look at these two components individually, before considering them together. For the sake of simplicity it would be convenient if both the force and impedance ratios were independent of frequency, and invariant with the type of floor surface. This would allow a simple ranking of the impact sound insulation using the normalized (or standardized) impact sound pressure levels. Unfortunately, we will soon see that there is no simple outcome.

We first look at the force ratio, $10\lg(F^2_{rms(h)}/F^2_{rms(r)})$ where the mean-square force for the real impact corresponds to the force applied by the heel of an adult female walking in a pair of high-heeled shoes as measured by Watters (1965). It has previously been noted that the force depends on the contact stiffness of the surface upon which the impact is made. The calculated force is shown in Fig. 3.34 for two different walking surfaces: hardwood floor and thin carpet on a concrete slab. The first point to note is that the force from the ISO tapping machine is higher than with footsteps. A high input force from the ISO tapping machine is used to ensure that the impact sound pressure levels are well-above the background noise for both laboratory and field measurements. The next point is that the force ratio varies with frequency. For thin carpet on a concrete slab, the force ratio rises with a particularly steep gradient above 125 Hz which indicates that the force spectrum for the ISO tapping machine hammer is significantly different to this particular footstep. This leads on to a particular issue with some soft floor coverings; non-linearity with high input forces.

As a consequence of the ISO tapping machine applying a high input force, the response of some soft floor coverings is non-linear (Lindblad, 1968, 1983). This occurs where the contact stiffness of the soft floor covering depends upon the force. Therefore, when the ISO tapping machine is used to measure the impact sound insulation of floors with soft floor coverings, the results are only representative of impacts that are the same as those from the ISO tapping machine hammer. Lindblad's (1968) investigation into non-linearity used a modified ISO tapping machine to give three different drop-heights, 40, 12.6, and 4 mm; the lower the drop-height, the lower the force. For linear systems, each successive decrease in drop-height would correspond to a 5 dB reduction in the normalized impact sound pressure level for each frequency band. This indicates whether the contact stiffness of a soft floor covering can be considered as

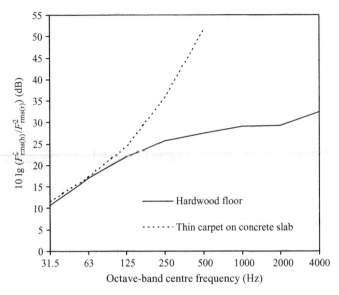

Figure 3.34

Mean-square force ratio for the ISO tapping machine relative to the force applied by the heel of an adult female walking in a pair of high-heeled shoes. Measured data are reproduced with permission from Watters (1965).

a linear or a non-linear spring. The results from two different soft floor coverings (a) and (b) on a concrete slab are shown in Fig. 3.35. Covering (a) shows a general trend of successive 5 dB reductions in the low-frequency range; the differences become larger at higher frequencies but the curve shapes are generally similar across the frequency range. Covering (b) shows distinct non-linearity. The different curve shapes indicate that the contact stiffness is effectively increasing with increasing force and behaves as a non-linear (hardening) spring.

Non-linearity complicates matters because the non-linear response of a soft floor covering may be relevant to some real impacts, but irrelevant to others. This makes it difficult to optimize and rank-order the performance of soft coverings to provide impact sound insulation against a range of impact sources (Lindblad, 1983). Another issue with soft floor coverings is that the measured improvement of impact sound insulation also depends upon the mass and contact area of the ISO tapping machine hammer (Vér, 1971); these are not necessarily representative of the mass and contact area of footsteps.

We will now look at the impedance ratio, $\mathrm{Re}\{1/(Z_{dp} + Z_h)\}/\mathrm{Re}\{1/(Z_{dp} + Z_r)\}$. This ratio will only be constant if the hammer impedance, Z_h, is equal to the impedance of the real impact source, Z_r for footsteps. The impedance of the human walker results from the combination of footwear and the mechanics of the human body. For footsteps we are interested in the impedance of the heel of the foot or shoe as it lands upon the floor. Watters (1965) and Scholl (2001) carried out heel-impedance measurements as part of their investigations into the tapping machine. Watters (1965) measured the heel-impedance of an adult female (weight of at least 45 kg) in a pair of high-heeled shoes. Scholl (2001) measured the heel-impedance of two adult males when barefoot and in flat-heeled shoes. These measurements are shown in Fig. 3.36 when the male or female subjects are standing. The measurements indicate a wide range of heel-impedances for different walkers and different footwear, particularly in the low- and mid-frequency ranges.

(a)

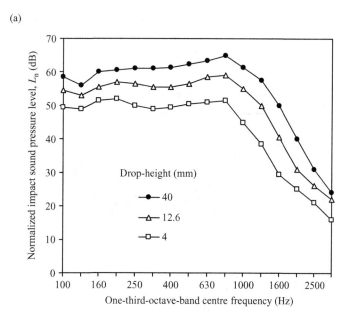

(b)

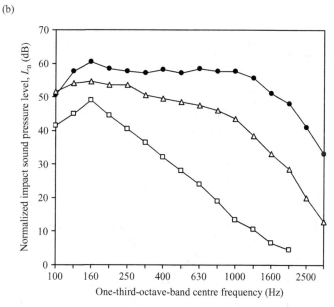

Figure 3.35

Normalized impact sound pressure levels measured using different drop-heights (40, 12.6, and 4 mm) on a modified ISO tapping machine. (a) Floor: 0.8 mm vinyl glued to a resilient backing of 2 mm vinyl foam on 160 mm concrete floor slab. (b) Floor: 1.1 mm vinyl glued to a resilient backing of 2 mm felt on 160 mm concrete floor slab. Measured data are reproduced with permission from Lindblad (1968).

In the same way that the tapping machine hammer can be represented as a lump mass, we can represent the heel-impedance using lump elements of mass, $Z = i\omega m$, and stiffness, $Z = k/i\omega$ (Watters, 1965). Figure 3.37 clearly shows that the impedance of a 0.5 kg mass used for the ISO tapping machine hammer does not represent the range of measured

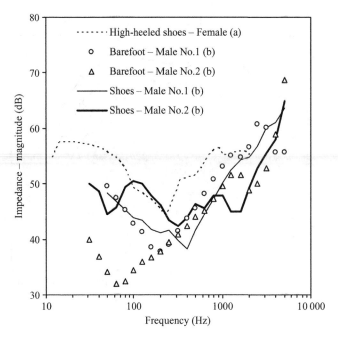

Figure 3.36

Measured heel-impedances for male and female walkers with different footwear. Measured data are reproduced with permission from (a) Watters (1965) and (b) Scholl (2001).

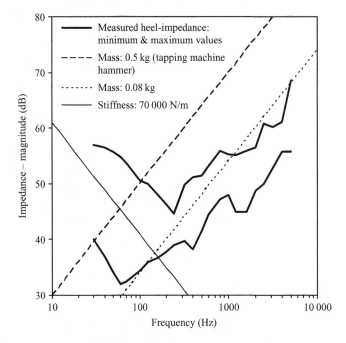

Figure 3.37

Range of measured heel-impedances from Fig. 3.36 compared with mass and stiffness impedances. Measured data are reproduced with permission from Watters (1965) and Scholl (2001).

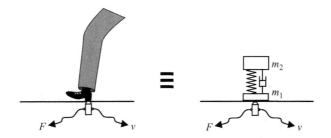

Figure 3.38

Illustration of the mass–spring–mass system used to represent the heel-impedance 'seen' by the floor.

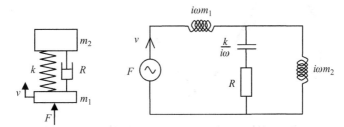

Figure 3.39

Equivalent electrical circuit used to calculate the input impedance of a mass–spring–mass system representing the heel-impedance of a walker in shoes.

heel-impedances. Therefore for footsteps we can say that $Z_h \neq Z_r$. However, the heel-impedance can be approximately represented by a stiffness of 70 000 N/m in the low-frequency range, and a mass of 0.08 kg in the mid-frequency range; these values are purely indicative because of the wide range in measured heel-impedances. These lump elements illustrate the general trend that is apparent in the individual measurements (Fig. 3.36) where the impedance in terms of $20 \lg|Z|$ initially decreases by 6 dB per octave to a minimum value before increasing by 6 dB per octave.

These lump elements of mass and stiffness can be used to create a mass–spring–mass system that is representative of the heel-impedance 'seen' by the floor as shown in Fig. 3.38 (Scholl, 2001; Warnock, 1983). The input impedance at the heel is calculated using an equivalent electrical circuit (see Fig. 3.39) by applying a force to the mass, m_1, that impacts upon the floor,

$$Z = \frac{F}{v} = i\omega m_1 + \frac{i\omega m_2 \left(\frac{k}{i\omega} + R\right)}{i\omega m_2 + \frac{k}{i\omega} + R} \tag{3.108}$$

A minimum value in the input impedance occurs at the resonance frequency of this mass–spring–mass system; we need this resonance to correspond to the minimum value that occurs in the heel-impedance spectrum. However, the spring needs to be highly damped (R term) with an equivalent loss factor of 0.8 to avoid very low input impedance at this resonance frequency. The input impedance below the resonance frequency is primarily determined by the spring stiffness although there will also be a peak due to an anti-resonance of the mass–spring–mass system. Above the resonance frequency the input impedance is primarily determined by the mass, m_1, that impacts upon the floor, hence we assign a value of 0.08 kg to m_1. Mass m_2

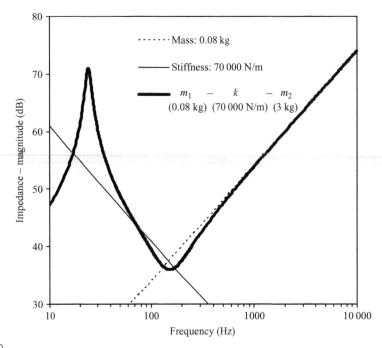

Figure 3.40

Impedance of a mass–spring–mass system used to represent the heel-impedance.

is then chosen so that the mass–spring–mass resonance frequency occurs close to the point where the impedance of the 0.08 kg mass equals the impedance of the 70 000 N/m stiffness. By using a value of 3 kg for m_2, the mass–spring–mass resonance frequency is 151 Hz and the peak due to the anti-resonance is shifted well-below 50 Hz to a frequency of 24 Hz. The impedance for this mass–spring–mass system is shown in Fig. 3.40. It is clear that if the mass of the tapping machine hammer were to be replaced by a hammer that was effectively a mass–spring–mass system, then an attempt could be made to simulate an idealized heel-impedance. Warnock (1983, 2000) effectively implemented a hammer based on a mass–spring–mass system in a machine that was used to simulate the force pulse generated by footsteps. As with real footsteps, the force levels from this machine were low, which meant that background noise often affected measurement of the impact sound pressure level.

We now have a mass–spring–mass system for a heel-impedance that can be used to look at the impedance ratio. This ratio is shown in Fig. 3.41 for four different plates commonly used for the upper surface of a floor or floating floor. The concrete and screed plates have relatively high impedances, hence the ratio is approximately 0 dB across the building acoustics frequency range because Z_h and Z_r have negligible effect. The chipboard and OSB plates have much lower impedances, so the frequency-dependent values of Z_h and Z_r have a significant effect, and cause the ratio to vary with frequency. For many floor surfaces, Z_{dp} will also vary with frequency in the low- and mid-frequency ranges where the assumption of an infinite plate is not appropriate (Section 2.8.3). This will cause the impedance ratio to become a more complex function of frequency that is specific to particular types of floors with specific dimensions.

To summarize, we have seen that both the force ratio and the impedance ratio can be quite complex functions of frequency with large variations that depend upon the type of floor surface.

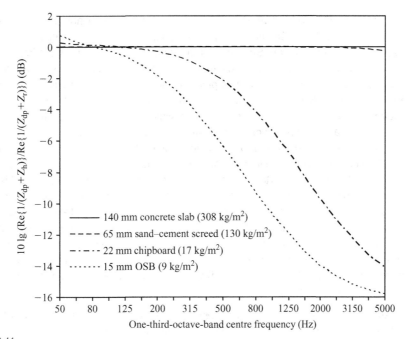

Figure 3.41

Impedance ratio for four different plates.

Hence, the ratio of the power inputs will also be a relatively complex function of frequency. This allows a conclusion to be drawn relating to one specific impact source, footsteps. The implication is that the normalized (or standardized) impact sound pressure level in individual frequency bands will not always correctly rank order all types of floor (with or without floor coverings) in terms of their impact sound insulation against footsteps. However, by calculating a single-number quantity from the individual frequency bands we will soon see that good correlation can be achieved with subjective assessment of impact sound insulation in dwellings. Similarly, the improvement of impact sound insulation in individual frequency bands will not always correctly rank order all types of floor covering in terms of their impact sound insulation against footsteps.

The advantages and disadvantages of the ISO tapping machine were known before it was adopted in the Standards. Subsequent studies to investigate potential improvements have generally produced one of the following options: (1) change the ISO tapping machine to simulate one particular type of impact, such as footsteps or a heavy impact from a child jumping, or, (2) keep the ISO tapping machine but use a rating system that combines the frequency band levels to produce a single-number quantity that correlates well with the subjective evaluation of impact sound insulation.

3.6.3.4 Modifying the ISO tapping machine

Proposals to change the ISO tapping machine have generally been to make modifications to the hammer impedance so that it is more representative of footsteps, or to change the impact source completely. Regarding the former, it has been suggested that a resilient material (acting as a spring) could be inserted between the face of the metal hammer and the floor (e.g. see Gerretsen, 1976; Scholl, 2001; Schultz, 1975). This modification to the hammer impedance has been incorporated into the relevant Standard for laboratory measurement of the improvement of impact sound insulation due to floor coverings on lightweight floors (ISO 140 Part 11). Springs can either be attached to the ISO tapping machine hammers or a resilient material can be placed on the floor underneath the hammers. This approach is based on representing the heel-impedance by a mass–spring–mass system. However, it assumes that the floor covering is a rigid plate with a sufficiently high mass that the lower mass in the mass–spring–mass system can be omitted, i.e. m_1 in Fig. 3.39 (Scholl, 2001). The resilient material significantly reduces the impact sound pressure level in the mid- and high-frequency ranges, therefore low background noise levels are important to ensure accurate measurements. A practical issue in using a resilient material is that its dynamic properties must not vary with time (either during the test or after repeated use) and temperature. An advantage of the steel hammers of the ISO tapping machine hammers is that they are generally hard wearing, although their diameter and curvature must be checked periodically. Note that the proposals described above to use a resilient material are different to the rubber coating that was previously in old versions of the measurement Standard (now superseded) which was to prevent the hammers damaging fragile floor coverings (ISO 140 Part 6:1978).

Other investigations have looked at completely different impact sources to the ISO tapping machine. Based on the work by Watters (1965), a mechanical machine was built to simulate the impact made by the heel of a shoe with interchangeable shoe types, although unwanted mechanical noise and vibration limited its application (Josse, 1970). To represent heavy, soft impacts on floors, research in Japan has investigated the use of rubber balls and tyres; these are discussed in Section 3.6.4.

3.6.3.5 Rating systems for impact sound insulation

The alternative to changing the ISO tapping machine has been to make alterations to the rating system that combines levels from individual frequency bands to produce a single-number quantity. The aim being that it should correlate well with subjective evaluation of impact sound insulation. This approach should not be viewed as 'correcting' features associated with the tapping machine such as the effects of hammer impedance and non-linearity with some soft floor coverings. However, rating systems that place emphasis on the low-frequency range may fortuitously avoid some of the problems with non-linearity that tend to be more apparent in the mid- and high-frequency ranges.

A rating system can form a relationship between objective measurements with a standardized impact source, and the subjective evaluation of impact sound insulation. We expect this to be a complex relationship, partly because of the variety of structure-borne sound sources in dwellings that range from footsteps to washing machines, but also due to the complexity in the subjective evaluation of impact sound. It is reasonable to assume that there will be a relationship between acceptable impact sound pressure levels and background noise due to a masking effect. However, there are also indications that impact noise in dwellings tends to disturb the occupant by startling them, which becomes important for sleep disturbance (Raw and Oseland, 1991). Although footsteps on floors and stairs are often listed as an impact source that is heard (Grimwood, 1997), there is uncertainty as to whether annoyance from footsteps is linked to the loudness, or to the fact that they are detectable (Gerretsen, 1976). It is also reasonable to assume that average subjective evaluations will vary between different countries and cultures. We are therefore expecting rather a lot from a rating system that essentially relates impact sound pressure levels using the ISO tapping machine to a subjective rating via a 'black box' of complex relationships.

There have been several proposals for rating systems (e.g. see Choudhury and Bhandari, 1972; Fasold, 1965; Gerretsen, 1976). Some countries have implemented their own rating systems, but here we will only discuss the rating method described in ISO 717 Part 2. This Standard describes calculation of the single-number quantities, $L_{n,w}$, $L'_{n,w}$, and $L'_{nT,w}$ along with spectrum adaptation terms, C_I, $C_{I,50-2500}$ for one-third-octave-bands, and $C_{I,63-2000}$ for octave-bands.

We start by looking at the link between objective and subjective evaluation of impact sound insulation. Field measurements by Bodlund (1985) in Sweden were used to compare different rating methods with subjective evaluations of the impact sound insulation in dwellings. This data set comprised 14 different groups of housing with 22 different construction types (concrete and timber joist floors with a variety of floor coverings).

The subjective rating used a seven-grade scale, where one was "Quite unsatisfactory" and seven was "Quite satisfactory". For each single block of multi-storey apartment houses, or residential block with a particular construction, the mean subjective score and the mean single-number quantity for the impact sound insulation was used rather than attempting to relate each individual score from an interviewee to a specific sound insulation measurement. We have to accept that it is difficult to relate objective and subjective ratings on an individual basis, but that relationships can be found by grouping the objective and subjective ratings to calculate mean values.

The correlation coefficient, r, between the mean weighted normalized impact sound pressure level, $<L'_{n,w}>$, and the mean subjective score was found to be 75% with the following straight-line relationship,

$$\langle L'_{n,w} \rangle = 80.6 - 5.48 X \tag{3.109}$$

where X is the mean subjective score (1–7).

Hence $L'_{n,w}$ (and also $L'_{nT,w}$) can be considered as adequate descriptors when we need to define performance standards for impact sound insulation in the field. However, to improve the correlation, Bodlund proposed an alternative to the rating curve in ISO 717 Part 2; this consisted of a straight line between the 50 and 1000 Hz one-third-octave-bands with a positive gradient of 1 dB per one-third-octave-band. Bodlund's rating curve is significantly different to the ISO rating curve that is used to calculate $L_{n,w}$, $L'_{n,w}$, or $L'_{nT,w}$ as can be seen from Fig. 3.42. The ISO rating curve tends to emphasize the insulation in the mid- and high-frequency ranges. Bodlund placed greater emphasis on the low-frequency range, but used the procedure described in ISO 717 Part 2 to shift the reference curve to determine a new single-number quantity, denoted here as $L'_{B,w}$. This emphasis on low frequencies is necessary because of the low-frequency content of common impacts such as footsteps.

Footsteps on floors that have a soft floor covering, and/or a floating floor tend to generate the highest sound pressure levels in the low-frequency range (Bodlund, 1985; Warnock, 2000). Although there are wide variations between different walkers with different footwear (refer back to Fig. 3.23), the general trend from a male walker in shoes with hard-heels can be seen in measurements from Warnock (2000). These data were collected in the laboratory with one

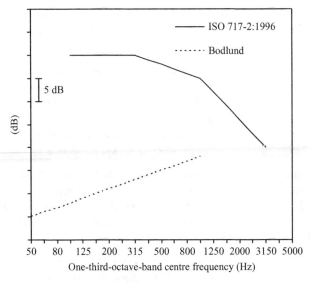

Figure 3.42

Rating curves used to calculate single-number quantities for impact sound insulation. Permission to reproduce extract from ISO 717-2 is granted by BSI on behalf of ISO.

male walker in one pair of shoes walking on different floor constructions. The maximum time-weighted sound levels generated by the walker in the receiving room below are shown in Fig. 3.43. Without a soft floor covering or a floating floor, the timber floors gave significantly higher levels than the concrete slab below 100 Hz. For both timber and concrete floors with a soft floor covering or a floating floor, the timber floors had significantly higher levels than the concrete slab below 50 Hz. Above 50 Hz there was a wide range of sound pressure levels from the different floor constructions and it is difficult to identify a trend.

Returning to Bodlund's work, the mean single-number quantity, $\langle L'_{B,w} \rangle$, calculated using Bodlund's rating curve which emphasized the low-frequency range showed stronger correlation with the mean subjective score ($r = 87\%$) than $L'_{n,w}$. The results are shown in Fig. 3.44 and give the following straight-line relationship,

$$\langle L'_{B,w} \rangle = 86.3 - 5.53\,X \tag{3.110}$$

Bodlund found that when the mean subjective score was less than 4.4, at least 20% of the interviewees rated the performance with a score less than 3. A mean subjective score of 4.4 corresponded to 51% of interviewees rating the performance with a score higher than 4. Bodlund proposed that mean subjective scores less than 4.4 should be deemed as unsatisfactory. Hence the straight-line relationships described above can be used to help identify suitable performance standards for impact sound insulation; although consideration should be given as to whether subjective scores from one country (in this case, Sweden) can be applied to other countries.

ISO 717 Part 2 does not directly implement Bodlund's rating system; however, it can be used to assess the single-number quantities that are defined in this Standard. ISO 717 Part 2 defines spectrum adaptation terms, C_I, or $C_{I,50-2500}$ such that when they are added to $L_{n,w}$, $L'_{n,w}$, or $L'_{nT,w}$, the resulting single-number quantity is equal to the energetic sum of the one-third-octave-band

(a) Without a soft floor covering or floating floor

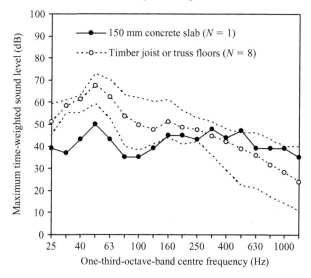

(b) With a soft floor covering or floating floor

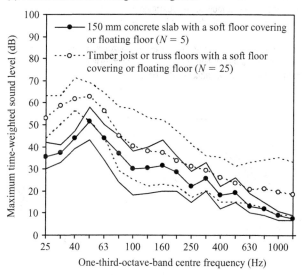

Figure 3.43

Maximum time-weighted sound levels (time constant, $\tau = 35\,\text{ms}$) in the receiving room from a male (approximate weight of 83 kg) walking in leather soled shoes with a hard rubber tip on the heels. Measured data are shown for floors with and without a soft floor covering or floating floor. Each graph shows the average of N measurements. When $N > 1$, the minimum and maximum levels from the set of measurements are shown using the same line style (solid or dotted lines) as the average values but without symbols. Measured data are reproduced with permission from Warnock (2000) and the National Research Council of Canada.

values (L_n, L'_n, or L'_{nT}) minus 15 dB. Hagberg (1996) has assessed this use of the spectrum adaptation term by using Bodlund's rating system; this was done on the basis that if an alternative single-number quantity, such as $L'_{n,w} + C_{I,50-2500}$, has a strong correlation with $L'_{B,w}$, then it is reasonable to assume that it will also have a strong correlation with subjective evaluation.

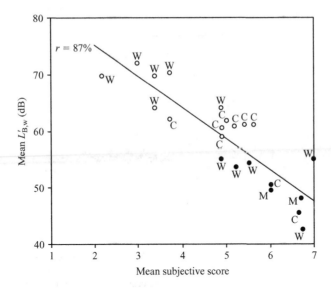

Figure 3.44

Correlation between objective and subjective assessment of the impact sound insulation, where the subjective score and $L'_{B,w}$ are both mean values for a single block of multi-storey apartment houses or a residential block. The subjective rating used a seven-grade scale where 1 = Quite unsatisfactory and 7 = Quite satisfactory. Data are reproduced with permission from Bodlund (1985).
Legend: o, separating floors in multi-storey apartment houses; •, horizontal impact sound insulation for floors in attached houses and multi-storey apartment houses; W, timber joist floor; C, concrete floor; M, mixed floor structures (concrete ground floor and lightweight second floor in two-storey attached dwellings).

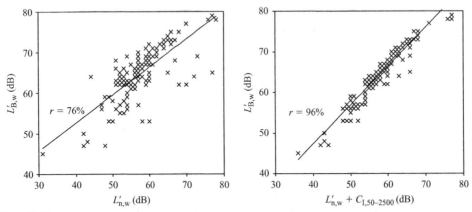

Figure 3.45

Correlation between $L'_{n,w}$ and $L'_{n,w} + C_{I,50\text{-}2500}$ with Bodlund's single-number quantity, $L'_{B,w}$. Data are reproduced with permission from Hagberg (1996).

Hagberg used 146 measurements to investigate the correlation between $L'_{n,w}$ and $L'_{B,w}$, as well as $L'_{n,w} + C_{I,50\text{-}2500}$ and $L'_{B,w}$; the results are shown in Fig. 3.45. For the straight-line relationship between $L'_{n,w}$ and $L'_{B,w}$ the correlation coefficient, r, is 76%. There is significantly higher correlation ($r = 96\%$) between $L'_{n,w} + C_{I,50\text{-}2500}$ and $L'_{B,w}$ due to emphasis on the low-frequency range. The correlation between $L'_{n,w} + C_I$ and $L'_{B,w}$ is slightly lower ($r = 90\%$) because $L'_{n,w} + C_I$ does not use one-third-octave-bands below 100 Hz.

The average value of $C_{I,50-2500}$ is -3 dB for concrete floors with a typical range of -11 to $+1$ dB, compared to an average of $+2.5$ dB for timber floors with a typical range of -2 to $+13$ dB (Rindel and Rasmussen, 1996). Rindel and Rasmussen note that there are potential problems in using high negative values of the spectrum adaptation term. These typically occur with bare concrete floors, or concrete floors with non-resilient floor coverings for which $L'_{n,w} + C_{I,50-2500}$ will not adequately account for the relatively high impact sound pressure levels that occur at high frequencies. However $L'_{n,w} + C_{I,50-2500}$ can still be used to make a stronger link between the objective and subjective rating of any floor that has a resilient floor covering or floating floor.

3.6.3.6 *Concluding discussion*

Despite all the criticisms that can be (and have been) made about the ISO tapping machine over the years, its use in building regulations has certainly improved the impact sound insulation of the building stock and identified constructions with poor impact sound insulation. From a regulatory point of view, there are pragmatic reasons to maintain the use of the ISO tapping machine and rely on a rating system to make the link between the subjective and objective ranking of impact sound insulation in dwellings. This approach ensures that any historical database of measurements does not become redundant because it allows measurement data to be re-processed with any new rating procedure. The ISO tapping machine also allows accurate measurements in the field by producing sound pressure levels that are usually well-above background noise levels. This is essential for regulations that require field tests to demonstrate compliance with a performance standard. For regulations on impact sound insulation in dwellings the choice of single-number quantity depends on what the regulation is aiming to achieve and how the regulation is implemented and enforced. As with airborne sound insulation this means that different countries may need to use different single-number quantities.

For engineers involved in floor design and the measurement of impact sound insulation (as well as manufacturers designing flooring elements) it is necessary to be aware of what can, and what cannot be inferred from measurements using the ISO tapping machine. It is therefore useful to have a general understanding of the tapping machine and the background behind the rating system.

Regulations often specify performance standards to be achieved in field measurements using the ISO tapping machine. In this case the situation is relatively straightforward. Laboratory measurements using the ISO tapping machine can be used to aid design decisions and can be incorporated into models that include flanking transmission for impact sound insulation (see Section 5.4.2). However, the situation can be different with bespoke designs; for example when establishing a suitable level of impact sound insulation below a dance studio or a sports hall in a multi-storey building. The existence of the ISO tapping machine and the ISO 717 Part 2 rating system make it convenient to use this source to specify the required level of impact sound insulation and to demonstrate compliance with measurements in the finished building. This may not always be appropriate. When assessing the ability of different floors to provide insulation against specific impact sources by using the ISO tapping machine, it is necessary to be aware of its limitations. The two main issues are the effect of the hammer impedance on the power input in comparison with the impedance of the real impact source, and the existence of non-linearity with some soft floor coverings. Over the years, some floating floors and floor coverings have inevitably been designed to take advantage of the physics of the ISO tapping machine, rather than to attenuate real impact sources. In some cases it can be beneficial to take laboratory measurements of the impact sound insulation with more than one impact source, for example with the ISO tapping machine and a heavy impact source such as the ISO rubber ball.

3.6.4 Heavy impact sources

Impacts from people on floors can generally be categorized as either light, hard impacts (e.g. footsteps in hard heeled shoes), or heavy, soft impacts (e.g. children running and jumping, adults exercising, footsteps in bare feet). We have previously discussed the fact that the ISO tapping machine does not simulate any specific impact source, but that the ISO rating system provides a link between the subjective and objective rating of impact sound insulation in dwellings. In previous surveys of impact sound insulation in dwellings (e.g. Bodlund, 1985) it is reasonable to assume that both light and heavy impacts occurred on a daily basis, so the subjective assessments are likely to have considered both types of impact. Hence, by using the ISO tapping machine and a rating system that gives a single-number quantity that correlates well with the subjective assessment, then to some unknown extent, both light and heavy impacts are taken into account. However, an issue arises purely because of this unknown extent, and the fact that the ISO tapping machine is not well-suited to ranking the impact sound insulation of floors against specific heavy impacts. In countries such as Japan, hard-heeled shoes are not worn in dwellings and the majority of dwellings are of lightweight construction. Heavy, soft impacts on lightweight timber or steel frame floors tend to give rise to higher impact sound pressure levels than heavyweight concrete floors; although there are inevitably exceptions due to the many different types of lightweight floor construction. To allow measurement of heavy, soft impacts, Japan has developed two impact sources, these are the rubber ball and the bang machine (see Fig. 3.46).

The bang machine consists of a tyre dropped from a height of 0.9 m. The test method and the bang machine specification are described in Japanese Industrial Standard JIS A 1418-2. It is well-suited to laboratory measurements, but less suitable for field tests as it produces heavy blows that could cause minor damage to decorated dwellings. In addition, it is less convenient to transport than the ISO tapping machine.

During the development of the rubber ball, different versions were produced to modify the characteristic of the impact force (Tachibana et al., 1998). The final specification for the ISO rubber ball is described in the relevant Standard (ISO 140 Part 11). The ISO rubber ball is a hollow sphere of 30 mm thick silicone rubber with an outer diameter of 180 mm. It has a weight of approximately 2.5 kg and is dropped from a height of 1 m above the floor, measured from the lower surface of the ball.

During the development of heavy impact sources, there was a wide range of equipment that could accurately measure peak, impulse, or maximum time-weighted sound levels. This is in contrast to the early development period for the tapping machine when it was not possible to accurately measure impulses (Schultz, 1981). This led to the design of a tapping machine that produced a continuous signal from a train of hammer impacts. However, when people provide a subjective assessment of footsteps and assess the ability of background noise to mask the noise from footsteps, the A-weighted peak, impulse ($\tau = 35\,\text{ms}$), or maximum time-weighted (Fast, $\tau = 125\,\text{ms}$) sound level generally gives better correlations than the equivalent continuous sound level (Ford and Warnock, 1974; Hamme, 1965; Olynyk and Northwood,

Figure 3.46

Heavy impact sources. Left: Rubber ball – manually dropped from a height of 1 m. Right: Bang machine. Photo provided by Dr J. Yoshimura at the Kobayasi Institute of Physical Research, Japan.

1965, 1968; Warnock, 1983). Nowadays the maximum time-weighted sound level, peak sound level, and impulse sound levels are all clearly defined as different descriptors (IEC 61672 Part 1); however, there is sometimes ambiguity about their usage in the past.

When a multi-modal system (e.g. a room) is excited by an impulse, the first peak (or trough) in the instantaneous measured response will not necessarily have the largest magnitude (Brüel, 1987). For this reason the maximum time-weighted sound level is determined during a time interval starting from just before the impact, to a time when the response has decayed to a negligible level. For each impulse that is produced from a single drop of the ISO rubber ball, the maximum time-weighted (Fast) sound level, L_{Fmax}, is measured in the receiving room. This is averaged for a number of excitation positions and stationary microphone positions to give the impact sound pressure level, $L_{i,Fmax}$ (ISO 140 Part 11). Some proposals for short test methods to measure the impact sound pressure level from heavy impact sources have used a single stationary microphone position, often near the centre of the floor at mid-height in the room; however, measured data indicates significant spatial variation in the sound field in the low- and mid-frequency ranges (Broch, 1983; Warnock, 2000). It is therefore necessary to measure at several different microphone positions to calculate the spatial average level.

Unlike the impact sound pressure level measured with the ISO tapping machine, the level measured with the rubber ball is not normalized to the absorption area, or standardized to the

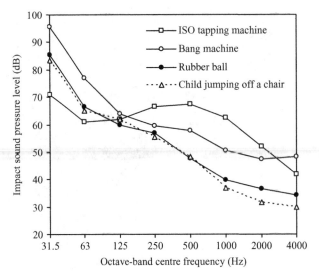

Figure 3.47

Impact sound pressure levels in a receiving room underneath a timber floor due to the ISO tapping machine, bang machine, rubber ball, and a child jumping off a chair. Measured data are reproduced with permission from Tachibana *et al.* (1998).

reverberation time of the receiving room (ISO 140 Part 11). Normalization or standardization is essential for the comparison of tests made with the ISO tapping machine, and to allow calculation of the impact sound pressure level in different rooms from laboratory test data. However, peak sound levels from an individual impact tend not to be significantly affected by the absorption area or reverberation time of the receiving room (Hamme, 1965, Ford and Warnock, 1974, Schultz, 1975). The same tendency has been assumed for maximum time-weighted (Fast) levels. Hence, normalization or standardization is not carried out. Another factor is that the reverberation times in laboratory receiving rooms are not excessively long or short; typically being between 1 and 2 s (ISO 140 Part 1) which also minimizes variation between laboratories.

Figure 3.47 allows comparison of the impact sound pressure levels from a timber floor with the ISO tapping machine, bang machine, rubber ball, and a child jumping off a chair (Tachibana *et al.*, 1998). In a similar way to the ISO tapping machine, the bang machine and the rubber ball do not simulate one specific type of heavy impact, although the rubber ball can give similar impact sound pressure levels to a child jumping off a chair. With heavy impacts on lightweight floors, the impact sound pressure levels are usually highest in frequency bands below 100 Hz. For this particular floor, the spectrum shape for the ISO tapping machine is distinctly different to the spectrum for heavy impacts. When trying to compare different floor constructions in terms of their impact sound insulation specifically against heavy, soft impacts it will be more reliable to use a heavy impact source.

Heavy impact sources are useful measurement tools for lightweight floors. However, because of the complexity in determining their input force, and the use of a maximum time-weighted sound level, there is no simple method for the prediction of the impact sound pressure level as there is with the ISO tapping machine.

3.6.5 Other measurement issues

Although most issues relating to the ISO tapping machine have been covered, there are some specific practical issues, such as the effect of time dependency on the impact sound pressure level and the effect of dust and dirt under the tapping machine hammers. These are discussed here along with more general aspects such as the size of the test specimen and static load.

3.6.5.1 Background noise correction

See discussion in Section 3.5.4.1.

3.6.5.2 Converting to octave-bands

Impact sound insulation descriptors in one-third-octave-bands can be converted to octave-bands using the three one-third-octave-bands that form the octave-band,

$$X_{OB} = 10 \lg \left(\sum_{n=1}^{3} 10^{X_{TOB,n}/10} \right) \tag{3.111}$$

where X represents L_n, L_{nT}, etc.

The improvement of impact sound insulation is converted using:

$$\Delta L_{OB} = -10 \lg \left(\frac{1}{3} \sum_{n=1}^{3} 10^{-\Delta L_{TOB,n}/10} \right) \tag{3.112}$$

3.6.5.3 Time dependency

After the tapping machine is switched on, the physical properties of some floor surfaces and floor coverings change with time. It is usually a matter of minutes before the properties stabilize, although in some cases, a steady-state is never reached and the sound pressure levels continue to vary with time. The relevant Standards (ISO 140 Parts 6 & 7) require the sound pressure level to have reached a steady-state before measurements are taken; and if a steady-state is not reached, then an appropriate measurement period should be established.

Time dependency occurs with some soft floor coverings. During each impact the tapping machine hammers can increase or decrease the contact stiffness and damping properties of the soft floor covering directly underneath the hammers. Hence the power input from the tapping machine changes with time due to variation in the physical properties of the floor covering. This can occur with materials such as carpet (Michelsen, 1982) or felt-backed PVC (Bodlund and Jonasson, 1983). An example is shown in Fig. 3.48 for carpet on a concrete slab where the sound pressure level was measured using successive 5 s samples (Michelsen, 1982). After the tapping machine was switched on, the sound pressure levels initially increase with time in the low- and mid-frequency range before stabilizing after ≈4 min.

Time dependency also occurs when the tapping machine causes damage to the surface of the floor during the test. This can occur with plate materials where the hammers cause indentations, or where a fragile surface finish breaks up due to the force of the impacts. Fournier and Val (1963) quote an example of a concrete floor with a fragile crust on the surface. The tapping machine hammers damaged the crust of the floor causing it to crack and break over a period

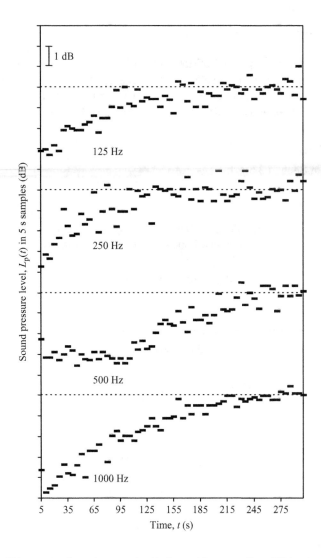

Figure 3.48

Variation of the one-third-octave-band sound pressure level in the receiving room with an ISO tapping machine on top of a carpet on a concrete floor slab. Consecutive samples were measured with an integration time of 5 s, starting at the point that the tapping machine was switched on. Measured data are reproduced with permission from Michelsen (1982).

of a few minutes. Significant damage from the tapping machine is less common when building materials have been designed to pass durability tests. However, the floor surface should be checked before and after the measurement for signs of damage that could have affected the results. In addition the sound pressure level shown on the analyser should be monitored during the test for any signs of a change.

Another factor that can vary the impact sound insulation over longer periods of time is the floor temperature; this is relevant when floors have an underfloor heating system. A significant effect can occur with soft floor coverings comprising foams and plastic coverings (Bodlund and Jonasson, 1983) where an increase in the floor temperature changes the stiffness and damping properties of the covering. This results in a change to the impact sound pressure level.

3.6.5.4 Dust, dirt, and drying time

Building work is rarely a clean process, and impact sound insulation tests in the field are not always carried out before the floors have been swept of wood shavings, sand, etc. This is relevant to measurements with the ISO tapping machine because excessive amounts of dust and dirt underneath the hammers can affect the results. The preceding discussion on the tapping machine indicated that the contact area between the hammer and the floor, as well as the contact stiffness affects the force pulse delivered by the hammer. Hence, dust and dirt can change the power input from the tapping machine into the floor. Once the tapping machine is switched on, any effect will initially change with time as the floor vibration causes the dust and dirt to move around. An indication of the effect of dust and dirt on two different floors is shown in Fig. 3.49. At the point that the tapping machine was switched on, the layer of dust and dirt was approximately 2 mm deep. Its effect is to reduce the sound pressure level in the mid- and high-frequency range. The effect on the single-number quantity can be significant with some floors, such as the concrete slab, but negligible for others.

As with airborne sound insulation, the impact sound insulation can vary depending on how much the test element has dried out. For a walking surface such as a floating floor screed, the contact stiffness changes as it dries out; this affects the power input from the ISO tapping machine. It is also worth noting that the stiffness and damping properties of adhesive used to fix floor tiles or other coverings can also change as it dries.

3.6.5.5 Size of test specimen

With soft floor coverings, such as carpet or vinyl flooring, it is appropriate to take measurements using small samples that are large enough to support the ISO tapping machine (ISO 140 Parts 8 & 11). However, the results can differ depending on whether the covering is loose-laid or if a fixing adhesive is used (Bodlund and Jonasson, 1983).

Floating floors need to cover the complete surface of the reference floor (ISO 140 Parts 8 & 11). The intention is that the measurement should only consider structure-borne sound transmission from the tapping machine into the reference floor via the floating floor. If a small area of floating floor is used, unwanted flanking transmission can occur via the uncovered area of the reference floor (lightweight or heavyweight). The floating floor that is excited by the tapping machine not only transmits structure-borne sound via the resilient material into the reference floor but also radiates sound into the source room (i.e. the room containing the tapping machine) which, in turn, excites the uncovered area of the reference floor. Using only a small area of floating floor tends to reduce the measured improvement in the sound pressure level in the mid- and high-frequency ranges. A full-size floating floor is also necessary to ensure that the driving-point mobility (which determines the power input from the tapping machine) is representative of the full-size floor. This may not occur with small areas of flooring that have a much lower modal density than the full-size floating floor.

3.6.5.6 Static load

Floating floors that provide impact sound insulation usually incorporate a resilient material. Therefore the static load that occurs *in situ* (such as from furniture or office equipment) can significantly alter the performance of the floating floor. This is usually more relevant to lightweight floating floors than heavier ones such as screeds.

(a) 140 mm concrete floor slab

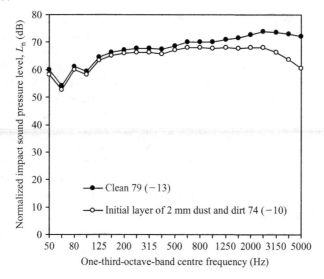

(b) Timber floating floor on a 140 mm concrete floor slab

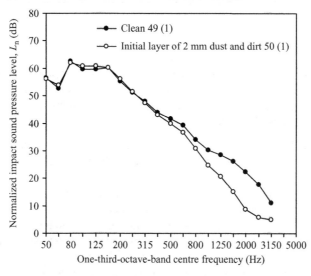

Figure 3.49

Effect of dust and dirt on the impact sound insulation measured with the ISO tapping machine. The rating for the single-number-quantity (ISO 717 Part 2) is shown as $L_{n,w}(C_I)$ on the legend. Measured data from Hopkins are reproduced with permission from BRE Trust.

In current measurement Standards, simulation of normal furnishing is defined by a uniformly distributed load of 20 to 25 kg/m² (ISO 140 Parts 8 & 11). Note that in a previous version of the Standard (now superseded), a loading of 100 kg/m² was recommended (ISO 140 Part 8:1978). On lightweight floating floors, 20 to 25 kg/m² typically reduces ΔL_w by up to 2 dB, whereas 100 kg/m² can reduce ΔL_w by up to 6 dB. It is therefore important to carry out impact sound insulation tests with the static load relevant to the end use.

A static load is most commonly provided by using an array of concrete blocks. These lump masses need to be placed directly on the floor. If they are placed on small strips of resilient material they effectively become an array of mass–spring resonators that can alter the response of the floating floor; this interferes with the required assessment of static load.

3.6.5.7 Excitation positions

The power input from the ISO tapping machine partly depends on the driving-point mobility of the floor. This varies across the surface of the floor so it is necessary to average several different excitation positions. The driving-point mobility on ribbed or spatially periodic plates with beams can differ between excitation positions on top of the ribs or beams compared to those in-between them. When the positions of the ribs or beams are not known and a sufficient number of random positions are used, this is partly overcome by positioning the tapping machine at an angle of 45° to the ribs/beams (ISO 140 Parts 6 & 7). This is also beneficial because this type of plate can have a strong modal response between the ribs or beams; hence, aligning the tapping machine with the beams could bias the average.

When measuring impact sound insulation in the field it is also useful to be aware of floors that show a significant decrease in vibration with distance across the floor (Section 2.7.7), because flanking transmission via connected walls may depend on the position of the tapping machine.

3.7 Rain noise

In some countries consideration of rain noise inside buildings is only necessary for moderate rainfall, whilst in others it is relevant to long rainy seasons with torrential downpours. As with impact sound insulation on floors, standardizing an impact source for rainfall, requires a degree of pragmatism. An artificial rainfall source in the laboratory needs to be linked to natural rainfall, but it also needs to generate sufficiently high levels to allow measurements that are unaffected by background noise.

3.7.1 Power input

To gain an overview of relevant parameters we start by looking at a raindrop as a structure-borne sound source. The impact of a drop of water upon a surface can be considered in two phases (Petersson, 1995). In the initial impact phase the mass of the drop remains unchanged and there is rapid deceleration. This is followed by a flow phase in which the drop 'breaks open' and the mass of the drop decreases. A falling drop is assumed to be initially spherical but as it travels through the air its shape becomes distorted, although its volume is assumed to remain constant. The force applied upon impact will depend on its distorted shape; however, as its exact shape is uncertain it is necessary to adopt idealized drop shapes. For a drop shape described by a paraboloid, the force pulse can be described by (Jagenäs and Petersson, 1986; Suga and Tachibana, 1994):

$$F(t) = \begin{cases} \rho_w \pi r^2 v_0^2 \left(1 - \frac{3v_0 t}{8r}\right) & \text{for } 0 \le t \le \frac{8r}{3v_0} \\ 0 & \text{for all other } t \end{cases} \tag{3.113}$$

where ρ_w is the density of water, r is the radius of the initially spherical drop, and v_0 is the drop velocity in the flow phase.

(a) Paraboloid drop shape

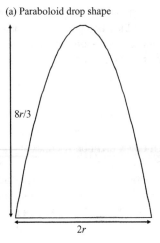

(b) Force pulse from a single drop

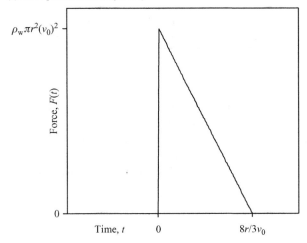

Figure 3.50

Idealized raindrop.

The paraboloid drop and its associated triangular shape force pulse from Eq. 3.113 are shown in Fig. 3.50. The Fourier transform of the force pulse (Eq. 3.113) is used to give the energy spectrum, $|F(f)|^2$, from which the FFT line spectrum can be combined to give one-third-octave or octave bands.

It is assumed that each raindrop falls upon a dry surface. In practice we are not really interested in the first few raindrops on a dry roof, but from the sound radiated during steady rainfall. This means there will be a layer of water over its surface and as most roof elements are sloped and non-porous there will be some water flowing down the surface. The kinetic energy of a falling drop can be transferred to an existing layer of water on the surface; this broadens the force pulse with an increase in the force applied at low frequencies and a reduction at high frequencies (Petersson, 1995). The effect of an existing layer of water is inherent in a measurement. However, it is not included in the above calculation of the force pulse; we simply note that

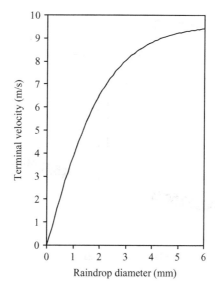

Figure 3.51

Terminal velocities for different raindrop diameters.

this introduces some additional uncertainty. A comparison of measurement and prediction in Section 4.5 indicates that the effect can sometimes be considered as negligible.

For a falling drop, equilibrium between the external forces acting upon it results in a terminal velocity, v_T. For simplicity these external forces can be taken as gravity and aerodynamic drag. Raindrops falling vertically in a calm atmosphere can be assumed to have reached their terminal velocity before hitting the ground. To calculate the force pulse (Eq. 3.113) it can be assumed that v_0 equals v_T for which the terminal velocity of a raindrop can be calculated from the empirical equation (Best, 1950):

$$v_T = 9.58 \left\{ 1 - \exp\left[-\left(\frac{D}{1.77} \right)^{1.147} \right] \right\}$$ (3.114)

where the raindrop is assumed to be spherical with diameter, D in mm.

Raindrops with diameters larger than 1 mm can be classed as rain whereas smaller drops are generally referred to as drizzle. Raindrop diameter depends upon the rainfall rate, temperature, and humidity. In temperate climates there is rarely any need to consider drop diameters larger than 5 to 6 mm; larger drops than this will break up into smaller drops on their way down. Terminal velocities are shown in Fig. 3.51 and generally increase with increasing raindrop diameter. The terminal velocity tends towards a plateau for large drops because the aerodynamic drag force increases due to the distorted drop shape.

For 2 and 5 mm drop diameters the minimum fall heights needed to achieve terminal velocity, when starting from an initial velocity of zero, are approximately 7 and 15 m respectively (e.g. see McLoughlin et al., 1994). Height limitations in the laboratory mean that arranging fall heights for artificial raindrops to reach terminal velocity is not always practical. Therefore the average fall velocity is usually lower and is determined by measurement or calculation. Drops generated from pressurized nozzles have an initial velocity, so a shorter fall height may achieve terminal velocity.

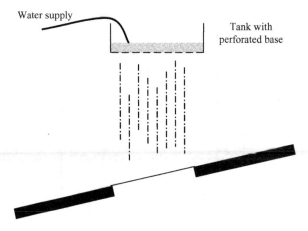

Figure 3.52

Measurement set-up using a water tank to provide a median drop diameter within specified tolerances.

In the laboratory, artificial rain can be generated with a single drop diameter. There will inevitably be some variation; hence, tolerances are given on the median drop diameter from the artificial rain source (ISO 140 Part 18). A tank of water with a perforated base forms a suitable source as shown in Fig. 3.52. The number of drops, N, that fall upon the excitation area in 1 s can then be calculated from the rainfall rate, R_r, in mm/h using:

$$N = \frac{R_r}{3\,600\,000} \frac{3}{4\pi r^3}$$

(3.115)

Natural rainfall has a statistical distribution of raindrop diameters. This can be modelled using the exponential distribution given by Marshall and Palmer (1948) in terms of $n(D)$ in $mm^{-1}\,m^{-3}$,

$$n(D) = 8000 \exp\left(-4.1DR_r^{-0.21}\right)$$

(3.116)

where D is the raindrop diameter in mm and R_r is the rainfall rate (i.e. rain intensity) in mm/h.

Figure 3.53 shows distributions of raindrop diameters for typical rainfall rates. This emphasizes the paucity of large diameter raindrops in the distribution. However, it is not the volume of water that is relevant, it is the power input into the structure; and this tends to be dominated by the small fraction of larger diameter drops that apply high forces due to their high terminal velocities (Ballagh, 1990). The Marshall–Palmer distribution of drop diameters can be used to calculate the number of drops, $N(D)$, with drop diameters between D and $D + \delta D$ (in mm) that fall upon the excitation area in 1 s,

$$N(D) = n(D)v_T\delta D$$

(3.117)

As with the ISO tapping machine, the impedance of an impacting body can affect the power input; although the effect is generally much less significant with a rain drop. Rather than treat the rain drop as a lump mass like the tapping machine hammer, a flow impedance is used to describe the raindrop impedance (Petersson, 1995).

$$Z_{drop} = \rho_w\pi r^2 v_T$$

(3.118)

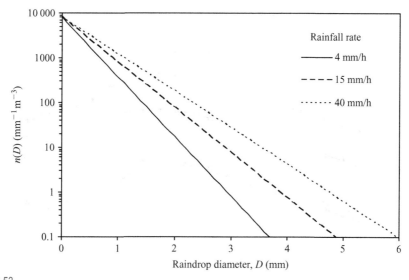

Figure 3.53

Marshall–Palmer distribution of raindrop diameters for different rainfall rates.

For artificial rainfall with single diameter drops, the power input into a plate with a driving-point impedance, Z_{dp} that is real-valued (i.e. an infinite plate) is calculated using:

$$W_{in} = NS_e|F(f)|^2 \frac{1}{Z_{dp} + Z_{drop}}$$ (3.119)

where N equals the number of drops that fall upon the excitation area, S_e in 1 s.

For natural rainfall the distribution can be described by minimum and maximum drop diameters, D_{min} and D_{max}, with a chosen diameter step (δD). The power input is calculated by summing the individual power inputs from Eq. 3.119 for each diameter in the distribution using $N(D)$ instead of N.

The measurement Standard for rain noise (ISO 140 Part 18) is based upon the rain type classifications for intense and heavy rain in IEC 60721-2-2. The latter Standard can be used to define moderate, intense, and heavy rain based on upper limits for the rainfall rate, drop diameter, and fall velocity. Hence moderate, intense, and heavy rain are defined as having rainfall rates of 4, 15, and 40 mm/h respectively, median drop diameters of 1, 2, and 5 mm respectively, and fall velocities of 2, 4, and 7 m/s respectively. These can be used to define artificial rain for use in the laboratory where drops with a median drop diameter often fall at velocities that are lower than their terminal velocity. It is logical to try and compare these with natural rain. However, it is not possible to definitively describe natural rain; the rainfall rate varies during a period of rainfall; often with an initially high rate followed by lower rates. Hence the duration and the rainfall rate are important, not just the depth in millimetres that falls over 1 h. For comparative purposes we can create one example of natural rainfall (there are many possibilities) by using the same rainfall rates as the artificial rain, the Marshall–Palmer model for the drop diameter distribution, and terminal velocities calculated using Eq. 3.114. For temperate climates we will assume $D_{min} = 0$ mm, $D_{max} = 5$ mm, and $\delta D = 0.1$. Other examples of natural rain can be

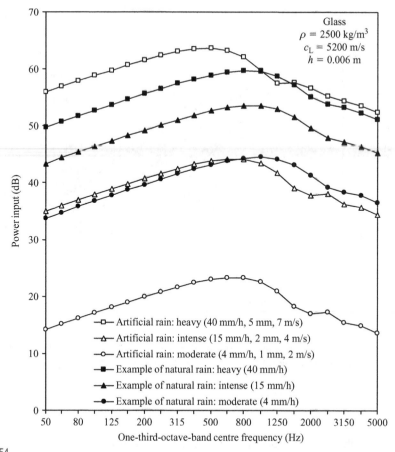

Figure 3.54

Predicted power inputs for artificial and natural rain falling upon a 1 m² excitation area on 6 mm glass.

created by changing D_{min} and D_{max}; note that reducing D_{max} can significantly reduce the power input and therefore requires justification against real rainfall statistics for any design work.

Figure 3.54 shows the power input (Eq. 3.119) for rainfall on 6 mm glass using the above descriptions of artificial and natural rain. For other plates with a frequency-independent driving-point impedance (i.e. infinite plates) and negligible drop impedance, the curves will have a similar shape. As with other impulses, the force spectrum is flat up to a cut-off frequency above which the power input decreases with increasing frequency. For one-third-octave-bands this results in an increasing power input up to the cut-off frequency which occurs in the mid- or high-frequency range. There is some similarity between artificial heavy and the natural heavy rain because the power input tends to be dominated by larger diameter drops. Using $D_{max} = 5$ mm means that the natural intense and natural moderate rainfall have higher power inputs than their artificial counterparts. For this particular example of natural rain, using artificial heavy and intense rain could give a reasonable indication of the range from moderate through to heavy natural rain. However, with some roof elements it is only possible to measure with heavy artificial rain in the laboratory in order that the sound pressure levels are well-above the background noise level.

3.7.2 Radiated sound

Two methods are described in the relevant Standard (ISO 140 Part 18) to determine the sound intensity or sound power radiated from underneath the test element; sound pressure level measurements in a reverberant room or sound intensity measurements.

For homogeneous elements the radiated power from the test element is proportional to the excitation area. The excitation area used in the measurements may not cover the entire surface hence it is useful to present values in terms of the sound intensity level, L_I. For sound pressure level measurements in a reverberant room this is calculated using:

$$L_I = L_p + 10 \lg \left(\frac{V}{T} \right) - 10 \lg \left(\frac{S_e}{S_0} \right) - 14\,\text{dB} \tag{3.120}$$

where V is the volume of the receiving room, T is the reverberation time in the receiving room, S_e is the excitation area, and reference area, $S_0 = 1\,\text{m}^2$.

For sound intensity measurements,

$$L_I = L_{In} + 10 \lg \left(\frac{S_m}{S_e} \right) \tag{3.121}$$

where L_{In} is the temporal and spatial average normal sound intensity level over the measurement surface.

Using Eq. 3.120 or 3.121, the sound power radiated by a homogeneous element when excited over its entire surface area can be calculated using:

$$L_W = L_I + 10 \lg \left(\frac{S}{S_0} \right) \tag{3.122}$$

where S is the surface area of the test element.

3.7.3 Other measurement issues

This section discusses the comparison of measurements from different roof elements and the application of measured data to *in situ* installations.

Example rain noise measurements on 6 mm float glass and an Insulating Glass Unit (IGU) formed from two panes of 6 mm glass are shown in Fig. 3.55. For the 6 mm glass there is a small peak in the sound power at the critical frequency. The critical frequency peak is only just discernible for the IGU, but there is a significant peak at the mass–spring–mass resonance frequency. Roof elements formed from single sheets of metal, plastic, or glass often have their critical frequency in the mid- or high-frequency range. Below the critical frequency, the radiation efficiency depends on the baffle orientation, boundary conditions, and the plate dimensions. If these factors differ significantly between measurements on different elements and/or they differ from *in situ*, it makes comparisons between different elements more awkward unless the effect of these differences can be modelled (see Section 4.5).

As with measurements of airborne or impact sound insulation in transmission suites, total loss factor measurement also needs consideration for rain noise measurements on solid homogenous plates. With rain excitation, the sound power radiated by a solid homogeneous plate is

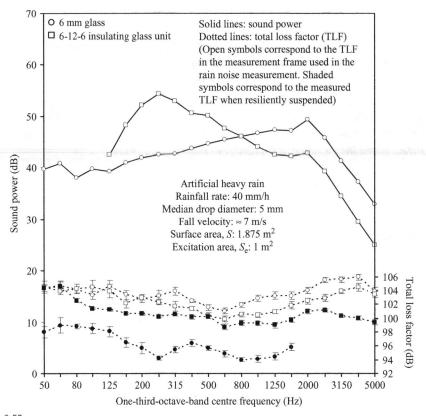

Figure 3.55

Radiated sound power from artificial heavy rain on glazing using sound intensity measurements according to ISO 15186 Part 1. Test element dimensions: $L_x = 1.5$ m, $L_y = 1.25$ m. 6 mm glass: $\rho_s = 15$ kg/m^2, $f_c = 2021$ Hz, mounted in a wooden frame with putty. 6-12-6 insulating glass unit (air filled): $f_{msm} = 200$ Hz, mounted in the same wooden frame but without putty. Both elements were orientated at an angle of 30° for water drainage. All total loss factors were determined from structural reverberation time measurements using MLS shaker excitation. All values are shown with 95% confidence intervals (values for resiliently suspended 6 mm glass are only shown below the critical frequency). Measured data from Hopkins are reproduced with permission from BRE and DfES.

dependent on the total loss factor of the plate. For this reason, damping compounds are sometimes applied to lightweight metal roofs to increase the internal loss factor, and therefore the total loss factor in order to reduce the radiated power (note that commonly used metals tend to have very low internal loss factors). If the mounting conditions change the total loss factor, this will also change the radiated sound power. To give an indication of the effect of the mounting, the total loss factors were measured once in the measurement frame used for the rain noise measurements and once when resiliently suspended. For the resiliently suspended 6 mm glass there are no structural coupling losses, only internal and radiation losses. The internal and radiation losses are low so the total loss factor increases significantly once it is installed in the frame where the coupling losses predominantly determine the total loss factor. For the 6 mm glass that forms one side of the IGU, the increase is much smaller because each pane of the IGU is tightly coupled to the other; hence, the structural coupling losses of each pane are already quite high when it is resiliently suspended.

Note that for laminated glass the bending stiffness and internal loss factor vary significantly with surface temperature. During the measurement the radiated sound power is therefore affected by the water temperature as well as room temperature. This affects the comparison of measurements as well as use of the measurement to determine the radiated sound power *in situ*. Compared to float glass of the same thickness this makes it possible for laminated glass to radiate more sound power at some frequencies and less sound power at others; this will depend on its temperature.

3.8 Reverberation time

In reverberant sound fields encountered in typical rooms, the decay curves vary throughout the space and it is necessary to determine the spatial average reverberation time. In Section 1.2.6.3 we looked at idealized decay curves in rooms; these were straight lines or smooth curves that were unaffected by random fluctuations or background noise. The ability to gain good estimates of the reverberation time from measured decay curves is determined by the combination of the acoustic system under test, the measurement procedure, signal processing, and evaluation of the decay curve.

There are two main methods that are used to determine the reverberation time in spaces or the structural reverberation time on structures; the interrupted noise method, and the integrated impulse response method (ISO 3382). The derivations in this section tend to refer to sound pressure as this is the most common application; but they are equally applicable to vibration. For accurate sound insulation measurements in rooms, we typically need to determine the reverberation time in seconds to two decimal places. For structural reverberation times the decays are much shorter and three decimal places are often required.

3.8.1 Interrupted noise method

The interrupted noise method uses random noise to create a steady-state level. After a steady-state has been achieved, the excitation is stopped and the subsequent decay of the sound pressure level is recorded. An example decay curve is shown in Fig. 3.56. Compared to the idealized decay curve previously shown in Fig. 1.19, the curve is not smooth; it is characterized by random fluctuations due to the random nature of the excitation signal. As the decaying signal level gets closer to the background noise level, the slope of the decay curve is altered by the background noise and it is no longer representative of the actual decay. When this occurs within the evaluation range, it prevents accurate determination of the reverberation time. Any significant effects on T_X due to background noise can be avoided by using a steady-state level that is at least $15 + X$ dB above steady background noise, preferably more, because background noise is not always steady. For example, the steady-state level should be at least 30 dB above background noise for T_{15} or at least 45 dB above background noise for T_{30}. The starting point for the evaluation range is 5 dB below the steady-state level, and therefore the end point will be at least 10 dB above the background noise level.

If measurement of a decay curve is repeated using exactly the same microphone and loud-speaker positions, then each time that the excitation is stopped, the modes will all have random phases and amplitudes. The interaction between the decaying modes will therefore vary with each measurement, which will result in different decay curves. For this reason, a single decay curve measured with interrupted noise is not particularly useful; hence, we need to take more

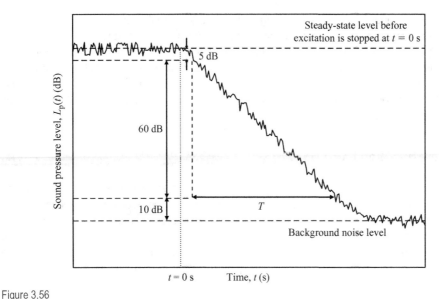

Figure 3.56

Example decay curve measured using the interrupted noise method.

than one measurement at each position to calculate an average value. This average can be calculated in two ways; we can either arithmetically average the reverberation times from each curve, or ensemble average the decay curves. Both are acceptable, but the latter is preferable because it results in smoother decay curves. A smooth curve is beneficial when determining the reverberation time, T_X, because in order to carry out linear regression over a range of X dB it is necessary to clearly identify the starting point and the end point that define the evaluation range. If there are large random fluctuations it can be difficult to identify the starting point that is 5 dB below the initial level and the end point that is $X + 5$ dB below the initial level. Ensemble averaging the decay curves is also preferable because the ensemble average of an infinite number of curves gives the same decay curve as the integrated impulse response for that point in the room (Schroeder, 1965). An infinite number of measurements is clearly out of the question, but by averaging several decay curves we obtain a single, much smoother, decay curve. This requires each individual decay curve, $L_p(t)$, to be synchronized at the time when the excitation is stopped. The ensemble average decay curve, $L_{p,av}(t)$, can then be calculated from the N individual decay curves using the following:

$$L_{p,av}(t) = 10 \lg \left(\frac{1}{N} \sum_{n=1}^{N} 10^{L_{p,n}(t)/10} \right) \qquad (3.123)$$

To determine the spatial average reverberation time from different source and receiver positions, it is acceptable to arithmetically average the reverberation times, or to ensemble average the decay curves. The benefits of the latter are the same as have just been discussed.

3.8.2   Integrated impulse response method

When an acoustic system is excited by a Dirac delta function the resulting response is the impulse response of that system, $h(t)$ (Section 1.2.2). An acoustic system is defined such that it includes the space or structure under test as well as the chain of measurement equipment. In practice, $h(t)$ is the signal received by the analyser from the measurement transducer, e.g. sound pressure from a microphone in a room or acceleration from an accelerometer on a wall. For a linear acoustic system, the impulse response completely describes that system; hence, the output signal from the system can be calculated from any known input signal that excites it.

The integrated impulse response method involves generating an impulse, and recording or sampling the response of an acoustic system to this impulse. Due to its infinite height and infinitely narrow width, it is not possible to create a Dirac delta function in practice. However, it is possible to create an impulse of sufficiently short duration that can represent the Dirac delta function. On structures, it is possible to generate an impulse with a hammer blow. In rooms an impulse can be generated with a gunshot from a starting pistol, balloon bursts, handclaps, or noise bursts via a loudspeaker. Although these impulse sources have been used in rooms for many years, not all of them are omnidirectional and are able to generate a flat spectrum at a sufficiently high level whilst avoiding very high crest factors (ratio of peak to rms). The latter can cause problems due to the limitations of the detector in the analyser. All of these problems can be avoided when the system to be measured is linear and time-invariant (LTI). The required impulse response can then be measured with swept-sine signals (Müller and Massarani, 2001) or a signal referred to as a Maximum Length Sequence (MLS) (Section 3.9).

The integrated impulse response method was introduced by Schroeder (1965) and involves signal processing of an impulse response using reverse-time integration. Using this method gives the same decay curve that would be determined by averaging an infinite number of decay curves from measurements using the interrupted noise method. This results in a decay curve that truly represents the characteristics of the acoustic system. In addition, a single measurement gives a decay curve without random fluctuations, which increases the accuracy of reverberation time calculations. We recall that for interrupted noise measurements, more than one measurement needs to be taken at each receiver position to reduce the effect of random variations.

From Schroeder (1965), the proof for the integrated impulse response method is based around interrupted noise measurements made with stationary white noise as the input signal, $x(t)$. We assume that the input signal was switched on at some time in the past, represented by time $t = -\infty$, and that the acoustic system reaches a steady state before $t = 0$. The input signal is interrupted at $t = 0$ in order to perform the decay measurement. The acoustic system under test has an impulse response, $h(t)$, hence the output signal, $y(t)$, at the receiver position is given by the convolution integral,

$$y(t) = \int_{-\infty}^{0} h(t - u)x(u)\mathrm{d}u \tag{3.124}$$

Replacing $t - u$ by a variable, w, gives,

$$y(t) = \int_{t}^{\infty} h(w)x(t - w)\mathrm{d}w \tag{3.125}$$

Due to the random nature of white noise, it is necessary to average all realizations of the input signal (also referred to as an ensemble average) to find the expected value of the squared output signal,

$$E[y^2(t)] = \int_{t}^{\infty} \int_{t}^{\infty} h(t_1)h(t_2)E[x(t - t_1)x(t - t_2)]\mathrm{d}t_1\mathrm{d}t_2 \tag{3.126}$$

$E[x(t-t_1)x(t-t_2)]$ is the expected value of the product of the signal with itself at another point in time; this equals the auto-correlation function. Hence, substituting the auto-correlation function for white noise from Eq. 3.5 into Eq. 3.126 gives

$$E[y^2(t)] = N \int_t^\infty h^2(w)\mathrm{d}w \tag{3.127}$$

where N is the power spectral density.

Equation 3.127 shows that the squared impulse response acquired from a single measurement can be integrated to give the expected value of the squared decay, $E[y^2(t)]$ from a large number of interrupted noise measurements.

Practical implementation of Eq. 3.127 involves recording or sampling the impulse response, and then playing it backwards into an analyser where the signal is squared and then integrated. The direction of signal analysis is from the end of the decay to the start of the decay; hence, this process is referred to as reverse-time integration or backwards integration. Effectively, the integration starts at the end of the squared signal. From this starting point, each sample of the decaying signal is replaced by the sum of the mean-square sound pressure of that sample and the summed mean-square sound pressure from all previous samples back to the starting point. This results in a new decay curve of level versus time, the logarithmic integrated impulse response. For a decay curve normalized to the total energy of the impulse response, the process of reverse-time integration gives the logarithmic integrated impulse response as

$$L_p(t) = 10\lg\left(\frac{\int_t^\infty h^2(w)\mathrm{d}w}{\int_0^\infty h^2(w)\mathrm{d}w}\right) = 10\lg\left(\frac{\int_\infty^t h^2(w)\mathrm{d}(-w)}{\int_0^\infty h^2(w)\mathrm{d}w}\right) \tag{3.128}$$

In practice, the tail end of any decaying signal is buried beneath the background noise. Therefore as reverse-time integration cannot actually start at $t=\infty$, it is necessary to find a starting point which gives a decay curve that is not significantly altered by the presence of background noise (Lundeby et al., 1995; Nilsson, 1992; Vorländer and Bietz, 1994). The effect of different starting points on the logarithmic integrated impulse response can be calculated using an idealized impulse response. An idealized squared impulse response with an exponential decay defined by the reverberation time is given by

$$h^2(t) = \exp\left(\frac{-6t\ln 10}{T}\right) \tag{3.129}$$

It is assumed that during the impulse measurement there is steady background noise with a mean-square pressure, p_{noise}^2. As an example, if we have a background noise level that is 40 dB below the level at $t=0$, then $p_{\text{noise}}^2 = 10^{-40/10}$.

The measured signal consists of the idealized impulse response combined with the background noise. Hence in Eq. 3.128 we need to replace $h^2(w)$ with $h^2(w) + p_{\text{noise}}^2$, and replace the integration limit, ∞, with t_{max}. For the idealized impulse response this gives the logarithmic integrated impulse response as

$$L_p(t) = \begin{cases} 10\lg\left(\dfrac{10^{-6t/T} - 10^{-6t_{\text{max}}/T} + \dfrac{6p_{\text{noise}}^2\ln 10}{T}(t_{\text{max}} - t)}{1 - 10^{-6t_{\text{max}}/T} + \dfrac{6p_{\text{noise}}^2 t_{\text{max}}\ln 10}{T}}\right) & \text{for } 0 \le t \le t_{\text{max}} \\[4mm] -\infty \text{ dB} & \text{for } t \ge t_{\text{max}} \end{cases} \tag{3.130}$$

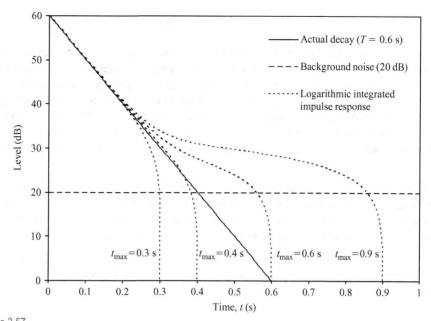

Figure 3.57

Example decay curves measured using the integrated impulse response method.

Examples of the logarithmic integrated impulse response calculated using Eq. 3.130 are shown in Fig. 3.57 for different values of t_{max}. The actual decay curve has a reverberation time of 0.6 s and has been shifted so that it starts at a level of 60 dB. The background noise level is 20 dB. The later part of the measured decay becomes shorter or longer depending upon the value of t_{max}. In this example, the level of the actual decay is equal to the background noise level at $t = 0.4$ s. When $t_{max} = 0.4$ s the logarithmic integrated impulse response between $t = 0$ and 0.4 s is only slightly affected by the presence of background noise. Hence, despite the fact that we can only estimate the actual background noise, suitable values of t_{max} can readily be found that allow calculation of the reverberation time.

The approach used to determine t_{max} depends on the availability of the background noise level during the impulse response measurement (ISO 3382). For steady background noise this level can usually be estimated from the horizontal tail of the impulse response where the decaying signal is well-below the background noise. As indicated in the example, the optimum value for t_{max} is at the time where the actual decay curve equals the background noise. In practice, we do not know the actual decay curve as this is the very reason that we are carrying out the measurement in the first place. In addition, the fluctuations that exist in real background noise mean that we cannot always obtain a good estimate from a short sample at the tail end of the impulse response. For these reasons, the procedures used to determine the optimum value tend to be based on iterative processes that are automated in the measurement equipment (Lundeby et al., 1995; ISO 3382).

As with the interrupted noise method, different source and receiver positions are needed to determine the spatial average reverberation time.

3.8.3 Influence of the signal processing on the decay curve

With the interrupted noise method, the signal that is processed by the analyser passes through the required filters before going to the detector. In the detector the signal is squared, and then integrated over a specific time interval using either linear or exponential averaging to determine an average value. Averaging is necessary to reduce the fluctuations in the squared signal, however if the averaging times are too long, it is not possible to accurately calculate the gradient of the decay curve. In the following discussion we are not concerned with the squaring of the signal, hence all references to the detector will relate to its function as an averaging device.

With the interrupted noise method, both the filters and the averaging device can alter the shape of the decay curve. With the integrated impulse response method, the signal is filtered and squared before using reverse-time integration to calculate the decay curve; hence, with this method it is only the filters that can alter the shape of the decay curve.

The effect of the detector and the filters on the decay curve is thoroughly described by Jacobsen (1987) and the same approach is used here by looking at the effect of the detector and the filters separately.

3.8.3.1 Effect of the detector

The effect of the averaging device on the decay curve is determined by convolving the impulse response of the detector, $d(t)$, with the squared signal from an interrupted noise measurement, $x(t)$. This equals the output from the analyser, $y(t)$, when we assume that it is only the detector that affects the measured decay. Convolution of these two continuous time signals, $d(t)$ and $x(t)$, is denoted as $d(t)*x(t)$ and is defined by the convolution integral,

$$y(t) = \int_{-\infty}^{\infty} d(u)x(t-u)du = \int_{-\infty}^{\infty} d(t-u)x(u)du \qquad (3.131)$$

where u is a dummy time variable.

The squared noise signal is interrupted at $t=0$ and results in an idealized exponential decay. This signal is defined using Eq. 3.129 to give

$$x(t) = \begin{cases} 1 & \text{for } t < 0 \\ \exp\left(\dfrac{-6t \ln 10}{T}\right) & \text{for } t \geq 0 \end{cases} \qquad (3.132)$$

where T is the actual reverberation time.

The impulse response for a linear averaging device with an integration (averaging) time, T_{int}, is

$$d(t) = \begin{cases} \dfrac{1}{T_{int}} & \text{for } 0 < t < T_{int} \\ 0 & \text{for all other } t \end{cases} \qquad (3.133)$$

For linear averaging, convolution gives the following decay curve,

$$
L_{\mathrm{p}}(t) = \begin{cases} 10\lg\left(1 - \dfrac{t}{T_{\mathrm{int}}} + \dfrac{T - T\exp\left(\dfrac{-6t\ln 10}{T}\right)}{6T_{\mathrm{int}}\ln 10}\right) & \text{for } 0 \leq t < T_{\mathrm{int}} \\[4mm] 10\lg\left(\exp\left(\dfrac{-6t\ln 10}{T}\right)\right)\left(\exp\left(\dfrac{6T_{\mathrm{int}}\ln 10}{T}\right) - 1\right)\dfrac{T}{6T_{\mathrm{int}}\ln 10} & \text{for } t > T_{\mathrm{int}} \end{cases}
$$

$$(3.134)$$

For exponential averaging, the impulse response is a decaying exponential curve that weights the average towards the latter part of the time sample. The detector therefore has its own decay time, T_{detector}, which is defined as

$$
T_{\mathrm{detector}} = 6\tau\ln 10 \tag{3.135}
$$

where τ is the time constant. When exponential averaging is implemented using an RC circuit (resistor capacitor), τ is equal to RC.

The impulse response for an exponential averaging device is

$$
d(t) = \begin{cases} \dfrac{1}{\tau}\exp\left(-\dfrac{t}{\tau}\right) & \text{for } 0 < t < \infty \\[3mm] 0 & \text{for } -\infty < t < 0 \end{cases} \tag{3.136}
$$

for which convolution with the idealized exponential decay gives the following decay curve,

$$
L_{\mathrm{p}}(t) = \begin{cases} 10\lg\left[\exp\left(\dfrac{-6t\ln 10}{T}\right)\left(\dfrac{T}{T - 6\tau\ln 10} - \dfrac{6\tau\ln 10}{T - 6\tau\ln 10}\exp\left(\dfrac{6t\ln 10}{T} - \dfrac{t}{\tau}\right)\right)\right] & \\ \qquad\qquad\qquad\qquad\qquad\qquad\qquad\qquad \text{for } \tau \neq \dfrac{T}{6\ln 10} \\[4mm] 10\lg\left[\exp\left(\dfrac{-6t\ln 10}{T}\right)\left(1 + \dfrac{6t\ln 10}{T}\right)\right] & \\ \qquad\qquad\qquad\qquad\qquad\qquad\qquad\qquad \text{for } \tau = \dfrac{T}{6\ln 10} \end{cases}
$$

$$(3.137)$$

Figure 3.58 shows the effect of linear and exponential averaging on the idealized decay using Eqs 3.134 and 3.137. All the decay curves have been shifted so that they start at 60 dB. To assess the effect of the detector on an idealized exponential decay, the decaying sound pressure level is plotted against the normalized time, t/T, so that it is applicable to any reverberation time, T. If the effect of the detector on the idealized decay is negligible, the decay will have the same gradient as the actual decay over the time interval, $0 \leq t/T \leq 1$. The figures show that the detector causes curvature of the decay curve, starting at the beginning of the decay. For some of the curves produced by exponential averaging this is easier to see if a straight edge is placed against them. The extent of the curvature over the time interval, $0 \leq t/T \leq 1$, depends on the properties of the linear or exponential averaging device.

The relevant Standards for reverberation time measurement (ISO 354 and ISO 3382) place requirements on the detector in terms of T_{int} or τ, but also require that the evaluation shall start 5 dB below the initial sound pressure level at $t = 0$.

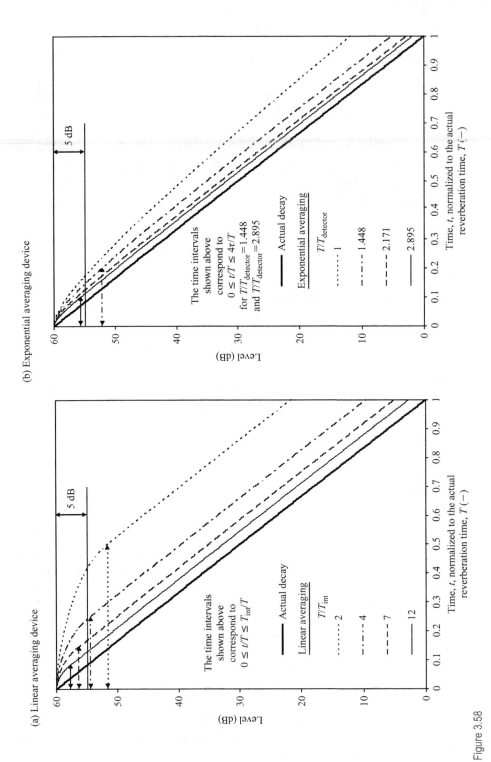

Figure 3.58

Effect of linear and exponential averaging devices on the decay curve.

With linear averaging, the detector causes significant curvature in the initial part of the decay curve, $0 \leq t < T_{int}$. However, when $t \geq T_{int}$, the decay curves shown in Fig. 3.58a have the same slope as the actual decay curve. The evaluation range therefore needs to exclude the time interval, $0 \leq t < T_{int}$, with the requirement that $T/T_{int} \geq 4$ (Jacobsen, 1987). Typical requirements in the relevant Standards are that T_{int} shall be less than $T/12$ or less than $T/7$. Combining these requirements with an evaluation range that starts 5 dB below the level at $t = 0$ means that there is no need to refer to a specific time at which the evaluation should begin. This can be seen in Fig. 3.58a where the requirement for an initial 5 dB drop excludes the initial non-linear part of the decay that is caused by the detector for both $T/T_{int} = 7$ and $T/T_{int} = 12$.

With exponential averaging, the detector can also cause curvature of the decay curve. Figure 3.58b shows that the decay curve for $T/T_{detector} = 1$ is significantly different to the actual decay curve. Jacobsen (1987) proposed that the evaluation range should exclude the time interval, $0 \leq t < 4\tau$ (equivalent to $0 \leq t < 0.29 T_{detector}$), provided that $T/T_{detector} > 2$. Typical requirements in the relevant Standards are that the time constant, τ, of an exponential averaging device shall be less than, but as close as possible to (a) $T/20$ (where $\tau < T/20$ is equivalent to $T/T_{detector} > 1.448$) or (b) $T/30$ (where $\tau < T/30$ is equivalent to $T/T_{detector} > 2.171$). An alternative requirement (used in ISO 354:1985 that has now been superseded) is that the equivalent averaging time of an exponential averaging device, 2τ, shall be less than, but as close as possible to $T/20$. The condition $2\tau < T/20$ is equivalent to $T/T_{detector} > 2.895$. By using an evaluation range that starts 5 dB below the level at $t = 0$, the time interval, $0 \leq t \leq 4\tau$ is excluded when $T/T_{detector} > 2.895$.

The time intervals that need to be excluded for linear and exponential averaging can be compared with the arrival time for the sound wave that travels directly from the sound source to the microphone, and for the first reflected wave that arrives at the microphone. We are interested in the decay of the reverberant field rather than propagation of the direct field. For this reason it is necessary to exclude the time interval in which the last sound that was produced by the source (i.e. just before it was switched off at $t = 0$) travels directly from the sound source to the microphone. In addition, we want to exclude the time interval before the arrival of the first reflected wave. This is because the mean-square pressure does not approximate an exponential decay until the arrival of the first-order reflections (Vorländer, 1995). As the first reflected wave always arrives later than the direct wave, it is the arrival of the first reflected wave that determines the time interval that should be excluded. By assuming a diffuse field, the mean free path can be used to estimate the average path length from the source to the microphone with a single reflection from a room surface. This assumes that the distance from the source to the reflection point on the surface, and the distance from the microphone to the reflection point on the surface are both equal to half the mean free path. The time taken for sound to travel a distance equal to the mean free path is approximately 7 ms in a 50 m³ room and approximately 11 ms in a 200 m³ room. For the majority of rooms, the initial 7 or 11 ms will be contained within the time intervals that are already excluded by using a starting point that is 5 dB below the level at $t = 0$ and by satisfying the requirements on the detector in the relevant Standards (ISO 354 and ISO 3382). Hence, there is no need to exclude an additional time interval based upon the position of the source and the microphone in relation to the room surfaces.

3.8.3.2 *Effect of the filters*

A filter has its own impulse response; hence, a filter also has its own decay time. To accurately measure a decaying sound or vibration signal that is sent through a band-pass filter, the decay time of the filter must be shorter than the actual reverberation time that is to be measured. This is necessary to ensure that we measure the reverberation time of the acoustic system and not the reverberation time of the filter itself (sometimes referred to as ringing of the filter). The normal measurement method is referred to as forward-filter analysis. This name is used to distinguish it from reverse-filter analysis where the signal is played backwards through the filter to reduce its influence on the decay curve.

3.8.3.2.1 Forward-filter analysis

For one-third-octave and octave-band filters that satisfy the relevant Standard (IEC 61260) the impulse response is not symmetrical along the time axis. This can be seen from the impulse responses for 6th order IIR Butterworth filters in Fig. 3.59. The asymmetry means that the filter responds relatively quickly to an input signal in comparison to the time it takes for the response of the filter to decay. This is a desirable feature for sound and vibration level measurements, but it is not always ideal for reverberation time measurements. When a sinusoidal signal is sent through a filter (where the frequency of the sinusoid is within the bandwidth of the filter) there will be a short time delay of $\approx 1/B$ s before the filter responds to this sinusoid and outputs a sinusoid with the same amplitude. The time interval on the decay curve, $0 \leq t < 1/B$, can therefore be ignored; note that when $BT > 12$, this time interval occurs within the initial 5 dB drop. We now need to look at how the relatively slow decay time of the filter impulse response can affect the measured decay curve.

Filtering is implemented by convolving the impulse response of a filter, $f(t)$, with the impulse response of an acoustic system, $x(t)$. The resulting impulse response is then processed using reverse-time integration as described in Section 3.8.2 on the integrated impulse response method.

The impulse response of a filter will vary depending upon the type of filter (e.g. Butterworth) and its implementation in analogue or digital form. With digital implementation of a filter, its impulse response can simply be determined by convolution with a unit impulse.

To define an idealized impulse response of an acoustic system it is convenient to use an exponentially decaying sinusoid because this will give a decay curve that is a straight line. The frequency of this sinusoid, f, is assumed to be equal to the band centre frequency of the filter. This impulse response can therefore be considered as representing a single decaying mode that lies within the filter passband. In practice this could occur on structures or in spaces in the low-frequency range. However, there is usually more than one decaying mode in a band, and, as we have seen in Section 1.2.6.3.2, this can result in decay curves that are not straight lines. Therefore, it is easier to draw general conclusions on the effect of the filter by using a single decaying sinusoid; so the idealized impulse response of the acoustic system is defined as

$$x(t) = \begin{cases} \cos(2\pi\, ft)\exp\left(\dfrac{-3t \ln 10}{T}\right) & \text{for } t > 0 \\ 0 & \text{for all other } t \end{cases} \tag{3.138}$$

where T is the actual reverberation time.

The effect of the filter on the decay curve depends upon the filter bandwidth and the actual reverberation time. For this reason, it is useful to calculate decay curves for different values of BT, where the time scale on the x-axis is normalized to the actual reverberation time (Jacobsen, 1987). The filter bandwidth, B, can be estimated using the band centre frequency, f, where $B = 0.23f$, for one-third-octave-bands and $B = 0.707f$ for octave-bands.

(a) One-third-octave-band filter

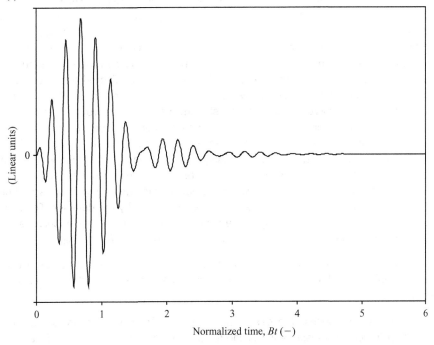

(b) Octave-band filter

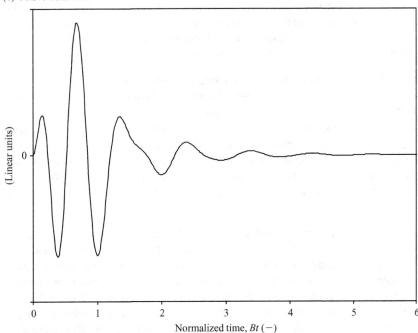

Figure 3.59

Impulse response of one-third-octave-band and octave-band filters.

Figure 3.60 shows forward-filter analysis for one-third-octave-band filters. The plot of the decay curve associated with the filter itself is shown alongside the plot of decay curves for the idealized impulse response with values of BT between 1 and 32. From the impulse response of the filter (Fig. 3.59), one would not expect its decay curve to be a straight line; in fact, the decay curve contains a number of prominent ripples over the 60 dB decay range. When the actual reverberation time is very short these ripples also appear in the filtered decay curve of the exponentially decaying sinusoid as can be seen in Fig. 3.60 when $BT = 1$. As BT increases from 1 to 32, the decay curve tends towards a straight line. We have previously seen that the detector affected only the initial part of the decay curve; here we see that filtering not only introduces curvature in the initial part of the decay curve but that it can also distort the main part of the decay curve. When $BT > 16$ the main part of the decay curve is straight with minor curvature in the initial 5 dB of the decay (Jacobsen, 1987). This minor curvature is unimportant with interrupted noise measurements because evaluation of the decay curve starts 5 dB below the level at $t = 0$ in order to minimize the curvature introduced by the detector. To show the effect of the filter, the full 60 dB decay range is shown. However, the decay curve is almost always evaluated within the initial 35 dB drop to calculate T_{10}, T_{15}, T_{20}, or T_{30}.

For octave-bands the decay curve associated with the filter has different ripples. However, the same requirement that $BT > 16$ to give straight decays is applicable to octave-band measurements.

For electroacoustic purposes, the design of band pass filters is primarily based on their attenuation at frequencies inside and outside the passband, rather than their effect on decay measurements (IEC 61260). The requirement that $BT > 16$ is appropriate for most one-third-octave and octave-band filters designed according to this Standard. However, this is quite a strict requirement as it is based on the idealized situation of a single decaying sinusoid. In addition, the impulse response depends upon the individual filter, so manufacturer's information for the specific filter should be used whenever it is available.

In room acoustics measurements, the integrated impulse response method is often used to determine the Early Decay Time (EDT) from the initial 10 dB of the decay. This method avoids any distortion caused by the detector, but significant curvature in the initial part of the decay curve caused by filtering needs to be avoided by ensuring that $BT > 16$. This requirement is usually satisfied in spaces used for music performance because octave-bands are used (i.e. large bandwidths) and the spaces tend to have quite long reverberation times. For sound insulation measurements in both the field and the laboratory the $BT > 16$ requirement is stricter than is necessary for the accuracy that is required; hence $BT > 8$ can be adopted instead (ISO 3382).

Figure 3.61 shows the minimum reverberation times that can be measured under the requirement that $BT > 16$ and $BT > 8$. We can look at the $BT > 8$ requirement in the context of one-third-octave-band measurements in furnished dwellings. It is reasonable to assume that furnished rooms have reverberation times of approximately 0.5 s. For reverberation times ≥ 0.35 s, the requirement that $BT > 8$ is always satisfied between 100 and 3150 Hz and it is possible to measure much lower reverberation times in the mid- and high-frequency ranges. If the requirement is not satisfied, the errors incurred by using forward-filter analysis depend upon the evaluation range and the shape of the actual decay curve; hence, they are not easily quantifiable. Errors can usually be avoided by using reverse-filter analysis that will be discussed in the next section or by using octave-band measurements.

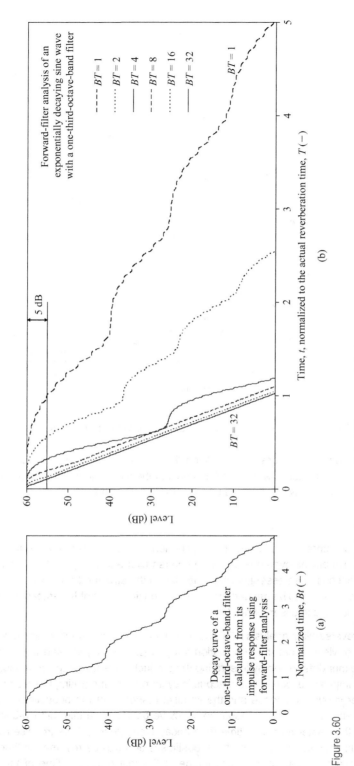

Figure 3.60

Effect of the filter on reverberation time measurement using forward-filter analysis. (a) Decay curve derived from the impulse response of the filter. (b) Decay curves for an exponentially decaying sinusoid for different values of *BT*.

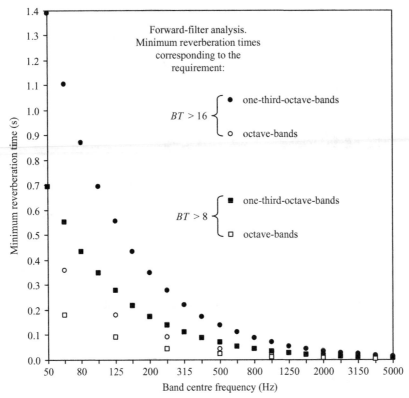

Figure 3.61

Requirements for the minimum reverberation times that can be measured using forward-filter analysis.

For the measurement of structural reverberation times it is useful to work in terms of loss factors, hence the requirement that $BT > 16$ means that the maximum measurable loss factor is 0.03161 for one-third-octave-bands and 0.09716 for octave-bands.

3.8.3.2.2   Reverse-filter analysis

Any problems encountered in satisfying the requirement that $BT > 16$ or $BT > 8$ for forward-filter analysis can usually be overcome by using reverse-filter analysis (Jacobsen and Rindel, 1987). This approach is often essential when measuring the structural reverberation times of building elements. It may also be needed for reverberation times in highly damped rooms such as recording studios (Rasmussen *et al.*, 1991).

To understand reverse-filter analysis it is useful to look at the convolution process from a qualitative point of view. From the convolution integral (Eq. 3.131) we see that the output from the filter is either determined by integrating the product of the filter impulse response and the reverse-time impulse response of the acoustic system or by integrating the product of the reverse-time filter impulse response and the impulse response of the acoustic system. We have already noted that the impulse response of octave-band or one-third-octave-band filters is asymmetric with a faster response time than decay time. Hence, if the impulse response of the filter (or the impulse response of the acoustic system) can be reversed in time before the convolution process, we can take advantage of the fast response time of the filter to

326

measure shorter reverberation times than with forward-filter analysis. Reverse-time analysis is implemented on analysers by storing the measured impulse response of the acoustic system, and then playing it backwards through the filters.

Figure 3.62 shows reverse-filter analysis for one-third-octave-band filters; this can be compared to forward-filter analysis previously shown in Fig. 3.60. Forward-filter analysis delays the signal, whereas reverse-filter analysis results in a negative time delay. However, this does not prevent determination of the reverberation time, which is determined using relative time, rather than absolute times. The decay curve of the filter is significantly shorter than with forward-filter analysis and does not contain prominent ripples. For this reason, the main unwanted effect of reverse-filter analysis is curvature in the initial part of the decay curve. When $BT > 4$ there is some curvature in the initial 5 dB of the decay, but after this the decay curve is straight (Jacobsen, 1987). Although this example uses one-third-octave-bands, the requirement that $BT > 4$ equally applies to octave-band measurements.

Another benefit of reverse-filter analysis is that the distortion of the decay curve caused by an exponential averaging device is reduced. The requirement on the detector that $T/T_{detector} > 2$ can be changed to $T/T_{detector} > 0.25$ when the evaluation range starts 5 dB below the level at $t = 0$ (Jacobsen and Rindel, 1987).

Figure 3.63 shows the minimum reverberation times that can be measured under the requirement that $BT > 4$. This requirement corresponds to a maximum measurable loss factor of 0.12643 for one-third-octave-bands and 0.38864 for octave-bands.

3.8.4 Evaluation of the decay curve

Measured decay curves are rarely perfect straight lines; hence, they need to be evaluated in such a way as to minimize errors in the calculated reverberation time. Evaluation is usually automated and carried out by the analyser to give a straight line that best fits the data in the chosen evaluation range. The simplest form of linear regression, and the most commonly used, is the least-squares method which is described in general textbooks on statistics. An alternative method to calculate the reverberation time is to use a weighting function. This can reduce the effect of random fluctuations in the decay curve at both ends of the evaluation range. Vigran and Sørdal (1976) show that symmetric weighting functions can give a low variance for the spatial average reverberation time, with negligible change to the mean value. To use a weighting function, $W(z)$, the reverberation time, T_X, is calculated from the decay curve using (Vigran and Sørdal, 1976)

$$T_X = \frac{60}{X} \int_{-\infty}^{\infty} W\left(\frac{L_p(t) - L_{-5\,dB}}{L_{-(X+5)dB} - L_{-5\,dB}}\right) dt \tag{3.139}$$

where $L_{-5\,dB}$ is the starting point of the evaluation range (5 dB below the initial sound pressure level), and $L_{-(X+5)dB}$ is the end point.

An example of a symmetric weighting function is a triangular function described by

$$W(z) = \begin{cases} 4z & \text{for } 0 \leq z \leq 0.5 \\ 4(1-z) & \text{for } 0.5 < z \leq 1 \\ 0 & \text{for all other } z \end{cases} \tag{3.140}$$

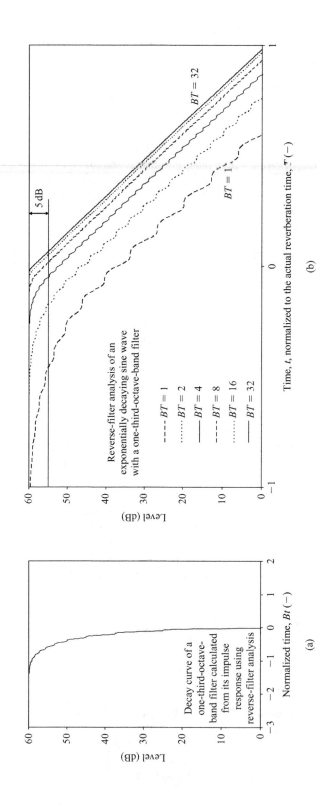

Figure 3.62

Effect of the filter on reverberation time measurement using reverse-filter analysis. (a) Decay curve derived from the impulse response of the filter. (b) Decay curves for an exponentially decaying sinusoid for different values of *BT*.

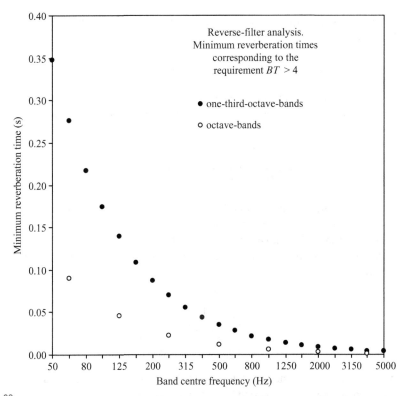

Figure 3.63

Requirement for the minimum reverberation times that can be measured using reverse-filter analysis.

3.8.5 Statistical variation of reverberation times in rooms

For interrupted noise measurements, the reverberation time varies between different measurement positions as well as between repeated measurements at the same position. For the integrated impulse response method, only a single measurement is needed at each position; so it is only the spatial variation within the room that needs to be considered. In rooms with diffuse fields, the standard deviation for these sources of variation can be calculated for the situation where interrupted noise measurements are taken using a point source driven with a Gaussian white noise signal (Davy *et al.*, 1979). In a diffuse field the decay curve is a straight line. However, the random nature of white noise means that the measured decay curve will have random fluctuations about this straight line; the approach taken by Davy *et al.* (1979) to determine the reverberation time is to fit a straight line to the decay curve using the least-squares method. It is assumed here that the filters and the detectors have no effect on the shape of the decay curve within the evaluation range.

For the spatial variation of the reverberation time, the standard deviation, $\sigma_{T,s}$, is (Davy *et al.*, 1979)

$$\sigma_{T,s} = \sqrt{\frac{720T}{B_s X^3} \left(\frac{10}{\ln 10}\right)^2 F\left(\frac{\ln 10}{10}X\right)} \qquad (3.141)$$

329

where X is the evaluation range (dB) used to calculate the reverberation time, B_s is the statistical bandwidth of the filter, and the function $F(\ )$ is

$$F(z) = 1 - \frac{3}{z}(1 + \exp(-z)) - \frac{12}{z^2}\exp(-z) + \frac{12}{z^3}(1 - \exp(-z)) \tag{3.142}$$

For the interrupted noise method using an exponential averaging device with its own decay time, $T_{detector}$, the standard deviation, $\sigma_{T,r}$, for repeated measurements at the same microphone position is (Davy et al., 1979).

$$\sigma_{T,r} = \sqrt{\frac{720T}{B_s X^3}\left(\frac{10}{\ln 10}\right)^2 F\left(\frac{\ln 10}{10}\frac{T}{T_{detector}}X\right)} \tag{3.143}$$

For the interrupted noise method using a linear averaging device there are a larger number of variables that determine the standard deviation. For a device which averages N samples during an integration time, T_{int}, with a time interval, a, between the start of successive averages ($a \geq T_{int}$), the standard deviation, $\sigma_{T,r}$, for repeated measurements at the same microphone position is (Davy et al., 1979)

$$\sigma_{T,r} = \sqrt{\frac{\left(\dfrac{10}{\ln 10}\right)^2\left(\dfrac{1}{B_s T_{int}} + \dfrac{2}{N}\right)}{\dfrac{X^3}{720aT} + \dfrac{X^2}{4T^2} + \dfrac{10aX}{T^3}}} \tag{3.144}$$

Octave-bands will have lower standard deviations than one-third-octave-bands, although the latter are most commonly used in sound insulation measurements.

The filter bandwidth used in the equations is the statistical bandwidth of the specific type of filter. For Butterworth filters this can be calculated from the effective bandwidth using Eq. 3.9 (Davy and Dunn, 1987). However, the statistical bandwidth is not always known or available; hence, an estimate can be made by using the band centre frequency, f, which gives a bandwidth of $0.23f$, for one-third-octave-bands and $0.707f$ for octave-bands.

For the interrupted noise method with N_r measurements at each of N_s different microphone positions, the total standard deviation in seconds can be estimated using:

$$\sigma_T = \sqrt{\frac{1}{N_s}\left(\sigma_{T,s}^2 + \frac{\sigma_{T,r}^2}{N_r}\right)} \tag{3.145}$$

For the integrated impulse response method, the total standard deviation in seconds can be estimated using:

$$\sigma_T = \sqrt{\frac{\sigma_{T,s}^2}{N_s}} \tag{3.146}$$

The spatial variation in terms of $\sigma_{T,s}$ is shown in Fig. 3.64 for one-third-octave-bands with a reverberation time of 1 s. This illustrates how the standard deviation decreases with increasing frequency and with increasing evaluation range. However, it should not be inferred that measurements should always use the largest evaluation range (i.e. T_{60}) in order to minimize the standard deviation. This model only applies to the diffuse field situation where there is a straight line decay. For non-diffuse fields where the decay curve is not a straight line, T_{10}, T_{15},

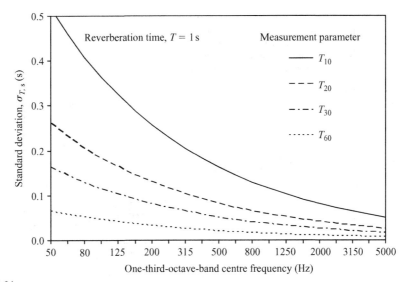

Figure 3.64

Standard deviation for the spatial variation of the reverberation time with different evaluation ranges.

or T_{20} needs to be used to relate the reverberant sound pressure to the sound power radiated into a room (Section 1.2.7.5.2); note that $\sigma_{T,s}$ calculated in this section is only applicable to diffuse fields.

To assess the total standard deviation for interrupted noise measurements we will take an example using exponential averaging where the time constant, τ, of an exponential averaging device is required to be less than, but as close as possible to $T/30$; to satisfy this requirement we will use $\tau = 0.99\ T/30$. For parallel filter measurement of all frequency bands (using either linear or exponential averaging), the choice of averaging time should be based on an estimate of the smallest reverberation time to be measured. To look at the effect of increasing the number of repeated measurements at each microphone position we will consider a 20 dB evaluation range. With parallel filter measurements used for sound insulation, T_{20} can be used for measurements over the entire building acoustics frequency range (ISO 140 Parts 3 & 4). From Fig. 3.65 it is clear that there is little to be gained by taking large numbers of repeat measurements, hence $N_r = 2$ is commonly used for reverberation time measurements that are intended for sound insulation calculations.

The standard deviation in seconds (Eqs 3.145 and 3.146) increases with increasing reverberation time. Hence, it may be inferred that more measurements are needed for longer reverberation times. However, to calculate the airborne or impact sound insulation, we need to know the total standard deviation in decibels that corresponds to the correction terms, $10\ \lg(T/T_0)$ and $10\ \lg(A/A_0)$; this can be estimated from the normalized standard deviation using

$$\sigma_{T(\text{dB})} \approx 4.34\frac{\sigma_T}{\overline{T}} \tag{3.147}$$

where $\overline{T}$ is the mean value of the reverberation time.

Examples of $\sigma_{T(\text{dB})}$ are shown in Figs 3.66 and 3.67 for the interrupted noise method and the integrated impulse response method respectively. This indicates that the standard deviation in

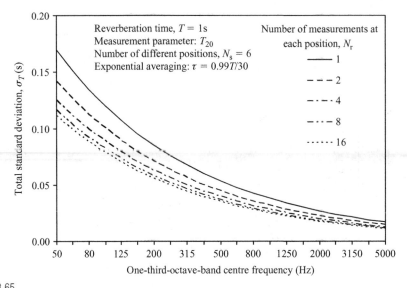

Figure 3.65

Total standard deviation for the interrupted noise method (exponential averaging) with different numbers of repeat measurements at each position.

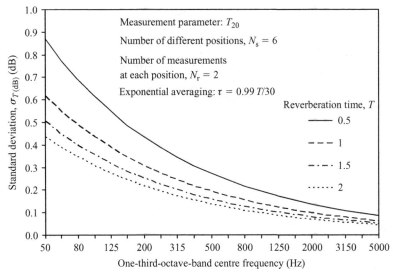

Figure 3.66

Standard deviation in decibels for the interrupted noise method using exponential averaging.

decibels increases with decreasing reverberation time and that the integrated impulse response method has slightly lower standard deviations than the interrupted noise method.

The standard deviations only apply to diffuse fields; hence the values in the low-frequency range will not be applicable to many typical rooms. Davy *et al.* (1979) note that Eqs 3.141, 3.143, and 3.144 tend to overestimate the standard deviation in the low-frequency range. Davy (1988) used

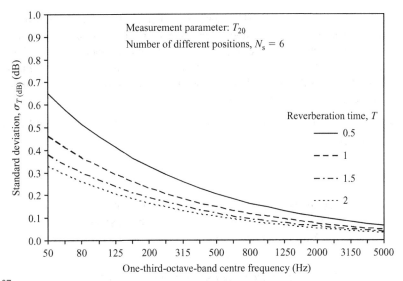

Figure 3.67

Standard deviation in decibels for the integrated impulse response method.

measurements in reverberant chambers (100–600 m³) to derive empirical correction factors to account for this overestimation based on the statistical modal overlap. The correction factors were derived using T_{30} where the decay curve could be considered as linear over the evaluation range; hence they are suited to reverberant chambers rather than typical rooms.

The choice of a suitable probability distribution for the spatial variation of reverberation time is less well-defined than for the spatial variation of mean-square sound pressure. Bodlund (1976) has shown that either a gamma or a normal (Gaussian) probability distribution could be assumed for diffuse fields in empty rooms with hard surfaces. However, the presence of absorption (particularly when concentrated on one surface) makes it harder to identify a single, simple probability distribution that is appropriate.

3.9 Maximum Length Sequence (MLS) measurements

When airborne sound insulation is measured using broad-band noise, the sound pressure level to be measured in the receiving room is not always sufficiently high above the background noise. In these situations it is necessary to increase the power output from the loudspeaker; this can be done by limiting the broad-band noise to an individual frequency band. However, the increase in level may still not be sufficient. This problem can be overcome by using a signal that is commonly referred to as MLS (Schroeder, 1979). This allows the measurement of signals (e.g. sound pressure, vibration) at low levels, as well as reverberation times in the presence of high background noise. MLS measurements are well-suited to reverberation time measurements as they determine the impulse response of an acoustic system; this facilitates use of the integrated impulse response method as well as reverse-filter analysis (Section 3.8.3.2.2).

The signal processing for MLS measurements is more complex than with broad-band noise measurements. Modern analysers usually automate the MLS measurement process so that it is not essential (although it is beneficial) to have an in-depth knowledge of the signal processing.

To reap the benefits of using MLS with sound insulation measurements, it is necessary to accept longer measurement times, restrict changes in the environmental conditions during the measurement, and prohibit the use of moving microphones and loudspeakers. This section starts with an overview of MLS in its application to sound insulation measurements and then reviews the limitations of the method. These limitations are due to the requirement for the system under test to be linear and time-invariant.

More details on the generation and processing of MLS signals can be found in Rife and Vanderkooy (1989).

3.9.1 Overview

MLS is a deterministic signal; in other words, it is exactly repeatable. This is in contrast to white noise, which is a random signal and is therefore described by its statistics. However, like white noise, the shape of the MLS frequency spectrum is white. An MLS signal theoretically has a crest factor of unity and this low value is beneficial because it gives high signal-to-noise ratios.

An MLS signal consists of a periodic binary sequence where each value in the sequence is either $+1$ or -1 and is equally spaced in time. One MLS period contains $2^n - 1$ values where n is an integer and describes the order of the sequence. For example, an MLS period of order $n = 15$ comprises a sequence of $2^{15} - 1$ values. The values in the sequence are generated from digital shift registers (Schroeder, 1979). The length of the MLS period, T_{MLS}, in seconds, is calculated by dividing the number of values in the sequence, by the sampling rate of the clock, f_{clock}, that is used to generate the sequence,

$$T_{\text{MLS}} = \frac{2^n - 1}{f_{\text{clock}}} \tag{3.148}$$

To send the MLS signal to a loudspeaker, it is converted to an analogue signal of bi-polar pulses. Part of an MLS sequence and its analogue form is shown in Fig. 3.68. The MLS sequence of $+1$ and -1 values can be viewed as a sequence of impulses with different polarities.

The MLS measurement process is used to determine the impulse response of a system, $h(t)$, where the system is defined as the combination of the measurement equipment and the acoustic system that is being measured. For a linear and time-invariant system with an impulse response, $h(t)$, the input signal $x(t)$ is related to the output signal, $y(t)$, by the convolution of $x(t)$ and $h(t)$.

$$y(t) = x(t) * h(t) \tag{3.149}$$

The impulse response is determined from the cross-correlation function, $R_{xy}(\tau)$ between the input and the output signals. Cross-correlation describes the relationship between the input and output signals in terms of the time displacement, τ, between these two signals after the output signal has been modified by the measurement equipment and the room and/or structure under test. It is particularly useful in extracting signals from noise where the input signal is related to the output signal. This is the situation in which MLS can be used to overcome the problem of high background noise. We recall that the MLS signal is deterministic; hence, we know everything about the signal that leaves the analyser to allow us to correlate it with the signal received back at the analyser from the microphone. The repeatable nature of the MLS signal means that when averaging the response of a linear and time-invariant system to an MLS signal it will be the same each time, whereas unwanted background noise will be uncorrelated

(a)

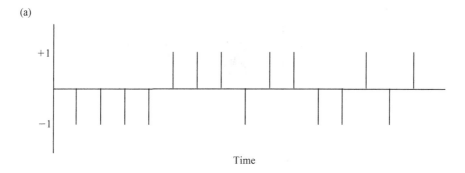

(b)

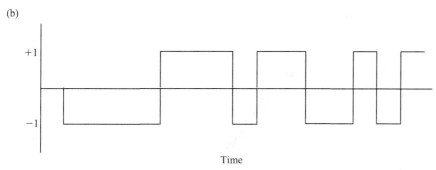

Figure 3.68

Part of an MLS period: (a) comprising a sequence of impulses with different polarities and (b) the analogue version of the MLS signal.

at different points in time. Cross-correlation can also be carried out with output signals of random noise by using Fast Fourier Transform (FFT) techniques. However, the random nature of the output signal means that more averages are needed than with MLS, and time-windowing is necessary because unlike MLS the output signal is not periodic.

The cross-correlation function is determined from the auto-correlation function of the input, $R_{xx}(\tau)$, convolved with the impulse response, and is given by,

$$R_{xy}(\tau) = R_{xx}(\tau) * h(\tau) \qquad (3.150)$$

For the MLS input signal, the auto-correlation function, $R_{xx}(\tau)$, is (almost) the Dirac delta function, $\delta(\tau)$. This allows us to make use of the fact that convolution of an impulse with the impulse response of the system will reproduce the impulse response of that system. Hence the impulse response, $h(\tau)$, is equal to the cross-correlation function, $R_{xy}(\tau)$.

$$R_{xy}(\tau) = \delta(\tau) * h(\tau) = h(\tau) \qquad (3.151)$$

For MLS signals, $R_{xy}(\tau)$ can be calculated using the Fast Hadamard Transform (FHT), an efficient cross-correlation algorithm. The measured sequence of impulses that are received from the microphone occur at different times, and have different polarities due to the $+1$ and -1 values used to create the sequence. The Hadamard transformation takes all the impulse responses and shifts them so that they are all aligned to occur at the same time with the

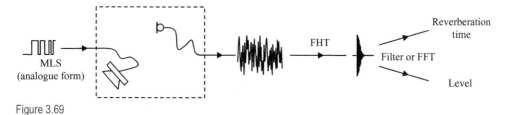

Figure 3.69

Schematic diagram of MLS measurement.

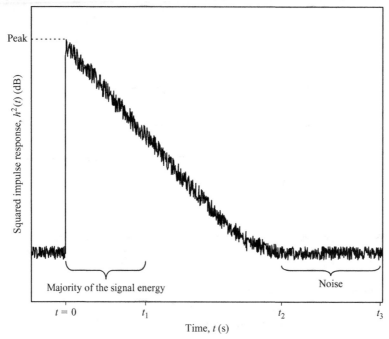

Figure 3.70

Squared impulse response of an acoustic system.

same polarity. This results in the impulse response, $h(t)$. The impulse response is then pro-cessed to give the spectrum or the decay curve. An outline of the MLS measurement process is shown in Fig. 3.69.

For level measurements, the squared impulse response, $h^2(t)$ shown in Fig. 3.70 is used to calculate the energy that is due only to the signal whilst avoiding the parts that are affected by background noise. The mean-square signal level, E_{signal}, is calculated by integrating $h^2(t)$ between 0 and t_1. Between t_1 and t_2, the signal level is affected by background noise and between t_2 and t_3, there is only background noise. Note that E_{signal} does not account for the signal energy embedded in the noise between t_1 and t_2. However, this will not cause errors as long as t_1 is chosen such that the signal energy between t_1 and t_2 is negligible compared to the energy between 0 and t_1. The requirement in the relevant Standard (ISO 18233) is that $t_1 \geq T/3$, where $T/3$ corresponds to the 20 dB down point from the peak of the squared impulse response. To quantify this error we use an idealized squared impulse response with an exponential decay that is calculated from the reverberation time of the system, T.

$$h^2(t) = X \exp\left(\frac{-6t \ln 10}{T}\right) \tag{3.152}$$

where X is the peak value of the squared impulse response.

Note that in describing features of the MLS measurement process, we need to refer to the reverberation time of the system, T. For measurements where the loudspeaker is in the same room as the microphone, use of the term reverberation time is unambiguous. When we measure the signal in the receiving room with the loudspeaker in a different room, we can still measure a reverberation time from the impulse response, but this particular reverberation time is not used in sound insulation calculations.

The estimate of the total signal energy is obtained by integrating the squared impulse response from 0 to t_1. The total signal energy is determined by integrating from 0 to T. Therefore the total signal energy is underestimated by ΔE in decibels, where,

$$\Delta E = 10 \lg\left(1 - \exp\left(\frac{-6t \ln 10}{T}\right)\right) \tag{3.153}$$

For $t_1 = T/3$, ΔE is $-0.04\,\text{dB}$. This error is sufficiently small that it is reasonable to assume that the majority of the energy is contained in the initial part of the squared impulse response between 0 and $T/3\,\text{s}$.

The sound pressure level, $L_{p,\text{MLS}}$, at a single microphone position in either the source or the receiving room is calculated from the squared impulse response using:

$$L_{p,\text{MLS}} = 10 \lg\left(\frac{W_0 \displaystyle\int_0^{t_1} h^2(t)dt}{C_{\text{ref}}}\right) \tag{3.154}$$

where W_0 is a constant describing the signal power per unit bandwidth of the MLS signal, and C_{ref} is an arbitrary reference value.

The background noise level is estimated by integrating the energy in the squared impulse response between t_2 and t_3.

$$L_{\text{background,MLS}} = 10 \lg\left(\frac{\dfrac{1}{t_3 - t_2}\displaystyle\int_{t_2}^{t_3} h^2(t)dt}{C_{\text{ref}}}\right) \tag{3.155}$$

To gain an improved estimate of the sound pressure level, $L_{p^*,\text{MLS}}$, this background noise can be 'removed' from the signal energy that is contained between 0 and t_1.

$$L_{p^*,\text{MLS}} = 10 \lg\left(\frac{\displaystyle\int_0^{t_1}\left[W_0 h^2(t)\right] - \left[\dfrac{1}{t_3 - t_2}\displaystyle\int_{t_2}^{t_3} h^2(t)dt\right]dt}{C_{\text{ref}}}\right) \tag{3.156}$$

With broad-band noise signals, a correction can be made for the background noise using Eq. 3.75. This correction must not be used with MLS measurements as account has already been taken of the background noise in the MLS measurement process.

By using several stationary microphone positions in the source and receiving rooms, the spatial average sound pressure levels L_{p1} and L_{p2} can be calculated and used to calculate the level difference needed to determine the airborne sound insulation.

The reverberation time is determined from the integrated impulse response in the same way as with other impulse measurements, such as a pistol shot, by the application of reverse-time integration (Section 3.8.2).

The length of the MLS period, T_{MLS}, should be equal to, and preferably longer than the reverberation time of the system being measured, T. This ensures that the MLS period will excite every resonance in the system (Bjor, 1995). It is usually necessary to average the measured impulse responses from several MLS periods to provide the required signal-to-noise ratio. Hence, this minimum period length also avoids problems with the tails of the measured impulse responses overlapping into the next MLS period, known as time aliasing. It therefore appears that we need to know the result of the measurement, i.e. the reverberation time, before we can set the MLS period length on the analyser to actually measure the reverberation time. This is not a problem in practice because the maximum reverberation time that is typically encountered in rooms, particularly in dwellings, is less than 8 s. Therefore the order of the MLS sequence and the sampling rate can be set to give a valid period length for the majority of sound insulation measurements.

The MLS measurement process uses one MLS period to excite the system so that it is in a steady state before using subsequent MLS periods to make the measurement. In contrast to measurements with broad-band noise, the MLS measurement process determines the background noise during the measurement period, which is the period of time that we are actually interested in. Both stationary and transient background noise energy is uniformly distributed in time over the squared impulse response. This 'smearing' of transient background noise energy over time means that MLS effectively shows a degree of immunity to unwanted transients (Rife and Vanderkooy, 1989).

For level measurements it is useful to assess the improvement in the signal-to-noise ratio, ΔSNR, that is gained by using the MLS measurement process compared to stationary random noise such as broad-band noise commonly used for sound insulation measurements (Vorländer and Kob, 1997). To do this the MLS signal is considered in two different ways: firstly using the MLS measurement process, and secondly by treating the MLS signal as a stationary noise signal and using it for measurements in the same way as with broad-band noise. By taking advantage of the fact that MLS is deterministic, several MLS periods can be synchronized and averaged to improve the signal-to-noise ratio. Between these MLS periods there will be correlation between the impulse responses, but there will be no correlation between the background noise during these periods. By averaging N MLS periods, the estimated improvement in the signal-to-noise ratio in decibels by using MLS compared to broad-band noise is (Vorländer and Kob, 1997; ISO 18233):

$$\Delta \text{SNR} = 10 \lg \left(\frac{N T_{MLS}}{t_1} \right) \qquad (3.157)$$

Hence doubling the number of MLS periods in the averaging process will increase the signal-to-noise ratio by 3 dB.

As noted previously, the majority of the signal energy is contained in the initial part of the squared impulse response. Hence for the squared impulse response derived from the MLS measurement process it is more appropriate to describe the signal-to-noise ratio for the MLS measurement process using the peak-to-mean-square noise ratio (Vorländer and Kob, 1997). The peak of the impulse response is indicated in Fig. 3.70, along with the portion of the signal used to estimate the mean-square noise.

The estimated improvement in the signal-to-noise ratio in decibels by using MLS compared to broad-band noise, i.e. the ratio of the peak-to-mean-square noise ratio for MLS, to the signal-to-noise ratio for broad-band noise, is (Vorländer and Kob, 1997):

$$\Delta \text{SNR} = 10 \lg \left(\frac{6 T_{\text{MLS}} \ln 10}{T} \right) = 10 \lg \left(\frac{T_{\text{MLS}}}{T} \right) + 11.4 \, \text{dB} \qquad (3.158)$$

Because T_{MLS} must at least equal T, the minimum value of ΔSNR is 11.4 dB. A typical value of T_{MLS} to cover the building acoustics frequency range is about 8 s, so with a reverberation time of 1 s, ΔSNR is 20.4 dB. Such large improvements in the signal-to-noise ratio mean that the MLS measurement process allows measurement of the airborne sound insulation in circumstances where high background noise would otherwise prevent the use of broad-band noise. MLS measurements can therefore be attempted in situations where the signal-to-noise ratio for broad-band noise measurements would be close to 0 dB.

3.9.2 Limitations

There are many advantages to using the MLS measurement process for sound insulation measurements, but there are limitations that need consideration in any measurement. These stem from the requirement that the system being measured must be linear and time-invariant. The relevant elements in this system are the chain of measuring equipment, the environmental conditions in which sound pressure level measurements are taken, and all the physical elements, such as the walls, floors, or ground, that affect sound propagation and transmission from the loudspeaker to the microphone.

In airborne sound insulation measurements the main factor that affects linearity is distortion from the loudspeaker and/or the amplifier when they are driven too strongly. Non-linearity is identifiable by large spikes in the tail of the impulse response between time t_2 and t_3. This problem is straightforward to solve by reducing the level of the output signal, and, where necessary, increasing the number of averages to achieve the required signal-to-noise ratio. Hence, it is the time-invariant requirement, rather than linearity, that is the more important issue for sound insulation measurements (Bjor, 1995; Svensson and Nielsen, 1999; Vorländer and Kob, 1997). In contrast, measurements using broad-band noise are relatively immune to time-variance and there are fewer problems with time-variance if the impulse response is determined using a swept-sine signal instead of MLS.

MLS measurement relies on the Hadamard transformation to shift the measured impulse responses so that they all occur at the same time with the same polarity. Therefore, any variable that changes sound propagation or transmission between the loudspeaker and the microphone is a potential source of error if it changes the impulse response of the system during the measurement. Common causes of such time-variance are air movement, variation in temperature, variation in humidity, moving loudspeakers, and moving microphones. Two types of time-variance need consideration: variation within an MLS period (intra-periodic) and variation between MLS periods (inter-periodic). For sound insulation measurements we typically use MLS periods that are a few or several seconds in length, hence both types are relevant. Time-variance has the potential to change the frequency, amplitude, and the phase of a wave. However, it only takes a small change to the speed of sound to introduce significant phase shifts; hence, this tends to be the more important factor (Vorländer and Kob, 1997). Both air movement and variations in temperature change the speed of sound. The resulting phase shift reduces the correlation between the impulse response of the system and the MLS signal, causing a decrease in the measured sound pressure level.

There tend to be fewer problems due to time-variance in the laboratory compared to the field because there is usually more control over the environmental conditions. However, it is good practice to monitor the environmental conditions in both the laboratory and the field during measurements with long averaging times. It is also beneficial to avoid using longer measurement times than are actually needed to achieve the required signal-to-noise ratio.

3.9.2.1 *Temperature*

An idealized model for the effect of temperature changes in a room during a measurement is described by Vorländer and Kob (1997). This theoretical model for single frequencies is useful in illustrating important features associated with a linear change in temperature. Three single frequencies are considered here: 100, 500, and 1000 Hz for which the actual reverberation time for each frequency is chosen to be 1 s. The inter-periodic temperature increase is chosen to be 0.0033°C per MLS period. Hence, by assuming an MLS period length that is equal to the reverberation time, and measuring over 600 periods, there will be a total increase of 2°C during a 10 min measurement. The calculated decay curves are shown in Fig. 3.71 alongside the actual decay of the system. For sound insulation measurements, the reverberation time

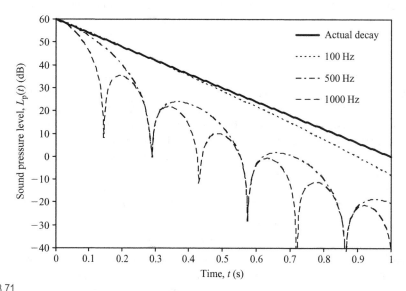

Figure 3.71

Predicted decay curves for MLS measurements at three single frequencies when there is a change in temperature during the measurement. The actual reverberation time at each frequency without temperature drift should be 1 s.

is often calculated from the first 35 dB decrease in the decay curve. During this time, the change in temperature can cause the initial decay rate to be slower than the later decay rate. A change in temperature tends to cause larger changes in the measured reverberation time at higher, rather than lower frequencies. These features have been observed in practice where measurements have been made in rooms undergoing changes in temperature (Bradley, 1996). An important feature identified by the Vorländer and Kob model is the minima in the decay curves. As with most single frequency models, it is reasonable to assume that the predicted minima will be deeper than would occur with frequency band measurements that are used in practice. However, the time at which the first minimum occurs in the decay curve can be used to establish guidance for the maximum allowable change in temperature, $\Delta\theta_{max}$ in °C, during a measurement (Vorländer and Kob, 1997). For both level and reverberation time measurements this results in an equation of the form,

$$\Delta\theta_{max} \leq \frac{X}{fT} \tag{3.159}$$

where T is the reverberation time.

Hence, the allowable change in temperature is smallest with long reverberation times and/or at high frequencies. The guidance in the relevant Standard (ISO 18233) defines the maximum allowable change in temperature over the measurement period as $X = 1300$ for level measurements, and $X = 200$ for reverberation time measurements using the decay range from 0 to 30 dB.

3.9.2.2 Air movement

Air movement occurs indoors due to HVAC systems, fans, or open windows. However, these sources are usually controllable; hence the main issue is with outdoor measurements of sound pressure levels in the presence of wind. In the measurement of façade sound insulation, the sound pressure level is measured outside the building; hence we are interested in the effect of wind on sound propagation from the loudspeaker to the microphone and to the test element. The wind changes the speed of sound and introduces phase shifts, whether the wind occurs in gusts or in more steady conditions.

Façade sound insulation measurements in a variety of wind conditions show that both the external sound pressure level and the signal-to-noise ratio are reduced by the presence of wind (Horvei *et al.*, 1998). The effect is particularly significant in the high-frequency range. It therefore tends to be important with façade elements, such as windows and ventilators for which the high-frequency range is often important in determining the single-number quantity for the façade sound insulation. Field tests from Horvei *et al.* (1998) have been used as an empirical basis upon which to set the limits on wind speeds in the relevant Standard (ISO 18233). This defines suitable wind conditions for external measurements by monitoring the wind speed close to the loudspeaker to ensure that average wind speeds are less than 4 m/s and the wind speed during gusts is less than 10 m/s.

3.9.2.3 Moving microphones

Sound pressure level and reverberation time measurements using broad-band noise can be made with moving microphones and, in the laboratory, moving loudspeakers. Whilst this is suitable for broad-band noise, any moving equipment such as microphones, loudspeakers, or diffusers, causes problems in satisfying the time-invariant requirement for MLS measurements.

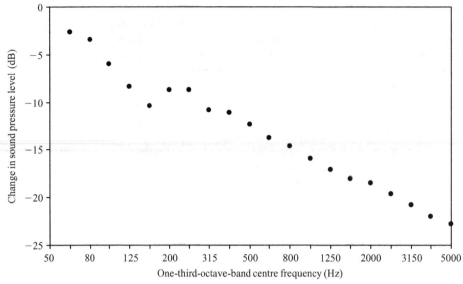

Figure 3.72

Measured change in the sound pressure level when using a rotating boom (sweep radius of 1 m and a 64 s sweep period) instead of stationary microphone positions with MLS measurements. Measured in a 206 m³ reverberation room with an MLS period of order $n = 15$, $T_{MLS} = 1.491$ s, 1000 periods used for averaging. Measured data are reproduced with permission from Weise and Schmitz (2000).

To determine spatial average sound pressure levels or reverberation times it is common to use microphones on a rotating boom rather than a number of stationary microphone positions. With MLS measurements, movement of the microphone introduces both intra-periodic and inter-periodic time-variance as the impulse response changes in magnitude and phase within an MLS period and between MLS periods. This time-variance reduces the correlation between the input and the output signals. The result is a reduction in the measured signal level compared to the level that would be measured by using stationary positions along the path of the rotating microphone (Bietz *et al.*, 1997; Weise and Schmitz, 2000). The reduction is dependent upon the sweep radius of the rotating microphone as well as frequency. Examples of measured data are shown in Fig. 3.72 for a 1 m sweep radius that is commonly used in airborne sound insulation measurements where there is no synchronization between the sweep time and the MLS period (Weise and Schmitz, 2000). This shows the general trend that the reduction is largest at high frequencies. However, for airborne sound insulation we are interested in a sound pressure level difference rather than absolute levels. It is possible to gain a reasonable estimate of the level difference between the source and receiving rooms if exactly the same sweep radius and sweep period are used in both rooms (Bietz *et al.*, 1997). In practice, the significant reduction in level at high frequencies along with the potential errors in setting up identical rotating booms means that the use of moving microphones with MLS is not a viable option (Weise and Schmitz, 2000). Hence, only stationary microphones are considered for MLS measurements.

3.10 Sound intensity

Instantaneous sound intensity is a vector quantity, and is equal to the product of the sound pressure and the particle velocity. We are usually interested in measuring the time-averaged intensity in stationary sound fields; this vector quantity describes the net flow of sound energy passing through a unit area that lies normal to the measurement surface. This can be used to calculate the sound power radiated by various surfaces in rooms.

If we were to measure the intensity of a propagating plane wave in an anechoic environment, this could be done simply by using measurements of the time-averaged mean-square sound pressure (Eq. 1.19). In practice we want to measure sound intensity in a variety of different sound fields in and around buildings, often where the radiated sound is reflected back from other surfaces and where there are other sound sources present. For example, measurement of sound intensity radiated by a separating wall in a reverberant room where there is flanking transmission from the surrounding walls and floors. In these situations, a single microphone cannot be used to give the magnitude and direction of the intensity, it is necessary to determine both the sound pressure and the particle velocity.

To illustrate the issues pertaining to sound intensity measurement in the presence of reflected waves it is simplest to look at the superposition of two plane waves travelling in opposite directions (Fahy, 1989). In a one-dimensional interference field, the temporal and spatial variation of the sound pressure is described by:

$$p(x, t) = [\hat{p}_+ \exp(-ikx) + \hat{p}_- \exp(ikx)] \exp(i\omega t) \tag{3.160}$$

where $\hat{p}_+$ and $\hat{p}_-$ are the amplitudes of the two waves.

Equation 3.160 can be re-written in terms of spatially varying amplitude and phase terms, $\hat{p}(x)$ and $\phi(x)$ as:

$$p(x, t) = \hat{p}(x) \exp[i(\omega t + \phi(x))] \tag{3.161}$$

The particle velocity can now be determined from Euler's equation,

$$u(x, t) = -\frac{1}{\rho_0} \int \frac{\partial p}{\partial x} \, dt \tag{3.162}$$

which gives:

$$u(x, t) = \frac{i}{\omega \rho_0} \frac{\partial p}{\partial x} = \frac{i}{\omega \rho_0} \left[\frac{d\hat{p}(x)}{dx} + i\frac{d\phi(x)}{dx} \hat{p}(x) \right] \exp[i(\omega t + \phi(x))] \tag{3.163}$$

The instantaneous sound intensity (sometimes called energy flux density), $I(x, t)$, in stationary sound fields is given by the product of sound pressure (Eq. 3.161) and particle velocity (Eq. 3.163). It has two components, the active intensity, I_A, and the reactive intensity, I_R,

$$I(x, t) = p(x, t)u(x, t) = I_A + iI_R \tag{3.164}$$

Active intensity describes the net flow of sound energy and is proportional to the phase gradient where,

$$I_A = -\frac{[\hat{p}(x)]^2}{\omega \rho_0} \frac{d\phi(x)}{dx} \cos^2(\omega t + \phi(x)) \tag{3.165}$$

It is the active component that is of most importance, and for stationary sound fields we need the time-averaged value. The time-averaged active intensity is proportional to the phase gradient, and is given by:

$$\langle I_A \rangle_t = -\frac{[\hat{p}(x)]^2}{2\omega \rho_0} \frac{d\phi(x)}{dx} \tag{3.166}$$

Reactive intensity describes non-propagating energy that is moving back and forth, and is proportional to the gradient of mean-square pressure,

$$I_R = -\frac{1}{4\omega\rho_0}\frac{d[\hat{p}(x)]^2}{dx}\sin[2(\omega t + \phi(x))]$$

(3.167)

The instantaneous reactive intensity can take non-zero values, but the time-averaged reactive intensity, $\langle I_R\rangle_t$, is zero. Quantifying the reactive intensity is of limited practical use. However, its existence helps to explain the difficulties that occur when trying to measure the active component. In a propagating plane wave there is only active intensity. In the one-dimensional interference field described above, there would only be reactive intensity if there was no dissipation of sound energy (i.e. no sound absorption). This does not occur in reality as there will always be some dissipation of energy, which will result in an active component. In three-dimensional space it is clear that there will also be a reactive sound field in reverberant rooms, but because surfaces always absorb some fraction of the incident sound intensity, there will also be an active component. The sound field is highly reactive in the nearfield of a vibrating plate or very close to a sound source, but it fades quite rapidly with distance which allows active intensity to be measured quite close to a vibrating surface, often between 0.1 and 0.3 m from the surface. For sound insulation, we need to measure the active intensity in sound fields which have various degrees of reactivity. The word 'active' is commonly omitted and we simply refer to intensity.

A thorough overview of sound intensity in theory and in practice can be found in the book by Fahy (1989).

3.10.1   p–p sound intensity probe

To measure sound intensity there are two main types of probe: the p–p and the p–u probe. The p–p probe comprises two pressure microphones and uses a finite-difference approximation to estimate the particle velocity. The p–u probe uses direct measurement of the pressure and the particle velocity. The most common type is the p–p probe; this typically has two microphones in a face-to-face configuration, separated by a solid spacer of length, d (see Fig. 3.73). We will focus on the measurement of active intensity using a p–p probe. This probe comprises two microphones, No.1 and No. 2, which measure the sound pressures, p_1 and p_2 respectively. Positive intensity is defined for a plane wave propagating in the x-direction from microphone No.1 to No. 2.

For a wave propagating in the x-direction, the particle velocity, u_x, is related to the pressure gradient, $\partial p/\partial x$, by Euler's equation (Eq. 3.162). The acoustic centres of the two microphones are separated by a distance, d, hence the pressure gradient can be determined using a finite-difference approximation,

$$\frac{\partial p}{\partial x} = \frac{p_2 - p_1}{d}$$

(3.168)

The microphones are equally spaced about the point at which the intensity estimate is made, so the sound pressure at the mid-point between the microphones is

$$p = \frac{1}{2}(p_1 + p_2)$$

(3.169)

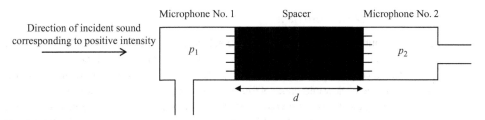

Figure 3.73

Sound intensity measurement: p–p probe.

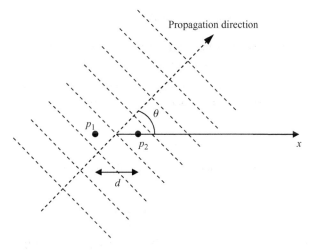

Figure 3.74

Orientation of a p–p sound intensity probe in a propagating plane wave field.

Therefore the time-averaged sound intensity in the x-direction, I_x, can be determined using

$$I_x = \langle pu_x \rangle_t = \left\langle \frac{1}{2}(p_1 + p_2)\frac{1}{\rho_0 d} \int (p_2 - p_1)\, dt \right\rangle_t \qquad (3.170)$$

A p–p probe measures the net intensity component along the axis of the probe, not the complete vector. The importance of this fact becomes clearer if we consider a single plane wave propagating at an angle, θ, to the x-axis (see Fig. 3.74). The probe axis is aligned with the x-axis. For a wave with intensity, I, the probe measures the component, $I \cos\theta$. For $\theta < 90°$, the intensity component along the probe axis is $10\lg(\cos\theta)$ dB below the level at $\theta = 0°$; at $60°$ the intensity component along the axis of the probe is 3 dB below the level corresponding to $0°$. At $\theta = 90°$ the wave propagates perpendicular to the probe axis, and there is no intensity component along the probe axis.

To determine the three-dimensional intensity vector it is necessary to measure in three mutually perpendicular directions to give the resultant intensity vector, $\mathbf{I}$,

$$\mathbf{I} = I_x\mathbf{i} + I_y\mathbf{j} + I_z\mathbf{k} \qquad (3.171)$$

345

3.10.1.1   *Sound power measurement*

To determine the sound power of a radiating object or surface, it is necessary to define a measurement surface, S_M, that encloses a volume around the object or surface. The basis for this lies in Gauss's divergence theorem; this concerns the fact that a field enclosed within a volume can only change by flow into or out of the volume. Application of this theorem means that the sound intensity inside a specified volume is determined by the stationary signal emitted by the sound source(s) enclosed within this volume, and by the sound intensity that flows into the volume due to stationary signals from sound sources outside it (if any). Hence the net sound power, W_{net}, of the source(s) enclosed within the volume can be found by using Gauss's theorem to transform a volume integral into a surface integral over the measurement surface,

$$W_{net} = \int_V \nabla.\mathbf{I}\,dV = \int_S \mathbf{I}.\mathbf{n}\,dS \qquad (3.172)$$

where $\mathbf{n}$ is the unit normal vector of the measurement surface that encloses the volume, with the vector pointing out of the volume enclosed by this surface.

The axis of the p–p probe must therefore be perpendicular to the measurement surface so that the intensity component being measured is normal to this surface. This component is denoted as I_n, and is given by

$$I_n = \mathbf{I}.\mathbf{n} \qquad (3.173)$$

The temporal and spatial average value over the measurement surface gives the normal sound intensity level, L_{In}, as

$$L_{In} = 10\lg\left(\frac{|I_n|}{I_0}\right) \qquad (3.174)$$

where $I_0 = 10^{-12}$ W/m^2. Note that the magnitude is used to give an unsigned value of I_n. This means that L_{In} does not give information on the direction of the measured intensity; I_n would be positive for a net intensity component flowing out of the measurement surface and negative when flowing into the surface.

3.10.1.1.1   *Measurement surfaces*

Example measurement surfaces are shown in Fig. 3.75. For the wall there are narrow strips around the edge of the measurement surface on the side walls, floor, and ceiling; it is important that these strips radiate insignificant sound power. We are only trying to measure one source (in this case, a wall) hence these strips are unwanted sound sources that are enclosed within the volume by the measurement surface. This is not usually an issue in the laboratory where flanking transmission is suppressed. In the field, measurements are almost always taken in the presence of flanking. Therefore, the sound power radiated by the test element should be more than 10 dB higher than the total sound power radiated by these narrow strips (ISO 15186 Part 2). If this is not satisfied then the sound radiated by these strips needs to be reduced by covering them with an additional lining. However, it is difficult to get an accurate measurement of the sound power radiated by these narrow strips, particularly when they are only ≈ 0.1 m wide, so a single scan along each strip is usually considered sufficient to estimate their sound power (ISO 15186 Part 2). Another problem can occur in the field when these strips absorb significant sound power. In this situation, the measured sound power will underestimate the actual value. When the strips are highly absorbent, they need to be covered with a lining that makes the combination of the lining and the wall/floor construction reflective.

(a) (b)

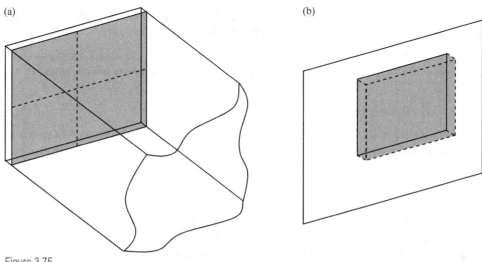

Figure 3.75

Measurement surface, S_M (shaded area) used to determine the net sound power radiated by (a) a wall in a room and (b) a window flush with the façade. The dashed lines indicate how the measurement surface could be divided into sub-areas for scanning measurements (e.g. four sub-areas for the wall and five sub-areas for the window).

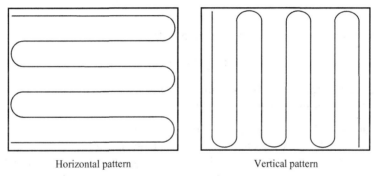

Horizontal pattern   Vertical pattern

Figure 3.76

Scanning patterns for intensity measurements.

3.10.1.1.2 Discrete point and scanning measurements

Measurements for sound insulation purposes are taken with stationary random signals, so intensity measurements can either be taken at discrete points on the measurement surface, or by scanning the probe across the surface. Both methods are commonly used in the laboratory, but scanning is usually more practical in the field.

For scanning measurements it is common to split the measurement surface into a number of sub-areas. The area or sub-area is scanned at a constant speed between 0.1 to 0.3 m/s, using both a horizontal and vertical scanning pattern (see Fig. 3.76); the arithmetic average of the measured intensity from the two patterns is usually deemed to be acceptable when the difference between the two patterns is less than 1.0 dB (Jonasson, 1991, 1993; ISO 15186 Parts 1 & 2). A suitable distance between the lines in the scanning pattern depends on the variation in the intensity over the measurement surface; this distance is usually the same as the distance of the probe from the surface (ISO 15186 Parts 1 & 2).

Calculation of the net sound power for discrete point and scanning measurements is described in the relevant Standards along with other checks on the sound intensity measurement.

3.10.1.2 Error analysis

To assess the errors involved in intensity measurement with a p-p probe it is simplest to look at single frequencies under the assumption that the results are generally applicable to frequency bands.

The finite-difference approximation introduces a systematic error in both the pressure at the mid-point and in the pressure gradient. These errors combine to give a systematic error in the intensity, and depend upon the sound field that is being measured. As before, we will assume a plane wave propagating at an angle, θ, to the x-axis (Fig. 3.74). In a plane wave field where the intensity is measured in the x-direction, the normalized error, $e_{FD}(I_x)$, due to the finite-difference approximation is (Fahy, 1989; Pavić, 1977):

$$e_{FD}(I_x) = \frac{I_x - I}{I} \approx -\frac{1}{6}\left(kd \cos\theta\right)^2 \tag{3.175}$$

where I_x is the value measured using the finite-difference approximation and I is the actual value.

Normalized errors in linear values are converted to decibels using

$$10 \lg\left(1 + e(I_x)\right) \tag{3.176}$$

Hence, as $e_{FD}(I_x)$ is negative, the finite-difference approximation causes the measured intensity level to be an underestimate of the actual value.

For spherical waves from a point source, Eq. 3.175 also applies where $kr \gg 1$, i.e. in the far-field (Fahy, 1989). Another source of particular interest is a vibrating plate such as a wall or floor; however, for finite plates it is difficult to quantify the error, even more so with non-homogeneous, isotropic plates. This does not cause problems in practice if the probe is at least 0.1 m away from the vibrating surface (ISO 15186 Parts 1, 2, & 3).

There is also a measurement error due to phase-mismatch in the equipment. For the propagating plane wave shown in Fig. 3.74, the phase difference between the two points is $kd \cos\theta$. However, there will also be an unwanted phase difference, $\pm\phi_{PM}$, between the two measurement channels. This occurs because for each channel there will inevitably be minor differences between the microphones and other hardware or signal processing equipment. Therefore the impulse response for each measurement channel will have a slightly different phase response. The total phase difference measured in a propagating plane wave field will be $kd \cos\theta \pm \phi_{PM}$, hence the normalized error, $e_{PM}(I_x)$, due to phase mismatch is

$$e_{PM}(I_x) = \frac{I_x - I}{I} \approx \frac{\pm\phi_{PM}}{kd \cos\theta} \tag{3.177}$$

where I_x is the value measured with phase-mismatch but without any other errors.

From Eq. 3.177 it is seen that phase-mismatch errors are smallest when the axis of the probe is aligned with the propagation direction.

With increasing frequency and increasing microphone spacing, d, the magnitude of the normalized error for the finite-difference approximation increases, whereas the magnitude of the normalized error for the phase-mismatch decreases. The combined normalized error due to the finite-difference error and phase-mismatch is

$$e(I_x) \approx -\frac{1}{6}\left(kd\cos\theta\right)^2 \pm \frac{\phi_{PM}}{kd\cos\theta} \qquad (3.178)$$

Equation 3.178 can be used to assess what length of microphone spacer is needed to cover the different parts of the building acoustics frequency range. It can be seen that there will be some cancellation of the individual errors when the phase-mismatch is positive. Hence the error is calculated for both positive and negative phase-mismatch. To make an assessment of suitable spacer lengths it is necessary to set a tolerable error; this can be taken as a normalized error of ±5%, which corresponds to limits of ±0.2 dB (Fahy, 1989).

For the band centre frequencies we will now assume the following phase-mismatch values: $\phi_{PM} = \pm0.05°$ between 20 and 250 Hz, and $\phi_{PM} = \pm f/5000°$ between 315 and 6300 Hz. These would satisfy the relevant requirement for a Class 1 p–p probe according to IEC 1043. As the error is also angle-dependent we will consider a plane wave propagating at angles of 0° and 60° to the x-axis. The combined normalized errors are shown in Fig. 3.77 for two common spacer lengths, 50 and 12 mm. With a 50 mm spacer the combined error is within the ±0.2 dB limits between 50 and 500 Hz; above 500 Hz it is the finite-difference error that causes the large error. With a 12 mm spacer the combined error is just within the ±0.2 dB limits between

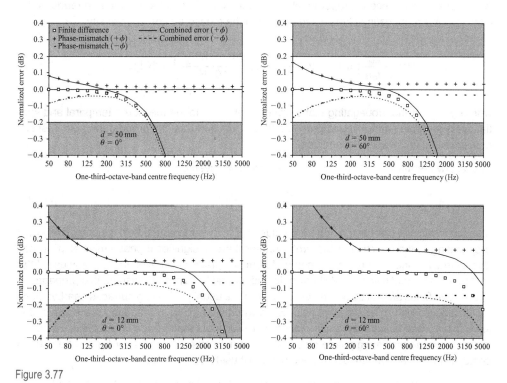

Figure 3.77

Normalized errors for a p–p probe in a propagating plane wave (± 0.2 dB corresponds to a normalized error of 5%).

200 and 2000 Hz. However, another important error needs to be considered, the effects of scattering and diffraction from the probe itself. Fortuitously for half-inch microphones with a 12 mm spacer in a face-to-face configuration, these effects compensate for the finite-difference error; this means that the 12 mm spacer can be used up to 10 000 Hz (Jacobsen et al., 1998). Hence it is possible to cover the entire building acoustics frequency range using a 50 and a 12 mm spacer. From Fig. 3.77 it is clear that if the phase mismatch could be accurately corrected, then measurements could be made over the entire building acoustics frequency range with a single 12 mm spacer (Jacobsen, 1991; Ren and Jacobsen, 1992); some intensity systems incorporate phase correction to achieve this.

The intensity probe is usually placed in an unknown sound field where the actual phase difference is ϕ. The measured phase difference will therefore be $\phi \pm \phi_{PM}$. In order to assess the validity of intensity measurements, it is useful to compare the measured phase difference to the corresponding phase difference in an idealized sound field. For this purpose, we use the phase difference in the direction of a propagating plane wave ($\theta = 0°$), for which the phase difference is kd. Now we can look at the ratio of the phase difference in the direction of a propagating plane wave to the measured phase difference. For reasons that will shortly become apparent, this ratio in decibels is referred to as the surface pressure-intensity indicator, F_{pI}, (and sometimes as the pressure-intensity index or field indicator),

$$F_{pI} = 10 \lg \left| \frac{kd}{\phi \pm \phi_{PM}} \right| \tag{3.179}$$

As the phase difference is proportional to the intensity, F_{pI} can be re-written in terms of the intensity for a propagating plane wave relative to the measured normal intensity,

$$F_{pI} = 10 \lg \left| \frac{\frac{\langle p^2 \rangle_{t,s}}{\rho_0 c_0}}{\langle I_n \rangle_t} \right| = L_p - |L_{In}| + 10 \lg \left(\frac{(20 \times 10^{-6})^2}{10^{-12}} \right) - 10 \lg (\rho_0 c_0) \tag{3.180}$$

where $\langle p^2 \rangle_{t,s}$ for the propagating plane wave is set equal to the measured temporal and spatial average mean-square sound pressure that is measured by the probe during the intensity measurement in the actual sound field.

The last two terms in Eq. 3.180 cancel out, and F_{pI} is given by,

$$F_{pI} = L_p - |L_{In}| \tag{3.181}$$

Measurements will only give a non-zero value of F_{pI} when the sound field is not a plane wave propagating in the direction of the probe and/or there is significant phase-mismatch. It may appear that this indicator is too far removed from any practical situation to be of any use. After all, the actual sound field is rarely known. In and around buildings we assume that propagating plane wave fields are not common because there are usually reflecting surfaces nearby; these result in some kind of reverberant field. Even when we have a propagating plane wave, it is likely to be propagating at an angle other than $\theta = 0°$; note that kd was only used in Eq. 3.179 to represent the maximum phase difference. The next step is to see how F_{pI} can be used in conjunction with another measurement to make it useful in practice.

The term 'residual intensity' is used to describe the intensity value that is reported by an analyser purely due to phase-mismatch. When F_{pI} is measured with the probe orientated in a

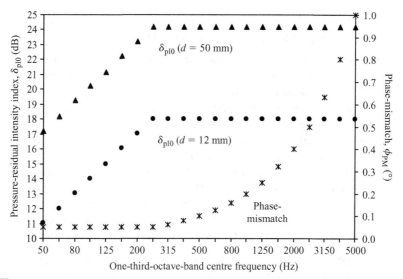

Figure 3.78

Example phase-mismatch errors for a p–p probe and the corresponding pressure-residual intensity indices for 50 and 12 mm spacers.

field of uniform sound pressure such that $\phi = 0$, the actual intensity is zero and the analyser only measures the residual intensity. In this situation, F_{pI} is referred to as the pressure-residual intensity index, δ_{pI0}, for a specific spacer distance.

$$\delta_{pI0} = 10 \lg \left| \frac{kd}{\pm \phi_{PM}} \right| \tag{3.182}$$

The pressure-residual intensity index is measured in a small cavity specially designed for the purpose. In this cavity the microphone diaphragms are exposed to the same sound pressure but they are not separated by the solid spacer; the spacer distance is merely entered into the analyser for calculation of δ_{pI0} according to Eq. 3.182.

For a specific spacer distance the pressure-residual intensity index should ideally be as large as possible. Figure 3.78 shows the pressure-residual intensity index for spacer lengths of 50 and 12 mm corresponding to phase-mismatch errors of $\phi_{PM} = \pm 0.05°$ between 50 and 250 Hz, and $\phi_{PM} = \pm f/5000°$ between 315 and 5000 Hz.

Both F_{pI} and δ_{pI0} use a propagating plane wave as a point-of-reference, hence the value, $\delta_{pI0} - F_{pI}$, can be used to give a measure of the error caused by phase-mismatch in the actual sound field,

$$\delta_{pI0} - F_{pI} = 10 \lg \left| \frac{\phi \pm \phi_{PM}}{\pm \phi_{PM}} \right| \tag{3.183}$$

Using $\delta_{pI0} - F_{pI}$ allows requirements to be set for negligible phase-mismatch error; the higher the value of $\delta_{pI0} - F_{pI}$, the lower the error. For precision or engineering grade accuracy, the error in the measured intensity level is within ± 0.5 dB when $\delta_{pI0} - F_{pI} > 10$ dB. For survey grade accuracy, it is within ± 1 dB when $\delta_{pI0} - F_{pI} > 7$ dB. These requirements are sometimes described by introducing another term, the dynamic capability index, L_d (ISO 9614 Part 1),

$$L_d = \delta_{pI0} - K \tag{3.184}$$

where K is the bias error factor. K is 7 dB for precision or engineering grade accuracy, or 10 dB for survey grade accuracy. The requirement can then be stated as $L_d > F_{pl}$.

Having established a requirement for negligible phase-mismatch error by using $\delta_{pl0} - F_{pl}$, an additional requirement can be set using F_{pl} by itself. This can indicate when a sound field is reactive and sufficiently different to a propagating plane wave as to render the measurement invalid. For intensity measurement of building elements in the laboratory, the requirement in the relevant Standards depends upon whether the surface that is being measured with the intensity probe is reflective or absorbent. The requirement is $F_{pl} \leq 10$ dB for sound reflecting test elements and $F_{pl} \leq 6$ dB for sound absorbing test elements (ISO 15186 Parts 1 & 3). The requirement is more stringent for sound absorbing elements because the intensity probe measures net flow of sound energy through unit area. If the surface of the test element that is being measured has a much higher absorption area than the total absorption area of the room, measurement of the net sound intensity will underestimate the sound power actually transmitted by the test element, and overestimate the airborne sound insulation (Roland et al., 1985; van Zyl et al., 1987).

When F_{pl} is too high there are two common causes: the sound field is too reverberant and/or the probe is too close to the element such that it is in the nearfield. These problems can often be solved by adding sound absorbent material in the space where the measurements are being carried out, and by moving the probe several centimetres further away from the test element. This highlights an advantage of splitting the measurement surface into sub-areas when using scanning sound intensity measurements. It is often easier to gain reliable measurements on some sub-areas than others; using sub-areas sometimes avoids having to re-measure the entire measurement surface.

Although it is not necessary to have a defined receiving room for sound intensity measurements, they are often carried out in the same reverberant receiving room that is used for standard sound insulation tests. Therefore it is useful to have a rule-of-thumb that indicates how much absorption is needed in this room to satisfy the requirement, $F_{pl} \leq 10$ dB. This can be determined using the definition of F_{pl} in Eq. 3.180. It is assumed that there is a diffuse field in a room where waves are incident from all possible angles upon a perfectly reflecting and rigid test element (Section 1.2.7.1.1). The sound pressure level very close to the surface of the element will be 3 dB higher than the temporal and spatial average sound pressure level in the central zone of the receiving room, $\langle p_2^2 \rangle_{t,s}$, hence

$$\langle p^2 \rangle_{t,s} = 2 \langle p_2^2 \rangle_{t,s} \tag{3.185}$$

The normal intensity, $\langle I_n \rangle_t$ can be estimated from the power transmitted into the room by the test element (Eq. 3.36) by assuming that all the transmitted sound is radiated by the test element, therefore

$$\langle I_n \rangle_t \approx \frac{\langle p_2^2 \rangle_{t,s}}{4 \rho_0 c_0} \frac{A}{S} \tag{3.186}$$

where S is the area of the test element and A is the absorption area in the room.

Assuming negligible phase-mismatch, an estimate of F_{pl} can now be calculated from Eqs 3.180, 3.185 and 3.186 giving

$$F_{pl} \approx 9 + 10 \lg \left(\frac{S}{A} \right) \tag{3.187}$$

This gives the rule-of-thumb that $S/A < 1.25$ is needed to satisfy the requirement $F_{pl} \leq 10\,dB$. For $S = 10\,m^2$, the receiving room will therefore need $A > 8\,m^2$, so additional absorbent material (e.g. foam or mineral wool) usually has to be added in both the field and the laboratory. Rectangular blocks of absorbent material are a practical option in both the field and the laboratory as they can be stacked on top of each other. These blocks can be placed a short distance away from walls and on top of the floor as long as these surfaces are not being measured with the intensity probe.

For $\approx 10\,m^2$ test elements in rooms without additional absorbent material (field or laboratory), F_{pl} is typically between 5 and 15 dB over the building acoustics frequency range.

3.11 Properties of materials and building elements

The availability of measured dynamic or acoustic properties facilitates the choice of materials at the design stage. They are particularly useful in situations where consideration is being given to substituting one material or element in a construction with a different one.

3.11.1 Airflow resistance

Airflow resistance can be used to predict sound absorption by, and sound transmission through porous materials. The fundamental parameters describing resistance to airflow are given in Section 1.3.2.1.2.

Two methods to determine the airflow resistance are described in ISO 9053. The most common method is the direct airflow method (Brown and Bolt, 1942). This uses an air supply (or vacuum) to produce a pressure drop across a specimen of porous material for which there is no temporal variation under steady-state conditions. Two measurements are required: the pressure drop between the two faces of the test specimen, and the volumetric airflow rate. To ensure a simple representative relationship between pressure and velocity for acoustic purposes it is necessary to ensure there is laminar (rather than turbulent) airflow. An example test cell is shown in Fig. 3.79; the dimensions and other details are described in ISO 9053.

An alternative method is the alternating airflow method (Venzke et al., 1972). This requires a device to produce a slowly alternating airflow in order to measure the alternating component of pressure in a volume enclosed by a sample of porous material. Equipment includes a 2 Hz calibrating pistonphone (non-standard apparatus), a 2 Hz alternating airflow device, and a condenser microphone to measure the alternating pressure. The direct airflow method is simpler to implement and tends to be more widely used.

Airflow resistance usually varies with linear airflow velocity (Brown and Bolt, 1942). Hence the airflow resistance is typically measured with a low linear airflow velocity; $0.5 \times 10^{-3}\,m/s$ is recommended in ISO 9053. This corresponds to the particle velocity in a plane wave with a sound pressure level of 80 dB.

Care needs to be taken when installing test specimens to ensure that there are no air gaps between the specimen and the test cell. For fibrous materials there can be a wide variation in the airflow resistance between samples taken from a single sheet, as well as between different sheets. A sufficiently large number of test specimens need to be measured to determine the average value. In addition, each test specimen should be weighed, and its thickness in the test

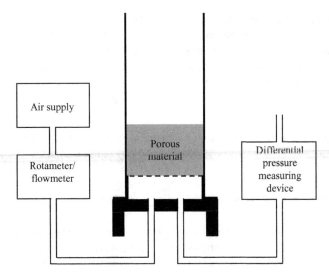

Figure 3.79

Basic measurement set-up for the airflow resistance of porous materials using the direct airflow method.

cell must be measured. This is particularly important if the intention is to establish a relationship between airflow resistance and bulk density; it is not always appropriate to assume a single value for all samples using the bulk density quoted by the manufacturer.

3.11.2 Sound absorption

Measuring the sound absorption is relevant to both rooms and cavities. Only a brief overview of the measurements is given here because there are detailed descriptions in the Standards themselves, or in other books on room acoustics or sound absorption (e.g. Mechel, 1989/1995/1998).

3.11.2.1 Standing wave tube

The standing wave tube can be used to determine the normal incidence sound absorption coefficient (ISO 10534 Parts 1 & 2). It is particularly useful for comparing small samples of material during product development or in quality control. It can also be used to assess absorbent material placed at the ends of long cavities in the low-frequency range where the normal incidence absorption coefficient is relevant to the axial modes. However, in rooms there is a wide range of angles of incidence and values at normal incidence are of limited use. If the absorber is locally reacting then the statistical sound absorption coefficient can be calculated using the measured impedance from the standing wave tube. However for many locally reacting materials, empirical laws can be used to calculate the normal incidence and statistical absorption coefficients (Section 1.3.5.2.1).

3.11.2.2 Reverberation room

A reverberation room typically has non-parallel walls, hard reflective surfaces, and diffusers to give a sound field that approximates a diffuse field in the steady-state, and during the decay process. Reverberation room measurements can be used to determine the sound absorption coefficient or absorption area of an absorber (ISO 354). Due to the restrictions on what can be tested in a standing wave tube, the absorption values determined using the reverberation room method have greater practical application, particularly to the calculation of reverberation times *in situ*.

The absorption area of the reverberation room excluding air absorption is calculated from Eqs 1.83 and 1.97, which gives

$$A = A_T - A_{air} = \frac{24V \ln 10}{c_0 T} - 4\,mV \tag{3.188}$$

Reverberation time measurements are carried out with and without the absorber in the room. Equation 3.188 is then used to calculate the absorption area in both cases to give the absorption areas, A_{with} and $A_{without}$. The absorption area of the test specimen, $A_{absorber}$ is then calculated from

$$A_{absorber} = A_{with} - A_{without} \tag{3.189}$$

The sound absorption coefficient, α_s, for an absorber placed over a wall or floor in the room is

$$\alpha_s = \frac{A_{absorber}}{S} \tag{3.190}$$

where S is the area of the wall or floor that is covered by the absorber.

Note that whilst α_s can be compared with predicted values for the statistical sound absorption coefficient, α_{st}, it is possible for α_s to take values that are larger than unity. This occurs due to diffraction at, or absorption by, the edges of the absorber. Diffraction occurs because sound waves that impinge upon the wall or floor effectively see two very different impedances adjacent to each other; the hard reflective wall/floor surface and the absorber. Edge effects from the finite size of the absorber need to be considered along with a decaying sound field that can be significantly different to a diffuse field due to the presence of the absorber(s). To allow a fair comparison of different absorbers the relevant Standard (ISO 354) therefore prescribes the surface area of a plane absorber, its position in the room, and mounting conditions.

3.11.3 Dynamic stiffness

Vibration isolation between plates is usually achieved by incorporating resilient elements; for example a resilient layer in a floating floor or wall ties in a cavity masonry wall. These resilient elements can often be treated as simple springs that are described by their dynamic stiffness; it is worth noting that there is not usually a simple relationship between dynamic stiffness and static stiffness. In the laboratory, the dynamic stiffness can be quantified using mass–spring systems by connecting the resilient element to one or more masses. The test set-up is arranged so that each mass acts as a simple lump mass that does not support wave motion in the frequency range that contains the resonance frequency of the mass–spring system.

When resilient elements are installed as part of a wall or floor it is assumed that they will undergo small displacements. Hence they will act as linear springs obeying Hooke's law with a linear relationship between stiffness and displacement. Whilst this is a reasonable assumption for airborne sources, it is not always appropriate for structure-borne sound sources (e.g. heavy impacts on a floor). To simplify the comparison of dynamic stiffness for different resilient elements, and to give values that can be used in prediction models, it is useful to quantify the

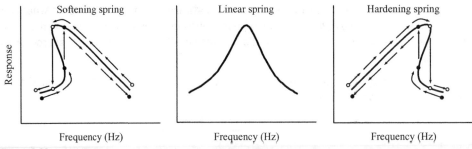

Figure 3.80

Mass–spring system: Frequency response with linear and non-linear (softening and hardening) springs. For the non-linear springs, the arrows indicate the response curve that occurs when the excitation frequency is either increased or decreased.

dynamic stiffness when they act as linear springs. Unfortunately, many resilient elements act as non-linear springs with increasing or decreasing stiffness with increasing input force; these are described as hardening or softening springs respectively. In addition, it is not always possible to identify the region that can be considered as linear. Figure 3.80 shows the frequency response at and near resonance for linear and non-linear springs in a mass–spring system. For a linear spring the peak of the curve occurs at the resonance frequency. Non-linear springs are more complicated because at some frequencies the theoretical curve has more than one value. For swept sine excitation, the response in these unstable regions jumps between different points on the curve depending upon whether the excitation frequency is increasing or decreasing.

Difficulties in measuring the resonance frequency of non-linear springs can be overcome by measuring the resonance frequency with very low input forces at a number of different force levels within a defined range. When these input forces are sufficiently low, linear regression can be used to determine a resonance frequency at zero input force (Pritz, 1987). This resonance frequency is then used to calculate the dynamic stiffness corresponding to zero input force. Whilst this is a step removed from reality, it is a pragmatic way of dealing with non-linear springs. It is also quite adequate for calculating the resonance frequencies of walls or floors that can be represented as mass–spring systems (see Sections 4.3.5.1 and 4.4.4).

In lightweight walls and floors, spring-like elements such as resilient bars, channels, or hangars are often used to provide isolation. The dynamic properties of these elements can be affected by the static load; for example a ceiling formed by resilient channels supporting one or two layers of plasterboard of different weights. In addition, these elements cannot always be treated as simple springs, and measurement of their dynamic properties usually requires more complex measurement procedures (Brunskog and Hammer, 2002).

3.11.3.1 Resilient materials used under floating floors

A test set-up to determine the dynamic stiffness of resilient materials used under floating floors is specified in ISO 9052 Part 1. This Standard is intended for materials subjected to static loads in the range 0.4 to 4 kPa whilst the actual test uses a static load of 2 kPa; this range makes it suitable for many floating floors such as those made from concrete screed or layers of sheet material. It is mainly intended for the comparison of resilient materials used in continuous layers (not small individual resilient mounts) under floating floors. However the measured dynamic stiffness can be used to estimate the mass–spring resonance frequency of floating floors and

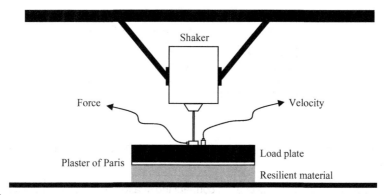

Figure 3.81

Measurement set-up for the dynamic stiffness of resilient materials.

wall linings as well as to predict the sound insulation at higher frequencies. Table A3 in the Appendix has some examples of measured dynamic stiffness values.

No pre-loading of the resilient material is applied before the measurement, i.e. applying a static load on top of the material for a certain time. Pre-loading can change the dynamic stiffness such that it is no longer representative of the situation when it is actually installed under a floating floor (Metzen, 1996).

3.11.3.1.1 Measurement

The simplest test set-up uses a mass–spring system as shown in Fig. 3.81. Measurement of the mass–spring resonance frequency can be used to calculate the dynamic stiffness (Cremer et al., 1973). A sample of resilient material is placed on a heavy rigid base. This is covered with a thin sheet of waterproof film before applying a layer of plaster of Paris to account for any surface irregularities and ensure excitation over the entire surface of the sample. For measurements according to ISO 9052 Part 1, the dimensions of this sample are 200 × 200 mm. A steel load plate with the same dimensions is then bedded down onto the plaster of Paris, and left to set. The mass of the load plate is 8 kg, which corresponds to a 2 kPa static load. The load plate must act as a lump mass, therefore the fundamental bending mode of the load plate must be well-above the mass–spring resonance frequency; this mode will usually be above 1500 Hz for a 200 × 200 × 26 mm steel load plate.

For porous samples (e.g. mineral wool, open-cell foams) it is important that the air is free to move in and out of the sides of the sample during the test; therefore the waterproof film and plaster of Paris must not cover the sides of the sample. The effect of air contained within a porous material on the calculation of dynamic stiffness will be discussed shortly. For non-porous samples (e.g. closed-cell foams) there is no air movement in and out of the sample. However, in the test set-up there can be air movement via the joint along the perimeter of the non-porous sample and the heavy rigid base; hence it is necessary to seal this joint with petroleum jelly (ISO 9052 Part 1).

A force transducer is used to measure the input force, and an accelerometer is positioned adjacent to the force transducer to measure the vertical vibration of the load plate. The excitation force must be applied to the centre of the load plate so that there is only a vertical component to the vibration. To prevent lateral and rotational forces being applied to the force

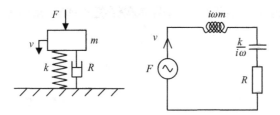

Figure 3.82

Mass–spring system representing the measurement set-up for the dynamic stiffness of resilient materials and its equivalent electrical circuit.

transducer, a drive rod or stinger is used to connect the shaker to the force transducer (Mitchell and Elliott, 1984). This must provide lateral flexibility whilst providing very high stiffness in the axial direction; piano wire often forms a suitable stinger for dynamic stiffness measurements.

The force and acceleration signals are taken to a dual-channel FFT analyser to calculate autospectra, and the Frequency Response Function that corresponds to the driving-point mobility, Y_{dp}. Calibration of the measurement system is carried out using a freely suspended mass (ISO 7626 Part 2). The actual system under test is likely to be non-linear with various different damping mechanisms depending upon the material. However, because low force levels are used it is still helpful to look at a linear mass–spring system with idealized damping. This can be used to calculate the driving-point mobility from the equivalent circuit shown in Fig. 3.82, which yields:

$$Y_{dp} = \frac{v}{F} = \frac{1}{i\omega m + \frac{k}{i\omega} + R} \tag{3.191}$$

where m is the mass of the load plate and k is the spring stiffness (alternatively they may represent the mass per unit area and stiffness per unit area). For a linear system with a viscous damper, the damping constant, R, is related to the loss factor by:

$$R = \eta\sqrt{km} \tag{3.192}$$

From Eq. 3.191, the magnitude of the driving-point mobility will be largest at the frequency where,

$$i\omega m + \frac{k}{i\omega} = 0 \tag{3.193}$$

which is defined as the resonance frequency, f_{ms}, of the mass–spring system,

$$f_{ms} = \frac{1}{2\pi}\sqrt{\frac{k}{m}} \tag{3.194}$$

The resonance frequency is determined at a number of different input force levels within a defined range; 0.1 to 0.4 N if $s' \leq 50\,\text{MN/m}^3$, and 0.2 to 0.8 N if $s' > 50\,\text{MN/m}^3$ (ISO 9052 Part 1). Linear regression can then be used to find the resonance frequency at zero input force. For each input force, the resonance frequency can be found from the magnitude and/or the phase of the driving-point mobility. The idealized linear mass–spring system described by Eq. 3.191 gives an indication of the important features in the driving-point mobility (see Fig. 3.83). At the resonance frequency, the magnitude reaches a peak and the phase is 0°. For lightly damped springs, the resonance frequency can be identified from the peak in the magnitude of the driving-point mobility or the peak in the vibration spectrum. High damping can

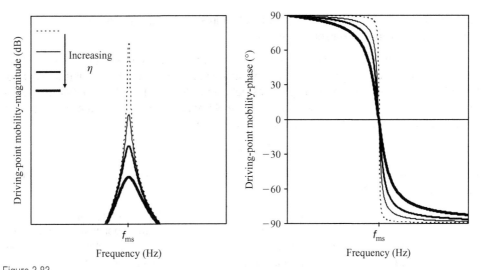

Figure 3.83

Driving-point mobility for a linear mass–spring system in the vicinity of the resonance frequency.

make it difficult to discern this peak and it is more accurate to identify the resonance frequency from the phase of the driving-point mobility. When measuring phase it is worth being aware that phase shifts can be introduced by charge amplifiers, and by changing the accelerometer orientation by 180°. In addition the charge amplifier may not have a flat phase response with frequency. To obtain the velocity signal from the accelerometer, a flat (but possibly phase-shifted) response can usually be obtained at low frequencies by taking the acceleration signal from the charge amplifier, and using post-processing to integrate the signal to give the velocity.

The excitation signal can be a sinusoid (often using an automated sine sweep), white noise, or some form of impulse (e.g. instrumented force hammer, MLS signal). ISO 9052 Part 1 states that sinusoidal signals form the reference method in the case that there is dispute over different resonance frequencies obtained with different signals. As it is quite common to measure slightly different resonance frequencies with different signals, there is little motivation to use anything other than a sinusoidal signal for the purpose of Standardization. With a force hammer it can be difficult to generate a flat force spectrum with a low input force; the force levels typically needed for a flat spectrum can drive the system into non-linear response.

3.11.3.1.2 Calculation of dynamic stiffness

The relevant parameter for resilient materials is the dynamic stiffness per unit area; this is specific to the thickness of material under test. To make the link between dynamic stiffness and the resonance frequency at zero input force requires consideration of any air movement in and out of the sample during the test. This needs to be related to the *in situ* situation where the resilient material is installed as a continuous layer under a floating floor, and the air only moves within the resilient material.

The dynamic stiffness of porous resilient materials is effectively determined by two springs connected in parallel; one spring representing the skeletal frame of the material, and another spring representing the air contained within the material that surrounds the skeletal frame

(Cremer *et al.*, 1973). These springs add together in series to give the dynamic stiffness per unit area of the installed resilient material, s', in N/m^3.

If the lateral airflow resistivity of the material is not too high, the small size of the test sample will allow air to move freely in and out of its sides during the measurement. So in this test set-up, the stiffness of the air within the sample does not come into play. Therefore the test only gives an estimate of the dynamic stiffness for the skeletal frame; this is referred to as the apparent dynamic stiffness per unit area, s'_t, and is calculated using

$$s'_t = 4\pi^2 \rho_s f_{ms}^2 \qquad (3.195)$$

where ρ_s is the mass per unit area of the load plate and f_{ms} is the mass–spring resonance frequency at zero input force.

For a porous resilient material *in situ*, the air contained within it acts as a spring with dynamic stiffness per unit area, s'_a. For most porous resilient layers used under floating floors, the mass–spring resonance frequency occurs in the low-frequency range or below it. At these low frequencies it can be assumed that there is isothermal compression of the air. The dynamic stiffness per unit area of the air is calculated from

$$s'_a = \frac{K}{\phi d} \qquad (3.196)$$

where K is the bulk compression modulus of air (for isothermal compression, $K = P_0 = 1.013 \times 10^5$ Pa), ϕ is the porosity of the porous resilient material, and d is the thickness of the sample under the load plate.

For porous resilient materials with lateral airflow resistivity in the range, $10 \leq r < 100$ kPa.s/m^2 (ISO 9052 Part 1), the dynamic stiffness per unit area of the installed resilient material, s', can be calculated from

$$s' = s'_t + s'_a \qquad (3.197)$$

Figure 3.84 shows s'_a for typical porosities and thicknesses of resilient material. Uncertainty in estimates of the porosity usually results in a negligible change to s'_a, but it is important to note that the stiffness of the contained air often forms a significant percentage of s'. For layers of mineral wool typically used under floating floors, s'_a is often between 25% and 250% of s'.

For non-porous materials or porous materials with very high lateral airflow resistivity, $r \geq 100$ kPa.s/m^2, there is no need to include the s'_a term, hence $s' = s'_t$ (ISO 9052 Part 1).

For porous resilient materials with lateral airflow resistivity, $r < 10$ kPa.s/m^2, then $s' = s'_t$ only if $s'_a \ll s'_t$, otherwise s' cannot be quoted as satisfying the Standard (ISO 9052 Part 1). For some reconstituted foams, $r < 10$ kPa.s/m^2, but $s'_a \ll s'_t$ is not satisfied; yet the test can still yield useful estimates by including the s'_a term (Hall *et al.*, 1996).

Having quantified the dynamic stiffness with this test set-up it would be ideal if the internal loss factor could be calculated from the same measurement; for example, using the 3 dB bandwidth of the resonance peak. However, the measured loss factor is only likely to be relevant to this particular test set-up. For porous materials, the measured loss factor has two components: damping due to the skeletal frame, and damping due to airflow in and out of the sides of the sample. The latter is often significant; hence the loss factor measured in the test set-up may not be relevant to the *in situ* situation where the lateral dimensions of the layer are usually much larger. For non-porous materials it is possible to rank order the loss factor measured with

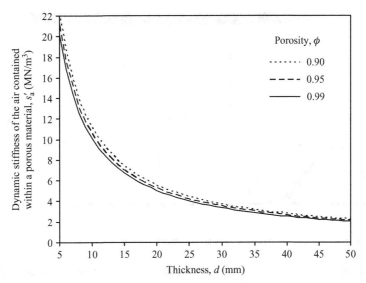

Figure 3.84

Dynamic stiffness per unit area for the air contained within a porous material.

different materials, but it should be noted that the use of petroleum jelly around the perimeter can also change the measured loss factor.

This test set-up only provides a single value of dynamic stiffness and provides no information on frequency dependence. Other measurements show that the dynamic Young's modulus and loss factor of the skeletal frame (measured in a vacuum) are independent of frequency for mineral wool (Pritz, 1986), but frequency-dependent for some foams (Pritz, 1994). For practical calculations it is rarely necessary to measure frequency-dependent values when it is only the low-frequency range that is of interest. It is usually sufficient just to be aware that the dynamic stiffness of some materials will be frequency-dependent.

3.11.3.2 *Wall ties*

Wall ties in masonry cavity walls are rarely simple rectangular strips or cylinders; they usually have a drip in the centre that is formed by twists, notches, or kinks. This allows water to drip down rather than pass between the leaves of a cavity wall. Differences between these twists, notches, or kinks, as well as the material properties, cross-sectional dimensions, and cavity depth give rise to different values of dynamic stiffness. Table A4 in the Appendix has some examples of measured dynamic stiffness values.

3.11.3.2.1 *Measurement*

The dynamic stiffness of wall ties can be determined by casting each end of a wall tie into a concrete cube, and measuring the axial mass–spring–mass resonance frequency (Craik and Wilson, 1995). The dynamic stiffness determined by the measurement is specific to the spacing, X mm between the cubes; this spacing corresponds to the depth of the wall cavity in which the tie is to be used. Wall ties with a highly resilient material at the centre may have a dynamic stiffness that is only determined by this material, so the stiffness will be independent of the

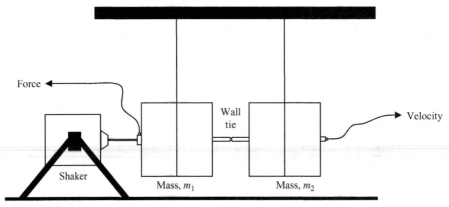

Figure 3.85

Measurement set-up to determine the transfer mobility for the dynamic stiffness of wall ties.

cavity depth (Wilson, 1992); these ties are sometimes used in high performance cavity walls in studios.

Standard concrete cube moulds can be used to make nominally identical 100 mm cubes using a concrete mix that rigidly holds the wall tie (Hall and Hopkins, 2001). The test set-up is shown in Fig. 3.85. To allow free vibration of the mass–spring–mass system, each cube is supported along its centre line by a loop of cord.

The measurement equipment is the same as described for resilient materials used under floating floors. Excitation is applied to the centre point on the outer surface of the cube. The force transducer measures the input force on cube mass, m_1, with an accelerometer to measure vibration in the horizontal direction on cube mass, m_2. The force and acceleration signals are taken to a dual-channel FFT analyser to calculate the Frequency Response Function that corresponds to the transfer mobility, Y_{tr}. Wall ties tend to act as non-linear springs, some of which are hardening and others are softening; an example of a wall tie acting as a hardening spring is shown in Fig. 3.86 (Hopkins et al., 1999). Problems with measuring non-linear springs can be overcome by applying low input forces and extrapolating to a resonance frequency at zero force input. Therefore it is still useful to look at a linear mass–spring–mass system with idealized damping using the equivalent circuit shown in Fig. 3.87; this gives the driving-point mobility as

$$Y_{dp} = \frac{v_1}{F} = \frac{i\omega m_2 + \frac{k}{i\omega} + R}{\left[i\omega m_1 \left(i\omega m_2 + \frac{k}{i\omega} + R\right)\right] + \left[i\omega m_2 \left(\frac{k}{i\omega} + R\right)\right]} \tag{3.198}$$

and the transfer mobility as

$$Y_{tr} = \frac{v_2}{F} = \frac{\frac{k}{i\omega} + R}{\left[i\omega m_1 \left(i\omega m_2 + \frac{k}{i\omega} + R\right)\right] + \left[i\omega m_2 \left(\frac{k}{i\omega} + R\right)\right]} \tag{3.199}$$

The magnitude of the transfer mobility is largest at the frequency where the denominator of Eq. 3.199 is smallest. This gives the resonance frequency, f_{msm}, of the mass–spring–mass system as

$$f_{msm} = \frac{1}{2\pi} \sqrt{\frac{k}{\left(\frac{m_1 m_2}{m_1 + m_2}\right)}} \tag{3.200}$$

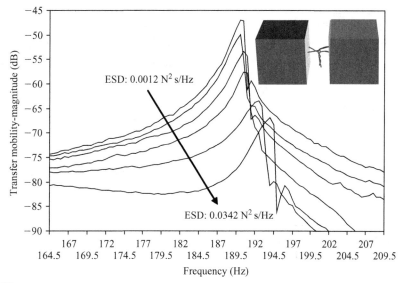

Figure 3.86

Measured transfer mobility of a mass–spring–mass system with a butterfly wall tie. Impulse excitation from an instrumented force hammer is applied at various excitation levels (shown in terms of the Energy Spectral Density, ESD). Measured data are reproduced with permission from Hopkins (1999).

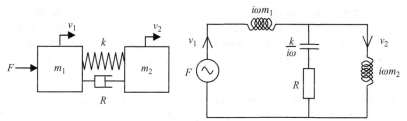

Figure 3.87

Mass–spring–mass system representing the measurement set-up for the dynamic stiffness of wall ties and its equivalent electrical circuit.

An anti-resonance can be found in the driving-point mobility where the numerator of Eq. 3.198 is smallest. This occurs at the frequency,

$$\frac{1}{2\pi}\sqrt{\frac{k}{m_2}}$$

(3.201)

The magnitude and phase of the driving point or transfer mobility for an idealized linear mass–spring–mass system (Eqs 3.198 and 3.199) are shown in Fig. 3.88. The peak in the magnitude and/or the phase passing through 0° can be used to identify the axial mass–spring–mass resonance frequency. It is often necessary to use the phase when the resonance is highly damped.

The aim is to excite only the axial mass–spring–mass resonance. However, twisting or oblique resonances may be excited with some wall ties, so it is necessary to identify the axial resonance with an additional measurement (Hopkins et al., 1999; Wilson, 1992). If there is doubt as to which peak in the transfer mobility corresponds to the axial resonance, then the single

363

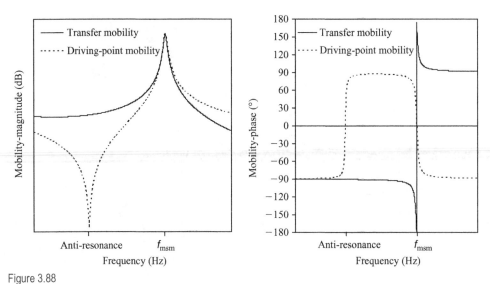

Figure 3.88

Transfer mobility and driving-point mobility for a linear mass–spring–mass system in the vicinity of the resonance frequency.

accelerometer on cube mass, m_2, can temporarily be replaced by two accelerometers. These are positioned near the edges of the cube at equal distances from the centre point. The accelerometers are aligned along the centre line of the cube in the horizontal plane, and then in the vertical plane in a subsequent measurement. At the axial mass–spring–mass resonance frequency the phase difference between these two accelerometers in both the horizontal and vertical planes will be zero; at any other resonance frequency the phase difference will be non-zero.

The resonance frequency is determined at a number of different input force levels in the range, 0.01 to 0.1 N (Hopkins *et al.*, 1999). Linear regression is then used to find the resonance frequency at zero input force.

3.11.3.2.2 *Calculation of dynamic stiffness*

From Eq. 3.200, the dynamic stiffness of a wall tie, $s_{X\,mm}$, in N/m for a cavity depth of X mm is

$$s_{X\,mm} = 2\pi^2 f_{msm}^2 m_{av} \qquad (3.202)$$

where f_{msm} is the mass–spring–mass resonance frequency at zero input force, and the average cube mass, m_{av}, is calculated from the following:

$$m_{av} = \frac{(m_1 + m_2 + m_{tie}) - m_{tie}}{2} \qquad (3.203)$$

where the term in brackets corresponds to the complete test specimen.

3.11.3.3 *Structural reverberation time*

Structural reverberation times for bending wave vibration on walls and floors are usually short. In addition, many building elements have low modal density and evaluation of the decay curves can be more awkward than with rooms. To relate the decay time to the reverberant vibration level it is usually necessary to use T_{10}, T_{15}, or T_{20} and measure to three decimal places.

To measure a smooth decay curve that is unaffected by the reverberation time of the filters it is generally best to use the integrated impulse response method with reverse-filter analysis (Sections 3.8.2 and 3.8.3.2.2). The response of the structure is measured using an accelero-meter. The impulse needed to excite bending waves can be provided by a single hammer hit from a plastic-headed hammer (e.g. see Craik, 1981) or by using a signal such as MLS from a shaker pushed up against the wall or floor (e.g. see Meier and Schmitz, 1999).

The measured reverberation time can be used to calculate the total loss factor. This is the sum of the internal loss factor and all the coupling loss factors (radiation and structural losses). If the mounting conditions for the structure are carefully arranged it is possible to use the total loss factor to estimate the internal loss factor, or the coupling loss factor due to sound radiation (see Sections 3.11.3.4 and 3.11.3.9).

As noted in Section 2.6.3, the internal loss factor of building materials can be non-linear if measured with high-vibration amplitudes. The response is generally expected to be linear for coupling loss factors between masonry/concrete elements that are connected at rigid junctions. However, non-linearity could potentially occur with certain junctions; for example, where there are resilient connections and the resilient material or resilient connecting element effectively acts as a non-linear spring. There is some evidence from measurements on a masonry wall in a transmission suite that hammer excitation gives higher total loss factors than MLS shaker excitation (Meier and Schmitz, 1999). This may possibly be attributed to non-linearity, but it has not yet been proved explicitly. Hammer excitation of walls and floors has been used for many years and the results generally correspond to theoretical predictions. It is possible that state-of-the-art measurement technology has exposed an issue that was previously hidden by difficulties in correctly evaluating the decay curves. Whatever the reason for these discrepancies, it is best to avoid driving a structure into non-linear response and to try and measure structural decays using vibration levels that are representative of the situation in practice. For lightweight and heavyweight walls excited with a hammer, it is possible to cause lateral deflections that are unrepresentative of those induced by airborne excitation. In contrast, the impact sound insulation of floors is usually measured with hammer excitation from the ISO tapping machine. It is worth noting that non-linearity of the total loss factor has not yet been identified as an issue with tapping machine measurements on bare concrete floors. To avoid any problem, a cautious approach is to use MLS shaker excitation instead of a plastic-headed hammer. However, the latter is quicker and more convenient than the former. If measurements are regularly needed on a specific type of element in the laboratory (e.g. masonry walls) it is worth investing time to compare both methods to see if they can be considered as equivalent within the range of measurement uncertainty.

3.11.3.4 Internal loss factor

For structure-borne sound transmission on plates and beams, the internal loss factor quantifies the damping due to the conversion of vibrational energy into heat energy. We will only look at measuring the internal loss factor for bending waves as these are the most important. Two methods are generally used to determine the internal loss factor. If the material comes in sheet form, measuring the structural reverberation time is usually the most convenient approach. Alternatively, short, narrow strips of material can be used to form beams; these are excited by a shaker and the loss factor can then be determined at the various resonance frequencies of the beams.

For measurements on plates, the structural reverberation time can be used to determine the total loss factor in frequency bands over the building acoustics frequency range. To minimize errors in the evaluation of the decay curves, there should ideally be at least five bending modes in the frequency bands of interest. If the coupling losses due to structural connections and sound radiation are negligible, the total loss factor provides an estimate of the internal loss factor. If the loss factor does not vary significantly with frequency, an average loss factor can be determined from a chosen frequency range below the critical frequency. To avoid problems with non-linearity, low levels with MLS shaker excitation can be applied.

For sheet materials such as plasterboard, chipboard, or OSB the plate can be suspended vertically by two loops of resilient cord. This resilient suspension minimizes the coupling losses, but the radiation losses from the plate are inherent in the measurement. Fortunately, homogeneous plates without stiffeners do not radiate particularly efficiently below the critical frequency. For most sheet materials the critical frequency is in the high-frequency range; this gives a relatively wide frequency range below the critical frequency that can be used to estimate the internal loss factor. It may be necessary to avoid the frequency band immediately below the critical frequency because of high radiation losses. These losses usually cause a very steep slope in the early part of the decay curve. Note that trying to estimate the radiation efficiency and subtract the calculated radiation losses from an unbaffled plate with free boundaries is highly likely to introduce errors. Below the critical frequency the decay curves are usually reasonably straight and T_{15} or T_{20} can be calculated accurately. Figure 3.89 shows example data from sheet material below the critical frequency. The internal losses for these particular materials can generally be considered as being independent of frequency. An arithmetic average value can therefore be calculated from frequency bands below the critical frequency; this usually

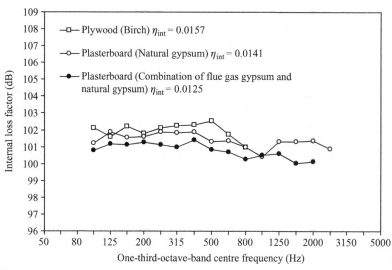

Figure 3.89

Internal loss factor of sheet materials determined from structural reverberation time measurements on plates. The legend contains arithmetic average values from the frequency range below the critical frequency. Measured data from Hopkins are reproduced with permission from ODPM and BRE.

gives a reasonable estimate for use in prediction models over the building acoustics frequency range.

For masonry blocks it is useful to measure the internal losses when the blocks are mortared together as they are in full-size walls. Unfortunately their weight often makes it difficult to structurally isolate the test element to minimize the coupling losses. One possibility is to create a beam of several mortared blocks and determine the loss factor at the resonance frequencies of the beam (Kuhl and Kaiser, 1952). Another possibility is to measure the structural reverberation time on full-size walls that are structurally isolated, so that the total loss factor is primarily determined by internal and radiation losses (Craik and Barry, 1992; Craik and Osipov, 1995). Structural isolation can be achieved by building a free-standing, full-size wall on top of a resilient material. However for structural stability it is necessary to have some support at points along the sides of the wall; contact at these points also needs to be made via a resilient material to minimize these unwanted structural coupling losses.

For laminated materials such as laminated glass, the internal loss factor usually varies with temperature. For this reason it is convenient to take measurements on short strips of material inside a small temperature-controlled chamber (Yoshimura and Kanazawa, 1984). The dimensions of the strip are chosen so that it acts as a beam with bending modes at or near the frequencies of interest. For the comparison of laminated glass products, the strip dimensions are standardized to 25 mm wide × 300 mm long (ISO/PAS 16940). This method can also be used with other materials, not just laminates.

The strip is excited at the mid-point using random noise from a shaker and the input impedance is measured at the excitation point using an impedance head (see Fig. 3.90). FFT analysis is then used to determine a transfer function corresponding to the input impedance. The resulting input impedance spectrum contains peaks at the various resonance frequencies of the strip. The frequency resolution must be sufficiently fine to accurately determine the resonance frequencies and their 3 dB down points. The strip is usually lightly damped so it is important to ensure that the transfer function accurately reproduces the resonance peaks (Randall, 1987). The internal loss factor can be calculated at each resonance frequency, f_i, using

$$\eta_{\text{int},i} = \frac{\Delta f_{3\,\text{dB},i}}{f_i} \qquad (3.204)$$

where $\Delta f_{3\,\text{dB},i}$ is the half-power bandwidth.

In comparison with measuring structural reverberation times on a plate, this method using beams is advantageous because it allows estimates of the internal loss factor near and above the critical frequency. However, it only gives internal loss factors at specific frequencies, so it is less efficient when there is significant variation with frequency. It is important to note that measured values for the internal loss factor depend on the mode shape; this means that there can be significant differences between values determined on beams and plates (Dunn et al., 1983). When the internal loss factor varies significantly with frequency, better estimates can be found by using both methods to cover the building acoustics frequency range. Figure 3.91 shows an example of the internal loss factor for laminated glass using both methods; note that the internal loss factor increases significantly with increasing frequency and increasing temperature.

3.11.3.5 Quasi-longitudinal phase velocity

In Section 2.3 we saw that the phase velocity for bending, transverse shear, and torsional waves can be related to the phase velocity for quasi-longitudinal waves, c_L. This makes the

(a) Test set-up

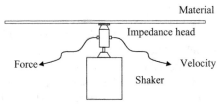

(b) Example spectrum for the input impedance

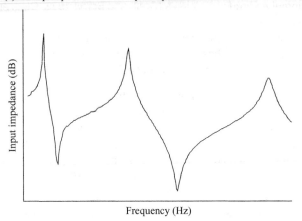

Figure 3.90

Measurement of the internal loss factor using short strips of material acting as beams.

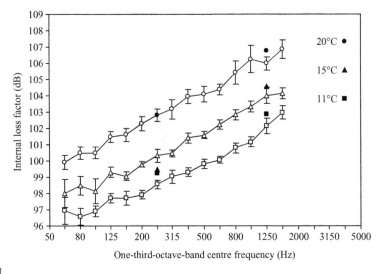

Figure 3.91

Measured internal loss factor of 6.4 mm laminated glass (3 mm–0.4 mm PVB–3 mm) at three different surface temperatures. Open symbols connected by lines correspond to measurement of the structural reverberation time on a plate shown with 95% confidence intervals. Shaded symbols correspond to individual resonance frequencies measured on short beams (average from two beams). The critical frequency is estimated to lie in the 2000 Hz one-third-octave-band; measurements on the plate are only shown below this frequency. Measured data from Hopkins are reproduced with permission from BRE.

quasi-longitudinal phase velocity a very useful property. Table A2 in the Appendix has some examples of measured values.

In principle it is possible to estimate the quasi-longitudinal phase velocity by identifying the critical frequency from the coincidence dip in the measured sound reduction index of a plate. In practice, many plates are orthotropic to some degree, the coincidence dip is not always easy to identify and the lowest point in the coincidence dip is not always close to the critical frequency; hence this method is prone to error. For small homogeneous specimens (e.g. solid masonry blocks) the quasi-longitudinal phase velocity can be measured using ultrasound (BS 1881-203:1986) or determined from modal analysis (Maysenhölder and Horvatic, 1998). The latter usefully allows estimates of both the Young's modulus and Poisson's ratio. In practice it is usually the properties of an entire wall or floor that are of interest, rather than the properties of individual blocks in a masonry wall. An impulse from a hammer can be used to excite quasi-longitudinal waves, and measure the time-of-flight across a beam or a plate (Craik, 1982a). To excite only quasi-longitudinal waves requires access to the end of a beam, or the edge of a plate. This can usually be arranged in the laboratory, but not always in the field. The advantage of this method is that measurements can be carried out over the length of a beam, or over the length and width of a plate; hence for modular elements such as masonry walls, the effect of the mortar joints can be included in the measurement. It can also be used for homogeneous beams or plates that are either isotropic or orthotropic. However, it is not appropriate for sandwich plates (e.g. plasterboard – rigid foam – OSB) or laminates (e.g. laminated glass) for which it is less meaningful to consider quasi-longitudinal waves, as well as being more difficult to excite them.

The measurement set-up is shown in Fig. 3.92. Two accelerometers are positioned along a measurement line separated by a distance, d. These must be aligned in the same direction to respond to in-plane vibration. An impulse from a hammer is used to excite quasi-longitudinal waves in the plane of the test element. The signals from the accelerometer are sent to a dual-channel analyser (or oscilloscope) to measure signal voltage against time. Quasi-longitudinal phase velocities are usually greater than 1400 m/s, hence the time resolution of the analyser and the distance between accelerometers (ideally >1 m) must be chosen to give sufficient accuracy (±5% is usually acceptable). The only part of the signal that is of interest is the initial rising slope, after this the signal is affected by waves reflected from the boundaries, some of which may have been converted into other wave types. The quasi-longitudinal wave velocity is calculated from

$$c_L = \frac{d}{\Delta t} \tag{3.205}$$

where d is the distance between accelerometers and Δt is the time between nominally identical points on the initial rising slope of the response.

Measured values from beams, $c_{L,b}$, and plates, $c_{L,p}$, can be used to calculate Young's modulus using Eqs 2.20 and 2.21 respectively.

Measurements can either be taken *in situ* or on isolated elements in the laboratory. For heavyweight elements such as masonry/concrete walls and floors it is convenient to take measurements *in situ*. There is rarely any need for them to be decoupled at the boundaries, other than to allow excitation along one edge for isotropic plates or two edges for orthotropic plates. For lightweight elements it is necessary to make a distinction between c_L measured on individual materials (e.g. a sheet of chipboard, a timber batten) and on combinations of materials

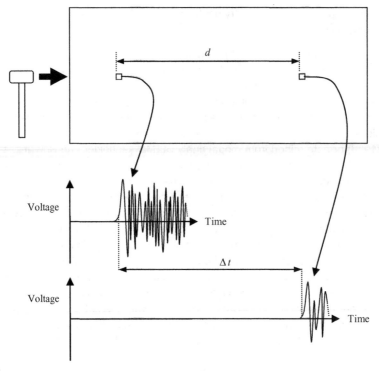

Figure 3.92

Measurement of the quasi-longitudinal wave velocity.

(e.g. sheets of chipboard screwed to timber battens to form a floating floor). Most lightweight walls or floors are formed from lightweight sheet material connected to beams. These beams effectively act as stiffeners; therefore the quasi-longitudinal phase velocity is usually different in the two orthogonal directions. To quantify c_L for individual materials, it is necessary to measure them in isolation. Lightweight sheet materials (e.g. plasterboard) can be suspended from resilient cord for the measurement. For heavier elements such as timber joists, it is usually sufficient to rest them on a resilient material on the floor. Note that problems can occur if they are placed directly onto a rigid surface (e.g. a concrete floor) because structure-borne waves may also be excited in this other structure that will affect the measurement.

3.11.3.6 Bending phase velocity

For homogeneous thin plates, the simplest option to determine the bending phase velocity is to calculate it from measurements of the quasi-longitudinal phase velocity. However, there are times when it is not possible to gain access to the edge of the element, and/or the plate acts as a thick plate at the frequencies of interest, and/or the element is non-homogeneous (e.g. hollow blocks). In these situations the frequency-dependent bending phase velocity can be determined by measuring the phase difference between two point on a plate (Nightingale et al., 2004; Rindel, 1994; Roelens et al., 1997). These measurements are more complex and tend to be required on reverberant plates which can make it difficult to accurately determine the phase difference of the propagating wave. In addition these plates are not always large enough to take measurements that are out of the nearfield. For these reasons it is not always possible to measure values across the entire building acoustics frequency range and the results can be highly dependent upon the signal processing.

3.11.3.7 Bending stiffness

The bending stiffness can be determined with small strips of material acting as beams; this uses the same test set-up as for internal loss factor measurements (Section 3.11.3.4). When the stiffness varies with temperature, this measurement can be arranged inside a small temperature-controlled chamber. The peaks in the input impedance spectrum correspond to the resonance frequencies, f_i. The bending stiffness for a plate of the material can be calculated at each resonance frequency from these beam measurements (Yoshimura and Kanazawa, 1984; ISO/PAS 16940)

$$B_{p,i} = \rho_s \left(\frac{\pi L^2 f_i}{2C_i^2} \right)^2 \tag{3.206}$$

where L is the length of the beam, and C_i is a constant depending on the resonance frequency. ($C_1 = 1.87510$, $C_2 = 4.69410$, $C_3 = 7.85476$: NB Above the third resonance frequency it can be difficult to achieve sufficient accuracy using this approach.)

3.11.3.8 Driving-point mobility

Driving-point mobility of structures is needed to calculate the power input for point force excitation of plates and beams. This is particularly relevant to the power input from the ISO tapping machine into a floor. Whilst the driving-point mobility can be predicted for simple homogeneous plates, it is not always possible for non-homogeneous plates, plates with low modal density, or plates connected to a framework of beams. The driving-point mobility can also be used to estimate the modal density, the loss factors of individual modes, and mode frequencies.

Driving-point mobility is given by the ratio of the velocity to the excitation force at the point of excitation. This is measured using dual-channel FFT analysis to give the required Frequency Response Function. For thin plates with a low mass per unit area, the force and velocity can be measured at the same point using an impedance head. Many plates that form building elements are too heavy and stiff to use impedance heads and require use of a separate accelerometer and a force transducer. It is not particularly convenient to use excitation from a shaker because of the difficulty in fixing the force transducer to the wall or floor (ISO 7626 Part 2). Hence the force transducer is usually mounted in an instrumented force hammer (ISO 7626 Part 5). The bandwidth of the impulse is affected by the tip attached to the force transducer as well as the hammer mass. On most building elements, a plastic tip (rather than rubber or metal) can be used to give a sufficiently flat force spectrum up to ≈ 1000 Hz. Calibration of the measurement system is carried out using a freely suspended mass as described in ISO 7626 Part 2.

To measure driving-point mobility, the excitation point and the accelerometer are ideally positioned exactly opposite each other, on either side of the plate. However, for floors and walls there may only be access to one side, so the accelerometer can only be positioned adjacent to the excitation point. Even when there is access to both sides, the inconvenience and the positional errors that occur in fixing an accelerometer on the opposite side of a large wall or floor make it preferable to use the same side. By measuring the acceleration at a position adjacent to the excitation point we are effectively measuring the transfer mobility between two points. However, when the bending wavelength is large compared to the distance, d, between the accelerometer and the excitation point (i.e. $k_B d \ll 1$), a good estimate of the driving-point mobility can be obtained. State-of-the-art accelerometers are sufficiently small that the distance between the excitation point and the centre of the accelerometer can usually be kept to $d \leq 20$ mm. For most plates that form walls and floors in buildings this will allow measurements over the low- and mid-frequency range.

Mobility measurement using FFT analysis outputs both real and imaginary parts. It is usually the real part that is of interest as this relates to power input and to the modal density. The imaginary part and magnitude may be affected by the contact stiffness of the material at the excitation point acting as a spring (Eq. 3.97).

To minimize phase errors from the charge amplifiers, no integration of the accelerometer signal takes place before FFT processing. Therefore, the analyser gives accelerance which is converted to mobility through a single integration (i.e. division by $i\omega$). Peaks in the spectrum of the driving-point mobility correspond to resonant modes; therefore if the modal density and damping are low, deep troughs (anti-resonances) will occur in the spectrum between the resonant peaks. Coherence values at anti-resonances are often low due to the weak response signal being contaminated with noise. This is usually apparent at low frequencies where the mode spacing is large. However, it is the response at the resonant peaks that is of most interest. If the FFT lines are used to calculate the average mobility in a frequency band containing one or more modal peaks, then the low coherence at anti-resonances usually has negligible effect on the frequency-average mobility. Frequency Response Function H_1 (Channel A: Force; Channel B: Acceleration) is usually used to give the optimum estimate because most noise is in the acceleration signal, rather than the force signal.

Spatial averaging is carried out by arithmetically averaging the linear values of the driving-point mobility. The standard deviation and 95% confidence interval can also be calculated; these can help decide whether a simple infinite plate model for the driving-point mobility would be adequate over part of the building acoustics frequency range. It is often worth storing all the individual measurements and only carrying out the averaging after checking each individual measurement for errors. For this reason it is useful to store a repeat measurement at each position. Errors can occur due to the operator not producing a single 'clean' impulse from the hammer hit. This can occur on crumbly, dusty, or powdery surfaces that can also cause weak accelerometer fixing.

Frequency-averaging to give the mobility (real part) in one-third-octave or octave-bands is needed for calculations of the power input or to determine the modal density. This is calculated by averaging the discrete FFT lines in each frequency band using the following:

$$\langle \text{Re}\{Y_{dp}\}\rangle_f = \frac{1}{f_u - f_l} \int_{f_l}^{f_u} \text{Re}\{Y_{dp}\} df \tag{3.207}$$

where f_l and f_u are the lower and upper limits of the frequency band respectively.

Estimating the modal density by counting modal peaks in the response of an isolated plate is generally prone to error; there may be modes with the same frequency, modes very close together, or modes that are not discernable when the damping is high. For homogenous plates, better estimates can often be calculated from the spatial average driving-point mobility (real part). (This is not appropriate for plates with attached beams where the driving-point mobility is significantly different above the beams, compared to between the beams.) The modal density for a frequency band with centre frequency, f, is (Clarkson and Pope, 1981; Cremer *et al.*, 1973)

$$n(f) = 4m\langle \mathrm{Re}\{Y_{\mathrm{dp}}\}\rangle_{f,s} \qquad (3.208)$$

where m is the mass of the plate or beam (kg) and $\langle\ \rangle_{f,s}$ indicates a frequency and spatial average value.

Loss factors of individual modes can either be calculated from the half-power bandwidth (Section 2.8) or with a more accurate approach from modal analysis such as using vector diagrams (e.g. see White, 1982). Mode frequencies can be identified from the peaks in the driving-point mobility. However, unless the plate is uncoupled and isolated from other structures, it can be difficult to relate these to the local mode frequencies that assume idealized boundary conditions.

In existing buildings, information is not always available on the thickness or density of walls and floors. If a simple length or mass measurement is not possible then the driving-point mobility can be measured in order to estimate either the thickness or the density (one of them must be known). This also requires a reasonable estimate for the quasi-longitudinal phase velocity. If the plate is homogenous then the driving-point mobility at high frequencies will tend towards that for an infinite plate and Eq. 2.190 can be used to estimate either the thickness or the density. One application of this approach is with concrete ground floors where there is no access underneath. However, caution is needed when a concrete slab has been cast directly onto the ground/hardcore because the slab will not necessarily act independently of what lies beneath it.

3.11.3.9 *Radiation efficiency*

Accurate prediction of the radiation efficiency is not possible for all plates and for all boundary conditions; unfortunately, reliance on measurement will not always provide the solution. This is mainly because radiation efficiency is dependent upon the type of excitation (e.g. mechanical point force excitation, bending waves impinging upon a boundary, plane sound waves). It is also specific to the plate dimensions and the arrangement and type of stiffening elements; hence measurements on one plate are not always transferable to other plate sizes.

Two relatively simple methods of measuring the radiation efficiency are considered here: one uses measurement of plate vibration and radiated sound power, the other estimates the radiation efficiency from measurement of the structural reverberation time. There are also more complex methods such as acoustical holography (Villot *et al.*, 1992).

Radiation efficiency makes the link between the radiated sound power and the mean-square bending wave velocity on the plate (Eq. 2.198). For this reason it is applicable to resonant transmission rather than non-resonant transmission; these terms are introduced in Section 4.3.1. With airborne excitation, non-resonant transmission tends to be the dominant sound transmission mechanism at frequencies well-below the critical frequency; hence, measurement of plate vibration and radiated sound power tends to result in an overestimate of the radiation efficiency (Macadam, 1976). As it is usually the radiation efficiency below the critical frequency that is of most interest, it is necessary to use mechanical excitation.

The radiation efficiency is usually measured with the intention of using it in an SEA model. For this reason we ideally want to excite the plate modes using statistically independent excitation forces. This should provide equipartition of modal energy and no correlated modal response (Section 4.2). However, statistically independent excitation forces are not easily provided. It is much more convenient to use point force excitation from an electrodynamic shaker. The disadvantage of mechanical excitation at a point is that it gives a correlated modal response and depending on the excitation point, some modes will be excited and not others (Fahy, 1985). Point excitation also causes the vibration at the excitation point to be higher than the spatial average plate vibration. To try and minimize these effects it is necessary to average the plate response by taking measurements at a number of different excitation positions. Below the critical frequency there will also be sound radiation from the nearfield at the excitation point. This can be calculated as described in Section 2.9.7. Fortunately, the power radiated by the nearfield is often negligible compared to power radiated by the bending modes. In principle, the sound power radiated by the nearfield on homogeneous plates could be calculated and extracted from the measured sound power; in practice this is prone to introducing errors due to the uncertainty in the plate impedance and input force.

The measurement set-up requires baffles around the perimeter of the plate to be arranged so that they are representative of *in situ*. If the plate is not baffled, there will be a degree of cancellation between pressures on opposite sides of the plate. Below the critical frequency the absence of baffles tends to reduce the radiation from plate modes acting as corner or edge radiators.

For a test element installed in a transmission suite, the radiation efficiency is conveniently determined by exciting the wall with a shaker in one room (source room) and measuring the radiated sound pressure in the other room (receiving room). This avoids measuring the self-generated noise of the shaker. By measuring the plate vibration along with the reverberation time in the receiving room, the radiated sound power and the mean-square plate velocity can be used to calculate the radiation efficiency from Eq. 2.198. As with the measurement of $\Delta R_{\text{resonant}}$ (Section 3.5.1.2.2) a signal such as MLS can be used to overcome problems in measuring low sound pressure levels in the receiving room. When the shaker is simply pushed up against the wall, the power input may vary between excitation positions. $L_{\text{pv},T}$ can then be calculated according to Eq. 3.49, and the radiation efficiency can be calculated from the following:

$$10 \lg \sigma = L_{\text{pv},T} + 20 \lg \left(\frac{20 \times 10^{-6}}{10^{-9}} \right) - 10 \lg T_0 + 10 \lg \left(\frac{6V \ln 10}{S \rho_0^2 c_0^3} \right) \qquad (3.209)$$

It is usually the radiation efficiency below and in the vicinity of the critical frequency that is of most interest. The radiation efficiency above the critical frequency tends towards unity for homogeneous isotropic plates. When measuring at frequencies above the critical frequency and above the thin plate limit, the measured radiation efficiency sometimes becomes significantly lower than unity; this is due to thick plate effects in combination with point excitation. The resulting values are not representative of the radiation efficiency under airborne excitation.

Example measurements on a solid homogeneous masonry wall with mechanical point force excitation are shown in Fig. 3.93. These measurements with point excitation indicate that the radiation efficiency can be slightly higher than unity at and above the critical frequency, even though predictions for masonry/concrete plates with airborne excitation are usually better when unity is assumed (Section 2.9.4.3).

For plates with high radiation efficiencies, very low internal losses, and a critical frequency above the frequency range of interest, it is possible to estimate the radiation efficiency using

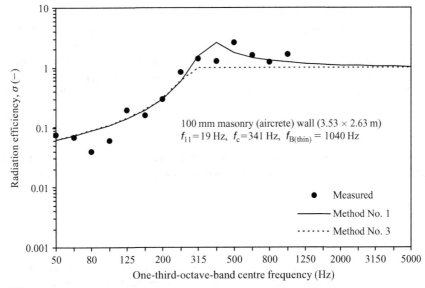

Figure 3.93

Measured radiation efficiency for a 100 mm masonry wall in a transmission suite along with predicted values (method Nos 1 & 3) assuming simply supported boundaries and baffles in the same plane as the wall. Measured data from Hopkins are reproduced with permission from BRE Trust.

structural reverberation time measurements to give the total loss factor. An example application is for plates with high critical frequencies that are bonded to a porous material for which the radiation efficiency into the porous material is often much higher than into air (Craik *et al.*, 2000). The total loss factor is determined from measurements of the structural reverberation time as described in Section 3.11.3.3. The plate needs to be isolated at its boundaries so that coupling losses are negligible. If the internal loss factor is known (or can be measured), this can be subtracted from the measured total loss factor to estimate the coupling loss factor due to sound radiation from both sides of the plate (note that this is twice the value given by Eq. 4.21). If the radiation efficiency is the same for both sides of the plate, then the result is simply divided by two. If they are different (i.e. one side radiates into a porous material and the other into air) then the estimate of the higher radiation efficiency into the porous material may be improved by subtracting the estimated coupling loss factor for radiation into air on one side.

3.12 Flanking transmission

Measurements relating to flanking transmission for either airborne or impact sound insulation can be considered in the following categories:

(1) Quantifying flanking sound transmission via building elements that form a sin-gle flanking path between rooms such as suspended ceilings and access floors (laboratory measurements).
(2) Quantifying flanking sound transmission for individual flanking paths across a junction where there is more than one possible flanking path (laboratory measurements).

(3) Quantifying flanking transmission between two or more connected elements to provide input data for SEA or SEA-based prediction models to calculate the *in situ* performance (mainly laboratory measurements, but sometimes made in the field).
(4) Quantifying the sound insulation provided by combinations of separating and flanking elements by simulating the *in situ* construction in a flanking laboratory.
(5) Identifying and quantifying the effect of flanking transmission on the sound insulation that is achieved *in situ*. This provides information that can be used to try and improve the sound insulation.

This section looks at measurements that fit into these five categories. The basic theory needed to arrange such measurements in the laboratory, as well as to apply and interpret the results is covered in Chapters 4 and 5.

3.12.1 Flanking laboratories

We recall that laboratory measurement of direct sound transmission across a building element requires a transmission suite with suppressed flanking transmission. In a similar way, laboratory measurement of flanking transmission requires a facility and/or measurement technique in which all flanking paths can be suppressed other than the flanking path(s) of interest. This increases the complexity of the test set-up. It also means that the type of laboratory facility and/or measurement technique may differ for different types of construction.

3.12.1.1 *Suspended ceilings and access floors*

Laboratory facilities for suspended ceilings and access floors are described in the relevant Standards (ISO 140 Parts 9 & 12). The basic principle is that all sound is transmitted between the two rooms via the ceiling/floor and the plenum (see Fig. 3.94). The dividing wall must have sufficiently high sound insulation that there is no significant transmission across it; the dividing wall is usually isolated from the suspended ceiling or access floor by a resilient material to minimize vibration transmission between them. One side wall and both end walls of the plenum are lined with highly absorbent material.

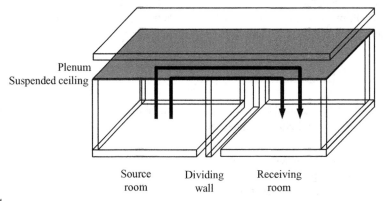

Figure 3.94

Outline sketch of a transmission suite used to measure airborne flanking transmission via a suspended ceiling (or access floor if viewed upside down).

376

Suspended ceilings and access floors are usually considered as introducing a single flanking path between two rooms. However, in terms of an SEA model there is more than one sound transmission path by which airborne or impact sound is transmitted between these two rooms. These paths can often be described by a plate–cavity–plate system (see Sections 4.3.5 and 5.3.1.1.1) if the sound field in the plenum can be considered as reverberant (i.e. not too highly damped). The transmission paths involve the sound field in the cavity and any structural coupling between the plates where they meet along the line of the dividing wall. The sound insulation measurement is specific to the test set-up with the plenum/floor/ceiling dimensions and plenum absorption that are used in the laboratory. The only normalization that can be carried out is to the absorption area in the receiving room.

The normalized flanking level difference, $D_{n,f}$, for suspended ceilings or access floors is (ISO 140 Parts 9 & 12):

$$D_{n,f} = D - 10 \lg \left(\frac{A}{A_0} \right) \tag{3.210}$$

where the reference absorption area, A_0, is $10\,\text{m}^2$. Note that for suspended ceilings this is normally quoted as $D_{n,c}$ but it is calculated in the same way as Eq. 3.210.

The normalized flanking impact sound pressure level, $L_{n,f}$, for access floors (ISO 140 Part 12) is given by:

$$L_{n,f} = L_p + 10 \lg \left(\frac{A}{A_0} \right) \tag{3.211}$$

3.12.1.2 Other flanking constructions and test junctions

Requirements for the laboratory facilities and measurements on other flanking constructions and test junctions are described in the relevant Standards (ISO 10848). The type of facility that is required depends on whether sound pressure level measurements are being used to determine $D_{n,f}$ and $L_{n,f}$ (as with suspended ceilings and access floors), or vibration measurements are being used to determine parameters such as the coupling loss factor or vibration reduction index for subsequent use in SEA or SEA-based models.

Sound pressure level measurements can be useful in quantifying flanking transmission paths where the test junction is formed from plates that are not well-suited to inclusion in SEA or SEA-based models. Examples include plates with a significant decrease in vibration level with distance, plates where the vibration field on each element is not a reverberant bending wave field, and plates where vibration transmission cannot be directly related to measurement of the lateral surface velocity on each element. Measurements in a laboratory facility are relatively simple to arrange when the elements that form the test junction lie in a single plane as previously seen in Fig. 3.94 for suspended ceilings or access floors. This set-up is appropriate when the separating wall or floor has negligible influence on transmission between the flanking elements that lie in the same plane; this could apply to some façade elements. However, with many wall and floor junctions the separating and flanking elements are perpendicular to each other and they all play a role in vibration transmission across the junction; hence shielding of one or more elements is required as indicated in Fig. 3.95. To measure transmission via any specific flanking path the connection at the junction must remain unchanged so that its role in the vibration transmission process is unaltered. It is therefore necessary to shield one or more elements to prevent them from being directly excited by airborne sound in the source room and to significantly reduce the sound radiated into the receiving room. Airborne sound

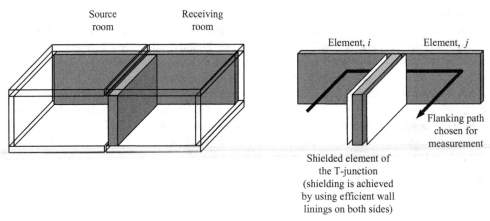

Figure 3.95

Illustration of shielding used to measure airborne sound transmission via one specific flanking path. In this example the test junction installed in the laboratory is a T-junction. The walls of the junction could be lightweight or heavyweight construction and different to each other.

insulation measurements can then be used to determine the normalized flanking level difference (Eq. 3.210) for each specific flanking path. With structure-borne excitation from the ISO tapping machine it is only necessary to shield elements in the receiving room if it can be assumed that the airborne sound generated in the source room does not cause an unwanted flanking path. This can be used to determine the normalized flanking impact sound pressure level (Eq. 3.211) for each specific flanking path.

The use of structure-borne excitation and vibration measurements is well-suited to elements with a reverberant bending wave field and no significant decrease in vibration with distance. This allows use of vibration measurements to determine parameters such as the coupling loss factor or vibration reduction index that can be used in SEA or SEA-based models. For such measurements it is not necessary to have a laboratory with reverberant rooms commissioned for sound pressure level measurement. This allows some flexibility in choosing where measurements are carried out. However, the way in which a test junction is installed in a laboratory facility affects the flow of vibrational energy between the test elements as well as the way in which those elements transmit energy into the laboratory structure.

There are many different types of test junction and many possible ways of installing a junction within a laboratory. For this reason, additional measurements are needed to assess flanking transmission between elements of the test junction via the laboratory structure. To make a full assessment of flanking transmission via all possible paths would require a prediction model of the test junction and the laboratory; however if such a model is available there is no need to make measurements. It is possible to overcome this problem if it can be assumed that the laboratory structure can be described by SEA plate subsystems. An assessment can then be made based on principles of energy flow from SEA (Section 4.2). As an example of this we consider a junction of three walls installed as a free-standing element in a laboratory as shown in Fig. 3.96. The aim of the measurement is to determine the velocity level difference, $D_{v,ij}$ between two elements i and j with excitation of element, i. If elements i and j are positioned on the same side of the vibration break in the laboratory floor there will be an unwanted flanking path between the separating wall and the flanking wall via the floor. For this unwanted flanking path to have negligible effect on $D_{v,ij}$, when element, i, is excited, there must be

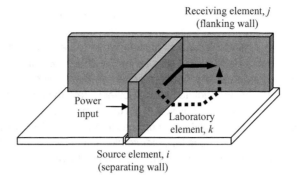

Receiving element, j
(flanking wall)

Power
input

Laboratory
element, k

Source element, i
(separating wall)

Figure 3.96

Example assessment of flanking transmission due to connections between the test junction and the laboratory. All test elements and laboratory elements are assumed to act as SEA subsystems. The solid arrow indicates the flanking path of interest for which vibration measurements are being used to give the velocity level difference, $D_{v,ij}$. The dotted arrow indicates an unwanted flanking path via laboratory element, k.

a positive net flow of vibrational energy from receiving element, j, to laboratory element, k. This assumes that elements i, j, and k all act as SEA plate subsystems and there is only bending wave motion. Hence when element i is excited by a structure-borne sound source, element, j must have higher modal energy than element, k; this can be written as

$$\frac{E_j}{n_j(f)B} > \frac{E_k}{n_k(f)B} \tag{3.212}$$

where E is energy (Eq. 2.237), $n(f)$ is the bending wave modal density (Eq. 2.139), and B is the bandwidth of the frequency band.

Equation 3.212 can now be written in terms of a velocity level difference, $D_{v,jk}$, between elements j and k that is measured during excitation of element i. This gives the flanking criterion as

$$D_{v,jk} + 10 \lg \left(\frac{\rho_{s,j} n_k(f)}{\rho_{s,k} n_j(f)} \right) > 0\,\text{dB} \tag{3.213}$$

This is more conveniently written in terms of the quasi-longitudinal phase velocity or critical frequency of the plate elements to give the following inequalities (ISO 10848 Part 1)

$$D_{v,jk} + 10 \lg \left(\frac{\rho_{s,j} c_{L,j}}{\rho_{s,k} c_{L,k}} \right) > 0\,\text{dB} \quad \text{or} \quad D_{v,jk} + 10 \lg \left(\frac{\rho_{s,j} f_{c,k}}{\rho_{s,k} f_{c,j}} \right) > 0\,\text{dB} \tag{3.214}$$

The above approach is not appropriate when each element in the test junction is connected to laboratory walls/floors that are too small to be considered as SEA plate subsystems.

Constructions can be built into flanking laboratories in different ways. One extreme is where all elements in the test junction are rigidly connected to the laboratory structure on all sides to form two (or more) box-shaped rooms. The other is where each element in the test junction is isolated from the laboratory structure with resilient materials or is supported on small laboratory floor elements that are all isolated from each other by vibration breaks. Neither option is ideal for all types of test element. For heavyweight elements the total loss factor of each element plays an important role in determining the vibration transmission between the elements. The effect of the total loss factor for each element on the coupling loss factor or vibration reduction index is discussed later with examples in Section 5.2.3.

3.12.2   Ranking the sound power radiated from different surfaces

In completed buildings where the sound insulation needs to be improved, it is necessary to identify which room surfaces need remedial treatment, such as a wall lining or an independent ceiling. Vibration and sound intensity measurements are useful diagnostic tools for this purpose. For both measurements the radiated sound power is determined for the various surfaces that face into the receiving room. In many situations these measurements can be used to rank order the sound power radiated by different surfaces in the room. This allows remedial treatments to be prescribed for the surfaces that radiate high sound power levels, and an estimate can be made of the potential improvement in the sound insulation.

3.12.2.1   Vibration measurements

For bending wave vibration, the sound power radiated by a surface is calculated from the temporal and spatial average mean-square velocity over the surface, $\langle v^2 \rangle_{t,s}$, using the following:

$$W = S \rho_0 c_0 \sigma \langle v^2 \rangle_{t,s} \tag{3.215}$$

where σ is the radiation efficiency.

Due to spatial variation of plate vibration over the plate surface (Section 2.7), a sufficiently large number of accelerometer positions are needed to calculate the spatial average value. Even for masonry walls in the high-frequency range, a single position can be up to 10 dB higher than the spatial average level due to material imperfections. Therefore the potential for errors is large if only a single accelerometer position is used to estimate the radiated power.

The radiation efficiency that is needed for Eq. 3.215 can be calculated for homogeneous isotropic plates when the critical frequency is known (Section 2.9.4). Above the critical frequency it can usually be assumed that the radiation efficiency is unity. The critical frequency of a solid masonry/concrete plate tends to occur in the low-frequency range, so a large part of the building acoustics frequency range can often be covered with this simple assumption. In contrast this will only cover a few frequency bands in the high-frequency range for many lightweight walls and floors.

The limitations of this method are that it cannot be applied to homogeneous walls or floors unless the critical frequency can be estimated, nor to walls and floors that do not have a reverberant bending wave field. This is awkward because construction details in an existing building that has been decorated can be difficult to ascertain. In addition it assumes that non-resonant transmission below the critical frequency is insignificant compared to resonant transmission; this can be assessed from the theory and examples in Section 4.3.1. A more practical limitation is that attaching accelerometers to walls and ceilings in the field tends to mark or damage decorated surfaces, and they cannot be attached to floors with fixed coverings such as carpets. All the above limitations can be overcome by using sound intensity.

3.12.2.2   Sound intensity

Sound intensity is a powerful tool with which to assess flanking transmission in buildings, and it avoids the limitations of vibration measurements in quantifying the radiated sound power (Villot and Roland, 1981). However, measurements over the complete building acoustics frequency range can take several hours if the majority of receiving room surfaces need to be measured and reactive sound fields make it difficult to take valid measurements. Field and laboratory measurements almost always require additional absorption to be introduced into the receiving room. In the field it is rarely possible to measure every single frequency band between 100 and 3150 Hz whilst satisfying the requirements on F_{pl} and $\delta_{pl0} - F_{pl}$ (ISO 15186 Part 2). Even in the laboratory there are often single bands in the low- or high-frequency range that are difficult to measure (usually $\leq$125 Hz and $\geq$2000 Hz). Depending on the time available for the field test it is not uncommon to leave the test site with valid measurements and positive intensity readings in less than half the frequency bands over the building acoustics frequency range for the majority of room surfaces. Fortunately it is possible to identify solutions to most flanking transmission problems without needing measured data in every single frequency band (e.g. see Carman and Fothergill, 1990; Hongisto, 2001).

In the field, scanning is easier to carry out than discrete point measurements. The sound power for each surface is calculated from the temporal and spatial average normal sound intensity level, L_{In}, and the measurement surface area, S_M, using

$$L_W = L_{In} + 10\lg S_M \qquad (3.216)$$

At the outset it is important to establish which surfaces are likely to be the dominant radiating surfaces; there will not always be enough time to measure every surface in a room. A short measurement on all surfaces with the intensity probe will indicate which ones are likely to have the highest sound power, and on which surfaces it will be difficult to satisfy the requirements on F_{pl} and $\delta_{pl0} - F_{pl}$. For each surface, the most practical option is to slowly scan across a single diagonal line from corner to corner (Pettersen et al., 1997).

Scanning speeds are usually between 0.1 and 0.3 m/s (ISO 15186 Part 2), hence long measurement times needed to scan all the walls and floors can be problematic in the field. It is unwise to increase the scanning speed, but time savings can sometimes be made by increasing the distance between the lines in the scanning pattern. With elements such as windows and doors this is not appropriate because there are often leaks at edges and over the surface so the sound radiation is rarely uniform. However for masonry/concrete elements where the air paths have been removed by sealing the surface with a bonded surface finish (e.g. plaster), it is often possible to double the distance between the lines in the scanning pattern without significant loss of accuracy.

Both during and after the measurements, it is useful to check whether all the dominant radiating surfaces have been measured. This is relatively simple if it can be assumed that there is a diffuse sound field in the receiving room. Reverberation times are needed to calculate the total absorption area, and the spatial average sound pressure level from the central zone of the receiving room. Essentially we need to check whether the room energy due to sound radiated from surfaces measured with the intensity probe is equal to the room energy calculated from the reverberant sound pressure level measurement. Note that the latter accounts for all sound transmission. To carry out this check it is necessary to account for higher energy density near the room boundaries by using the Waterhouse correction to estimate the room energy from the reverberant sound pressure level; this is more important for the low-frequency range. The result can be expressed as an energy level difference, ΔE, in decibels given by

$$\Delta E = \left(10\lg\left(\sum_{i=1}^{N} 10^{L_{Wi}/10}\right) + 10\lg\left(\frac{4}{A}\right)\right) - \left(L_p + 10\lg\left(1 + \frac{S_T\lambda}{8V}\right)\right) \qquad (3.217)$$

where L_{Wi} is the sound power level for radiating surface i. Note that ΔE is the same as K_2 defined in an informative annex of ISO 15186 Part 2.

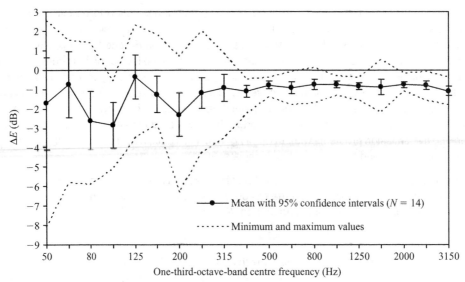

Figure 3.97

ΔE values from a flanking laboratory with 14 different masonry test constructions. The receiving rooms used for the measurements had volumes of $\approx 50\,\text{m}^3$ with one separating element and one, two, or three flanking elements. Measured data from Hopkins are reproduced with permission from ODPM and BRE.

The factors that affect interpretation of ΔE are measurement uncertainty in the sound pressure, sound intensity, and reverberation time. In addition there are errors due to the assumption of a diffuse field and the associated fact that the Waterhouse correction is only usually a good estimate for empty box-shaped rooms with a minimum volume of $50\,\text{m}^3$. However, when ΔE equals 0 dB and the intensity measurements satisfy the requirements on F_{pl} and $\delta_{pl0} - F_{pl}$ it is reasonable to assume that all the dominant radiating surfaces have been measured accurately. Negative values of ΔE are an indication that not all the radiated sound has been accurately measured using sound intensity. Positive values of ΔE can occur in the low-frequency range due to a non-diffuse sound field which invalidates the Waterhouse correction and increases the measurement uncertainty.

Example values of ΔE from masonry/concrete test constructions in a flanking laboratory are shown in Fig. 3.97. It is usually easier to measure under laboratory conditions than in the field; hence it will rarely be possible to achieve $\Delta E = 0$ dB over a wide frequency range in the field. Based on these measurements, an acceptable range for field measurements can be taken as $-2 < \Delta E \le 0$ dB. It can sometimes be difficult to achieve values in this range with lightweight constructions (ISO 15186 Part 2). This is partly due to the complexity of sound radiation from lightweight walls and floors that form flanking elements. When lightweight walls and floors are excited along one junction line, there is often a decrease in vibration level with distance away from the junction; this gives a corresponding decrease in the radiated sound power (Nightingale, 1996). This causes some measurement problems. Firstly, the volume enclosed by the measurement surface for the separating wall may contain strips of flanking wall/floor that radiate significant sound power; if so, additional linings will be required to reduce their sound radiation. Secondly, it may be difficult to accurately quantify the sound power that is radiated by a flanking element if there is a rapid decrease in the radiated sound intensity level with increasing distance from the excited junction.

Field measurements are often taken in furnished buildings and it is not always possible to fix additional linings to the strongly radiating surfaces that are close to the measurement surface. If the airborne sound insulation is predominantly determined by a separating element with relatively low sound insulation then it can be difficult to measure radiation from flanking walls and floors. A common example is an internal floor ($R_w \approx 40\,dB$) that needs to be upgraded to a separating floor in flat conversions. Quantifying the flanking transmission from the existing walls with sound intensity measurements may not be possible until the sound insulation of the floor has been upgraded.

Building façades are often formed from several elements that are adjacent to each other (e.g. wall, infill panels, curtain walling, window, door). Before using sound intensity as a diagnostic tool to identify the dominant radiating surface(s) it may be necessary to shield the other surfaces using additional linings (e.g. see Vermeir et al., 1996). Care needs to be taken in shielding the various elements because the dominant sound source will sometimes be the frame or gap in-between the elements.

It is important to note that sound intensity measurements quantify sound radiation from separating and flanking elements; they do not automatically quantify the sound transmitted along a certain transmission path. If we consider sound transmission between two rooms in terms of the paths Dd, Fd, Df, and Ff (previously defined in Fig. 3.14) it is clear that the sound power radiated by a separating or flanking element can result from more than one transmission path. However, if the test construction is arranged in a similar way to the laboratory set-ups in Figs 3.94 and 3.95 and there is only one dominant flanking path, then measurements can indeed be used to quantify transmission via this path.

Two examples are now used to illustrate the ranking of sound power from different surfaces. They also indicate the influence of flanking transmission at levels of airborne sound insulation commonly required in building regulations. The overall sound insulation is not only determined by the separating wall or floor; there are many transmission paths involving the separating and flanking elements that determine the sound insulation in situ. The first example compares the radiated sound power from intensity and vibration measurements. This is shown in Fig. 3.98 for a masonry separating wall in a masonry building. The intensity measurements show that the separating and flanking walls radiate similar levels of sound power over a wide frequency range and at some frequencies the flanking wall radiates more than the separating wall. Above the critical frequency of each wall, the sound intensity and vibration measurements are in close agreement. Below the critical frequency there will be errors in the vibration measurements; these are mainly due to the predicted radiation efficiency and partly due to non-resonant transmission below 100 Hz. For heavyweight walls and floors (without lightweight linings) the required insight into the flanking transmission can often be gained from either measurement method. For lightweight walls and floors, reliance is placed on sound intensity measurements due to the complexity in predicting the radiated sound from vibration measurements. The second example in Fig. 3.99 shows the sound powers for a separating timber floor and flanking walls in a timber frame building. In this example ΔE generally falls in the range $-2 < \Delta E \leq 0\,dB$; note that the remaining flanking wall that contained the entrance door was not scanned due to a lack of time. At low frequencies it was only the separating floor (surface 1) that gave positive intensity satisfying the requirements on F_{pl} and $\delta_{pl0} - F_{pl}$. In the mid-frequency range, the sound power radiated by the separating floor was approximately equal to the combined sound power radiated by all the flanking walls (surfaces 2 to 6).

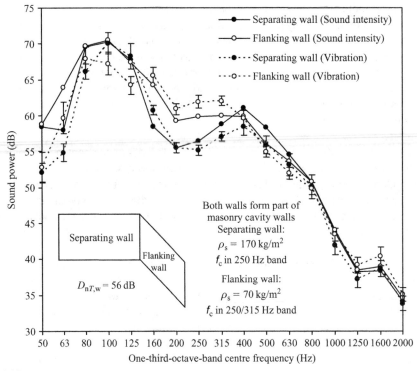

Figure 3.98

Example measurement of sound power radiated by separating and flanking walls in a heavyweight construction. Comparison of measurements using sound intensity and vibration. Sound power derived from vibration measurements is shown with the 95% confidence intervals for the spatial average velocity level. Measured data from Hopkins are reproduced with permission from ODPM and BRE.

3.12.3   Vibration transmission

Vibration transmission can be quantified using measurement of structural intensity or the velocity level difference; the latter is the simpler method. These measurements are usually required to calculate coupling parameters for use in an SEA or SEA-based prediction model. The two most widely used parameters are the coupling loss factor and the vibration reduction index. As these parameters have their basis in SEA, the basic assumptions made in SEA (Section 4.2) need to be satisfied by the elements being measured.

3.12.3.1   Structural intensity

Structural intensity measurement can be used to quantify the net structural power; sometimes called structure-borne sound power. This gives an insight into the net energy flow that lies 'underneath' the reverberant vibration field. Bending waves tend to be the most important wave type for sound insulation in buildings and it is unusual to need measurements of in-plane wave intensity. For this reason only structural intensity measurement of bending wave vibration is considered here. As with sound intensity, it is the active component of intensity that is useful rather than the reactive component. The magnitude and direction of the active intensity vectors changes with time, hence the time-averaged value for stationary vibration signals is needed for most practical purposes. In contrast to sound intensity measurement, we cannot place a

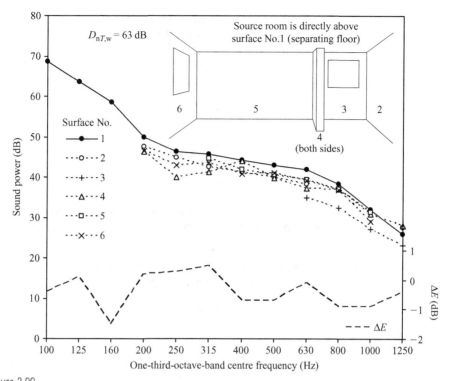

Figure 3.99

Example measurements of sound power radiated by a separating floor and flanking walls in a lightweight construction using sound intensity. Measured data from Hopkins and Turner are reproduced with permission from ODPM and BRE.

structural intensity probe at a point in space where the energy flow takes place; this lies within the structure. However, by taking measurements on the surface of a structure, the results can be related to energy flow within its cross-section. Structural intensity is therefore quoted in terms of the net power per unit width (W/m) across a line on the surface.

Structural intensity measurement for bending waves on beams or plates with a two-accelerometer probe was formulated and implemented by Noiseux (1970). The assumptions in this approach limit its validity to propagating bending waves on thin plates where nearfields are negligible. A more rigorous approach to structural intensity measurement that avoids the assumption of measuring in the far-field was developed by Pavić (1976). This requires intensity probes that comprise arrays of four accelerometers for beams and eight accelerometers for plates. However, as the size of the array increases beyond two accelerometers, errors in the signal processing and the positioning of the accelerometers tends to reduce the advantage of using a large array to give a full description of the structural intensity (Bauman, 1994). For practical measurements on building structures it is usually sufficient to use a two-accelerometer probe as shown in Fig. 3.100 (Kruppa, 1986).

By deriving the structural intensity for bending wave motion in the x-direction on a thin homogeneous isotropic plate, the result can be simply adapted to thin beams. We will use the sign conventions and variables for bending waves on thin beams and plates defined in Section 2.3.3 for positive power flow in the positive x-direction. Referring back to this bending wave theory,

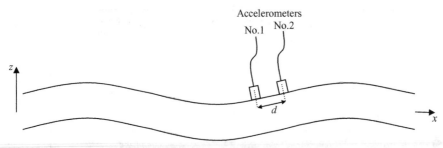

Figure 3.100

Structural intensity measurement (bending wave motion): a–a probe.

we see that it is most practical to formulate all variables in terms of the lateral displacement, η; this displacement is relatively large and easily determined from acceleration measurements using accelerometers.

The time-averaged structural intensity in the x-direction, I_x, has two components: a force component, I_{xF}, and a moment component, I_{xM} (Noiseux, 1970) such that,

$$I_x = I_{xF} + I_{xM} \tag{3.218}$$

where the force component is

$$I_{xF} = \langle Q_x v_z \rangle_t = \left\langle B_p \frac{\partial}{\partial x} \left(\frac{\partial^2 \eta}{\partial x^2} + \frac{\partial^2 \eta}{\partial y^2} \right) \left(\frac{\partial \eta}{\partial t} \right) \right\rangle_t \tag{3.219}$$

and the moment component which comprises bending and twisting moment contributions, I_{xB} and I_{xT}, is

$$I_{xM} = I_{xB} + I_{xT} = \langle M_{xy} \omega_y + M_{xx} \omega_x \rangle_t = \left\langle -B_p \left(\frac{\partial^2 \eta}{\partial x^2} + v \frac{\partial^2 \eta}{\partial y^2} \right) \left(\frac{\partial^2 \eta}{\partial t \partial x} \right) + B_p (1 - v) \frac{\partial^2 \eta}{\partial x \partial y} \left(-\frac{\partial^2 \eta}{\partial t \partial y} \right) \right\rangle_t \tag{3.220}$$

Hence the time-averaged structural intensity in the x-direction is

$$I_x = B_p \left\langle \frac{\partial}{\partial x} \left(\frac{\partial^2 \eta}{\partial x^2} + \frac{\partial^2 \eta}{\partial y^2} \right) \frac{\partial \eta}{\partial t} - \left(\frac{\partial^2 \eta}{\partial x^2} + v \frac{\partial^2 \eta}{\partial y^2} \right) \frac{\partial^2 \eta}{\partial t \partial x} - (1 - v) \frac{\partial^2 \eta}{\partial x \partial y} \frac{\partial^2 \eta}{\partial x \partial y} \right\rangle_t \tag{3.221}$$

The same approach applies to the structural intensity in the y-direction, I_y, which yields:

$$I_x = B_p \left\langle \frac{\partial}{\partial y} \left(\frac{\partial^2 \eta}{\partial x^2} + \frac{\partial^2 \eta}{\partial y^2} \right) \frac{\partial \eta}{\partial t} - \left(\frac{\partial^2 \eta}{\partial y^2} + v \frac{\partial^2 \eta}{\partial x^2} \right) \frac{\partial^2 \eta}{\partial t \partial y} - (1 - v) \frac{\partial^2 \eta}{\partial x \partial y} \frac{\partial^2 \eta}{\partial t \partial x} \right\rangle_t \tag{3.222}$$

Hence the resultant intensity vector, $\mathbf{I}$, is

$$\mathbf{I} = I_x \mathbf{i} + I_y \mathbf{j} \tag{3.223}$$

where the magnitude is

$$|\mathbf{I}| = \sqrt{I_x^2 + I_y^2} \tag{3.224}$$

and the direction in terms of the angle, θ, from the x-axis is

$$\theta = \arctan \frac{I_y}{I_x} \tag{3.225}$$

For a thin beam we only need to consider the structural intensity in the x-direction. This is found from Eq. 3.221 by replacing B_p with B_b and removing all terms involving the y-direction to give:

$$I_x = B_b \left\langle \frac{\partial^3 \eta}{\partial x^3} \frac{\partial \eta}{\partial t} - \frac{\partial^2 \eta}{\partial x^2} \frac{\partial^2 \eta}{\partial t \partial x} \right\rangle_t \tag{3.226}$$

Having formulated the equations that describe the structural intensity, we now look at how a two-accelerometer probe can be used to estimate this intensity. To simplify the measurement procedure, Noiseux (1970) noted that for a plane bending wave field in the absence of nearfields, the force component is equal to the moment component. Noiseux's proposal for measurements in plane wave fields was to measure only the moment component and assume that $I_{xF} = I_{xM}$. However, the moment component contains the angular velocity term, $\partial^2 \eta / \partial x \partial y$, which is obtained by using four accelerometers (Pavić, 1976). To avoid using more than two accelerometers, Noiseux introduced a modified moment component, I'_{xM}, which, using the notation in Chapter 2 is

$$I_{xM} \approx I'_{xM} = \left\langle \frac{M_{xy} - M_{yx}}{1 + \nu} \omega_y \right\rangle_t = -B_p \left\langle \left(\frac{\partial^2 \eta}{\partial x^2} + \frac{\partial^2 \eta}{\partial y^2} \right) \frac{\partial^2 \eta}{\partial t \partial x} \right\rangle_t \tag{3.227}$$

The term in brackets can be simplified for plane waves in the free-field, where

$$\frac{\partial^2 \eta}{\partial x^2} + \frac{\partial^2 \eta}{\partial y^2} = k_B^2 \eta \tag{3.228}$$

This gives a displacement, whereas we normally measure acceleration. So, if we assume that the signal is a sinusoid of frequency, ω, or a narrow band of noise with a centre frequency, ω, then,

$$\eta = \frac{1}{(i\omega)^2} \frac{\partial^2 \eta}{\partial t^2} \tag{3.229}$$

Hence the modified moment component in Eq. 3.227 can be written in the form:

$$I'_{xM} = \frac{B_p k_B^2}{\omega^2} \left\langle \frac{\partial^2 \eta}{\partial t^2} \frac{\partial^2 \eta}{\partial t \partial x} \right\rangle_t \tag{3.230}$$

This requires measurement of the time-averaged value of the product of the lateral acceleration and the angular velocity. Noiseux proposed use of two accelerometers: one to measure the lateral acceleration and the other to measure the angular acceleration. Since this time, single transducers have become available that can output both lateral and angular acceleration; these can work well in practice (e.g. see Bauman, 1994). However, the modified moment component also allows a two-accelerometer probe to be used where the accelerometers are positioned side-by-side, and separated by a distance, d between their centres (see Figs 3.100 and 3.101). Using this approach allows the angular velocity to be determined using a finite-difference approximation. This has similarities to sound intensity measurement with a p–p probe where a finite-difference approximation is used to determine the particle velocity; hence we will refer to an a–a probe for structural intensity. This approach conveniently allows use of a two-channel analyser that is capable of sound intensity measurement (Rasmussen and Rasmussen, 1983). It also allows analysis of the measurement errors in a similar way to sound intensity.

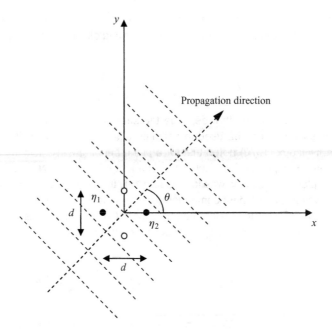

Figure 3.101

Orientation of two a–a probes in a propagating plane wave field to measure structural intensity in the x- and y-directions.

3.12.3.1.1   a–a structural intensity probe

An a–a probe comprises two accelerometers: No.1 and No.2, for which the associated lateral displacements are η_1 and η_2 respectively (Fig. 3.101). For positive intensity, a plane bending wave propagates in the x-direction from accelerometer No.1 to No.2. The accelerometers are equally spaced about the point for which the intensity estimate is made, so the lateral acceleration at the mid-point between the accelerometers is

$$\frac{\partial^2 \eta}{\partial t^2} = \frac{1}{2}\left(\frac{\partial^2 \eta_1}{\partial t^2} + \frac{\partial^2 \eta_2}{\partial t^2}\right) \tag{3.231}$$

and the angular velocity is estimated using a finite-difference approximation,

$$\frac{\partial^2 \eta}{\partial t \partial x} = \frac{1}{d}\left(\frac{\partial \eta_2}{\partial t} - \frac{\partial \eta_1}{\partial t}\right) = \frac{1}{d}\int\left(\frac{\partial^2 \eta_2}{\partial t^2} - \frac{\partial^2 \eta_1}{\partial t^2}\right)dt \tag{3.232}$$

Hence, the time-averaged structural intensity in the x-direction is

$$I_x = 2I'_{xM} = \frac{2B_p k_B^2}{\omega^2}\left\langle \frac{1}{2}\left(\frac{\partial^2 \eta_1}{\partial t^2} + \frac{\partial^2 \eta_2}{\partial t^2}\right)\frac{1}{d}\int\left(\frac{\partial^2 \eta_2}{\partial t^2} - \frac{\partial^2 \eta_1}{\partial t^2}\right)dt\right\rangle_t \tag{3.233}$$

The probe measures the intensity component along its axis; hence I_y is measured by rotating the probe by 90°.

For sound intensity measurements made with a p–p probe, the sound pressure signals are processed according to Eq. 3.170, in either one-third-octave or octave-bands. The similarity between Eqs 3.170 and 3.233 means that structural intensity can be measured using a dual-channel analyser designed for sound intensity (Rasmussen and Rasmussen, 1983). Each channel needs to be calibrated to measure acceleration in decibels re 20×10^{-6} m/s², and the spacing, d, must be set on the analyser to the accelerometer spacing. This gives the structural intensity level in decibels re 10^{-12} W/m when the intensity level from the analyser is corrected by adding:

$$10 \lg \left(\frac{\rho_0 \sqrt{B_p \rho_s}}{\pi f} \right) \text{ for plates} \tag{3.234}$$

and

$$10 \lg \left(\frac{\rho_0 \sqrt{B_b \rho_l}}{\pi f} \right) \text{ for beams.} \tag{3.235}$$

3.12.3.1.2 Structural power measurement

Measurement of net structural power with an a–a probe is defined in a similar way to sound power, but using one-dimension rather than two. Instead of using a surface integral, a line integral is used to calculate the net structural power. A measurement line is chosen that encloses the source that is of interest; we are usually interested in the net structural power transmitted across a junction of plates and/or beams. Examples of measurement lines on the receiving plate are shown in Fig. 3.102.

The axis of the a–a probe must be perpendicular to the measurement line so that the intensity component, I_n, that is being measured is normal to this line, and is defined by Eq. 3.173. For N equally spaced measurement points along a measurement line of length, L, the net structural

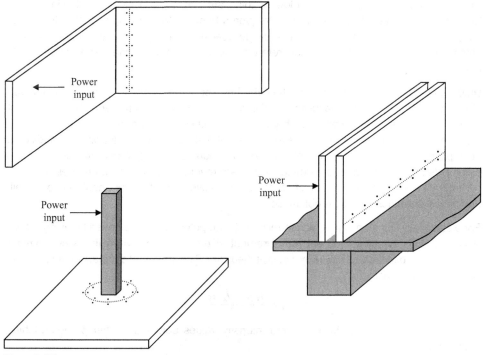

Figure 3.102

Examples of measurement lines used to determine the net structural power.

power, W_{net}, is

$$W_{net} = \frac{L}{N} \sum_{i=1}^{N} I_{n,i} \qquad (3.236)$$

3.12.3.1.3 Error analysis

Structural intensity is usually used to determine the power flowing across a line (i.e. measuring I_x or I_y) or to determine the vector intensity (i.e. measuring I_x and I_y). In a similar way to sound intensity measured with a p–p probe, there will be errors due to the finite-difference approximation and phase-mismatch. We will assume a plane bending wave propagating at an angle, θ, to the x-axis as previously shown in Fig. 3.101.

The normalized errors, $e_{FD}(I_x)$ and $e_{FD}(I_y)$, due to the finite-difference approximation are:

$$e_{FD}(I_x) \approx -\frac{1}{6}(k_B d \cos \theta)^2 \quad \text{and} \quad e_{FD}(I_y) \approx -\frac{1}{6}(k_B d \sin \theta)^2 \qquad (3.237)$$

The normalized errors, $e_{PM}(I_x)$ and $e_{PM}(I_y)$, due to phase-mismatch are:

$$e_{PM}(I_x) \approx \frac{\pm \phi_{PM}}{k_B d \cos \theta} \quad \text{and} \quad e_{PM}(I_y) \approx \frac{\pm \phi_{PM}}{k_B d \sin \theta} \qquad (3.238)$$

The combined normalized error due to the finite-difference error and phase-mismatch is calculated in the same way as described in the section on sound intensity. This combined error can be used to define an accelerometer spacing on a plate or beam of specific material and thickness by choosing a tolerable error (e.g. $\pm 5\%$). The magnitude of the structural intensity will only be valid for thin plates, i.e. below the thin plate limit, $f_{B(thin)}$. This means that in contrast to the p–p sound intensity probe, all valid frequency bands in the building acoustics frequency range can often be covered by using a single spacing. For a probe with $|\phi_{PM}| < 0.4°$ and measurements taken on masonry/concrete walls or floors, an appropriate spacing is usually between 50 and 100 mm.

Unlike sound intensity, there is no well-established set of indicators that can be used to validate a structural intensity measurement. For sound intensity measurements the reactivity of the sound field can be reduced by adding absorbent material into the room. In buildings there is little or no possibility of reducing the reactivity of the vibration field by introducing additional absorption at the boundaries of walls, floors, or columns. Hence structural intensity measurements are always taken on structures that are reverberant to various degrees. On very reverberant structures, the phase difference of the active intensity component may be small, and phase-mismatch may be problematic.

For a stationary vibration signal, phase-mismatch can potentially be removed by using probe-switching. This involves a repeat measurement at each position when the accelerometer positions have been rotated by 180°, i.e. switched over. The improved estimate of the intensity is then calculated using

$$I_x = \frac{I_{x,0°} - I_{x,180°}}{2} \qquad (3.239)$$

where $I_{x,0°}$ and $I_{x,180°}$ are the measured intensity values before and after probe-switching respectively.

If the phase-mismatch is negligible, probe-switching should give intensity of the same magnitude but opposite sign. If the phase-mismatch is greater than the actual phase difference in the vibration field the intensity will not change sign after probe-switching. (In principle this can be done with a sound intensity probe, but it is often easier to ensure accurate repositioning in two dimensions rather than three.) When switching the accelerometers it should be ensured that the fixing strength is the same (particularly when using bees wax), and that the accelerometer's axis of minimum transverse sensitivity is not aligned at a different angle relative to the probe axis.

Commercial sound intensity probes come with matched pairs of microphones to minimize phase-mismatch errors. In contrast, individual accelerometers for structural intensity probes are usually taken off the shelf. Problematic phase-mismatch (typically $|\phi_{PM}| > 0.4°$) can often be avoided by using accelerometers with consecutive serial numbers. However, the phase difference between two accelerometers should always be checked, even if probe-switching is to be used. It can be measured by rigidly fixing one accelerometer on top of the other, and then fixing the pair of accelerometers on top of an electrodynamic shaker so they are exposed to the same vibration signal. Dual-channel FFT analysis can then be used to determine the phase from the Frequency Response Function.

Measurement using an a–a probe is only valid for plane bending waves in the absence of nearfields. However, it is difficult to quantify the error and offer definitive guidance on appropriate distances from junctions or structural discontinuities. On masonry/concrete elements a distance of 0.3 to 0.6 m from a junction is usually adequate for the building acoustics frequency range (Craik *et al.*, 1995; Kruppa, 1986). On lightweight elements such as plasterboard walls, the probe will often be close to stud connections. However, it is difficult to generalize due to the frequency-dependent effects of different screw spacings and different types of studs (e.g. heavy thick timber or light thin steel).

Above the thin plate limit (but below the fundamental quasi-longitudinal mode) the calculated error in the measured magnitude of the structural intensity is an overestimate of the actual intensity by more than 20% (Maysenhölder, 1990). However, when it is only the direction of the intensity vector that is of interest on homogeneous isotropic plates, it should still be possible to gain reasonable estimates above the thin plate limit.

When measuring two-dimensional intensity vectors, the finite-difference and phase-mismatch errors combine to give an absolute error, $\varepsilon(\theta)$, in radians for the propagation angle, θ, where

$$\varepsilon(\theta) = \left| \arctan \left(\frac{\sin\theta \left(1 + \dfrac{\pm\phi_{PM}}{k_B d\,\sin\theta} - \dfrac{1}{6}(k_B d\,\sin\theta)^2 \right)}{\cos\theta \left(1 + \dfrac{\pm\phi_{PM}}{k_B d\,\cos\theta} - \dfrac{1}{6}(k_B d\,\cos\theta)^2 \right)} \right) \right| - \theta \tag{3.240}$$

3.12.3.1.4 *Visualizing net energy flow*

Structural intensity measurements can be used to visualize net energy flow across the surface of a plate. Typically, a square grid is drawn over the plate surface so that the measurement points are defined by the intersections of the grid lines. This results in a grid of active intensity vectors over the surface.

Interpreting a grid of measured vectors is not always intuitive, even when the measurement errors are insignificant. At a specific point, or along a line on the surface, there can be an outgoing or incoming flow of vibrational energy; referred to as a source or a sink respectively. Sources and sinks are usually the features of interest. However, individual bending wave

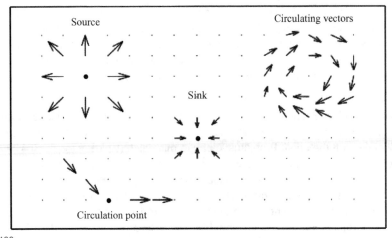

Figure 3.103

Idealized examples of a source, a sink, and circulating vectors.

modes on a plate give rise to a circulatory flow of active intensity vectors that sometimes make it difficult to identify these features (Cuschieri, 1991; Noiseux, 1970). Fortunately, strongly circulating vector fields are not usually evident when the frequency band is sufficiently wide, the response is multi-modal, and the plate has sufficiently high damping. One-third-octave-band measurements in the low-frequency range on masonry/concrete plates may show some circulation (see Kruppa, 1986). Note that circulatory active intensity due to individual modes also occurs in two- or three-dimensional sound fields (Fahy, 1989). When discussing vector fields it is useful to introduce another term, circulation point; this is defined as a grid point at which the energy flow is simply passing through (Maysenhölder and Schneider, 1989). These points are of no specific interest, but it is important to be aware of them as they may cause confusion in the identification of a source or a sink. Idealized examples of a source, sink, and circulating vectors are shown in Fig. 3.103. To identify these features it is necessary to make sure that the grid has a sufficient number of points; if the grid spacing is too coarse, some features may be interpreted incorrectly.

The sound insulation problem under investigation may involve an airborne or a structure-borne sound source. In general, the latter is more likely to result in useful measurements of structural intensity. Grid measurements on walls or floors that face into a source room excited by an airborne source tend not to be particularly informative (Kruppa, 1986). This is primarily due to difficulty in measuring active intensity in a highly reactive vibration field.

An example is now used to look at some of the features that can be encountered in practice with structural intensity measurements on a masonry construction (Hopkins, 2000). The test construction used for the measurements is shown in Fig. 3.104. In this set-up the separating wall is excited directly by a structure-borne sound source that transmits vibration to the flanking wall across the corner junction. Measurements were taken before and after the introduction of a window opening into the flanking wall. Two different grids were used on the flanking wall; a relatively coarse grid of 13×7 positions (entire flanking wall) and a more detailed grid of 12×15 positions over the area around the opening. One-third-octave-band measurements were taken with probe switching to remove the phase error, and a probe spacing of 50 mm. The structure-borne sound source was a plastic headed hammer. This source was used to

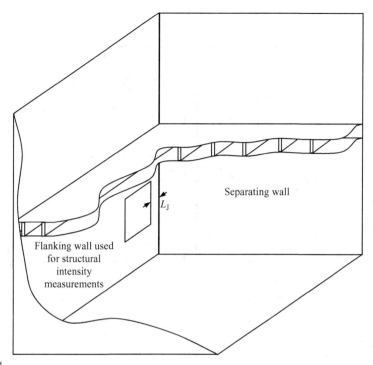

Figure 3.104

Sketch of the test construction used for structural intensity measurements. Flanking wall: 100 mm masonry wall ($L_x = 4$ m, $L_y = 2.4$ m, $\rho_s = 70$ kg/m^2, $c_L = 2370$ m/s, solid aircrete blocks with mortar on each side, 13 mm plaster finish). Window opening: 0.9×1.2 m, $L_J = 0.25$ m.

excite the separating wall at many different positions during a 20 s period with ≈ 5 hammer hits per second. A selection of the structural intensity plots are shown in Fig. 3.105.

An example of disordered vectors is seen in the 50 Hz band for the wall with an opening; note that this could potentially occur on walls or floors without openings. This seemingly random array of vectors is shown to emphasize the importance of using a sufficiently large number of measurement positions when trying to identify the position of a source.

The flanking wall is far from being an ideal plate. The lintel forms a discontinuity, and with a window opening there are only narrow strips of wall near the edges of the opening. When measurements are used to quantify the net structural power transmitted from the separating wall to the flanking wall, the measurement line is usually 0.3 to 0.6 m from a junction. This gives a limited number of measurement positions underneath and above the opening. The modified moment component and the assumption that the force component equals the moment component is not valid near the free edges of an opening. However, the magnitude of the intensity in the direction perpendicular to the junction is not significantly in error (Hopkins, 1997). The measured vectors generally point in a direction parallel to the free boundaries of the opening as would be expected for the true intensity vectors.

Examples of well-ordered vector flow can be seen for the wall with and without an opening. The vectors either head towards the upper wall boundary or towards the left wall boundary. The discontinuity formed by the lintel limits the length of the upper boundary over which the vectors can head towards the in-line junction that leads to the flanking wall on the first floor.

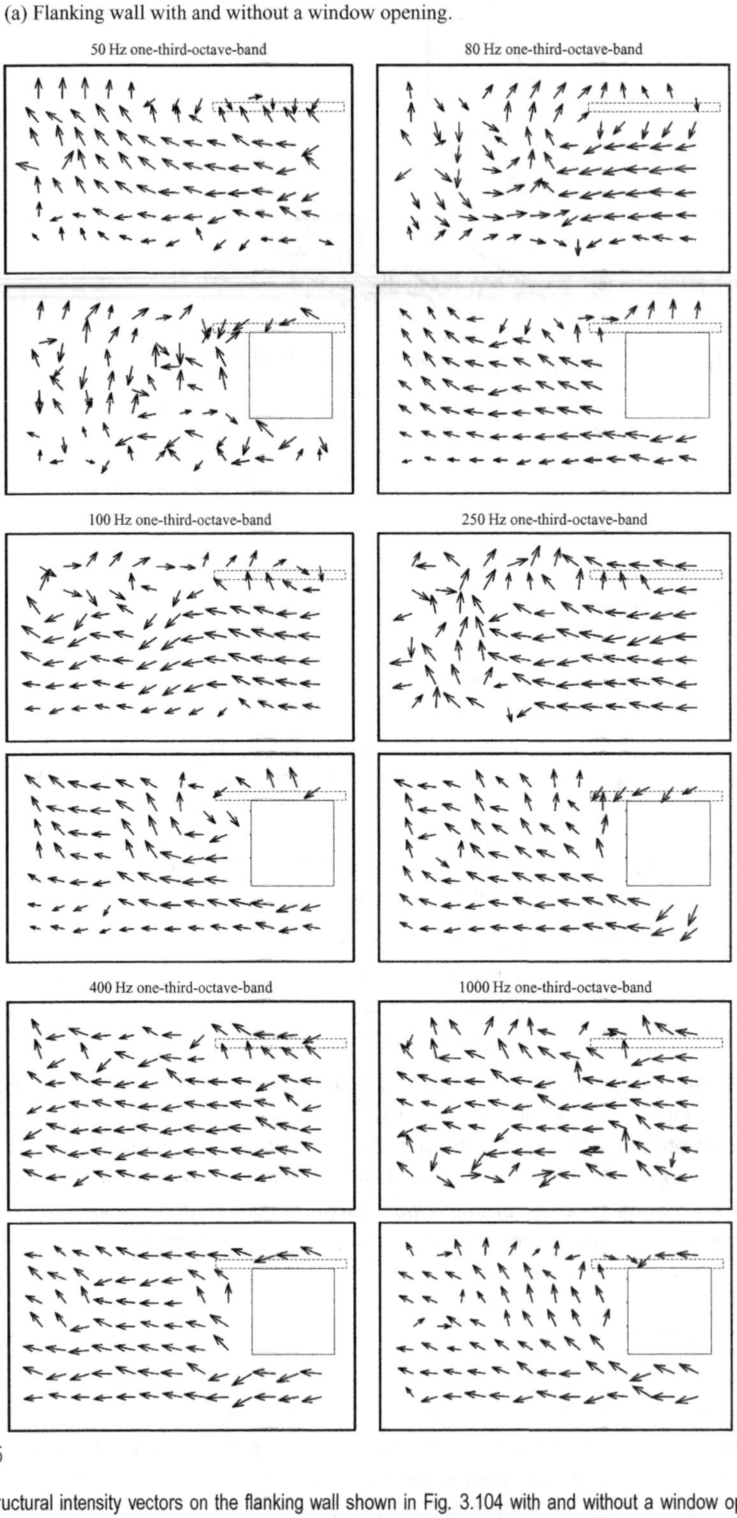

(a) Flanking wall with and without a window opening.

Figure 3.105

Measured structural intensity vectors on the flanking wall shown in Fig. 3.104 with and without a window opening (vector magnitude in decibels). Measured data are reproduced with permission from Hopkins (2000).

(b) Flanking wall with a window opening (detailed grid)

250 Hz one-third-octave-band

400 Hz one-third-octave-band

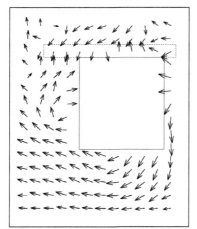

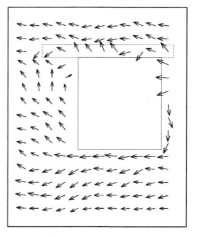

Figure 3.105

(*Continued*)

3.12.3.1.5 *Identifying construction defects*

High levels of sound insulation are often achieved by isolating two parts of a construction. For example a floating floor is isolated from the floor base by a resilient layer, and the leaves of a cavity wall are sometimes isolated from each other over their surface (i.e. no wall ties, lintels, etc.) and only connected at the foundations. Defects in the construction can remove this isolation and significantly reduce the sound insulation. A common example is the bridging of a floating screed floor via a gap or hole in the resilient layer; screed fills the hole and makes a rigid connection between the screed and the floor base. Vibration levels by themselves often give little or no clue in detecting the connection point on reverberant plates. Similarly, placing a pressure microphone or intensity probe at successive positions on a grid near the surface of the plate is unlikely to identify the connection point. Structural intensity offers the possibility of tracking down structural connections between plates.

A procedure to identify the location of a point source, such as a rigid connection on an isotropic, homogeneous plate is described by Maysenhölder and Schneider (1989). The first step is to divide the rectangular plate into four equal areas and measure the intensity vector at the centre of each of these areas. This gives a rough indication of the region containing the source. Subsequent measurements can then be used to more accurately identify its position. An example to identify the position of a single rigid connection between the leaves of a masonry cavity wall is shown in Fig. 3.106 (Maysenhölder and Schneider, 1989).

Structural intensity measurements have also been used to track down a bridged screed in a factory where there was excessive transmission of vibration from a structure-borne sound source (Sorainen and Rytkönen, 1989). However, an airborne and a structure-borne sound source may give similar patterns of vectors (Rasmussen and Rønnedal, 1992). This makes it necessary to look at structural intensity measurements alongside other measurements (or predictions of the sound transmission paths) before coming to a conclusion. As with any measurement of structural intensity on reverberant plates, it is important to check that the phase error is low and the results are not affected by the residual intensity (Craik *et al.*, 1996).

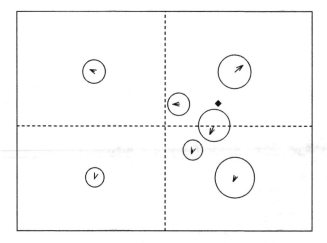

Figure 3.106

Structural intensity measurements used to identify the position of a single sound bridge (indicated by ◆) between two leaves of a masonry cavity wall. Shaker excitation was used on one leaf with structural intensity measurements on the other leaf to determine the frequency-average intensity (500 to 2000 Hz). Intensity vectors are shown with arrows and vibration levels are shown with circles for which the highest vibration levels have the largest diameter. Measured data are reproduced with permission from Maysenhölder and Schneider (1989).

Success in identifying a connection partly depends on the width of the frequency band(s) or frequency-average that is used to determine the structural intensity vectors. This is usually determined by trial and error because the frequency range in which there is significant transmission via the connection is not usually known *a priori*. However, one-third-octave-bands may be too narrow for this purpose. For point connected masonry/concrete plates, a bandwidth between 500 and 2000 Hz wide can be beneficial in identifying sources (Maysenhölder and Schneider, 1989; Craik *et al.*, 1996).

3.12.3.2 *Velocity level difference*

The velocity level difference is used to calculate vibration transmission between two elements (e.g. walls, floors, columns) across a junction or some other type of connection. For the purpose of calculating the vibration reduction index, its measurement is described in the relevant Standard (ISO 10848 Part 1).

One element is treated as the source element, which is excited by an airborne or a structure-borne sound source; the other element is treated as the receiving element (see Fig. 3.107). For connected walls and floors between a source and receiving room, there are many possible permutations of source and receiving element. For this reason, the velocity level difference, $D_{v,ij}$, is defined as the difference between the velocity level on source element, i, and receiving element, j. Each element can therefore be numbered to simplify their identification.

At the outset it is important to establish the aim of the measurement. Usually it is to quantify the difference between the bending wave vibration on elements, i and j; hence it is the velocity perpendicular to the surface of each element that is required. Excitation of element, i, is arranged to ensure that only bending waves are excited; this is most conveniently done with a structure-borne sound source. Only element, i, can be the source element, so when using airborne excitation via the sound field in a room, it is not possible for elements i and j to

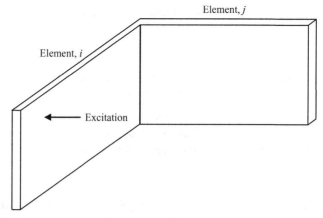

Figure 3.107

Measurement of velocity level difference between two walls connected across an L-junction.

face into the same room. When the bending waves on the source or receiving plate impinge upon the element boundaries and the junction, there is often some conversion to in-plane waves. In this measurement, transmission via the junction is effectively treated as a black box; bending waves are measured on the source and receiving elements, yet transmission across the junction involving wave conversion is treated as an unknown process.

If the intention is to determine a spatial average velocity level difference over the surface of each element with a structure-borne sound source, then it is important to ensure that there is no significant decrease in vibration with distance across the element (Section 2.7.7). This is particularly important for lightweight walls and floors because the measurements may be specific to a particular layout of the frame and the sheets/boards; as well as specific to particular excitation and measurement points.

Bending waves are usually of most interest, and it is rarely necessary to measure the velocity level difference for in-plane wave motion on the source and/or receiving element. This is fortunate because interpretation of such measurements can be complex. The accelerometer does not distinguish between transverse shear and quasi-longitudinal motion, and the presence of bending waves can affect the measurement of in-plane motion (Craik, 1998).

Measurements are usually taken in one-third-octave or octave-bands over the building acoustics frequency range.

3.12.3.2.1 Stationary excitation signal and fixed power input

For structure-borne excitation of a wall or floor, a random noise signal can be sent to an electrodynamic shaker that is pushed up against the element; this is only possible if the power input is the same each time the shaker is moved to a different source position. In practice, it is awkward and time-consuming to measure the power input; although it can be done with structural intensity or by using a force transducer between the shaker and the element. It is simpler for a floor, because the ISO tapping machine can be considered as a structure-borne sound source that provides a fixed power input.

For a stationary signal from either a structure-borne or airborne sound source, $D_{v,ij}$ is calculated from

$$D_{v,ij} = L_{v,i} - L_{v,j} \tag{3.241}$$

where $L_{v,i}$ and $L_{v,j}$ are the temporal and spatial average velocity levels determined from different source positions and different accelerometer positions for each source position.

The temporal and spatial average value, standard deviation and 95% confidence interval can be calculated from Eqs 3.18, 3.19, and 3.20 respectively. Corrections for background noise can be made in the same way as for sound pressure measurements using Eq. 3.75.

3.12.3.2.2 Impulse excitation

An impulse can be generated using a hammer weighing ≈0.5 kg with a hard plastic head; this provides a reasonably flat spectrum over the building acoustics frequency range, and avoids excessive damage to the surface. One approach is to repeatedly hit the wall during a defined measurement period. Multiple hits have the advantage of incorporating many different source positions (Craik, 1982b). The decay of the signal in each filter band depends on the reverberation time of the filter as well as the element. Masonry/concrete elements tend to have short reverberation times, so to maintain a steady signal it is usually necessary to use at least two hits per second. An alternative approach is to use a single impulse from a hammer hit. A single hit is more suitable than multiple hits when the measurer is likely to cause vibration whilst moving around the floor and cause an increased level of vibration on the receiving element. An additional consideration for lightweight floors is that the mass of the measurer and the measurement equipment on the floor must not affect the velocity level, or induce a static deflection that alters vibration transmission via the junction under test.

A dual-channel analyser and a pair of accelerometers are needed to measure the vibration on elements, i and j simultaneously. This is because of the variation in the energy spectral density between each individual hammer hit. The term 'excitation' will now be used to refer to either a single hit at different positions, or multiple hits over the surface. To account for the spatial variation in vibration and ensure that different modes are excited, M different excitations are applied with N different pairs of accelerometer positions for each excitation. For excitation, m, and pair of accelerometer positions, n, the velocity level difference $(D_{v,ij})_{m,n}$ is determined using

$$(D_{v,ij})_{m,n} = 10\lg \left(\frac{\int_{0}^{T_{int}} v_i^2(t)\mathrm{d}t}{\int_{0}^{T_{int}} v_j^2(t)\mathrm{d}t} \right)_{m,n} \tag{3.242}$$

where T_{int} is the integration time for linear averaging.

Level differences from different excitations and different accelerometer positions are arithmetically averaged to give $D_{v,ij}$ according to:

$$D_{v,ij} = \frac{1}{MN} \sum_{m=1}^{M} \sum_{n=1}^{N} (D_{v,ij})_{m,n} \tag{3.243}$$

The excitation is applied after $t = 0$, and the resulting signal must have decayed to a negligible level (e.g. decreased by 60 dB) before $t = T_{int}$. This requirement needs to be balanced against the need to avoid excessive averaging of background noise after the excitation has ceased;

so T_{int} must not occur too long after the end of the excitation. With a single hit on highly damped elements, T_{int} may only need to be 2 or 3 s. If so, it is better to automate the process. A computer can be used to start and stop the analyser, and an audible tone can be emitted to let the measurer know when the integration is starting and finishing. The frequency of this tone must be well-above the highest frequency band being measured.

Although $D_{v,ij}$ for an impulsive source is defined in terms of a ratio of velocities, it is usually more accurate to use the acceleration signal from the charge amplifiers. This avoids potential distortion of the impulse due to phase non-linearities in the charge amplifiers. There is no need to convert each acceleration signal to velocity because $D_{v,ij}$ is formed from a ratio.

In the light of non-linearities relating to structural reverberation times (Section 3.11.3.3), it is prudent to pay attention to the strength of the hammer hits and the resulting vibration levels. This issue can be avoided by measuring $(D_{v,ij})_{m,n}$ using MLS shaker excitation at low levels. On masonry/concrete elements, the MLS signal may need a red noise spectrum to give a relatively flat acceleration spectrum on the source element. MLS is also advantageous when the background level and/or the velocity level difference is high.

In Section 2.7 it was seen that there can be significant spatial variation of vibration over the surface of a plate. Examples of standard deviations for $D_{v,ij}$ between solid masonry walls measured with impulse excitation are shown in Fig. 3.108. A generalized curve is included on the figure because the curve shape can usually be described using just four points. The assumption of a normal (Gaussian) probability distribution for the velocity level difference is reasonable for solid masonry walls when there is no significant decrease in vibration level with distance (Hopkins, 2000). The 95% confidence interval (Eq. 3.20) can therefore be calculated from the standard deviation in decibels.

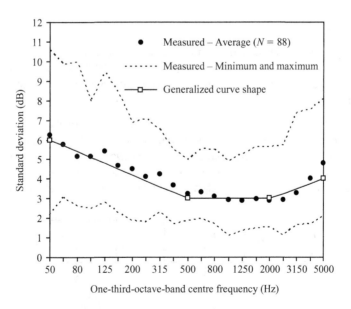

Figure 3.108

Measured standard deviations for the velocity level difference, $D_{v,ij}$ (hammer or MLS shaker excitation). Average, minimum and maximum values calculated from 88 measurements on different combinations of solid masonry walls with surface areas of $\approx 10 m^2$, with and without plaster, and with mass per unit areas in the range 70 to 430 kg/m^2. Measured data from Hopkins are reproduced with permission from ODPM and BRE.

3.12.3.2.3 Excitation and accelerometer positions

Accelerometers need to be used at a number of random positions over the surface of each element to account for the spatial variation in vibration. However, at these positions, the direct field of the excitation and any nearfields close to the boundaries must be negligible (Sections 2.7.5 and 2.7.2). Suitable distances from the excitation point or from the boundaries will vary for each element as well as with frequency. In practice it is awkward to prescribe different distances for different elements at different frequencies. It is therefore necessary to compromise and choose distances based upon typical elements as well as the practicalities of the measurement (Pedersen, 1993). In addition, the excitation positions must be chosen to ensure excitation of the different modes and so that the direct field is negligible at the junction or other connection under test. The values quoted in ISO 10848 Part 1 are representative of distances in common use. The distance between the accelerometer and the excitation point is at least 1 m, the distance between the accelerometer and the boundaries is at least 0.25 m, and the distance between accelerometer positions is at least 0.5 m. The distance between the excitation position and the junction is at least 1 m, and is at least 0.5 m from the other boundaries.

As with spatial sampling of sound pressure in rooms, the inclusion of correlated samples on a plate should also be avoided. As a rule-of-thumb for a two-dimensional diffuse bending wave field it can be assumed that accelerometer positions should be separated from each other by a distance, $d \geq \lambda_B/2$. However a wide variety of plate materials and thicknesses are encountered in practice, the majority of which do not have diffuse bending wave fields over the entire building acoustics frequency range. A minimum distance between accelerometer positions of 0.5 m is a pragmatic choice because the direct field near the excitation point limits the available area in which accelerometers can be fixed.

3.12.3.3 Coupling Loss Factor, η_{ij}

Two methods are particularly convenient to determine the coupling loss factor: the first uses the velocity level difference and the total loss factor (Craik, 1982b), the second uses structural intensity (Craik, 1995). In this section it is more appropriate to refer to SEA subsystems rather than building elements.

The method using the velocity level difference is the most basic form of experimental SEA. We restrict our attention to two connected subsystems, i and j, for which we want to determine the coupling loss factor, η_{ij}. A thorough introduction to the two-subsystem SEA model shown in Fig. 3.109 is given in Section 4.2. This model is based on a blinkered view of the building because these two subsystems will rarely be isolated from other subsystems. As an example,

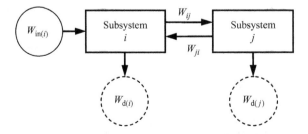

Figure 3.109

Two-subsystem SEA model.

consider a floor connected to a wall; they will inevitably be connected to other walls or columns for structural support. However, we can account for these other subsystems by considering the coupling losses from our two subsystems of interest to these other subsystems. Effectively, these other subsystems are treated purely as places of energy dissipation. This is reasonable when these other subsystems do not form flanking transmission paths between the two subsystems of interest. This can often be arranged in a laboratory, but field measurements will usually be affected by other flanking paths.

From Fig. 3.109, a power input, $W_{in(i)}$, is applied to subsystem i from which some power is dissipated through internal losses and coupling losses to other subsystems, and some is transmitted to subsystem j. For subsystem i the dissipated and transmitted powers are $W_{d(i)}$ and W_{ij} respectively. The transmitted power from subsystem i to subsystem j is

$$W_{ij} = \omega \eta_{ij} E_i \tag{3.244}$$

This transmitted power can also be calculated from the perspective of subsystem j, because the power transmitted from i to j must be the same as the total power dissipated by subsystem j. This is described using the total loss factor of subsystem j,

$$W_{ij} = \omega \eta_j E_j \tag{3.245}$$

Assuming that there is negligible power flow back from subsystem j to i, the coupling loss factor is calculated by equating Eqs 3.244 and 3.245 to give

$$\eta_{ij} = \frac{E_j}{E_i} \eta_j \tag{3.246}$$

Hence the coupling loss factor in decibels can be calculated from the velocity level difference, $D_{v,ij}$, and the total loss factor of subsystem j using

$$L_{\eta_{ij}} = L_{\eta_j} - \left[D_{v,ij} + 10 \lg \left(\frac{m_i}{m_j} \right) \right] \tag{3.247}$$

where m_i and m_j are the mass of subsystems i and j respectively.

An alternative method of determining the coupling loss factor is to use structural intensity to quantify the net power, $W_{net,ij}$ transmitted from i to j. From the two-subsystem SEA model,

$$W_{net,ij} = W_{ij} - W_{ji} = \omega \eta_{ij} E_i - \omega \eta_{ji} E_j \tag{3.248}$$

hence the coupling loss factor is

$$\eta_{ij} = \frac{W_{net,ij} + \omega \eta_{ji} E_j}{\omega E_i} \tag{3.249}$$

In practice, the power returning from subsystem j to i is usually negligible, hence

$$\eta_{ij} \approx \frac{W_{net,ij}}{\omega E_i} \tag{3.250}$$

and the coupling loss factor in decibels can be calculated using

$$L_{\eta_{ij}} = 10 \lg \left(\frac{W_{net,ij}}{10^{-12}} \right) - [L_{v,i} + 10 \lg (\omega m_i)] \tag{3.251}$$

An advantage with structural intensity is that it is possible to make use of the probe directivity to distinguish between different transmission paths, thus avoiding problems with flanking transmission via other subsystems. As an example, it can be used to quantify the coupling loss factor between two leaves of a cavity wall via the foundations when there is another connection between the leaves, such as a lintel above a window.

Measured coupling loss factors between plates are not only specific to the type of junction, but also the properties of both plates, including the area of the source plate and the junction length. This can be seen by referring back to the coupling loss factor definition in Eq. 2.154. Hence care needs to be taken in the comparison of measured coupling loss factors from different laboratories/buildings and when using them in SEA models of other constructions. Measurements on beams are restricted in a similar way.

3.12.3.4 Vibration Reduction Index, K_{ij}

The vibration reduction index, K_{ij}, is defined as (EN 12354 Part 1)

$$K_{ij} = \overline{D_{v,ij}} + 10 \lg \left(\frac{L_{ij}}{\sqrt{a_i a_j}} \right) \tag{3.252}$$

where L_{ij} is the junction length between elements i and j, a is the absorption length calculated from the measured structural reverberation time (Eq. 2.148) and the direction-averaged velocity level difference, $\overline{D_{v,ij}}$, is

$$\overline{D_{v,ij}} = \frac{D_{v,ij} + D_{v,ji}}{2} \tag{3.253}$$

The reasons behind this definition of the vibration reduction index and the reason for using a direction-averaged velocity level difference are discussed in Section 5.4 along with the SEA-based model in which they are used.

References

Andres, H.G. (1965/1966). Über ein Gesetz der räumlichen Zufallsschwankung von Rauschpegeln in Räumen und seine Anwendung auf Schalleistungsmessungen, *Acustica*, **16**, 279–294.

Anon. (1996). *Microphone handbook, Vol. 1: Theory*, Brüel & Kjær.

Ballagh, K.O. (1990). Noise of simulated rainfall on roofs, *Applied Acoustics*, **31**, 245–264.

Bauman, P.D. (1994). Measurement of structural intensity: analytic and experimental evaluation of various techniques for the case of flexural waves in one-dimensional structures, *Journal of Sound and Vibration*, **174** (5), 677–694.

Bendat, J.S. and Piersol, A.G. (2000). *Random data: analysis and measurement procedures*, John Wiley & Sons, Inc., New York. ISBN: 0471317330.

Best, A.C. (1950). Empirical formulae for the terminal velocity of water drops falling through the atmosphere, *Quarterly Journal of the Royal Meteorological Society*, **76**, 302–311.

Bietz, H., Bethke, G. and Schmitz, A. (1997). Effects of time variances using MLS techniques in building acoustics for measuring sound insulation and reverberation time, *Proceedings of Internoise 97*, Budapest, Hungary, 1429–1432.

Bjor, O-H. (1995). M-sequence for building acoustic measurements, *Proceedings of the Institute of Acoustics*, **17** (5), 101–109.

Bodlund, K. (1976). Statistical characteristics of some standard reverberant sound field measurements, *Journal of Sound and Vibration*, **45** (4), 539–557.

Bodlund, K. (1985). Alternative reference curves for evaluation of the impact sound insulation between dwellings, *Journal of Sound and Vibration*, **102** (3), 381–402.

Bodlund, K. and Jonasson, H.G. (1983). Measurement of the impact sound insulation improvement, *Proceedings of Internoise 83*, Edinburgh, Scotland, 619–622.

Bradley, J.S. (1996). Optimizing the decay range in room acoustics measurements using maximum-length-sequence techniques, *Journal of the Audio Engineering Society*, **44** (4), 266–273.

Bradley, J.S. and Chu, W.T. (2002). Errors when using façade measurements of incident aircraft noise, *Proceedings of Internoise 2002*, Dearborn, MI, USA.

Broch, J.T. (1983). On low frequency impact sound insulation measurements, *Proceedings of Internoise 83*, Edinburgh, Scotland, 1115–1118.

Brown, R.L. and Bolt, R.H. (1942). The measurement of flow resistance of porous acoustic materials, *Journal of the Acoustical Society of America*, **13** (4), 337–344.

Brüel, P.V. (1987). Measurement of impulsive signals, *Proceedings of Internoise 87*, Beijing, China, 1339–1342.

Brunskog, J. and Hammer, P. (2002). Measurement of the acoustic properties of resilient, statically tensile loaded devices in lightweight structures, *Building Acoustics*, **9** (2), 99–137.

Brunskog, J. and Hammer, P. (2003). The interaction between the ISO tapping machine and lightweight floors, *Acta Acustica*, **89**, 296–308.

BS 1881-203:1986 Testing concrete – Part 203: Recommendations for measurement of velocity of ultrasonic pulses in concrete, British Standards Institution.

Carman, T.A. and Fothergill, L.C. (1990). Investigation of sound transmission in buildings using the sound intensity technique, *Proceedings of the Institute of Acoustics*, **12** (5), 1–8.

Choudhury, N.K.D. and Bhandari, P.S. (1972). Impact noise rating of resilient floors, *Acustica*, **26**, 135–140.

Clarkson, B.L. and Pope, R.J. (1981). Experimental determination of modal densities and loss factors of flat plates and cylinders, *Journal of Sound and Vibration*, **77** (4), 535–549.

Cook, R.K., Waterhouse, R.V., Berendt, R.D., Edelman, S. and Thompson Jr. M.C. (1955). Measurement of correlation coefficients in reverberant sound fields, *Journal of the Acoustical Society of America*, **27** (6), 1072–1077.

Cops, A., Minten, M. and Myncke, H. (1987). Influence of the design of transmission rooms on the sound transmission loss of glass – Intensity versus conventional method, *Noise Control Engineering Journal*, **28** (3), 121–129.

Craik, R.J.M. (1981). Damping of building structures, *Applied Acoustics*, **14**, 347–359.

Craik, R.J.M. (1982a). The measurement of the material properties of building structures, *Applied Acoustics*, **15**, 275–282.

Craik, R.J.M. (1982b). The measurement of structure-borne sound transmission using impulsive sources, *Applied Acoustics*, **15**, 355–361.

Craik, R.J.M. (1990). On the accuracy of sound pressure level measurements in rooms, *Applied Acoustics*, **29**, 25–33.

Craik, R.J.M. (1992). The influence of the laboratory on measurements of wall performance, *Applied Acoustics*, **35**, 25–46.

Craik, R.J.M. (1998). In-plane wave propagation in buildings, *Applied Acoustics*, **53** (4), 273–289.

Craik, R.J.M. and Barry, P.J. (1992). The internal damping of building materials, *Applied Acoustics*, **35**, 139–148.

Craik, R.J.M. and Osipov, A.G. (1995). Structural isolation of walls using elastic interlayers, *Applied Acoustics*, **46**, 233–249.

Craik, R.J.M. and Wilson, R. (1995). Sound transmission through masonry cavity walls, *Journal of Sound and Vibration*, **179** (1), 79–96.

Craik, R.J.M., Ming, R. and Wilson, R. (1995). The measurement of structural intensity in buildings, *Applied Acoustics*, **44**, 233–248.

Craik, R.J.M., Wilson, R. and Ming, R. (1996). The location of building defects using structural intensity, *Building Acoustics*, **3** (4), 217–231.

Craik, R.J.M., Tomlinson, D. and Wilson, R. (2000). Radiation into a porous medium, *Proceedings of the Institute of Acoustics*, **22** (2), 383–388.

Cremer, L. (1976/1977). Die Problematik der Korperschallprufung von Decken, *Acustica*, **36** (3), 173–183.

Cremer, L., Heckl, M. and Ungar, E.E. (1973). *Structure-borne sound*, Springer-Verlag. ISBN: 0387182411.

Cuschieri, J.M. (1991). Experimental measurement of structural intensity on an aircraft fuselage, *Noise Control Engineering Journal*, **37** (3), 97–107.

Davy, J.L. (1988). The variance of decay rates at low frequencies, *Applied Acoustics*, **23**, 63–79.

Davy, J.L. and Dunn, I.P. (1987). The statistical bandwidth of Butterworth filters, *Journal of Sound and Vibration*, **115** (3), 539–549.

Davy, J.L., Dunn, I.P. and Dubout, P. (1979). The variance of decay rates in reverberation rooms, *Acustica*, **43**, 12–25.

Dunn, S.E., Stevens, K.K. and Uhlar, D.A. (1983). Considerations in applying damping coating material properties determined from beam tests to the control of plate vibration problems, *Proceedings of Internoise 83*, Edinburgh, Scotland, 541–544.

Fahy, F.J. (1985). *Sound and structural vibration. Radiation, transmission and response*, Academic Press, London. ISBN: 0122476700.

Fahy, F.J. (1989). *Sound intensity*, Elsevier Science Publishers Ltd. ISBN: 1851663193.

Fasold, W. (1965). Untersuchungen über den Verlauf der Sollkurve für den Trittschallschutz im Wohnungsbau, *Acustica*, **15**, 271–284.

Ford, R.D. and Warnock, A.C.C. (1974). Impact noise on floors, *Technical Paper No. 417*, National Research Council Canada, June 1974.

Fournier, D. and Val, M. (1963). Anomalies relevées dans les mesures de transmission de bruits d'impacts á l'aide de machines á frapper normalisées, *Proceedings of a Colloquium on Noise in Dwellings and Offices*, Marseille, France, September 1963.

Gerretsen, E. (1976). A new system for rating impact sound insulation, *Applied Acoustics*, **9**, 247–263.

Gerretsen, E. (1990). How indifferent is the sound reduction index of a wall to its boundary conditions? *Proceedings of Internoise 90*, Gothenburg, Sweden, 47–50.

Gösele, K. (1956). Trittschall – Entstehung und Dämmung, *VDI-Berichte "Schall und Schwingungzen in Festkörpen"*, **8**, 23–28.

Gösele, K. (1961). Luftschalldämmung, *Proceedings of the Third International Congress on Acoustics*, Vol. II. Elsevier, 989–1000.

Goydke, H. and Fischer H.-W. (1983). On impact sound level measurements uncertainties caused by properties variations of the standard tapping machine, *Proceedings of Internoise 83*, Edinburgh, Scotland, 1123–1126.

Goydke, H., Siebert, B.R.L. and Scholl, W. (2003). Considerations on the evaluation of uncertainty values of building acoustic single-number quantities, *Proceedings of Euronoise 2003*, Naples, Italy.

Grimwood, C. (1997). Complaints about poor sound insulation between dwellings in England and Wales, *Applied Acoustics*, **52** (3/4), 211–223.

Guy, R.W. and Mulholland, K.A. (1979). Some observations on employing a panel's cill and reveal to enhance noise reduction, *Applied Acoustics*, **3**, 377–388.

Guy, R.W. and Sauer, P. (1984). The influence of sills and reveals on sound transmission loss, *Applied Acoustics*, **17**, 453–476.

Guy, R.W., De Mey, A. and Sauer, P. (1985). The effect of some physical parameters upon the laboratory measurements of sound transmission loss, *Applied Acoustics*, **18**, 81–98.

Hagberg, K. (1996). Acoustic requirements based on ISO/DIS 717, *NKB Committee and Work Reports 1996:02*, The Nordic Committee for Building Regulations (NKB). (9515307813)

Hall, R. and Hopkins, C. (2001). Dynamic stiffness of wall ties used in masonry cavity walls: measurement procedure, *BRE Information Paper IP3/01*, BRE, Watford, England. ISBN: 1860814611.

Hall, R., Bougdah, H. and Mackenzie, R.K. (1996). A method for predicting $L'_{nT,w}$ for lightweight floating floors comprising low density flexible polyurethane foam on concrete supporting floors, *Building Acoustics*, **3** (2), 105–117.

Hamme, R.N. (1965). Sound transmission through floor-ceiling structures: I. Evolution of new impact test method, *Report No. IBI-1-I, November 1965*. Geiger and Hamme Laboratories, Michigan, USA.

Heckl, M. and Seifert, K. (1958). Untersuchungen über den Einfluss der Eigenresonanzen der Messräume auf die Ergebnisse von Schalldämmessungen, *Acustica*, **8**, 212–220.

Hongisto, V. (2001). Technical note: a case study of flanking transmission through double structures, *Applied Acoustics*, **62**, 589–599.

Hopkins, C. (1997). Energy flow measurement between walls containing a window aperture, *Proceedings of the Institute of Acoustics*, **19** (8), 261–266.

Hopkins, C. (2000). Structure-borne sound transmission between coupled plates, PhD thesis, Department of Building Engineering and Surveying, Heriot-Watt University, Edinburgh, Scotland.

Hopkins, C., Wilson, R. and Craik, R.J.M. (1999). Dynamic stiffness as an acoustic specification parameter for wall ties used in masonry cavity walls, *Applied Acoustics*, **58**, 51–68.

Horvei, B., Olsen, H. and Ustad, A. (1998). Nordtest project on MLS based method for sound insulation measurements, *SINTEF Report No. STF40 A98008. SINTEF Telecom and Informatics*, Trondheim, Norway.

ISO 140-6:1978 Acoustics – Measurement of sound insulation in buildings and of building elements – Part 6: Laboratory measurements of impact sound insulation of floors, International Organization for Standardization.

ISO 140-8:1978 Acoustics – Measurement of sound insulation in buildings and of building elements – Part 8: Laboratory measurement of the reduction of transmitted impact noise by floor coverings on a standard floor. International Organization for Standardization.

ISO 354:1985 Acoustics – Measurement of sound absorption in a reverberation room, *International Organization for Standardization*.

Jacobsen, F. (1987). A note on acoustic decay measurements, *Journal of Sound and Vibration*, **115** (1), 163–170.

Jacobsen, F. (1991). A simple and effective correction for phase mismatch in intensity probes, *Applied Acoustics*, **33**, 165–180.

Jacobsen, F. and Rindel, J.H. (1987). Letter to the editor. Time reversed decay measurements, *Journal of Sound and Vibration*, **117** (1), 187–190.

Jacobsen, F., Cutanda, V. and Juhl, P.M. (1998). A numerical and experimental investigation of the performance of sound intensity probes at high frequencies, *Journal of the Acoustical Society of America*, **103**, 953–961.

Jagenäs, A. and Petersson, B.A.T. (1986). The water drop as a structural acoustic source, *Proceedings of Internoise 1986*, USA, 349–352.

Jonasson, H. (1991). Measurements of sound reduction index with intensity technique. *Nordtest Project 746–88, Technical Report SP 1991:23*, Swedish National Testing Institute SP.

Jonasson, H.G. (1993). Sound intensity and sound reduction index, *Applied Acoustics*, **40**, 281–293.

Jonasson, H. and Carlsson, C. (1986). Measurement of sound insulation of windows in the field. Nordtest Project 556-85, *Technical Report SP-RAPP 1986:37*, Swedish National Testing Institute SP.

Josse, R. (1970). Une machine destinée à reproduire fidèlement les bruits de pas pour l'étude du comportement réel des revêtements de sol, *Cahiers du CSTB No. 924 January/February 1970*, Centre Scientifique et Technique du Batiment, France.

Kihlman, T. (1970). Sound transmission in building structures of concrete, *Journal of Sound and Vibration*, **11** (4), 435–445.

Kihlman, T. and Nilsson, A.C. (1972). The effects of some laboratory designs and mounting conditions on reduction index measurements, *Journal of Sound and Vibration*, **24** (3), 349–364.

Kruppa, P. (1986). Measurement of structural intensity in building constructions, *Applied Acoustics*, **19**, 61–74.

Kuhl, W. and Kaiser, H. (1952). Absorption of structure-borne sound in building materials without and with sand-filled cavities, *Acustica*, **2** (1), 179–188.

Leppington, F.G. (1996). Acoustic radiation from plates into a wedge-shaped fluid region: application to the free plate problem, *Proceedings of the Royal Society*, **452**, London, 1745–1764.

Leppington, F.G., Heron, K.H., Broadbent, E.G. and Mead, S.M. (1987). Resonant and non-resonant acoustic properties of elastic panels. II. The transmission problem, *Proceedings of the Royal Society*, **A412**, London, 309–337.

Lindblad, S.G. (1968). Impact sound characteristics of resilient floor coverings: a study on linear and nonlinear dissipative compliance, *Division of Building Technology*, Lund Institute of Technology, Sweden.

Lindblad, S.G. (1983). Non-linearity problems connected with impact sound insulation measurements, *Proceedings of Internoise 83*, Edinburgh, Scotland, 611–614.

Lubman, D. (1971). Spatial averaging in sound power measurements, *Journal of Sound and Vibration*, **16** (1), 43–58.

Lubman, D. (1974). Precision of reverberant sound power measurements, *Journal of the Acoustical Society of America*, **56** (2), 523–533.

Lubman, D., Waterhouse, R.V. and Chien, C. (1973). Effectiveness of continuous spatial averaging in a diffuse sound field, *Journal of the Acoustical Society of America*, **53** (2), 650–659.

Lundeby, A., Vigran, T.E., Bietz, H. and Vorländer, M. (1995). Uncertainties of measurements in room acoustics, *Acustica*, **81**, 344–355.

Lyon, R.H. (1969). Statistical analysis of power injection and response in structures and rooms, *Journal of the Acoustical Society of America*, **45**, 545–565.

Lyon, R.H. and DeJong, R.G. (1995). *Theory and application of Statistical Energy Analysis*, Butterworth-Heinemann, MA, USA. ISBN: 0750691115.

Macadam, J.A. (1976). The measurement of sound radiation from room surfaces in lightweight buildings, *Applied Acoustics*, **9** (2), 103–118.

Marshall, J.S. and Palmer, W.Mck. (1948). The distribution of raindrops with size, *Journal of Meteorology*, **5**, 165–166.

Maysenhölder, W. (1990). Rigorous computation of plate-wave intensity, *Acustica*, **72**, 166–179.

Maysenhölder, W. and Horvatic, B. (1998). Determination of elastodynamic properties of building materials, *Proceedings of Euronoise 98*, Munich, Germany, 421–424.

Maysenhölder, W. and Schneider, W. (1989). Sound bridge localization in buildings by structure-borne sound intensity measurements, *Acustica*, **68**, 258–262.

McLoughlin, J., Saunders, D.J. and Ford, R.D. (1994). Noise generated by simulated rainfall on profiled steel roof structures, *Applied Acoustics*, **42**, 239–255.

Mechel, F.P. (1989/1995/1998). *Schallabsorber Band I, II and III*, Stuttgart: S. Hirzel Verlag. ISBN: 3777604259/ISBN: 3777605727/ISBN: 3777608092.

Meier, A. and Schmitz, A. (1999). Application of total loss factor measurements for the determination of sound insulation, *Building Acoustics*, **6** (2), 71–84.

Meier, A., Schmitz, A. and Raabe, G. (1999). Inter-laboratory test of sound insulation measurements on heavy walls. Part 2 – Results of main test, *Building Acoustics*, **6** (3/4), 171–186.

Metzen, H.A. (1996). Estimation of the reduction in impact sound pressure level of floating floors from the dynamic stiffness of insulation layers, *Building Acoustics*, **3** (1), 33–53.

Michelsen, N. (1982). Repeatability of sound insulation measurements, *Technical Report No.36*, Danish Acoustical Laboratory (now DELTA Acoustics), Denmark.

Mitchell, L.D. and Elliott, K.B. (1984). How to design stingers for vibration testing of structures, *Sound and Vibration Magazine*, April 1984, 14–18.

Müller, S. and Massarani, P. (2001). Transfer-function measurement with sweeps, *Journal of the Audio Engineering Society*, **49**, 443–471.

Nightingale, T.R.T. (1996). Acoustic intensity as a tool for assessing sound isolation and flanking transmission in lightweight building constructions, *Proceedings of Internoise 96*, Liverpool, England, 2685–2690.

Nightingale, T.R.T., Halliwell, R.E. and Pernica, G. (2004). Estimating in-situ material properties of a wood joist floor: Part 1 – Measurements of the real part of bending wavenumber, *Building Acoustics*, **11** (3), 175–196.

Nilsson, E. (1992). Decay processes in rooms with non-diffuse sound fields. *Report TVBA-1004*, Doctoral dissertation, Lund Institute of Technology, Sweden.

Noiseux, D.U. (1970). Measurement of power flow in uniform beams and plates, *Journal of the Acoustical Society of America*, **47** (1/2), 238–247.

Olesen, H.S. (2002). Laboratory measurement of sound insulation in the frequency range 50 Hz to 160 Hz – A Nordic intercomparison, *Nordtest Technical Report 489* (www.nordtest.org). ISSN: 0283-7234.

Olynyk, D. and Northwood, T.D. (1965). Subjective judgments of footstep-noise transmission through floors, *Journal of the Acoustical Society of America*, **38**, 1035–1039.

Olynyk, D. and Northwood, T.D. (1968). Assessment of footstep noise through wood-joist and concrete floors, *Journal of the Acoustical Society of America*, **43** (4), 730–733.

Pavić, G. (1976). Measurement of structure borne wave intensity, Part I: Formulation of the methods, *Journal of Sound and Vibration*, **49** (2), 221–230.

Pavić, G. (1977). Measurement of sound intensity, *Journal of Sound and Vibration*, **51** (4), 533–545.

Pedersen, D.B. (1993). Measurement of vibration attenuation through junctions of building structures. *Nordtest Project 967-91, DTI No. 260 2 8011*, Danish Technological Institute, Aarhus, Denmark. (8777563042)

Pedersen, D.B., Roland, J., Raabe, G. and Maysenhölder, W. (2000). Measurement of the low-frequency sound insulation of building components, *Acustica*, **86**, 495–505.

Petersson, B.A.T. (1995). The liquid drop impact as a source of sound and vibration, *Building Acoustics*, **2** (4), 585–623.

Pettersen, O.K.Ø., Olsen, H. and Vigran, T.E. (1997). On spatial sampling using the scanning intensity technique, *Applied Acoustics*, **50** (2), 141–153.

Pritz, T. (1986). Frequency dependence of frame dynamic characteristics of mineral and glass wool materials, *Journal of Sound and Vibration*, **106** (1), 161–169.

Pritz, T. (1987). Non-linearity of dynamic modulus of mineral wool slabs, *Proceedings of Internoise 87*, Beijing, China, 1367–1370.

Pritz, T. (1994). Dynamic Young's modulus and loss factor of plastic foams for impact sound insulation, *Journal of Sound and Vibration*, **178** (3), 315–322.

Randall, R.B. (1987). *Frequency analysis*, Brüel & Kjær. ISBN: 8787355078.

Rasmussen, B. and Rønnedal, P. (1992). Use of structural intensity for source localization in building acoustics, *Proceedings of the 14th ICA Congress*, Beijing, China.

Rasmussen, B., Rindel, J.H. and Henriksen, H. (1991). Design and measurement of short reverberation times at low frequencies in talks studios, *Journal of the Audio Engineering Society*, **39** (1/2), 47–57.

Rasmussen, P. and Rasmussen, G. (1983). Intensity measurements in structures, *Proceedings of the 11th ICA Congress*, Paris, France, 231–234.

Raw, G.J. and Oseland, N.A. (1991). Subjective response to noise through party floors in conversion flats, *Applied Acoustics*, **32**, 215–231.

Ren, M. and Jacobsen, F. (1992). A simple technique for improving the performance of intensity probes, *Noise Control Engineering Journal*, **38** (1), 17–25.

Rife, D.D. and Vanderkooy, J. (1989). Transfer-function measurement with maximum-length sequences, *Journal of the Audio Engineering Society*, **37** (6), 419–444.

Rindel, J.H. (1994). Dispersion and absorption of structure-borne sound in acoustically thick plates, *Applied Acoustics*, **41**, 97–111.

Rindel, J.H. and Rasmussen, B. (1996). Some consequences of including low frequencies in the evaluation of floor impact sound, *Proceedings of the Third Joint Meeting of the Acoustical Society of America and the Acoustical Society of Japan*, Honolulu, USA, December 1996.

Roelens, I., Nuytten, F., Bosmans, I. and Vermeir, G. (1997). In-situ measurement of the stiffness properties of building components, *Applied Acoustics*, **52** (3/4), 289–309.

Roland, J. (1995). Adaptation of existing test facilities to low frequency measurements, *Proceedings of Internoise 95*, Newport Beach, USA, 1113–1116.

Roland, J., Martin, C. and Villot, M. (1985). Room to room transmission: what is really measured by intensity? *Proceedings of 2nd International Congress on Acoustic Intensity*, Senlis, France, 539–545.

Schmitz, A., Meier, A. and Raabe, G. (1999). Inter-laboratory test of sound insulation measurements on heavy walls. Part 1 – Preliminary test, *Building Acoustics*, **6** (3/4), 159–169.

Scholl, W. (2001). Impact sound insulation: The standard tapping machine shall learn to walk!, *Building Acoustics*, **8** (4), 245–256.

Schroeder, M.R. (1965). New method of measuring reverberation time, *Journal of the Acoustical Society of America*, **37**, 409–412.

Schroeder, M.R. (1979). Integrated-impulse method measuring sound decay without using impulses, *Journal of the Acoustical Society of America*, **66** (2), 497–500.

Schultz, T.J. (1975). A proposed new method for impact noise tests, *Proceedings of Internoise 75*, Sendai, Japan, 343–350.

Schultz, T.J. (1981). Impact noise testing and rating – 1980, *Report Number NBS-GCR-80-249 prepared for the National Bureau of Standards*, Department of Commerce, Washington, DC, U.S.A by Bolt Beranek and Newman Inc.

Sorainen, E. and Rytkönen, E. (1989). Determination of propagation paths of vibration in the floor of a control room using vibration intensity measurements, *Applied Acoustics*, **26**, 1–7.

Suga, H. and Tachibana, H. (1994). Sound radiation characteristics of lightweight roof constructions excited by rain, *Building Acoustics*, **1** (4), 249–270.

Svensson, U.P. and Nielsen, J.L. (1999). Errors in MLS measurements caused by time-variance in acoustic systems, *Journal of the Audio Engineering Society*, **47**, 907–927.

Tachibana, H., Tanaka, H., Yasuoka, M. and Kimura, S. (1998). Development of new heavy and soft impact source for the assessment of floor impact sound insulation of buildings, *Proceedings of Internoise 98*, Christchurch, New Zealand.

Timoshenko, S.P. and Goodier, J.N. (1970). *Theory of elasticity*, McGraw-Hill, New York. ISBN: 0070858055.

Utley, W.A. and Pope, C.N. (1973). The measurement of damping in large panels, *Applied Acoustics*, **6**, 143–149.

van den Eijk, J. (1972). Sound insulation between dwellings: correction to 10.log S/A or to 10.log T/0.5? *Applied Acoustics*, **5**, 305–307.

van Zyl, B.G., Erasmus, P.J. and Anderson, F. (1987). On the formulation of the intensity method for determining sound reduction indices, *Applied Acoustics*, **22**, 213–228.

Venzke, G., Behr, R. and Deicke, H. (1972). Erweiterte Möglichkeiten zur Messung des Strömungswiderstandes von porösen Schicten, *Acustica*, **26**, 141–146.

Vér, I.L. (1971). Impact noise isolation of composite floors, *Journal of the Acoustical Society of America*, **50** (1/4), 1043–1050.

Vermeir, G., Bosmans, I. and Mees, P. (1996). Laboratory analysis of the airborne sound transmission through a prefabricated structural glazing, *Proceedings of Internoise 96*, Liverpool, England, 1795–1800.

Vian, J.P. and Drouin, C. (1977). Etude d'une nouvelle methode de mesure des bruits d'impact, *CSTB Report No.2.77.003 April 1977*, Centre Scientifique et Technique du Batiment, France.

Vigran, T.E. and Sørdal, S. (1976). Comparison of methods for measurement of reverberation time, *Journal of Sound and Vibration*, **48**, 1–13.

Villot, M. and Roland, J. (1981). Measurement of sound powers radiated by individual room surfaces using the acoustic intensity method, *Proceedings of International Congress on Recent Developments in Acoustic Intensity Measurement*, Senlis, France, 153–159.

Villot, M., Chavériat, G. and Roland, J. (1992). Phonoscopy: an acoustical holography technique for plane structures radiating in enclosed spaces, *Journal of the Acoustical Society of America*, **91** (1), 187–195.

Vorländer, M. (1995a). Survey test methods for acoustic measurements in buildings, *Building Acoustics*, **2** (1), 377–389.

Vorländer, M. (1995b). Revised relation between the sound power and the average sound pressure level in rooms and consequences for acoustic measurements, *Acustica*, **81**, 332–343.

Vorländer, M. and Bietz, H. (1994). Comparison of methods for measuring reverberation time, *Acustica*, **80**, 205–215.

Vorländer, M. and Kob, M. (1997). Practical aspects of MLS measurements in building acoustics, *Applied Acoustics*, **52** (3/4), 239–258.

Warnock, A.C.C. (1983). Floor impact noise and foot simulators, *Proceedings of Internoise 83*, Edinburgh, Scotland, 1127–1130.

Warnock, A.C.C. (1998). Floor research at NRC Canada, *Proceedings of COST Workshop on Building Acoustics "Acoustic performance of medium-rise timber buildings"* December 1998, Dublin, Ireland.

Warnock, A.C.C. (2000). Investigation of use of the tire impact machine as standard device for rating impact sound transmission of floors, *NRC Report NRCC-44301*, National Research Council Canada.

Watters, B.G. (1965). Impact-noise characteristics of female hard-heeled foot traffic, *Journal of the Acoustical Society of America*, **37** (4), 619–630.

Weise, W. (2003). Measurement uncertainties for sound field levels in rooms, *Building Acoustics*, **10** (4), 281–287.

Weise, W. and Schmitz, A. (2000). Rotating microphones and MLS in building acoustics, *Acustica*, **86**, 62–69.

White, R.G. (1982). Vibration testing (Chapter 27). In White, R.G. and Walker, J.G. (eds.), *Noise and vibration*, Ellis Horwood, 713–753. ISBN: 0853125023.

Wilson, R. (1992). Sound transmission through double walls, PhD thesis, Department of Building Engineering and Surveying, Heriot-Watt University, Edinburgh, Scotland.

Yoshimura, J. (2006). Personal communication, Kobayasi Institute of Physical Research, Japan.

Yoshimura, J. and Kanazawa, J. (1984). Influence of damping characteristics on the transmission loss of laminated glass, *Proceedings of Internoise 84*, Honolulu, USA, 589–592.

Direct sound transmission

4.1 Introduction

Direct sound transmission across a single building element occurs where the element is excited by an airborne or a structure-borne sound source on one side, and radiates sound from the other side without any flanking transmission.

This chapter looks at predicting and interpreting features that describe the airborne and impact sound insulation of various building elements. Splitting up a building into its component parts (e.g. floating floor, concrete base floor, ceiling finish, flanking walls, wall linings) is a convenient way to approach calculation of the sound insulation; it also suits the way that designers and manufacturers approach the design of a building. In many cases this approach is well-suited to prediction using Statistical Energy Analysis (SEA); hence this is discussed at the beginning of the chapter. For some types of building elements this approach allows insight into sound transmission mechanisms from which decisions can be made on ways to improve the sound insulation. However, it is not suited to all types. The inability of prediction models to deal with every single type of building element indicates why laboratory measurements are so important in providing information at the design stage. At the same time we find that prediction models illustrate the inherent limitations of many laboratory measurements; in some cases the most useful information lies somewhere between the two.

4.2 Statistical energy analysis

SEA is a framework of analysis for predicting the transmission of sound and vibration in built-up structures by using a statistical approach with energy as the primary variable. SEA was introduced in the 1960s and is a well-established engineering tool used in construction, ship, automobile, and aerospace industries (Lyon and DeJong, 1995). It is introduced in this chapter because if the concepts are grasped in the context of direct sound transmission, it is easier to apply them to flanking transmission in the next chapter. It is important to note that SEA provides a framework for the analysis of complex systems; the classical theories of sound and vibration transmission can be, and usually are, incorporated within this framework.

The origins of SEA reside in a linear system comprised of two 'weakly' coupled oscillators excited by independent broadband random noise (Lyon and Maidanik, 1962; Scharton and Lyon, 1968). Considering the temporal average energy for each oscillator, E_1 and E_2, the net energy flow between oscillators is proportional to the difference in the uncoupled energies of the oscillators. Net power transfer takes place from the oscillator with higher energy to the oscillator with lower energy, and can be expressed using K_1 and K_2 as (temporarily undefined) coupling terms:

$$W_{\text{net},12} = K_1(E_1 - K_2 E_2) \qquad (4.1)$$

409

This approach is then extended from two oscillators, to two sets of oscillators that are coupled together. It is now appropriate to refer to the oscillators as modes and to introduce the term 'subsystem' to represent one set of oscillators. A subsystem comprises a similar group of modes each with similar modal energy for the form of excitation, where the subsystem response is determined in frequency bands. Between the modes of a subsystem it is assumed that there is no correlation and no significant transfer of energy. For this to hold true, the subsystems must be 'weakly' coupled with power input from statistically independent excitation forces. With this more general approach for subsystems 1 and 2, we have $K_1 = \omega \eta_{12}$ (which introduces a coupling loss factor, η_{12}) and $K_2 = n_1/n_2$ (which introduces a ratio of modal densities). This gives the following relationship between two coupled subsystems; commonly referred to as the consistency relationship,

$$\frac{\eta_{12}}{n_2} = \frac{\eta_{21}}{n_1} \tag{4.2}$$

Power flowing from 1 to 2 can now be written as

$$W_{12} = \omega \eta_{12} E_1 \tag{4.3}$$

where ω is the angular frequency.

From Eqs 4.1–4.3 the net power flow can now be described using only coupling loss factors as

$$W_{\text{net},12} = W_{12} - W_{21} = \omega \eta_{12} E_1 - \omega \eta_{21} E_2 \tag{4.4}$$

For a frequency band with a bandwidth, $\Delta \omega$, and a centre frequency, ω, the modal energy of a subsystem is defined as the total subsystem energy divided by the mode count, N, in that band. Equation 4.4 can therefore be written in terms of the modal energies as

$$W_{\text{net},12} = \omega \eta_{12} n_1(\omega) \Delta \omega \left(\frac{E_1}{n_1(\omega) \Delta \omega} - \frac{E_2}{n_2(\omega) \Delta \omega} \right) = \omega \eta_{12} N_1 \left(\frac{E_1}{N_1} - \frac{E_2}{N_2} \right) \tag{4.5}$$

This shows that the net power transferred between two coupled subsystems is proportional to the difference in their modal energies. The thermal analogy is a useful way of describing this process (Lyon and DeJong, 1995). For each subsystem, the modal energy can be considered as acoustic temperature, such that there will be heat (energy) flow from subsystems of high temperature (high modal energy) to those of low temperature (low modal energy). This provides the basis upon which systems with more than two subsystems can be studied when the following assumptions are satisfied (Hodges and Woodhouse, 1986; Lyon and DeJong, 1995):

1. Statistically independent excitation forces.
2. Equal probability of modes occurring in a certain frequency range.
3. Equipartition of modal energy in a subsystem, and incoherent modal response between modes in the coupled subsystems.
4. 'Weak' (or 'light') coupling between subsystems.

These four points are discussed below.

The requirement for statistically independent excitation forces is due to the SEA assumption that when individual subsystems are coupled together, the modal vibrations must be uncorrelated so that a linear relationship can be used between net power transfer and modal energies using local modes. If the excitation causes the modal response to be coherent then the requirement for equipartition of modal energy will not be satisfied. Statistically independent excitation forces can be realized using rain-on-the-roof excitation (also called delta-correlated excitation); this is defined as unity magnitude, random phase, multi-point excitation over the entire subsystem. In measurements it is common to use point excitation for reasons of practicality. When point excitation is applied at the anti-node of two or more modes that are sufficiently close in frequency, correlation exists between the modal responses, thus violating one of the assumptions in SEA (Fahy, 1970). However, physical experiments on structures indicate that using point excitation and averaging the response from a number of randomly chosen excitation points can be used to approximate statistical independence between the modes (Bies and Hamid, 1980). Hence sound and vibration measurements using point sources are almost always averaged from a number of different source positions.

For real spaces and structures, mode frequencies and mode shapes are rarely the precise entities that analytic solutions and deterministic models (such as finite element methods) would lead us to believe. There will be uncertainty in describing the modes of real spaces or structures due to uncertainty in the dimensions, material properties, and particularly for building structures, the quality of workmanship. Hence it is reasonable to assume that there is equal probability of the mode frequencies falling within a certain frequency range. Embracing the issue of uncertainty when describing modes is a liberating step as it allows use of the statistical modal density. We no longer have to be concerned by the fact that for a given room, wall, or floor, there are large numbers of modes to deal with, for which we are unable to accurately quantify the mode frequency or mode shape. It is often assumed that uncertainty only occurs at 'high frequencies'. However, from the viewpoint of a structure-borne sound wave, buildings tend to be relatively imprecise, highly variable structures and we can often assume that there is uncertainty in the modal description over most of the building acoustics frequency range.

Equipartition of modal energy in a subsystem means that every mode has equal energy, which, as we rarely know otherwise, is often assumed; partly because we have already assumed broadband statistically independent excitation forces. This form of broadband excitation also tends to satisfy the requirement for incoherent modal response between modes in the coupled subsystems.

For predictive SEA, 'weak' coupling can (to some extent) be considered as occurring when the local mode behaviour of an uncoupled subsystem is hardly changed when it is coupled to the other subsystems such that energy flow can be related to the local modal energies. However, 'weak' coupling between subsystems is an awkward criterion about which there has been much debate. A review of the literature by James and Fahy (1994) suggests that there is confusion between the validation of the fundamental SEA equation (Eq. 4.1), the use of SEA with wave theory calculation of the coupling loss factor, and the requirements necessary to use experimental SEA. This has led to different definitions of 'weak' coupling depending upon the model under consideration. James and Fahy (1994) concluded that 'weak' coupling definitions that were created to assess the validity of Eq. 4.1 are of little or no use.

4.2.1 Subsystem definition

We can now start to define SEA subsystems that form parts of a building. These are either space subsystems (e.g. rooms, cavities) or structural subsystems (e.g. plates, beams).

Subsystems are defined by their ability to store modal energy. Therefore, the boundaries of a subsystem must cause reflections so that the sound or vibration field is reverberant for the specific wave type considered in the subsystem. Reflections occur when there is an impedance change at a boundary. Hence for space subsystems where there is only one wave type, the surfaces that define a room or cavity usually define the subsystem. For structural subsystems it can be slightly more complex. Although bending waves are of primary importance for sound radiation, in-plane waves can be important for structure-borne sound transmission. As these waves have different modal energies, they need to be represented as separate subsystems. For example, a plate can be represented by three subsystems using a separate subsystem for bending, transverse shear, and quasi-longitudinal waves. Conversion between these wave types at a junction can therefore be included in an SEA model using coupling loss factors from one subsystem to another. The subsystem boundaries may vary depending upon the wave type under consideration. This is relevant to structural subsystems where junction impedances are very different for bending waves compared to in-plane waves (Fahy, 1974). Fortunately for rigidly connected masonry/concrete plates, the visible boundaries of a wall or floor that face into a room usually give a reasonable demarcation of a subsystem for either bending, transverse shear, or quasi-longitudinal waves (Craik, 1998). A plate or beam can be represented by one subsystem for each wave type, although there are many situations where it is only necessary to consider bending waves and a single subsystem will be sufficient. This is generally the case for direct transmission described in this chapter. In the next chapter we will consider separate subsystems for each of the three wave types on a plate.

In Section 2.7.7 examples were given of non-reverberant plates that had a significant decrease in vibration with distance. This does not automatically mean that SEA cannot be used; it may still be suitable when the excitation is distributed over the plate surface (i.e. airborne excitation) although it does indicate that other models should be considered. If this decrease is only due to high internal damping for bending wave motion, a rule of thumb for the maximum dimension of a plate subsystem, L_{max}, is (Lyon and DeJong, 1995)

$$L_{max} < \frac{c_{g,B}}{\omega \eta_{int}} \qquad (4.6)$$

When there is a significant decrease in vibration with distance, use of SEA will depend on the type of excitation and the most important transmission mechanisms. Structural excitation at a single point or along a line on a plate may indicate a significant decrease in vibration across plates that are highly damped, non-homogeneous or spatially periodic. For airborne or multi-point structural excitation (i.e. rain-on-the-roof) over the plate surface it may still be possible to include the plate as a single subsystem in an SEA model. Alternatively it may be necessary to model a single plate as more than one subsystem to account for these losses over distance.

4.2.2 Subsystem response

SEA gives the temporal and spatial average response of a subsystem in terms of its energy. By assuming a statistical description for each subsystem, the subsystem response represents the ensemble average of 'similar' subsystems with physical parameters drawn from statistical distributions. Hence it is not necessary to know the exact geometry of a room when representing it as a space subsystem, just the volume. This approach makes SEA an attractive form of analysis from an engineering viewpoint. It also simplifies interpretation of the results because the predicted subsystem energies usually have a smooth variation with frequency. However, this has implications for the comparison of SEA predictions against measured data. To validate SEA models we ideally need to compare the prediction against the average value calculated from measurements on a number of 'similar' constructions because SEA does not predict the response of an individual system with specific modal features.

Ideally we would like to calculate the variance as well as the average response. Unfortunately, calculation of the variance for SEA systems is only possible to a fairly limited extent (Lyon and DeJong, 1995) although research continues to address this issue (e.g. see Langley and Cotoni, 2004). For airborne sound insulation the variance is mainly determined by the variation in the coupling loss factors and the spatial variation of the sound pressure level. For impact sound insulation, the power input also varies due to variation in the driving-point impedance over the plate surface. Whilst we cannot predict the variance, there are some ideal conditions that give reasonable agreement between an SEA prediction and the measured result from a single construction. To minimize spatial variance of the response and to ensure that resonant transmission occurs under damping control of the modes we require 'high' mode counts and 'high' modal overlap in the frequency band of interest. Quantifying suitably 'high' values depends on the system under analysis.

For plates and/or beams that are coupled together at a junction, a wave approach can be used to calculate the transmission coefficients; these are needed for calculation of the coupling loss factors (Section 5.2). Assuming that the junction has been correctly modelled, it is useful to know the requirements on the subsystems such that these apply to the ensemble with a low variance. Computational and physical experiments indicate that the wave transmission coefficients only give accurate estimates when the geometric mean of the modal overlap factors, M_{av}, is at least equal to unity (Fahy and Mohammed, 1992; also see Clarkson and Ranky, 1984; Davies and Wahab, 1981). The geometric mean of the modal overlap factors for subsystems i and j is given by

$$M_{av} = \sqrt{M_i M_j} \qquad (4.7)$$

For coupled plates, an additional condition for each plate subsystem is that there should be at least five modes in the frequency band, $N_s \geq 5$ (Fahy and Mohammed, 1992). In the low- and mid-frequency ranges, many plate and beam subsystems in buildings have zero, fractional, or low mode counts in one-third-octave-bands. Uncertainty in predicting the mode frequencies gives rise to uncertainty as to which bands are under damped modal control, and which (if any) are not. For this reason it is more meaningful to use the statistical mode count. We will adopt the condition $N_s \geq 5$ as a quantitative definition of the term 'multi-modal'. There are many plates in buildings (particularly masonry/concrete walls and floors) that are not multi-modal for bending modes in the low- and mid-frequency ranges. For structural coupling between plates that is calculated using a wave approach, SEA can still be used if the empirical conditions, $M_{av} \geq 1$ and $N_s \geq 5$ are not met for each plate subsystem, but the levels of uncertainty may be large. These conditions tend to be overly restrictive in the application of SEA to buildings where masonry/concrete plates are not multi-modal, but have high total loss factors (Craik et al., 1991). Section 5.2.3 contains examples and further discussion relating to these conditions for vibration transmission between plates.

In contrast to many plates in buildings, rooms often form multi-modal subsystems in the low- and mid-frequency range. Therefore coupling between a room and a plate may involve resonant transmission between a multi-modal space subsystem and a plate subsystem with a fractional mode count. In addition the space subsystem may have high modal overlap compared to the plate subsystem with a value well-below unity. In such situations, reasonable estimates can still be achieved when $N_s \geq 1$ for the plate subsystem and $M_{av} \geq 1$; examples are given in Section 4.3.1.

4.2.3 General matrix solution

SEA requires knowledge of the dissipative subsystem losses, the coupling losses between subsystems and the actual or nominal power input into the subsystem(s). The losses are described using loss factors; these give the fraction of energy transferred per radian cycle. Three loss factors are defined: internal (dissipative) subsystem losses (η_{ii}), coupling losses between subsystems (η_{ij}), and total subsystem losses (η_i). These have already been introduced as the internal loss factor, the coupling loss factor, and the total loss factor in Chapters 1 and 2. The internal loss factor accounts for energy that is converted to heat; it can also be used to account for energy that is transferred to parts of the system that are not included in the SEA model. The coupling loss factor accounts for energy transferred to another subsystem. The total loss factor of a subsystem is the sum of its internal loss factor and all the coupling loss factors from that subsystem (Eq. 1.106).

A two-subsystem model with a single power input illustrates the principles of energy flow between subsystems (see Fig. 4.1). This is the simplest situation of relevance to sound insulation. It could represent power input from a loudspeaker in a source room, with the source and receiving rooms as subsystems 1 and 2 respectively; the sound could be transmitted between the rooms via an aperture, ventilator, or porous material. Alternatively it could represent power input from the ISO tapping machine into a plate (subsystem 1) representing a floor that radiates sound into the receiving room (subsystem 2). From conservation of energy, the power balance equations for subsystems 1 and 2 are

$$W_{in(1)} + \omega\eta_{21}E_2 = \omega\eta_{11}E_1 + \omega\eta_{12}E_1 \tag{4.8}$$

$$\omega\eta_{12}E_1 = \omega\eta_{22}E_2 + \omega\eta_{21}E_2 \tag{4.9}$$

When the power input and the loss factors are known, Eqs 4.8 and 4.9 can be solved to find the subsystem energies. However, as we almost always deal with more than two subsystems,

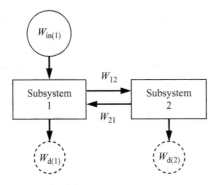

Figure 4.1

Two-subsystem SEA model. A power input, $W_{in(1)}$, is applied to subsystem 1. Transmitted power between subsystems is denoted by W_{12} and W_{21}. Dissipated power (through internal losses and coupling losses to other subsystems) is denoted by $W_{d(1)}$ and $W_{d(2)}$.

the power balance equations are generalized into a matrix solution for N subsystems to give the general SEA matrix solution,

$$
\begin{bmatrix}
\sum_{n=1}^{N} \eta_{1n} & -\eta_{21} & -\eta_{31} & \cdots & -\eta_{N1} \\
-\eta_{12} & \sum_{n=1}^{N} \eta_{2n} & -\eta_{32} & & \\
-\eta_{13} & -\eta_{23} & \sum_{n=1}^{N} \eta_{3n} & & \\
\vdots & & & \ddots & \\
-\eta_{1N} & & & & \sum_{n=1}^{N} \eta_{Nn}
\end{bmatrix}
\begin{bmatrix}
E_1 \\ E_2 \\ E_3 \\ \vdots \\ E_N
\end{bmatrix}
=
\begin{bmatrix}
\dfrac{W_{in(1)}}{\omega} \\[2mm]
\dfrac{W_{in(2)}}{\omega} \\[2mm]
\dfrac{W_{in(3)}}{\omega} \\[2mm]
\vdots \\[2mm]
\dfrac{W_{in(N)}}{\omega}
\end{bmatrix}
\tag{4.10}
$$

where η_{ij} is the coupling loss factor from subsystem i to j, and η_{ii} is the internal loss factor for subsystem i.

For general discussions it is convenient to simplify Eq. 4.10 into the form

$$
[\eta]\{E\} = \left\{ \frac{W_{in}}{\omega} \right\}
\tag{4.11}
$$

where $[\eta]$ is the square matrix of loss factors, $\{E\}$ is the column matrix for energy, and $\left\{ \frac{W_{in}}{\omega} \right\}$ is the column matrix for power input terms.

The energy matrix contains the unknown subsystem energies in which we are interested. We are able to fill the loss factor matrix and the power input matrix with predicted (or measured) values. It is unlikely that every subsystem will be connected to every other subsystem, so some of the coupling loss factors in the loss factor matrix will be zero. The power input matrix will only usually have one row that is non-zero because with sound insulation measurements we only usually have one power input into a subsystem; e.g. a loudspeaker in the source room, or the ISO tapping machine on a floor.

The subsystem energies are determined using

$$
\{E\} = [\eta^{-1}] \left\{ \frac{W_{in}}{\omega} \right\}
\tag{4.12}
$$

where $[\eta^{-1}]$ is the inverse of the loss factor matrix.

In some situations there will be more than one coupling loss factor between two subsystems; the individual coupling loss factors can then be added together to give a single value for use in the matrix solution. An example of this could occur with two rooms separated by a solid plate containing an aperture and a small area of porous material. In this case there will be three non-resonant coupling loss factors between the rooms; one for non-resonant (mass law) transmission across the plate, one for transmission through the aperture, and one for transmission through the porous material.

4.2.4 Converting energy to sound pressures and velocities

Solving the general SEA matrix gives subsystem energies, which can then be converted into more practical variables such as sound pressure or velocity.

For space subsystems, energy is converted to a mean-square pressure using Eq. 1.154. In Section 1.2.8.1 we discussed use of the Waterhouse correction to account for higher energy density near the boundaries of reverberant rooms. Unlike real spaces, there is no spatial variation of energy within an SEA space subsystem; it simply represents the spatial average reverberant energy. To compare measured and predicted sound pressure levels in a room, the Waterhouse correction can either be accounted for in the measured level or in the SEA predicted level. In practice, the correction is rarely used because it is only significant in the low-frequency range where other uncertainties in the SEA model are usually much larger. Typically, we are interested in the sound pressure level difference, D, between rooms which can be calculated from the energy ratios between space subsystems, i and j, using

$$D = 10 \lg \left(\frac{p_i^2}{p_j^2} \right) = 10 \lg \left(\frac{E_i}{E_j} \right) + 10 \lg \left(\frac{V_j}{V_i} \right) \tag{4.13}$$

Equation 4.13 allows subsequent calculation of all the descriptors commonly used to describe airborne sound insulation such as R, D_n, D_{nT}.

For structural subsystems, energy is converted to a mean-square velocity using Eq. 2.237. Energy ratios between structural subsystems, i and j, can be converted to a velocity level difference, $D_{v,ij}$, using

$$D_{v,ij} = 10 \lg \left(\frac{v_i^2}{v_j^2} \right) = 10 \lg \left(\frac{E_i}{E_j} \right) + 10 \lg \left(\frac{m_j}{m_i} \right) \tag{4.14}$$

Note that $D_{v,ij}$ is usually defined with i as the source subsystem, hence when i is not the source subsystem this should be stated for clarity.

When validating an SEA model with measurements on space and structural subsystems it is sometimes necessary to look at the ratio of sound pressure to vibration; in this situation it is simplest to convert all the measured variables to energy, and to work purely in terms of the energy level difference.

4.2.5   Path analysis

Whilst sound pressure or velocity are usually the variables of interest for each subsystem, they do not give an immediate insight into how sound and vibration is transmitted from one subsystem to another along the various transmission paths. With SEA it is possible to use path analysis to assess the relative importance of one transmission path compared to another (Craik, 1996). The ability to carry out path analysis alongside the matrix solution makes SEA a powerful tool with which to make design decisions.

We start by considering a simple, ordered SEA system as shown in Fig. 4.2a, focusing on only the first three subsystems. The power balance equations can be written in such a way that we exclude the power input term; this allows us to use the energy in the source subsystem as the reference point in the path analysis,

$$\omega \eta_{12} E_1 = \omega \eta_2 E_2 \tag{4.15}$$

$$\omega \eta_{23} E_2 = \omega \eta_3 E_3 \tag{4.16}$$

(a)

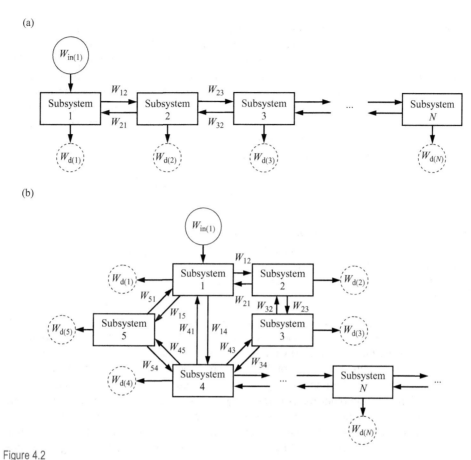

(b)

Figure 4.2

Example SEA systems: (a) Simple, ordered system and (b) more realistic/complex system.

Equations 4.15 and 4.16 can now be combined to find the energy ratio, E_1/E_3, due to energy flowing along the transmission path, $1 \rightarrow 2 \rightarrow 3$,

$$\frac{E_1}{E_3} = \frac{\eta_2 \eta_3}{\eta_{12} \eta_{23}} \tag{4.17}$$

Hence for any system with power injected into subsystem 1, the energy ratio between subsystem 1 and subsystem N for transmission along the chain of subsystems, $1 \rightarrow 2 \rightarrow 3 \rightarrow \cdots \rightarrow N$, is

$$\frac{E_1}{E_N} = \frac{\eta_2 \eta_3 \ldots \eta_N}{\eta_{12} \eta_{23} \ldots \eta_{(N-1)N}} \tag{4.18}$$

Note that Eqs 4.13 and 4.14 can be used to convert the energy ratio for each path to sound pressure or velocity ratios.

The path with the lowest energy ratio is the strongest path, and the path with the highest energy ratio is the weakest path. It is simplest to convert the energy ratio into an energy level difference in decibels by taking 10 times the logarithm (base 10).

Equation 4.18 may give the impression that path analysis only applies to a very simple ordered SEA system (Fig. 4.2a), but it equally applies to real systems with more complex connections between subsystems. An example is given in Fig. 4.2b where there are many transmission paths of potential interest. For example, to assess the strength of different transmission paths between subsystems 1 and 4 we could start by comparing E_1/E_4 for the following three paths: (1) $1 \to 4$, (2) $1 \to 2 \to 3 \to 4$, and (3) $1 \to 5 \to 4$. It is also useful to combine different paths to give E_1/E_4 for a specific combination of paths. As we usually look at energy level differences in decibels, the energy level difference due to transmission between subsystem 1 and subsystem N along P different paths can be calculated from

$$10 \lg \left(\frac{E_1}{E_N} \right)_{\substack{\text{Due to} \\ P \text{ paths}}} = -10 \lg \left(\sum_{p=1}^{P} \left(\frac{E_1}{E_N} \right)_p^{-1} \right)$$

(4.19)

There are other permutations that we have not yet considered; although paths cannot re-enter the subsystem that contains the source, they can revisit other subsystems. Using the system in Fig. 4.2b, two examples for E_1/E_4 would be $1 \to 5 \to 4 \to 5 \to 4$ and $1 \to 2 \to 3 \to 2 \to 3 \to 4$. In practice, these paths are often insignificant compared to the paths that visit each subsystem only once.

In a complete building there are many transmission paths that determine the overall sound insulation and so the matrix solution is used to determine the distribution of energy between the subsystems. However, when we want to find ways of increasing the sound insulation it is useful to know whether there is a dominant transmission path. A dominant path is defined here as giving nominally the same energy level difference as the matrix solution; therefore the combination of all the other paths is relatively unimportant. We can then test out various changes to the dominant path that might lead to an increase in the sound insulation. Changes to one path will change the relative importance of the other paths; therefore the overall sound insulation then needs to be checked by re-calculating the sound insulation using the matrix solution.

4.3 Airborne sound insulation

This section starts by looking at the airborne sound insulation of a solid homogeneous isotropic plate as a basis from which to look at the wide range of building elements that are encountered in practice.

4.3.1 Solid homogeneous isotropic plates

A solid homogeneous isotropic plate represents the simplest type of wall or floor that is found in a building. Many wall and floor constructions are far more complex than this idealized form; yet an understanding of sound transmission via this simple plate is of fundamental importance as it is often used as a benchmark for comparison with more complex constructions. The main features of sound transmission can also be explained by considering a plate of finite thickness but infinite size. In practice, many problems in sound insulation design, prediction, and measurement revolve around the finite size of plates and their connections to other plates. By starting with an SEA model for a solid, homogeneous, finite size plate, it is easier to grasp the various concepts, and to extend the calculations to more complicated constructions that

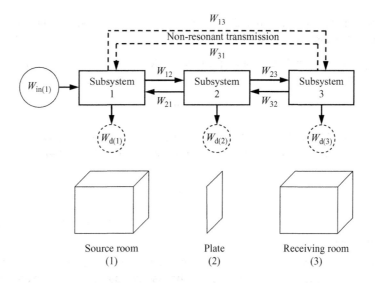

Figure 4.3

Three-subsystem SEA model for airborne sound transmission between a source room (subsystem 1) and a receiving room (subsystem 3) via a solid homogeneous plate (subsystem 2). Arrows with dashed lines represent non-resonant transmission, and arrows with solid lines represent resonant transmission.

are found in practice. However, we will make use of the infinite plate model to gain insights into angle-dependent sound transmission and to note the important links between finite and infinite plate models for airborne sound transmission.

For airborne sound transmission from a source room to a receiving room across a solid homogeneous plate, a three-subsystem SEA model is required as shown in Fig. 4.3. The plate subsystem only needs to represent bending waves because radiation from in-plane modes is insignificant. Note that by limiting the model to three subsystems we are ignoring flanking transmission and assuming that vibration can be transmitted from the plate to any connected structure, but not vice versa.

SEA is based on energy flow between groups of modes at resonance; this is referred to as resonant transmission. For airborne sound transmission across a plate it is also necessary to include non-resonant transmission between the space subsystems. In an SEA model this form of transmission bypasses the plate subsystem (see Fig. 4.3) even though the transmitted power is determined by the plate properties (Crocker and Price, 1969). We will now look at resonant and non-resonant transmission separately before combining them to calculate the overall sound insulation.

In deriving the SEA equations we will often refer to coupling between subsystems i and j rather than using the specific subsystem numbers (1, 2, and 3). As there are only three subsystems there is little scope for confusion and the generalization will make it easier for the reader to apply the equations to other SEA systems.

4.3.1.1 Resonant transmission

Resonant transmission concerns coupling between modes with resonance frequencies that fall within the frequency band of interest. In the three-subsystem model considered here, it occurs between room modes and plate bending modes. The SEA framework conveniently allows this to be described in a few short steps.

The sound power, W_{ij}, radiated from a plate subsystem, i, that is undergoing bending wave motion, into a room subsystem, j, is

$$W_{ij} = \omega \eta_{ij} E_i \qquad (4.20)$$

Substituting Eqs 2.237 and 2.198 into Eq. 4.20 gives the coupling loss factor from a plate to a room in terms of the frequency-average radiation efficiency (Section 2.9.4) as

$$\eta_{ij} = \frac{\rho_0 c_0 \sigma}{\omega \rho_s} \qquad (4.21)$$

The consistency relationship (Eq. 4.2) can now be used to calculate the coupling loss factor from a room to a plate, η_{ji}.

We can use SEA path analysis to calculate a resonant sound reduction index, R_R, for the resonant transmission path $1 \rightarrow 2 \rightarrow 3$. This is only the resonant component of the sound reduction index that we determine using standard sound insulation measurements in a transmission suite. Assuming that ρ_0 and c_0 are the same in both rooms, Eqs 4.17 and 4.13 give the resonant sound reduction index as

$$R_R = 10 \lg \left(\frac{\eta_2 \eta_3}{\eta_{12} \eta_{23}} \frac{V_3}{V_1} \right) + 10 \lg \left(\frac{S}{A} \right) \qquad (4.22)$$

Using the consistency relationship with Eqs 1.59 and 2.139 for the room and plate modal densities respectively, Eq. 4.22 can be rewritten in terms of the plate properties as

$$R_R = 10 \lg \left(\frac{2\pi^2 h c_L \rho_s^2 f^3 \eta}{\sqrt{3} \rho_0^2 c_0^4 \sigma^2} \right) \qquad (4.23)$$

where η is the total loss factor of the plate. Note that R_R only increases at 9 dB/octave if η and σ are independent of frequency.

As an example we consider a plate of 6 mm glass. We will assume that the glass is installed in a filler wall in a transmission suite and that the edges of the glass pane are embedded in putty held within a wooden frame. Resonant transmission is under damping control of the plate so we need to determine the total loss factor of the plate from its individual loss factors. The internal loss factor for glass is usually between 0.003 and 0.006; this is a property of the glass and does not include any dissipative loss mechanisms related to the putty. The coupling losses are due to sound radiation and structural coupling. The coupling loss factor for sound radiation is calculated using Eq. 4.21; note that because the plate radiates into both the source room and the receiving room, the sum of the coupling losses due to radiation will be twice this value. We assume that the plate boundaries can be considered as being simply supported and that the filler wall represents an infinite rigid baffle; the radiation efficiency under these assumptions is calculated using method no.1 (refer back to 6 mm glass in Fig. 2.65a). At this point we hit a complication because the structural coupling losses for this particular junction are difficult to calculate accurately. This is partly because the properties of the putty change as it dries out, and partly because the junction detail is not simple to model even when the properties of the putty are fixed. Hence, we accept that whilst we can estimate some parts of

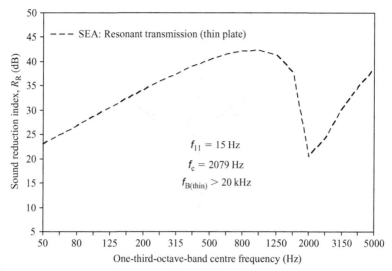

Figure 4.4

Resonant transmission across a plate of 6 mm glass. *Plate properties: $L_x = 1.5$ m, $L_y = 1.25$ m, $h = 0.006$ m, $\rho_s = 15$ kg/m^2,* $c_L = 5200$ m/s, $\nu = 0.24$, total loss factor $\eta = 0.024$.

the total loss factor, we cannot estimate all of them. In such cases the total loss factor can be calculated from measurements of the structural reverberation time. Measured total loss factors are usually frequency-dependent, but to simplify matters for this example it is reasonable to use a single frequency-average value.

The sound reduction index for resonant transmission is shown in Fig. 4.4. We can only consider resonant transmission when the fundamental mode is below the frequency range of interest. The thin plate limit is well-above 20 kHz, which conveniently means that thin plate theory can be used across the entire building acoustics frequency range. As the total loss factor depends upon the specific installation (and is usually frequency-dependent) it is only appropriate here to make a brief qualitative assessment of the general trends. Starting at the lowest frequency band, the sound insulation increases with frequency, then begins to level off as it approaches the critical frequency. A minimum value is reached very close to the critical frequency, after which the sound insulation begins to increase with frequency once more.

4.3.1.2 *Non-resonant transmission (mass law)*

Non-resonant transmission between two spaces separated by a plate is quantified using a non-resonant transmission coefficient, τ_{NR}. This gives a non-resonant sound reduction index, R_{NR}, for the non-resonant transmission path $1 \rightarrow 3$. We will shortly look at quantifying the transmission coefficient using infinite and finite plate models, but the first step is to determine the coupling loss factor.

In a three-dimensional space with a diffuse sound field, the sound power incident upon a plate of surface area, S, is calculated from Eq. 3.34, giving

$$W = SI = E \frac{c_0}{d_{mfp}} \frac{S}{S_T} = \frac{Ec_0 S}{4V} \tag{4.24}$$

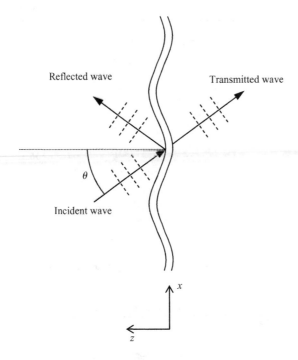

Figure 4.5

Plane wave incident upon an infinite plate along with the reflected and transmitted waves.

Therefore the sound power, W_{ij}, transmitted between two space subsystems i and j, across a plate subsystem with a non-resonant transmission coefficient, τ_{NR}, is

$$W_{ij} = \frac{E_i c_0 S}{4 V_i} \tau_{NR} = \omega \eta_{ij} E_i \tag{4.25}$$

which gives the coupling loss factor,

$$\eta_{ij} = \frac{c_0 S}{4 \omega V_i} \tau_{NR} \tag{4.26}$$

The consistency relationship (Eq. 4.2) can be used to calculate the coupling loss factor, η_{ji}, in the reverse direction.

4.3.1.2.1 Infinite plate theory

For an infinite plate, non-resonant transmission is based on the assumption that it simply acts as a limp mass with no stiffness. For this reason it is also referred to as mass law transmission, or forced transmission. It is assumed that the infinite plate lies in the xy plane and that a plane wave is incident upon the plate at an angle, θ, as shown in Fig. 4.5. For this infinite plate we only need to focus on the x- and z-dimensions, so the incident sound wave can be described by

$$p_i(x, z) = \hat{p}_i \exp(-ikx \sin \theta) \exp(ikz \cos \theta) \tag{4.27}$$

where the time dependence, $\exp(i\omega t)$, has been excluded for brevity.

Therefore the reflected wave is

$$p_r(x, z) = \hat{p}_r \exp(-ikx \sin \theta) \exp(-ikz \cos \theta) \tag{4.28}$$

and the transmitted wave is

$$p_t(x, z) = \hat{p}_t \exp(-ikx \sin\theta) \exp(ikz \cos\theta) \tag{4.29}$$

Continuity of sound pressure at $z = 0$ gives the total pressure acting on the plate as

$$\hat{p} = \hat{p}_i + \hat{p}_r - \hat{p}_t \tag{4.30}$$

The plate acts as a limp mass, so from Newton's third law the total pressure equals the inertial reaction from the plate,

$$\hat{p} = i\omega\rho_s\hat{v} \tag{4.31}$$

where ρ_s is the mass per unit area of the plate.

At $z = 0$, continuity requires that the lateral plate velocity equals the z-component of the particle velocity on each side of the plate,

$$\hat{v} = \hat{u}_i + \hat{u}_r = \hat{u}_t \tag{4.32}$$

Relating the particle velocities to the sound pressures (Eq. 1.16) therefore gives

$$\hat{v} = (\hat{p}_i - \hat{p}_r)\frac{\cos\theta}{\rho_0 c_0} = \hat{p}_t\frac{\cos\theta}{\rho_0 c_0} \tag{4.33}$$

From Eqs 4.30, 4.31, and 4.33, the angle-dependent transmission coefficient, $\tau_{NR,\theta}$, equals the ratio of the transmitted sound power to the incident sound power,

$$\tau_{NR,\theta} = \frac{W_t}{W_i} = \left|\frac{\hat{p}_t}{\hat{p}_i}\right|^2 = \frac{1}{1 + \left(\dfrac{\omega\rho_s \cos\theta}{2\rho_0 c_0}\right)^2} \tag{4.34}$$

Hence non-resonant transmission is lowest at normal incidence ($0°$) and increases with increasing angle of incidence. Equation 4.34 can be used to calculate a diffuse field transmission coefficient, τ_d, for an infinite plate. When one side of the plate is exposed to a diffuse sound field, all angles of incidence are equally probable and the plane waves that are incident upon the plate will have equal intensity, I. We therefore consider the incident sound as radiating from a hemisphere that encloses, and is centred around a unit area on the plate. The incident and transmitted intensities can then be determined from the component of the incident plane wave intensity that is normal to the plate surface, $I \cos\theta$, by integrating over the element of the solid angle, $d\Omega = \sin\theta \, d\theta \, d\phi$. Hence the total transmitted intensity, I_t, is determined from

$$I_t = \int_\Omega \tau_\theta I \cos\theta \, d\Omega = \int_0^{2\pi} \int_0^{\pi/2} \tau_\theta I \cos\theta \sin\theta \, d\theta \, d\phi \tag{4.35}$$

The total incident intensity can be found in the same way. By considering a unit area on the plate, the diffuse field transmission coefficient, $\tau_{NR,d}$, is defined as the ratio of the total transmitted intensity to the total incident intensity. The incident plane wave intensity is a constant, and τ_θ does not vary with ϕ, hence the integration simplifies to give,

$$\tau_{NR,d} = \frac{\displaystyle\int_0^{\pi/2} \tau_\theta \cos\theta \sin\theta \, d\theta}{\displaystyle\int_0^{\pi/2} \cos\theta \sin\theta \, d\theta} = \int_0^{\pi/2} \tau_\theta \sin 2\theta \, d\theta \tag{4.36}$$

The sound field in the low-frequency range is not usually diffuse for typical rooms and will not contain all angles of incidence. In addition, with transmission suite measurements it is common to mount a plate within a niche (Section 3.5.1.3.3); therefore the plate may be shielded from some angles of incidence near 90°. An empirical adjustment is often quoted which changes the upper integration limit in the numerator and denominator of Eq. 4.36 from 90° to 78° (Vér and Holmer, 1988). Other empirical values that are sometimes quoted are 75° and 80°. Here we will also use 78° to define field incidence, and the field incidence transmission coefficient, $\tau_{NR,f}$.

Whilst the transmission coefficient is convenient for SEA calculations, it is simpler to compare the transmission loss in decibels for different types of non-resonant transmission using

$$R_{NR} = 10\lg\left(\frac{1}{\tau_{NR}}\right) \tag{4.37}$$

Therefore the transmission loss for non-resonant transmission is

$$R_{NR,\theta^\circ} = 10\lg\left(1 + \left(\frac{\omega\rho_s\cos\theta}{2\rho_0 c_0}\right)^2\right) \tag{4.38}$$

for a single angle of incidence, which gives

$$R_{NR,0^\circ} = 10\lg\left(1 + \left(\frac{\omega\rho_s}{2\rho_0 c_0}\right)^2\right) \tag{4.39}$$

for normal incidence, and

$$R_{NR,d} = R_{NR,0^\circ} - 10\lg(0.23R_{NR,0^\circ}) \tag{4.40}$$

for diffuse incidence (Vér and Holmer, 1988) assuming $R_{NR,0^\circ} > 15$ dB, otherwise use numerical integration of Eq. 4.36, and

$$R_{NR,f} = R_{NR,0^\circ} - 5\,\text{dB} \tag{4.41}$$

for field incidence (Vér and Holmer, 1988) assuming $R_{NR,0^\circ} > 15$ dB, otherwise use numerical integration of Eq. 4.36.

Figure 4.6 shows non-resonant sound reduction indices for an infinite plate of 6 mm glass. These vary significantly with the angle of incidence. A single angle of incidence is relevant to façade sound insulation measurements when a loudspeaker is directed towards the façade. In practice there will usually be a range of angles of incidence, whether it is from a reverberant sound field within a room, or an external environmental noise source. For this reason we are almost always interested in an angular average value, such as for diffuse or field incidence. As field incidence excludes angles close to grazing which have very low transmission loss, the field incidence values are higher than for diffuse incidence.

4.3.1.2.2 Finite plate theory

For airborne sound transmission across finite plates, non-resonant transmission describes transmission due to bending modes that have their resonance frequencies outside the frequency band of interest (Sewell, 1970). Looking back at Fig. 2.63 we find that individual modes with resonance frequencies below the critical frequency will have a higher radiation efficiency at frequencies above their resonance frequency, than actually at their resonance frequency. So when the frequency band of interest is below the critical frequency, radiation from modes with

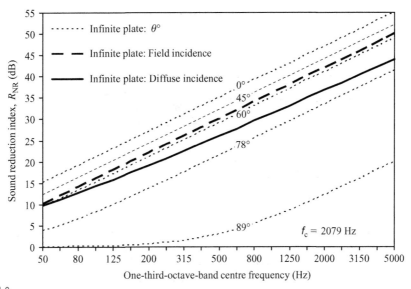

Figure 4.6

Non-resonant transmission across an infinite plate of 6 mm glass ($\rho_s = 15\,\text{kg/m}^2$, $c_L = 5200\,\text{m/s}$).

resonance frequencies within the band can be lower than radiation from modes with resonance frequencies outside the band that have been excited 'off-resonance'. For the former modes, the modal response is under damping control; for the latter modes, it is predominantly under mass control. Non-resonant transmission can therefore be considered as being unaffected by the plate damping.

In contrast to infinite plate theory, finite plate theory considers the plate to have both mass and stiffness. Whilst the mass per unit area is still important in quantifying the non-resonant transmission, the role of bending modes means that it also depends on the plate dimensions and the critical frequency. For a finite plate, the non-resonant transmission coefficient below the critical frequency is given by (Leppington et al., 1987)

$$\tau_{NR} = \left(\frac{2\rho_0}{\rho_s k (1 - \mu^{-4})}\right)^2 \left\{ \begin{array}{l} \ln(k\sqrt{S}) + 0.16 - U(L_x/L_y) \\[2mm] + \dfrac{1}{4\mu^6} \left[\begin{array}{l} (2\mu^2 - 1)(\mu^2 + 1)^2 \ln(\mu^2 - 1) \\ + (2\mu^2 + 1)(\mu^2 - 1)^2 \ln(\mu^2 + 1) \\ - 4\mu^2 - 8\mu^6 \ln\mu \end{array} \right] \end{array} \right\} \tag{4.42}$$

where

$$\mu = \sqrt{\frac{f_c}{f}} \tag{4.43}$$

and $U(L_x/L_y)$ is a function of the rectangular shape of the plate

$$U\left(\frac{L_x}{L_y}\right) = U\left(\frac{L_y}{L_x}\right) = \frac{1}{2\pi}\left(\frac{L_x}{L_y} + \frac{L_y}{L_x}\right) \ln\left[1 + \left(\frac{L_x}{L_y}\right)^2\right]$$

$$- \left(0.5 + \frac{L_x}{\pi L_y}\right) \ln\left(\frac{L_x}{L_y}\right) - \frac{\ln 2}{\pi} - \frac{2}{\pi}\int_{L_x/L_y}^{1} \frac{\arctan t}{t}\,dt \tag{4.44}$$

425

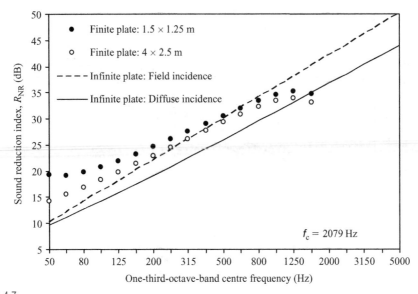

Figure 4.7

Non-resonant transmission across finite and infinite plates of 6 mm glass ($\rho_s = 15\,\text{kg/m}^2$, $c_L = 5200\,\text{m/s}$).

Plates that form windows, doors, walls, and floors usually have aspect ratios in the range, $1/3 \leq L_x/L_y \leq 3$, and satisfactory estimates for the transmission coefficient can be found by ignoring the $U(L_x/L_y)$ term in Eq. 4.42.

Using 6 mm glass as an example we can now compare the non-resonant sound reduction index for finite plates with infinite plates. This is done using SEA path analysis for the non-resonant transmission path $1 \rightarrow 3$. For a finite glass plate, a reasonable size to consider is $1.5 \times 1.25\,\text{m}$, but to illustrate the effect of plate size, we will also use rather unrealistic dimensions of $4 \times 2.5\,\text{m}$. The non-resonant sound reduction indices are shown in Fig. 4.7. For the finite plates, the smaller plate has higher values than the larger plate, and both finite plates have significantly higher values than an infinite plate assuming diffuse incidence. As the frequency approaches the critical frequency, the non-resonant sound reduction index for the finite plates starts to decrease. Above the critical frequency, the non-resonant transmission coefficient is undefined for the finite plates.

The sound reduction index for field incidence increases by 6 dB per doubling of frequency. This is steeper than the average slope for each of the finite plates although at some frequencies the finite plate curves have similar values to the field incidence curve. Field incidence assumes that angles of incidence are restricted to being equally probable between 0° and 78°. Its name unfortunately suggests that this assumption is always valid in the field (i.e. *in situ*); which it is not. The assumption simply gives fortuitous agreement with measured data for particular plate sizes with particular critical frequencies, usually with particular mounting conditions in a niche and usually where non-resonant transmission dominates in the low- and mid-frequency ranges and the sound field in the rooms is far from being diffuse. Whilst the infinite plate formulae for diffuse and field incidence are very useful for quick calculations and illustrative purposes, they do not describe all the features of non-resonant transmission that relate to finite size plates (Leppington *et al.*, 1987). Unless stated, all calculations of non-resonant transmission in SEA models will use finite plate theory to calculate the non-resonant transmission from this point onwards.

Note that when modelling more complex plates (e.g. plates that are non-homogeneous, periodic, or profiled) using an infinite plate approach it is necessary to link the models to measurements that will be made on finite plates in practice. For convenience this is often done by restricting the range for the angles of incidence in the same way as with field incidence.

Unlike with finite plate theory, non-resonant transmission across infinite plates is defined above the critical frequency. This implies that if the plate damping is sufficiently high such that resonant transmission is negligible at frequencies well-above the critical frequency, then non-resonant transmission will dominate at these high frequencies. It is unusual to find homogeneous plates in buildings that are this highly damped in the building acoustics frequency range, and using infinite plate theory for non-resonant transmission above the critical frequency is not usually appropriate.

4.3.1.3 Examples

We will now compare predictions of the sound reduction index with transmission suite measurements for different solid plates. For these particular laboratory measurements it is reasonable to assume that the test elements are surrounded by a rigid infinite baffle, and that the plate boundaries are simply supported. Glass and plasterboard plates are used to illustrate non-resonant transmission and the dip in the sound reduction index at the critical frequency. Masonry walls are used to show the importance of resonant transmission across the entire building acoustics frequency range. These have a plaster finish to remove any transmission via air paths through the blocks and/or the mortar joints. In contrast to the plates of glass and plasterboard, the masonry walls have low mode counts and low modal overlap. To ensure accurate prediction of the resonant transmission, the measured total loss factors are used in the SEA models. By using the measured total loss factor we are accounting for all the coupling losses from the plate to the laboratory structure, but not including the laboratory structure in the model. For the masonry walls the measured total loss factor is approximately described by $\eta = 0.01 + 0.3f^{-0.5}$.

For the plasterboard and masonry walls the measurements were made in a transmission suite with reverberation times between 1 and 2 s, and source and receiving room volumes of 130 and 115 m^3 respectively. The modal overlap factors in such rooms (refer back to Fig. 1.24) will be much larger than for most masonry/concrete plates. It is therefore reasonable to assume that it is the geometric mean of the modal overlap factors for the plate and the source room (or receiving room) that is relevant when assessing the SEA prediction of resonant transmission. Note that the source room is used for these calculations of M_{av} (Eq. 4.7) because the two rooms are similar and the small difference is not important here.

SEA path analysis is used to predict the sound reduction index for the resonant path $1 \rightarrow 2 \rightarrow 3$, and the non-resonant path $1 \rightarrow 3$. The matrix solution is used to give the overall transmission due to resonant and non-resonant transmission; however, in these particular examples, combining the resonant and non-resonant transmission paths using Eq. 4.19 gives the same result. As we are interested in a wide range of plate thicknesses and materials in buildings, the important plate parameters are included on the figures: statistical mode count, modal overlap factor, fundamental mode frequency, critical frequency, and the thin plate limit (bending waves).

4.3.1.3.1 Glass

A sheet of 6 mm glass has already been used as an example in this chapter (refer back to Figs 4.4 and 4.7). Figure 4.8a shows the predicted sound reduction index for the resonant and

(a)

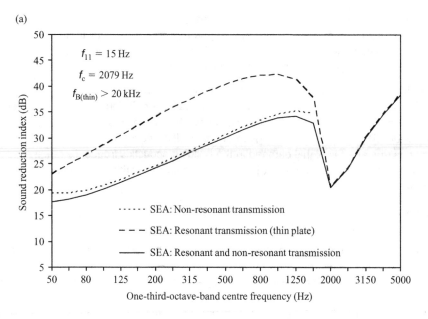

(b)

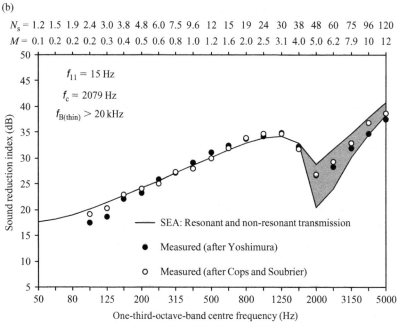

Figure 4.8

Airborne sound insulation of 6 mm glass: (a) predicted non-resonant and resonant transmission using SEA and (b) comparison of measurements and SEA. Upper x-axis labels show the predicted statistical mode count and modal overlap factor for the plate in each frequency band. (*Note*: M_{av} is not shown here because measurements were taken in two different laboratories with different volumes and reverberation times.) Measurements according to ISO 140 Part 3 and niche detail according to ISO 140 Parts 1 and 3. *Plate properties*: $L_x = 1.5$ m, $L_y = 1.25$ m, $h = 0.006$ m, $\rho_s = 15$ kg/m^2, $c_L = 5200$ m/s, $\nu = 0.24$, total loss factor $\eta = 0.024$ (frequency-average value from structural reverberation time measurements on similar test elements). Measured data are reproduced with permission from Cops and Soubrier (1988) and Yoshimura (2006).

non-resonant paths separately as well as in combination. Below the critical frequency, non-resonant transmission generally dominates over resonant transmission and the overall sound insulation is mainly determined by non-resonant transmission. At and above the critical frequency there is only resonant transmission.

In the vicinity of the critical frequency it is notoriously difficult to accurately predict the radiation efficiency (even when $N_s \geq 30$ and $M \geq 3$). This can partly be overcome by calculating lower and upper limits for the radiation efficiency using method no. 2 (Section 2.9.4.2). SEA calculations are therefore carried out twice; once with the lower limit and once with the upper limit. Measured data usually lies within the shaded area between the two limits. Comparison of the combination of resonant and non-resonant transmission with measurements is shown in Fig. 4.8b (Cops and Soubrier, 1988; Yoshimura, 2006). Close agreement below the critical frequency confirms that non-resonant transmission for a finite plate gives a better estimate than diffuse or field incidence for an infinite plate (refer back to Fig. 4.7).

4.3.1.3.2 Plasterboard

A partition formed from 12.5 mm plasterboard without a frame is used to provide an example of a plate with a high statistical mode count over the entire frequency range. The structural coupling losses from the plasterboard to the laboratory structure are negligible; hence the total loss factor equals the sum of the internal loss factor plus the two radiation coupling loss factors (i.e. both sides).

Figure 4.9a allows an assessment of the resonant transmission path. Below the critical frequency, there is a large difference between the SEA prediction for resonant transmission and the measured values. This indicates that there is a much more important sound transmission mechanism than resonant transmission at these frequencies. As we have previously seen with the 6 mm glass plate, this is non-resonant transmission. At and above the critical frequency where there is only resonant transmission, the measured values lie within the range predicted for the radiation efficiency using method no. 2 (Section 2.9.4.2). In Fig. 4.9b the agreement between measurements and the three different predictions for non-resonant transmission below the critical frequency confirms that non-resonant transmission is the dominant mechanism. The SEA prediction uses finite plate theory to determine the non-resonant transmission coefficient. Infinite plate theory for a diffuse field tends to overestimate the sound transmission measured in practice; field incidence fortuitously gives a reasonable estimate, but there is no firm basis on which it can be applied to plates of all sizes. Combining resonant and non-resonant transmission (finite plate theory) gives good agreement between measurements and the SEA prediction over the entire building acoustics frequency range as shown in Fig. 4.9c.

4.3.1.3.3 Masonry wall (A)

Masonry wall (A) is a 115 mm aircrete wall with a 13 mm plaster finish on one side (see Fig. 4.10). As ρ and c_L for plaster and aircrete are similar this allows the prediction to assume a 128 mm thick solid homogeneous plate. Measurements on the plastered wall are used to give c_L and the total loss factor. In Section 4.3.8.1 we will discuss the effect of a bonded surface finish such as plaster in more detail.

Below the critical frequency, resonant transmission dominates over non-resonant transmission. This can be seen by the fact that the curve for resonant transmission is very similar to the curve for the combination of resonant and non-resonant transmission. Due to the low mode count, the

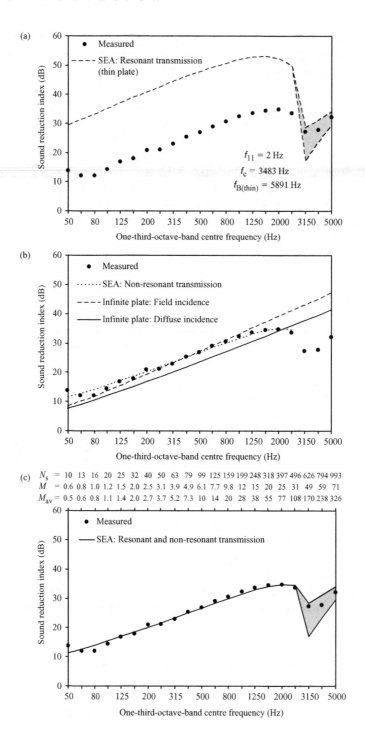

Figure 4.9

Measured and predicted airborne sound insulation of 12.5 mm plasterboard. Upper x-axis labels show the predicted statistical mode count and modal overlap factor for the plate and M_{av} (plate and room). Measurements according to ISO 15186 Part 3 (50–100 Hz) and ISO 15186 Part 1 (125–5000 Hz). *Plate properties*: $L_x = 3.53$ m, $L_y = 2.63$ m, $h = 0.0125$ m, $\rho_s = 10.8$ kg/m^2, $c_L = 1490$ m/s, $\nu = 0.3$, $\eta_{int} = 0.0141$. Measured data from Hopkins are reproduced with permission from ODPM and BRE.

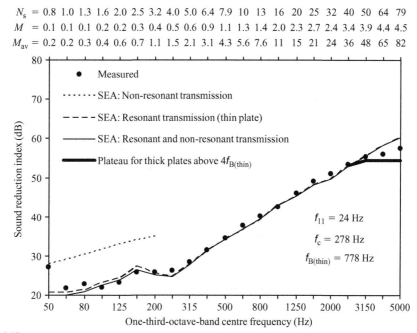

$$N_s = 0.8\ 1.0\ 1.3\ 1.6\ 2.0\ 2.5\ 3.2\ 4.0\ 5.0\ 6.4\ 7.9\ 10\ 13\ 16\ 20\ 25\ 32\ 40\ 50\ 64\ 79$$
$$M = 0.1\ 0.1\ 0.1\ 0.2\ 0.2\ 0.3\ 0.4\ 0.5\ 0.6\ 0.9\ 1.1\ 1.3\ 1.4\ 2.0\ 2.3\ 2.7\ 2.4\ 3.4\ 3.9\ 4.4\ 4.5$$
$$M_{av} = 0.2\ 0.2\ 0.3\ 0.4\ 0.6\ 0.7\ 1.1\ 1.5\ 2.1\ 3.1\ 4.3\ 5.6\ 7.6\ 11\ 15\ 21\ 24\ 36\ 48\ 65\ 82$$

Figure 4.10

Measured and predicted airborne sound insulation of a 115 mm masonry wall (solid aircrete blocks) with a 13 mm lightweight plaster finish (one side). Upper x-axis labels show the predicted statistical mode count and modal overlap factors for the plate and M_{av} (plate and room). Measurements according to ISO 15186 Part 3 (50–100 Hz) and ISO 15186 Part 1 (125–5000 Hz). *Plate properties:* $L_x = 3.53$ m, $L_y = 2.63$ m, $h = 0.128$ m, $\rho_s = 71$ kg/m², $c_L = 1820$ m/s, $\nu = 0.2$, measured total loss factor. The plateau is calculated using material properties corresponding to the plate thickness: $c_L = 1920$ m/s and an internal loss factor $\eta_{int} = 0.0125$. Measured data from Hopkins are reproduced with permission from ODPM and BRE.

radiation efficiency has been calculated using method no. 4 (Section 2.9.4.4). For $0.2 \leq M_{av} < 1$ the measured and predicted values show close agreement; however if many similar walls were measured one would expect to see a wide range of values due to the low mode counts and low modal overlap. Hence there will often be individual bands where the differences are large. In this particular example, the largest difference occurs in the 50 Hz band.

At and above the critical frequency where $N_s \geq 4$ and $M_{av} \geq 1$ there is close agreement between measurements and predictions up to a frequency of $4f_{B(thin)}$. Above $4f_{B(thin)}$ the measured values start to level-off to form a plateau; the prediction of this plateau is discussed in Section 4.3.1.4.

4.3.1.3.4 Masonry wall (B)

Masonry wall (B) is a 100 mm dense aggregate masonry wall with a 13 mm plaster finish on one side (see Fig. 4.11). Compared to the plaster, the dense aggregate wall has a high mass per unit area and is much stiffer; hence the plaster can be ignored in calculating the plate thickness and density for the SEA model.

Below the critical frequency the mode count $N_s < 3$, so the radiation efficiency has been calculated using method no. 4 (refer back to Section 2.9.4.4 and Fig. 2.65d). If we assume that non-resonant theory for finite plates still gives a better estimate than the infinite plate theory, we find that resonant transmission dominates over non-resonant transmission. Below the critical frequency where $0.2 \leq M_{av} < 1$ the agreement between measurement and prediction

N_s = 0.6 0.7 0.9 1.2 1.4 1.8 2.3 2.9 3.6 4.6 5.8 7.3 9.2 12 14 18 23 29 36 46 58
M = 0.1 0.1 0.1 0.1 0.2 0.2 0.3 0.5 0.6 0.8 0.9 0.9 1.1 1.2 1.4 1.6 1.9 1.9 2.4 3.3 3.9
M_{av} = 0.2 0.2 0.2 0.4 0.5 0.7 1.0 1.4 2.0 2.9 3.8 4.8 6.5 8.7 12 16 22 26 38 56 77

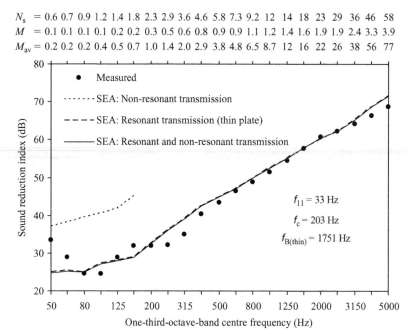

Figure 4.11

Measured and predicted airborne sound insulation of a 100 mm masonry wall (solid dense aggregate blocks) with a 13 mm lightweight plaster finish (one side). Upper x-axis labels show the predicted statistical mode count and modal overlap factors for the plate and M_{av} (plate and room). Measurements according to ISO 15186 Part 3 (50–100 Hz) and ISO 15186 Part 1 (125–5000 Hz). *Plate properties:* $L_x = 3.53$ m, $L_y = 2.63$ m, $h = 0.1$ m, $\rho_s = 200$ kg/m², $c_L = 3200$ m/s, $v = 0.2$, measured total loss factor. Measured data from Hopkins are reproduced with permission from ODPM and BRE.

is generally similar to when $M_{av} \geq 1$. To make a thorough assessment of the uncertainty we cannot rely on a single measurement, we would need to measure a number of similar walls.

Above the critical frequency where $N_s \geq 1$ and $M_{av} \geq 1$ there is close agreement between measurement and prediction. The thin plate limit lies within the high-frequency range; this provides more evidence that it is reasonable to use thin plate theory up to $4f_{B(thin)}$.

4.3.1.3.5 Masonry wall (C)

For masonry wall (C) we look at a thicker version of masonry wall (B). This is a 215 mm dense aggregate masonry wall with a 13 mm plaster finish on each side (see Fig. 4.12); as before, the plaster is ignored in calculating the plate thickness and density.

Assuming simply supported boundaries, the fundamental mode frequency is 70 Hz. This is above the lowest one-third-octave-band considered in the building acoustics frequency range, and close to the critical frequency. For this reason, only resonant transmission is predicted and shown above the fundamental mode.

In the low-frequency range where $0.1 \leq M_{av} < 1$, the agreement between measurement and prediction is not significantly worse than in the mid- and high-frequency ranges where $N_s \geq 1$ and $M_{av} \geq 1$. Although errors are incurred by using thin plate theory between $f_{B(thin)} \leq f < 4f_{B(thin)}$ they are not large enough to warrant changing to thick plate theory. Above $4f_{B(thin)}$ the curve reaches a plateau due to thickness resonances; these are discussed in the following section.

N_s = 0.3 0.3 0.4 0.5 0.7 0.9 1.1 1.3 1.7 2.2 2.7 3.4 4.3 5.4 6.7 8.6 11 13 17 22 27
M = 0.04 0.03 0.03 0.1 0.1 0.1 0.2 0.2 0.3 0.4 0.3 0.3 0.4 0.4 0.5 0.5 0.7 0.9 1.1 1.2 1.4
M_{av} = 0.1 0.1 0.1 0.3 0.4 0.5 0.7 1.0 1.3 2.0 2.2 2.6 3.8 5.2 6.6 8.9 13 18 26 34 46

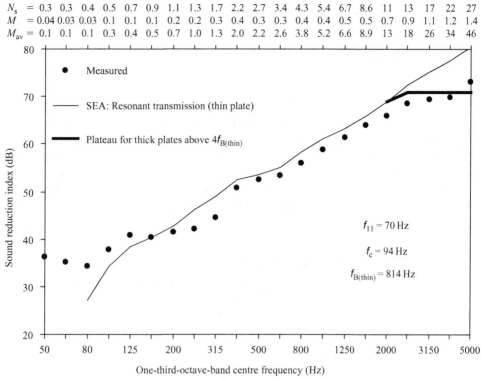

Figure 4.12

Measured and predicted airborne sound insulation of a 215 mm masonry wall (solid dense aggregate blocks) with a 13 mm lightweight plaster finish (each side). Upper x-axis labels show the predicted statistical mode count and modal overlap factor for the plate and M_{av} (plate and room). Measurements according to ISO 15186 Part 3 (50–100 Hz) and ISO 15186 Part 1 (125–5000 Hz). *Plate properties:* $L_x = 3.53$ m, $L_y = 2.63$ m, $h = 0.215$ m, $\rho_s = 430$ kg/m², $c_L = 3200$ m/s, $\nu = 0.2$, measured total loss factor. The plateau is calculated using material properties corresponding to the plate thickness: $c_L = 4000$ m/s and an internal loss factor $\eta_{int} = 0.01$. Measured data from Hopkins are reproduced with permission from ODPM and BRE.

4.3.1.4 Thin/thick plates and thickness resonances

In the low-frequency range, thick masonry/concrete plates sometimes have their fundamental bending mode above the critical frequency. In such cases it is not suitable to calculate non-resonant transmission below the critical frequency using finite plate theory. However, resonant transmission can still be calculated at and above the fundamental mode using thin plate bending wave theory.

For most solid masonry/concrete walls the thin plate limit for bending waves falls in the mid- or high-frequency range. In Section 2.3.3.1 we discussed this thin plate limit in terms of a 10% difference in the phase velocity between pure bending waves on thin plates and bending waves on thick plates. Whilst it is referred to as a 'limit', thin plate theory does not instantly break down at a specific frequency. For direct airborne sound insulation across a solid homogeneous plate, thin plate theory can often be used up to a frequency of $4f_{B(thin)}$ (Ljunggren, 1991). We will now treat $4f_{B(thin)}$ as a limit whilst acknowledging that it is not quite so clear-cut in practice. Errors from using thin plate theory in the range $f_{B(thin)} \leq f < 4f_{B(thin)}$ are usually less than 3 dB. This is often tolerable due to the uncertainty in predicting the total loss factor. Above $4f_{B(thin)}$ the airborne sound insulation effectively stops increasing with frequency and reaches a plateau

Figure 4.13

Dilatational wave motion corresponding to thickness resonance on a plate.

with dips due to thickness resonances across the plate (Ljunggren, 1991). It can be useful to visualize a longitudinal wave in air impinging upon a thick infinite plate at normal incidence (Vér, 1992); and this wave exciting longitudinal waves across the plate thickness that efficiently radiate sound into the air on the other side of the plate. In practice, we are interested in an incident sound field that is diffuse and where the thickness modes are described by dilatational waves (see Fig. 4.13). The mode frequencies for these thickness modes are calculated using,

$$f_r = \frac{rc_D}{2h} \tag{4.45}$$

where r takes positive integer values 1, 2, 3, etc.

The phase velocity for dilatational waves, c_D, is (Timoshenko and Goodier, 1970)

$$c_D = \sqrt{\frac{\lambda + 2\mu}{\rho}} \tag{4.46}$$

where Lamé's constants, μ and λ, are

$$\mu = G = \frac{E}{2(1 + v)} \tag{4.47}$$

$$\lambda = \frac{vE}{(1 + v)(1 - 2v)} \tag{4.48}$$

and E and v correspond to the material properties of the plate for longitudinal wave motion in the thickness direction.

This sound transmission mechanism is not included within the SEA framework as there are rarely more than two thickness modes in the high-frequency range for masonry/concrete plates. In practice, the thickness resonances don't always appear as distinct dips in the sound reduction index. A plateau therefore provides a reasonable estimate for the sound reduction index and can be estimated according to (Ljunggren, 1991)

$$R_{\text{plateau}} = 20 \lg \left(\frac{\rho c_D}{4\rho_0 c_0} \right) + 10 \lg \left(\frac{\eta_{\text{int}}}{0.02} \right) \tag{4.49}$$

If $4f_{B(\text{thin})}$ falls within the frequency range of interest, a smooth transition into the plateau region can be achieved using R calculated from thin plate theory where $R < R_{\text{plateau}}$, and changing over to R_{plateau} in frequency bands where $R \geq R_{\text{plateau}}$. Due to increasing attenuation of the dilatational waves with increasing frequency it is likely that the plateau will only extend a few octaves above $4f_{B(\text{thin})}$ before starting to increase with frequency once more (Ljunggren, 1991). It is unusual to need to predict the sound insulation above 5000 Hz so for typical solid masonry/concrete walls there is no need to consider the region beyond the plateau.

Using this approach with masonry/concrete plates gives measured values of R that are typically within ± 3 dB of the plateau region. Most thin surface finishes (such as plaster or render) that are bonded to a thick plate will have negligible effect on the plateau region. Note that if the

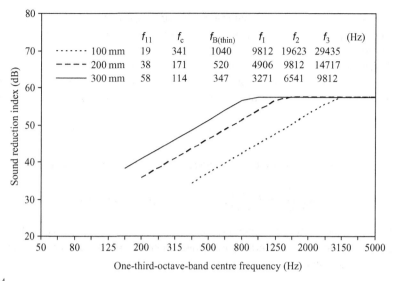

Figure 4.14

SEA predictions above the critical frequency (resonant transmission only) using thin plate theory combined with the plateau region. *Plate properties*: $L_x = 4\,\text{m}$, $L_y = 2.5\,\text{m}$, $h = 0.1/0.2/0.3\,\text{m}$, $\rho_s = 80/160/240\,\text{kg/m}^2$, $c_L = 1900\,\text{m/s}$, $\nu = 0.2$, internal loss factor $\eta_{\text{int}} = 0.0125$, total loss factor $\eta = 0.0125 + f^{-0.5}$.

solid plate is orthotropic, and the thin plate theory uses an effective bending stiffness, then the effective Young's modulus will be different to the Young's modulus that applies to the dilatational modes in the thickness direction.

We have already seen an example of a thick plate in Fig. 4.12 for a 215 mm dense aggregate masonry wall. The first two thickness modes of this wall are $f_1 = 9607\,\text{Hz}$ and $f_2 = 19\,215\,\text{Hz}$. Although f_1 is well above the building acoustics frequency range it still affects the top of the high-frequency range.

For thermal purposes, some low-density walls are used in a wide range of thicknesses. Figure 4.14 shows an example of how the plateau region potentially forms an upper limit for the airborne sound insulation in the high-frequency range. For the 200 and 300 mm plates the first thickness resonance, f_1, lies within the building acoustics frequency range.

4.3.1.5 Infinite plates

So far we have mainly focused on finite plates. It might seem that further analysis of infinite plates would be of little benefit; this is not the case. We will soon see that above the critical frequency, the infinite plate formulae yield the same sound reduction index as for finite plates. In addition, very large walls and floors that face into large spaces can be modelled as infinite plates. Infinite plates also provide a convenient way of assessing the effect of different angles of incidence. Whilst we usually need to know the sound reduction index for a diffuse incidence sound field, there are some occasions where the incident sound field inside or outside a building is highly directional.

We have already looked at non-resonant transmission across an infinite plate acting as a limp mass. To complete the analysis we need to calculate the transmission coefficient when this infinite plate has mass, bending stiffness, and damping.

As in Section 4.3.1.2.1 we assume that the infinite plate lies in the xy plane and that a plane wave is incident upon the plate at an angle, θ (refer back to Fig. 4.5). Therefore the sound pressure for the incident, reflected, and transmitted waves can be taken from Eqs 4.27–4.29. At this point it is necessary to make a brief return to the equation of motion for bending waves on an isotropic plate. This is to account for the force per unit area applied by the sound pressure, $p(x, y, t)$, that drives the plate into motion. The equation of motion (Eq. 2.87) therefore becomes

$$-\frac{\partial Q_x}{\partial x}\,dx\,dy - \frac{\partial Q_y}{\partial y}\,dy\,dx + p(x, y, t) = \rho_s dx\,dy\frac{\partial^2 \eta}{\partial t^2} \tag{4.50}$$

which gives the wave equation for bending waves on a thin homogeneous plate as

$$B_p\left(\frac{\partial^4 \eta}{\partial x^4} + 2\frac{\partial^4 \eta}{\partial x^2 \partial y^2} + \frac{\partial^4 \eta}{\partial y^4}\right) + \rho_s\frac{\partial^2 \eta}{\partial t^2} = p(x, y, t) \tag{4.51}$$

Note that to be consistent with Chapter 2 we will temporarily use η as the lateral displacement of the plate. Once we have derived the transmission coefficient we will continue to use it to represent loss factors again.

The sound pressure that drives the plate is

$$p(x, y, t) = \hat{p}\exp(-ik_x x)\exp(-ik_y y)\exp(i\omega t) \tag{4.52}$$

and the plate displacement must take the same form, hence

$$\eta(x, y, t) = \hat{\eta}\exp(-ik_x x)\exp(-ik_y y)\exp(i\omega t) \tag{4.53}$$

Substituting Eq. 4.53 into the wave equation (Eq. 4.51) gives

$$B_p[(k_x^2 + k_y^2)^2 - k_B^4] = \frac{p}{\eta} \tag{4.54}$$

It is now convenient to define a surface impedance for the plate, Z_p, as the ratio of the complex sound pressure that drives the plate to the complex plate velocity,

$$Z_p = \frac{p}{v} = \frac{p}{i\omega\eta} \tag{4.55}$$

Hence Eqs 4.54 and 4.55 give the surface impedance as

$$Z_p = \frac{B_p}{i\omega}[(k_x^2 + k_y^2)^2 - k_B^4] \tag{4.56}$$

Returning to the derivation of sound transmission, we are restricting our attention to the bending wave that propagates in the positive x-direction, so we can set $k_y = 0$. In the x-dimension, the sound pressure and plate velocity must have the same spatial dependence, hence $k_x = k\sin\theta$ and the surface impedance simplifies to

$$Z_p = \frac{B_p}{i\omega}(k^4\sin^4\theta - k_B^4) \tag{4.57}$$

Ignoring time dependence, Eq. 4.57 can now be used to describe the plate velocity in terms of the sound pressure acting on the plate at $z = 0$,

$$v = \hat{v}\exp(-ikx\sin\theta) = \frac{(\hat{p}_i + \hat{p}_r - \hat{p}_t)}{Z_p}\exp(-ikx\sin\theta) \tag{4.58}$$

At $z=0$ there must be continuity for the z-component of the particle velocity on both sides of the plate (Eq. 4.32). This gives the relationship between the plate velocity and the sound pressures described by Eq. 4.33. The angle-dependent transmission coefficient for an infinite isotropic plate with mass and stiffness, $\tau_{\infty,\theta}$, can now be determined from Eqs 4.33 and 4.58, yielding

$$\tau_{\infty,\theta} = \frac{W_t}{W_i} = \left|\frac{\hat{p}_t}{\hat{p}_i}\right|^2 = \frac{1}{\left|1 + \dfrac{Z_p \cos\theta}{2\rho_0 c_0}\right|^2} \tag{4.59}$$

To include the effect of damping in the calculation of $\tau_{\infty,\theta}$ we use the loss factor, η, and calculate the surface impedance using $B_p(1+i\eta)$ instead of B_p. An infinite wall or floor clearly has no boundaries at which it can lose energy via coupling losses to other walls and floors, so the loss factor represents the internal loss factor for the plate material.

An example of the angle-dependent sound reduction index is shown in Fig. 4.15a for a 150 mm thick infinite plate with the properties of cast in situ concrete. From Eq. 4.59 we see that when $\theta=90°$, the sound reduction index will be 0 dB at all frequencies; this is due to trace matching between the incident sound wave and the bending wave (Section 2.9.2). Whilst this angle can potentially occur, it is of more interest to look at an angle very close to 90° to see how much the sound reduction index differs from that for 90°; for $\theta=89°$ we see that there is a dip very close to the critical frequency, above which the level climbs steeply. At any frequency above the critical frequency there will always be an angle of incidence at which trace matching (also called coincidence) occurs. For each angle of incidence ($\theta>0°$) this results in a coincidence dip in the sound reduction index; such dips can be seen in Fig. 4.15a where $f>f_c$. The depth of the coincidence dip is determined by the damping. At frequencies well above the coincidence dip, each curve tends towards a slope that increases by 18 dB per doubling of frequency.

When $\theta=0°$ there is no trace matching and the sound reduction index is the same as for an infinite plate acting as a limp mass (Eq. 4.39) at all frequencies. When $\theta>0°$ for $f<f_c$ the surface impedance (Eq. 4.57) is primarily determined by the mass per unit area because $k<k_B$, hence

$$Z_p \approx \frac{B_p}{i\omega}k_B^4 = i\omega\rho_s \tag{4.60}$$

Using this surface impedance to determine the angle-dependent transmission coefficient (Eq. 4.59) gives the same equation as when the infinite plate acts as a limp mass (Eq. 4.34). Well-below the critical frequency, we therefore find that sound transmission is only determined by the mass per unit area.

To determine the sound reduction index for diffuse incidence, integration of the transmission coefficient (Eq. 4.59) is carried out in the same way as for non-resonant transmission (Eq. 4.36). This gives a smooth curve up to frequencies just above the critical frequency. At higher frequencies, the curve is full of peaks and troughs due to the coincidence dips. To give a smooth curve above the critical frequency, the sound reduction index for diffuse incidence is calculated using (Cremer, 1942)

$$R_\infty = R_{NR,0°} + 10\lg\left(\frac{f}{f_c}-1\right) + 10\lg\eta - 2\,\text{dB} \quad \text{for } f>f_c \tag{4.61}$$

The sound reduction index for diffuse incidence is shown in Fig. 4.15b for the same 150 mm thick infinite plate. Above the critical frequency, the grey parts of the lowest three curves indicate where curve fitting was used to connect the smooth curve given by Eq. 4.61 to the value calculated by integration of the angle-dependent transmission coefficient.

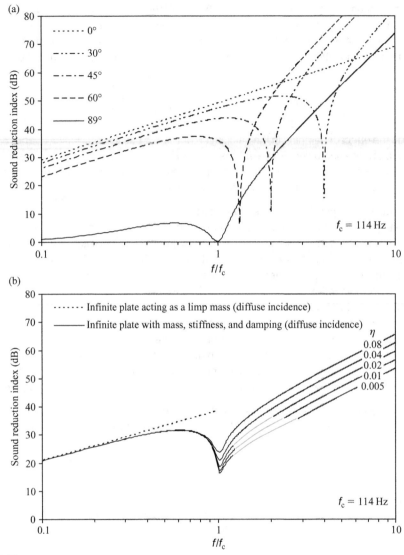

Figure 4.15

Predicted airborne sound insulation for a 150 mm thick infinite plate: (a) angle-dependent sound reduction index for an infinite plate (mass, stiffness, and damping) with the properties of cast *in situ* concrete and (b) sound reduction index (diffuse incidence) for an infinite plate with different loss factors. *Plate properties:* $h = 0.15$ m, $\rho_s = 330$ kg/m^2, $c_L = 3800$ m/s, $\nu = 0.2$, $\eta_{int} = 0.005$.

Figure 4.15b shows that well-below the critical frequency the infinite plate acts as a limp mass and that the plate stiffness and damping have no effect on the sound reduction index. As the frequency approaches the critical frequency, the effect of stiffness in the surface impedance becomes more important and causes the curve to sag down below the diffuse incidence mass law. This sagging curve leads to a dip at the critical frequency; the lowest frequency at which coincidence occurs. At and above the critical frequency, the sound reduction index for diffuse incidence is under damping control. Referring back to the single angles of incidence in

Fig. 4.15a it is clear that angular weighting carried out by integration of the angle-dependent transmission coefficient gives a curve that is primarily determined by the coincidence dips. It is these dips that are under damping control. At frequencies well-above the critical frequency the sound reduction index for diffuse incidence increases by 3 dB per doubling of the loss factor. For a frequency-independent loss factor, the curves well-above the critical frequency tend towards a slope that increases by 9 dB per doubling of frequency.

We can now make a link between the sound reduction index for a finite and an infinite plate. For sound radiation from bending modes on finite plates, modes are classified as corner, edge, or surface radiators depending on the frequency (Section 2.9.3). Below the critical frequency the radiation from these modes depends on the plate boundaries; this is because most of the sound power is radiated by corners or edges. Above the critical frequency, the plate modes are all surface radiators and the perimeter length of the plate does not affect the radiation efficiency. So although the existence of plate boundaries defines the modes, these boundaries do not effect the sound radiation by these modes above the critical frequency. For this reason, the sound insulation of finite plates above the critical frequency can be modelled using a plate without boundaries, an infinite plate. The link that remains to be made is how the damping affects the mean-square velocity of a finite plate and the equivalent infinite plate. For infinite plates the only losses are material losses; hence the loss factor that is used to incorporate damping is the internal loss factor. For finite plates we must also consider the coupling losses because the plate is usually coupled to other parts of the structure (e.g. to other plates). So to model a finite plate above the critical frequency using the infinite plate equation, we need to replace the internal loss factor with the total loss factor of the finite plate. Above $4f_{B(thin)}$ the plateau region for thick plates is applicable.

An example to illustrate the equivalence of a finite and an infinite plate above the critical frequency is shown in Fig. 4.16 that compares measurement and prediction for a 150 mm concrete slab. SEA is used to calculate resonant transmission for the finite plate above the critical frequency assuming a radiation efficiency of unity. The finite and the infinite plate predictions are calculated for two different damping scenarios: one where the damping equals the internal loss factor and the other where the damping equals the total loss factor of the finite plate that was measured. Figure 4.16 shows that above the critical frequency in each scenario, the finite plate SEA prediction tends towards the infinite plate prediction. The measurements only agree with the finite plate SEA prediction and the infinite plate prediction that use the total loss factor, not the internal loss factor. Note that the measured data does not have a deep dip at the critical frequency like the infinite plate theory. A shallow or unidentifiable dip is a feature of masonry/concrete plates that leads to the use of radiation efficiencies of unity at and above the critical frequency.

Unlike the internal loss factor, the total loss factor for masonry/concrete plates that are connected to other plates is usually frequency-dependent (Section 2.6.5). Therefore the slope of the sound reduction index well-above the critical frequency will often differ from 9 dB per doubling of frequency.

There is no equivalence between finite and infinite plates below the critical frequency. When modelling finite plates using SEA we need to consider both non-resonant and resonant transmission. The relative importance of these paths to each other has previously been illustrated with a number of examples in Section 4.3.1.3. These show that resonant or non-resonant (mass law) transmission may dominate below the critical frequency. This is in contrast to

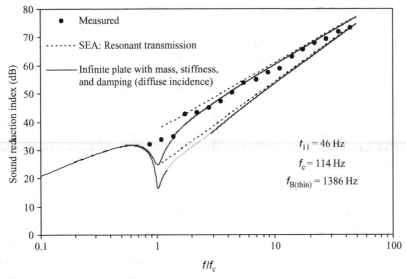

Figure 4.16

Sound reduction index (diffuse incidence) for a 150 mm concrete slab. Comparison of predicted data (finite and infinite plates) with measurements. For the predicted data, the upper curves are predictions using the measured total loss factor, $\eta \approx 1.83 f^{-0.62}$, the lower curves are predictions using the internal loss factor, $\eta_{int} = 0.005$. Measurements according to ISO 140 Part 3. *Plate properties:* $L_x = 3.4$ m, $L_y = 3.3$ m, $h = 0.15$ m, $\rho_s = 330$ kg/m², $c_L = 3800$ m/s, $\nu = 0.2$. Measured data are reproduced with permission from ODPM and BRE.

infinite plate theory where the surface impedance is determined by the mass term below the critical frequency; this implies that there will only be mass-law (non-resonant) transmission.

For infinite plates, the field incidence transmission coefficient for an infinite plate acting as a limp mass assumes that all angles of incidence are between 0° and 78°. In Section 4.3.1.2.2 we noted that although this sometimes provides fortuitous agreement with measured data it is a less robust approach than using finite plate theory for non-resonant transmission. A reduced range of angles is sometimes assumed in transmission suite measurements due to the mounting of the test element within a niche. It is therefore of interest to assess the effect of a reduced range of incident angles on the coincidence dip. To do this it is convenient to use the ranges 0–90°, 0–78°, and 0–68° for a sheet of 6 mm glass as shown in Fig. 4.17. In this example the effect of reducing the range of angles of incidence is to shift the coincidence dip upwards by ≈100 Hz. This serves as a reminder of the uncertainty associated with predictions in the vicinity of the critical frequency when the incident sound field cannot be considered as diffuse.

4.3.1.6 Closely connected plates

It is usually clear when part of a wall or floor can be modelled as a single plate. It is not so obvious when there are two or more plates that are very closely connected together over their surfaces with an air gap of a millimetre or so. A common example occurs in lightweight walls and floors when layers of board (e.g. plasterboard, plywood, chipboard) are closely connected together with nails, screws, or dabs of adhesive. Other examples include sheets of metal that are screwed or glued to timber doors. Modelling such multi-layer plates is complex and it is

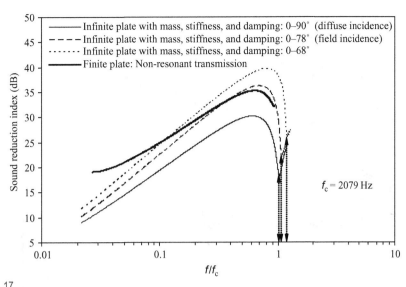

Figure 4.17

Comparison of the sound reduction indices for 6 mm glass with different ranges for the angle of incidence. The frequency of the coincidence dip is indicated by arrows. *Plate properties: $L_x = 1.5$ m, $L_y = 1.25$ m, $h = 0.006$ m, $\rho_s = 15$ kg/m^2, $c_L = 5200$ m/s, $\nu = 0.24$, total loss factor $\eta = 0.024$.*

not usually possible to describe all of their behaviour by a solid plate, a laminated plate, or a plate–cavity–plate system. In addition, variation in workmanship and uniformity of materials means that the width of the air gap between two plates can differ considerably between similar constructions. This usually means that design decisions concerning substitution of one type of plate in a wall or floor for another type are reliant upon laboratory measurements.

The existence of a very narrow air gap sometimes causes a mass–spring–mass resonance (see Section 4.3.5.1). Its presence is apparent with closely connected plates that are used in some, but not all, lightweight wall and floor constructions (Warnock, 2000). This may partly be due to non-uniform air gaps and variation in the strength of other sound transmission mechanisms.

For solid homogeneous plates we have already noted the difficulty in predicting excitation and radiation at the critical frequency. When two layers of board are closely connected together with screws, nails, or dabs of adhesive, the depth and width of the critical frequency dip often changes. However, for boards with the same material properties the dip itself does not usually shift down to a lower frequency as it would for a solid homogeneous plate with the combined thickness of the two individual boards (Sharp, 1978; Sharp and Beauchamp, 1969). Figure 4.18 shows measurements where the critical frequency dip in the 3150 Hz band does not move to a different band when two plates are closely connected together. This can be a desirable design feature if it keeps the critical frequency dip at the top of the building acoustics frequency range. By spot bonding boards with different critical frequencies it is sometimes possible to remove any visible dip at either of the critical frequencies (Matsumoto *et al.*, 2006). At frequencies well-below the lower critical frequency and well-below any mass–spring–mass resonance, a rough estimate for the non-resonant transmission can be found by assuming that the closely connected plates act as a solid plate with the combined mass per unit area of the individual plates. Resonant transmission is highly dependent on the type of connections between the plates.

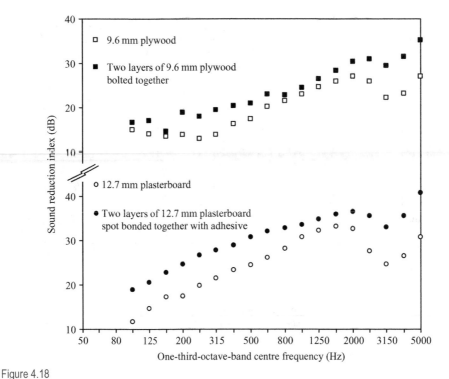

Figure 4.18

Comparison of measured sound reduction indices for single plates and two plates closely connected together. Measured data are reproduced with permission from Sharp and Beauchamp (1969) and Sharp (1978).

For thin sheet materials that are closely connected, the internal losses tend to be frequency dependent and higher than for the individual plates; however, the effect is highly dependent upon the type of connection, air space, and the plates that are connected together (Trochidis, 1982). For two layers of 12.5 mm plasterboard that are screwed together, an increase in the internal loss factor of up to 50% can occur below the critical frequency. Internal losses near the critical frequency are more difficult to measure accurately, but they are still important because they affect the depth of the critical frequency dip.

4.3.2 Orthotropic plates

Orthotropic plates have two critical frequencies; one in each of the two orthogonal directions. Many plate materials are orthotropic to some degree. It is therefore important to identify when a plate can simply be modelled as an isotropic plate, and when it is necessary to take full account of their orthotropic nature. As a rule of thumb, if the critical frequencies are only a few one-third-octave-bands apart then the plate can often be treated as an isotropic plate using the effective bending stiffness; the errors are insignificant compared to the uncertainty in the radiation efficiency and the critical frequencies themselves. When the critical frequencies are separated by an octave or more, the effect on the sound reduction index is more significant. Infinite plate theory is now used as the starting point to investigate the effect of the two critical frequencies.

4.3.2.1 Infinite plate theory

Infinite plate theory for thin homogeneous isotropic plates with mass, stiffness, and damping can be extended to orthotropic plates (Heckl, 1960, 1981). The bending wave equation for an orthotropic plate forced into motion by sound pressure is given by

$$B_{p,x}\frac{\partial^4 \eta}{\partial x^4} + 2B_{p,eff}\frac{\partial^4 \eta}{\partial x^2 \partial y^2} + B_{p,y}\frac{\partial^4 \eta}{\partial y^4} + \rho_s\frac{\partial^2 \eta}{\partial t^2} = p(x,y,t) \tag{4.62}$$

assuming that $B_{p,xy} \approx B_{p,eff}$ (Section 2.3.3.2).

Substituting the plate displacement (Eq. 4.53) into the wave equation (Eq. 4.62) gives the surface impedance for an orthotropic plate as

$$Z_p = i\omega\rho_s\left[1 - \frac{B_{p,x}k_x^4}{\omega^2 \rho_s} - 2\frac{B_{p,eff}k_x^2 k_y^2}{\omega^2 \rho_s} - \frac{B_{p,y}k_y^4}{\omega^2 \rho_s}\right] \tag{4.63}$$

For an isotropic plate we focussed purely on the x-direction; this clearly cannot be done for an orthotropic plate. To determine the constants, k_x and k_y, we maintain use of the angle, θ, as the angle of incidence from the plate normal, and introduce ϕ as the angle between the x-axis and the projection of the wave number for the incident wave onto the $k_x k_y$ plane (see Fig. 4.19). Hence,

$$k_x = k\sin\theta\cos\phi$$
$$k_y = k\sin\theta\sin\phi \tag{4.64}$$

The surface impedance can now be found in terms of the bending wave number in the x- and y-directions by substituting Eq. 4.64 into Eq. 4.63,

$$Z_p = i\omega\rho_s\left[1 - \left(\frac{\cos^2\phi}{k_{B,x}^2} + \frac{\sin^2\phi}{k_{B,y}^2}\right)^2 k^4\sin^4\theta\right] \tag{4.65}$$

Using the same approach as for an isotropic infinite plate, the angle-dependent transmission coefficient for an infinite orthotropic plate with mass and stiffness can now be found by inserting the surface impedance (Eq. 4.65) into Eq. 4.59. To determine the diffuse incidence transmission coefficient, the integration over θ and ϕ is carried out according to (Heckl, 1960)

$$\tau_{\infty,d} = \frac{2}{\pi}\int_0^{\pi/2}\int_0^1 \frac{d(\sin^2\theta)d\phi}{\left|1 + \frac{Z_p\cos\theta}{2\rho_0 c_0}\right|^2} \tag{4.66}$$

The effect of damping is included in the surface impedance by using $B_{p,x}(1+i\eta)$ and $B_{p,y}(1+i\eta)$ to calculate the bending wave numbers. The integration in Eq. 4.66 gives a smooth curve up to frequencies just above the highest critical frequency, depending on the damping. Above this the curve is full of peaks and troughs due to the coincidence dips. To give a smooth curve above the highest critical frequency, the sound reduction index can be calculated using Eq. 4.61 for an infinite isotropic plate by replacing f_c with the effective critical frequency, $f_{c,eff}$.

Figure 4.20 compares an isotropic concrete plate to orthotropic versions of the same plate that have been created by increasing the stiffness in the x-dimension. The damping used in the calculation corresponds to a total loss factor that is representative of a finite-sized concrete slab connected to other masonry/concrete plates. Below the lowest critical frequency, the infinite plate acts as a limp mass and the prediction is made using the diffuse incidence mass

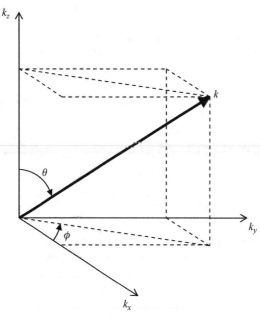

Figure 4.19

Angles relating the wave number of the incident sound wave to k_x, k_y, and k_z.

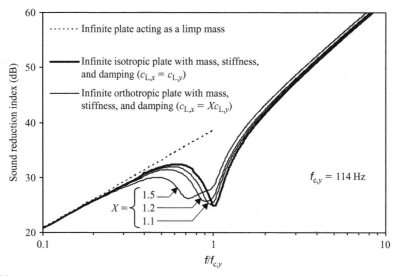

Figure 4.20

Sound reduction index (diffuse incidence) for a 150 mm thick infinite plate (mass, stiffness, and damping) modelled as being isotropic and orthotropic. The isotropic plate is based on the properties of cast *in situ* concrete. Orthotropic plates are created by keeping $c_{L,y}$ and the other properties the same as the isotropic plate, but by increasing $c_{L,x}$ to give different orthotropic plates. *Plate properties:* $h = 0.15$ m, $\rho_s = 330$ kg/m², $\nu = 0.2$, total loss factor $\eta = 0.005 + f^{-0.5}$. Isotropic plate: $c_L = c_{L,x} = c_{L,y} = 3800$ m/s.

law (Eqs 4.36 and 4.40). For 10% or 20% differences between $c_{L,x}$ and $c_{L,y}$ the coincidence dip is seen to broaden slightly. Such differences are quite common for masonry/concrete plates. When predicting the sound insulation it is reasonable to treat these orthotropic plates as isotropic by using the effective bending stiffness. This is in contrast to a 50% difference where there are two distinct critical frequency dips at $f_{c,x} = 76\,\text{Hz}$ and $f_{c,y} = 114\,\text{Hz}$; this feature will not be predicted when modelling the plate with an effective bending stiffness. The extended critical frequency region is defined by the lowest and highest critical frequencies. At and above the lowest critical frequency, the sound reduction index is affected by the plate damping. Above the highest critical frequency, the plate acts as an isotropic plate with a bending stiffness equal to the equivalent bending stiffness.

Many sheet materials used in lightweight walls and floors have low damping, and the coincidence dip is quite prominent. Whilst some sheet materials are significantly orthotropic, most sheets (whether isotropic or orthotropic) are fixed to a framework of studs, battens, or joists. It is often the frame that causes large changes to the bending stiffness so that they act as orthotropic plates. For the moment we will ignore the complexity of including the frame. As before, we consider an infinite isotropic plate, and then increase the stiffness in the x-dimension to make it orthotropic. The example in Fig. 4.21 assumes that the isotropic plate has similar properties to plasterboard. For differences in the stiffness that are larger than 50% there is a characteristic extended critical frequency region. For relatively small changes, such as 20%, the short slope in this critical frequency region runs in the opposite direction. It has previously been mentioned that it is difficult to accurately predict sound insulation at the critical frequency; the assumption that a plate is perfectly isotropic is one of the reasons.

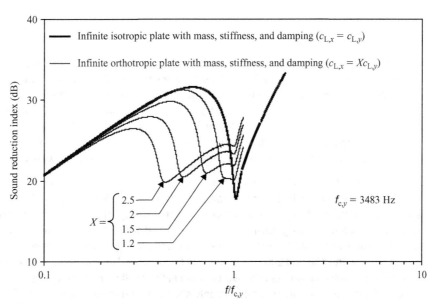

Figure 4.21

Sound reduction index (diffuse incidence) for a 12.5 mm thick infinite plate (mass, stiffness, and damping) modelled as being isotropic and orthotropic. The isotropic plate is based on the properties of plasterboard. Orthotropic plates are created by keeping $c_{L,y}$ and the other properties the same as the isotropic plate, but by increasing $c_{L,x}$ to give different orthotropic plates. Plate properties: $h = 0.0125\,\text{m}$, $\rho_s = 10.8\,\text{kg/m}^2$, $\nu = 0.3$, internal loss factor $\eta_{\text{int}} = 0.0141$. Isotropic plate: $c_L = c_{L,x} = c_{L,y} = 1490\,\text{m/s}$.

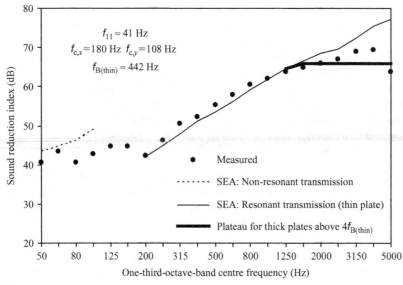

Figure 4.22

Sound reduction index for an orthotropic 240 mm masonry wall (solid blocks) with a 5 mm plaster finish (each side). Measurements according to ISO 140 Part 3. *Plate properties:* $L_x = 3.96$ m, $L_y = 3.00$ m, $h = 0.24$ m, $\rho_s = 430$ kg/m², $c_{L,x} = 1500$ m/s, $c_{L,y} = 2500$ m/s, $\nu = 0.2$, measured total loss factor. The plateau is calculated using material properties corresponding to the plate thickness: $c_L = 2500$ m/s and an internal loss factor $\eta_{int} = 0.01$. Measured data are reproduced with permission from Schmitz *et al.* (1999).

4.3.2.2 Masonry/concrete plates

Masonry walls and concrete floors with embedded reinforcement are usually orthotropic rather than isotropic. For cast *in situ* concrete floors and solid block/brick walls (horizontal and vertical joints mortared), the bending stiffness is not usually more than 20% higher in one direction than the other. However walls and floors can also be built from beams and/or blocks that are rigidly bonded together in one direction, but only touch or slot together in the other direction. The orthotropic nature of these plates is often more pronounced, but for many plates it is possible to use the effective bending stiffness to predict the sound insulation as if the plate were isotropic.

Examples for solid block walls with all joints mortared have already been seen in Section 4.3.1.3. Here we look at an example of a highly orthotropic masonry wall from Schmitz *et al.* (1999). The blocks are slotted together in the horizontal *x*-direction and are only rigidly connected with mortar in the vertical *y*-direction. This gives a bending stiffness that is 178% higher in the *y*-direction than in the *x*-direction. Figure 4.22 shows the measured and predicted data. The predicted critical frequencies are in the 100 and 200 Hz one-third-octave-bands. Non-resonant and resonant transmission are predicted using the effective bending stiffness; this gives an effective critical frequency of 139 Hz. Resonant transmission is only shown at frequencies above the effective critical frequency because of the complexity in predicting the radiation efficiency when there are two widely spaced critical frequencies, relatively few plate modes, and laboratory surfaces that form perpendicular baffles to the wall. Instead, the non-resonant transmission is shown; this gives a rough indication of the maximum sound reduction index that can occur at these low frequencies. Above the effective critical frequency, the predicted resonant transmission is in close agreement with the measurements up until $4f_{B(thin)}$. Above this frequency the plateau

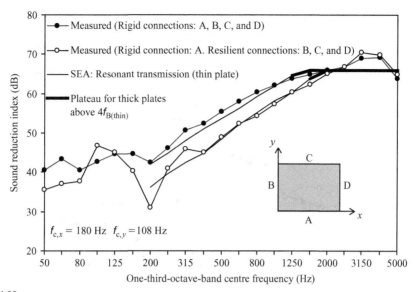

Figure 4.23

Orthotropic 240 mm masonry wall with rigid connections (mortar) or resilient connections (mineral fibre and sealant) at the plate boundaries. Measurements according to ISO 140 Part 3. Plate properties are described in Fig. 4.22. Measured data are reproduced with permission from Schmitz *et al.* (1999).

for the thick plate provides a reasonable estimate. The first thickness resonance occurs in the 5000 Hz one-third-octave-band for which the measurement shows a distinct dip.

The general shape of the extended critical frequency region corresponds to infinite plate theory (diffuse incidence) for orthotropic plates. However, even when the damping is incorporated using the total loss factor, this theory significantly underestimates the sound reduction index both in the critical frequency region and below it.

For the orthotropic masonry wall discussed above, all four boundaries were fixed with rigid mortar to the laboratory structure. Most walls in buildings are rigidly connected on at least two sides. The effect of reducing the number of rigidly connected boundaries and replacing them with resilient connections is to reduce the coupling losses from the plate. For isotropic plates, it is not particularly relevant which rigid connection is replaced by a resilient connection; however, the effect can be more marked with orthotropic plates. This can be seen in Fig. 4.23 by comparing the masonry wall with rigid connections to the same wall with resilient connections on three sides (Schmitz *et al.*, 1999). Compared to the rigidly connected wall, the sound reduction index is much lower between the highest critical frequency ($f_{c,x}$) and $4f_{B(thin)}$. This is due to the reduction in the total loss factor and is confirmed by the agreement between the measurements and the prediction for resonant transmission using the measured total loss factors. Concerning the critical frequencies, it is worth considering what would happen if we didn't know that the wall was orthotropic. The measured sound reduction index for the wall with resilient connections would then lead us to believe that there was only one critical frequency in the 200 Hz band; and we would infer from this that the wall was isotropic. With the resilient connections, the average coupling loss factor for the two plate boundaries that run parallel to the *x*-direction is higher than the average of those that run parallel to the *y*-direction. In combination with the low modal density and the low total loss factor, this causes the coincidence dip for $f_{c,x}$ to be

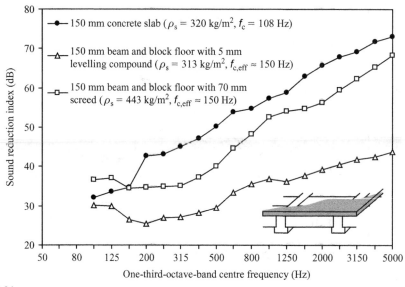

Figure 4.24

Comparison of a solid homogeneous concrete floor slab with an orthotropic beam and block floor (different surface finishes). Both blocks and beams are solid, and the only rigid material that bonds them together is the surface finish. The measured total loss factors for these three floors are within 2 dB of each other. Measurements according to ISO 140 Part 3. Measured data are reproduced with permission from ODPM and BRE.

more prominent than $f_{c,y}$. This indicates a degree of complexity with orthotropic plates when applying a laboratory measurement to *in situ* where the boundary conditions are different.

Some orthotropic plates are built from beams and blocks without any material such as mortar to rigidly bond them together within the cross-section of the plate. This occurs with some types of beam and block floor. These floors commonly have a surface finish such as a screed or levelling compound which means that they are only rigidly bonded together across their surface. For such floors, the assumption of a homogeneous orthotropic plate to predict the airborne sound insulation is not always appropriate. An example is shown in Fig. 4.24 comparing measurements on a solid homogenous concrete floor slab to one specific type of beam and block floor (solid beams and blocks) with different thicknesses of surface finish.

4.3.2.3 Masonry/concrete plates containing hollows

Many blocks, bricks, and slabs are not homogeneous and isotropic due to the hollows inside them. Hollows reduce the mass per unit area, cause orthotropic or anisotropic behaviour, and form cavities that may support one-, two-, or three-dimensional sound fields within the plate.

In concrete floor slabs with circular hollows, the sound field in the hollows usually has negligible effect on the direct transmission of airborne sound across the slab (Vinokur, 1995). Resonant transmission across heavyweight concrete slabs dominates over most of the building acoustics frequency range. The main effect of the hollows is to give a different bending stiffness in different directions compared to a solid slab, and to change the coupling losses to the supporting walls.

Apart from hollows with large volumes inside diaphragm walls, the cavities inside many masonry bricks and blocks are usually so small that they do not support modes over the

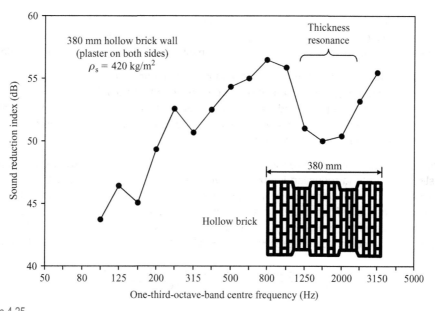

Figure 4.25

Measured sound reduction index for a hollow brick wall. Measurements according to ISO 140 Part 3. Measured data are reproduced with permission from Lang (1993).

majority of the building acoustics frequency range. Common examples are perforated bricks with narrow hollows that are used to achieve high levels of thermal insulation. There are a wide variety of designs for which the sound insulation is highly variable. This is mainly due to a deep dip at the lowest thickness resonance that typically occurs in the mid- or high-frequency range (e.g. see Fringuellino and Smith, 1999; Scholl and Weber, 1998). An example is shown in Fig. 4.25 (Lang, 1993). Unlike thickness resonances with solid masonry/concrete walls (Section 4.3.1.4), the frequency of the lowest thickness resonance is more difficult to predict but can be identified by measurements on single blocks (Weber and Bückle, 1998).

4.3.2.4 Profiled plates

Profiled plates are mainly used to increase the structural strength of thin sheet materials. Profiles make the bending stiffness of the plate much higher in the direction along the raised profile, than in the perpendicular direction. This results in different critical frequencies in the two orthogonal directions, sometimes remarkably different. Consequently, it is useful to estimate the critical frequencies of profiled plates from the bending stiffness for each direction. The bending stiffness calculation is described in Section 2.3.3.2.1. Profiled sheet materials are commonly encountered as metal cladding used for facades and roofing. As an example, consider a flat sheet of 0.65 mm thick steel with a critical frequency of $\approx$19 kHz. When this plate is given a trapezoidal profile commonly used for cladding, the lowest critical frequency may be as low as 150 Hz in the direction of the raised profiles. In the orthogonal direction, the critical frequency will be higher than for the flat sheet (i.e. >19 kHz). For these orthotropic plates, the extended critical frequency region covers the majority of the building acoustics frequency range.

When the two critical frequencies are widely spaced (but fall within the building acoustics frequency range) it is not useful to predict the sound insulation by modelling the profiled plate as an isotropic plate using the effective bending stiffness. This would give a single coincidence dip within the extended critical frequency region. However, this approach can be used to estimate the sound insulation at frequencies above the highest critical frequency of the profiled plate.

Below the lowest critical frequency, the sound reduction index for profiled plates with a low mass per unit area (e.g. metal cladding) can be assumed to be dominated by non-resonant transmission. An infinite plate model assuming diffuse incidence will overestimate the mass-law transmission; hence it is necessary to assume field incidence, or to use finite plate theory to calculate the non-resonant transmission.

In the extended critical frequency region, the model for an infinite orthotropic plate can be used to estimate the sound reduction index. However, using Eq. 4.66 underestimates the sound reduction index in the lower part of the extended critical frequency region. This is because diffuse incidence for infinite plates does not adequately represent non-resonant transmission across finite plates. To overcome this, the following approximation based on Eq. 4.66 can be used in the extended critical frequency region, $\text{Min}(f_{c,x}, f_{c,y}) < f < \text{Max}(f_{c,x}, f_{c,y})$ (Heckl, 1960)

$$\tau_{\infty,d} \approx \frac{\rho_0 c_0}{\pi \omega \rho_s} \frac{\text{Min}(f_{c,x}, f_{c,y})}{f} \left[\ln \left(\frac{4f}{\text{Min}(f_{c,x}, f_{c,y})} \right) \right]^2 \qquad (4.67)$$

This approximation provides a relatively smooth link from the field incidence mass law (infinite plate) below the lowest critical frequency to the infinite orthotropic plate model. Although the approximation ignores the effect of damping, this error can be considered as negligible in the context of other assumptions. More important is the assumption in the bending stiffness calculations for the profiled plate (Section 2.3.3.2.1) that allows the plate to be represented by

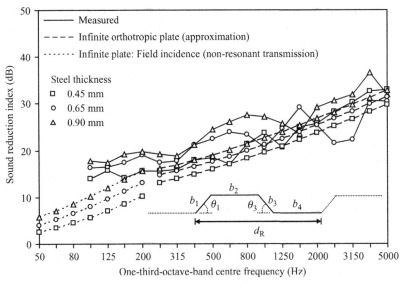

Figure 4.26

Measured sound reduction indices for profiled steel cladding of different thicknesses. Measurements according to ISO 140 Part 3. *Plate properties:* $L_x = 2.4$ m, $L_y = 3$ m, $\rho = 7800$ kg/m³, c_L(steel) $= 5270$ m/s, $v = 0.28$ (*Note:* x- and y-directions are defined so that the raised profiles run parallel to the x-direction). *Profile:* $b_1 = b_3 = 46.1$ mm, $b_2 = b_4 = 95$ mm, $\theta_1 = \theta_3 = 49.4°$, $\theta_2 = \theta_4 = 0°$, $d_R = 250$ mm. For all three thicknesses, $f_{c,x} = 216$ Hz, $f_{c,y} > 15$ kHz. Measured data are reproduced with permission from Lam and Windle (1995).

a flat infinite orthotropic plate. This assumption concerns the bending wavelength being much larger than the repetition distance, d_R, of the profile. When this assumption is not satisfied, the shape of the profile within the repetition distance will determine the mode shapes, and the radiation efficiency of these modes.

For profiled metal cladding with trapezoidal profiles, it is often found that $\lambda_{B,y} < d_R$, over most of the building acoustics frequency range; it is only in the low- and mid-frequency ranges that $\lambda_{B,eff} > d_R$. The effect of this can be seen in Fig. 4.26. This compares the infinite orthotropic plate approximation (Eq. 4.67) in the extended critical frequency region with measurements on different thickness plates with the same profile. Differences between measurements and predictions in the extended critical frequency region are generally less than 5 dB; even though it is only in the low- and mid-frequency ranges that $\lambda_{B,eff} > d_R$. However, one should be aware that the sound insulation of profiled plates is characterized by fluctuations that depend on the shape of the profile (Cederfeldt, 1974; Lam and Windle, 1995). Figure 4.27 gives an indication of the variation between different profiles for a single thickness of plate material. The infinite plate approximation takes no account of the plate modes that are determined by the shape of the profile. It is resonant transmission via these plate modes that causes significant dips in the sound reduction index at frequencies above the lowest critical frequency (see Fig. 4.27). To some extent it is possible to predict these fluctuations by using finite element methods to calculate the plate modes (Lam and Windle, 1995).

4.3.3 Low-frequency range

The definition of the sound reduction index gives the impression that the measured sound insulation of any test element can be described in such a way that it is independent of the source and receiving room, and can be referenced to the element area such that the result can be used for different sizes of the same test element. In the low-frequency range, simple scaling in terms of size is not always possible because non-resonant (mass law) transmission for finite plates depends on the plate dimensions, and resonant transmission can be highly variable when the rooms and/or the plates have low modal density. The assumption that the sound reduction index is independent of the source and receiving room tends to break down somewhere in the low-frequency range; most noticeably below 100 Hz. It is in this range that statistical descriptions of sound and vibration fields and the coupling between these fields no longer satisfy some of the assumptions made in SEA. On the basis that these fields can no longer be described in a statistical sense, it is assumed that the mode shapes and mode frequencies can be calculated exactly using deterministic models. Analytic descriptions of the modal sound and vibration fields or finite element methods can then be used to couple together the modes of the source and receiving rooms via a separating wall or floor (Gagliardini et al., 1991; Kropp et al., 1994; Maluski and Gibbs, 2000; Osipov et al., 1997; Pietrzyk and Kihlman, 1997). Deterministic models have mainly been used for a solid rectangular plate between two rectangular rooms to shed light on some of the factors that cause large variations in low-frequency airborne sound insulation for nominally identical separating walls, floors, or windows.

Parametric studies with deterministic models have been used to determine the sound reduction index of a solid homogeneous isotropic plate when placed between different sizes of room; these cover a range of room dimensions that can typically be found in dwellings. The results show that the sound reduction index in the low-frequency range is highly dependent on the source and receiving room dimensions (i.e. their modes) as well as the test element itself (Kropp et al., 1994; Osipov et al., 1997; Pietrzyk and Kihlman, 1997). This occurs whether

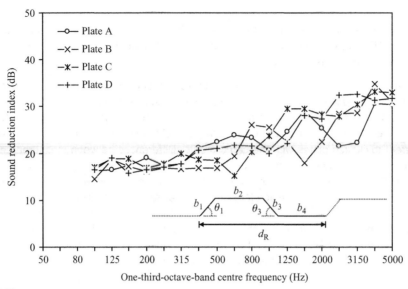

Figure 4.27

Measured sound reduction indices for 0.65 mm profiled steel cladding with different profiles. Measurements according to ISO 140 Part 3. *Plate properties: $L_x = 2.4$ m, $L_y = 3$ m, $\rho = 7800$ kg/m^3, c_L(steel) = 5270 m/s, $\nu = 0.28$. Profile for plate A: $b_1 = b_3 = 46.1$ mm, $b_2 = b_4 = 95$ mm, $\theta_1 = \theta_3 = 49.4°$, $\theta_2 = \theta_4 = 0°$, $d_R = 250$ mm ($f_{c,x} = 216$ Hz, $f_{c,y} \approx 20$ kHz). Profile for plate B: $b_1 = b_3 = 39.4$ mm, $b_2 = b_4 = 57$ mm, $\theta_1 = \theta_3 = 62.8°$, $\theta_2 = \theta_4 = 0°$, $d_R = 150$ mm ($f_{c,x} = 210$ Hz, $f_{c,y} \approx 21$ kHz). Profile for plate C: $b_1 = b_3 = 62.7$ mm, $b_2 = b_4 = 95$ mm, $\theta_1 = \theta_3 = 61.4°$, $\theta_2 = \theta_4 = 0°$, $d_R = 250$ mm ($f_{c,x} = 134$ Hz, $f_{c,y} \approx 21$ kHz). Profile for plate D: $b_1 = b_3 = 46.1$ mm, $b_2 = 95$ mm, $b_4 = 20$ mm, $\theta_1 = \theta_3 = 49.4°$, $\theta_2 = \theta_4 = 0°$, $d_R = 175$ mm ($f_{c,x} = 246$ Hz, $f_{c,y} \approx 21$ kHz). Measured data are reproduced with permission from Lam and Windle (1995).*

the dominant transmission mechanism is resonant or non-resonant (mass law) transmission. A study on a solid plate (30 kg/m^2) across which there was only non-resonant (mass law) transmission, gave a standard deviation of 3–5 dB with a range of approximately 15 dB for the sound reduction index (one-third-octave-bands) in the low-frequency range (Kropp *et al.*, 1994). Similar studies on resonant transmission across solid concrete plates gave higher standard deviations with values up to 10 dB (Osipov *et al.*, 1997; Pietrzyk and Kihlman, 1997). A transmission mechanism that was not included in these particular studies was the mass–spring–mass resonance frequency; this usually occurs in the low-frequency range with plate–cavity–plate constructions and is also likely to cause high standard deviations.

These parametric studies confirm earlier observations (Heckl and Seifert, 1958) that low-frequency airborne sound transmission tends to be lower between rooms with identical dimensions than between rooms with different dimensions. This can be attributed to strong coupling between identical room modes via the separating wall or floor. It is relatively easy to visualize this for identical axial modes in each room that are perpendicular to the separating wall or floor. For identical rooms a quick assessment of potentially adverse dips in the airborne sound insulation can be made by calculating the mode frequencies for the first few axial modes; dips may be exacerbated if they coincide with the mass–spring–mass resonance frequency of the separating wall or floor. However, it is not just identical room dimensions that cause low sound insulation. There can be other combinations of plate modes and room modes that are particularly well-coupled and result in dips in the sound insulation (Osipov *et al.*, 1997; Pietrzyk and Kihlman, 1997).

In the laboratory, accurate low-frequency measurements with good reproducibility can be achieved by using sound intensity as discussed in Section 3.5.1.1.1. This allows a fairer comparison of test elements. However, it does not overcome the difficulty in applying such measurements to estimate the airborne sound insulation *in situ* when the same element is installed between rooms with different room dimensions. For simple test elements such as solid plates there is the potential to predict the airborne sound insulation using deterministic models. However, this is not always straightforward. Using a single deterministic model assumes that it is possible to accurately determine the eigenfrequencies and eigenfunctions for a specific pair of rooms and a specific separating element. This takes no account of the uncertainty in describing their modal features. The modal overlap is not usually high, and local mode frequencies of the rooms and the plate(s) tend to be widely spaced apart. For this reason, any deterministic model will output a sound pressure level difference between the source and the receiving rooms that is characterized by peaks and troughs. Uncertainty in one mode frequency in the model can shift a peak or trough into an adjacent one-third-octave-band. Using only one deterministic model takes no account of any uncertainty. It assumes that the material properties of the wall and its boundary conditions are known in sufficient detail to accurately calculate the mode frequency and mode shape; this is rarely the case. It also assumes that the impedances of the room boundaries are known exactly, room temperatures are fixed, and that the source room will only be excited at one or more specific loudspeaker positions; this is rarely, if ever, the case. Comparisons of single deterministic models with a laboratory measurement often show good agreement, but there are almost always differences of up to 10 dB in an individual one-third-octave-band (Kropp *et al.*, 1994; Maluski and Gibbs, 2000; Osipov *et al.*, 1997). To use a deterministic model as a design tool it is necessary to account for uncertainty in the variables that determine the modal response (e.g. dimensions, damping boundary conditions, temperature). This can potentially be done using Monte-Carlo simulations, but the computation time may only be justifiable in critical applications such as the design of recording studios.

With attached dwellings it is quite common to have rooms with nominally identical dimensions on either side of the separating wall or floor. Theoretically this is the least favourable room layout from an acoustic point of view whilst it is the most practical from a construction viewpoint. There are a number of reasons why the adverse effects of identical room dimensions on sound insulation are not always apparent, identifiable, or predictable in the field. Firstly, adverse effects will not always be detected by field sound insulation measurements. These measurements are primarily intended for sound fields that can be considered as diffuse. Sound pressure levels are usually measured in the central zone of the room at certain minimum distances from the room boundaries. In small rooms with relatively few room modes, these positions tend to be on, or near the nodal planes of these modes. Therefore it is possible to measure a low spatial average sound pressure level in the source room, receiving room, or both rooms. Secondly, there is the question as to which sound reduction index (if any) can be empirically adjusted to calculate the sound insulation between rooms with identical dimensions in the field. A laboratory measurement on a separating element with low modal density will only apply to the size of test element and boundary conditions used in the laboratory; this assumes that the effect of the laboratory receiving room can be removed by using intensity measurements and effectively converting this room into a duct (Section 3.5.1.1.1). Thirdly, occupied rooms in different dwellings tend to be filled with large items of furniture in different ways; hence the room modes are unlikely to be identical in practice. The issue of identical axial modes may be more relevant to buildings such as hotels, where rooms are furnished in exactly the same way in a symmetrical fashion about the separating wall. Fourthly, the temperature is not usually exactly the same in each room so the mode frequencies will differ in each room; changing the temperature in either the source or receiving room by at least a few degrees centigrade can change the sound pressure level difference in a single one-third-octave-band by at least a few decibels in the low-frequency range (e.g. see Scholes, 1969). Fifthly, there are many different types of constructions used for separating walls and floors for which there are different sound transmission mechanisms dominating in the low-frequency range. Empirical corrections to the sound reduction index of the separating element have previously been attempted by splitting constructions into lightweight and heavyweight categories (Gibbs and Maluski, 2004). To predict the airborne sound insulation *in situ* where flanking transmission is present there are relatively few simple, quick, and practical options other than to use SEA or SEA-based models and make an allowance for uncertainty in the low-frequency range. This allowance can be made on an empirical basis from previous measurements, or with Monte-Carlo simulations using deterministic models.

4.3.4 Membranes

Roofs in commercial buildings and very large spaces such as arenas are sometimes formed from membranes using materials such as PTFE or ETFE. Each membrane typically has a mass per unit area less than $2\,kg/m^2$ and is under tension. For membranes it is not bending wave motion but wave motion due to tension of the membrane. The phase velocity, c_m, is independent of frequency and is given by (Morse and Ingard, 1968)

$$c_m = \sqrt{\frac{T}{\rho_s}}$$

(4.68)

where T is the tension per unit length around the edge of the membrane (N/m).

The modes of a membrane stretched over a rectangular frame are given by (Morse and Ingard, 1968)

$$f_{p,q} = \frac{c_m}{2}\sqrt{\left(\frac{p}{L_x}\right)^2 + \left(\frac{q}{L_y}\right)^2}$$

(4.69)

where p and q take positive integer values 1, 2, 3, etc.

A reasonable estimate for the airborne sound insulation of non-porous single layer membranes can be calculated using infinite plate theory for an infinite plate acting as a limp mass (Weber and Mehra, 2002). Practical roof constructions usually comprise two membranes separated by an air gap; however it is not suitable to predict the performance by assuming a plate–cavity–plate system.

The airborne sound insulation in the low-frequency range can be increased by loading a membrane with additional weights, and then increasing the tension. To reduce adverse effects in the mid-frequency range this membrane is then used as part of a double layer with a wide air gap as shown in Fig. 4.28 (Hashimoto *et al.*, 1996).

4.3.5 Plate–cavity–plate systems

We will now use SEA models to look at airborne sound transmission across a plate–cavity–plate system such as a cavity wall or floor. As with solid plates we will only consider bending wave

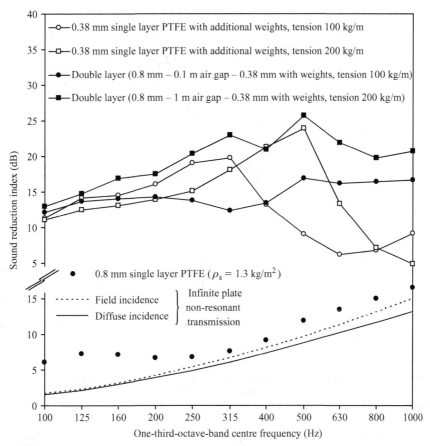

Figure 4.28

Measured sound reduction indices for different membrane roof elements. For the single layer membrane, infinite plate theory (non-resonant transmission) is shown for comparison with the measurement. Measured data are reproduced with permission from Hashimoto *et al.* (1996).

motion. The five-subsystem SEA model for a plate–cavity–plate system is shown in Fig. 4.29. Note that by limiting the model to five subsystems we are ignoring flanking transmission by assuming that vibration is transmitted from each plate to the rest of the building or laboratory structure, but not vice versa. Vibration can be transmitted between the two plates via structural connections as well as via the sound field in the cavity.

Structural connections often form an important transmission path. Examples of structural connections between plates include frameworks of beams, foundations, wall ties, and other plates that form flanking walls or floors. In some cases the structural coupling can simply be introduced as a coupling loss factor between the plates, in other cases it is necessary to expand the model by adding other beam and plate subsystems.

The sound field in the cavity is affected by the absorption within it, or around its boundaries. Cavities will generally be referred to as being empty (this includes cavities with an absorbent material around the perimeter of the cavity), or as being partly or fully filled with porous material (partly filled refers to filling across the cavity depth).

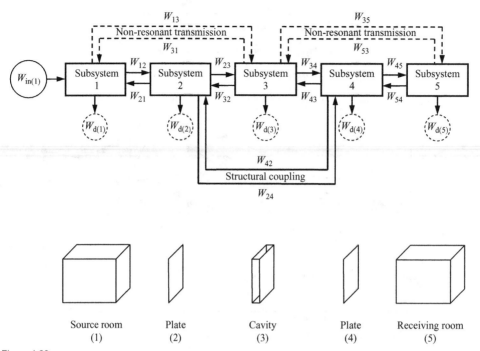

Figure 4.29

Five-subsystem SEA model for airborne sound transmission between a source room and a receiving room across a plate–cavity–plate system. Arrows with dashed lines represent non-resonant transmission, and arrows with solid lines represent resonant transmission.

Plate–cavity–plate systems superficially appear to be simple; they very rarely are. One single type of prediction model (based on either finite or infinite plates, deterministic or statistical approaches) is not usually able to accurately predict all important features of sound transmission for even quite simple cavity walls, floors, and windows. For design work, neither measurement nor prediction on their own will solve all problems. It is usually only in combination that sufficient insight into the transmission mechanisms can be gained. For laboratory measurements it is useful to have a model that can give some basic insight into the transmission process. This can be used to decide how the plate–cavity–plate system should be mounted in the laboratory to give a measurement that will be relevant *in situ*.

The advantage of SEA lies in the fact that structural coupling between the plates and flanking transmission are often important with cavity walls and floors *in situ*. Its disadvantages become more apparent when these two features are absent. SEA is not only useful for quantifying the airborne sound insulation of plate–cavity–plate systems in the laboratory and the field but also in qualitative discussions of the transmission mechanisms. Plates and cavities in buildings usually support modal behaviour over the majority of the building acoustics frequency range; but not always with high mode counts. Alternative models to SEA that are based on infinite plate theories or numerical methods can also be used to model a wide variety of plate–cavity–plate systems; although when compared with measurements the errors are similar to SEA. To be of practical use, plate–cavity–plate models based on infinite plates usually need to be modified in some way to deal with finite plate size (Villot *et al.*, 2001). The errors that occur with infinite plate models can be attributed to modal behaviour that is not considered in the model. In choosing an appropriate model we sometimes find that the mode counts in frequency bands are too low or too high to suit one single model over the entire building acoustics frequency range. SEA is by no means suited to modelling every plate–cavity–plate system over this frequency range; however, the process of trying to make a system fit an SEA model usually sheds light on the key features that need to be considered in the design.

This section uses three examples of plate–cavity–plate systems to form a basis from which the various transmission mechanisms can be discussed. Before we look at these examples it is necessary to consider another type of non-resonant transmission between two rooms connected via a plate–cavity–plate system. This is modelled by assuming that the gas in the cavity acts as a spring and the plates act as lump masses. The result is a mass–spring–mass resonance frequency that often falls in the low-frequency range. Plate–cavity–plate models based on infinite plates very usefully show that below the mass–spring–mass resonance frequency the two plates effectively act as a single plate with the combined mass per unit area and thickness of the two individual plates (Beranek and Work, 1949; Fahy, 1985; London, 1950). Hence the mass–spring–mass resonance frequency can be used to estimate the frequency below which the cavity wall acts as a single plate, and above which it acts as a plate–cavity–plate system.

4.3.5.1 Mass–spring–mass resonance

In the low-frequency range there is a non-resonant transmission mechanism across a cavity wall or floor that can be modelled by treating the gas contained within an empty cavity as a spring and each plate as a lump mass. It is assumed that the plates forming the cavity are non-porous or have a sufficiently high airflow resistivity that the gas does not enter the plates when it undergoes compression. For any gas in an enclosed cavity with a depth, L_z, the dynamic stiffness per unit area of the gas, s'_g, is

$$s'_g = \frac{K}{L_z} = \frac{\gamma P}{L_z} \tag{4.70}$$

where K is the bulk compression modulus of the gas.

The mass–spring–mass resonance frequency has already been calculated using an equivalent circuit in Section 3.11.3.2.1. To use Eq. 3.200 with a plate–cavity–plate system it is only necessary to replace the stiffness with the stiffness per unit area, and each mass with the mass per unit area. This gives

$$f_{msm} = \frac{1}{2\pi} \sqrt{\frac{s'_g}{\left(\frac{\rho_{s1}\rho_{s2}}{\rho_{s1} + \rho_{s2}}\right)}} \tag{4.71}$$

where ρ_{s1} and ρ_{s2} are the mass per unit area of plates 1 and 2 respectively.

For air-filled cavities (adiabatic compression, $\gamma = 1.4$), the dynamic stiffness per unit area of the enclosed air, s'_a, is given by

$$s'_a = \frac{1.4P_0}{L_z} = \frac{\rho_0 c_0^2}{L_z} \tag{4.72}$$

so the mass–spring–mass resonance frequency for air at 20°C is

$$f_{msm} = 60 \sqrt{\frac{\rho_{s1} + \rho_{s2}}{\rho_{s1}\rho_{s2}L_z}} \tag{4.73}$$

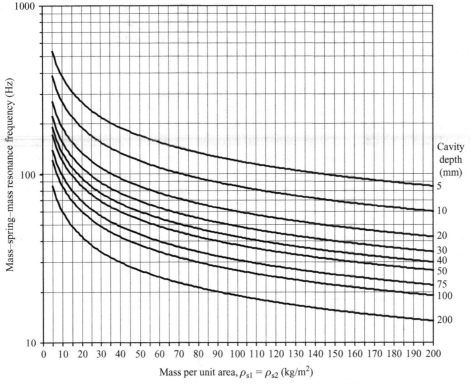

Figure 4.30

Mass–spring–mass resonance frequencies for a plate–cavity–plate system where the cavity is filled with air, and both plates have the same mass per unit area.

Figure 4.30 shows calculated resonance frequencies for air-filled cavities (Eq. 4.73) where both plates have the same mass per unit area.

When a cavity is partly filled across its depth with a porous material that has a porosity close to unity (e.g. mineral wool), then Eq. 4.73 can also be used to estimate the resonance frequency. If the porous material has a low porosity, the cavity depth, L_z, can be replaced by an effective cavity depth, $L_{z,eff}$, given by

$$L_{z,eff} = (L_z - d) + \phi d \qquad (4.74)$$

where d is the thickness of the porous material and ϕ is the porosity of the porous material.

For fully filled cavities the plates can be strongly coupled together by the porous material and the plate may need to be modelled as a sandwich panel. If so, the simple lumped element model will no longer be appropriate. However, in some cases an estimate of the resonance frequency can be calculated if the dynamic stiffness of the porous material is considered as the combined stiffness of the enclosed air and the skeletal frame of the material, s'. This can be measured in the same way as for resilient materials used under floating floors if the static load does not significantly alter the dynamic stiffness (Section 3.11.3.1). The mass–spring–mass resonance frequency can then be calculated using Eq. 4.71 by replacing s'_g with s'. In some cases the dynamic stiffness of the skeletal frame may be negligible and the resonance frequency will only be determined by the air contained within the porous material. This occurs with very low-density mineral wool used in fully filled cavity walls for which it is necessary to reconsider the assumption of adiabatic compression for the enclosed air. For cavities filled with fibrous materials, adiabatic compression is only a reasonable assumption in the high-frequency range. The mass–spring–mass resonance usually occurs in the low-frequency range where isothermal compression ($\gamma = 1.0$) occurs due to heat conduction by the fibres, hence

$$f_{msm} = 51 \sqrt{\frac{\rho_{s1} + \rho_{s2}}{\rho_{s1} \rho_{s2} L_z}} \tag{4.75}$$

Laboratory measurements on simple plate–cavity–plate systems (i.e. no framework other than around the perimeter of the plate) show that the mass–spring–mass resonance for a cavity that is fully filled with a fibrous material tends to occur at a lower frequency than with an empty cavity (e.g. see Nightingale, 1999). For fully filled cavities it is possible to treat the porous material as an equivalent gas and to calculate the bulk compression modulus for Eq. 4.70 by assuming an appropriate model for the porous material (Allard, 1993). However, there are limits as to how useful it is to pursue more accurate calculations of the mass–spring–mass resonance frequency. Many plates and cavities support modal behaviour in the vicinity of the resonance frequency; therefore the lumped element model assuming simple masses and a spring becomes less appropriate. Lightweight walls and floors, as well as some lightweight linings on heavyweight wall and floor bases, require a frame to support the sheet material. The type of frame, stud spacing, and screw spacing can change the resonance frequency by constraining and stiffening the sheet material; this means it can no longer be treated as a lump mass. Measurements on plasterboard cavity walls indicate that the resonance frequency is often highest for timber studs with closely spaced screws and lowest for light steel studs with widely spaced screws; reducing the stud spacing tends to increase the frequency at which resonance occurs (Quirt *et al.*, 1995).

Cavity dimensions differ widely between various plate–cavity–plate systems. When the mass–spring–mass resonance frequency is below the lowest cavity mode there is usually a marked dip in the airborne sound insulation due to the resonance. It is hard to generalize when it occurs above the lowest cavity mode. In this situation the air is not acting as a lump spring and the mass–spring–mass resonance frequency can be regarded as a transition point from single plate behaviour to plate–cavity–plate behaviour. However, this may still introduce a significant dip, a step, or change in slope in the airborne sound insulation curve.

When designing plate–cavity–plate systems, the aim is usually to choose the cavity depth and the mass per unit area of the plates so that the mass–spring–mass resonance falls well-below the important frequency range. One good reason to do this is due to a lack of models that can accurately predict the frequency range over which the dip will occur, and how deep any dip will be. The risk of including the resonance frequency within the important frequency range is usually only taken when measurements are available to assess the adverse effects (if any).

Some example laboratory measurements on plate–cavity–plate systems with distinct mass–spring–mass resonance dips are shown in Fig. 4.31. The dip is typically less than 8 dB when compared to the level in the adjacent one-third-octave-bands.

Modelling the air in the cavity as a spring can potentially be incorporated into an SEA model over a wide frequency range by coupling the two plates together using the air stiffness (Brekke, 1981; Craik and Wilson, 1995). The stiffness of the air spring will decrease with increasing cavity depth which can result in an increase in the airborne sound insulation. This mechanism can be included as a form of structural coupling that directly couples plate subsystems 2 and

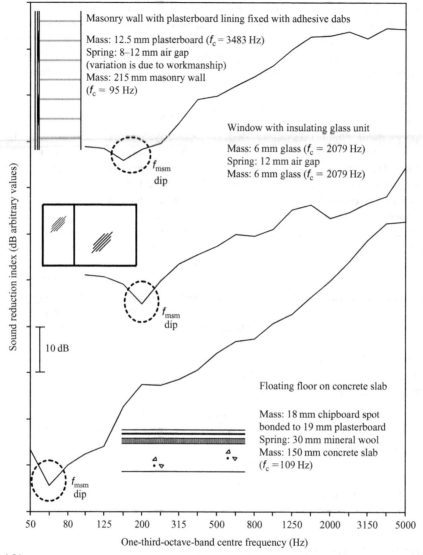

Figure 4.31

Examples of plate–cavity–plate systems with mass–spring–mass resonance frequencies in the building acoustics frequency range. Measured data are reproduced with permission from ODPM and BRE.

4 together. Unfortunately it is difficult to identify when an air spring is, or is not appropriate for a particular plate–cavity–plate system. In addition it relies on a semi-empirical approach to determine the coupling loss factor (Brekke, 1981; Craik and Wilson, 1995). These factors make it rather awkward to use, hence it is not considered in the following sections.

4.3.5.1.1 Helmholtz resonators

To reduce the adverse effect of the mass–spring–mass resonance on the sound insulation, attempts have been made to reduce the depth of the associated dip using arrays of Helmholtz

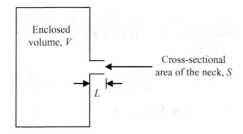

Figure 4.32

Helmholtz resonator.

resonators within the plate–cavity–plate system. These are tuned to resonate and absorb sound energy at or near the mass–spring–mass resonance frequency. For windows with secondary glazing, resonators can be placed around the perimeter of the cavity within the supporting cavity wall to avoid obstructing the line of sight (Enger and Vigran, 1985). For cavity walls or wall linings the resonators can be incorporated into one or both of the plates on the side that faces into the cavity (Mason and Fahy, 1988; Warnock, 1993).

The cross-section through a typical Helmholtz resonator is shown in Fig. 4.32. At frequencies where the wavelength is much larger than the dimensions of the resonator, the slug of air contained within the neck acts as a lump mass and the larger volume of air in the cavity acts as a spring. Hence its resonance frequency is given by the mass–spring resonance,

$$f = \frac{1}{2\pi}\sqrt{\frac{k}{m}} = \frac{c_0}{2\pi}\sqrt{\frac{S}{V(L + \Delta L)}} \tag{4.76}$$

where ΔL is an end correction to define the effective mass of the slug of air (Ingard, 1953). For a slit-shaped neck with rectangular cross-section, $\Delta L \approx 0.4\,S^{0.5}$ (Junger, 1975).

There are two main reasons why resonators are not widely used. Both theory and practice show that if the resonators are tuned at or near the mass–spring–mass resonance frequency it is possible to remove the dip in the sound insulation, or at least reduce its depth. However, this is usually at the expense of slightly reducing the sound insulation at other frequencies. There are also practical issues. If workmanship is poor, the use of resonators may introduce air paths or structural coupling (i.e. bridging) which can negate any improvement in the sound insulation at the mass–spring–mass resonance.

An alternative approach to try and reduce the mass–spring–mass resonance dip is through active noise control (e.g. see Jakob and Möser, 2003; Kårekull, 2004). In practice, the cost and complexity is often prohibitive for use in buildings.

4.3.5.2 Using the five-subsystem SEA model

This section shows how the five-subsystem SEA model can be used to help make design decisions. Three examples are used to illustrate the basic principles and common assumptions from which it should be clear how changes to the design of a building element can sometimes be made on a qualitative basis without the need for calculations. Details concerning the calculation of structural coupling, and sound transmission into and out of the cavity are included after the examples so that the initial focus is on the application and interpretation of the SEA model.

As with the three-subsystem SEA model for solid plates, we can calculate sound reduction indices for the various transmission paths. There is one purely non-resonant path, $1 \rightarrow 3 \rightarrow 5$, for which the sound reduction index is

$$R_{1 \rightarrow 3 \rightarrow 5} = 10 \lg \left(\frac{\eta_3 \eta_5}{\eta_{13} \eta_{35}} \frac{V_5}{V_1} \right) + 10 \lg \left(\frac{S}{A} \right) \tag{4.77}$$

Resonant transmission paths include $1 \rightarrow 2 \rightarrow 3 \rightarrow 4 \rightarrow 5$ and $1 \rightarrow 2 \rightarrow 4 \rightarrow 5$ for which path analysis gives

$$R_{1 \rightarrow 2 \rightarrow 3 \rightarrow 4 \rightarrow 5} = 10 \lg \left(\frac{\eta_2 \eta_3 \eta_4 \eta_5}{\eta_{12} \eta_{23} \eta_{34} \eta_{45}} \frac{V_5}{V_1} \right) + 10 \lg \left(\frac{S}{A} \right) \tag{4.78}$$

and

$$R_{1 \rightarrow 2 \rightarrow 4 \rightarrow 5} = 10 \lg \left(\frac{\eta_2 \eta_4 \eta_5}{\eta_{12} \eta_{24} \eta_{45}} \frac{V_5}{V_1} \right) + 10 \lg \left(\frac{S}{A} \right) \tag{4.79}$$

When there is more than one type of structural connection between the two plates that does not require modelling as a separate subsystem (e.g. wall ties and foundations), then Eq. 4.79 can be used for each connection separately or the individual structural coupling loss factors can be summed to give η_{24}; note that they need to be summed to calculate the general matrix solution (Eq. 4.10).

The sound reduction index for other paths can be calculated in the same way, such as those with a combination of resonant and non-resonant transmission, e.g. $1 \rightarrow 2 \rightarrow 3 \rightarrow 5$ and $1 \rightarrow 3 \rightarrow 4 \rightarrow 5$.

4.3.5.2.1 Windows: secondary glazing

This first example concerns two sheets of glass separated by a cavity depth of at least 50 mm. The model could form the basis for an internal observation window between two rooms in a building, or curtain walling, or domestic secondary glazing to improve the façade sound insulation. Section 4.3.12.3 will discuss the reasons why the five-subsystem SEA model is not appropriate for modelling insulating glass units, i.e. sealed units with cavity depths typically less than 20 mm. This example uses a rather wide observation window (3×1.25 m) with 6 or 10 mm glass on either side of a 50, 100, or 200 mm cavity; this is chosen so that the cavity has sufficient local modes in the low-frequency range. Note that for secondary glazing used with typical sizes of windows in the façades of dwellings, a lack of cavity modes often restricts the SEA model to the mid- and high-frequency ranges. The mass–spring–mass resonance is estimated to fall in or below the 100 Hz one-third-octave-band. Hence with only one cavity mode below the 100 Hz band we will only model at and above the 125 Hz band.

Each pane of glass is usually supported on one leaf of a cavity wall. For specialist applications such as a studio observation window, the cavity reveal lining will rarely be continuous across the cavity. It is therefore assumed that transmission via the reveal lining or the supporting cavity wall is insignificant so that the structural coupling path $1 \rightarrow 2 \rightarrow 4 \rightarrow 5$ can be ignored. In practice it is the air paths around the frame and the structural paths due to bridging across the reveal that limit the achievable sound insulation.

Figure 4.33 shows path analysis and the predicted sound reduction index using the matrix solution. In this example the four transmission paths shown on the graph can be summed using Eq. 4.19 to give approximately the same result as the matrix solution. Below the lower of the two critical frequencies the non-resonant path $1 \rightarrow 3 \rightarrow 5$ is generally the strongest, and

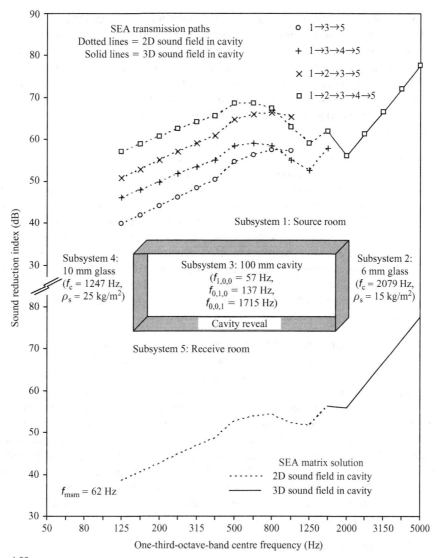

Figure 4.33

SEA model of a wide observation window with a 100 mm deep cavity between panes of 6 and 10 mm glass. Path analysis is shown in the top part of the figure and the matrix solution in the lower part. *Plate properties*: $L_x = 3$ m, $L_y = 1.25$ m, $c_L = 5200$ m/s, $\nu = 0.24$, radiation efficiency uses method No. 2 (lower limit only), total loss factor $\eta = 0.024 + \sum \eta_{ij}$, $\alpha = 0.02$ (for the plate surfaces facing into the cavity at and above $f_{0,0,1}$ in the cavity). *Cavity properties*: Empty cavity, $L_x = 3$ m, $L_y = 1.25$ m. Reveal properties: $\bar{\alpha}_P = 0.05$ (125–400 Hz), $\bar{\alpha}_P = 0.1$ (500–5000 Hz).

the resonant path $1 \rightarrow 2 \rightarrow 3 \rightarrow 4 \rightarrow 5$ is the weakest; but there is no dominant path. Above the highest critical frequency of the plates there is only the resonant path $1 \rightarrow 2 \rightarrow 3 \rightarrow 4 \rightarrow 5$. As there is no structural coupling, all paths involve the cavity. The cavity sound field is two dimensional below $f_{0,0,1}$ and covers a large part of the building acoustics frequency range; hence increasing the reveal absorption increases the sound reduction index. This can be seen by the slight step in the curve that occurs between 400 and 500 Hz where the absorption coefficient increases. It is not uncommon for reveal materials to have significantly different absorption coefficients in adjacent one-third-octave-bands and such steps may be more distinct in practice. In this example the critical frequency for the 10 mm glass occurs when the cavity sound field is two dimensional. Therefore increasing the reveal absorption can be used to try and reduce the depth of the critical frequency dip. The critical frequency for the 6 mm glass occurs when the sound field is three dimensional; hence the possibilities for significantly changing the absorption are more limited because it is the two panes of glass that form the largest surfaces in the cavity. Both critical frequencies are evident as dips in the resulting sound reduction index.

Figure 4.33 shows that paths involving non-resonant transmission into and out of the cavity play an important role in determining the overall sound insulation. For such constructions it is important to note that there is more than one approach to predict the non-resonant coupling loss factor between rooms and cavities. The calculations used to create Fig. 4.33 are based on the approach of Price and Crocker (1970). Below the first cross-cavity mode this approach shows no increase in the sound reduction index with increasing cavity depth; this is at odds with experimental evidence (e.g. see Brekke, 1981; Mulholland, 1971; Nightingale, 1999). An alternative SEA model to Price and Crocker for non-resonant transmission into and out of cavities is given by Craik (2003); this takes account of the cavity depth. Both of these approaches are described in Section 4.3.5.3. Due to the importance of non-resonant transmission for this particular plate–cavity–plate system we will now look at both of them.

Figure 4.34a uses the approach of Price and Crocker to compare the above result with different windows that have 6 mm glass on each side of a 50, 100, or 200 mm cavity (i.e. 6–50–6, 6–100–6, 6–200–6). Below the first cross-cavity mode the sound reduction index is the same for the three different cavity depths with 6 mm glass on each side. At and above the first cross-cavity mode they differ due to the three-dimensional nature of the sound field in the cavity; the cavity total loss factor changes as well as the coupling loss factors into and out of the cavity. Identical critical frequencies for both plates result in a deep dip in the sound reduction index at the critical frequency. Using different glass thicknesses such as 6–100–10 gives two shallower critical frequency dips. Compared to 6–100–6, the higher mass per unit area of the 10 mm glass in 6–100–10 reduces the strength of the non-resonant transmission paths $1 \rightarrow 3 \rightarrow 5$ and $1 \rightarrow 2 \rightarrow 3 \rightarrow 5$. This results in 6–100–10 having higher sound insulation than 6–100–6 below the critical frequency dip of the 10 mm glass.

Figure 4.34b uses the approach of Price and Crocker at and above the first cross-cavity resonance and the approach of Craik below the first cross-cavity resonance. For 6–50–6, 6–100–6, and 6–200–6 in the low- and mid-frequency ranges (where non-resonant transmission is dominant) there is a distinct increase in the sound reduction index with increasing cavity depth.

The significant differences between Fig. 4.34a and b below the first cross-cavity resonance indicate the difficulty in making design decisions with an SEA model when non-resonant transmission is dominant. As yet there is insufficient evidence to choose one or the other approach for all plate–cavity–plate systems. Fortunately, when there are dominant structural coupling paths the uncertainty in predicting the non-resonant path becomes less important. The difficulty in accurately predicting sound radiation at the critical frequency has already been noted with solid plates. This problem is exacerbated in lightweight plate–cavity–plate systems where the first cross-cavity resonance and the critical frequencies often occur within a few frequency bands of each other. At these frequencies, accurately predicting differences due to different cavity depths is not usually possible and reliance tends to be placed on laboratory measurements.

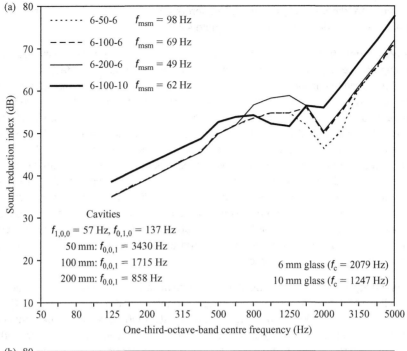

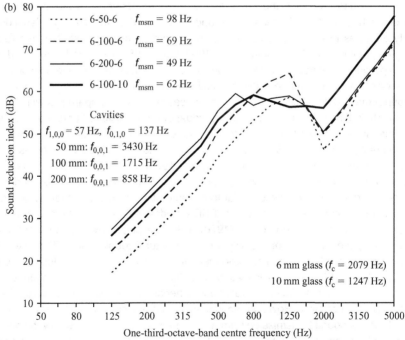

Figure 4.34

SEA model (matrix solution) for different combinations of 6 and 10 mm glass and different cavity depths (50, 100, and 200 mm): (a) using the approach of Price and Crocker to model non-resonant transmission into and out of the cavity and (b) using the approach of Price and Crocker at and above the first cross-cavity resonance, and the approach of Craik below the first cross-cavity resonance. Plate and cavity properties are given in Fig. 4.33.

Because there are different ways of modelling non-resonant transmission, it is useful to briefly discuss alternative models for plate–cavity–plate systems.

Alternative models (infinite plates): An alternative model for the plate–cavity–plate system uses an impedance approach that accounts for the above-mentioned effect of cavity depth. This is based on finite thickness plates and a finite depth cavity for a system of infinite extent (Beranek and Work, 1949). The model works well for sound at normal incidence when the cavity is partly or fully filled with a porous material. However, for sound transmission between rooms we are more interested in diffuse sound fields in the source room. The impedance model has been adapted to diffuse incidence for plate–cavity–plate systems with cavities that are either empty, partly filled, or fully filled with a porous material (Fahy, 1985; Novak, 1992; Ookura and Saito, 1978). Impedance models are ideally suited to cavities that are fully filled with absorbent material; it is then reasonable to assume that any sound waves propagating in the cavity (parallel to the plates) are so highly attenuated that there is no need to consider cavity modes. For empty cavities at frequencies above the lowest cavity mode (which often occurs in the low-frequency range) the errors tend to be larger than with fully filled cavities because the cavity modes are ignored.

Most plate–cavity–plate systems in buildings have cavity depths between 50 and 300 mm; for thinner cavities the plates are often tightly coupled together by the air in the cavity. For cavity depths between 50 and 300 mm we now consider existing laboratory measurements on systems with empty cavities and negligible structural coupling. These indicate that in any individual one-third-octave-band between the mass–spring–mass resonance and the first cross-cavity mode, there is the potential to increase the sound insulation by up to 10 dB by at least doubling the cavity depth as long as these cavity depths are within the range 50–300 mm (e.g. see Brekke, 1981; Mulholland, 1971; Nightingale, 1999; Quirt, 1982; Utley and Mulholland, 1968). For design purposes it is useful to identify such improvements. However, we need to consider the accuracy of impedance models (diffuse incidence sound field) at frequencies between the mass–spring–mass resonance and the first cross-cavity mode. Comparisons with laboratory measurements indicate that differences in individual one-third-octave-bands between predictions and measurements are typically up to 10 dB for empty cavities or cavities partly filled with fibrous materials, and up to 5 dB for cavities that are almost fully filled with fibrous materials (Nightingale, 1999; Novak, 1992; Ookura and Saito, 1978). Hence although impedance models indicate the correct trend of increasing sound insulation with increasing cavity depth, the error in predicting absolute values is similar to the potential improvement that is being sought. Similarly we should consider the accuracy of SEA models below the first cross-cavity mode. The original work by Price and Crocker (1970) was on a plate–cavity–plate system formed from aluminium plates and an empty cavity without structural coupling; the difference between one-third-octave-band SEA predictions and measurements of the sound reduction index was generally less than 2 dB. Almost all validation of the theory on building elements tends to be on plate–cavity–plate systems with structural coupling because most constructions require structural connections for stability. However, an indication of the accuracy can be found by looking at the prediction of sound transmission from a room into an empty cavity below the first cross-cavity mode where the cavity sound field is two dimensional. In such situations the differences between predictions and measurements for the cavity sound pressure level are typically up to 5 dB (Craik and Smith, 2000a; Hopkins, 1997; Wilson, 1992). The existing evidence does not seem to indicate that one model (SEA or impedance approach) will always be significantly 'better' than the other.

Neither SEA nor impedance models are perfectly suited to predicting the change in sound insulation due to increasing the cavity depth below the first cross-cavity mode. It is fortunate that structurally isolated plates forming cavity walls and floors are not common in buildings. For most plate–cavity–plate systems there will be structural coupling as well as flanking transmission, therefore errors in the non-resonant path of an SEA model often become less important due to the existence of many other transmission paths. To accurately quantify the effect of changing the cavity depth it is often necessary to use carefully designed laboratory measurements that avoid confounding factors such as niche effects, unwanted structural coupling, and changes in the total loss factor of the plates at different positions in the test aperture.

4.3.5.2.2 Masonry cavity wall

In this example a masonry cavity wall is modelled in three different scenarios. Although it is possible to create an SEA model for this wall in a specific laboratory or in a specific building we will deliberately chose an example that is a step removed from the reality *in situ* and in the laboratory. This allows us to look at features that are important to both. (Note that a comparison of SEA with measurements on a masonry cavity wall construction will be shown in Section 5.3.2.2 for the combination of direct and flanking transmission.)

Scenario (A) has no structural connections between the plates so that all sound transmission occurs via the cavity, scenario (B) has wall ties connecting the plates, and scenario (C) has wall ties and a foundation connecting the plates. As with SEA modelling of solid plates in Section 4.3.1.3, we use the total loss factor to model each plate as if it were connected to other walls and floors without actually including them in the model. This is done by treating the sum of these coupling loss factors (estimated to be $0.3f^{-0.5}$) as an internal loss factor. When modelling a separating wall *in situ* where the connected flanking walls and floors are included in the model there is no need to make this adjustment. Non-resonant transmission into and out of the cavity is modelled using the approach of Price and Crocker.

The predicted sound reduction indices using the SEA matrix solution are shown in Fig. 4.35a for scenarios A, B, and C. Path analysis is shown for scenarios A and C to help assess the strength of different paths by comparing them with the matrix solution in Fig. 4.35b.

Scenario A has a mass–spring–mass resonance frequency that is well-below 50 Hz and outside of the building acoustics frequency range. It is common to design masonry cavity walls to have resonance frequencies below the frequency range of interest. This is done to reap the benefits of higher sound insulation due to the separation between the wall leaves; we recall that below the resonance frequency a cavity wall effectively acts as a single solid wall. As all transmission paths involve the cavity, the sound reduction index will change when the cavity total loss factor is changed; hence the addition of absorption in the cavity will increase the sound reduction index (and vice versa). Below the first cross-cavity mode the sound field in the cavity is two dimensional so absorption could either be placed around the perimeter of the cavity, or the cavity could be partially or fully filled with absorbent material across its depth. In a transmission suite the absorption around the cavity perimeter may vary between laboratories unless specific material is introduced around the perimeter as part of the test element.

Scenario A gives an unrealistically high sound reduction index because there are no structural connections or flanking transmission. Note that above the critical frequency, the matrix solution gives the same sound reduction index as path $1 \rightarrow 2 \rightarrow 3 \rightarrow 4 \rightarrow 5$.

(a)

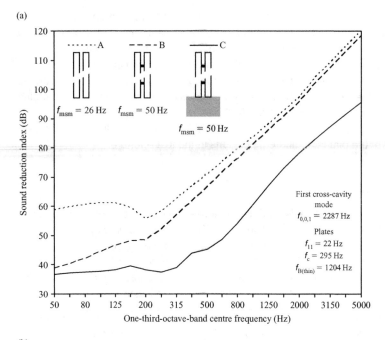

(b)

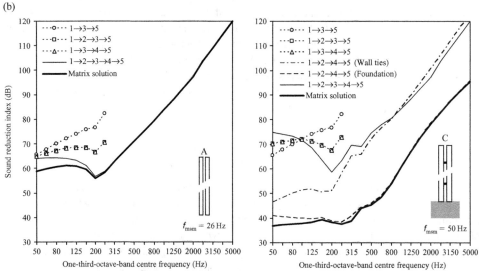

Figure 4.35

SEA model of a masonry cavity wall: (a) matrix solution and (b) path analysis for comparison with the matrix solution. *Plate properties:* $L_x = 4\,\text{m}$, $L_y = 2.5\,\text{m}$, $h = 0.1\,\text{m}$, $\rho_s = 140\,\text{kg/m}^2$, $c_L = 2200\,\text{m/s}$, $\nu = 0.2$, radiation efficiency uses method No. 3, total loss factor $\eta = 0.01 + 0.3\,f^{-0.5} + \sum \eta_{ij}$ (Note that $0.3f^{-0.5}$ is being used here to represent the sum of the structural coupling losses to connected walls and floors that are not included in the model.). *Cavity properties:* Empty cavity, $L_x = 4\,\text{m}$, $L_y = 2.5\,\text{m}$, $L_z = 0.075\,\text{m}$, $T = 0.3\,\text{s}$. *Wall ties:* 2.5 ties/m², $k = s_{75\text{mm}} = 2 \times 10^6\,\text{N/m}$. *Foundation:* $d_1 = 0.25\,\text{m}$, $d_2 = 0.6\,\text{m}$, $\rho = 2000\,\text{kg/m}^3$, $s'_{\text{soil}} = 1.96 \times 10^9\,\text{N/m}^3$, $\eta_{\text{soil}} = 0.96$.

The addition of wall ties gives scenario B where the sound reduction index is now significantly reduced in the low-frequency range. The combined mass–spring–mass resonance frequency of the air spring and the wall ties is 50 Hz so in practice there could be a slight dip in the sound insulation at this frequency. This mass–spring–mass transmission mechanism is not included in the SEA model; if important, it is easiest to include it as an empirical adjustment.

Scenario C is created by adding a common foundation (concrete) where the ground is stiff clay with stones. This reduces the sound reduction index across the entire building acoustics frequency range. Note that this foundation detail has deliberately been chosen to give strong foundation coupling so that scenarios A and C are indicative of the extremes. Different foundation details will be discussed in Section 4.3.5.4.3 and the wave theory used in these calculations will be given in Section 5.2.4.

Measurements on masonry cavity walls in the laboratory are not easy to interpret in terms of their performance *in situ*. We recall that with solid homogeneous walls we could convert a result from one laboratory to another laboratory by using the measured total loss factor (refer back to Section 3.5.1.3.2). With plate–cavity–plate systems there is more than one path involving resonant transmission; hence there are no simple conversions. The engineer is caught between a rock and a hard place; laboratory measurements are important because 'perfect' theoretical models do not exist, yet it is difficult to apply the laboratory measurement without the aid of a theoretical model. Various tactics have been used to try and overcome this difficulty. One possibility is to establish rule-of-thumb conversions from a specific mounting condition in the laboratory to a specific situation in buildings. This may be possible when masonry cavity walls are not rigidly connected to the foundations; they are sometimes built off resilient materials so the transmission path via the foundations may not be as important. In addition, structural coupling via the foundations may dominate on the ground floor of a multi-storey building but not several floors above it. In some transmission suites it is possible to build foundations below the aperture. Otherwise, if the aperture is sufficiently high a foundation can be built within the aperture and shielded with linings (Parmanen *et al.*, 1988). Another approach is to use a flanking laboratory to test the combination of the separating cavity wall and some of the flanking walls and floors to try and simulate the actual building. Note that transmission via the foundations can be affected by the underlying soil; hence there are limitations to building representative foundations in a transmission suite or a flanking laboratory.

4.3.5.2.3 Timber joist floor

This example looks at sound transmission through a basic timber joist floor. This commonly forms an internal floor within a dwelling, and its sound insulation often requires upgrading in a building that is being converted into flats. In this case an SEA model is used to form a basis upon which decisions can be made on how to improve the sound insulation.

Figure 4.36 shows the timber joist floor along with path analysis and the predicted sound reduction index using the matrix solution. The five-subsystem model has been used with a single subsystem to model transmission via all the floor cavities (subsystem 3) and an extra subsystem has been used to model structural coupling via all the joists (subsystem 6). The structural coupling between the chipboard/plasterboard and the joists is modelled by assuming point connections at screws/nails. Note that when modelling structure-borne sound excitation of lightweight walls and floors it may be necessary to model every cavity and every joist as an individual subsystem.

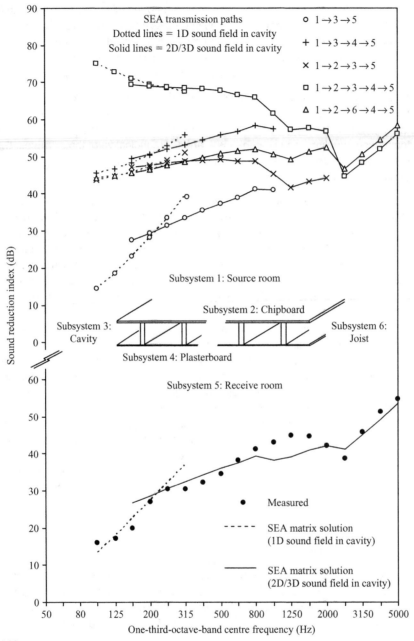

Figure 4.36

SEA model of a timber joist floor (matrix solution and path analysis). *Plate properties:* $L_x = 5.45$ m, $L_y = 4.15$ m, radiation efficiency uses method No. 2 (lower limit only). Chipboard: $h = 0.025$ m, $\rho_s = 19$ kg/m^2, $c_L = 2210$ m/s, $\nu = 0.3$. Plasterboard: $h = 0.0125$ m, $\rho_s = 9.7$ kg/m^2, $c_L = 1910$ m/s, $\nu = 0.3$. *Beam properties:* 225 mm deep timber joists spaced at 450 mm centres with screw/nail fixings at ≈ 500 mm centres (modelled as point connections). $L_x = 4.15$ m, $h_y = 0.05$ m, $h_z = 0.225$ m. Fundamental beam mode, $f_1 = 31$ Hz. *Cavity properties:* Empty cavity, $L_x = 4.15$ m, $L_y = 0.4$ m, $L_z = 0.225$ m. Cavity modes that demarcate one-, two-, and three-dimensional sound fields are $f_{1,0,0} = 41$ Hz, $f_{0,1,0} = 429$ Hz, $f_{0,0,1} = 762$ Hz. Mass–spring–mass resonance frequency: 50 Hz. Total loss factors are predicted from the sum of the coupling loss factors and the internal loss factor. Measured data from Hopkins are reproduced with permission from ODPM and BRE.

The mass–spring–mass resonance frequency is 50 Hz and the lowest cavity mode is 41 Hz, hence the plate–cavity–plate system is used across the entire building acoustics frequency range. Above 41 Hz the cavity has a one-dimensional sound field which changes to a two-dimensional field at 429 Hz, and to a three-dimensional field at 762 Hz. Although each element of the floor can support local modes, one-dimensional subsystems such as the cavity (below the 400 Hz band) and the beam have quite low mode counts; we can refer back to the examples for a similar cavity and joist in Figs 1.55 and 2.26 respectively. The lower part of Fig. 4.36 shows measured data for comparison with the SEA matrix solution. For the one-dimensional sound field, non-resonant transmission into and out of the cavity is predicted using the approach of Craik. For the two- and three-dimensional sound fields it is modelled using the approach of Price and Crocker. These approaches are described in Section 4.3.5.3. An abrupt transition between models is not usually appropriate, so the overlap between them is shown. The agreement between the measured and predicted sound insulation indicates that it is now appropriate to draw conclusions from the path analysis shown in the upper part of Fig. 4.36.

In the low- and mid-frequency range, non-resonant transmission via the path $1 \rightarrow 3 \rightarrow 5$ tends to dominate. Therefore increasing the airborne sound insulation would require increasing the mass per unit area of the chipboard or the plasterboard, and/or putting highly absorptive material in the cavity. However, the cavity dimensions also play a role because the sound field is one dimensional and there are relatively few modes. For this reason the low-frequency sound insulation in individual frequency bands can change significantly with different joist depth, spacing, and length; these vary depending on the structural requirements. The model is too crude to predict these changes accurately, but they can be observed in laboratory measurements on timber floors. Note that timber frame walls usually have much shallower cavities than these floors and the mass–spring–mass resonance often plays an important role in the low-frequency range.

In the high-frequency range, paths $1 \rightarrow 2 \rightarrow 3 \rightarrow 5$, $1 \rightarrow 2 \rightarrow 6 \rightarrow 4 \rightarrow 5$ and $1 \rightarrow 2 \rightarrow 3 \rightarrow 4 \rightarrow 5$ mainly determine the sound insulation. Sound transmission via paths that involve the sound field in the cavity could be reduced with absorption in the cavity. However, structural coupling via the joists along path $1 \rightarrow 2 \rightarrow 6 \rightarrow 4 \rightarrow 5$ will limit the maximum achievable sound insulation. Hence separating timber floors between dwellings also tend to require resilient devices such as hangers, bars, or channels to isolate the plasterboard from the joists thus reducing transmission via this structural path.

4.3.5.3 Sound transmission into and out of cavities

Resonant and non-resonant coupling loss factors for plates and rooms were introduced in Section 4.3.1. For sound radiation from a plate into a space, use of the radiation efficiency equations assumes that the plate is radiating into a free field. This free-field assumption also applies to non-resonant transmission between two spaces. For rooms these assumptions are appropriate. However, in comparison with rooms, cavity depths are very shallow. It is therefore questionable whether the free-field assumption is appropriate for cavities. For resonant transmission, comparisons of measured and predicted data indicate that the assumption is reasonable for all types of plates (e.g. see Craik and Smith, 2000a; Price and Crocker, 1970; Wilson, 1992). Some experiments also indicate that it is reasonable for non-resonant transmission across a plate into a cavity when the other plate that forms the cavity is sufficiently rigid that there is negligible interaction with the sound field in the cavity (Craik and Smith, 2000a). Examples of this include cavities between plasterboard and non-porous masonry/concrete plates. Note that for plate–cavity–plate systems where the resonant transmission path $1 \rightarrow 2 \rightarrow$ structural coupling $\rightarrow 4 \rightarrow 5$ is dominant, accurate modelling of transmission into and out of the cavity becomes less important. For lightweight cavity walls and floors, where neither plate can be considered rigid, an alternative approach can be used to predict non-resonant transmission from the cavity to the room and vice versa (Craik, 2003).

The approach that was originally used by Price and Crocker (1970) is described now because it is still a useful starting point for many models. For resonant transmission, the coupling loss factor from the plate to the cavity is calculated using Eq. 4.21. The coupling loss factor from the cavity to the plate can then be calculated with the consistency relationship (Eq. 4.2) by using the modal density of the cavity. For non-resonant transmission, the coupling loss factor is calculated from the room to the cavity using Eq. 4.26, and the consistency relationship is used for the reverse direction.

Cavities often have one- or two-dimensional sound fields in the low- and mid-frequency ranges where all wave motion is parallel to the plate surfaces. From Craik (2003), an alternative model to that of Price and Crocker (1970) can be used to predict non-resonant transmission via a plate that connects a cavity (i) to a room (j) below the first cross-cavity mode. This model assumes that the sound pressure associated with waves in the cavity (propagating parallel to the plate) cause the plate to move with a lateral velocity that is only determined by its mass per unit area, hence

$$\hat{v}^2 = \frac{\hat{p}_i^2}{\omega^2 \rho_s^2} \tag{4.80}$$

The phase velocity of the structural wave induced in the plate must follow that of the sound wave in the cavity, so from Eq. 2.198 the radiated power from the plate can be given as

$$W_{ij} = \sigma_{f_c} S \rho_0 c_0 \langle v^2 \rangle_{t,s} \tag{4.81}$$

where S is the surface area of the plate that faces into the cavity and σ_{f_c} is the radiation efficiency of the plate at the critical frequency (i.e. the frequency at which the phase velocities for sound in the cavity and the structural wave in the plate are equal).

Equations 4.3, 4.80, and 4.81 can now be used to determine the coupling loss factor from the cavity to the room using

$$W_{ij} = \omega \eta_{ij} E_i = \frac{\sigma_{f_c} S \rho_0 c_0}{\omega^2 \rho_s^2} \langle p_i^2 \rangle_{t,s} \tag{4.82}$$

hence, the coupling loss factor at frequencies below the first cross-cavity mode is given by (Craik, 2003)

$$\eta_{ij} = \frac{\rho_0^2 c_0^3 S \sigma_{f_c}}{\omega^3 V_i \rho_s^2} \tag{4.83}$$

Note: This assumes that ρ_0 and c_0 are the same in the cavity and the room.

The coupling loss factor for non-resonant transmission from the room to the cavity can be calculated from the consistency relationship. To cover the entire building acoustics frequency range it will sometimes be necessary to switch over to the Price and Crocker approach in the vicinity of the first cross-cavity mode.

4.3.5.4 *Structural coupling*

There is always more than one way to model structural coupling between two parts of a structure; and in the absence of a model there is usually more than one way to measure it. In any assessment of a plate–cavity–plate structure it is useful to consider two aspects of structural coupling separately. The first aspect concerns connections between two plates over their surface such as a framework, resilient mounts, or wall ties that connect the two plates together (i.e. connections that would exist whether the plate was modelled as being of finite or infinite size). The second aspect concerns the connections around the boundaries of a finite plate; these may be different in the laboratory and *in situ*.

4.3.5.4.1 Point connections between plates and/or beams

Point connections between parallel plates, between plates and beams, and between beams are very common. Plates can be connected together by wall ties, dabs of adhesive, resilient mounts, or bridged screeds. Plates and beams are often connected with screws, nails, or bolts. We start by modelling a single point connection. There is usually more than one connection so we will need to look at the assumptions made when using this model for many point connections. When point connections are closely spaced along a line then the appropriateness of the point connection model depends on the wavelength of the structure-borne sound waves.

Simple models to quantify the power flow across a single point connection make use of the driving-point mobility (or impedance) of the plates or beams, and the mobility (or impedance) of the point connection (Cremer *et al.*, 1973; Lyon and DeJong, 1995). These are conveniently calculated using equivalent circuits based on impedance or mobility (e.g. see Harris and Crede, 1976). The following approach is general and has been used for resilient mounts under floating floors (Vér, 1971), wall ties between masonry cavity walls (Craik and Wilson, 1995; Narang, 1994) and screw/nail connections between plasterboard and a timber frame (Craik and Smith, 2000b).

We assume that there is a propagating bending wave on a beam or plate (denoted as subsystem i) as shown in Fig. 4.37. This is coupled to another beam or plate (subsystem j) by a single point connection. The propagating bending wave velocity far from the point connection on plate i is v_0, but the presence of the point connection changes the velocity directly above this connection to v_i. This results in a velocity, v_j, directly above the connection on the other beam or plate. The axial force, F, on the connection is proportional to $v_i - v_j$. To determine the power transmitted by the point connection, the equivalent circuit in Fig. 4.37 yields the following relationships for F and v_j,

$$F = \frac{v_0}{Y_i + Y_j + Y_c}$$

$$v_j = FY_j = \frac{v_0 Y_j}{Y_i + Y_j + Y_c}$$

(4.84)

where Y_i and Y_j are the driving-point mobilities and Y_c is the mobility of the point connector.

The transmitted power is then given by

$$W_{ij} = \frac{1}{2} \mathrm{Re}\left\{ F v_j^* \right\} = \frac{v_0^2}{2} \frac{\mathrm{Re}\left\{ Y_j \right\}}{\left| Y_i + Y_j + Y_c \right|^2}$$

(4.85)

473

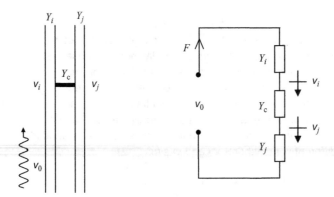

Figure 4.37

Point connection between two subsystems *i* and *j* and the equivalent circuit. Each subsystem could be a plate or a beam. The illustration shows a distinct gap between the two structures; this would apply to wall ties in a cavity wall, or a resilient mount between a floating floor and a base floor. However the model can also be used for screw, nail, or bolt connections between plates and/or beams for which this gap will be very small.

which can be written in terms of the coupling loss factor as

$$W_{ij} = \omega \eta_{ij} E_i = \omega \eta_{ij} m_i \frac{v_0^2}{2} \qquad (4.86)$$

where m_i is the mass of subsystem *i*.

For *N* identical point connections, equating Eqs 4.85 and 4.86 gives the coupling loss factor as

$$\eta_{ij} = \frac{N}{\omega m_i} \frac{\text{Re}\{Y_j\}}{|Y_i + Y_j + Y_c|^2} \qquad (4.87)$$

To calculate the coupling loss factor, the beams and plates can be modelled as being of infinite extent; this allows the impedances in Section 2.8.3 to be used to calculate the driving-point mobilities.

This approach assumes that the power transmitted across the connection is only due to forces and that any power transmitted by moments is negligible. It also assumes that each connection acts independently of the others; hence there must be no correlation in the vibration field above the connection points. Correlation could occur with an evenly spaced array of point connections. However, even if the intention is to construct a wall or floor with a perfectly periodic array of point connections, very few walls and floors are built so exactly and have spatially uniform material properties; therefore the assumption of uncorrelated points is often reasonable.

For a resilient point connection such as a resilient mount or a wall tie, it may be appropriate to model it as a simple linear spring for which

$$Y_c = \frac{i\omega}{k} \qquad (4.88)$$

where *k* is the dynamic stiffness of the point connection acting as a spring (N/m). Note that for wall ties, *k* is identical to s_{Xmm} as described in Section 3.11.3.2.

For rigid point connections such as screws, nails, or bolts, the stiffness can be assumed to be infinite and $Y_c = 0$.

For plates connected by springs across an air-filled cavity (such as cavity walls with ties, or floating floors on resilient mounts) a mass–spring–mass resonance frequency can be calculated for the air stiffness and the spring stiffness acting in parallel,

$$f_{msm} = \frac{1}{2\pi} \sqrt{\frac{s_a' + \frac{N}{S}k}{\left(\frac{\rho_{s1}\rho_{s2}}{\rho_{s1} + \rho_{s2}}\right)}} \tag{4.89}$$

where S is the plate area.

Below this mass–spring–mass resonance frequency the two plates effectively act as a single plate with the combined mass per unit area and thickness of the two individual plates.

Real point connections are not always simple: springs may be non-linear, structures can be offset from each other at the connection point, the effective contact area of point connections varies depending on the structural wavelength, and the connection may support wave motion across its length at high frequencies. These complexities become apparent in laboratory studies on isolated structures and result in more complex models or empirical solutions for specific connectors. In the absence of such information, the simplicity of the impedance/mobility approach is ideal for initial estimates.

Full-size walls and floors that are only rigidly or resiliently connected along their edges at equally spaced points to form L-, T-, and X-junctions are less common in buildings. For these junctions, other models are available that can be used to determine the coupling loss factors (Bosmans, 1998; Bosmans and Vermeir, 1997).

For a beam that is connected to a plate using screws, nails, or bolts the point connection model is appropriate when the structural wavelength is much smaller than the distance between the connections. When the wavelength is much larger than this distance, the junction between the beam and plate acts as a line connection. As a rule of thumb for bending waves on lightweight timber frame walls and floors, the transition from a line connection to individual point connections can be assumed to start when the screw/nail spacing is approximately equal to half a bending wavelength (Craik and Smith, 2000b). Hence the line connection model is sometimes needed in the low-frequency range with a changeover to the point connection model at higher frequencies. This simplification does not deal with the effective contact length of point connections which varies with frequency and makes it difficult to identify a transition frequency without using more complex models (Bosmans and Nightingale, 2001). When there is uncertainty in whether to assume line or point connections, it is simplest to model them both to give an indication of the potential range.

4.3.5.4.2 Line connections

Line connections occur between plates, and between plates and beams. Examples include lightweight walls and floors with a framework, and masonry/concrete cavity walls where there is a return at a wall boundary. Line junctions occur with lightweight cavity walls and floors where the plates on either side of the wall are connected to the same beam (see Fig. 4.38). For a line of closely spaced point connections (e.g. screws, bolts), the line connection model is appropriate when the spacing is much smaller than the bending wavelength (as discussed in Section 4.3.5.4.1).

Figure 4.38

Examples of line connections between plates in a plate–cavity–plate system.

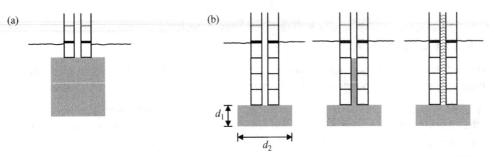

Figure 4.39

Examples of different foundation details for masonry/concrete cavity walls. The ground floor is not shown; this could be built in to each cavity leaf and either lie on the ground or be suspended above it. (a) Concrete deep trench fill: Trench is filled to within a few tens of centimetres of the ground surface. The trench may be up to 3.5 m below ground level. (b) Strip footing: Wall extends below the damp proof course to rest upon a strip concrete footing (e.g. $d_1 \approx 0.25\,\text{m}$, $d_2 \approx 0.6\,\text{m}$). Below ground level there are a number of different options: thicker or different blocks, wall ties, poured concrete infill, and thermal insulation within the cavity.

For line connections between parallel plates it is also possible to use an equivalent circuit approach in the same way as with point connections (Sharp, 1978). However it is usually necessary to adopt a wave approach. For line connections such as those shown in Fig. 4.38, a comprehensive overview for bending wave transmission along with tabulated values are given in the book by Craik (1996). It is sometimes necessary to consider both bending and in-plane waves for which other models are available (Craik and Smith, 2000b; Langley and Heron, 1990). Vibration transmission between connected plates is discussed in Section 5.2.

4.3.5.4.3 *Masonry/concrete walls: foundations*

Structural coupling via the foundations is often an important transmission path at the boundary of a masonry/concrete cavity wall. Different foundation details are often used for the same cavity wall depending on the soil conditions and the structural requirements; a few different foundation details are shown in Fig. 4.39. The variety and complexity of foundation details has implications for the interpretation of laboratory measurements on cavity walls.

To include structural coupling via the foundations into the five-subsystem SEA model, it is possible to rely purely on measurements. Structural intensity or vibration level differences can be used to determine the coupling loss factor as described in Section 3.12.3.3. Note that it is not always possible to use vibration level differences if there are other significant transmission paths such as those involving the cavity or wall ties. By measuring the structural coupling, the foundation detail is effectively treated as a black box; bending waves enter the black box on one leaf, and emerge on the other leaf of the cavity wall. Example measurements are shown

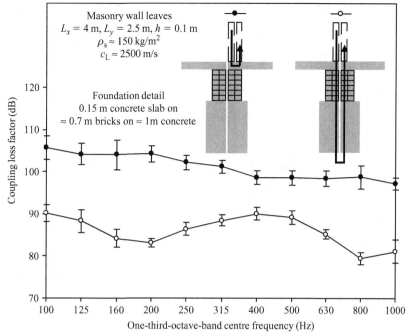

Figure 4.40

Measured coupling loss factors for vibration transmission via a split foundation (two separate foundations separated by a 50 mm gap), and on one side of the split foundation. Measurements were taken using structural intensity and are shown with 95% confidence intervals. Note that the coupling loss factor depends upon the junction length with the foundation and the plate area. Measured data from Hopkins are reproduced with permission from ODPM, BRE, and BRE Trust.

in Fig. 4.40 for a masonry cavity wall placed in two different positions on a split foundation in a flanking laboratory. Whilst measurements are essential for quantifying the performance of complicated foundations, reliance on measured data is not particularly desirable as there are many possible foundation details and many different types of soil. To give an indication of whether different soils are likely to be a significant factor it is necessary to use a theoretical model for vibration transmission via the foundation.

Complete models of many foundation details are made complicated by the fact that they are built from several components that don't form simple beams or plates, and that are highly damped by the soil. In addition, these components are not all rigidly bonded together because they must also provide thermal insulation and prevent ingress of moisture. However, a model for bending wave transmission via a simplified strip foundation is useful in highlighting the role of soil stiffness in this structural transmission path (Wilson and Craik, 1995). This model is described in Section 5.2.4 from which Fig. 4.41 shows the calculated coupling loss factors for three different soils (soil properties are given in Table 5.1). The peaks are due to the soil being modelled as a lump spring; in practice the soil is likely to be anisotropic so the coupling loss factors should be used to identify general trends rather than specific resonance peaks. The measured and predicted coupling in this section indicates that for masonry cavity walls there can be differences up to 10 dB with different foundation details. Earlier analysis in Section 4.3.5.2.2 indicates that the transmission path via the foundations

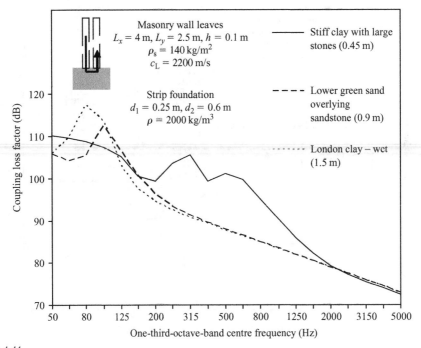

Figure 4.41

Predicted coupling loss factors for vibration transmission via a strip foundation with different soils.

is sufficiently strong that different foundation details and different soils are likely to account for some of the variation in airborne sound insulation between nominally identical cavity walls *in situ*.

4.3.5.4.4   *Lightweight cavity walls*

For many lightweight cavity walls, the complexity in predicting the structural coupling between the two wall leaves means that reliance is generally placed upon extensive laboratory testing (e.g. see Walker, 1993). As with laboratory measurements of masonry cavity walls, it is necessary to consider the absence or inclusion of structural coupling around the perimeter of a lightweight cavity wall. For laboratory measurements, the relevant Standard (ISO 140 Part 1) gives requirements on the laboratory structure that forms the test aperture. Previously there had been no specific requirements for the dimensions and properties of the materials that form the test aperture when testing lightweight cavity walls; during this time an inter-laboratory comparison of airborne sound insulation was carried out on such walls (Fausti *et al.*, 1999). A sample of the results are shown in Fig. 4.42 (Smith *et al.*, 1999). These indicate that structural coupling around the perimeter can significantly alter the airborne sound insulation over the majority of the building acoustics frequency range. It is worth noting that it is generally easier to achieve a high degree of isolation between wall leaves in the laboratory than *in situ*. Rigid connections between wall leaves around the perimeter of a wall or floor *in situ* may be necessary for fire stopping; these can significantly reduce the sound insulation unless materials are carefully chosen to minimize the structural coupling (Craik *et al.*, 1997).

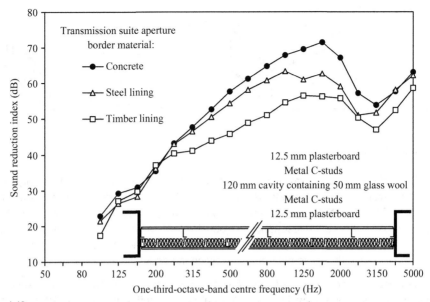

Figure 4.42

Airborne sound insulation of a lightweight cavity wall measured in laboratories with different border materials around the perimeter of the aperture. Measurements according to ISO 140 Part 3. Measured data are reproduced with permission from Smith *et al.* (1999).

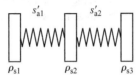

Figure 4.43

Mass–spring–mass–spring–mass system.

4.3.5.5 Plate–cavity–plate–cavity–plate systems

For a plate–cavity–plate–cavity–plate system modelled with lump masses and springs as shown in Fig. 4.43, there are two resonance frequencies to consider (Blevins, 1979)

$$f_{msmsm} = \frac{1}{2^{3/2}\pi}\sqrt{X \pm \sqrt{X^2 - 4s'_{a1}s'_{a2}\left(\frac{1}{\rho_{s1}\rho_{s2}} + \frac{1}{\rho_{s2}\rho_{s3}} + \frac{1}{\rho_{s1}\rho_{s3}}\right)}}$$

where

$$X = \frac{s'_{a1} + s'_{a2}}{\rho_{s2}} + \frac{s'_{a1}}{\rho_{s1}} + \frac{s'_{a2}}{\rho_{s3}} \qquad (4.90)$$

where the dynamic stiffness per unit area for each air-filled cavity is given by Eq. 4.72.

The lower resonance frequency is calculated using the negative sign in Eq. 4.90. Note that the frequency range between the two resonances can span one or several one-third-octave-bands.

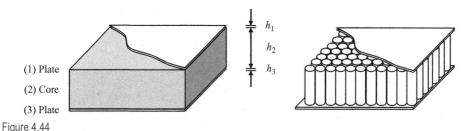

(1) Plate
(2) Core
(3) Plate

Figure 4.44

Example sandwich panels.

The SEA model for a plate–cavity–plate system can be extended to a plate–cavity–plate–cavity–plate system. An SEA model is appropriate above the higher of the two resonance frequencies. One application is triple glazing in studios which often has individual cavity depths of at least 100 mm; hence the resonance frequencies tend to be in or below the low-frequency range. With increasing frequency, transmission paths involving structural coupling between the plates will start to dominate over paths involving the sound fields in the cavities.

For three closely spaced sheets of glass it is only possible to draw limited conclusions from laboratory measurements in which the individual cavity depths are at most 20 mm and the lowest resonance falls within the frequency range of interest. These only tend to show a discernible dip in the sound insulation at the lower resonance frequency and generally offer small improvements over what can be achieved with double glazing (Brekke, 1981; Quirt, 1983). However, it has been shown that the sound insulation between the resonance frequencies can be higher when the ratio of the lower to the higher resonance is much less than unity and the layout of the plates and cavities is not symmetrical about the central plate (Vinokur, 1990).

4.3.6 Sandwich panels

Sandwich panels are sometimes used to form internal walls, external walls, doors, or roof elements. They usually consist of a relatively lightweight core material with a plate (e.g. plasterboard, cement particle board, steel) bonded to this core on each side (see Fig. 4.44). A variety of core materials are used such as foam, mineral wool, or a cardboard honeycomb. Initial design considerations are usually given to the structural and/or thermal performance so the properties of the core may end up being isotropic, orthotropic, or anisotropic, with or without rigid connections between the two plates across the core.

Early work by Kurtze and Watters (1959) was concerned with identifying a type of plate that would have a high static stiffness and low weight, but without the high bending stiffness that causes resonant transmission to become as dominant as it does with solid plates. Their sandwich panel design was based on 'promoting' transverse shear motion rather than bending wave motion. This required a core with low shear stiffness, and thin and relatively stiff plates on either side. Wave motion on the panel was described by combining the panel bending stiffness and shear stiffness in parallel. The resulting motion is advantageous because of the non-dispersive nature of transverse shear waves. It extends the frequency range over which the phase velocity for wave motion on the panel is lower than the phase velocity in air. Therefore by moving the lowest coincidence frequency to higher frequencies, the panel can act as a limp mass over a larger part of the frequency range. Later work identified adverse resonance effects relating to thickness deformation of the core (Ford et al., 1967). In practice this limits the ability to

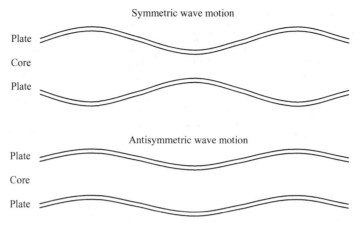

Symmetric wave motion

Plate

Core

Plate

Antisymmetric wave motion

Plate

Core

Plate

Figure 4.45

Sandwich panel – wave motion.

achieve the non-resonant (mass law) sound reduction index at all frequencies below coincidence. Sandwich panels are commonly used in automotive or aeronautic industries because their acoustic performance can usually be optimized for a specific noise source (e.g. see Makris et al., 1986; Moore and Lyon, 1991; Wen-chao and Chung-fai, 1998).

For porous core materials the sound reduction index of a sandwich panel can be calculated using infinite plate assumptions and Biot theory to model two compressional waves and a shear wave propagating in the core (Bolton et al., 1996; Lauriks et al., 1992).

For a non-porous core with propagating shear and dilatational waves, resonant transmission across a sandwich panel can be modelled in terms of symmetric and antisymmetric wave motion on the panel (Dym et al., 1974, 1976; Moore and Lyon, 1991). Symmetric and antisymmetric motions, correspond to dilatational and bending wave motion respectively (see Fig. 4.45). Symmetric motion involves thickness deformation of the core. This gives rise to a dilatational resonance that usually causes an adverse dip in the sound reduction index. For a sandwich panel where both plates are identical, the dilatational resonance frequency, f_d, is given by (Moore and Lyon, 1991)

$$f_d = \frac{1}{2\pi} \sqrt{\frac{\left[\dfrac{2(\lambda_2 + 2\mu_2)}{h_2}\right]}{\rho_1 h_1 + \dfrac{\rho_2 h_2}{6}}} \qquad (4.91)$$

where subscript 1 refers to either plate (assuming $h_1 = h_3$ and $\rho_1 = \rho_3$), and subscript 2 refers to the core for which μ_2 and λ_2 are Lamé's constants given by Eqs 4.47 and 4.48.

Reasonable estimates for the dilatational resonance frequency can also be determined from the measured dynamic stiffness (or Young's modulus) of the core by calculating the mass–spring–mass resonance frequency; i.e. replacing s'_g with s' of the core in Eq. 4.71.

Below f_d, an estimate for the sound reduction index can be found by assuming only non-resonant (mass law) transmission. This can be calculated from the mass per unit area of the sandwich panel using either the diffuse or field incidence mass law for an infinite plate (Section 4.3.1.2.1).

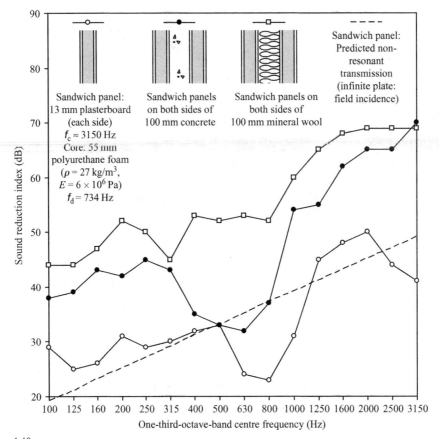

Figure 4.46

Measured sound reduction index for a sandwich panel and two other constructions formed using these panels. Predicted values for non-resonant (mass law) transmission are shown as an estimate for the single sandwich panel below the dilatational resonance, f_d. Measured data are reproduced with permission from Homb et al. (1983).

Measurements from Homb *et al.* (1983) on a sandwich panel in various configurations are shown in Fig. 4.46. The adverse effect of the dilatational resonance at f_d is apparent for the single sandwich panel; there is a deep dip below the infinite plate theory for non-resonant (mass law) transmission. With many lightweight sandwich panels this resonance occurs in the mid- or high-frequency ranges (e.g. see Jones, 1981). However, these panels are usually combined to form one or both leaves of a cavity wall, or connected to other plates to form a multi-layer construction. Figure 4.46 shows two examples of this; in some constructions the adverse effect of the resonance dip can be avoided, in others its effect will still be evident.

4.3.7   Composite sound reduction index for several elements

In Chapter 5 we will look at the combination of direct and flanking transmission that involves vibration transmission between connected plates. The situation is often simpler for direct sound transmission across partitions that are formed from more than one element. Some partitions such as façade walls or corridor walls contain a number of different elements; e.g., the wall itself along with any windows, doors, holes, gaps, apertures, or ventilators. These elements are often exposed to the same sound field, which is assumed to be diffuse. For walls, windows, and doors it can usually be assumed that there is no significant exchange of vibrational energy between them that alters their individual performance. It can also be assumed that there will be no interaction between holes, gaps, apertures, and ventilators that are widely spaced apart; the more important factor is their position in the wall (e.g. in the middle, at an edge, or in a corner). Regardless of whether the sound reduction index of each element is measured or predicted, they can be combined to give a single composite sound reduction index for the wall.

For N elements each with a sound reduction index, R_n, and an area, S_n, the composite sound reduction index is

$$R_{total} = -10 \lg \left(\frac{1}{\sum_{n=1}^{N} S_n} \sum_{n=1}^{N} S_n 10^{-R_n/10} \right)$$
(4.92)

Note that Eq. 4.92 does not imply that all sound reduction indices can be scaled to different areas. For example, the measured sound reduction index for a $1.5 \times 1.25\,\mathrm{m}$ sheet of $6\,\mathrm{mm}$ glass is not relevant when assessing a $0.1 \times 0.3\,\mathrm{m}$ sheet to be used as a vision panel in a door. For all plates it is worth calculating the fundamental bending mode and the critical frequency to help decide whether using a test result for a different size of test element is appropriate for the dominant sound transmission mechanism.

Section 4.3.10 contains calculation of the sound reduction index for slit-shaped and circular apertures for which the opening area needs to be used in Eq. 4.92. As these are idealized models, the opening area of each aperture is well-defined and it is clear that the sound reduction index only applies to specific aperture dimensions. For practical purposes, measurements on holes, gaps, slits, and ventilators are not normalized to an opening area and are almost always described using $D_{n,e}$ rather than R.

4.3.8 Surface finishes and linings

Surface finishes and linings not only increase or decrease the radiated sound from a base wall or floor, they can also alter its basic properties (e.g. mass, stiffness, damping, airflow resistance). A bonded surface finish can usually be considered as an integral part of the plate that forms the wall or floor. In contrast, linings are often considered as a 'bolt-on' component that is interchangeable; hence laboratory measurements of the sound reduction improvement index are commonly quoted for linings. This does not imply that these measurements apply to all other base walls and floors; the result is usually specific to one type of base wall or floor.

4.3.8.1 Bonded surface finishes

Many walls and floors have a bonded finish over the entire surface such as plaster, render, or screed. These typically have a minimum thickness of $5\,\mathrm{mm}$ and can alter the airflow resistivity, internal loss factor, the mass per unit area, and bending stiffness of the base plate. Such changes subsequently affect the radiation and structural coupling losses to and from the plate.

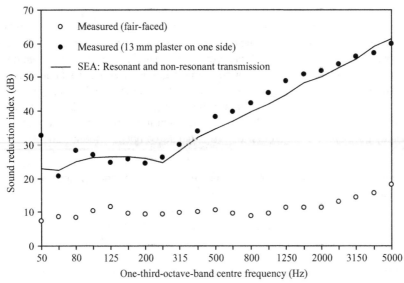

Figure 4.47

Effect of a bonded surface finish on a highly porous plate. 100 mm masonry wall (expanded clay blocks) with a 13 mm lightweight plaster finish (one side). Measurements according to ISO 140 Part 3. *Plate properties: $L_x = 3.53$ m, $L_y = 2.63$ m, $h = 0.1$ m, $\rho_s = 77$ kg/m², $c_L = 2330$ m/s, $\nu = 0.2$, measured total loss factor.* Fair-faced blocks have an airflow resistivity, $r = 6400$ Pa.s/m². Plaster finish: $\rho_s = 10$ kg/m². Measured data from Hopkins are reproduced with permission from ODPM and BRE.

A common bonded surface finish on a masonry wall is plaster or render. Most solid masonry walls with plaster or render on one side (or both sides) can be adequately modelled as a homogeneous plate (refer back to the examples in Section 4.3.1.3). In some cases, the surface finish changes the bending stiffness and the total loss factor that is measured on the base plate; hence these properties need to be used in any model. For a porous masonry wall, a bonded surface finish can effectively remove the non-resonant transmission path via the pores. An example is shown in Fig. 4.47; this fair-faced wall will later be modelled as a porous plate in Section 4.3.9.2. With a plaster finish on one side, this highly porous wall can be modelled as a solid homogeneous plate using SEA in the same way as for other masonry walls. This suggests that despite the porous interface with the air on the fair-faced side of the wall, the estimate for the radiation efficiency (method no. 3 in Section 2.9.4.3) is still reasonable over the building acoustics frequency range.

When using transmission suite measurements to assess any differences between a fair-faced masonry wall and the same wall with a bonded finish it is very useful to measure the total loss factor. A surface finish can cause changes to the internal and coupling loss factors that result in either an increase or a decrease in the total loss factor over various parts of the building acoustics frequency range.

Measurements on 100 and 200 mm brick walls (1600 kg/m³) indicate that the addition of plaster increases the quasi-longitudinal phase velocity (Craik and Barry, 1992). This increase also occurs with other walls built from solid masonry blocks (densities up to 2000 kg/m³) and causes the plastered wall to have a lower critical frequency than when it is fair faced. This shift in the critical frequency is sometimes observed in laboratory measurements.

4.3.8.2 *Linings*

Linings are used on both separating and flanking elements (e.g. floating floors, thermal wall linings on masonry/concrete walls, suspended ceilings). Most linings consist of a plate with a low mass per unit area (compared to the base wall or floor) and a critical frequency in the high-frequency range. This plate is isolated from the base wall or floor by an empty cavity, or a cavity that is partially or fully filled with an absorbent and/or resilient material. The plate usually needs to be supported by the base element with structural connections across this cavity. This general description fits the five-subsystem SEA model that was previously used to model a plate–cavity–plate system in Section 4.3.5. This model is appropriate for simple wall linings such as plasterboard which is screwed to timber battens and screwed to a solid masonry/concrete wall or floor. The five-subsystem model is used in the same way as for a timber joist floor by adding a sixth subsystem to model structural coupling via the battens (Wilson and Craik, 1996). Many linings use resilient structural connections that are too complicated to model without measuring properties relating to their dynamic stiffness. Reliance therefore tends to be placed upon laboratory measurement of the sound reduction improvement index.

Some linings have no structural connections to the base wall or floor; these are commonly referred to as independent linings. A frame is needed to support the independent lining so the perimeter of this frame will be connected to the flanking walls and floors; structural coupling between the lining and the base wall or floor therefore depends on these flanking elements. In laboratory measurements the boundary conditions are usually arranged to avoid vibration transmission via such a path. *In situ* this structural coupling is usually more important when the flanking walls and floors are lightweight (e.g. timber frame walls/floors) rather than heavyweight (e.g. concrete walls/floors).

Linings are formed from plates with a high critical frequency to try and take advantage of their low radiation efficiency at frequencies well-below the critical frequency. This is based on the assumption that all sound radiation by the lining will be due to resonant transmission and that transmission paths involving non-resonant (mass law) transmission via the cavity will be unimportant. This is often the case when the cavity is filled with highly absorbent material and the base wall or floor has a high mass per unit area. However, the air in the cavity or any resilient connections across the cavity can act as a spring element. This results in a mass–spring–mass resonance frequency (Sections 4.3.5.1 and 4.3.5.4.1) at which more sound can be radiated than from the base wall or floor without the lining. The addition of linings tends to increase the sound insulation in the mid- and high-frequency ranges whilst reducing it near the mass–spring–mass resonance. For this reason the design of linings usually starts with the aim of getting the mass–spring–mass resonance frequency to be well-below the lowest frequency of interest. However, constraints on the cavity depth and available resilient materials/connectors often results in the resonance frequency falling within the low-frequency range. Whilst this is an important part of the building acoustics frequency range, exact resonance frequencies and their adverse effects are not easily predicted for lightweight linings (refer back to the discussion in Section 4.3.5.1). Hence laboratory testing is usually needed to hone this aspect of the design.

The sound reduction improvement index is usually measured with airborne excitation to give ΔR, but it can also be measured with mechanical excitation to give $\Delta R_{Resonant}$ as previously described in Section 3.5.1.2.2. For the majority of flanking walls and floors that are built from masonry/concrete it is only resonant transmission via the lining that is relevant. This applies to the situation where the base wall or floor is only excited by structure-borne sound or mechanical excitation. ΔR includes non-resonant (mass law) transmission below the critical frequency

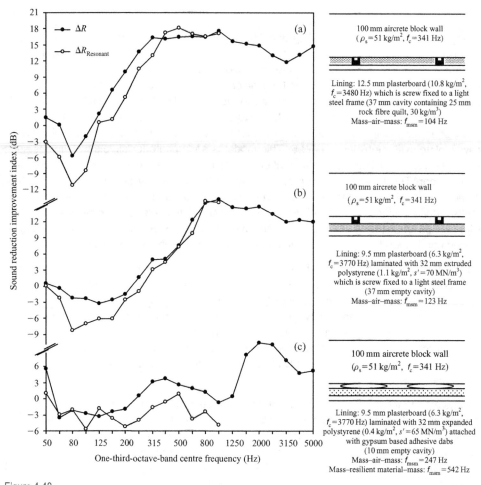

Figure 4.48

Measured sound reduction improvement index for three different wall linings on the same solid masonry wall. Measurements used to determine ΔR were made according to ISO 140 Part 3. Note that $\Delta R_{\text{Resonant}}$ is only shown below the thin plate limit. Measured data from Hopkins are reproduced with permission from BRE Trust.

and if the base wall or floor is porous, non-resonant transmission through the pores. Below the critical frequency of non-porous base walls and floors, $\Delta R_{\text{Resonant}}$ will usually be lower than ΔR. For porous base walls and floors, $\Delta R_{\text{Resonant}}$ may be lower than ΔR across a wider frequency range.

The measured sound reduction improvement indices for three different wall linings on the same masonry wall are shown in Fig. 4.48. Note that whilst negative values for the sound reduction improvement index are usually due to mass–spring–mass resonances, other peaks, and troughs can occur near the critical frequencies of the base wall and the lining. In the high-frequency range a dip or plateau often occurs due to the high-radiation efficiency of the lining at and above its critical frequency. The critical frequency of the base wall and the mass–spring–mass resonance are usually both in the low-frequency range. The mass–spring–mass resonance may only have a distinguishable dip when it is well separated from the critical

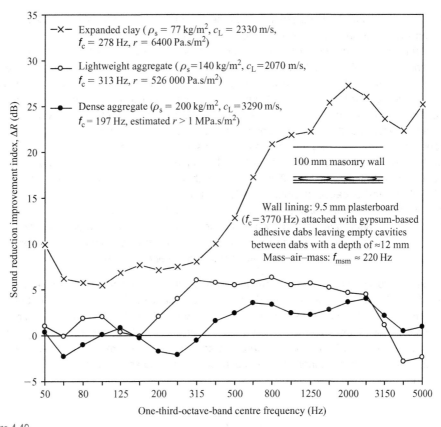

Figure 4.49

Measured sound reduction improvement index for one type of wall lining on three different 100 mm solid masonry walls. Measurements used to determine ΔR were made according to ISO 140 Part 3. Measured data from Hopkins are reproduced with permission from ODPM and BRE.

frequency of the base wall. Examples (a) and (b) have a single mass–spring–mass resonance frequency due to the air in the cavity; these are well-below the critical frequency of the base wall. Example (c) has two resonance frequencies; one associated with the air in the cavity (close to the critical frequency of the base wall) and the other due to the resilient material acting as a spring. Note that any calculated mass–spring–mass resonance frequency only gives a rough indication of the frequency bands in which the sound reduction index of the wall is actually reduced by the lining. Mechanical excitation tends to emphasize the dip at the mass–spring–mass resonance. This aspect is worth noting for point force excitation such as with impacts on walls and floors, or some types of machinery/equipment. However, highly negative values at the resonance frequency are not usually observed when a flanking wall or floor is excited along its boundaries by structure-borne sound waves transmitted from other connected plates.

The measured sound reduction improvement indices for the same wall lining on three different masonry walls are shown in Fig. 4.49. These walls all have very different airflow resistivities and mechanical impedances. This indicates that one type of lining can have significantly different sound reduction improvement indices depending on the base wall or floor. Different base elements usually have different impedances and this will change vibration transmission between the lining and the base plate across the structural connections; this can be seen by referring back to the theory for point connections in Section 4.3.5.4.1. Laboratory measurements are usually only taken on base walls or floors that have been sealed with paint or plaster, so non-resonant transmission due to porosity is usually excluded from the measurement. Outside of the laboratory, wall linings are usually applied to fair-faced walls and it is not appropriate to assume that the measurement of the sound reduction improvement index applies to any fair-faced wall. It has been shown that a highly porous masonry wall may reduce the depth of the mass–spring (air)–mass dip in the sound reduction index due to a lining with a cavity behind it (Heckl, 1981; Warnock, 1992). The majority of masonry walls do not have very low airflow resistivities, and it is difficult to confirm that this effect is always measurable when $r > 20\,000$ Pa.s/m^2 because of other variables that affect the level of sound insulation. Some laboratory measurements indicate that applying a bonded surface finish such as render onto a 200 mm dense aggregate masonry wall (2100 kg/m^3) to seal the wall before applying a plasterboard lining gives a lower sound reduction index near the mass–spring–mass resonance compared to when the base wall is fair faced (Mackenzie et al., 1988).

The sound reduction improvement index is only measured with a lining on one side of a base wall or floor. Estimates of the sound reduction index of the same base element with the same or different linings on each side can be calculated by simply adding ΔR for each lining to R for the base wall. Measurements on non-porous walls show that this approach can give good estimates (Warnock, 1991) although there is also evidence that simple addition of ΔR near the mass–spring–mass resonance should only be considered as a rough estimate (Fothergill and Alphey, 1989).

4.3.9   Porous materials (non-resonant transmission)

We now look at sound transmission through porous materials that form a partition, or form part of a wall or floor. Fundamental aspects of the theory relating to porous materials were introduced in Section 1.3.2, and provide the necessary background for this section.

The model that is described here gives the normal incidence sound reduction index, $R_{0°}$, for single sheets of homogeneous porous materials (Bies and Hansen, 1980; Schultz, 1988). For porous materials in long thin cavities, normal incidence is relevant because it applies to the axial modes. However, it is usually diffuse incidence sound fields that are of more interest. Fortunately, the sound reduction index, R, measured in the laboratory for these materials is often similar to $R_{0°}$ for normal incidence (Schultz, 1988). Hence $R_{0°}$ can be used as an estimate for R.

Three frequency ranges are used in the model, A, B, and C. Each range is defined in terms of the material thickness, d, relative to the wavelength of sound within the equivalent gas, λ_{pm} (Section 1.3.2.2). For fibrous materials and some foams this can be calculated from empirical equations such as those of Delany and Bazley (Section 1.3.2.2).

The calculations for frequency ranges A, B, and C are described below.

Frequency range A: $d < \lambda_{pm}/10$

In this frequency range the skeletal frame of the porous material has a low mass impedance, therefore the compressions and rarefactions of the air particles in the pores of the material cause the entire frame to move. This allows sheets of porous material to be modelled using a lumped parameter approach from electrical circuit theory. The normal incidence sound reduction index is (Schultz, 1988)

$$R_{0^\circ} = 10 \lg \left[1 + \frac{\frac{rd}{\rho_0 c_0} \left(\frac{2\pi f \rho_s}{\rho_0 c_0} \right)^2 \left(4 + \frac{rd}{\rho_0 c_0} \right)}{4 \left[\left(\frac{rd}{\rho_0 c_0} \right)^2 + \left(\frac{2\pi f \rho_s}{\rho_0 c_0} \right)^2 \right]} \right] \qquad (4.93)$$

where r is the airflow resistivity, d is the thickness of porous material, and ρ_s is the mass per unit area of the sheet of porous material.

Frequency range B: $\lambda_{pm}/10 \le d < \lambda_{pm}$

There is no specific model for this frequency range. Hence in range B, R_0 is determined by fitting a curve to the R_0 values that have been calculated for ranges A and C. A suitably smooth transition from range A to B to C can usually be achieved by using at least a third-order polynomial curve. When the errors in the curve fit are negligible, this smooth curve may be used to represent R_0 in ranges A, B, and C.

Range B can cover a large part of the building acoustics frequency range. To carry out the curve fitting it is sometimes necessary to calculate values that are below the 50 Hz one-third-octave-band in range A, and above the 5000 Hz one-third-octave-band in range C.

Frequency range C: $d \ge \lambda_{pm}$

In this range the frame can be considered as rigid and the concept of an equivalent gas can be used to represent sound propagation within the porous material. Hence the calculations require $Z_{0,pm}$ and k_{pm} for the equivalent gas. These may be calculated from empirical equations such as those of Delany and Bazley (Section 1.3.2.2).

A fraction of the sound energy that is incident upon the porous material will enter the material; the remaining fraction is reflected back from the entry surface. Once inside the porous material, the sound is attenuated as it propagates through the material towards the exit surface. At the exit surface, a fraction is reflected and the remaining fraction is transmitted; the subsequent travels of the reflected wave within the material are assumed to have an insignificant effect on the overall sound transmission.

The propagation loss, ΔL_P (dB) within the porous material is (Eq. 1.177)

$$\Delta L_P = \frac{20}{\ln 10} \left| \mathrm{Im}\{k_{pm}\} \right| d \qquad (4.94)$$

The entry/exit loss, ΔL_E (dB) at the entry or exit surface is determined from the reflection coefficient for the porous material, R_{pm}, using

$$\Delta L_E = -10 \lg \left(1 - \left| R_{pm} \right|^2 \right) \qquad (4.95)$$

where the reflection coefficient for normal incidence is (Eq. 1.74)

$$R_{pm} = \frac{\dfrac{Z_{0,pm}}{\rho_0 c_0} - 1}{\dfrac{Z_{0,pm}}{\rho_0 c_0} + 1} \qquad (4.96)$$

(a) Rock wool

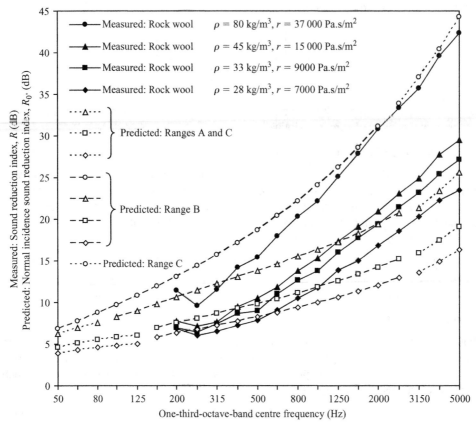

Figure 4.50

Sound reduction index for 100 mm thick sheets of mineral wool – comparison of measured and predicted data. Symbols used for the predicted values correspond to those used for the measurements. Measurements according to ISO 140 Part 3. Measured data from Hopkins are reproduced with permission from ODPM and BRE.

Therefore the entry/exit loss is

$$\Delta L_E = -10 \lg \left[1 - \frac{\left[\left(\mathrm{Re} \left\{ \frac{Z_{0,pm}}{\rho_0 c_0} \right\} \right)^2 + \left(\mathrm{Im} \left\{ \frac{Z_{0,pm}}{\rho_0 c_0} \right\} \right)^2 - 1 \right]^2 + 4 \left(\mathrm{Im} \left\{ \frac{Z_{0,pm}}{\rho_0 c_0} \right\} \right)^2}{\left[\left(1 + \mathrm{Re} \left\{ \frac{Z_{0,pm}}{\rho_0 c_0} \right\} \right)^2 + \left(\mathrm{Im} \left\{ \frac{Z_{0,pm}}{\rho_0 c_0} \right\} \right)^2 \right]^2} \right] \tag{4.97}$$

The normal incidence sound reduction index, R_{0°, can then be calculating using

$$R_{0^\circ} = \Delta L_P + 2\Delta L_E \tag{4.98}$$

4.3.9.1 Fibrous sheet materials

We will now look at some examples for transmission across 100 mm thick sheets of rock and glass wool (average fibre diameter $\approx 5\,\mu$m) in the direction of longitudinal airflow resistivity. Comparison of measured R from a transmission suite ($S \approx 2\,\mathrm{m}^2$) and predicted R_{0° are shown in Fig. 4.50. Although the model gives sound transmission at normal incidence, this

(b) Glass wool

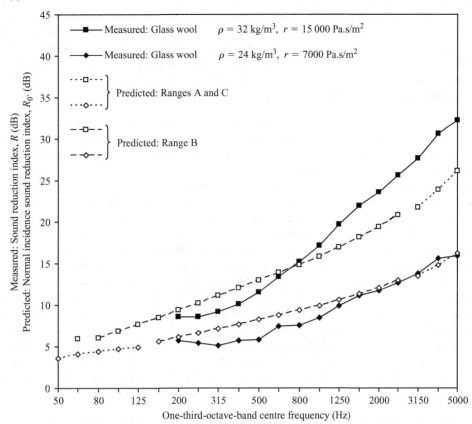

Figure 4.50

(*Continued*)

is a reasonable estimate of the sound reduction index with a diffuse incidence sound field. For 100 mm thick fibrous materials, the difference between measured R and predicted $R_{0°}$ is typically up to ±5 dB. Differences between rock and glass wool are not identified by this simple model. This would require more thorough models that also account for structural waves in the porous material (Biot theory); these waves become important for very thick layers. Such models for porous materials can be found in the book by Allard (1993).

4.3.9.2 Porous plates

Sound transmission across porous plates depends on the mass, stiffness, and damping of the plate as well as the airflow resistance, porosity, and other parameters that describe sound propagation through the pores.

Porous plates in buildings are most commonly encountered in the form of fair-faced masonry walls. For these walls there may also be air paths due to gaps or slits formed at the joints between the blocks. It is simplest to consider two types of masonry wall; those with mortar joints along all edges where all airflow is assumed to occur through the blocks, and those where distinct slits are created between blocks that only touch, or slot together. For the latter type, sound transmission through the slit can be measured if the slit is particularly complicated, or calculated for a simple straight slit (see Section 4.3.10.1). For both types, the air path introduces a non-resonant transmission path across the plate. The sound reduction index of a porous blockwork wall is therefore determined by sound transmission due to resonant transmission as well as three types of non-resonant transmission: (1) mass law, (2) slits at the block joints, and (3) sound propagation through the porous blocks.

We have already seen that non-resonant transmission through a porous material with a rigid frame can be calculated by treating the material as an equivalent fluid. This requires knowledge of the complex wavenumber and the characteristic impedance. For porous plates such as masonry walls it is necessary to either measure these parameters, calculate them by assuming an idealized geometry for the porous structure, or to use empirical models (Allard, 1993; Attenborough, 1993; Voronina, 1997; Wilson, 1997). Information on the microstructure of a masonry block from which to choose a model is rarely available. However, measurement of the airflow resistance on small samples cut from individual blocks is relatively quick and simple. This is measured in the same way as for other porous materials (Section 3.11.1). A model is then needed to link the airflow resistivity to the complex wave number and the characteristic impedance. Unless a specific empirical model is available it is possible to fall back on the empirical equations of Delany and Bazley (Section 1.3.2.2). Although these were based on measurements of fibrous materials with porosities close to unity, the resulting equations are semi-empirical due to the way in which the data was normalized. For this reason they have a slightly more general application to porous materials. (An example of this is the way that these equations were previously used to estimate the ground impedance for outdoor sound propagation models, even though ground porosities are often well-below unity. Nowadays they are rarely used because more accurate models have been developed for specific types of ground.)

Figure 4.51 compares the measured and predicted sound reduction index for two fair-faced masonry walls built from highly porous blocks (mortared along all edges). One wall is built from wood fibre aggregate blocks and the other with expanded clay blocks; both of which have airflow resistivities less than 20 000 Pa.s/m^2. The equivalent gas model gives a reasonable estimate of the sound reduction index with differences between measurements and predictions typically up to ± 5 dB; a similar accuracy to the previous examples shown for mineral wool.

Block density is not an indicator of airflow resistivity; it depends on the material and the manufacturing process. For this reason there are blocks with identical densities that have very different airflow resistivities as well as low-density blocks which have a high airflow resistivity.

Available data indicates that for design purposes, the sound reduction index of fair-faced masonry walls (up to 200 mm thick) may be considered in three main groups: (1) low airflow resistivity: $r \leq 20\,000$ Pa.s/m^2 where the sound reduction index may be estimated as described above; (2) intermediate airflow resistivity: $20\,000 < r \leq 300\,000$ Pa.s/m^2 where the sound reduction index can only be accurately determined by laboratory measurements on full-size walls; and (3) high airflow resistivity: $r > 300\,000$ Pa.s/m^2 where gaps at the mortar joints start to become more important than airflow through the blocks, (Richards, 1959; Warnock, 1992; Watters, 1959; Williamson and Mackenzie, 1971). For groups (1) and (2), the sound reduction index can often be increased above that of the fair-faced wall by using paint to seal one or both surfaces of the porous blocks. Note that bonded surface finishes such as plaster may not only change the airflow resistivity but other wall properties too (Section 4.3.8.1).

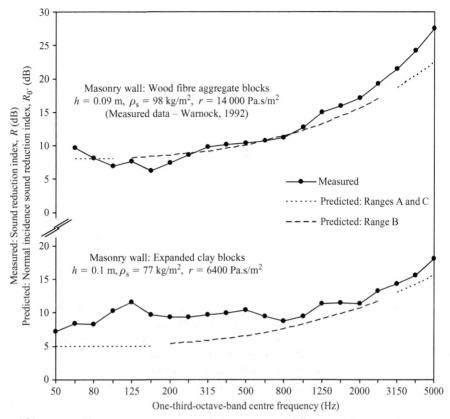

Figure 4.51

Sound reduction index for porous masonry blockwork walls – comparison of measured and predicted data. Measured data for wood fibre blocks according to ASTM E90 are reproduced with permission from Warnock (1992) and the National Research Council of Canada. Measured data for expanded clay blocks according to ISO 140 Part 3 from Hopkins are reproduced with permission from ODPM and BRE.

The sound reduction improvement index for a lining on a wall or floor is also affected by the porosity of the base wall or floor (Section 4.3.8.2).

4.3.9.3 *Coupling loss factor*

Sound transmission across a sheet of porous material or a porous plate is incorporated into the SEA framework as a non-resonant transmission mechanism between two spaces. The coupling loss factor between two rooms is calculated using Eq. 4.26 where S is the surface area of the porous sheet or porous plate. The non-resonant transmission coefficient is calculated from $R_{0°}$. The coupling loss factor between cavities or between a cavity and a room is determined using the same approach as in Section 4.3.1.2 for a one- or two-dimensional incident sound field.

4.3.10 Air paths through gaps, holes, and slits (non-resonant transmission)

Air paths tend to reduce the airborne sound insulation so sealing is an important issue with all wall and floor constructions. Slits can be found along unsealed edges of walls and floors and

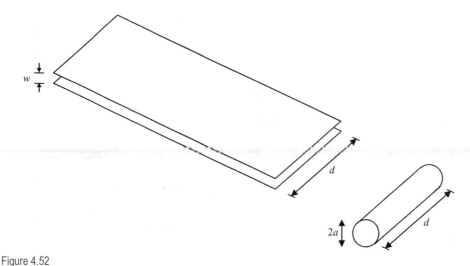

Figure 4.52

Slit-shaped and circular apertures.

between bricks, blocks, or sheet materials whereas circular apertures can be found as small drill holes, large holes formed for ventilation purposes, or pipes used for cables. To identify the importance of air paths it is useful to be able to estimate their effect on the sound insulation. For slit-shaped apertures with straight-edges or for circular apertures (see Fig. 4.52) there are approximate diffraction models to determine sound transmission for a plane wave impinging upon an aperture at normal incidence. These models also provide reasonable estimates for diffuse sound fields. For this reason they are useful in identifying the basic features that affect the transmission loss particularly when trying to identify possible leakage problems. In addition they show that the sound reduction index for the slit cannot simply be assumed to be 0 dB at all frequencies.

Real air paths often take tortuous routes through more complex cross-sections than a simple slit-shaped or circular aperture. Accurate prediction is fraught with difficulty for many real apertures, especially with resilient seals around windows and doors where the sound transmission is dependent upon their compression *in situ*. However, the models for simple apertures are useful in validating laboratory measurements before taking measurements on more complex apertures.

4.3.10.1 Slit-shaped apertures (straight-edged)

For a slit-shaped aperture in the middle of a large plate, or along an edge formed by perpendicular plates, the transmission coefficient is (Gomperts, 1964; Gomperts and Kihlman, 1967)

$$\tau = \frac{mK \cos^2(Ke)}{2n^2 \left\{ \frac{\sin^2(KX + 2Ke)}{\cos^2(Ke)} + \frac{K^2}{2n^2}[1 + \cos(KX)\cos(KX + 2Ke)] \right\}} \tag{4.99}$$

where the end correction, e, for a straight-edged slit (assuming a slit of infinite length in an infinite baffle) is

$$e = \frac{1}{\pi}\left(\ln\frac{8}{K} - 0.57722\right) \tag{4.100}$$

and $X = d/w$ (where d is the depth of the slit, w is the width of the slit), $K = kw$ (where k is the wave number), m is a constant for the incident sound field ($m = 8$ for a diffuse field or $m = 4$ for a plane wave at normal incidence), and n is a constant depending on the position of the slit ($n = 1$ for a slit in the middle of a plate such as along one or both vertical edges of a door, and $n = 0.5$ for a slit along an edge such as along the threshold of a door).

Equation 4.99 assumes cylindrical waves are radiated from the slit and that $w \ll \lambda$. This allows calculations over the building acoustics frequency range for most slits. Equation 4.99 also assumes that the slit is of infinite length. In practice, sound transmission via a slit can be considered to be independent of the slit length, l, when $l > \lambda$, and will depend only on its width and depth (Gomperts and Kihlman, 1967). To assess its suitability for typical slit lengths in the low-, mid- and high-frequency ranges we can look at the wavelength for the lowest band centre frequency in each range; $\lambda_{50Hz} = 6.86\,\text{m}$ (low), $\lambda_{200Hz} = 1.72\,\text{m}$ (mid), $\lambda_{1000Hz} = 0.34\,\text{m}$ (high). Hence it is suited to slits running along the longest dimension of most walls and floors over most of the building acoustics frequency range. However for most windows and doors it only tends to be appropriate in the mid- and high-frequency ranges; this is sufficient for most calculations. On the basis of comparisons with measurements, it can be assumed that the actual sound reduction index in the low-frequency range will be higher than the calculated value (Gomperts and Kihlman, 1967).

Maxima in the transmission coefficient occur at the resonance frequencies across the slit. For a plane wave at normal incidence, resonance frequencies occur when

$$d + 2e = z\frac{\lambda}{2} \tag{4.101}$$

where $z = 1, 2, 3$ etc.

Figure 4.53 shows the sound reduction index for various different slits that can be found in buildings using Eqs 4.99 and 4.37. These assume that the incident sound field is diffuse ($m = 8$). Note that the sound reduction index for an aperture can be negative at some frequencies and that dips occur at the resonance frequencies. Below the first resonance frequency the sound reduction index tends to be higher for a slit in the middle of a plate rather than along an edge. The deepest dip tends to occur at the first resonance frequency. The resonance dips become more prominent in the building acoustics frequency range as the slit depth increases from 6 to 200 mm.

The presence of apertures with prominent dips at the resonance frequencies needs to be borne in mind when scrutinizing sound insulation test results for indications of air paths. Plates with critical frequencies in the high-frequency range (e.g. glass, plasterboard) may have critical frequency dips at similar frequencies to resonance dips associated with the aperture.

The model uses normal incidence, but the resonance frequencies depend on the angle of incidence; for an incident angle of 45° the resonance frequency tends to be higher than for normal incidence (Mechel, 1986). Very few slits in buildings are straight-edged rectangular slits with perfectly uniform dimensions and it is diffuse incidence sound fields that are of most interest; hence it is rarely possible to accurately predict their resonance frequencies.

Laboratory measurements indicate that Eq. 4.99 is valid when $w < 0.3\lambda$ (Bodlund and Carlsson, 1989; Gomperts and Kihlman, 1967; Oldham and Zhao, 1993). These measurements also indicate that deep dips at the resonance frequencies predicted by Eq. 4.99 do not always occur in practice. This model assumes that the air is non-viscous; hence no account is taken

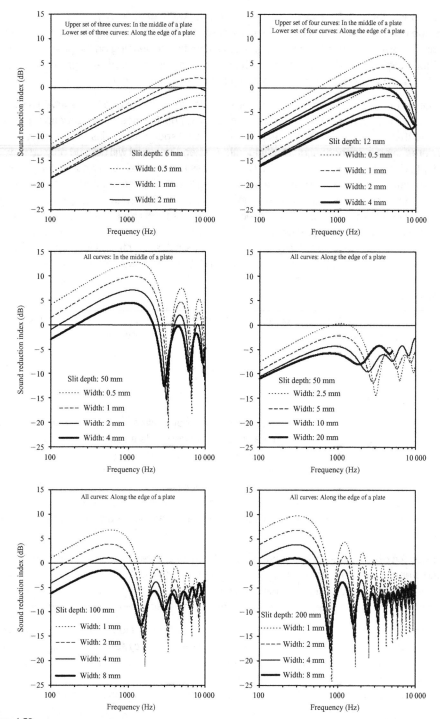

Figure 4.53

Predicted sound reduction index for different size slits.

of any damping mechanism. Viscous losses at resonance tend to be more important for long hairline cracks with widths less than a few millimetres. Validation of more accurate theories which incorporate viscous losses is awkward because of the variation between laboratory measurements; this is possibly due to the effect of different materials that form the slit, and the accuracy with which narrow slits can be created and measured (Lindblad, 1986). As a rule-of-thumb it can be assumed that the deep dips predicted by Eq. 4.99 at the resonances are unlikely to occur in practice when $d/w > 50$.

4.3.10.2 *Circular aperture*

Sound transmission through a circular aperture can be predicted by assuming infinitely thin, rigid pistons at each end of the circular aperture to simulate the motion of air particles at the entry and exit points of the aperture; it is then assumed that plane waves propagate inside the aperture (Wilson and Soroka, 1965).

The transmission coefficient for a plane wave at normal incidence on a circular aperture with radius, a, and depth, d, in an infinite baffle is (Wilson and Soroka, 1965)

$$\tau = \frac{4R_0}{4R_0^2[\cos{(kd)} - X_0 \sin{(kd)}]^2 + [(R_0^2 - X_0^2 + 1)\sin{(kd)} + 2X_0\cos{(kd)}]^2} \tag{4.102}$$

where R_0 and X_0 are the resistance and reactance terms for the radiation impedance of a piston in an infinite baffle given by

$$Z_{rad} = \rho_0 c_0 \pi a^2 (R_0(2ka) + iX_0(2ka)) \tag{4.103}$$

The resistance term, R_0 is given by

$$R_0(2ka) = 1 - \frac{2J_1(2ka)}{2ka} = \frac{(2ka)^2}{2 \times 4} - \frac{(2ka)^4}{2 \times 4^2 \times 6} + \frac{(2ka)^6}{2 \times 4^2 \times 6^2 \times 8} - \cdots \tag{4.104}$$

and the reactance term, X_0 is given by

$$X_0(2ka) = \frac{4}{\pi}\left(\frac{2ka}{3} - \frac{(2ka)^3}{3^2 \times 5} + \frac{(2ka)^5}{3^2 \times 5^2 \times 7} - \cdots\right) \tag{4.105}$$

For $2ka > 17.8$ the following approximations can be used (Wilson and Soroka, 1965)

$$R_0(2ka) = 1 - \frac{1}{ka}\sqrt{\frac{1}{\pi ka}}\sin\left(2ka - \frac{\pi}{4}\right)$$

$$X_0(2ka) = \frac{2}{\pi ka}\left[1 - \sqrt{\frac{1}{ka}}\sin\left(2ka + \frac{\pi}{4}\right)\right] \tag{4.106}$$

Although the derivation is based upon a plane wave at normal incidence it was originally validated against transmission suite measurements; for practical purposes it can be assumed to be equally applicable to reverberant or diffuse incidence sound fields (Wilson and Soroka, 1965). The assumption of plane waves propagating in one dimension within the aperture implies that $\lambda \gg 2a$. The upper frequency limit for a cylindrical tube can be taken as $2a/\lambda \approx 0.6$, but validation against measurements indicates that it still gives reasonable estimates even when

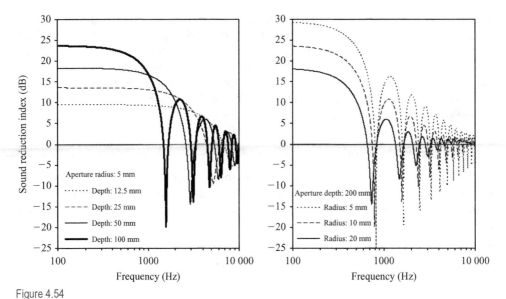

Figure 4.54

Predicted sound reduction index for different size circular holes.

$2a/\lambda$ is slightly greater than unity (Oldham and Zhao, 1993; Wilson and Soroka, 1965). Other models for oblique incidence allow for holes positioned at an edge or corner, and for the effect of sealing and filling the holes (Mechel, 1986).

In the far field, the directivity of the radiated sound from the aperture can be inferred from the directivity of a piston. When $\lambda \gg 2\pi a$ there is uniform radiation in all directions over a hemisphere, but the radiation becomes increasingly directional when $\lambda < 2\pi a$ with sound mainly radiated in the direction perpendicular to the surface of the circular aperture (Morse and Ingard, 1968).

Figure 4.54 shows the sound reduction index from Eqs 4.102 and 4.37 for different circular apertures that can be found in buildings. The sound reduction index below the first resonance (which often occurs in the low- and mid-frequency ranges) can be much higher than with typical slits (refer back to Fig. 4.53). However, in the same way as with slits, dips also occur in the sound reduction index at the resonance frequencies (Eq. 4.101). Deep dips can be predicted for long, small holes but they don't tend to occur in practice due to viscous losses of the air. A damping factor, D, to account for these losses at the resonance frequencies is (Gomperts and Kihlman, 1967; Oldham and Zhao, 1993)

$$D = \left(1 + \frac{\rho_0 \sqrt{2\mu\omega}\left(\dfrac{d + 0.5\pi a}{\pi a^3}\right)}{\dfrac{\rho_0\omega^2}{\pi c_0}}\right)^{-2} \tag{4.107}$$

where μ is the coefficient of viscosity for air ($1.56 \times 10^{-5}\,\text{m}^2/\text{s}$ at 20°C and 0.76 mHg).

The correction $-10\lg D$ in decibels is added to the sound reduction index at the resonance frequency (or each frequency band containing a resonance). This correction has been found to give good agreement with measurements at the first resonance frequency (Oldham and Zhao, 1993).

4.3.10.3 More complex air paths

Many air paths are formed by apertures that are neither slit shaped nor circular. Trying to apply idealized shapes of apertures to more complex shapes is rarely appropriate. For this reason, laboratory measurements are usually necessary to quantify their effect.

Particularly complicated slits can occur around window frames; these slits may have one or more internal bends, or the slits on either side of the frame may lead into a larger internal volume embodied within the frame. The resulting decrease in sound insulation is not always confined to the high-frequency range, it can also occur in the low- and mid-frequency ranges too. Measurements confirm that gaps around frames rarely act as simple rectangular slits, and sometimes act similarly to Helmholtz resonators if there are thin slits on either side that lead to a larger volume of air within the frame (Burgess, 1985; Lewis, 1979).

Electrical socket boxes in separating walls are often blamed for poor airborne sound insulation in the field, particularly when they are back to back; although this is not always justified. With lightweight walls, the adverse effects (if any) tend to occur in the mid- and high-frequency ranges and are highly dependent upon the wall construction, workmanship, and the type of socket box (Nightingale and Quirt, 1998; Royle, 1986).

Lining the interior surfaces of a slit-shaped aperture with a porous absorbent material will not necessarily increase the sound reduction index at all frequencies. For two slits with the same width and length, but where the interior of one slit has an absorbent lining, the sound reduction index for the latter may be higher at high frequencies when the absorption coefficient of the lining is high, but it can be lower at low frequencies (Bodlund and Carlsson, 1989).

4.3.10.4 Using the transmission coefficients

To determine transmission coefficients for one-third-octave or octave-bands, the values at single frequencies are averaged in each frequency band. Below the first resonance frequency, reasonable estimates can be found by taking the value at the band centre frequency. Assuming that we already have the sound reduction index for a wall or floor we can estimate the adverse effect of introducing apertures. This can be calculated using Eq. 4.92 with the predicted sound reduction index for a slit-shaped or circular aperture.

The transmission coefficients in this section are for single apertures in a baffle, not closely spaced arrays of slit-shaped or circular apertures that are sometimes used to form ventilation devices. When there are N circular apertures that are spaced apart by at least $\lambda/4$, the resulting sound reduction index is $10 \lg N$ lower than the sound reduction index for a single aperture (Morfey, 1969). Note that when calculating the effect of more than one aperture using Eq. 4.92 it is the sound reduction index for a single aperture that is used along with the total area of all the apertures.

The change to the sound reduction index, R, of a wall (area, S) due to the introduction of N apertures (each with an area, $S_{a,n}$ and a sound reduction index, $R_{a,n}$) can be described by ΔR, where

$$\Delta R = R_{\text{with apertures}} - R_{\text{without apertures}} = -10 \lg \left[\frac{S + \dfrac{1}{10^{-R/10}} \sum_{n=1}^{N} S_{a,n}(10^{-R_{a,n}/10})}{S + \sum_{n=1}^{N} S_{a,n}} \right] \quad (4.108)$$

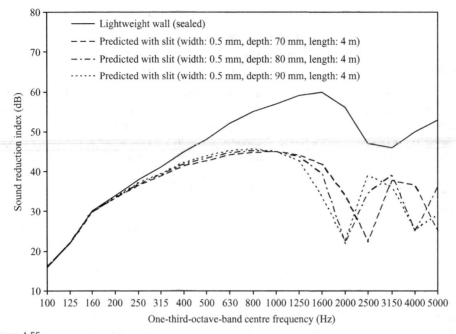

Figure 4.55

Predicted effect of a long, narrow slit-shaped aperture along one edge of a lightweight wall on the sound reduction index.

Note that ΔR is usually referred to as the sound reduction improvement index. For apertures this is a rather positive way of describing a generally negative effect.

Narrow slits along the top and/or bottom of a wall can go unnoticed visually. These sometimes occur at the top of heavyweight walls where they meet the roof or underside of a floor. They can also occur along unsealed tracks of the framework along the perimeter of lightweight walls. These slits may only be 0.5 mm wide but they are still capable of significantly reducing the sound insulation. This is illustrated by the example in Fig. 4.55. Note that gaps around the perimeter of lightweight walls do not usually form simple slit-shaped apertures, but the general trends shown in this example correspond to those observed in measurements (Royle, 1986). In practice the resonance dips are not usually quite as deep because of damping due to air viscosity.

An example showing the effect of circular apertures is shown in Fig. 4.56 for 50 small holes drilled through a 215 mm solid brick wall to give a total open area of 6637 mm^2 (Fothergill and Alphey, 1987). For each aperture, the first, second, third, and fourth resonances occur in the 800, 1600, 2500, and 3150 Hz bands respectively. These circular apertures are narrow and long, and the deep dips predicted by Eq. 4.102 (particularly at the first resonance) do not occur in practice because no account has been taken of viscous losses in the aperture. Using a damping correction (Eq. 4.107) in bands that contain resonances gives close agreement with the measurements. Of course the sound insulation is unlikely to be your primary concern if the neighbours start to drill 50 holes through your bedroom wall. A more important point is illustrated in the upper part of Fig. 4.56. This shows that with a single sheet of plasterboard on each side of the wall the effect of the holes in the brick wall is hardly identifiable; note that the holes were not covered up by the adhesive dabs used to fix the plasterboard. When a construction fails a field test and the airborne sound insulation in the mid- and high-frequency

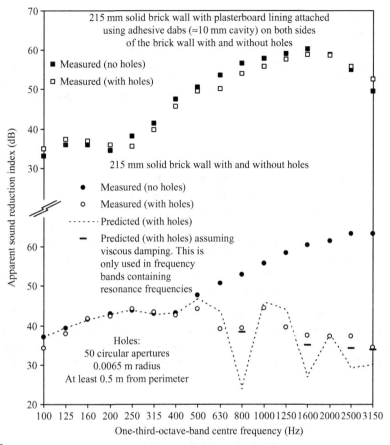

Figure 4.56

Measured and predicted effect of 50 small circular holes on the airborne sound insulation of a brick wall. Measurements according to ISO 140 Part 4. Measured data are reproduced with permission from Fothergill and Alphey (1987), ODPM, and BRE.

ranges is lower than expected, it is common to start looking for holes, slits, and gaps. This is a logical, and relatively simple, first step. However, most buildings contain plenty of hidden holes and discovery of a few of them may place unwarranted emphasis on poor workmanship as the cause of failure. Whilst poor workmanship is often to blame, this example provides a reminder that holes do not always have a significant effect on the overall sound insulation. In practice it is necessary to pay equal (if not more) consideration to flanking transmission as a possible cause of failure.

Sound transmission across an aperture is incorporated into the SEA framework as a non-resonant transmission mechanism between two spaces. Hence the coupling loss factor between these two spaces is calculated using Eq. 4.26 where S is the surface area of the aperture opening. Although transmission through a slit is predicted on the assumption that the slit is infinitely long, its actual opening area is used when calculating the coupling loss factor.

For continuous holes or slits across cavity walls such as those formed by pipes or box-shaped sections, the effect of the aperture can be calculated in the same way as for an aperture in a solid plate. However, the device/object that forms this aperture may also introduce structural coupling across the cavity wall. For holes or slits in one or both leaves of a plate–cavity–plate system, it may be appropriate to use an SEA model to determine sound transmission into and out of a reverberant cavity. This is not suitable when the slits or holes in each plate are opposite each other because the sound radiated by an aperture is highly directional. If there is an aperture in the plate that faces into the receiving room, use of the transmission coefficients requires that the cavity is large enough to support a three-dimensional sound field so that there is sound incident upon the aperture.

4.3.11   Ventilators and HVAC

Most ventilators have quite complex forms and incorporate absorbent material; hence it is necessary to rely on laboratory measurements to quantify their sound insulation. A few through-wall vents have very simple grilles and are not vastly different to the simple circular apertures discussed in Section 4.3.10.2; these can sometimes be adequately modelled as circular apertures in the low- and mid-frequency range (Ohkawa *et al.*, 1984). Equation 4.99 for slit-shaped apertures highlights an important point for the sound insulation of any ventilator; the effect of the incident sound field. The sound insulation depends upon whether a ventilator is positioned in the middle of a wall, along an edge, or in a corner.

Prediction of sound transmission from HVAC systems is thoroughly covered in industry guidance documents such as those by ASHRAE (Anon, 2003) and CIBSE (Leventhall *et al.*, 2002). These usually give reverberant sound pressure levels or sound power levels that can simply be combined with predictions of the sound transmitted by direct and flanking transmission in the building.

4.3.12   Windows

Many window constructions are formed from an insulating glass unit (IGU). These are also referred to as thermal glazing units, or double glazing, and consist of two panes of glass separated by a cavity (typically <20 mm) in a hermetically sealed unit. It is difficult to predict accurately the sound insulation of windows formed from an IGU; hence there is a dependence on laboratory measurements. In addition, the effect of seals around an openable window can only be assessed through measurement. However, most of the important features that determine the sound reduction index can be discussed in terms of the simpler models that have already been introduced in this chapter. These can often be used to help make design decisions alongside laboratory measurements.

4.3.12.1   Single pane

We start by considering a single pane of float glass. This can be considered as a solid homogenous isotropic plate for which its sound reduction index is determined by non-resonant and resonant transmission for a finite plate. An example for 6 mm glass has already been discussed in Section 4.3.1.3.1. Other types of glass such as wired, textured/patterned, or toughened glass have nominally identical material properties to float glass; although an average thickness will need to be assumed for textured/patterned glass. The main factors that cause significant changes in the measured sound reduction index are the boundary conditions and the niche in which the glass is mounted (refer back to Section 3.5.1.3.3 on the niche effect). Concerning resonant transmission, the radiation efficiency below the critical frequency can be calculated by assuming that the boundaries are simply supported or clamped (Section 2.9.4). Glass is mounted in many different types of frame; hence either boundary condition is possible, as well as various degrees of clamping in-between. The mounting can also affect the total loss factor by introducing edge damping (an additional internal loss) as well as changing the structural coupling losses from the glass. Non-resonant transmission usually dominates over resonant transmission below the critical frequency, so changes to the boundary conditions only tend to become apparent at frequencies near, at and above the critical frequency.

Non-resonant transmission across finite plates depends on plate size (Section 4.3.1.2.2); smaller plates tend to transmit less than larger plates. For this reason, it is possible for a perfectly airtight window formed from a large number of small glass panes to have higher sound insulation than a window with a single pane (identical area) at frequencies below the critical frequency. In practice this improvement may only be seen in the low-frequency range because of the increased length of slits/gaps around the frame with windows consisting of multiple panes that are not perfectly sealed.

4.3.12.2 Laminated glass

We now consider laminated glass. This is formed from two sheets of glass that are permanently bonded together by a relatively soft interlayer such as polyvinyl butyral (PVB) or polymethyl methacrylate (PMM). A laminate is advantageous because of the high damping that is achieved by constraining an interlayer between two plates, as well as a reduction in the bending stiffness compared to a solid plate. The internal loss factor of a laminate plate is high because of energy losses associated with shear deformation of the interlayer. For this reason they are useful for attenuating bending wave motion rather than in-plane wave motion. However, the material properties of laminate plates are not as simple as with most solid plates because both the bending stiffness and the internal loss factor vary with frequency and temperature. Therefore reliance tends to be placed upon measurement of these properties rather than trying to predict them from the individual properties of the interlayer and the plates. These measurements were discussed in Sections 3.11.3.4 and 3.11.3.7. An example of the frequency and temperature dependence of the internal loss factor for laminated glass was previously shown in Fig. 3.91. There can be large differences between laminates with different types of interlayer. They are often optimized to give the highest damping at a specific temperature and/or a specific frequency, as well as to shift the critical frequency up to higher frequencies where the adverse critical frequency dip may be more tolerable. With increasing frequency and/or increasing temperature over the building acoustics frequency range, the internal loss factor tends to increase, and the bending stiffness tends to decrease (Kerry and Ford, 1983; Yoshimura and Kanazawa, 1984).

The benefits of laminate glass compared to a single pane of glass with the same thickness are that the critical frequency is shifted to a higher frequency and the depth of the critical frequency dip is reduced by the higher damping. To take advantage of the higher internal losses at higher temperatures it is beneficial for the laminate pane in an IGU to face the side with the higher temperature; this is dependent on the climate and could be indoors or outdoors. Temperature effects for a single pane of laminate glass tend to be important near, at and above the critical frequency. This needs to be considered when comparing measured sound insulation data from the laboratory and the field, and when making comparisons between measurements on different laminates. An example of the effect of temperature on the sound reduction index of a single pane of laminate glass is shown in Fig. 4.57 (Yoshimura and Kanazawa, 1984).

Material property measurements on laminates sometimes give values at individual frequencies so it is necessary to interpolate between them in order to find values of the bending

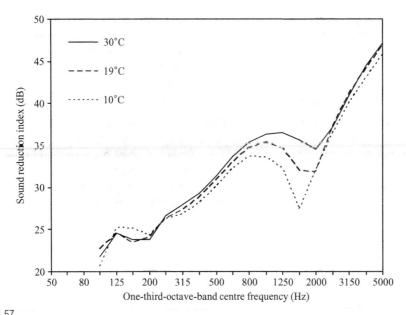

Figure 4.57

Effect of surface temperature on the measured sound reduction index of a 1.7 × 1.3 m pane of laminated glass (4 mm glass/1.1 mm interlayer/4 mm glass) in an aluminium frame. Measured data are reproduced with permission from Yoshimura and Kanazawa (1984).

stiffness and the internal loss factor for each one-third-octave-band centre frequency. With these values it is possible to estimate the sound reduction index by assuming that the plate is solid, homogenous, and isotropic as described in Section 4.3.1. Note that it is the total loss factor that needs to be used in the calculation and this will depend on the mounting in the frame. This approach to prediction generally gives good agreement with laboratory measurements (Ford, 1994; Yoshimura and Kanazawa, 1984).

4.3.12.3   Insulating glass unit (IGU)

The next step is to consider an IGU comprised of single or laminated glass panes. For these units the modal behaviour of the two panes is tightly coupled together by the air or other gas in the cavity as well as the structural connections (spacer bar and sealant) around the perimeter. It is therefore not possible to satisfy the SEA assumption that net power transferred between the two panes (or between each pane and the cavity) is proportional to the difference in their modal energies. Other prediction models are generally needed for these types of systems such as finite element methods, or methods based on multi-layered infinite structures (Villot et al., 2001). However, these units are so common in buildings it is useful to be able to identify the basic features that affect the sound reduction index. Two important features that cause dips in the sound reduction index are the mass–spring–mass resonance frequency and the critical frequencies of each pane.

The mass–spring–mass resonance frequency for air-filled units can be estimated using Eq. 4.72. For units filled with other gases (e.g. argon, sulphur hexafluoride) Eqs 4.70 and 4.71 can be used. Compared with air, a different gas may only shift the mass–spring–mass frequency into an adjacent one-third-octave-band. The mass–spring–mass resonance frequency for most IGUs occurs in the range 100–400 Hz. In this frequency range there are also vibrational modes of the unit (Pietrzko, 1999); these are better described as global modes of the unit due to the strong coupling between the individual panes. Therefore the individual panes of glass do not act as simple lump masses at the mass–spring–mass resonance and sound transmission is determined by a combination of modal behaviour and the air acting as a spring. Any change in the vibrational modes can change the mass–spring–mass resonance frequency and the depth of the dip. These global modes are altered by the structural connections at the perimeter of the unit as well as the dimensions of the unit (Rehfeld, 1997). This means that it is not always appropriate to assume similar performance from a nominally identical unit with different dimensions. Variation in the depth of the dip is exacerbated when the mass–spring–mass resonance lies in the low-frequency range because the sound reduction index may also be affected by the modes of the source and/or receiving room (Section 4.3.3). The depth of the mass–spring–mass resonance dip can be deep when the width and height of the IGU are identical; this can occur even when the panes have different thickness or one is a laminate (Michelsen, 1983). If we temporarily consider the local mode model for each individual pane with identical dimensions then we would expect the modal response of each pane to be characterized by degenerate modes at which there will be a strong response. This strong response will also be found in the global modes of the unit.

We now look at the frequency range in-between the mass–spring–mass resonance and the lowest critical frequency of an IGU. The most common range of cavity depths is between 6 and 16 mm. Over this range the panes are still tightly coupled together. The change in sound reduction index over this range of cavity depths is relatively small; hence tabulated sound reduction indices often quote average values for this range of cavities (e.g. see EN 12758). However, the type of spacer and sealant can still significantly change the sound reduction index in this frequency range due to its effect on the structural coupling around the perimeter of the panes (Gösele et al., 1977). Any advantage in choosing a unit with low structural coupling can be negated by the non-resonant transmission path from the source to the receiving room if there are gaps or slits around the frame of an openable window. If these airpaths are present they usually reduce the sound reduction index in the mid- and high-frequency ranges (e.g. see Michelsen, 1983). For argon-filled units, the phase velocities, and densities for air and argon are not vastly different, hence for most practical purposes the sound reduction index of these units is considered to be the same (Inman, 1994). However large differences occur when the gas is sulphur hexafluoride. This gas has a much lower phase velocity and a much higher density than air (see Table A1). The mass–spring–mass resonance frequency may not be significantly different to air-filled units, but coupling via the sound field in the cavity is significantly reduced. This leads to large increases in the sound insulation above the mass–spring–mass resonance frequency (Gösele et al., 1977; Kerry and Inman, 1986; Rückward, 1981). An example of the effect of different gas fills in an IGU is shown in Fig. 4.58 (Kerry and Inman, 1986).

If the IGU has panes with identical thickness there is usually a distinct dip at the critical frequency. For this reason most manufacturers produce units where one pane is thicker than the other to avoid a single critical frequency dip that is overly deep. Using panes with different thickness gives two shallower critical frequency dips. A single deep dip can also be avoided by using one laminate and one non-laminate pane; the higher internal damping of the laminate also being beneficial at the critical frequency. For units filled with sulphur hexafluoride, each plate radiates sound into the air in the room on one side, and into the gas in the cavity on the other. The plate therefore has a very different critical frequency on each side due to different phase velocities in air and in sulphur hexafluoride. The latter has a much lower phase velocity

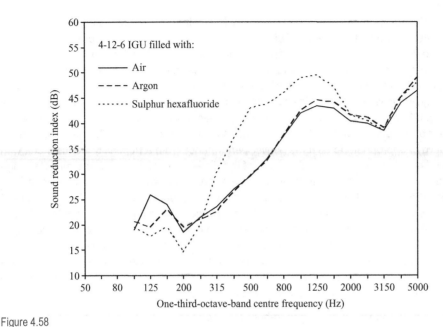

Figure 4.58

Effect of different gas fills on the measured sound reduction index of an IGU. Measurements according to ISO 140 Part 3. Measured data are reproduced with permission from Kerry and Inman (1986).

than air hence the critical frequency for radiation into the cavity is significantly lower than into air. These different critical frequencies are sometimes apparent as dips in the sound reduction index (Gösele *et al.*, 1977; Kerry and Inman, 1986).

4.3.12.4 Secondary/multiple glazing

Whilst different gas fills can be used to increase the sound insulation of an IGU in the mid- and high-frequency ranges, it is difficult to increase it in the low-frequency range without using secondary (or multiple) glazing. This requires two panes of glass separated by a wide air gap; note that one or both panes could be an IGU or a laminate. With a wide airspace of at least 50 mm the two panes are no longer tightly coupled. Most practical designs for secondary glazing use at least 6 mm thick glass and have at least a 50 mm cavity; hence the mass–spring–mass resonance frequency tends to be below 100 Hz.

If the two panes are structurally isolated from each other and there are no air paths around the frame, then at frequencies between the mass–spring–mass resonance and the first cross-cavity mode there is usually the potential to increase the sound insulation by up to 10 dB by at least doubling a cavity depth within the range 50–150 mm; so the resulting cavity depth is within the range 100–300 mm. When the panes and the cavity are large enough to support local modes, the five-subsystem SEA model for a plate–cavity–plate system can be used to aid design decisions. Note that any increase in sound insulation with increasing cavity depth may not be identified by an SEA model depending on how the non-resonant transmission is modelled (Sections 4.3.5.2.1 and 4.3.5.3). For practical window constructions any increase in the cavity depth beyond 300 mm tends to give negligible improvement due to the existence of flanking transmission, structural coupling, and air leakage (e.g. see Inman, 1994). Deep critical frequency dips can be avoided by using panes with different critical frequencies; this is seen in the example from Section 4.3.5.2.1.

Absorbent reveals in the cavity are used to reduce sound transmission via paths involving the cavity. Most reveal linings are only highly absorbent in the high-frequency range. This is usually above the first cross-cavity mode so the five-subsystem SEA model can be used to assess the effect of different reveal linings.

Infinite plate models for plate–cavity–plate systems show dips in the sound insulation due to the cross-cavity modes (Fahy, 1985; London, 1950). These dips are rarely (if ever) observed in one-third-octave-band measurements. However such models have led to designs of plate–cavity–plate systems with non-parallel plates on the basis that this will prevent or suppress any adverse effects due to cross-cavity modes. By placing one pane of glass at an angle to the other pane, the cavity depth will vary between $L_{z,min}$ and $L_{z,max}$. When $L_{z,max} = 3L_{z,min}$, laboratory measurements from Quirt (1982) indicate that the sound insulation is nominally the same as a rectangular cavity with a depth of $2L_{z,min}$; the sound insulation is lower than a rectangular cavity with a depth of $3L_{z,min}$; and the sound insulation is higher than a rectangular cavity with a depth of $L_{z,min}$. In practice this means that there is little to be gained from using non-parallel plates. Above the first cross-cavity mode this is apparent from the five-subsystem SEA model because the three-dimensional sound field in the cavity is assumed to be diffuse. Hence if the cavity volume and absorption area are unchanged, there will be no increase in sound insulation if it is only the shape of the cavity that is changed. It has also been shown that the sound insulation does not increase if the central pane of glass in a triple-glazed window is placed at an angle to the parallel outer panes (Rose, 1990).

4.3.13 Doors

Doors are sometimes formed from solid homogeneous plates, but more often from closely connected plates, sandwich panels, or plate–cavity–plate systems on a frame. There tends to be a dependence on laboratory measurements for all door constructions because the effectiveness of the seals around the perimeter is critically important in determining the achievable sound insulation. For compression seals (as opposed to drop-down or wipe seals) the performance may be specific to the type of door and door closer. The design aim with doors and doorsets is often to minimize the door mass and the required opening force whilst maximizing the sound insulation. This requires simultaneous consideration of sound transmission via the door and the seals (e.g. see Hongisto, 2000; Hongisto et al., 2000).

For doors without any seals, sound transmission via the gaps around the perimeter means that accurately predicting transmission via the door itself can become less important than predicting transmission via the gaps. For simple slit-shaped apertures between the door and the frame, the sound reduction index can be predicted as described in Section 4.3.10.1. An example is shown in Fig. 4.59 for a 12 mm thick glass door (such as in commercial buildings) with and without slit-shaped apertures at one door jamb or the threshold. Due to limitations of the slit model, the effect of the slits is only shown where $l > \lambda$. The slits cause a significant decrease in the sound reduction index in the mid- and high-frequency ranges. In this example the door is relatively thin so there are no resonance dips due to the slits within the building acoustics frequency range.

4.3.14 Empirical mass laws

In Section 4.3.1.2 we looked at non-resonant (mass law) transmission across infinite and finite plates for which the mass per unit area plays the lead role (for infinite plates) or a dominant role

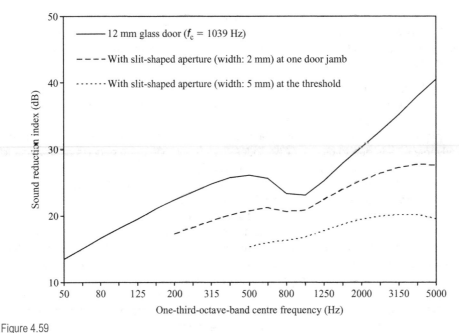

Figure 4.59

Predicted effect of slit-shaped apertures on the sound reduction index of a 12 mm thick glass door (0.8 × 2.0 m). The door jamb is assumed to be in the middle of a wall, rather than along an edge like the threshold.

(for finite plates) in determining the non-resonant transmission coefficient. On the basis that sound insulation is complex and should be simplified as far as possible, it has been the custom for many years to create empirical relationships that link the mass per unit area to the measured weighted sound reduction index. The concept is attractive because of its simplicity and if they are used to give rough estimates (say ±5 dB R_w) they can be very useful. However, looking back at the examples in this chapter it is clear that bending stiffness, damping (internal and total loss factors), plate size, thickness resonances, and porosity also play important roles in determining the sound reduction index; not all of these variables are simply related to the mass per unit area. For masonry walls this has led to different empirical mass laws for different types of blocks in different countries. There are now so many empirical mass laws in existence without traceability to specific types of block properties, surface finish, and boundary conditions that the concept has become devalued. An empirical mass law of any worth is limited to a specific end use. For example, interpolating between measured sound reduction indices for one type of masonry wall with different thicknesses, or producing a design curve for quick calculations where the sound reduction indices have been normalized to a stated total loss factor.

Before establishing an empirical mass law it is important to identify the dominant sound transmission mechanisms that primarily determine the weighted sound reduction index. For many solid masonry walls with high airflow resistivity, resonant transmission tends to be the dominant transmission mechanism over most of the building acoustics frequency range. For this reason the sound reduction index that is measured in the laboratory depends upon the total loss factor. This will vary depending on the material properties of the test element and the laboratory structure. Unfortunately the potential for normalizing measurements to a reference total loss factor over the entire building acoustics frequency range does not extend to all types of masonry walls (refer back to Sections 3.5.1.3.2 and 4.3.1.4). There is some logic in forming separate empirical mass laws for fair-faced masonry walls with high airflow resistivity, and masonry walls with a bonded surface finish (e.g. plaster). This is because a bonded surface finish tends to alter the bending stiffness as well as sealing the pores on the surface of a masonry wall. It may be possible to combine them into a single dataset afterwards.

For non-porous, homogeneous, isotropic solid plates it is much more convenient to create theoretical mass laws rather than using laboratory measurements (Gerretsen, 1999); this allows the total loss factor to be chosen in terms of its relevance to the field or the laboratory situation.

4.4 Impact sound insulation

For impact sound insulation it is simplest to consider heavyweight and lightweight floors separately. This is partly due to the complexity of many lightweight floor constructions and the use of more than one impact source to measure their impact sound insulation.

Heavyweight or lightweight base floors often use floor coverings to improve the impact sound insulation. These tend to be soft coverings or floating floors; both of which are usually least effective in the low-frequency range and most effective in the high-frequency range. Floor coverings are used on top of the base floor to increase the impact sound insulation by changing the power input from the impact source and/or isolating the walking surface from the base floor. The resulting improvement of impact sound insulation depends upon the impact source (e.g. ISO tapping machine, ISO rubber ball) and the base floor. As a first step it is better to reduce the power input into the base plate using a floor covering rather than trying to redesign the ceiling to reduce the radiated sound. The latter approach may lead to significant flanking transmission from the base plate into the flanking walls that subsequently radiate into the receiving room.

4.4.1 Heavyweight base floors

The simplest situation is where the ISO tapping machine excites a solid homogeneous isotropic plate that radiates sound into the receiving room. This is described by a two-subsystem SEA model where subsystem 1 is the plate and subsystem 2 is the receiving room (see Fig. 4.60). To simplify matters it is reasonable to assume that the power flow that returns to the plate from the room is negligible (i.e. $\eta_{21} = 0$). Recalling Eqs 4.8 and 4.9 and writing them in terms of the total loss factor gives

$$W_{\text{in}(1)} \approx \omega\eta_{11}E_1 + \omega\eta_{12}E_1 = \omega(\eta_{11} + \eta_{12})E_1 = \omega\eta_1 E_1 \qquad (4.109)$$

$$\omega\eta_{12}E_1 = \omega\eta_2 E_2 \qquad (4.110)$$

The energy in the receiving room is found by substituting Eq. 4.109 into Eq. 4.110 to give

$$E_2 = \frac{\eta_{12}}{\eta_1\eta_2} \frac{W_{\text{in}(1)}}{\omega} \qquad (4.111)$$

Equation 4.111 can now be used to give the required impact sound pressure level. The coupling loss factor from the plate to the room, η_{12}, is given by Eq. 4.21. The power input from the ISO tapping machine was discussed in detail in Section 3.6.3. Here we will take the simplest approach by assuming that there are short-duration impacts on a thick concrete floor slab. The hammer impedance is assumed to be negligible compared to the driving-point impedance of the plate, and the latter is assumed to equal that of an infinite plate. Hence the power input is

$$W_{\text{in}} = F_{\text{rms}}^2 \frac{2.3\rho c_L h^2}{(2.3\rho c_L h^2)^2 + (\omega m)^2} \approx \frac{3.9B}{2.3\rho c_L h^2} \qquad (4.112)$$

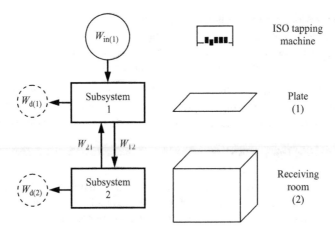

Figure 4.60

Two-subsystem SEA model for impact sound insulation of a solid floor.

which gives the normalized impact sound pressure level as

$$L_n = 10 \lg \left(\frac{\rho_0^2 c_0^2 \sigma}{\rho^2 h^3 c_L \eta} \right) + X \text{ dB} \tag{4.113}$$

where η is the total loss factor of the plate, $X = 78$ dB for one-third-octave-bands and $X = 83$ dB for octave-bands.

Figure 4.61 compares measurements on a 140 mm concrete floor slab with two predicted values; one prediction uses the infinite plate mobility (Eq. 4.113), and the other uses the measured driving-point mobility (50–1250 Hz). Below the critical frequency where $N_s < 1$ the differences between measurements and predictions are typically up to 5 dB. The use of measured mobility does not significantly improve the prediction in the low-frequency range where the mode count is low and the modal fluctuations are pronounced; this is partly due to uncertainty in the measured structural reverberation time and the predicted radiation efficiency. At and above the critical frequency there is close agreement between measurements and predictions. In the high-frequency range the difference often increases because the force spectrum is no longer flat and starts to tail off with increasing frequency (Section 3.6.3.2); this effect is often exacerbated in field measurements when the surface is not smooth and clean (Section 3.6.5.4). In addition, the thin plate assumption for the driving-point mobility starts to break down above the thin plate limit.

Heavyweight floors such as beam and block floors tend to be more complex to model due to their modular nature. Beam and block floors with a bonded surface finish sometimes allows their driving-point mobility to be approximated by an infinite homogeneous plate (refer back to Fig. 2.54). This does not automatically imply that they act as homogeneous plates in all respects. When these floors are bonded across their surface they do not always behave as homogeneous plates concerning their airborne sound insulation (refer back to Fig. 4.24). In addition there may be a significant decrease in vibration level with distance across a beam and block floor (refer back to Fig. 2.42). A comparison of the impact sound insulation for beam and block floors with a concrete floor slab is shown in Fig. 4.62. Although they have a similar performance to the concrete slab, the model for a homogeneous plate does not necessarily provide an adequate model for how these particular beam and block floors behave. There is no general rule that covers all types of modular floor. However, there are some designs of

N_s = 0.6 0.7 0.9 1.1 1.4 1.8 2.3 2.8 3.6 4.5 5.7 7.1 9.1 11 14 18 23 28 36 45 57

M = 0.4 0.2 0.1 0.2 0.2 0.2 0.2 0.2 0.3 0.4 0.5 0.7 0.7 0.8 1.0 1.1 1.1 1.1 1.6 1.9 2.1

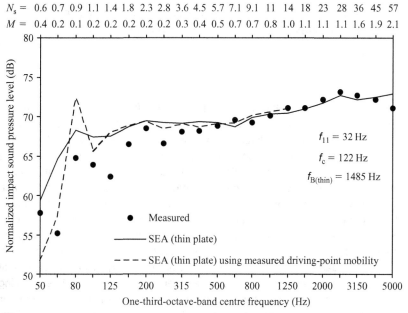

$f_{11} = 32\,\text{Hz}$

$f_c = 122\,\text{Hz}$

$f_{B(\text{thin})} = 1485\,\text{Hz}$

● Measured

——— SEA (thin plate)

– – – – SEA (thin plate) using measured driving-point mobility

Figure 4.61

140 mm concrete floor slab. Upper x-axis labels show the predicted statistical mode count and modal overlap factor for the plate in each frequency band. Measurements according to ISO 140 Part 6. *Plate properties: $L_x = 4.2$ m, $L_y = 3.6$ m, $h = 0.14$ m, $\rho_s = 345$ kg/m^2, $c_L = 3800$ m/s, $\nu = 0.2$, measured total loss factor (approximately $0.005 + 0.3f^{-0.5}$). Measured data from Hopkins are reproduced with permission from ODPM and BRE.

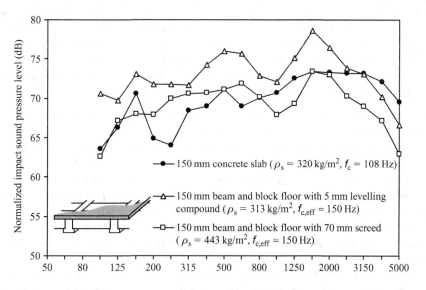

——●—— 150 mm concrete slab ($\rho_s = 320$ kg/m^2, $f_c = 108$ Hz)

——△—— 150 mm beam and block floor with 5 mm levelling compound ($\rho_s = 313$ kg/m^2, $f_{c,\text{eff}} \approx 150$ Hz)

——□—— 150 mm beam and block floor with 70 mm screed ($\rho_s = 443$ kg/m^2, $f_{c,\text{eff}} \approx 150$ Hz)

Figure 4.62

Comparison of a solid homogeneous concrete floor slab with an orthotropic beam and block floor (different surface finishes). The blocks and beams are solid, and the only rigid material that bonds them together is the surface finish. The measured total loss factors for these three floors are within 2 dB of each other. Measurements according to ISO 140 Part 6. Measured data are reproduced with permission from ODPM and BRE.

beam and block floor where the bonded finish results in a plate that can be modelled as being homogeneous and orthotropic (e.g. see Gerretsen, 1986; Patrício, 2001).

4.4.2 Lightweight base floors

For lightweight plates that form the walking surface of the floor, the force spectrum from the ISO tapping machine is significantly different to concrete floors (refer back to Fig. 3.32). Most lightweight floors are formed from plates and beams so the driving-point mobility is frequency-dependent and will differ depending on whether the tapping machine is above or in-between the beams. The application of SEA models to quantify the impact sound insulation tends to be limited; although these models are sometimes useful in a qualitative way. The impact sound insulation of timber joist floors without a resiliently mounted ceiling or floating floor can be predicted with more complex analytic and numerical models (Brunskog and Hammer, 2003 a, b). For such floors, these models indicate that significant increases in the impact sound insulation can be gained by increasing the joist depth (when there is absorbent material in the cavity), and increasing the mass per unit area of the ceiling by adding additional sheet material such as plasterboard (Brunskog and Hammer, 2003c). Lightweight separating floors between dwellings tend to be constructed from several layers of board materials and use isolating elements such as resilient hangers or resilient channels to support the ceiling. These aspects make it more difficult to predict the impact sound insulation.

Timber joist floor constructions vary around the world, and optimizing these constructions usually relies upon laboratory measurements. There can be significant variation in the impact sound insulation between nominally identical timber floors *in situ* (e.g. see Johansson, 2000) and laboratory measurements usefully identify aspects of the construction that cause some of this variation (e.g. see Fothergill and Royle, 1991; Warnock and Birta, 1998). In the low- and mid-frequency ranges, the impact sound insulation spectrum is often characterized by peaks and troughs. Many changes to the floor construction tend to enhance, reduce, or shift the frequency of these peaks and troughs. Whether these changes are of benefit usually depends on the rating system used to determine the single-number quantity.

In the absence of a simple model for lightweight floors it is still possible to interpret and make decisions from measurements by using SEA path analysis in a qualitative manner. An example using four timber joist floors is shown in Fig. 4.63. The walking surface is unchanged so the power input from the tapping machine can be taken as being the same for each floor. Floor A is the most basic construction to which changes can be made. Floor B is formed from A by adding mineral wool into the cavities to absorb sound. However, the change in the impact sound insulation is only a few decibels. We can either infer that the transmission path via the joists dominates over any paths involving the sound field in the cavity, or that the mineral wool is not very effective at absorbing sound in the cavity. If the former assumption is reasonable, then reducing the structural coupling between the joists and the ceiling should give a significant improvement. This is confirmed by floors C and D where the plasterboard is supported by resilient channels. This does not automatically imply that we can now remove the mineral wool from the cavities of floors C and D and achieve the same result. The resilient channels have significantly reduced the strength of the structural transmission path, so transmission paths via the sound field in empty cavities may now be of greater importance. Whilst resiliently suspended ceilings tend to improve the impact sound insulation in the building acoustics frequency range they sometimes reduce it below 50 Hz (Parmanen *et al.*, 1998).

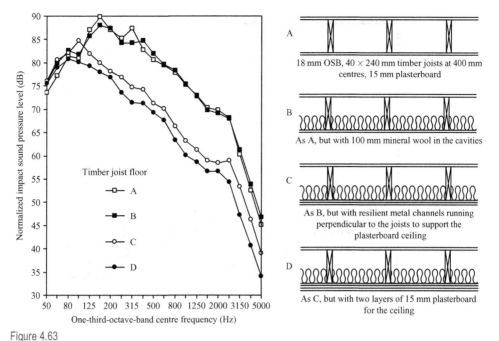

A

18 mm OSB, 40 × 240 mm timber joists at 400 mm centres, 15 mm plasterboard

B

As A, but with 100 mm mineral wool in the cavities

C

As B, but with resilient metal channels running perpendicular to the joists to support the plasterboard ceiling

D

As C, but with two layers of 15 mm plasterboard for the ceiling

Figure 4.63

Impact sound insulation of different timber joist floors using the ISO tapping machine. Measurements according to ISO 140 Part 6. Measured data from Hopkins and Hall are reproduced with permission from ODPM and BRE.

4.4.3 Soft floor coverings

Soft floor coverings are commonly found in the form of carpet or vinyl. In terms of the improvement of impact sound insulation, ΔL, their performance tends to differ depending on whether the base floor is heavyweight or lightweight.

4.4.3.1 Heavyweight base floors

The ISO tapping machine is used as the impact source on heavyweight floors. The improvement of impact sound insulation due to a soft floor covering on such a heavyweight floor can be determined using the contact stiffness of the spring-like floor covering that is 'seen' by the hammer of the tapping machine. This is appropriate because a soft floor covering usually has negligible effect on the total loss factor and bending stiffness of a heavyweight base floor. It can therefore be assumed that it only alters the force input. On heavyweight floors with or without a soft covering, the force pulse from each hammer gives rise to an under-critical oscillation (Section 3.6.3.1). The magnitude of the peak force from the tapping machine can be calculated with and without a soft covering to give $|F_n|_{\text{with}}$ and $|F_n|_{\text{without}}$ respectively. The improvement of impact sound insulation can then be calculated from (Lindblad, 1968; Vér, 1971)

$$\Delta L = 20 \lg \left(\frac{|F_n|_{\text{without}}}{|F_n|_{\text{with}}} \right) \tag{4.114}$$

Example force spectra for a 140 mm concrete floor slab with and without soft floor coverings are shown in Fig. 4.64. For covering No. 1, $E/d = 1.5 \times 10^{11}$ N/m³, which is indicative of a few millimetres of solid PVC. For covering No. 2, $E/d = 2.8 \times 10^8$ N/m³; this could potentially

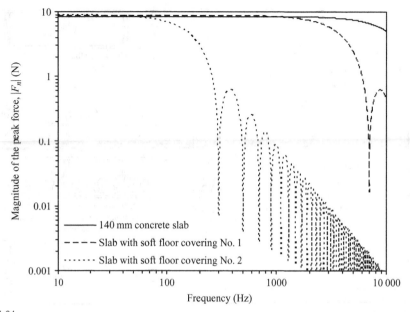

Figure 4.64

Force spectrum for the ISO tapping machine on a concrete floor slab with and without a soft floor covering.

represent some carpet or vinyl layers with a soft resilient backing. We will shortly discuss the difficulty in relating these force spectra to real soft floor coverings.

Whilst the force spectrum is initially flat at low frequencies, there is a cut-off frequency, f_{co}, above which the force decreases. For the concrete slab, the cut-off frequency (Eq. 3.102) due to the contact stiffness of the concrete (Eq. 3.97) is above the building acoustics frequency range ($f_{co} \approx 7000$ Hz). When the soft floor coverings are fixed to the slab, the contact stiffness of the covering (Eq. 3.98) gives the cut-off frequency as $f_{co} \approx 2300$ Hz for covering No. 1, and $f_{co} = 100$ Hz for covering No. 2. Below f_{co} the soft floor covering does not significantly alter the force input compared to the bare slab; hence it does not improve the impact sound insulation.

For soft floor coverings there are deep troughs in the force spectra above the cut-off frequency; these occur at frequencies, nf_{co} where $n = 3, 5, 7$, etc. These troughs occur because the model does not include the internal loss factor, η_{int}, of the soft floor covering. This damping can be included in the model by replacing E for the floor covering with $E(1 + i\eta_{int})$. However these loss factors are not usually known, and can be frequency dependent (Pritz, 1996). This is not particularly problematic because material damping is usually high enough for the troughs to be shallow; hence after averaging into one-third-octave or octave-bands there will rarely be any significant ripple in the curve.

Using the force spectrum we can now calculate the improvement of impact sound insulation from Eq. 4.114; this is shown in Fig. 4.65. The peaks in ΔL correspond to the troughs in the force spectra. If internal damping is incorporated using the internal loss factor, the curves will tend towards a straight slope of 12 dB/octave (equivalent to 40 dB/decade) for $f \geq f_{co}$. In practice, measurements of ΔL rarely show any ripple. Below f_{co}, ΔL is ≈ 0 dB. Hence by calculating f_{co} it is possible to estimate ΔL with two straight lines. This model assumes a linear spring for the soft floor covering and is useful in identifying general features of ΔL. However,

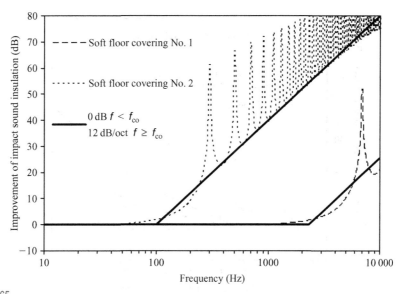

Figure 4.65

Improvement of impact sound insulation due to soft floor coverings on a 140 mm concrete floor slab.

we now need to return to an issue relating to the ISO tapping machine that was discussed in Section 3.6.3.3; namely, soft floor coverings acting as non-linear springs (Lindblad, 1968). The linear spring model results in a single characteristic frequency, f_{co}, above which there is a 12 dB/octave increase in impact sound insulation. However, the relatively high force from the ISO tapping machine causes a non-linear response from some soft floor coverings. This results in ΔL having more than one distinct slope. In practice, some materials have two or three slopes ranging from 5 to 22 dB/octave. Considering this fact alongside the finding that some coverings have a frequency-dependent Young's modulus (Pritz, 1996) means that reliance tends to be placed upon laboratory measurements of ΔL.

Increasing the thickness of a homogeneous soft covering decreases the contact stiffness, which in turn will reduce the cut-off frequency and subsequently increase ΔL. Whilst increasing the thickness is generally beneficial, materials used for soft coverings do not always show such simple relationships and there can sometimes be negligible increase in ΔL with increasing thickness.

For soft floor coverings on heavyweight floors, typical ranges of ΔL are shown in Fig. 4.66 for common materials. The dominant role of the contact stiffness in determining ΔL means that laboratory measurements of soft coverings usually give good estimates of ΔL when *in situ* for many other types of concrete base floor, such as ribbed or hollow core plank floors.

4.4.3.2 Lightweight base floors

For lightweight floors the impact source may be the ISO tapping machine or a heavy impact source such as the ISO rubber ball. With a soft floor covering the tapping machine tends to give minor improvements in the impact sound insulation for the low-frequency range, with more significant improvements in the mid- and high-frequency ranges. Heavy impact sources indicate much smaller improvements, often with little or no improvement in the low- and mid-frequency range. An example is shown in Fig. 4.67 for a carpet on a timber floor (Inoue *et al.*, 2006).

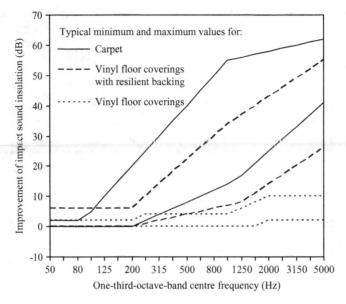

Figure 4.66

Typical ranges for the improvement of impact sound insulation using the ISO tapping machine for soft floor coverings on heavyweight floors.

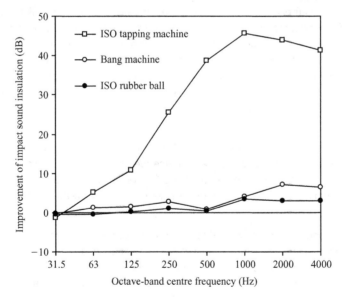

Figure 4.67

Improvement of impact sound insulation for a carpet on a timber floor (Reference floor No.1 from ISO 140 Part 11). Measurements according to ISO 140 Part 11. Measured data are reproduced with permission from Inoue *et al.* (2006).

4.4.4 Floating floors

Floating floors generally consist of a rigid walking surface 'floating' on a resilient material with no rigid connections to the surrounding walls at the edges of the floating floor. There are three

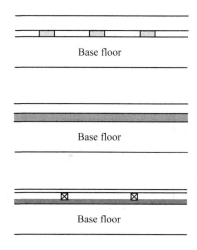

Base floor

Base floor

Base floor

Figure 4.68

Main types of floating floor.

main types as shown in Fig. 4.68; the connection between the walking surface and the base floor via the resilient material may be at individual points, continuous over their entire surface, or along lines (i.e. along battens/joists). These resilient connections result in a mass–spring type resonance associated with the floating floor. The performance of any floating floor relies on its isolation from the base floor; hence it is highly dependent on the quality of workmanship. Floating floors are easily bridged with rigid connections such as nailing through a lightweight floating floor, screed pouring through gaps in the resilient layer, or when walking surfaces are tightly butted up against the side walls.

4.4.4.1 Heavyweight base floors

When the base floor is much heavier than the floating floor, the mass–spring resonance frequency for the walking surface and the resilient layer modelled by a mass–spring system is approximately the same as for a mass–spring–mass system which includes the base floor. As with cavity walls the design aim is usually to have this resonance well-below the important frequency range as it tends to have an adverse effect on ΔL in the frequency band in which it falls (i.e. negative values of ΔL).

4.4.4.1.1 Resilient material as point connections

The SEA model described by Vér (1971) can be used to predict the improvement of impact sound insulation when the plate that forms the walking surface supports a reverberant bending wave field. This walking surface is connected to a heavyweight base floor by N resilient mounts with a dynamic stiffness, k. To reduce the number of subsystems in the model it is assumed that sound radiated into the cavity between the two plates is highly attenuated within the cavity by an absorbent material. It is further assumed that this material does not connect the two plates and transmit vibration between them; all transmission occurs via the mounts. This means that there is no need to include the cavity as a subsystem in the model. The number of subsystems can now be reduced to the bare minimum because we are only interested in the change in sound pressure level in the receiving room. This is equal to the change in vibration level of the base floor; hence we can also exclude the receiving room from the model. The result is a

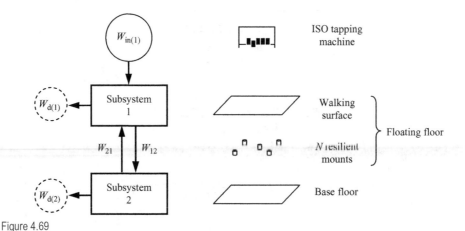

Figure 4.69

Two-subsystem SEA model for a floating floor connected by resilient mounts.

two-subsystem SEA model where subsystem 1 is the plate that forms the walking surface and subsystem 2 is the base floor (see Fig. 4.69).

This example shows the advantage of working in terms of energy and coupling loss factors because the basic SEA model is the same as the one that has just been used for impact sound insulation of a heavyweight base floor (Section 4.4.1). It is only the coupling loss factors and the type of subsystem that has changed; hence the energy in the base floor is given by Eq. 4.111. If the floating floor is now removed and the tapping machine is placed directly on the base floor, the energy of the base floor is found from

$$W_{in(2)} = \omega \eta_2 E_2 \qquad (4.115)$$

We can assume that the total loss factor of the base floor is not increased by coupling to the floating floor. This is reasonable for a heavyweight base floor that is connected to supporting walls on all sides because the structural coupling losses to these walls are usually much higher. The improvement of impact sound insulation is now given by the ratio of the base floor energy without the floating floor to with the floating floor. Hence combining Eqs 4.111 and 4.115 gives

$$\Delta L = 10 \lg \left(\frac{W_{in(2)}}{W_{in(1)}} \frac{\eta_1}{\eta_{12}} \right) \qquad (4.116)$$

Equation 4.116 can now be expanded and simplified with the aid of a few more assumptions. Firstly, the hammer impedance is assumed to be negligible compared to the driving-point impedance of each of the plate subsystems. It is also assumed that both plates can be modelled as infinite plates. The resilient mounts are modelled as springs for which vibration of the walking surface is assumed to be transmitted to the base floor only by forces, rather than by moments. The coupling loss factor, η_{12}, between the walking surface and the base floor is then given by Eq. 4.87 for which the assumptions were previously discussed in Section 4.3.5.4.1. The final simplifying assumption in calculating Eq. 4.87 is that the plate mobilities are sufficiently low that

$$|Y_1 + Y_2 + Y_c|^2 \approx |Y_c|^2 = \frac{\omega^2}{k^2} \qquad (4.117)$$

This gives the improvement of impact sound insulation as

$$\Delta L \approx 10\lg\left(\frac{2.3\rho_{s1}^2 c_{L1} h_1 \eta_1 S_1 \omega^3}{Nk^2}\right) \tag{4.118}$$

where k is the dynamic stiffness of each resilient mount (N/m) and N is the number of mounts. Subscript 1 indicates properties of subsystem 1, the plate that forms the walking surface of the floating floor.

Equation 4.118 is only appropriate above the mass–spring–mass resonance frequency of the system (Eq. 4.89). Below this frequency it is reasonable to assume that $\Delta L = 0$ dB. At the resonance frequency ΔL can be negative and is usually in the range $-10\,\text{dB} \leq \Delta L \leq 0\,\text{dB}$. The equation usefully shows that above the mass–spring–mass resonance frequency, ΔL increases at 30 dB/decade. The assumptions in the above derivation are suited to concrete plates, and have been shown to give good agreement with resilient mounts such as cork (Vér, 1971). For such floors this model identifies ways of improving the impact sound insulation; e.g., by reducing the number of mounts, using a thicker plate for the walking surface, or by trying to increase the total loss factor of the walking surface by increasing the internal damping of the plate.

For more general designs of floating floor it is necessary to work back through the various assumptions and revise the model accordingly. This is needed for floating floors where the walking surface has a much lower mass per unit area and there is an empty cavity without absorbent (such as for access floors used for cabling). Note that the model cannot be used for a continuous resilient layer simply by increasing the number of mounts because of the assumption that the vibration field at the connection points is uncorrelated. To make sure that consideration is given to other sound transmission mechanisms it is convenient to return to the five-subsystem model for a plate–cavity–plate system (Fig. 4.29); where subsystem 2 now forms the walking surface and subsystem 4 is the base floor. The first step is to reconsider the power input because more than one may now be necessary. For a structure-borne sound source such as the ISO tapping machine or machinery mounted on the floor, the power is injected directly into subsystem 2. However, for point excitation of a lightweight plate there may be significant sound power radiated by the nearfield (Section 2.9.7); this can be included as a power input directly into the cavity subsystem. In addition, structure-borne sources tend to generate some airborne noise from their mechanical parts, so it may be necessary to have another power input for sound power radiated into subsystem 1 (the room in which the structure-borne sound source is operating). For most walking surfaces with a low mass per unit area, calculation of the power input from the ISO tapping machine requires consideration of the interaction between the hammer and the walking surface. Rather than using Fourier transforms as in Section 3.6.3.1, estimates can be made using Eq. 4.112 but without making the approximation that the hammer impedance is negligible. The structural coupling loss factor for the resilient mounts can then be calculated according to Eq. 4.87 without making the approximation in Eq. 4.117. As the resilience of the mounts is increased, the path involving structural coupling between the plates will gradually become less important than the transmission path via the sound field in the cavity. Path analysis can then be used to try and optimize the design. For structural reasons the mounts may not be particularly resilient. The model can still be used for walking surfaces with rigid mounts by setting $Y_c = 0$, although the floor is no longer 'floating'.

If the walking surface of any type of floating floor does not completely cover the base floor then sound radiated by the walking surface back into the source room can subsequently excite the base floor which radiates sound into the receiving room. This transmission path, $2 \to 1 \to 4 \to 5$, is conveniently assessed using an SEA model and becomes more important with resilient layers that are dynamically soft. For lightweight floating floors it is usually more important in the mid- and high-frequency ranges near the critical frequency of the walking surface.

4.4.4.1.2   Resilient material over entire surface

One example of this type of floor is a screed on top of a continuous resilient layer. This kind of floating floor can be modelled according to Cremer (1952) where the walking surface and the base plate act as homogeneous infinite plates and the resilient layer acts as a series of closely spaced springs. The lump spring model that is used to transmit forces from the walking surface to the base floor means that wave motion in the resilient layer is not included. This approach allows use of the measured dynamic stiffness for the resilient layer (Section 3.11.3). It is further assumed that the walking surface and the base floor are formed from heavyweight plates with similar density and similar Young's modulus. In addition, the walking surface is taken to be thinner than the heavyweight base floor. Excitation by a point force (i.e. the hammer of the ISO tapping machine) excites a propagating bending wave on the walking surface that subsequently excites bending waves on the base floor via the resilient layer. The full derivation is given in Cremer et al., (1973) and uses two coupled bending wave equations to describe the motion of the two plates. Despite the complexity of the transmission process, the resulting equation is remarkably compact. Under the assumption that the hammer impedance is negligible compared to the driving-point impedance of the plates, the improvement of impact sound insulation is given by (Cremer et al., 1973)

$$\Delta L = 40 \lg \left(\frac{f}{f_{ms}} \right) \tag{4.119}$$

for which the resonance frequency (as with a mass–spring system) is given by

$$f_{ms} = \frac{1}{2\pi} \sqrt{\frac{s'}{\rho_s}} \tag{4.120}$$

where s' is the dynamic stiffness per unit area (Section 3.11.3.1) and ρ_s is the mass per unit area of the walking surface.

If the dynamic stiffness for a specific thickness is not known or a number of different materials are combined together, a rough estimate for N resilient layers on top of each other can be calculated from

$$s' = \left(\sum_{n=1}^{N} \frac{1}{s'_n} \right)^{-1} \tag{4.121}$$

This type of floor may use a lightweight plate to form the walking surface. In this case the hammer impedance may no longer be negligible compared to the driving-point impedance as the frequency increases. Account therefore needs to be taken of the reduction in power input above the limiting frequency, f_{limit} (Eq. 3.106). Assuming that the walking surface still acts as an infinite plate, this gives ΔL as (Cremer et al., 1973)

$$\Delta L = 40 \lg \left(\frac{f}{f_{ms}} \right) + 10 \lg \left[1 + \left(\frac{\omega m}{2.3 \rho c_L h^2} \right)^2 \right] \tag{4.122}$$

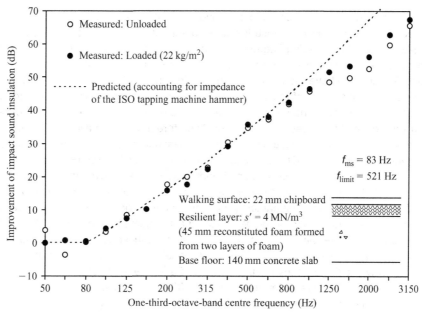

Figure 4.70

Improvement of impact sound insulation (ISO tapping machine) due to a floating floor (lightweight walking surface on continuous resilient layer) on a 140 mm concrete floor slab. Measurements according to ISO 140 Part 8. Measured data from Hopkins and Hall are reproduced with permission from BRE Trust.

where ρ, c_L, h correspond to the plate that forms the walking surface and m is the mass of the ISO tapping machine hammer (0.5 kg).

For a walking surface with a driving-point impedance, Z_{dp}, Eq. 4.122 can also be written in terms of the limiting frequency above which the power input starts to decrease with increasing frequency (Cremer *et al.*, 1973)

$$\Delta L = 40 \lg \left(\frac{f}{f_{ms}} \right) + 10 \lg \left[1 + \left(\frac{f}{f_{limit}} \right)^2 \right] \qquad (4.123)$$

Equations 4.119, 4.122, 4.123 for ΔL are valid above f_{ms} and have been found to give reasonable estimates in the frequency range: $f_{ms} < f < 4f_{ms}$ (Cremer *et al.*, 1973). When calculating single-number quantities this is usually the most important part of the building acoustics frequency range. Equation 4.119 indicates that ΔL increases by 40 dB/decade (recall that 30 dB/decade was predicted for a reverberant plate on individual resilient mounts). An example for a lightweight plate as the walking surface is shown in Fig. 4.70 to illustrate and aid discussion on a few points. For prediction purposes it is simplest to assume that $\Delta L = 0$ dB in all frequency bands below the band containing f_{ms}. A dip may occur in the frequency band containing f_{ms}, and sometimes in adjacent bands too. The uncertainty in predicting the frequency band with a negative ΔL is typically plus or minus one frequency band, as seen in this example. The depth of the mass–spring dip is highly variable; in addition the dip may become shallower, deeper, or remain unchanged when an additional static load is applied to the floor. A rule of thumb for the frequency band containing f_{ms} is that ΔL will fall in the range $-5\,\text{dB} \le \Delta L \le 0\,\text{dB}$. In the high-frequency range the resilient layer usually supports wave motion and the resilient layer can no longer be modelled as a simple lump spring element.

In practice, not all floating floors with continuous resilient layers have a single slope of 40 dB/decade. This is mainly because the walking surface and the base floor do not act as infinite plates and with increasing frequency the resilient layers no longer act as simple springs. The infinite plate assumption is now considered for a floating screed and a structural concrete floor slab. These often have similar material properties and the former is usually thinner than the latter by at least a factor of three. Structural slabs sometimes act as infinite plates when they span complete floors and show a noticeable decrease in vibration level with distance. Floating screeds tend to be cast in each room. Therefore they are smaller, and by their very nature they must be isolated from the walls at the edges; hence any structural coupling losses to these walls are negligible and the bending waves are reflected. If the screed has high internal damping, the amplitude of the propagating bending waves may have sufficiently decreased with distance that reflections from the boundaries can be neglected; hence it can be treated as an infinite plate. High internal damping has been found to occur with asphalt screeds for which Eq. 4.119 can give good agreement with measurements (Cremer et al., 1973). However, this is not the case with many other screeds such as those formed from sand–cement where the internal loss factor is low; these screeds usually act as finite plates with a reverberant bending wave field. For many floating screeds, Eq. 4.119 tends to overestimate ΔL and the frequency dependence is better described by a 30 dB/decade slope; hence the following empirical solution is commonly used (EN 12354 Part 2)

$$\Delta L = 30 \lg \left(\frac{f}{f_{ms}} \right) \qquad (4.124)$$

Single-number quantities predicted using Eq. 4.124 show good agreement with measurements on floating screeds (Metzen, 1996). However, in the mid- and high-frequency ranges the slope can be shallower than 30 dB/decade; sometimes tending towards a plateau as shown by the example in Fig. 4.71 (Gudmundsson, 1984a). This can make it harder to identify minor bridging defects with concrete screeds because a reduction in ΔL due to bridging tends to occur in the mid- and high-frequency ranges (Cremer et al., 1973; Villot and Guigou-Carter, 2003).

At low frequencies where the resilient layer does not support wave motion, it can be treated as a lump spring element. Resilient layers used under floating floors are typically between 5 and 50 mm thick for which most materials can be treated as a lump spring in the low-frequency range. Even when this is appropriate there are other factors that can make this spring less than simple. From Section 3.11.3.1 we recall that for porous resilient materials the dynamic stiffness is the combined stiffness of the skeletal frame and the air contained within it. For mineral wool, the dynamic stiffness and loss factor of the skeletal frame can vary non-linearly with static load (Gudmundsson, 1984b), and non-linearly with strain amplitude (Pritz, 1990). When the dynamic stiffness of a porous resilient layer is predominantly determined by the dynamic stiffness of the contained air, these non-linear effects for the skeletal frame tend to be less significant. However, there is a wide variety of resilient materials and there will inevitably be some materials that have the same dynamic stiffness in laboratory measurements but act as springs with a different dynamic stiffness under different walking surfaces. To keep this in perspective it is important to note that for most floating floor designs, we get a reasonable estimate of the frequency band that contains the mass–spring resonance dip and Eq. 4.119 or 4.124 gives an adequate estimate for ΔL in the low-frequency range. In the mid- and high-frequency ranges, reverberant bending wave fields on the walking surface and wave motion in the resilient layer require more complex models (Gudmundsson, 1984a). Modelling wave motion relies on properties of the resilient material that can be highly variable, are not usually available, and are not particularly easy to model (e.g. mineral wool is anisotropic).

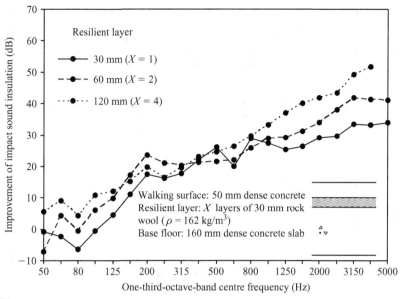

Figure 4.71

Improvement of impact sound insulation for a concrete screed on resilient layers of different thickness. Measurements according to ISO 140 Part 8. Measured data are reproduced with permission from Gudmundsson (1984).

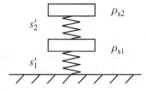

Figure 4.72

Mass–spring–mass–spring system on a rigid base.

If one floating floor is placed on top of another floating floor and there is a heavyweight base floor, a mass–spring–mass–spring system is formed (see Fig. 4.72). This double floating floor has two resonance frequencies given by

$$f_{msms} = \frac{1}{2^{3/2}\pi}\sqrt{X \pm \sqrt{X^2 - \frac{4s_1's_2'}{\rho_{s1}\rho_{s2}}}} \quad \text{where } X = \frac{s_1'}{\rho_{s1}} + \frac{s_2'}{\rho_{s1}} + \frac{s_2'}{\rho_{s2}} \quad (4.125)$$

Comparison of a single and a double floating floor is shown in Fig. 4.73 (Hopkins and Hall, 2006). A double floating floor can avoid the adverse dip in ΔL that occurs with a single floating floor at the mass–spring resonance frequency. However, because the double floating floor has two resonance frequencies, the steep increase in ΔL does not begin until frequencies above the higher resonance frequency.

4.4.4.1.3 Resilient material along lines

A floating floor formed by a lightweight plate rigidly connected to a series of parallel timber battens that rest upon a resilient layer is commonly referred to as a timber raft. Qualitatively

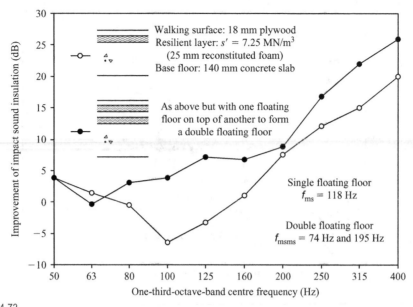

Figure 4.73

Improvement of impact sound insulation (ISO tapping machine) due to a single or double floating floor on a 140 mm concrete floor slab. Measurements according to ISO 140 Part 8. Measured data from Hopkins and Hall are reproduced with permission from BRE Trust.

it is useful to consider the five-subsystem SEA model that was previously used for a plate–cavity–plate system. For vibration transmission between the walking surface and the base plate this model indicates that it is not only necessary to consider structural coupling via the battens and resilient layer, but also via the sound field in the cavities. The latter path will be more important when the resilient layer is dynamically soft. Using an SEA model to quantify vibration transmission across a timber raft by treating the resilient layer as a lumped spring element has been found to be difficult (Stewart and Craik, 2000). Other complexities also occur: there is a different input power and different vibration transmission across the plate for hammers above and in-between battens; the timber raft tends to act as a spatially periodic plate with a decrease in vibration level with distance and there can be significant nearfield radiation into the cavity directly beneath the ISO tapping machine.

4.4.4.2 Lightweight base floors

The improvement of impact sound insulation for a floating floor on a lightweight base floor is generally specific to one type of base floor as there will usually be different sound transmission mechanisms. Certain aspects such as the resonance frequency and driving-point mobility can be used for basic design decisions, but calculation of ΔL is less amenable to simple models. For this reason, reliance is placed on laboratory measurements, particularly for heavy impact sources. Examples for the improvement of impact sound insulation with a timber raft on different base floors are shown in Fig. 4.74; note that this does not imply that similar trends exist for all lightweight base floors and floating floors. When impact sound insulation is critical in the low-frequency range, any conclusion about the efficacy of a floating floor not only depends on the base floor but also upon the excitation. In this particular example the differences between the base floors become particularly apparent in the mid- and high-frequency ranges.

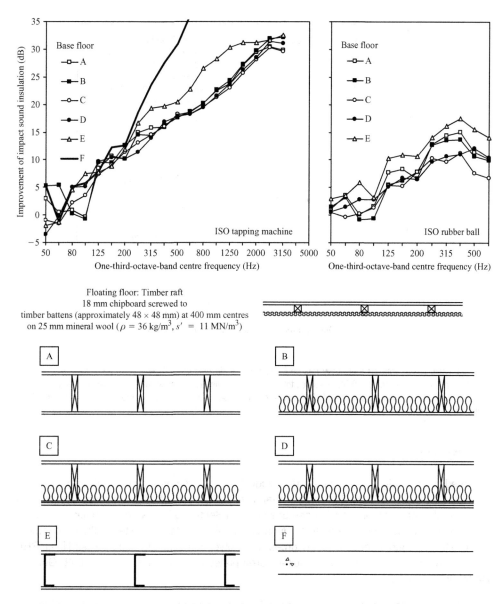

Figure 4.74

Improvement of impact sound insulation for a timber raft floating floor on different base floors measured using the ISO tapping machine and ISO rubber ball. Measurements according to ISO 140 Part 11 (lightweight base floors) and ISO 140 Part 8 (heavyweight base floor). Base floor D corresponds to Reference floor No.2 in ISO 140 Part 11: 18 mm OSB, 40 × 240 mm timber joists at 400 mm centres, 100 mm mineral wool, resilient metal channels running perpendicular to the joists, two layers of 15 mm plasterboard. Base floors A, B, and C use different components of base floor D. Base floor E is comprised of 22 mm chipboard, light steel C-joists and 15 mm plasterboard. Base floor F is a 150 mm concrete floor slab (ISO 140 Part 8 reference floor). Measured data from Hopkins and Hall are reproduced with permission from ODPM and BRE.

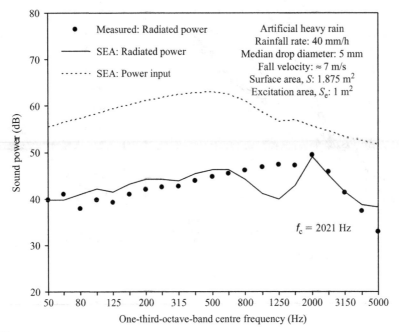

Figure 4.75

Radiated sound power from artificial heavy rain on 6 mm glass. *Plate properties:* $L_x = 1.5\,m$, $L_y = 1.25\,m$, $h = 0.006\,m$, $\rho_s = 15\,kg/m^2$, $c_L = 5350\,m/s$, $\nu = 0.24$, measured total loss factor, radiation efficiency calculated assuming plate lies in an infinite baffle. Plate orientated at an angle of 30° for water drainage. Sound intensity measurements according to ISO 15186 Part 1. Measured data from Hopkins are reproduced with permission from BRE.

4.5 Rain noise

The prediction of rain noise uses similar calculations to those already used for impact sound insulation on floors with the ISO tapping machine. Calculation of the power input for artificial and natural rainfall has previously been described in Section 3.7.1. Referring back to the SEA assumptions in Section 4.2, excitation from artificial rain is quite well-suited to the requirement for statistically independent excitation forces. The power input can therefore be used in an SEA model or other types of models for multi-layered plates (Guigou-Carter *et al.*, 2002; Guigou-Carter and Villot, 2003).

In a similar way to impact sound insulation for a homogenous plate (Section 4.4.1) we can use the same two-subsystem SEA model to calculate the reverberant sound pressure level in the room below the element in the roof that is being excited by rain (either artificial rain or an idealized model for natural rain). This can be converted to a sound power level for the element if required.

A comparison of measured and predicted sound power from artificial heavy rain on 6 mm glass is shown in Fig. 4.75. The power input from the artificial rain is also shown because the predicted decrease that initially occurs above the cut-off frequency for the force pulse is not observed in the measurements; this can be seen by the absence of any dip in the measurement between 1000 and 1600 Hz. However, in general there is close agreement between measurement and prediction. This suggests that any change to the power input due to drops falling on the existing layer of water that is running down the glass can reasonably be ignored in most practical calculations.

References

Allard, J.F. (1993). *Propagation of sound in porous media: modelling sound absorbing materials*, Elsevier Science Publishers Ltd., London and New York. ISBN: 185166887X.

Anon. (2003). *ASHRAE handbook – HVAC applications*. ASHRAE, USA. ISBN: 1931862230.

Attenborough, K. (1993). Models for the acoustical properties of air-saturated granular media, *Acta Acustica*, **1**, 213–226.

Beranek, L.L. and Work, G.A. (1949). Sound transmission through multiple structures containing flexible blankets, *Journal of the Acoustical Society of America*, **21** (4), 419–428.

Bies, D.A. and Hamid, S. (1980). In situ determination of loss and coupling loss factors by the power injection method, *Journal of Sound and Vibration*, **70** (2), 187–204.

Bies, D.A. and Hansen, C.H. (1980). Flow resistance information for acoustical design, *Applied Acoustics*, **13**, 357–391.

Blevins, R.D. (1979). *Formulas for natural frequency and mode shape*, Van Nostrand Reinhold, New York. ISBN: 0442207107.

Bodlund, K. and Carlsson, C.-A. (1989). Ljudläckage via springor och tätlister, *Technical Report SP 1989:28*, Swedish National Testing Institute SP. (9178481791)

Bolton, J.S., Shiau, N.-M. and Kang, Y.J. (1996). Sound transmission through multi-panel structures lined with elastic porous materials, *Journal of Sound and Vibration*, **191** (3), 317–347.

Bosmans, I. (1998). Analytic modelling of structure-borne sound transmission and modal interaction at complex plate junctions, PhD thesis, Katholieke Universiteit, Leuven, Belgium. ISBN: 9056821318.

Bosmans, I. and Nightingale, T.R.T. (2001). Modelling vibrational energy transmission at bolted junctions between a plate and a stiffening rib, *Journal of the Acoustical Society of America*, **109** (3), 999–1009.

Bosmans, I. and Vermeir, G. (1997). Diffuse transmission of structure-borne sound at periodic junctions of semi-infinite plates, *Journal of the Acoustical Society of America*, **101** (6), 3443–3456.

Brekke, A. (1981). Calculation methods for the transmission loss of single, double and triple partitions, *Applied Acoustics*, **14**, 225–240.

Brunskog, J. and Hammer, P. (2003a). The interaction between the ISO tapping machine and lightweight floors, *Acta Acustica*, **89**, 296–308.

Brunskog, J. and Hammer, P. (2003b). Prediction model for the impact sound level of lightweight floors, *Acta Acustica*, **89**, 309–322.

Brunskog, J. and Hammer, P. (2003c). Design possibilities for impact noise insulation in lightweight floors – A parameter study, *Proceedings of Euronoise 2003*, Naples, Italy.

Burgess, M.A. (1985). Resonator effects in window frames, *Journal of Sound and Vibration*, **103** (3), 323–332.

Cederfeldt, L. (1974). Sound insulation of corrugated plates – a summary of laboratory measurements. Report. No. 55, Division of Building Technology, Lund Institute of Technology, Sweden.

Clarkson, B.L. and Ranky, M.F. (1984). On the measurement of the coupling loss factor of structural connections, *Journal of Sound and Vibration*, **94** (2), 249–261.

Cops, A. and Soubrier, D. (1988). Sound transmission loss of glass and windows in laboratories with different room design, *Applied Acoustics*, **25**, 269–280.

Craik, R.J.M. (1996). *Sound transmission through buildings using statistical energy analysis*, Gower. ISBN: 0566075725.

Craik, R.J.M. (1998). In-plane wave propagation in buildings, *Applied Acoustics*, **53** (4), 273–289.

Craik, R.J.M. (2003). Non-resonant sound transmission through double walls using statistical energy analysis, *Applied Acoustics*, **64**, 325–341.

Craik, R.J.M. and Barry, P.J. (1992). The internal damping of building materials, *Applied Acoustics*, **35**, 139–148.

Craik, R.J.M. and Smith, R.S. (2000a). Sound transmission through double leaf lightweight partitions. Part I: Airborne sound, *Applied Acoustics*, **61**, 223–245.

Craik, R.J.M. and Smith, R.S. (2000b). Sound transmission through lightweight parallel plates. Part II: Structure-borne sound, *Applied Acoustics*, **61**, 247–269.

Craik, R.J.M. and Wilson, R. (1995). Sound transmission through masonry cavity walls, *Journal of Sound and Vibration*, **179** (1), 79–96.

Craik, R.J.M., Steel, J.A. and Evans, D.I. (1991). Statistical energy analysis of structure-borne sound transmission at low frequencies, *Journal of Sound and Vibration*, **144** (1), 95–107.

Craik, R.J.M., Nightingale, T.R.T. and Steel, J.A. (1997). Sound transmission through a double leaf partition with edge flanking, *Journal of the Acoustical Society of America*, **101** (2), 964–969.

Cremer, L. (1942). Theorie der Luftschalldämmung dünner Wände bei schrägem Einfall, *Akustische Zeitschrift*, **7**, 81–104.

Cremer, L. (1952). Theorie des Klopfschalles bei Decken mit schwimmendem Estrich, *Acustica*, **2**, 167–178.

Cremer, L., Heckl, M. and Ungar, E.E. (1973). *Structure-borne sound*, Springer-Verlag. ISBN: 0387182411.

Crocker, M.J. and Price, A.J. (1969). Sound transmission using statistical energy analysis, *Journal of Sound and Vibration*, **9** (3), 469–486.

Davies, H.G. and Wahab, M.A. (1981). Ensemble averages of power flow in randomly excited coupled beams, *Journal of Sound and Vibration*, **77**, 311–321.

Dym, C.L. and Lang, M.A. (1974). Transmission of sound through sandwich panels, *Journal of the Acoustical Society of America*, **56** (5), 1523–1532.

Dym, C.L., Ventres, C.S. and Lang, M.A. (1976). Transmission of sound through sandwich panels: A reconsideration, *Journal of the Acoustical Society of America*, **59** (2), 364–367.

Enger, J. and Vigran, T.E. (1985). Transmission loss of double partitions containing resonant absorbers, *Proceedings of the Institute of Acoustics*, **7** (2), 125–128.

Fahy, F.J. (1970). Energy flow between oscillators: special case of point excitation, *Journal of Sound and Vibration*, **11**, 481–483.

Fahy, F.J. (1974). Statistical energy analysis – a critical review, *Shock and Vibration Digest*, **6** (7), 14–33.

Fahy, F.J. (1985). *Sound and structural vibration. Radiation, transmission and response*, Academic Press, London. ISBN: 0122476700.

Fahy, F.J. and Mohammed, A.D. (1992). A study of uncertainty in applications of SEA to coupled beam and plate systems. Part 1: Computational experiments, *Journal of Sound and Vibration*, **158** (1), 45–67.

Fausti, P., Pompoli, R. and Smith, R.S. (1999). An intercomparison of laboratory measurements of airborne sound insulation of lightweight plasterboard walls, *Building Acoustics*, **6** (2), 127–140.

Ford, R.D. (1994). Predicting the sound reduction index of laminated glass, *Applied Acoustics*, **43**, 49–55.

Ford, R.D., Lord, P. and Walker, A.W. (1967). Sound transmission through sandwich constructions, *Journal of Sound and Vibration*, **5** (1), 9–21.

Fothergill, L.C. and Alphey, R.S. (1987). The effect on sound insulation of small holes through solid masonry walls, *Applied Acoustics*, **21**, 247–251.

Fothergill, L.C. and Alphey, R.S. (1989). The effect of wall linings on the sound insulation between dwellings, *Applied Acoustics*, **26**, 57–66.

Fothergill, L.C. and Royle, P. (1991). The sound insulation of timber platform floating floors in the laboratory and field, *Applied Acoustics*, **33**, 249–261.

Fringuellino, M. and Smith, R.S. (1999). Sound transmission through hollow brick walls, *Building Acoustics*, **6** (3/4), 211–224.

Gagliardini, L., Roland, J. and Guyader, J.L. (1991). The use of a functional basis to calculate acoustic transmission between rooms, *Journal of Sound and Vibration*, **145** (3), 457–478.

Gerretsen, E. (1986). Calculation of airborne and impact sound insulation between dwellings, *Applied Acoustics*, **19**, 245–264.

Gerretsen, E. (1999). Predicting the sound reduction of building elements from material data, *Building Acoustics*, **6** (3/4), 225–234.

Gibbs, B.M. and Maluski, S. (2004). Airborne sound level difference between dwellings at low frequencies, *Building Acoustics*, **11** (1), 61–78.

Gomperts, M.C. (1964). The "sound insulation" of circular and slit-shaped apertures, *Acustica*, **14** (1), 1–16.

Gomperts, M.C. and Kihlman, T. (1967). The sound transmission loss of circular and slit-shaped apertures in walls, *Acustica*, **18**, 144–150.

Gösele, K., Gösele, U. and Lakatos, B. (1977). Einfluß einer Gasfüllung auf die Schalldämmung von Isolierglasscheiben, *Acustica*, **38**, 167–174.

Gudmundsson, S. (1984a). Sound insulation improvement of floating floors. A study of parameters, *Report TVBA-3017*, Department of Building Acoustics, Lund Institute of Technology, Sweden. ISBN: 02818477.

Gudmundsson, S. (1984b). Transmission of structure-borne sound at various types of junctions with thin elastic layers. *Report TVBA-3016*, Department of Building Acoustics, Lund Institute of Technology, Sweden. ISBN: 02818477.

Guigou-Carter, C., Villot, M. and Horlaville, C. (2002). Study of simulated rainfall noise on roofs and glazings, *Proceedings of Forum Acusticum 2002*, Seville, Spain.

Guigou-Carter, C. and Villot, M. (2003). Study of simulated rainfall noise on multi-layered systems, *Proceedings of Euronoise 2003*, Naples, Italy.

Harris, C.M. and Crede, C.E. (1976). *Shock and vibration handbook*, McGraw-Hill. ISBN: 0070267995.

Hashimoto, N., Katsura, M., Nishikawa, Y., Katagihara, K., Torii, T. and Nakata, M. (1996). Experimental study on sound insulation of membranes with small weights for application to membrane structures, *Applied Acoustics*, **48** (1), 71–84.

Heckl, M. (1960). Untersuchungen an orthotropen Platten, *Acustica*, **10**, 109–115.

Heckl, M. (1981). The tenth Sir Richard Fairey Memorial lecture: Sound transmission in buildings, *Journal of Sound and Vibration*, **77** (2), 165–189.

Heckl, M. and Seifert, K. (1958). Untersuchungen über den Einfluss der Eigenresonanzen der Messräume auf die Ergebnisse von Schalldämmessungen, *Acustica*, **8**, 212–220.

Hodges, C.H. and Woodhouse, J. (1986). Theories of noise and vibration transmission in complex structures, *Reports on Progress in Physics*, **49**, 107–170.

Homb, A., Hveem, S. and Strøm, S. (1983). Lydisolerende konstruksjoner: Datasamling og beregningsmetode, *Norges byggforskningsinstitutt*. ISBN: 8253601875.

Hongisto, V. (2000). Sound insulation of doors – Part 1: Prediction models for structural and leak transmission, *Journal of Sound and Vibration*, **230** (1), 133–148.

Hongisto, V., Keränen, J. and Lindgren, M. (2000). Sound insulation of doors – Part 2: Comparison between measurement results and predictions, *Journal of Sound and Vibration*, **230** (1), 149–170.

Hopkins, C. (1997). Sound transmission across a separating and flanking cavity wall construction, *Applied Acoustics*, **52** (3/4), 259–272.

Hopkins, C. and Hall, R. (2006). Impact sound insulation using timber platform floating floors on a concrete floor base, *Building Acoustics*, **13** (4), 273–284.

Ingard, U. (1953). On the theory and design of acoustic resonators, *Journal of the Acoustical Society of America*, **25** (6), 1037–1061.

Inman, C. (1994). A practical guide to the selection of glazing for acoustic performance in buildings, *Acoustics Bulletin (UK Institute of Acoustics)*, **19** (5), 19–24.

Inoue, K.*, Yasuoka, M.** and Tachibana, H.*** (2006). Personal communication, *Nihon University,** Tokyo University of Science, ***University of Tokyo, Japan.

Jakob, A. and Möser, M. (2003). Active control of double-glazed windows. Part I: Feedforward control. Part II: Feedback control, *Applied Acoustics*, **64**, 163–196.

James, P. and Fahy, F.J. (1994). Weak coupling in statistical energy analysis, *ISVR Technical Report No. 228 February 1994*, Institute of Sound and Vibration (ISVR), UK.

Johansson, C. (2000). Field measurements of 170 nominally identical timber floors – A statistical analysis, *Proceedings of Internoise 2000*, Nice, France.

Jones, R.E. (1981). Field sound insulation of load-bearing sandwich panels for housing, *Noise Control Engineering*, **16** (2), 90–105.

Junger, M.C. (1975). Helmholtz resonators in load-bearing walls, *Noise Control Engineering*, **4** (1), 17–25.

Kårekull, O. (2004). Active control of light-weight double panel partitions, *Technical Report SP 2004:02*, Swedish National Testing Institute SP.

Kerry, G. and Ford, R.D. (1983). Temperature effects on the sound insulation of laminated glass, *Proceedings of the 11th ICA congress*, Paris, France, 191–194.

Kerry, G. and Inman, C. (1986). The problems of using single figure indices to describe the acoustic performance of traditional and newly developed windows, *Proceedings of the Institute of Acoustics*, **8** (4), 151–156.

Kropp, W., Pietrzyk, A. and Kihlman, T. (1994). On the meaning of the sound reduction index at low frequencies, *Acta Acustica*, **2**, 379–392.

Kurtze, G. and Watters, B.G. (1959). New wall design for high transmission loss or high damping, *Journal of the Acoustical Society of America*, **31**(6), 739–748.

Lam, Y.W. and Windle, R.M. (1995). Noise transmission through profiled metal cladding. Part I: Single skin measurements. Part II: Single skin SRI prediction, *Building Acoustics*, **2** (1), 341–376.

Lang, J. (1993). Measurement of flanking transmission in outer walls in test facilities, *Applied Acoustics*, **40**, 239–254.

Langley, R.S. and Cotoni, V. (2004). Response variance prediction in the statistical energy analysis of built-up systems, *Journal of the Acoustical Society of America*, **115**, 706–718.

Langley, R.S. and Heron, K.H. (1990). Elastic wave transmission through plate/beam junctions, *Journal of Sound and Vibration*, **143** (2), 241–253.

Lauriks, W., Mees, P. and Allard, J.F. (1992). The acoustic transmission through layered systems, *Journal of Sound and Vibration*, **155** (1), 125–132.

Leppington, F.G., Heron, K.H., Broadbent, E.G. and Mead, S.M. (1987). Resonant and non-resonant acoustic properties of elastic panels. Part II: The transmission problem, *Proceedings of the Royal Society, London*, **A412**, 309–337.

Leventhall, G., Tucker, P. and Oldham, D. (2002). *Noise and vibration control for HVAC, CIBSE Guide B5*. The Chartered Institution of Building Services Engineers, London, UK. ISBN: 1903287251.

Lewis, P.T. (1979). Effect of frame construction on the sound insulation of unsealed windows, *Applied Acoustics*, **12**, 15–24.

Lindblad, S.G. (1968). Impact sound characteristics of resilient floor coverings: A study on linear and nonlinear dissipative compliance, Division of Building Technology, Lund Institute of Technology, Sweden.

Lindblad, S. (1986). Sound transmission through slits, *Proceedings of Internoise 86*, Cambridge, USA, 403–406.

Ljunggren, S. (1991). Airborne sound insulation of thick walls, *Journal of the Acoustical Society of America*, **89** (5), 2338–2345.

London, A. (1950). Transmission of sound through double walls, *Journal of the Acoustical Society of America*, **22**, 605–615.

Lyon, R.H. and DeJong, R.G. (1995). *Theory and application of statistical energy analysis*, Butterworth-Heinemann, MA, USA. ISBN: 0750691115.

Lyon, R.H. and Maidanik, G. (1962). Power flow between linearly coupled oscillators, *Journal of the Acoustical Society of America*, **34** (5), 623–639.

Mackenzie, R., Ma, B., Wilson, R. and Stewart, M. (1988). The effect of cement render upon the sound insulation of dry lined concrete block party wall, *Proceedings of the Institute of Acoustics*, **10** (8), 21–28.

Makris, S.E., Dym, C.L. and MacGregor Smith, J. (1986). Transmission loss optimization in acoustic sandwich panels, *Journal of the Acoustical Society of America*, **79** (6), 1833–1843.

Maluski, S.P.S. and Gibbs, B.M. (2000). Application of a finite-element model to low-frequency sound insulation in dwellings, *Journal of the Acoustical Society of America*, **108** (4), 1741–1751.

Mason, J.M. and Fahy, F.J. (1988). The use of acoustically tuned resonators to improve the sound transmission loss of double-panel partitions, *Journal of Sound and Vibration*, **124** (2), 367–379.

Matsumoto, T., Uchida, M., Sugaya, H. and Tachibana, H. (2006). Technical note: Development of multiple drywall with high sound insulation performance, *Applied Acoustics*, **67**, 595–608.

Mechel, F.P. (1986). The acoustic sealing of holes and slits in walls, *Journal of Sound and Vibration*, **111** (2), 297–336.

Metzen, H.A. (1996). Estimation of the reduction in impact sound pressure level of floating floors from the dynamic stiffness of insulation layers, *Building Acoustics*, **3** (1), 33–53.

Michelsen, N. (1983). Effect of size on measurements of the sound reduction index of a window or a pane, *Applied Acoustics*, **16**, 215–234.

Moore, J.A. and Lyon, R.H. (1991). Sound transmission loss characteristics of sandwich panel constructions, *Journal of the Acoustical Society of America*, **89** (2), 777–791.

Morfey, C.L. (1969). Acoustic properties of openings at low frequencies, *Journal of Sound and Vibration*, **9** (3), 357–366.

Morse, P.M. and Ingard, K.U. (1968). *Theoretical acoustics*, McGraw-Hill, New York. ISBN: 0691084254.

Mulholland, K.A. (1971). Sound insulation measurements on a series of double plasterboard panels with various infills, *Applied Acoustics*, **4**, 1–12.

Narang, P.P. (1994). Sound bridging by wall ties in cavity brick walls: Theory and experiment, *Journal of Sound and Vibration*, **174** (2), 169–180.

Nightingale, T.R.T. (1999). Preliminary results of a systematic study of sound transmission through a cavity wall assembly, *Canadian Acoustics*, **27** (3), 58–59.

Nightingale, T.R.T. and Quirt, J.D. (1998). Effect of electrical outlet boxes on sound insulation of a cavity wall, *Journal of the Acoustical Society of America*, **104** (1), 266–274.

Novak, R.A. (1992). Sound insulation of lightweight double walls, *Applied Acoustics*, **37**, 281–303.

Ohkawa, H., Tachibana, H. and Koyasu, M. (1984). Sound transmission characteristics of building elements with small openings for ventilation purposes, *Proceedings of Internoise 84*, Honolulu, USA, 603–606.

Oldham, D.J. and Zhao, X. (1993). Measurement of the sound transmission loss of circular and slit-shaped apertures in rigid walls of finite thickness by intensimetry, *Journal of Sound and Vibration*, **161** (1), 119–135.

Ookura, K. and Saito, Y. (1978). Transmission loss of multiple panels containing sound absorbing materials in a random incidence field, *Proceedings of Internoise 78*, San Francisco, USA, 637–642.

Osipov, A., Mees, P. and Vermeir, G. (1997). Low-frequency airborne sound transmission through single partitions in buildings, *Applied Acoustics*, **52** (3/4), 273–288.

Parmanen, J., Heinonen, R. and Sivonen, V. (1988). Kaksinkertaisen tiiliseinän ääneneristävyys. Äänen sivutiesiirtymä perustuksen kautta, *Research Note 919*, VTT, Technical Research Centre of Finland. ISBN: 9513832392.

Parmanen, J., Sipari, P. and Uosukainen, S. (1998). Sound insulation of multi-storey houses – Summary of Finnish impact sound insulation results, *Proceedings of COST workshop on building acoustics "Acoustic performance of medium-rise timber buildings"* December 1998, Dublin, Ireland.

Patrício, J. (2001). Can beam-block floors be considered homogeneous panels regarding impact sound insulation? *Building Acoustics*, **8** (3), 223–236.

Pietrzko, S.J. (1999). Vibrations of double wall structures around mass–air–mass resonance, *Proceedings of the 137th meeting of the Acoustical Society of America joined with the 2nd convention of the European Acoustics Association (Forum Acusticum) 1999*, Berlin, Germany.

Pietrzyk, A. and Kihlman, T. (1997). The sensitivity of sound insulation to partition location – case of heavyweight partitions, *Proceedings of Internoise 97*, Budapest, Hungary, 727–730.

Price, A.J. and Crocker, M.J. (1970). Sound transmission through double panels using statistical energy analysis, *Journal of the Acoustical Society of America*, **47** (1/3), 683–693.

Pritz, T. (1990). Non-linearity of frame dynamic characteristics of mineral and glass wool materials, *Journal of Sound and Vibration*, **136** (2), 263–274.

Pritz, T. (1996). Dynamic Young's modulus and loss factor of floor covering materials, *Applied Acoustics*, **49** (2), 179–190.

Quirt, J.D. (1982). Sound transmission through windows I. Single and double glazing, *Journal of the Acoustical Society of America*, **72** (3), 834–844.

Quirt, J.D. (1983). Sound transmission through windows II. Double and triple glazing, *Journal of the Acoustical Society of America*, **74** (2), 534–542.

Quirt, J.D., Warnock, A.C.C. and Birta, J.A. (1995). Sound transmission through gypsum board walls: sound transmission results, *NRC-CNRC Internal Report IRC-IR-693*, October 1995, National Research Council Canada.

Rehfeld, M. (1997). Low frequency behaviour of double glazings in laboratories, *Proceedings of Internoise 97*, Budapest, Hungary, 743–746.

Richards, R.L. (1959). Unpublished test report: Sound transmission loss and air flow resistance measurements, Riverbank Acoustical Laboratories, Illinois, USA.

Rose, K. (1990). *BBC Engineering – Guide to acoustic practice*, Second Edition, BBC Engineering, UK. ISBN: 0563360798.

Royle, P. (1986). The effect of leakage on the sound insulation of plasterboard constructions, *Proceedings of the Institute of Acoustics*, **8** (4), 183–188.

Rückward, W. (1981). Sound insulation of laminated insulating glazings, *Proceedings of Internoise 81*, Amsterdam, The Netherlands, 435–438.

Scharton, T.D. and Lyon, R.H. (1968). Power flow and energy sharing in random vibration, *Journal of the Acoustical Society of America*, **43** (6), 1332–1343.

Schmitz, A., Meier, A. and Raabe, G. (1999). Inter-laboratory test of sound insulation measurements on heavy walls. Part 1: Preliminary test, *Building Acoustics*, **6** (3/4), 159–169.

Scholes, W.E. (1969). A note on the repeatability of field measurements of airborne sound insulation, *Journal of Sound and Vibration*, **10** (1), 1–6.

Scholl, W. and Weber, L. (1998). Einfluβ der Lochung auf die Schalldämmung und Schall-Längsdämmung von Mauersteinen. Ergebnisse einer Literaturauswertung, *Bauphysik*, **20** (2), 49–55.

Schultz, T.J. (1988). Chapter 15, Wrappings, enclosures, and duct linings. In Beranek, L.L. (ed.), *Noise and vibration control*, Institute of Noise Control Engineering, Washington, DC, 476–511. ISBN: 0962207209.

Sewell, E.C. (1970). Transmission of reverberant sound through a single-leaf partition surrounded by an infinite rigid baffle, *Journal of Sound and Vibration*, **12**, 21–32.

Sharp, B.H. (1978). Prediction methods for the sound transmission of building elements, *Noise Control Engineering*, **11** (2), 53–63.

Sharp, B.H. and Beauchamp, J.W. (1969). The transmission loss of multilayer structures, *Journal of Sound and Vibration*, **9** (3), 383–392.

Smith, R.S., Pompoli, R. and Fausti, P. (1999). An investigation into the reproducibility values of the European inter-laboratory test for lightweight walls, *Building Acoustics*, **6** (3/4), 187–210.

Stewart, M.A. and Craik, R.J.M. (2000). Impact sound transmission through a floating floor on a concrete slab, *Applied Acoustics*, **59**, 353–372.

Timoshenko, S.P. and Goodier, J.N. (1970). *Theory of elasticity*, McGraw-Hill, New York. ISBN: 0070858055.

Trochidis, A. (1982). Körperschalldämpfung mittels Gas-oder Flüssigkeitsschichten, *Acustica*, **51** (4), 201–212.

Utley, W.A. and Mulholland, K.A. (1968). The transmission loss of double and triple walls, *Applied Acoustics*, **1**, 15–20.

Vér, I.L. (1971). Impact noise isolation of composite floors, *Journal of the Acoustical Society of America*, **50** (1/4), 1043–1050.

Vér, I.L. (1992). Chapter 9, Interaction of sound waves with solid structures. In Beranek, L.L. and Vér, I.L. (eds.), *Noise and vibration control engineering*, J. Wiley and Sons, New York, 245–366. ISBN: 0471617512.

Vér, I.L. and Holmer, C.I. (1988). Chapter 11, Interaction of sound waves with solid structures. In Beranek, L.L. (ed.) *Noise and vibration control*, Institute of Noise Control Engineering, Washington, DC, 270–361. ISBN: 0962207209.

Villot, M. and Guigou-Carter, C. (2003). Modeling in building acoustics: an overview, *Proceedings of Euronoise 2003*, Naples, Italy.

Villot, M., Guigou, C. and Gagliardini, L. (2001). Predicting the acoustical radiation of finite size multi-layered structures by applying spatial windowing on infinite structures, *Journal of Sound and Vibration*, **245** (3), 433–455.

Vinokur, R.Y. (1990). Transmission loss of triple partitions at low frequencies, *Applied Acoustics*, **29**, 15–24.

Vinokur, R.Y. (1995). Sound insulation by concrete panels with cylindrical circular hollows, *Applied Acoustics*, **45**, 131–138.

Voronina, N. (1997). An empirical model for rigid frame porous materials with high porosity, *Applied Acoustics*, **51** (2), 181–198.

Walker, K.W. (1993). 20+ years of sound rated partition design, *Sound and Vibration Magazine, July*, 14–21.

Warnock, A.C.C. (1991). Sound transmission through concrete blocks with attached drywall, *Journal of the Acoustical Society of America*, **90** (3), 1454–1463.

Warnock, A.C.C. (1992). Sound transmission through two kinds of porous concrete blocks with attached drywall, *Journal of the Acoustical Society of America*, **92** (3), 1452–1460.

Warnock, A.C.C. (1993). Sound transmission through slotted concrete blocks with attached gypsum board, *Journal of the Acoustical Society of America*, **94** (5), 2713–2720.

Warnock, A.C.C. (2000). Airborne and impact sound insulation of joist floor systems: A collection of data, *NRC-CNRC Internal Report NRCC-44210, 2000*, National Research Council Canada.

Warnock, A.C.C. and Birta, J.A. (1998). Summary report for consortium on fire resistance and sound insulation of floors: Sound transmission class and impact insulation class results, *IRC Report IR-766*, National Research Council Canada.

Watters, B.G. (1959). Transmission loss of some masonry walls, *Journal of the Acoustical Society of America*, **31** (7), 898–911.

Weber, L. and Bückle, A. (1998). Schalldämmung von Lochsteinen neue Erkenntnisse, *Bauphysik*, **20** (6), 239–245.

Weber, L. and Mehra, S-R. (2002). Luftschalldämmung und Akustiche Materialeigenschaften von Folien und Membranen, *Zeitschrift für Lärmbekämpfung*, **49** (4), 129–136.

Wen-chao, H. and Chung-fai, N. (1998). Sound insulation improvement using honeycomb sandwich panels, *Applied Acoustics*, **53** (1–3), 163–177.

Williamson, J.J. and Mackenzie, R.K. (1971). Sound insulation of lightweight concrete, *Build International, July/August*, 244–252.

Wilson, R. (1992). Sound transmission through double walls, PhD thesis, Department of Building Engineering and Surveying, Heriot-Watt University, Edinburgh, Scotland.

Wilson, D.K. (1997). Simple, relaxational models for the acoustical properties of porous media, *Applied Acoustics*, **50** (3), 171–188.

Wilson, R. and Craik, R.J.M. (1995). Sound transmission via the foundation of a cavity wall, *Building Acoustics*, **2** (4), 569–583.

Wilson, R. and Craik, R.J.M. (1996). Sound transmission through dry lined walls, *Journal of Sound and Vibration*, **192** (2), 563–579.

Wilson, G.P. and Soroka, W.W. (1965). Approximation to the diffraction of sound by a circular aperture in a rigid wall of finite thickness, *Journal of the Acoustical Society of America*, **37** (2), 286–297.

Yoshimura, J. (2006). Personal communication, Kobayasi Institute of Physical Research, Japan.

Yoshimura, J. and Kanazawa, J. (1984). Influence of damping characteristics on the transmission loss of laminated glass, *Proceedings of Internoise 84*, Honolulu, USA, 589–592.

Combining direct and flanking transmission

5.1 Introduction

Flanking transmission is omnipresent in buildings and its effects are not confined to any particular part of the building acoustics frequency range. In fact it is not uncommon for the flanking structure to radiate similar or higher sound power levels than the separating wall or floor itself.

Figure 5.1 shows field measurements of airborne sound insulation for solid brick separating walls in dwellings with a variety of different flanking walls and floors. The variation is not only due to the different flanking constructions but also due to variation in workmanship, material properties, room dimensions, and measurement uncertainty. Sound insulation distributions of single-number quantities tend to be skewed towards lower values of sound insulation. The lower bound is only limited by the ingenuity of the builder to alter sound transmission paths and introduce new transmission paths through the quality of workmanship as well as by substitution of different materials to those that were specified. The wide range of results for the nominally identical separating wall illustrates the importance of understanding how the combination of direct and flanking transmission determines the *in situ* performance.

This chapter looks at general principles of predicting direct and flanking transmission based on the well-established use of Statistical Energy Analysis (SEA) models for predicting sound transmission in buildings (Craik, 1996; Cremer, *et al.*, 1973; Gerretsen, 1979, 1986; Kihlman, 1967). It starts by looking at vibration transmission between plates connected at a junction. An example building plan showing some plate junctions is shown in Fig. 5.2. The principles apply to both lightweight and heavyweight constructions. However many lightweight walls and floors do not act as simple reverberant subsystems, hence the examples will focus on heavyweight walls and floors, and issues relating to low mode counts and low modal overlap. The chapter then moves on to look at the general application of SEA models and SEA-based models (EN 12354) to the prediction of sound insulation.

Flanking transmission significantly increases the complexity of prediction. With so many transmission paths there is the potential to predict what appears to be the correct sound insulation *in situ* by using a model that does not adequately describe the actual sound transmission process. It is equally possible to inadvertently compensate for errors in one part of a model with invalid assumptions in another part of the model. For this reason, emphasis is placed on the many assumptions and limitations that are involved in the prediction of flanking transmission. These need to be kept firmly in perspective otherwise it is all too easy to disregard the possibility of success despite all evidence to the contrary. A degree of pragmatism is required. If the aim is to consistently predict the sound insulation between a specific pair of rooms in a specific building to an accuracy of $\pm 1\,dB$ in each frequency band across the entire building acoustics frequency range, then disappointment will generally follow. Realistic aims are to estimate the average sound insulation for many similar constructions and to gain sufficient insight into

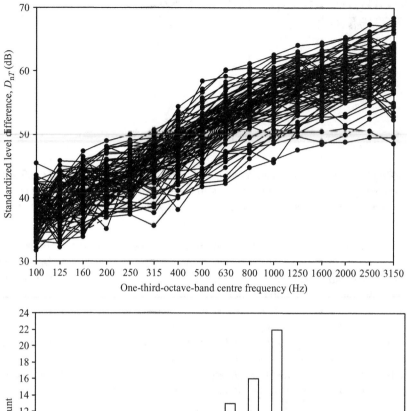

Figure 5.1

Field measurements from ninety-one solid brick separating walls (≈215 mm brick wall with ≈13 mm plaster finish on both sides) with different flanking constructions. The external flanking walls are masonry cavity walls but with different materials, cavity depths and wall ties; other flanking walls and floors varied. Room volumes were between 25 and 60 m³. Measured data are reproduced with permission from ODPM & BRE.

important sound transmission mechanisms that will allow design decisions to be made, and help identify solutions to sound insulation problems in existing buildings.

5.2   Vibration transmission across plate junctions

When predicting vibration transmission across junctions of walls and floors there is more than one model that can be used, each of which has limitations in its application to real buildings.

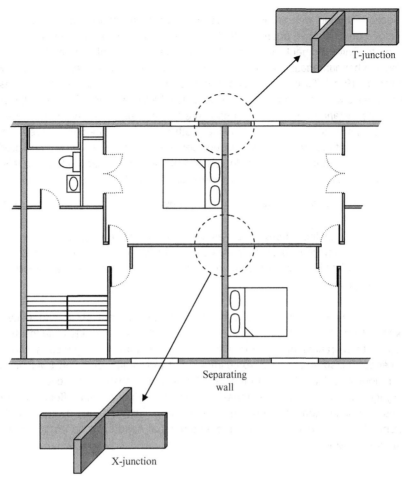

Figure 5.2

Example plate junctions in buildings.

Prediction is simplest when each plate that forms a wall or floor can be modelled as a solid, homogeneous, isotropic plate using thin plate theory. Whilst this is often a reasonable assumption there are many walls and floors that require more complex models. Just as with direct sound transmission in Chapter 4, idealized models are beneficial in providing an insight into the process of vibration transmission and they form a useful benchmark against which more complex constructions can be compared.

Predictions using a wave approach based on semi-infinite plates are commonly used under the assumption that all the plates have diffuse vibration fields. This assumption is not always appropriate, so use is also made of modal approaches and numerical methods. With any approach it is usually possible to model the junction so that either there is only bending wave motion on each plate, or bending and in-plane wave motion is possible on each plate. The output from any of these calculations can be converted into coupling parameters that can be used in SEA or SEA-based models.

In this section we start with an overview of the wave approach that considers only bending waves, move on to a wave approach that considers both bending and in-plane waves, and finally look at Finite Element Methods (FEM) as one example of a numerical method. (For examples of other numerical approaches, see Cuschieri, 1990; Guyader et al., 1982; Rébillard and Guyader, 1995.) To predict sound transmission between adjacent rooms it is usually sufficient to use a wave approach that only considers bending waves. For non-adjacent rooms that are some distance apart (i.e. with structure-borne sound transmission across a few or several junctions) it is necessary to consider both bending and in-plane waves. To gain an insight into some of the more awkward issues concerning the measurement and prediction of vibration transmission it is necessary to compare these different approaches in this section; this is done using SEA and numerical experiments with FEM.

Plate junctions in buildings can generally be described as one of four idealized types:

1. a rigid junction where the plates are rigidly connected together
2. a rigid junction where the plates are all rigidly connected to a beam/column
3. a resilient junction where one or more plates are connected via a resilient material
4. hinged junctions.

In this chapter the focus is on rigid junctions because these are applicable to many junctions of brick/block/concrete walls and floors in buildings. Idealized models of junctions are very useful tools but in practice there is more than one way of forming a rigid connection between plates; hence measurements are sometimes used to quantify vibration transmission. This can be fraught with difficulty due to the variation in workmanship that occurs at junctions of nominally identical walls and floors, measurement uncertainty and the effect of different plate properties and dimensions on vibration transmission. The examples using FEM, SEA, and measurements in this section illustrate the features that need to be considered when relying purely on measurements.

5.2.1   Wave approach: bending waves only

Junctions of walls and floors can often be described by an X, T, L or in-line junction as shown in Fig. 5.3. In the following derivations it is assumed that the plates forming these junctions are solid, homogeneous, isotropic plates modelled with thin plate theory.

In the same way that airborne sound insulation is described using the sound reduction index; we can make use of a transmission coefficient to describe structure-borne sound transmission. However with plate junctions there are more permutations for the source plate, i, and the receiver plate, j. The transmission coefficient, τ_{ij}, is therefore defined as the ratio of the power, W_{ij}, that is transmitted across the junction to plate j, to the power, $W_{inc,i}$, that is incident on the junction on plate i,

$$\tau_{ij} = \frac{W_{ij}}{W_{inc,i}}$$ (5.1)

which in decibels gives a transmission loss, TL_{ij}, as

$$TL_{ij} = 10 \lg \left(\frac{1}{\tau_{ij}} \right)$$ (5.2)

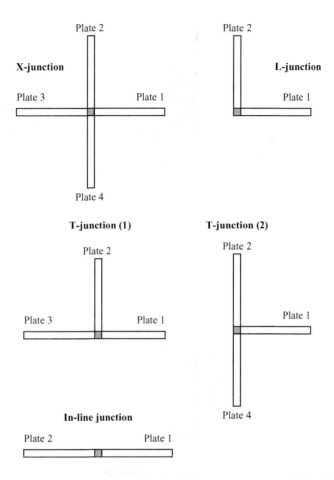

Figure 5.3

Common plate junctions. The hatched box represents the junction beam used for the purpose of modelling vibration transmission with a wave approach.

For use in SEA models, the coupling loss factor, η_{ij}, is calculated from the transmission coefficient, τ_{ij}, as described in Section 2.6.4.

5.2.1.1 Angular averaging

The wave approach considers a plane wave on plate i that impinges upon a junction at a specific angle of incidence, θ, as shown in Fig. 5.4. This gives an angle-specific transmission coefficient, $\tau_{ij}(\theta)$. In practice we are not interested in the transmission coefficient at a specific angle but in the angular average transmission coefficient. In a diffuse vibration field it is assumed that all angles of incidence are equally probable. As the intention is to use the transmission coefficient within the framework of SEA, we note that the SEA assumption of equipartition of modal energy on each plate relates to the assumption that the incident energy is uniformly distributed in angle.

The intensity that is incident upon the junction length, L_{ij}, is proportional to $L_{ij} \cos \theta$, hence the power transmitted from plate i to plate j is (Cremer, 1973)

$$W_{ij} = \int_{-\pi/2}^{\pi/2} \tau_{ij}(\theta) L_{ij} \cos \theta \, \mathrm{d}I(\theta) \tag{5.3}$$

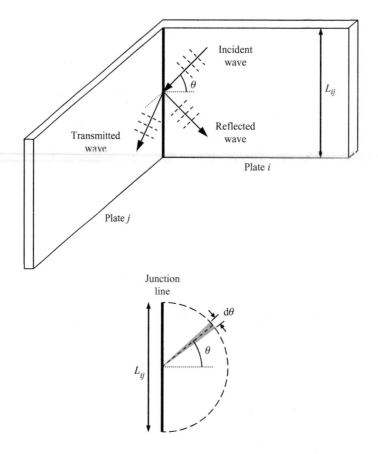

Figure 5.4

Plane wave incident at an angle, θ, upon a junction, giving rise to a reflected and a transmitted wave.

As the incident energy is uniformly distributed in angle, the intensity $dI(\theta)$ associated with $d\theta$ (see Fig. 5.4) is given by

$$dI(\theta) = \frac{c_{g,i}E_i}{S_i}\frac{d\theta}{2\pi}$$ (5.4)

Equations 5.3 and 5.4 now give the transmitted power as

$$W_{ij} = \frac{c_{g,i}E_iL_{ij}}{\pi S_i}\int_0^{\pi/2}\tau_{ij}(\theta)\cos\theta\,d\theta$$ (5.5)

and the angular average transmission coefficient, τ_{ij}, is therefore given by

$$\tau_{ij} = \int_0^{\pi/2}\tau_{ij}(\theta)\cos\theta\,d\theta$$ (5.6)

The angular average transmission coefficient is weighted towards values at normal incidence. For this reason, normal incidence transmission coefficients are sometimes used to give rough estimates of the angular average value; the reason is not that it is common for all incident waves to be normally incident upon a plate junction.

To reduce the number of calculations it is useful to note a relationship between the angular average transmission coefficients, τ_{ij} and τ_{ji}. For a bending wave that is incident upon a junction of homogeneous isotropic plates where only reflected and transmitted bending waves are generated at the junction, the SEA consistency relationship (Eq. 4.2) links τ_{ij} to τ_{ji} using

$$\tau_{ij} = \tau_{ji}\sqrt{\frac{h_i c_{L,i}}{h_j c_{L,j}}} = \tau_{ji}\sqrt{\frac{f_{c,j}}{f_{c,i}}} \tag{5.7}$$

5.2.1.2 Angles of incidence and transmission

In calculating the transmission coefficient over the range of possible angles of incidence it is necessary to relate the angle of incidence to the angle of transmission. From Snell's law of refraction, the incident, reflected and transmitted waves must have the same spatial dependence along the junction line in terms of their wavenumber. The angle of incidence on plate i is therefore related to the angle of transmission on plate j by

$$k_i \sin\theta_i = k_j \sin\theta_j \tag{5.8}$$

where k is the wavenumber on each plate; in this section we are only concerned with bending waves but it equally applies to quasi-longitudinal or transverse shear waves.

The angle of transmission can only take real (not complex) values. Therefore when $k_i > k_j$ there must be a cut-off angle, θ_{co}. For angles of incidence greater than the cut-off angle there is no transmitted wave and the transmission coefficient is zero. From Eq. 5.8 this cut-off angle is given by

$$\theta_{co} = \arcsin\left(\frac{k_j}{k_i}\right) \tag{5.9}$$

In this section we only consider bending waves and because the frequency-dependence of the bending wavenumber is the same for all thin plates the cut-off angle does not vary with frequency. However, when both bending and in-plane waves are considered (Section 5.2.2), the cut-off angle can be frequency-dependent.

5.2.1.3 Rigid X, T, L, and in-line junctions

In many buildings there is a degree of symmetry between adjacent rooms; hence some of the plates that form the junction will be identical in terms of their thickness and material properties. This conveniently reduces the number of variables that are needed to calculate the transmission coefficients. Referring to the plate junctions in Fig. 5.3, the following assumptions apply:

1. X-junction: plates 1 and 3 are identical, and plates 2 and 4 are identical
2. T-junction (1): plates 1 and 3 are identical
3. T-junction (2): plates 2 and 4 are identical.

This allows the transmission coefficients to be calculated using the variables χ and ψ (Cremer et al., 1973)

$$\chi = \frac{k_{B2}}{k_{B1}} = \sqrt[4]{\frac{\rho_{s2}B_1}{\rho_{s1}B_2}} = \sqrt{\frac{h_1 c_{L1}}{h_2 c_{L2}}} = \sqrt{\frac{f_{c2}}{f_{c1}}} \tag{5.10}$$

$$\psi = \frac{B_2 k_{B2}^2}{B_1 k_{B1}^2} = \frac{h_2 c_{L2}\rho_{s2}}{h_1 c_{L1}\rho_{s1}} = \frac{\rho_{s2}f_{c1}}{\rho_{s1}f_{c2}} \tag{5.11}$$

The wave approach used to model an incident bending wave that results in only reflected and transmitted bending waves is thoroughly covered in the book by Craik (1981); this also contains tabulated data for bending wave transmission across different junctions of plates and beams. In this section a short qualitative discussion is given along with the main equations needed to calculate the transmission coefficients for the most common junctions; X, T, L, and in-line junctions.

To model the rigid junction the plates are connected by a junction beam (see Fig. 5.3). This beam has no mass, does not support wave motion and has a rigid cross-section. We can now define the conditions that result in an incident bending wave only giving reflected and transmitted bending waves (not in-plane waves). This occurs when the junction beam is simply supported (sometimes referred to as being pinned) so that it cannot undergo displacement, but is free to rotate. Hence when a bending wave is incident upon the junction, the junction beam can 'transfer' rotation to the other plates, this will only give rise to bending waves. The junction beam is used to define equilibrium and continuity conditions that must be satisfied for this to occur. The equilibrium conditions ensure that the sum of the moments acting on the junction beam is zero. The continuity conditions ensure that the displacement of the beam and each plate along their line of connection is zero, and that the rotation of the beam and the plates is equal. The resulting transmission coefficients are independent of frequency; this simplifies their calculation as well as simplifying further calculations in SEA or SEA-based models.

For the junctions of perpendicular plates shown in Fig. 5.3 it is convenient to refer to 'transmission around a corner' and 'transmission across a straight section'. To calculate the transmission coefficients we assume that there is an incident bending wave on plate 1. For the X- and the L-junction, any of the plates can be chosen as plate 1 to calculate the transmission coefficients. For T-junctions, transmission around the corner depends on the relative orientation of the other plates and we need to refer to T-junctions (1) and (2); it is not necessary to make this distinction for transmission across the straight section of a T-junction.

Using the plate numbering system in Fig. 5.3 with an incident bending wave on plate 1, transmission around the corner of any rigid X- , T- or L-junction is given by (Craik, 1981, 1996).

If $\chi \geq \sin\theta$, then

$$\tau_{12}(\theta) = \frac{0.5 J_1 J_2 \psi \cos\theta \sqrt{\chi^2 - \sin^2\theta}}{(J_2\psi)^2 + \chi^2 + J_2\psi \left(\sqrt{1 + \sin^2\theta}\sqrt{\chi^2 + \sin^2\theta} + \sqrt{1 - \sin^2\theta}\sqrt{\chi^2 - \sin^2\theta}\right)} \tag{5.12}$$

else if $\chi < \sin\theta$, then

$$\tau_{12}(\theta) = 0$$

where the constants J_1 and J_2 depend on the junction. For X-junctions, $J_1 = 1$ and $J_2 = 1$. For T-junction (1), $J_1 = 2$ and $J_2 = 0.5$. For T-junction (2), $J_1 = 2$ and $J_2 = 2$. For L-junctions, $J_1 = 4$ and $J_2 = 1$.

For an incident bending wave on plate 1, transmission across a straight section of a rigid X-junction or the straight section of rigid T-junction (1) is given by (Craik, 1981, 1996).

If $\chi \geq \sin\theta$, then

$$\tau_{13}(\theta) = \frac{0.5\chi^2 \cos^2\theta}{(J_3\psi)^2 + \chi^2 + J_3\psi \left(\sqrt{1 + \sin^2\theta}\sqrt{\chi^2 + \sin^2\theta} + \sqrt{1 - \sin^2\theta}\sqrt{\chi^2 - \sin^2\theta}\right)} \tag{5.13}$$

else if $\chi < \sin\theta$, then

$$\tau_{13}(\theta) = \frac{\cos^2\theta}{2 + \dfrac{(J_3\psi)^2 C^2}{\chi^4} + \dfrac{2J_3\psi C}{\chi^2}\sqrt{1 + \sin^2\theta}}$$

where $C = \sqrt{\chi^2 + \sin^2\theta} + \sqrt{\sin^2\theta - \chi^2}$

and the constant J_3 depends on the junction. For X-junctions, $J_3 = 1$. For T-junction (1), $J_3 = 0.5$.

For an in-line junction, the normal incidence transmission coefficient gives estimates within 1 dB of the angular average value when $\chi \geq 1$. When $\chi < 1$, calculation in the reverse direction can be carried out first and then Eq. 5.7 can be used to give the value in the other direction (Craik, 1996). For a bending wave on plate 1 at normal incidence upon the junction line, transmission across a rigid in-line junction is given by (Cremer et al., 1973)

$$\tau_{12} \approx \tau_{12}(0°) = \left[\frac{2(1 + \chi)(1 + \psi)\sqrt{\chi\psi}}{\chi(1 + \psi)^2 + 2\psi(1 + \chi^2)} \right]^2 \tag{5.14}$$

5.2.1.3.1   Junctions of beams

The calculations for plate junctions in Section 5.2.1.3 can be adapted to beams. Calculations for junctions of beams are simpler because all waves are at normal incidence, so only the normal incidence transmission coefficient is used and no angular averaging is needed. The variables χ and ψ are calculated from Eqs 5.10 and 5.11 using the bending wavenumbers for beams, the beam bending stiffness and the mass per unit length.

Note that timber studs and joists that form beams in timber-frame constructions may need to be modelled as hinged rather than rigid junctions (Craik and Galbrun, 2005).

5.2.2   Wave approach: bending and in-plane waves

Compared to the wave approach for only bending waves, the inclusion of in-plane waves increases the complexity of the solution, as well as subsequent computation of the transmission coefficients. Major steps in applying the wave approach to bending and in-plane wave generation at plate junctions were made in the work of Cremer et al. (1973) and Kihlman (1967). Subsequent increases in computing power aided further advances in application and development of the theory with other types of plate junctions (Craven and Gibbs, 1981; Gibbs and Gilford, 1976; Langley and Heron, 1990; Wöhle et al., 1981). In general there are two aspects being considered simultaneously; the physics of the problem, and numerical computation. There are advantages in approaching a derivation from one or the other viewpoint as it is difficult to do justice to both aspects simultaneously. The derivation in this section only gives a brief overview of the theory to try and simplify subsequent computation. It is based on the notation and derivation given by Mees and Vermeir (1993) and Bosmans (1998); this makes a clear link back to the theory given by Cremer et al. (1973).

An example junction is shown in Fig. 5.5 where there are four plates connected by a junction beam. The derivation is general and equally applies to two, three, or four plates at any angle to each other. Hence it can be used for L, T, X and in-line junctions. The coordinate systems used

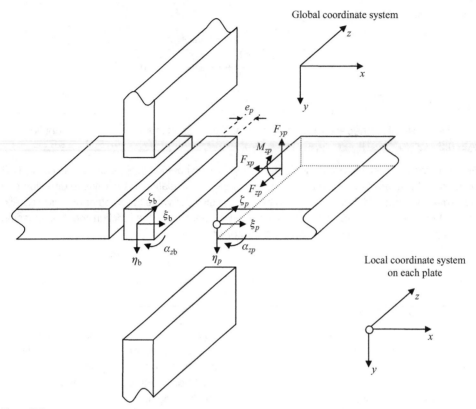

Figure 5.5

Exploded view of a X-junction indicating the variables used to describe wave motion on each plate (subscript p) and on the junction beam (subscript b). The z-axis of the global coordinate system is aligned along the centre line of the junction beam.

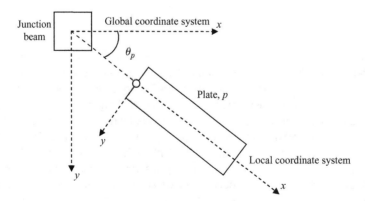

Figure 5.6

Local and global coordinate system for each plate connected to the junction beam.

to describe each plate and their position relative to the junction beam are shown in Fig. 5.6. Wave motion on each plate is described using a local coordinate system where each plate lies in the xz plane. Each plate that forms the junction is then connected to the junction beam using the global coordinate system.

The plates are assumed to be solid, homogeneous, and isotropic, and are modelled using thin plate theory. The model uses a junction beam to connect the plates together, but it does not represent a physical part of the real junction. The junction beam has no mass, does not support wave motion and has a rigid cross-section. This beam is free to rotate and to undergo displacement in the three coordinate directions; this allows generation of in-plane waves at the junction. With each plate there is an offset, e_p, from the junction beam. For rigid junctions this offset can be set to zero; note that masonry/concrete walls or floors can be rigidly connected together in more than one way, and there is no clear way of exactly defining any offset. However, it is included here because the offset does have a physical meaning when altering this basic model to account for a junction beam with mass and stiffness that supports wave motion, resilient layers between one or more plates and the junction beam, or a hinged junction (see Bosmans, 1998; Mees and Vermeir, 1993).

Incident waves can be bending, quasi-longitudinal, or transverse shear waves that travel in the negative x-direction towards the junction at $x = 0$ (local coordinates). For any incident wave with unit amplitude that impinges upon the junction at an angle of incidence, θ_i, the general form for the wave is

$$\exp(ik_i x \cos\theta_i)\exp(-ik_i z \sin\theta_i) \tag{5.15}$$

where k_i is the wavenumber.

This gives the general form for transmitted waves as

$$T\exp(-ik_x x)\exp(-ik_i z \sin\theta_i) \tag{5.16}$$

where T is the complex amplitude.

Note that the time-dependence, $\exp(i\omega t)$, has been excluded for brevity, and that although the z-dependence is needed to derive the plate and beam parameters at the junction ($x = 0$) it is omitted wherever it eventually cancels out in the final set of equations.

5.2.2.1 Bending waves

The general equations for bending wave motion are given in Section 2.3.3. However to accommodate the number of subscripts that are needed here, the subscript notation is slightly different. The variables are shown in Fig. 5.5 for one of the plates that form the junction. The subscript p identifies each plate that forms the junction. For lateral displacement, η, associated with bending wave motion, there is rotation about the z-axis by an angle, α_z, given by

$$\alpha_{zp} = \frac{\partial\eta}{\partial x} \tag{5.17}$$

a bending moment per unit width,

$$M_{zp} = -B\left(\frac{\partial^2\eta}{\partial x^2} + v\frac{\partial^2\eta}{\partial z^2}\right) \tag{5.18}$$

and a shear force per unit width,

$$F_{yp} = -B\left[\frac{\partial^3\eta}{\partial x^3} + (2-v)\frac{\partial^3\eta}{\partial x\partial z^2}\right] \tag{5.19}$$

5.2.2.1.1 Incident bending wave at the junction

For an incident bending wave on plate $p = 1$ that is described by the general form in Eq. 5.15 and impinges upon the junction at $x = 0$, the displacement, rotation (Eq. 5.17), bending moment (Eq. 5.18) and shear force (Eq. 5.19) are given by

$$\eta_{1i} = 1 \tag{5.20}$$

$$\alpha_{1i} = ik_{B1} \cos\theta_i \tag{5.21}$$

$$M_{z1i} = B_1 k_{B1}^2 (\cos^2\theta_i + \nu_1 \sin^2\theta_i) \tag{5.22}$$

$$F_{y1i} = iB_1 k_{B1}^3 \cos\theta_i [\cos^2\theta_i + (2 - \nu_1)\sin^2\theta_i] \tag{5.23}$$

5.2.2.1.2 Transmitted bending wave at the junction

For the transmitted bending wave propagating on plate p in the positive x-direction, the displacement is given by

$$\eta_p(x, z) = [T_{BNp} \exp(-ik_{Bpx1}x) + T_{Np} \exp(-ik_{Bpx2}x)] \exp(-ik_i z \sin\theta_i) \tag{5.24}$$

where T_{BNp} (Bending wave/Nearfield) and T_{Np} (Nearfield) are complex amplitudes.

Solutions to the wave equation lead to either real or imaginary values for the wavenumber, k_{Bpx1}, depending on the angle of incidence; but only give imaginary values for the wavenumber, k_{Bpx2}. To interpret Eq. 5.24 in terms of propagating bending waves and nearfields we use Snell's law (Eq. 5.8) to relate the angle of incidence to the angle at which waves are transmitted. Snell's law gives a cut-off angle, θ_{co} (Eq. 5.9) for the angle of incidence; at larger angles there is no transmitted propagating wave. Therefore the first term in the square brackets of Eq. 5.24 corresponds to a propagating bending wave with complex amplitude, T_{BNp}, when $\theta_i \leq \theta_{co}$ but changes to a nearfield when $\theta_i > \theta_{co}$. The second term corresponds to a nearfield with complex amplitude, T_{Np}, regardless of the angle of incidence. Substituting the general form for a transmitted wave into the bending wave equation gives the wavenumbers as follows:

$$\text{If } \theta_i \leq \theta_{co}, \text{ then } k_{Bpx1} = \sqrt{k_{Bp}^2 - k_i^2 \sin^2\theta_i}$$

$$\text{else, if } \theta_i > \theta_{co} \text{ then } k_{Bpx1} = -i\sqrt{k_i^2 \sin^2\theta_i - k_{Bp}^2} \tag{5.25}$$

$$k_{Bpx2} = -i\sqrt{k_{Bp}^2 + k_i^2 \sin^2\theta_i} \tag{5.26}$$

The displacement, rotation, bending moment and shear force for the transmitted bending wave at $x = 0$ can now be written in terms of T_{BNp} and T_{Np},

$$\eta_p = T_{BNp} + T_{Np} \tag{5.27}$$

$$\alpha_{zp} = -ik_{Bpx1} T_{BNp} - ik_{Bpx2} T_{Np} \tag{5.28}$$

$$M_{zp} = B_p T_{BNp}(k_{Bpx1}^2 + \nu_p k_i^2 \sin^2\theta_i) + B_p T_{Np}(k_{Bpx2}^2 + \nu_p k_i^2 \sin^2\theta_i) \tag{5.29}$$

$$F_{yp} = -iB_p k_{Bpx1} T_{BNp}[k_{Bpx1}^2 + (2 - \nu_p)k_i^2 \sin^2\theta_i]$$

$$- iB_p k_{Bpx2} T_{Np}[k_{Bpx2}^2 + (2 - \nu_p)k_i^2 \sin^2\theta_i] \tag{5.30}$$

To simplify computation and calculations of wave intensities it is useful to rename the various terms in Eqs 5.27–5.30 as constants $C1$ and $C2$ (Bosmans, 1998)

$$\begin{bmatrix} \eta_p \\ \alpha_{zp} \\ M_{zp} \\ F_{yp} \end{bmatrix} = \begin{bmatrix} C1_{\eta p} & C2_{\eta p} \\ C1_{\alpha zp} & C2_{\alpha zp} \\ C1_{Mzp} & C2_{Mzp} \\ C1_{Fyp} & C2_{Fyp} \end{bmatrix} \begin{bmatrix} T_{BNp} \\ T_{Np} \end{bmatrix} \tag{5.31}$$

5.2.2.2 In-plane waves

Bending wave motion could be dealt with in isolation, but the in-plane wave motion is linked to both quasi-longitudinal and transverse shear waves. The in-plane displacements, ξ and ζ, are shown in Fig. 5.5, these are related by two coupled equations of motion (Cremer *et al.*, 1973)

$$\frac{E}{1-v^2}\left(\frac{\partial^2 \xi}{\partial x^2} + v\frac{\partial^2 \zeta}{\partial x \partial z} \right) + \frac{E}{2(1+v)}\left(\frac{\partial^2 \xi}{\partial z^2} + \frac{\partial^2 \zeta}{\partial x \partial z} \right) - \rho\frac{\partial^2 \xi}{\partial t^2} = 0 \tag{5.32}$$

$$\frac{E}{1-v^2}\left(\frac{\partial^2 \zeta}{\partial z^2} + v\frac{\partial^2 \xi}{\partial x \partial z} \right) + \frac{E}{2(1+v)}\left(\frac{\partial^2 \zeta}{\partial x^2} + \frac{\partial^2 \xi}{\partial x \partial z} \right) - \rho\frac{\partial^2 \zeta}{\partial t^2} = 0 \tag{5.33}$$

From Cremer *et al.* (1973) the wave fields are described using a potential, Φ, and a stream function, Ψ,

$$\Phi(x, z) = [\Phi_+ \exp(ik_L x \cos \theta) + \Phi_- \exp(-ik_L x \cos \theta)] \exp(-ik_L z \sin \theta) \tag{5.34}$$

$$\Psi(x, z) = [\Psi_+ \exp(ik_T x \cos \theta) + \Psi_- \exp(-ik_T x \cos \theta)] \exp(-ik_T z \sin \theta) \tag{5.35}$$

from which the in-plane displacements are given by

$$\xi = \frac{\partial \Phi}{\partial x} + \frac{\partial \Psi}{\partial z} \tag{5.36}$$

$$\zeta = \frac{\partial \Phi}{\partial z} - \frac{\partial \Psi}{\partial x} \tag{5.37}$$

and the equations of motion are

$$\Delta \Phi + k_L^2 \Phi = 0 \text{ for quasi-longitudinal waves} \tag{5.38}$$

$$\Delta \Psi + k_T^2 \Psi = 0 \text{ for transverse shear waves} \tag{5.39}$$

where Δ is the harmonic operator.

The in-plane waves give rise to a normal force, F_x, and an in-plane shear force, F_z, (see Fig. 5.5) which are related to the in-plane displacements by (Timoshenko and Woinowsky-Krieger, 1959)

$$F_x = \frac{Eh}{1-v^2}\left(\frac{\partial \xi}{\partial x} + v\frac{\partial \zeta}{\partial z} \right) \tag{5.40}$$

$$F_z = Gh\left(\frac{\partial \xi}{\partial z} + \frac{\partial \zeta}{\partial x} \right) \tag{5.41}$$

5.2.2.2.1 Incident quasi-longitudinal wave at the junction

A unit amplitude quasi-longitudinal wave on plate $p = 1$, that is described by the general form in Eq. 5.15 is given by

$$\Phi(x) = \exp(ik_i x \cos \theta_i) \exp(-ik_i z \sin \theta_i) \tag{5.42}$$

When this wave impinges upon the junction at $x = 0$, the displacements (Eqs 5.36 and 5.37), normal force (Eq. 5.40) and in-plane shear force (Eq. 5.41) are given by

$$\xi_{1i} = ik_i \cos \theta_i \tag{5.43}$$

$$\zeta_{1i} = -ik_i \sin \theta_i \tag{5.44}$$

$$F_{x1i} = \frac{-E_1 h_1}{1 - \nu_1^2} k_i^2 (\cos^2 \theta_i + \nu_1 \sin^2 \theta_i) \tag{5.45}$$

$$F_{z1i} = 2G_1 h_1 k_i^2 \cos \theta_i \sin \theta_i \tag{5.46}$$

5.2.2.2.2   Incident transverse shear wave at the junction

A unit amplitude transverse shear wave on plate $p = 1$ described by the general form in Eq. 5.15 is given by

$$\Psi(x) = \exp(ik_i x \cos \theta_i) \exp(-ik_i z \sin \theta_i) \tag{5.47}$$

When this wave impinges upon the junction at $x = 0$, the displacements (Eqs 5.36 and 5.37), normal force (Eq. 5.40) and in-plane shear force (Eq. 5.41) are given by

$$\xi_{1i} = -ik_i \sin \theta_i \tag{5.48}$$

$$\zeta_{1i} = -ik_i \cos \theta_i \tag{5.49}$$

$$F_{x1i} = 2G_1 h_1 k_i^2 \cos \theta_i \sin \theta_i \tag{5.50}$$

$$F_{z1i} = G_1 h_1 k_i^2 (\cos^2 \theta_i - \sin^2 \theta_i) \tag{5.51}$$

5.2.2.2.3   Transmitted in-plane waves at the junction

For the transmitted in-plane waves propagating on plate p in the positive x-direction, the displacements are given by

$$\Phi_p(x, z) = T_{Lp} \exp(-ik_{Lpx} x) \exp(-ik_i z \sin \theta_i) \tag{5.52}$$

$$\Psi_p(x, z) = T_{Tp} \exp(-ik_{Tpx} x) \exp(-ik_i z \sin \theta_i) \tag{5.53}$$

where T_{Lp} and T_{Tp} are complex amplitudes.

Snell's law gives the cut-off angle for the quasi-longitudinal and transverse shear waves so their wavenumbers are given as follows:

$$\text{If } \theta_i \leq \theta_{co}, \text{ then } k_{Lpx} = \sqrt{k_{Lp}^2 - k_i^2 \sin^2 \theta_i}$$

$$\text{else, if } \theta_i > \theta_{co} \text{ then } k_{Lpx} = -i\sqrt{k_i^2 \sin^2 \theta_i - k_{Lp}^2} \tag{5.54}$$

$$\text{If } \theta_i \leq \theta_{co}, \text{ then } k_{Tpx} = \sqrt{k_{Tp}^2 - k_i^2 \sin^2 \theta_i}$$

$$\text{else, if } \theta_i > \theta_{co} \text{ then } k_{Tpx} = -i\sqrt{k_i^2 \sin^2 \theta_i - k_{Tp}^2} \tag{5.55}$$

The displacements, normal force and in-plane shear force for the transmitted in-plane wave at the junction ($x = 0$) can now be written in terms of T_{Lp} and T_{Tp},

$$\xi_p = -ik_{Lpx}T_{Lp} - ik_i \sin \theta_i T_{Tp} \tag{5.56}$$

$$\zeta_p = -ik_i \sin \theta_i T_{Lp} + ik_{Tpx}T_{Tp} \tag{5.57}$$

$$F_{xp} = \frac{-E_p h_p T_{Lp}}{1 - v_p^2}(k_{Lpx}^2 + v_p k_i^2 \sin^2 \theta_i) - 2G_p h_p k_{Tpx} k_i \sin \theta_i T_{Tp} \tag{5.58}$$

$$F_{zp} = -2G_p h_p k_{Lpx} k_i T_{Lp} \sin \theta_i - G_p h_p T_{Tp}(k_i^2 \sin^2 \theta_i - k_{Tpx}^2) \tag{5.59}$$

As with a transmitted bending wave it is also useful to define constants C1 and C2; from Eqs 5.56–5.59 these can be defined as (Bosmans, 1998)

$$\begin{bmatrix} \xi_p \\ \zeta_p \\ F_{xp} \\ F_{zp} \end{bmatrix} = \begin{bmatrix} C1_{\xi p} & C2_{\xi p} \\ C1_{\zeta p} & C2_{\zeta p} \\ C1_{Fxp} & C2_{Fxp} \\ C1_{Fzp} & C2_{Fzp} \end{bmatrix} \begin{bmatrix} T_{Lp} \\ T_{Tp} \end{bmatrix} \tag{5.60}$$

5.2.2.2.4 Conditions at the junction beam

Up till this point the focus has been on wave motion on each plate. To calculate the required transmission coefficients for an incident wave on plate 1, the derived equations must satisfy equilibrium and continuity conditions at the junction beam. The equilibrium conditions in Eqs 5.61–5.64 ensure that the sum of forces acting on the junction beam equals zero. To do this it is necessary to make use of the offset, e_p, as well as the angle, θ_p, between each plate and the junction beam in the global coordinate system. The continuity conditions in Eqs 5.65–5.68 ensure continuity between plate and beam motion (displacement and rotation) along the line that they are connected together; these equations are needed for each plate that forms the junction.

The following set of equations are solved by converting them into matrix format and using a matrix inversion to give the complex amplitudes (T_{BNp}, T_{Np}, T_{Lp}, T_{Tp}) on each plate as well as the junction beam parameters (ξ_b, η_b, ζ_b, α_{zb}). For use in SEA calculations it is sufficient to carry out the calculations at the band centre frequencies:

$$\sum_p (F_{xp} \cos \theta_p - F_{yp} \sin \theta_p) + (F_{x1i} \cos \theta_1 - F_{y1i} \sin \theta_1) = 0 \tag{5.61}$$

$$\sum_p (F_{xp} \sin \theta_p + F_{yp} \cos \theta_p) + (F_{x1i} \sin \theta_1 + F_{y1i} \cos \theta_1) = 0 \tag{5.62}$$

$$-\sum_p M_{zp} + \sum_p e_p F_{yp} + (e_1 F_{y1i} - M_{z1i}) = 0 \tag{5.63}$$

$$\sum_p F_{zp} + F_{z1i} = 0 \tag{5.64}$$

$$\xi_p(+\xi_{1i} \text{ if } p = 1) = \xi_b \cos \theta_p + \eta_b \sin \theta_p \tag{5.65}$$

$$\eta_p(+\eta_{1i} \text{ if } p = 1) = -\xi_b \sin \theta_p + \eta_b \cos \theta_p + e_p \alpha_{zb} \tag{5.66}$$

$$\zeta_p(+\zeta_{1i} \text{ if } p = 1) = \zeta_b \tag{5.67}$$

$$\alpha_{zp}(+\alpha_{z1i} \text{ if } p = 1) = \alpha_{zb} \tag{5.68}$$

5.2.2.2.5 Transmission coefficients

The transmission coefficient for any incident wave on plate 1 that is transmitted to plate p at a specific angle of incidence is calculated from the ratio of the wave intensities in the x-direction using

$$\tau(\theta_i) = \frac{I_{xp}(\theta_i)}{I_{x1i}(\theta_i)} \tag{5.69}$$

Note that Eq. 5.69 is also used to calculate transmission coefficients from an incident wave to a different wave type on the same plate, as well as to the same wave type on the same plate; although the latter are referred to as reflection coefficients.

For a unit amplitude wave incident upon the junction, the x-direction intensities are (Cremer et al., 1973)

$$I_{Bx1i}(\theta_i) = B_1 \omega k_{B1}^3 \cos \theta_i \quad \text{for an incident bending wave} \tag{5.70}$$

$$I_{Lx1i}(\theta_i) = 0.5 \rho_{s1} \omega^3 k_{L1} \cos \theta_i \quad \text{for an incident quasi-longitudinal wave} \tag{5.71}$$

$$I_{Tx1i}(\theta_i) = 0.5 \rho_{s1} \omega^3 k_{T1} \cos \theta_i \quad \text{for an incident transverse shear wave} \tag{5.72}$$

For the transmitted waves on plate p, the x-direction intensities are (Cremer et al., 1973)

$$I_{Bxp}(\theta_i) = \frac{1}{2} \text{Re}\{F_{yp}(-i\omega\eta_p)^* + M_{zp}(i\omega\alpha_{zp})^*\} \quad \text{for transmitted bending waves} \tag{5.73}$$

$$I_{LTxp}(\theta_i) = \frac{1}{2} \text{Re}\{F_{xp}(-i\omega\xi_p)^* + F_{zp}(-i\omega\zeta_p)^*\} \quad \text{for transmitted in-plane waves} \tag{5.74}$$

where * denotes the complex conjugate.

The transmitted wave intensities can now be re-written in terms of the complex amplitudes determined from the matrix solution as (Bosmans, 1998)

$$I_{Bxp}(\theta_i) = \frac{-\omega}{2} \text{Im}\{C1_{Fyp} - C1_{Mzp} C1_{\alpha zp}^*\} |T_{BNp}|^2 \tag{5.75}$$

$$I_{Lxp}(\theta_i) = \frac{-\omega}{2} \text{Im}\{C1_{Fxp} C1_{\xi p}^* + C1_{Fzp} C1_{\zeta p}^*\} |T_{Lp}|^2 \tag{5.76}$$

$$I_{Txp}(\theta_i) = \frac{-\omega}{2} \text{Im}\{C2_{Fxp} C2_{\xi p}^* + C2_{Fzp} C2_{\zeta p}^*\} |T_{Tp}|^2 \tag{5.77}$$

The angular average transmission coefficients are calculated using Eq. 5.6. Transmission coefficients usually fluctuate rapidly with incident angle; hence it is important to use a fine angular resolution. This is in contrast to the wave approach for only bending waves where the transmission coefficients tend to vary more slowly with incident angle. Each coupling loss factor between the plates is calculated from Eq. 2.154.

5.2.2.2.6 Application to SEA models

To incorporate bending, quasi-longitudinal, and transverse shear waves into an SEA model; each wave type requires its own subsystem. An example is shown in Fig. 5.7 using a T-junction. Some transmission coefficients (and hence coupling loss factors) between different wave types may be zero. However, it is necessary to include three subsystems for each plate because even if wave conversion at the junction does not directly connect two subsystems together, there can be transmission paths involving other subsystems that do connect them together. If there is equipartition of modal energy between quasi-longitudinal and transverse shear modes it is possible to use a single in-plane subsystem to represent both quasi-longitudinal and transverse

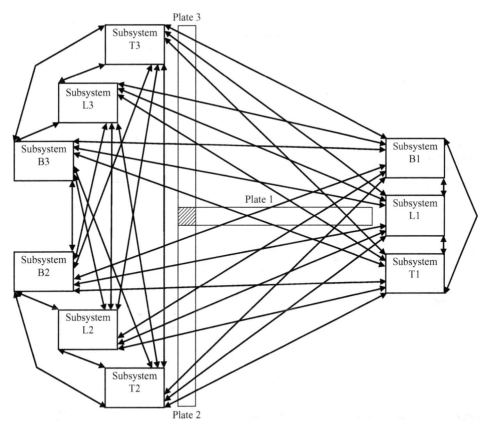

Figure 5.7

Nine-subsystem SEA model for three plates forming a T-junction. Subsystems for bending (B), quasi-longitudinal (L), and transverse shear (T) wave motion are used for each plate. Double headed arrows are used to indicate all possible couplings between subsystems; depending on the junction detail, some of the coupling loss factors will be zero.

shear waves (Craik and Thancanamootoo, 1992). As this is not usually known *a priori* and requires calculation of the same transmission coefficients, this approach is not often used unless it is beneficial in simplifying analysis of the results.

The wave approaches discussed in this section assume semi-infinite, thin plates. At this point we want to use SEA subsystems to represent reverberant vibration fields associated with each type of in-plane wave on finite plates. Wavelengths of in-plane waves are much longer than those for bending waves (refer back to Fig. 2.2) and the fundamental quasi-longitudinal and transverse shear modes tend to occur in the mid-frequency range. We can only consider use of an SEA model where there are reverberant vibration fields; this means that inclusion of subsystems for in-plane waves will only usually be appropriate in the mid- and high-frequency ranges above the fundamental in-plane mode frequencies. Now it is useful to look at the thin plate limit for bending waves on masonry/concrete walls in terms of the frequencies at which the fundamental in-plane modes occur. Figure 5.8 shows these fundamental mode frequencies for typical wall dimensions and quasi-longitudinal phase velocities that correspond to masonry/concrete. For 100 mm plates the thin plate limit is much higher than the fundamental in-plane mode frequencies. For 200 mm plates we find that at frequencies where these plates

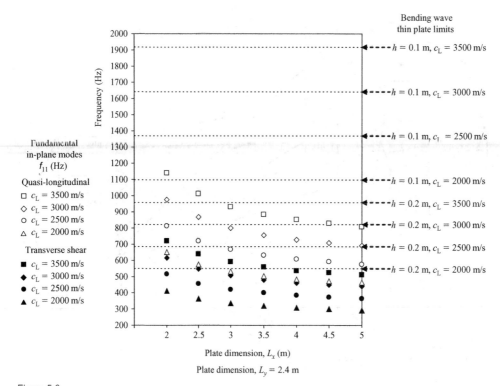

Figure 5.8

Comparison of thin plate limits with fundamental in-plane mode frequencies for a plate with dimensions L_x and L_y.

can start to be considered as being multi-modal for bending waves, the confounding features of non-diffuse in-plane wave fields need to be considered at the same time as thick plate bending wave theory. As previously noted in practical calculations of airborne sound insulation (Section 4.3.1.4), the thin plate limit is not an absolute limit and different approaches can be taken. The simplest approach is to continue to use thin plate bending wave theory along with the transmission coefficients derived for thin plates; comparisons with measurements often show that this is perfectly adequate (e.g. see Kihlman, 1967). An alternative is to continue to use the thin plate transmission coefficients but to use the group velocity for thick plates when calculating coupling loss factors from the bending wave subsystem (Craik and Thancanamootoo, 1992). A more complex alternative is to derive transmission coefficients that take account of shear deformation and rotatory inertia (McCollum and Cuschieri, 1990). A purely theoretical consideration of incorporating in-plane waves into SEA models of masonry/concrete buildings tends to be rather pessimistic of the outcome (e.g. see de Vries et al., 1981). However, comparisons of measurements and predictions in masonry/concrete buildings indicate reasonable agreement for vibration transmission across a single junction as well as long distance transmission across many junctions (Craik and Thancanamootoo, 1992; Kihlman, 1967; Roland, 1988).

5.2.2.3 Example: Comparison of wave approaches

It is now useful to compare the two wave approaches. We will look at a T-junction of masonry walls that represents a solid separating wall (plate 1) rigidly connected to two solid flanking walls (plates 2 and 3).

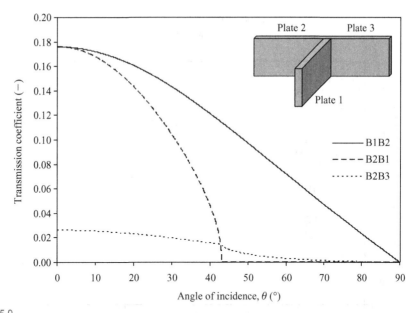

Figure 5.9

Angle-dependent transmission coefficients between three plates that form a T-junction ($h_1 = 0.215\,\text{m}$, $h_2 = h_3 = 0.1\,\text{m}$, $\rho = 2000\,\text{kg/m}^3$, $c_L = 3200\,\text{m/s}$, $\nu = 0.2$). These are calculated using the wave approach for bending waves only. The legend refers to BiBj which corresponds to an incident bending wave on plate i that results in a transmitted bending wave on plate j.

The wave approach for only bending waves gives angle-dependent transmission coefficients that are independent of frequency. These are shown in Fig. 5.9. For an incident wave on plate 2 that is transmitted to plate 1, $k_{B2} > k_{B1}$, and there is a cut-off angle, $\theta_{co} = 43°$ (Eq. 5.9) above which no bending wave is transmitted and the transmission coefficient is zero. As plates 2 and 3 are the same, this cut-off angle would also occur for an incident wave on plate 3 that is transmitted to plate 1.

The wave approach for bending and in-plane waves gives angle-dependent transmission coefficients that vary with frequency, and there are more transmission coefficients to consider. For this reason it is simpler to present the results by grouping all the transmission and reflection coefficients together. As an example; for an incident bending wave on plate 1, conservation of energy ensures that

$$\tau_{B1B1} + \tau_{B1B2} + \tau_{B1B3} + \tau_{B1L1} + \tau_{B1L2} + \tau_{B1L3} + \tau_{B1T1} + \tau_{B1T2} + \tau_{B1T3} = 1 \qquad (5.78)$$

where B, L, and T indicate bending, quasi-longitudinal, and transverse shear waves respectively. Note that τ_{B1B1} would normally be referred to as a reflection coefficient.

This grouping makes it possible to use a single graph to show the transmission and reflection coefficients at a single frequency when one wave type, on one plate, is incident upon the junction. Examples are shown in Fig. 5.10 for an incident bending wave. On these graphs it is the vertical distance between the lines that gives the transmission coefficient. At 100 Hz (Fig. 5.10 a and b), wave conversion from bending to in-plane waves mainly occurs below 10°, whereas above 10° the majority of the incident intensity is only reflected or transmitted as bending waves. The trends are similar at 1000 Hz (Fig. 5.10 c and d) but there is wave

(a) Incident bending wave on plate 1 at 100 Hz

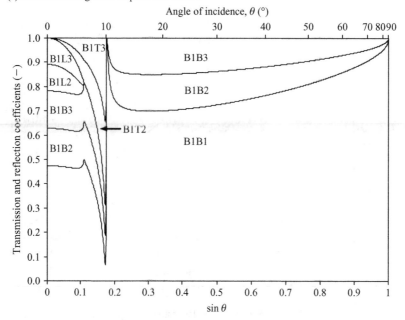

(b) Incident bending wave on plate 2 at 100 Hz

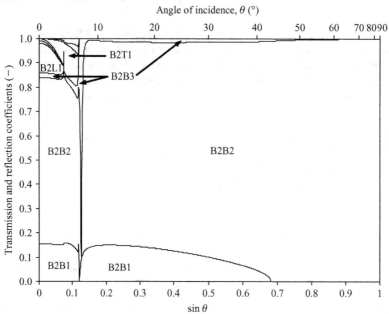

Figure 5.10

Angle-dependent transmission coefficients between three plates that form a T-junction ($h_1 = 0.215$ m, $h_2 = h_3 = 0.1$ m, $\rho = 2000$ kg/m³, $c_L = 3200$ m/s, $\nu = 0.2$). These are calculated using the wave approach for bending and in-plane waves. The descriptors refer to BiBj / BiLj / BiTj which correspond to an incident bending wave (B) on plate i that results in a transmitted bending wave (B) / quasi-longitudinal wave (L) / transverse shear wave (T) on plate j.

(c) Incident bending wave on plate 1 at 1000 Hz

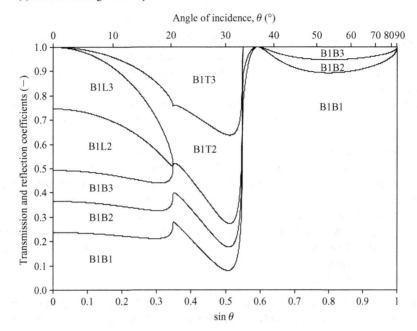

(d) Incident bending wave on plate 2 at 1000 Hz

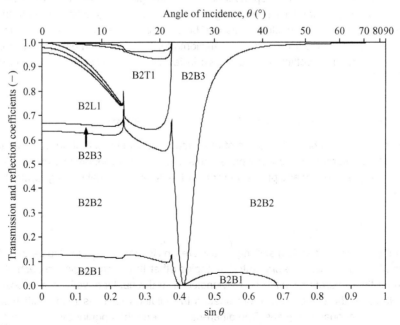

Figure 5.10

(*Continued*)

conversion over a slightly wider range of incident angles. At both 100 and 1000 Hz, we note that τ_{B2B3} has high values over a relatively narrow range of angles.

The features relating to these transmission coefficients can be considered in the context of the equivalent angle of incidence for an individual mode (Section 2.5.3). This indicates that with typical wall dimensions there is a limited range of angles at which waves impinge upon the plate boundaries, particularly in the low- and mid-frequency ranges (refer back to bending modes for a 215 mm wall in Fig. 2.30). Hence wave conversion from bending to in-plane waves may be predicted to occur at angles of incidence that are not present on a finite size plate. Fortunately, the uncertainty in describing wave motion on real walls and floors means that consideration of precise angles of incidence is not usually appropriate or necessary. When using these transmission coefficients with SEA, the assumption is made that there are diffuse vibration fields. Therefore it is the angular average transmission coefficient that is needed. This provides a reasonable estimate if there are a sufficient number of modes to approximately cover the range of angles of incidence. This makes a link to a requirement often used in SEA that there should be at least five modes in each frequency band ($N_s \geq 5$) to give estimates with low variance (Fahy and Mohammed, 1992). This requirement is not always satisfied for masonry/concrete walls and floors in the low- and mid-frequency ranges; this is discussed further in the examples in Section 5.2.3.

Angular average transmission coefficients are independent of frequency when the wave approach assumes that there are only bending waves. This is a useful feature to note because whilst this is the simpler and more convenient model, measurements on real walls and floors usually exhibit some frequency-dependence due to the generation of in-plane waves. This tends to be more apparent across the straight section of a T or X-junction rather than around the corner of an L, T, or X-junction. This can be seen in Fig. 5.11 by comparing the angular average bending wave transmission coefficients for the T-junction using the two different wave approaches; the coefficients are shown in terms of the transmission loss in decibels using Eq. 5.2.

5.2.2.4 Other plate junctions modelled using a wave approach

This section contains a brief discussion of other idealized junctions that are relevant to buildings. Note that automotive and aerospace industries also use models of wave transmission between connected beams and plates and some of these are applicable to lightweight building structures.

5.2.2.4.1 Junctions of angled plates

The wave approach for bending and in-plane waves in Section 5.2.2 can also be used for rigid junctions where the plates are orientated at angles other than right-angles to each other. It is common for the separating wall to be perpendicular to the flanking walls to give box-shaped rooms. However there are other room shapes in buildings such as those with hexagonal, triangular, or octagonal floor plans. The difference between the angular average transmission coefficients for perpendicular masonry/concrete walls in a T-junction when compared with a junction of the same three walls with angles between 60° and 135° can be up to 10 dB or more depending on the wall properties. Note that this does not usually result in similarly large differences for the airborne sound insulation between adjacent rooms in terms of a single-number quantity such as R_w or $D_{nT,w}$.

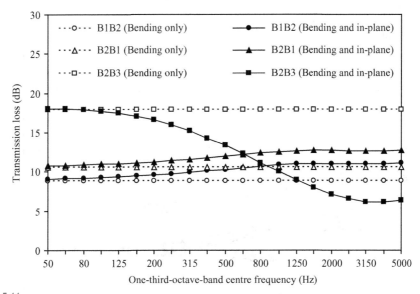

Figure 5.11

Angular average transmission coefficients (shown in decibels using the transmission loss, TL_{ij}) for bending waves transmitted across a T-junction ($h_1 = 0.215\,\text{m}$, $h_2 = h_3 = 0.1\,\text{m}$, $\rho = 2000\,\text{kg/m}^3$, $c_L = 3200\,\text{m/s}$, $\nu = 0.2$). Transmission coefficients were calculated using two wave approaches: one for bending waves only, and the other for bending and in-plane waves.

5.2.2.4.2 Resilient junctions

Examples of resilient plate junctions in buildings include a concrete floor slab resting on a continuous resilient layer on top of the supporting walls, as well as walls, floors or floating floors where a resilient material is used to form a vibration break or expansion joint. From Mees and Vermeir (1993) the wave approach for bending and in-plane waves can be extended to include an isotropic, homogeneous resilient layer at a plate junction. The resilient layer is treated as a spring hence it must be sufficiently thin that it does not support wave motion over the frequency range of interest.

Using the T-junction of masonry walls from Section 5.2.2.3, the transmission losses for the rigid T-junction can be compared to two examples of resilient T-junctions in Fig. 5.12. These indicate large changes in bending and in-plane wave transmission due to the resilient layers at the junction. For the resilient junction in Fig. 5.12b, the bending wave transmission loss corresponding to τ_{B2B3} in the mid- and high-frequency ranges is 0 dB; hence there is potential for significant flanking transmission via the two flanking walls. Both of the resilient junctions (Fig. 5.12 a and b) indicate the need to include in-plane wave subsystems in an SEA model. It is important to note that the insertion of resilient materials at junctions of walls and/or floors is not a panacea for increasing the sound insulation between rooms. Due to conservation of energy the use of resilient junctions results in a different distribution of energy between the space and structural subsystems compared to rigid junctions; hence resilient junctions can increase the sound insulation between some rooms in a building whilst decreasing it between others (Osipov and Vermeir, 1996). For this reason it is useful to be able to incorporate resilient junctions in a full SEA model of a building because calculating sound transmission between two specific rooms along a limited number of flanking paths will sometimes be misleading.

(a) Rigid junction

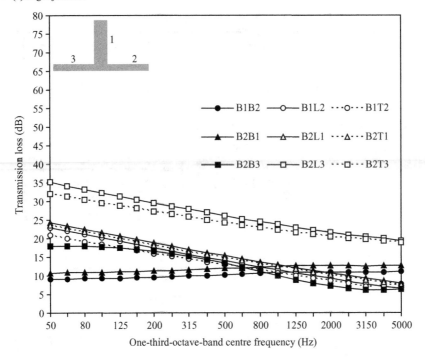

(b) Resilient junction (one resilient layer)

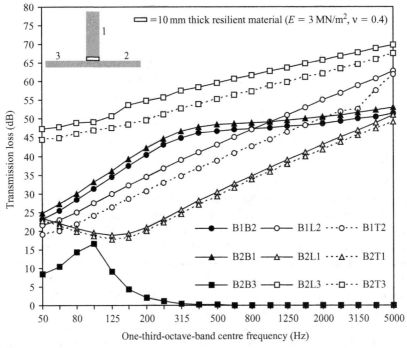

Figure 5.12

Rigid and resilient T-junctions. Angular average transmission coefficients (shown in decibels using the transmission loss, TL_{ij}) for an incident bending wave transmitted across a T-junction ($h_1 = 0.215$ m, $h_2 = h_3 = 0.1$ m, $\rho = 2000$ kg/m^3, $c_L = 3200$ m/s, $\nu = 0.2$). Values are calculated using a wave approach for both bending and in-plane waves. Note that B1L1, B1T1, B2L2, and B2T2 are not shown on the graph.

(c) Resilient junction (two resilient layers)

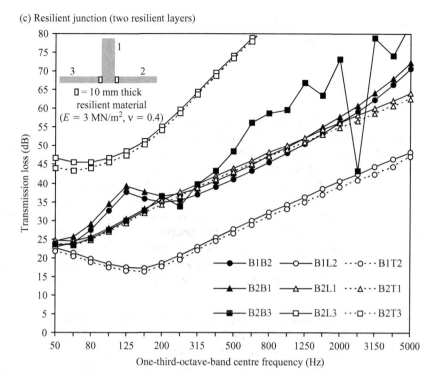

Figure 5.12

(*Continued*)

Comparisons of measurements and predictions on heavyweight walls and floors with resilient junctions show reasonable agreement when considered alongside the uncertainty in describing the properties of the resilient layer (Craik and Osipov, 1995; Pedersen, 1995). The Young's modulus of a resilient layer usually shows some dependence on both frequency and static load. In addition, the transmission loss can be highly sensitive to the Poisson's ratio of the layer (this is typically between 0.1 and 0.5), but less sensitive to its internal damping (Mees and Vermeir, 1993).

5.2.2.4.3 *Junctions at beams/columns*

The junction beam used in the wave models in Section 5.2 is not a physical beam at the junction. However, many buildings consist of a concrete or steel frame with walls built in-between the columns, these plate junctions can be modelled by using a junction beam with mass and stiffness that supports wave motion (Bosmans, 1998; Steel, 1994). Transmission between the beams (columns) and the floor slab can be predicted using an impedance approach (Craik, 1996; Ljunggren, 1985; Lyon and Eichler, 1964).

5.2.2.4.4 *Hinged junctions*

A hinged plate junction has negligible transfer of bending moments due to an incomplete line connection across the cross-section of one or more of the plates along the junction line. A model for a hinged junction is given by Mees and Vermeir (1993). This tends to be more relevant to plates which have not been fully connected over their cross-section due to poor workmanship.

Hinged junctions are less common than rigid or resilient plate junctions for masonry/concrete load-bearing walls in buildings.

5.2.3   Finite element method

The Finite Element Method (FEM) provides an example of how numerical models can be used to determine parameters for SEA or SEA-based models (e.g. see Simmons, 1991; Steel and Craik, 1994). The wave approaches in Sections 5.2.1 and 5.2.2 assume a diffuse vibration field on each plate. Whilst this assumption is convenient, it is not always appropriate for vibration transmission between finite walls and floors over the building acoustics frequency range. Numerical models allow modelling of finite plates and complex junction connections. However, they can be computationally intensive; hence wave approaches are often considered to be adequate for practical purposes when the junction is relatively simple. FEM is primarily used in this section to perform numerical experiments that illustrate issues relevant to finite plates with non-diffuse vibration fields and low modal overlap. These have implications for both the prediction and measurement of vibration transmission *in situ* and in the laboratory.

A brief discussion is included here on relevant features of FEM models. For a thorough overview the reader is referred to a text such as that by Zienkiewicz (1977) or Petyt (1998). The structure under analysis is discretized into elements that are connected at nodal points to form a mesh of elements; this mesh may be a simple rectangular grid for a homogeneous rectangular plate. In the dynamic analysis, the unknowns to be determined are the degrees of freedom (displacements and rotations) of all nodes in the mesh. The general equation of motion for linear systems under steady-state excitation by sinusoidal point forces is

$$[M]\left\{\frac{\partial^2 \hat{w}}{\partial t^2}\right\} + [C]\left\{\frac{\partial \hat{w}}{\partial t}\right\} + [K]\{\hat{w}\} = \{\hat{F}\} \qquad (5.79)$$

where w is the nodal displacement (complex), F is the applied force (complex), $[M]$ is the mass matrix, $[C]$ is the damping matrix, and $[K]$ is the stiffness matrix.

The displacement and force vectors are given by

$$\hat{w} = |\hat{w}|\exp(i\phi)\exp(i\omega t) \qquad (5.80)$$

$$\hat{F} = |\hat{F}|\exp(i\varphi)\exp(i\omega t) \qquad (5.81)$$

where ϕ is the phase shift due to damping, and φ is the applied phase shift to the input force.

Due to the harmonic time-dependence of the displacement and the force, Eq. 5.79 simplifies to

$$(-\omega^2[M] + i\omega[C] + [K])\{\hat{w}\} = \{\hat{F}\} \qquad (5.82)$$

There are many different types of finite elements. As FEM is mainly used at frequencies below the thin plate limit for bending waves, a plate element is usually chosen to simulate thin plate theory. The finite element dimensions required to achieve a certain degree of accuracy depends upon the frequency. This can be determined by successively reducing the element size until there is satisfactory convergence towards a solution; suitable element sizes are usually between $\lambda_B/3$ and $\lambda_B/6$.

Damping is often introduced into the model using the fraction of critical damping, the constant damping ratio, ζ_{cdr}. The relationship between the loss factor and the constant damping ratio is

$$\zeta_{cdr} = \frac{\eta}{2} \tag{5.83}$$

Whilst it is important to assign damping to the plates in a realistic way, it is not always straightforward. As an example, consider a FEM model of three walls connected to form a T-junction. In a building these walls will be connected to other walls and floors. It is therefore the total loss factor that is needed to calculate the constant damping ratio for the model rather than the internal loss factor. Whilst the total loss factor can be estimated (Section 2.6.5) it is not always possible to assign different damping to each wall in a model; fortunately the total loss factors are sometimes similar. The total loss factor for masonry/concrete walls and floors is usually much higher than the internal loss factor so care also needs to be taken that the damping used in the FEM model does not cause a significant decrease in vibration with distance across each plate (Section 2.7.7).

Around the plate boundaries there will need to be constrained nodes to account for connections to other plates that are not included in the FEM model; this is usually done by assuming simply supported boundaries. In a similar way to the wave approach, the nodes along the junction line can either be unconstrained to allow generation of both bending and in-plane waves, or simply supported so that only bending wave motion is considered.

As with laboratory measurements of the velocity level difference, excitation can be applied at single points. One advantage of using FEM is that rain-on-the-roof excitation can be applied to try and satisfy the SEA requirement for statistically independent excitation forces. This requires forces with unity magnitude and random phase to be applied at all the unconstrained nodes over the surface of the source plate. The output from the FEM model can be used to calculate the subsystem energy and power input that apply to an SEA model of the plate junction. These values correspond to individual frequencies whilst for practical purposes we require frequency bands; hence the energy or power at a number of frequencies can be averaged to give a representative value for the frequency band. As with the laboratory measurement, it is bending wave motion that is of interest on both source and receiving plates. FEM models usually output nodal displacements in all three coordinate directions, but it is the lateral displacement corresponding to bending wave motion that is used in subsequent calculations. At a sinusoidal frequency, ω, the energy associated with each plate is

$$E = m\langle v^2 \rangle_{t,s} = \omega^2 \left[\frac{1}{2} \sum_{n=1}^{N} m_n |\hat{w}_n|^2 \right] \tag{5.84}$$

where N is the number of unconstrained nodes on the plate, $\hat{w}$ is the peak lateral displacement and m is the mass associated with the node. (Note that in Chapter 2 we used $\hat{\eta}$ rather than $\hat{w}$ for bending wave displacement but η is now needed for loss factors.)

With single point excitation, consideration needs to be given to excluding displacement from nodes near the excitation point where the direct field dominates over the reverberant field (Section 2.7.5); this becomes more important when high damping is used in the FEM model. This makes it more convenient to use rain-on-the-roof excitation over all unconstrained nodes for which the vibration of the entire plate represents the reverberant energy stored by the plate subsystem.

The velocity at the node where the power is injected is complex and is related to the displacement by

$$\hat{v} = i\omega\hat{w} \qquad (5.85)$$

For single point excitation where the applied force only has a real component, the injected power is

$$W_{in} = \frac{1}{2}\text{Re}\{\hat{F}\hat{v}^*\} = \frac{-\omega}{2}\text{Re}\{\hat{F}\}\text{Im}\{\hat{w}\} \qquad (5.86)$$

For rain-on-the-roof excitation at P nodes the forces must be complex, so the injected power is

$$W_{in} = \frac{\omega}{2}\sum_{p=1}^{P}(\text{Im}\{\hat{F}\}\text{Re}\{\hat{w}\} - \text{Re}\{\hat{F}\}\text{Im}\{\hat{w}\})_p \qquad (5.87)$$

We now look at ways in which the plate energies can be used to determine parameters of more practical use. The simplest calculation is the velocity level difference between any two of the plates that form the junction. This is calculated from the plate energies using

$$D_{v,ij} = 10\lg\left(\frac{E_i}{E_j}\right) - 10\lg\left(\frac{m_i}{m_j}\right) \qquad (5.88)$$

where m_i and m_j are the mass of plates i and j respectively.

This numerical experiment with FEM has effectively simulated the physical experiment described in Section 3.12.3.3 to determine the coupling loss factor from the measured velocity level difference. The calculations and the assumptions are identical. It therefore follows that the vibration reduction index can also be calculated from the velocity level difference (Section 3.12.3.4). These calculations can be considered as basic forms of experimental SEA (ESEA).

A more involved form of ESEA makes use of the SEA power balance equations to determine the loss factors through inversion of an energy matrix (Lyon, 1975). This makes use of a general ESEA matrix that can be determined from the general SEA matrix (Eq. 4.10),

$$\begin{bmatrix} \sum_{n=1}^{N}\eta_{1n} & -\eta_{21} & -\eta_{31} & \cdots & -\eta_{N1} \\ -\eta_{12} & \sum_{n=1}^{N}\eta_{2n} & -\eta_{32} & & \\ -\eta_{13} & -\eta_{23} & \sum_{n=1}^{N}\eta_{3n} & & \\ \vdots & & & \ddots & \\ -\eta_{1N} & & & & \sum_{n=1}^{N}\eta_{Nn} \end{bmatrix} \begin{bmatrix} E_{11} & E_{12} & E_{13} & \cdots & E_{1N} \\ E_{21} & E_{22} & E_{23} & & \\ E_{31} & E_{32} & E_{33} & & \\ \vdots & & & \ddots & \\ E_{N1} & & & & E_{NN} \end{bmatrix}$$

$$= \begin{bmatrix} \dfrac{W_{in(1)}}{\omega} & 0 & 0 & \cdots & 0 \\ 0 & \dfrac{W_{in(2)}}{\omega} & 0 & & \\ 0 & 0 & \dfrac{W_{in(3)}}{\omega} & & \\ \vdots & & & \ddots & \\ 0 & & & & \dfrac{W_{in(N)}}{\omega} \end{bmatrix} \qquad (5.89)$$

where E_{ij} is the energy of subsystem i when the power is input into subsystem j.

Numerical experiments with FEM are carried out to fill in all the terms in the energy and power matrices. The energy matrix should be well conditioned if the system has been partitioned into suitable subsystems and $\eta_{ij} \ll \eta_{ii}$ (referred to as ESEA 'weak' coupling); so the energy terms on the diagonal (E_{11} to E_{NN}) should be significantly larger than off-diagonal terms. However, ill-conditioned matrices can still occur due to errors in the subsystem energies; this can result in negative coupling loss factors. Different approaches can be taken if this problem occurs. One possibility is to change the matrix layout used to calculate the coupling loss factors (see Lalor, 1990), an alternative approach is to use matrix-fitting routines to persuade/force the system to fit an SEA model (Clarkson and Ranky, 1984; Hodges *et al.*, 1987; Woodhouse, 1981). Failure of the matrix inversion to produce positive coupling loss factors does not prove that the system can or cannot be modelled using SEA. Assuming that it is only errors in the energies that cause the negative values, the system may still be suited to SEA. Fortunately, building elements such as masonry walls are quite highly damped when they are fully connected to other walls and floors; this tends to reduce the occurrence of negative coupling loss factors.

5.2.3.1 *Introducing uncertainty*

Using a deterministic model such as FEM does not allow statistical considerations to be ignored. With any physical construction there is uncertainty in the material properties and dimensions. For finite plates with fractional or generally low mode counts this becomes overtly obvious when predicting vibration parameters at individual frequencies and converting the result into one-third-octave or octave-bands. Comparisons of measurements and predictions often show that the peaks and troughs in the response do not always occur in the same frequency bands.

To take uncertainty into account it is possible to use a Monte Carlo technique to generate an ensemble of similar constructions. Each of these constructions can then be modelled using a deterministic approach so that individual results can be compared, or the ensemble of results can be analysed to give statistical parameters. The technique is based on random number generation to determine each variable based on a chosen statistical distribution. The exact statistical distribution is rarely known, but for physical properties it is reasonable to assume a normal (Gaussian) distribution, $N(\mu, \sigma)$. Note that this distribution is not bounded and the physical variables (e.g. dimensions, bending stiffness) can only be positive, but by confining its use to $\sigma/\mu < 0.3$ it is possible to avoid any significant errors (Keane and Manohar, 1993).

5.2.3.2 *Example: Comparison of FEM with measurements*

We will shortly compare SEA using wave approaches with FEM. Before this it is useful to compare FEM with measurements to confirm that it provides a reasonable model. This makes use of a physical experiment on five rigidly connected masonry walls that form a free-standing H-block on a large concrete floor (Hopkins, 2003a). The H-block is essentially two T-junctions joined by a single wall that is common to both of them. Each rectangular wall is only connected on two or three of its four sides, so the sum of its coupling loss factors is relatively low and the total loss factor is similar to the internal loss factor. This means that the modal fluctuations in the velocity level differences are easier to distinguish. It also simplifies matters because the same damping values can be used for all walls in the FEM model; note that for fully connected walls it is more complicated to account for different damping on each plate associated with bending and in-plane motion. The Monte Carlo approach described in Section 5.2.3.1 is used to create an ensemble of similar constructions based on the estimated uncertainty for all dimensions and material properties. FEM is used to model each construction in the ensemble to determine the ensemble average velocity level difference and 95% confidence interval. Finite elements for thin plate bending wave theory are used along with rain-on-the-roof excitation of bending wave motion on the chosen source plate. Velocity level differences represent lateral motion to each plate surface (i.e. bending wave motion).

Figure 5.13 compares measurements with two FEM predictions, one using a simply supported junction line and the other using an unconstrained junction line. The large modal fluctuations in the low-frequency range show that even when trying to account for the uncertainty, the predicted peaks and troughs can still end up in the adjacent frequency band to which they are measured. For transmission around the corner in the mid- and high-frequency ranges, the 95% confidence intervals for the measurements and the predictions overlap each other and either model for the junction line could be assumed to be appropriate. However, for transmission across the straight section in the high-frequency range, only the unconstrained junction line is

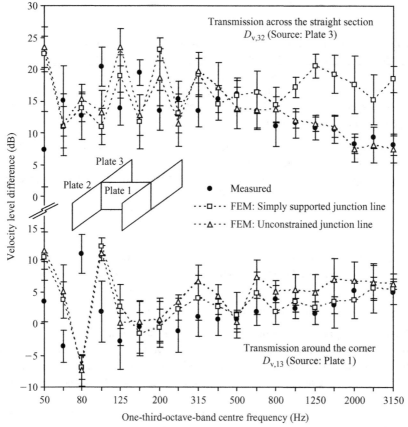

Figure 5.13

Comparison of FEM with measurements of vibration transmission between masonry walls. The H-block of five rigidly connected masonry walls is free-standing on a wide 300 mm thick concrete floor. Plate 1 is a 215 mm separating wall ($L_x = 4.5$ m, $L_y = 2.5$ m) and the other four plates are 100 mm flanking walls ($L_x = 3.6$–4.1 m, $L_y = 2.5$ m). All walls are built from solid dense aggregate blocks ($\rho = 2000$ kg/m³). The thin plate limit for bending waves, $f_{B(thin)}$, is in the 800 Hz band for plate 1 and the 2000 Hz band for plates 2 and 3. Predictions and measurements are both shown with 95% confidence intervals. Measured data from Hopkins are reproduced with permission from ODPM and BRE.

appropriate. At frequencies above the thin plate limits there is no indication that any significant errors occur by using only thin plate theory.

5.2.3.3 Example: Comparison of FEM with SEA (wave approaches) for isolated junctions

In Section 4.3.1 we looked at resonant coupling between the sound field in a room and bending wave motion on a plate. Even when the plates had relatively low mode counts and low modal overlap, there was reasonable agreement between measurements and SEA because the plates were coupled to rooms where the mode counts and modal overlap factors were much higher. For resonant coupling between masonry/concrete plates it is common for both subsystems to have low mode counts and low modal overlap. This section uses FEM and SEA to discuss statistical and deterministic views of vibration transmission with excitation of bending waves on one plate and comparison of the velocity level difference between plates (bending wave motion).

Here it is useful to look at FEM models for an ensemble of similar junctions where the walls have slightly different mode frequencies. This allows an insight into issues that occur with real masonry/concrete structures where the mode counts are low and there is uncertainty in predicting individual mode frequencies. There is more than one variable that can be altered to generate this ensemble; we will vary the L_x dimension that lies perpendicular to the junction line. Variation in L_x is created with a Monte Carlo approach by using random numbers drawn from a normal distribution, $N(\mu, \sigma)$. The mean value, $\mu = L_x$ will either be 3, 3.5 or 4 m and the standard deviation is chosen to be $\sigma = 0.25$ m. This gives a low degree of variation ($\sigma/\mu < 0.1$) so that the SEA prediction using the mean value for L_x is not significantly different to any individual junction with a different L_x value. The level of uncertainty introduced into this ensemble would not be unusual at the early design stage of a building. Even when the building plans and dimensions are finalized there will still be uncertainty in the material parameters that can change the mode frequencies. In these examples the ensemble is quite small; the ensemble has 10 members for 50–1000 Hz, and only 5 members for 1250–3150 Hz.

Figure 5.14a shows the velocity level difference for an L-junction where only bending waves are considered in the SEA and FEM models. The first point to note is the high level of variation in the individual members of the FEM ensemble; hence a single deterministic model is likely to be of little practical use even with relatively low levels of uncertainty. Individual members of the FEM ensemble show large fluctuations when $N_s < 5$ and $M_{av} < 1$. However, the arithmetic average of the FEM ensemble shows close agreement with the SEA model, particularly under the conditions that $M_{av} \geq 1$ and $N_s \geq 5$ (Fahy and Mohammed, 1992).

Figure 5.14b shows the same L-junction modelled with both bending and in-plane waves. The FEM ensemble average and SEA show similar frequency-dependence but the ensemble average does not tend towards the SEA solution quite as quickly as in Fig. 5.14a. The fundamental in-plane modes occur just as each bending wave subsystem becomes multi-modal ($N_s \geq 5$) and the frequency at which the FEM ensemble average and SEA converge is higher than in Fig. 5.14a.

For a T-junction we will only look at an unconstrained junction line because the comparison of FEM with measurements in Section 5.2.3.2 indicates that this is a more appropriate model for rigidly connected masonry walls in the high-frequency range. The predicted velocity level differences are shown in Fig. 5.15. As with the L-junction, there are large variations

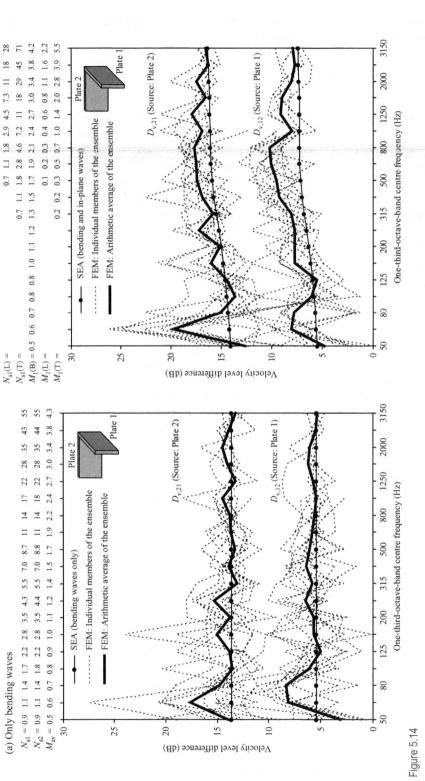

(a) Only bending waves

N_{s1} =	0.9	1.1	1.4	1.7	2.2	2.8	3.5	4.3	5.5	7.0	8.7	11	14	17	22	28	35	43	55
N_{s2} =	0.9	1.1	1.4	1.8	2.2	2.8	3.5	4.4	5.5	7.0	8.8	11	14	18	22	28	35	44	55
M_{av} =	0.5	0.6	0.7	0.8	0.9	1.0	1.1	1.2	1.4	1.5	1.7	1.9	2.2	2.4	2.7	3.0	3.4	3.8	4.3

(b) Bending and in-plane waves

N_{s1}(B) =	0.9	1.1	1.4	1.7	2.2	2.8	3.5	4.3	5.5	7.0	8.7	11	14	17	22	28	35	43	55
N_{s1}(L) =									0.7	1.1	1.8	2.9	4.5	7.3	11	18	28
N_{s1}(T) =					0.7	1.1	1.8	2.8	4.6	7.2	11	18	29	45	71
M_1(B) =	0.5	0.6	0.7	0.8	1.0	1.1	1.2	1.3	1.5	1.7	1.9	2.1	2.4	2.7	3.0	3.4	3.8	4.2
M_1(L) =									0.1	0.2	0.3	0.4	0.6	0.8	1.1	1.6	2.2
M_1(T) =					0.2	0.2	0.3	0.5	0.7	1.0	1.4	2.0	2.8	3.9	5.5

Figure 5.14

Predicted velocity level differences for an L-junction modelled with (a) only bending waves and (b) bending and in-plane waves. On (a) the upper x-axis labels show the predicted statistical mode counts for each plate and the geometric modal overlap factor in each frequency band. On (b) the upper x-axis labels only show the predicted statistical mode counts and modal overlap factor in each frequency band for plate 2 are similar and there are several permutations for the geometric modal overlap factor. Note that values are only shown above the fundamental local mode. Plate properties: L_{x1} = 4.0 m, L_{x2} = 3.5 m, L_{y1} = L_{y2} = 2.4 m, h_1 = h_2 = 0.1 m, ρ_{s1} = 140 kg/m², ρ_{s2} = 60 kg/m², c_{L1} = 2200 m/s, c_{L2} = 1900 m/s, ν = 0.2, loss factor $\eta = f^{-0.5}$. (This was used in the SEA and FEM models so that the losses were indicative of walls connected on all sides as they would be in situ. The modal overlap factors are calculated using this loss factor and should therefore be treated as minimum values because the total loss factor would also include the coupling loss factors from this junction.)

between individual members of the FEM ensemble. Below the fundamental in-plane modes where $M_{av} < 1$ and $N_s < 5$ for bending wave motion, the largest differences between ensemble average FEM values and SEA occur across the straight section of the T-junction. With low modal overlap and low mode counts there is a tendency for the SEA wave approach to overestimate the strength of the coupling loss factors. However, the differences between the ensemble average FEM values and SEA are reasonably low and the latter can be used for most practical purposes. Above the fundamental in-plane modes there are many permutations of M_{av} between subsystems of different wave types and this simple descriptor becomes less useful.

5.2.3.4 Example: Statistical distributions of coupling parameters

In the previous example there were large differences in vibration transmission between junctions with similar walls when the mode counts and the modal overlap were low. Information on the statistical distribution of coupling parameters is now needed to calculate the ensemble average and the standard deviation when such parameters are calculated from numerical experiments or measured on a set of similar junctions. Any probability density function is likely to be specific to certain systems, as well as specific to the choice of coupling parameter, the method by which the ensemble is created, type of excitation, type of subsystems, and the prediction or measurement method. Numerical experiments used to study various different dynamical systems indicate that probability density functions are not always described by a normal distribution; it has been found that coupled plates or beams often have right-skewed distributions if modal overlap factors are less than unity (Fahy and Mohammed, 1992; Hodges and Woodhouse, 1989; Manohar and Keane, 1994; Wester and Mace, 1999).

We now look at statistical distributions for an ensemble of junctions by taking the approach of Fahy and Mohammed (1992). An ensemble is created using a Monte Carlo approach with random numbers drawn from a normal distribution. An ensemble of 30 similar junctions of masonry walls is created in exactly the same way as in Section 5.2.3.3 by varying the L_x dimension of each plate and using a FEM model for each junction. We will focus on the low- and mid-frequency ranges where modal overlap is lowest and it is reasonable to model only bending wave motion by using a simply supported junction line with both L and T-junctions. Coupling loss factors are determined using ESEA as described in Section 5.2.3. These are then used to produce normal quantile plots. This type of plot indicates normal distributions when the majority of values in the ensemble lie along a straight line; any outliers will occur far from this line. If there are two distinctly different gradients for the ensemble, this indicates a skewed distribution. Figure 5.16 shows examples for an ensemble of random numbers drawn from a normal distribution; the lines are approximately straight but there are some fluctuations. These are intended for visual comparison with the normal quantile plots for the coupling loss factors shown in Fig. 5.17 for an L and a T-junction (Hopkins, 2002). When the modal overlap factor is less than unity for the coupled plates, the normal quantile plots for the linear coupling loss factors have distinctly non-normal, right-skewed distributions, and tend to have the widest spread of values. However, when plotted in decibels they tend towards straight lines; hence a practical simplification is to describe the linear coupling loss factor by a log-normal distribution. The mean, standard deviation and confidence intervals can then be calculated from the values in decibels. Upper and lower confidence intervals can subsequently be used in path analysis with SEA or SEA-based models to give an indication of the range of possible values.

The SEA matrix solution requires linear coupling loss factors, so a reverse transformation is needed to convert the mean and standard deviation from logarithmic to linear values. The

$N_{s1}(B) =$				0.4	0.6	0.7	0.9	1.1	1.4	1.8	2.2	2.8	3.5	4.4	5.6	6.9	8.9	11	14	18
$N_{s1}(L) =$														0.9	1.4	2.1	3.5	5.4	8.5	13
$N_{s1}(T) =$											0.8	1.3	2.2	3.4	5.3	8.7	14	21	34
$N_{s2}(B) =$	0.5	0.7	0.8	1.0	1.3	1.7	2.1	2.6	3.3	4.2	5.2	6.6	8.4	10	13	17	21	26	33
$M_1(B) =$				0.2	0.2	0.3	0.3	0.3	0.4	0.4	0.5	0.5	0.6	0.7	0.8	0.9	1.0	1.1	1.2	1.4
$M_1(L) =$														0.1	0.2	0.3	0.4	0.5	0.7	1.0
$M_1(T) =$											0.2	0.2	0.3	0.5	0.7	0.9	1.3	1.8	2.6
$M_2(B) =$	0.3	0.4	0.4	0.5	0.5	0.6	0.6	0.7	0.8	0.9	1.0	1.1	1.3	1.4	1.6	1.8	2.0	2.3	2.6

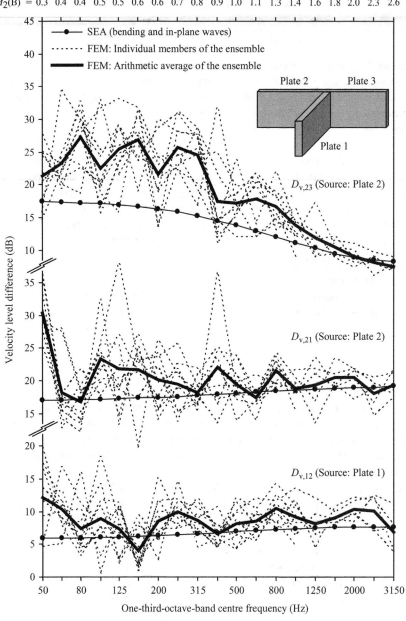

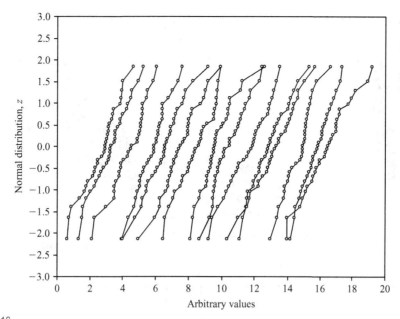

Figure 5.16

Normal quantile plot for 15 sets of 30 random numbers drawn from a normal distribution.

normal distribution $N(\mu, \sigma)$ has two parameters, μ and σ, and is referred to as a two-parameter distribution. The log-normal distribution is generally a three-parameter distribution $\Lambda(\tau, \mu_\Lambda, \sigma_\Lambda)$ where τ is the lowest possible value or threshold of the distribution. Assuming that the linear coupling loss factor can potentially be zero or infinitesimally small, then $\tau = 0$, so the two-parameter log-normal distribution $\Lambda(\mu_\Lambda, \sigma_\Lambda)$ can be used. The log-normal distribution of linear η_{ij} is now denoted by $\Lambda(\mu_\Lambda, \sigma_\Lambda)$, and the normal distribution of $\lg(\eta_{ij})$ by $N(\mu, \sigma)$. This gives the geometric mean, $\mu_\Lambda = 10^\mu$ and the geometric standard deviation, $\sigma_\Lambda = 10^\sigma$. The SEA matrix solution can now be used with μ_Λ representing the ensemble average coupling loss factor. Similar issues with non-normal distributions occur with the vibration reduction index because it is directly related to the coupling loss factor (see Section 5.4.1.1). When the modal overlap factor is greater than unity the spread of the coupling loss factors is much narrower and the normal distribution can be used to describe either the linear values or those in decibels.

Non-normal distributions with modal overlap factors less than unity were found by Fahy and Mohammed (1992) in numerical experiments using thin steel coupled plates. To date, this

←

Figure 5.15

Predicted velocity level differences for a T-junction modelled with bending and in-plane waves. The upper x-axis labels show the predicted statistical mode counts and modal overlap factor in each frequency band for plate 1 (all three wave types) and for plate 2 (bending waves). Note that these are only shown above the fundamental local mode. Values for plate 3 (bending waves) are similar to plate 2 (bending waves) and values for plates 2 and 3 (in-plane waves) are similar to plate 1 (in-plane waves). Plate properties: $L_{x1} = 4.0\,\text{m}$, $L_{x2} = 3.5\,\text{m}$, $L_{x3} = 3.0\,\text{m}$, $L_{y1} = L_{y2} = L_{y3} = 2.4\,\text{m}$, $h_1 = 0.215\,\text{m}$, $h_2 = h_3 = 0.1\,\text{m}$, $\rho_{s1} = 430\,\text{kg/m}^2$, $\rho_{s2} = \rho_{s3} = 200\,\text{kg/m}^2$, $c_L = 3200\,\text{m/s}$, $\nu = 0.2$, loss factor $\eta = f^{-0.5}$ (This was used in the SEA and FEM models so that the losses were indicative of walls connected on all sides as they would be *in situ*. The modal overlap factors are calculated using this loss factor and should therefore be treated as minimum values because the total loss factor would also include the coupling loss factors from this junction.)

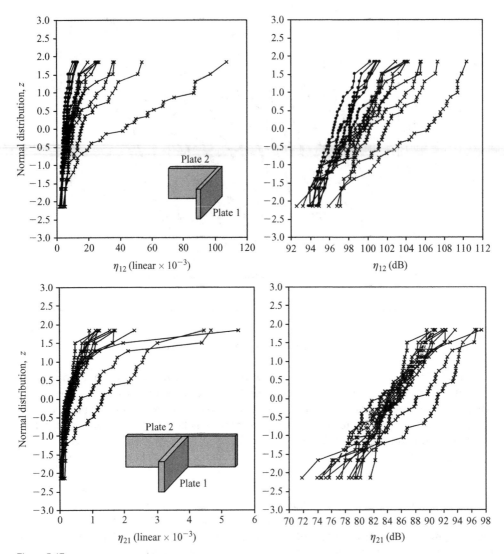

Figure 5.17

Normal quantile plots for the coupling loss factor (linear values and decibels) calculated using FEM (bending waves only) and ESEA. Fourteen curves are shown, each corresponding to a one-third-octave-band between 50 and 1000 Hz. Each curve is formed from an ensemble of 30 members. Curves denoted by –x– indicate that the modal overlap factor in the frequency band is less than unity for plate i and/or plate j for the relevant η_{ij}. L-junction plate properties: $L_{x1} = 4.0$ m, $L_{x2} = 3.5$ m, $L_{y1} = L_{y2} = 2.4$ m, $h_1 = h_2 = 0.1$ m, $\rho_{s1} = 140$ kg/m², $\rho_{s2} = 60$ kg/m², $c_{L1} = 2200$ m/s, $c_{L2} = 1900$ m/s, $\nu = 0.2$, loss factor $\eta = f^{-0.5}$. T-junction plate properties: $L_{x1} = 4.0$ m, $L_{x2} = 3.5$ m, $L_{x3} = 3.0$ m, $L_{y1} = L_{y2} = L_{y3} = 2.4$ m, $h_1 = 0.215$ m, $h_2 = h_3 = 0.1$ m, $\rho_{s1} = 430$ kg/m², $\rho_{s2} = \rho_{s3} = 60$ kg/m², $c_{L1} = 3200$ m/s, $c_{L2} = c_{L3} = 1900$ m/s, $\nu = 0.2$, loss factor $\eta = f^{-0.5}$.

feature has therefore been identified with lightly damped plates that have relatively high modal densities, as well as with highly damped plates with relatively low modal densities. Note that these findings apply to ensembles created using parameters drawn from a normal distribution. It is not possible to draw wide-ranging conclusions relating to all numerical or physical experiments. However, it emphasizes that more cautious conclusions should be drawn when using results from a single deterministic model or a single physical experiment when the modal

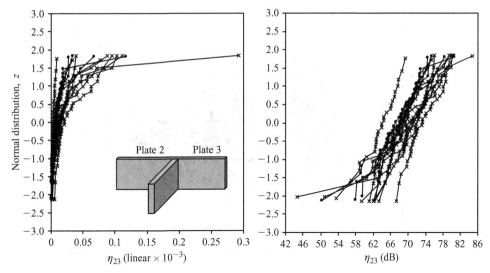

Figure 5.17

(*Continued*)

overlap factor of any plate in the junction is less than unity. This has implications for laboratory measurements used to determine coupling loss factors or the vibration reduction index with junctions of masonry/concrete walls (Hopkins, 1996). There are potential advantages in building free-standing wall constructions so that no laboratory facility is needed and the number of flanking transmission paths can be minimized. However, when these plates are unconnected on some sides they will have lower total loss factors, and hence lower modal overlap factors than *in situ*. Therefore the resulting coupling parameters measured in the laboratory are likely to be different *in situ*. Similar issues with low modal overlap occur when the walls and floors are connected by a resilient material at the junction.

Most masonry/concrete walls and floors are rigidly connected and it is more convenient to use the wave approach (bending waves only) to predict the coupling loss factor than to carry out numerical or physical experiments. SEA predictions using the wave approach tend to give reasonable estimates with low variance when the plates that form the junction satisfy the empirical conditions, $M_{av} \geq 1$ and $N_s > 5$ (Fahy and Mohammed, 1992) where M_{av} is defined in Eq. 4.7. For masonry/concrete plates these conditions are rarely satisfied in the low-frequency range, and sometimes in the mid-frequency range too. When they are not satisfied the wave approach tends to overestimate the coupling loss factor; this can be inferred from the examples in Section 5.2.3.3 where the velocity level difference from SEA is generally lower than from the FEM model. The modal overlap factor provides a simple, practical variable with which to define these conditions, but it is not necessarily the key variable. Some numerical experiments show it to be an inadequate indicator for rectangular plates (Wester and Mace, 1996). Using the same ensemble for the L-junction as in the previous numerical experiments, Fig. 5.18 indicates that there is no clear cut-off point and that for practical purposes the empirical conditions for masonry/concrete walls could be relaxed. By allowing for errors in an SEA prediction that are similar to those encountered from variation due to workmanship, Craik *et al.* (1991) propose more lenient conditions when using the wave approach (bending waves only) to calculate coupling loss factors in SEA; for a 5 dB error limit these are $M > 0.25$ and $N_s > 0.3$ for each plate.

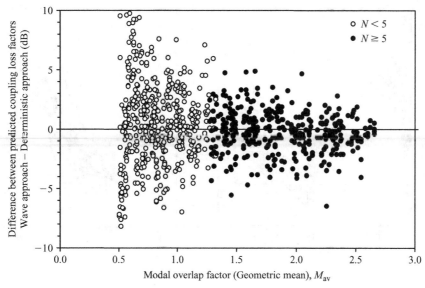

Figure 5.18

Comparison of coupling loss factors determined from the wave approach with those from the deterministic approach using FEM and ESEA. The same L-junction is used as described in Fig. 5.17. Each point corresponds to η_{12} or η_{21} for one of the 30 members of the ensemble and for one of the one-third-octave-bands between 50 and 1000 Hz.

5.2.3.5 Example: Walls with openings (e.g. windows, doors)

Flanking walls that are connected to a separating wall or floor usually contain openings such as windows and doors. If an opening is introduced into a wall close to a junction it can alter vibration transmission between the plates. The wave approach is well-suited to the calculation of transmission coefficients between plates without openings where a wave is incident upon the junction at a certain angle of incidence. It is less suited to plates with an opening close to the junction because simple assumptions about angles of incidence and transmission are no longer valid and any constraints on the displacement of the junction beam will vary along its length. It is therefore easier to take a modal view of vibration transmission.

Two aspects require consideration when modelling walls with an opening: the first is whether the wall can still be modelled as a single plate, and the second is how the vibration transmission is changed by the presence of an opening. Modelling of doors and windows needs to be considered on a case-by-case basis. Here we will focus on masonry/concrete walls with windows because this type of external flanking wall often forms part of important flanking paths.

The local mode shapes of a plate with an opening depend upon the boundary conditions of the opening. At this point we can refer back to Fig. 2.33 where a window opening was modelled with simply supported or free boundaries. For windows and doors, most permutations of these two boundary conditions are possible. For windows in external cavity masonry walls, the boundaries are not usually connected in a way that would justify the simply supported model; partly to avoid thermal bridging and ingress of moisture. An exception for doors and windows is a lintel that rigidly connects the inner to the outer leaf. When most or all boundaries of an opening are free (i.e. unconstrained), it is usually possible to model the wall as a single plate. If most or all of its boundaries are simply supported then the wall may act as one or more smaller plates that

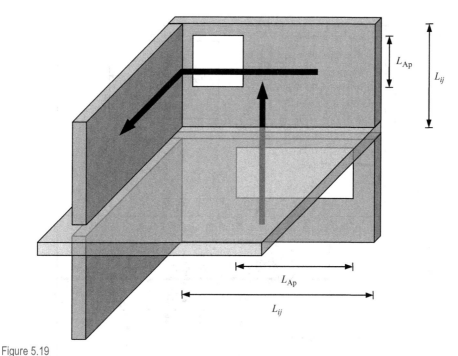

Figure 5.19

Window openings in flanking walls.

are separated by the opening. However, this will be limited to low frequencies where strips of wall between the wall boundaries and the boundaries of the opening cannot support vibration. Over the building acoustics frequency range, many masonry/concrete walls with openings can simply be modelled as a single plate with a reduced surface area due to the opening. Note that below the critical frequency, the boundary conditions of the opening can affect the radiation efficiency of the wall. As a rule-of-thumb, the frequency-average radiation efficiency still gives a reasonable estimate without any modification if the ratio of wall area to opening area is at least two and the boundaries of the opening are not simply supported.

We now consider the effect of an opening on the power flow between connected walls; this was illustrated with structural intensity measurements in Section 3.12.3.1.4. For a diffuse incidence vibration field on a junction, an opening close to the junction can be seen as reducing the length of the junction line (see Fig. 5.19); hence a simple estimate for the reduction in the coupling loss factor (in decibels) due to an opening is given by

$$10 \lg \left(\frac{L_{ij} - L_{Ap}}{L_{ij}} \right)$$

(5.90)

where L_{Ap} is the opening dimension that lies closest to and parallel to the junction line, L_{ij}. (Note that when using the vibration reduction index this estimate would be an increase instead of a reduction.)

As we have already established that the vibration field is not usually diffuse, we will use the results from the L-junction in Section 5.2.3.3 to assess the estimate in Eq. 5.90. Additional calculations with FEM and Monte Carlo methods are used to model an ensemble of similar

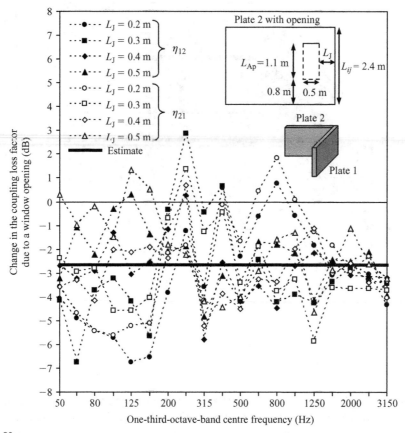

Figure 5.20

Predicted change in the coupling loss factor due to window openings in an L-junction. Plate properties: $L_{x1} = 4.0\,\text{m}$, $L_{x2} = 3.5\,\text{m}$, $L_{y1} = L_{y2} = 2.4\,\text{m}$, $h_1 = h_2 = 0.1\,\text{m}$, $\rho_{s1} = 140\,\text{kg/m}^2$, $\rho_{s2} = 60\,\text{kg/m}^2$, $c_{L1} = 2200\,\text{m/s}$, $c_{L2} = 1900\,\text{m/s}$, $\nu = 0.2$, loss factor $\eta = f^{-0.5}$. Opening: Top boundary is simply supported. All other boundaries are free.

L-junctions with a small window opening at different positions in the flanking wall. The effect of the opening is given by the difference between the ensemble average coupling loss factor with and without the opening. It is reasonable to assume that the effect of openings on vibration transmission will be most apparent when they are very close to the junction. Therefore the distance, L_J, between the edge of the opening and the junction line is varied from 0.2 to 0.5 m. Comparisons with laboratory measurements indicate that for an L-junction without an opening, a simply supported junction line is appropriate in the FEM model (Hopkins, 2003). However, when there is an opening close to the junction, there is better agreement with measurements when the junction line is unconstrained to allow generation of bending and in-plane waves. The L-junction used in this numerical example is kept within bounds that are similar to the construction validated with measurements (Hopkins, 2003b). For each member of the ensemble, ESEA is used to determine coupling loss factors by assuming that the L-junction can be represented by only two bending wave subsystems (i.e. one subsystem for each plate).

The predicted change in the coupling loss factors is shown in Fig. 5.20. These show large fluctuations about the estimated value in the low- and mid-frequency ranges. Measurements on similar walls with windows also show these large fluctuations; these are shown in Fig. 5.21

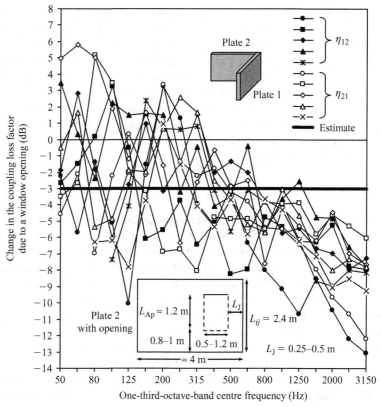

Figure 5.21

Measured change in the coupling loss factor due to window openings from five different L-junctions. Plate properties are similar to those in Fig. 5.20. Measured data from Hopkins are reproduced with permission from ODPM and BRE.

where the window openings were cut out of each flanking wall after they were built; this allowed measurements to initially be taken on the walls without an opening. In the mid- and high-frequency ranges where $L_J > \lambda_B/4$ the measurements indicate a larger reduction than given by the simple estimate or the FEM model. This is partly due to rain-on-the-roof excitation over the entire wall in FEM compared to measurements with point excitation at distances well away from the junction. Numerical and physical experiments on T-junctions also show large fluctuations from which it is difficult to assess whether the simple estimate is always appropriate.

As it is impractical and not particularly informative to calculate coupling loss factors using FEM for every combination of window and junction we need to consider how the simple estimate can be used with the wave approach. It is fortunate that there are many transmission paths between two rooms in a building. Therefore as L_{Ap} is typically half of L_{ij}, the additional uncertainty of ± 3 dB in this coupling loss factor can be tolerated as it will often have a small effect on the overall sound insulation. To assess the effect of the opening with an SEA model the wave approach can be used to calculate coupling loss factors with and without the simple estimate, or the openings can be completely ignored if they are small. Any decision to exclude a small wall from the model (because it contains an opening) should primarily be based around its inability to support bending modes. Care is needed to ensure that this will not exclude important flanking paths that cross more than one junction; the effect of excluding flanking walls or floors from an SEA model is shown with an example in Section 5.3.2.1. For most masonry/concrete constructions, the predicted change in the single-number quantity due to an opening will be a small increase in the sound insulation up to a few decibels. This small but beneficial aspect of windows in flanking walls has been identified in statistical analysis of many field airborne sound insulation measurements in flats (Sewell and Savage, 1987). If the windows are openable, consideration also needs to be given to the flanking path involving airborne sound transmitted between open windows in flanking walls on either side of the separating wall (Fothergill and Hargreaves, 1992; Kawai *et al.*, 2004).

5.2.3.6 *Example: Using FEM, ESEA, and SEA with combinations of junctions*

Numerical methods have been used in the previous examples to model isolated junctions consisting of plates with low mode counts and low modal overlap. For these isolated junctions, the combination of FEM, ESEA, and the Monte Carlo technique can be used to give ensemble average coupling loss factors. These are now incorporated in an SEA model of a larger construction formed from several different junctions. For this example we will look at seven coupled plates formed by connecting T-junctions and L-junctions as shown in Fig. 5.22. In this example we are getting closer to a situation of more practical interest, namely two adjacent rooms. The focus here is on the low-frequency range where $M_{av} < 1$ and $N_s < 5$ and only bending waves need consideration. Two SEA models are used for the construction; one using coupling loss factors from the wave approach and the other using ensemble average coupling loss factors from isolated junctions. The velocity level difference from the SEA models is compared with an ensemble of 30 similar constructions modelled with FEM. This ensemble is created by varying the L_x dimension of the plates in the same way as for the isolated junctions (Section 5.2.3.3); note that the L_x values for the three plates of one T-junction are applied to the other plates so that the perpendicular plates remain perpendicular to each other in each construction.

Predicted $D_{v,ij}$ values are shown in Fig. 5.23 (Hopkins, 2002). Figure 5.23a shows $D_{v,ij}$ between plates that share the same junction line; these are relevant to the measurement of coupling parameters (i.e. η_{ij} or K_{ij}) as well as to SEA-based models that only consider vibration transmission across one plate junction (see Section 5.4). Figure 5.23b shows $D_{v,ij}$ between plates that do not share the same junction line. The values are generally higher but act as a reminder that when predicting airborne or impact sound insulation between two adjacent rooms we will need to consider the role of all the walls and floors that form each room.

From Fig. 5.23 the SEA model using the wave approach tends to give a lower $D_{v,ij}$ than individual members of the ensemble; this was previously seen with the isolated junctions in Section 5.2.3.3. There is no clear advantage in using the ensemble average coupling loss factors; they

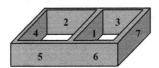

Figure 5.22

Seven coupled plates. Plate properties: $L_{x1} = L_{x4} = L_{x7} = 4.0\,\text{m}$, $L_{x2} = L_{x5} = 3.5\,\text{m}$, $L_{x3} = L_{x6} = 3.0\,\text{m}$, $L_y = 2.4\,\text{m}$, $h_1 = h_7 = 0.215\,\text{m}$, $h_2 = h_3 = h_4 = h_5 = h_6 = 0.1\,\text{m}$, $\rho_{s1} = \rho_{s7} = 430\,\text{kg/m}^2$, $\rho_{s4} = 140\,\text{kg/m}^2$, $\rho_{s2} = \rho_{s3} = \rho_{s5} = \rho_{s6} = 60\,\text{kg/m}^2$, $c_{L1} = c_{L7} = 3200\,\text{m/s}$, $c_{L4} = 2200\,\text{m/s}$, $c_{L2} = c_{L3} = c_{L5} = c_{L6} = 1900\,\text{m/s}$, $\nu = 0.2$, loss factor $\eta = f^{-0.5}$.

(a) Source and receiving plates connected together at a junction

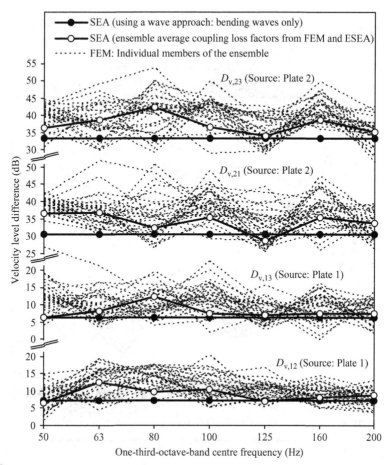

Figure 5.23

Predicted velocity level differences for the seven coupled plates.

can give a worse, similar, or better estimate than the wave approach. Note that these plates have low mode counts (local modes) and a numerical method has been used to model vibration transmission based on global modes of the isolated junctions. The results have then been used to derive coupling loss factors assuming that the transmission can be modelled on a local mode basis. The isolated junctions have then been combined to make a larger construction of seven coupled plates which has different global modes. The conclusion for masonry/concrete plates with low mode counts and low modal overlap is that SEA models incorporating coupling loss factors from either a single deterministic model or a single laboratory measurement on one junction, need to be treated with some caution. If the statistics of the ensemble are of primary importance, these can be calculated using numerical methods and Monte Carlo techniques. However, the time needed to create and compute the numerical models tends to be prohibitive. A simpler and more pragmatic approach to predictions for buildings is to use SEA with the wave approach and then make an allowance for uncertainty when interpreting the results and applying them to a single construction. For laboratory measurement of coupling parameters,

(b) Source and receiving plates that are not directly connected together at a junction

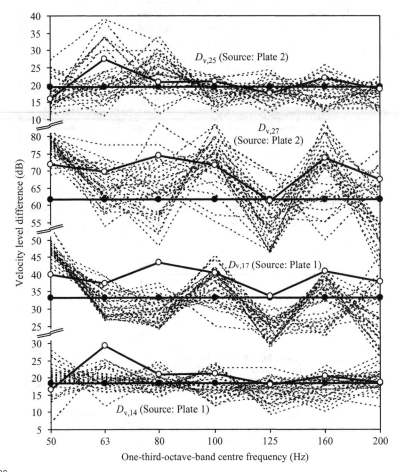

Figure 5.23

(*Continued*)

it is costly to build and measure an ensemble of similar junctions. However it is important to note that field measurements confirm the high levels of variation in the low-frequency range for junctions of masonry/concrete walls and floors (Craik and Evans, 1989).

5.2.4 Foundation coupling (Wave approach: bending waves only)

An important transmission path for masonry/concrete cavity walls occurs via the foundations (Sections 4.3.5.2.2 and 4.3.5.4.3). The following model from Wilson and Craik (1995) considers bending wave transmission between two plates on a strip foundation and highlights the role of soil stiffness in vibration transmission. In this model, bending waves on the source wall are assumed to cause rotation and displacement of the strip foundation; this is then resisted by the soil stiffness and the inertia of the foundation. As motion of the foundation causes both shearing and compression of the soil, the two relevant soil parameters are the shear stiffness per unit area, G_{soil}, and the compression stiffness per unit area, s'_{soil}. It is necessary to incorporate

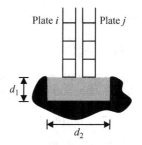

Figure 5.24

Cavity wall on a strip foundation.

damping in the compression stiffness using the loss factor, hence s'_{soil} is used in the form, $s'_{soil}(1 + i\eta_{soil})$. The shear stiffness per unit area can then be calculated from

$$G_{soil} = \frac{s'_{soil}(1 + i\eta_{soil})}{2(1 + \nu_{soil})} \tag{5.91}$$

where ν_{soil} is Poisson's ratio for soils (if unknown this can usually be assumed to be 0.25).

Two plates, i and j, are connected to a strip foundation as shown in Fig. 5.24. It is assumed that the strip foundation has a rectangular cross-section, a mass per unit length, ρ_l, and does not support bending or twisting motion. The incident bending wave on plate i has unit amplitude and is incident upon the strip foundation at an angle of incidence, θ_i, from a line on the plate that is perpendicular to the junction. Interaction of the incident wave with the foundation gives rise to a reflected bending wave on plate i with an amplitude, R_B, and a nearfield on plate i with amplitude, R_n. The angle of transmission is found from Snell's law (Section 5.2.1.2). When $(k_{B,i}/k_{B,j}) \sin \theta_i \leq 1$, a bending wave is transmitted on plate j with an amplitude, T_B, at an angle, θ_j, as well as a nearfield with amplitude T_n. Otherwise there is no transmitted bending wave, only a nearfield on plate j.

The following four equations can be solved in matrix form to give R, R_n, T, and T_n, for a specific angle of incidence, although it is only T that is needed to calculate the transmission coefficient (Wilson and Craik, 1995)

$$-R - R_n + T + T_n = 1$$

$$ik_{B,i} \cos \theta_i R + k_{n,i} R_n - ik_{B,j} \cos \theta_j T - k_{n,j} T_n = ik_{B,i} \cos \theta_i$$

$$[B_{p,i}k_{B,i}^2(\cos^2 \theta_i + \nu_i \sin^2 \theta_i) - ik_{B,i} \cos \theta_i(K_\phi - \omega^2 I_r)]R$$

$$- [k_{n,i}(K_\phi - \omega^2 I_r) + B_{p,i}(k_{n,i}^2 - \nu_i k_{B,i}^2 \sin^2 \theta_i)]R_n$$

$$+ B_{p,j}k_{B,j}^2(\cos^2 \theta_j + \nu_j \sin^2 \theta_j)T - B_{p,j}(k_{n,j}^2 - \nu_j k_{B,j}^2 \sin^2 \theta_j)T_n$$

$$= -ik_{B,i} \cos \theta_i(K_\phi - \omega^2 I_r) - B_{p,i}k_{B,i}^2(\cos^2 \theta_i + \nu_i \sin^2 \theta_i) \tag{5.92}$$

$$[iB_{p,i}k_{B,i}^3 \cos \theta_i(\cos^2 \theta_i + (2 - \nu_i) \sin^2 \theta_i) - (K_f - \omega^2 \rho_l)]R$$

$$- [B_{p,i}k_{n,i}(k_{n,i}^2 - (2 + \nu_i)k_{B,i}^2 \sin^2 \theta_i) + (K_f - \omega^2 \rho_l)]R_n$$

$$+ iB_{p,j}k_{B,j}^3 \cos \theta_j(\cos^2 \theta_j + (2 - \nu_j) \sin^2 \theta_j)T$$

$$- B_{p,j}k_{n,j}(k_{n,j}^2 - (2 + \nu_j)k_{B,j}^2 \sin^2 \theta_j)T_n$$

$$= K_f - \omega^2 \rho_l + iB_{p,i}k_{B,i}^3 \cos \theta_i(\cos^2 \theta_i + (2 - \nu_i) \sin^2 \theta_i)$$

where the nearfield wavenumber on each plate, k_n, is

$$k_n = \sqrt{k_B^2(1 + \sin^2 \theta)} \qquad (5.93)$$

and the effective stiffness, K_f, acting on a metre length of foundation due to compression of the soil along the two edges of the foundation, and shearing of the soil under the foundation is

$$K_f = 2d_1 s'_{soil}(1 + i\eta_{soil}) + d_2 G_{soil} \qquad (5.94)$$

(where the foundation dimensions, d_1 and d_2 are shown in Fig. 5.24)

and K_ϕ is the stiffness of the soil resisting rotation of the foundation,

$$K_\phi = \frac{G_{soil} d_1 d_2^2}{2} + \frac{d_2^3 s'_{soil}(1 + i\eta_{soil})}{12} \qquad (5.95)$$

and for the rectangular cross-section of the strip foundation, the mass moment of inertia, I_r, is

$$I_r = \frac{\rho_l}{12} \left(\frac{d_1}{d_2} + \frac{d_2}{d_1} \right) \qquad (5.96)$$

The transmission coefficient at a specific angle of incidence is given by

$$\tau_{ij}(\theta) = \frac{\rho_{s,j} c_{B,j} \cos \theta_j}{\rho_{s,i} c_{B,i} \cos \theta_i} |T|^2 \qquad (5.97)$$

The above calculations are repeated to give the angular average transmission coefficient, τ_{ij}, using Eq. 5.6. The coupling loss factor, η_{ij}, is calculated from τ_{ij}, using Eq. 2.154; for bending waves this is

$$\eta_{ij} = \frac{c_{B,i} L_{ij} \tau_{ij}}{\pi^2 f S_i} \qquad (5.98)$$

where L_{ij} is the length of the line connection between plate subsystems i and j.

From Wilson and Craik (1995), the compression stiffness and damping can be measured using the method of Briaud and Cepert (1990). This measurement is based on the assumption that the soil simply acts as a spring when a lump mass on top of the soil is excited by a vertical force. A trench is dug in the ground to the required depth with a relatively flat surface at its base. A concrete cuboid ($\approx 100\,kg$) is then cast onto this surface to form the lump mass. Measurement of the driving-point mobility of this cuboid can then be used to identify the mass–spring resonance frequency; this measurement is similar to that used for the dynamic stiffness of resilient materials (Section 3.11.3.1). The spring representing the soil can be highly damped and it is necessary to estimate this damping from measurements. For linear springs the ratio of the 3 dB bandwidth of the resonance peak to the resonance frequency equals the loss factor when $\eta < 0.3$. When $\eta \geq 0.3$ these values can only be treated as rough estimates because the ratio is no longer linearly related to the loss factor. Example data from three different soils are shown in Table 5.1. Note that the stiffness may vary significantly with depth below ground level.

5.3 Statistical energy analysis

The fundamental aspects of SEA that are needed to predict the combination of direct and flanking transmission have been covered in Section 4.2. Calculation of transmission coefficients for vibration transmission between idealized plate junctions has been discussed in Section 5.2; and

Table 5.1. Measured soil stiffness data (UK)

Type of soil	Depth below ground level (m)	Compression stiffness per unit area, s'_{soil} (N/m³)	Loss factor, $\eta_{soil}(-)$
Stiff clay with large stones*	0.45	1.96×10^9	0.96
Lower green sand overlying sandstone#	0.9	1.63×10^8	0.2
London clay – wet#	1.5	8.67×10^7	0.4

Measurements from *Wilson and Craik, and #Hopkins courtesy of BRE, ODPM, and BRE Trust.

these can be used to calculate the coupling loss factors. This section looks at how measured data can be included within the SEA framework and then uses some examples to illustrate the prediction of direct and flanking transmission.

5.3.1 Inclusion of measured data

The SEA framework is well-suited to the inclusion of laboratory measurements in cases where the internal, coupling, or total loss factor of a subsystem cannot be accurately predicted or where the airborne sound insulation of a building element is too complex to model.

5.3.1.1 Airborne sound insulation

To calculate transmission of airborne sound between adjacent rooms, laboratory measurements of airborne sound insulation can be included within the SEA framework. This is useful when direct transmission across an element is too complicated to model and the element itself does not affect any important flanking transmission paths. One example of this is a lightweight separating wall between adjacent rooms where a flanking path is formed by a continuous concrete floor. The SEA framework can be used to predict structure-borne sound transmission via the concrete floor and to incorporate the measured sound reduction index for direct sound transmission across the separating wall. Another example occurs with a building element such as a doorset, window or vent in a solid separating wall between adjacent rooms. It can often be assumed that any transmission of vibration between the separating wall and this type of element will have negligible effect on the overall sound insulation. Therefore the solid separating wall can simply be modelled, and laboratory measurements can be incorporated in the SEA model for the doorset, window or vent.

To use the measured sound reduction index of a separating element to calculate airborne sound transmission between two rooms, the element is not included as a subsystem in the SEA model. We deliberately ignore the mechanisms by which sound is transmitted across the element, such as non-resonant transmission, resonant transmission, mass–spring–mass resonances, dilatational wave motion; these are treated as a black box of unknowns. The room on each side of the separating element is included in the SEA model as a space subsystem and a non-resonant transmission mechanism is used to transmit sound between them. Hence we can convert the sound reduction index, R, into a coupling loss factor (Eq. 4.26) between two subsystems i and j using

$$\eta_{ij} = \frac{c_0 S}{4\omega V_i} 10^{-R/10} \tag{5.99}$$

where S is the area of the separating element.

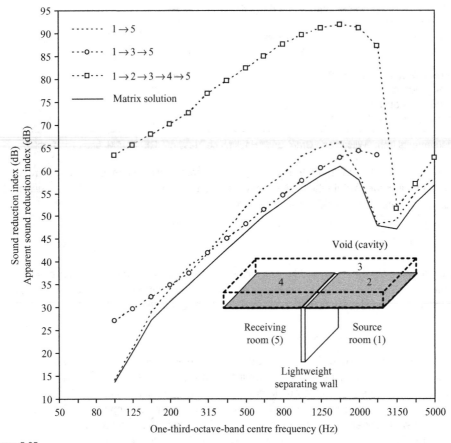

Figure 5.25

Predicted airborne sound insulation between two adjacent rooms with flanking transmission via the ceiling void. SEA model incorporates the measured sound reduction index for the lightweight separating wall. Rooms: $L_x = 4\,\text{m}$, $L_y = 3\,\text{m}$, $L_z = 2.5\,\text{m}$, $T = 0.5\,\text{s}$. Ceiling void: $L_x = 4\,\text{m}$, $L_y = 6\,\text{m}$, $L_z = 0.5\,\text{m}$, $T = 0.5\,\text{s}$. Ceiling: 12.5 mm plasterboard, $L_x = 4\,\text{m}$, $L_y = 3\,\text{m}$, $\rho_s = 10.8\,\text{kg/m}^2$, $c_L = 1490\,\text{m/s}$, $\eta_{\text{int}} = 0.0141$.

Similarly, the coupling loss factor for the element-normalized level difference is

$$\eta_{ij} = \frac{c_0}{4\omega V_i} 10^{-(D_{n,e} - 10\,\text{dB})/10} \tag{5.100}$$

The consistency relationship can be used to calculate the coupling loss factor in the reverse direction.

5.3.1.1.1 Example

This example is based on a lightweight separating wall with flanking transmission via a ceiling void as shown in Fig. 5.25. A laboratory measurement of the sound reduction index is used for the lightweight separating wall (e.g. plasterboard on light steel frame); hence this wall is not assigned to a subsystem. A five-subsystem model is used to determine the combination of direct and flanking transmission. Note that we are effectively using a plate–cavity–plate system as a flanking construction rather than a separating construction as in Section 4.3.5. Each plate for the ceiling is modelled as a single sheet of plasterboard (ignoring its supporting frame) and it is assumed to be uncoupled from the separating wall and the ceiling in the adjacent room. The radiation efficiency for the plasterboard is ideally calculated using method no. 2 to give an indication of lower and upper limits near the critical frequency; however, to simplify this example only the lower limit is used. The direct path, $1 \rightarrow 5$ corresponds to the sound reduction index of the separating wall, and the matrix solution corresponds to the apparent sound reduction index. Significant flanking transmission occurs via the non-resonant path $1 \rightarrow 3 \rightarrow 5$ below the critical frequency of the plasterboard, and via the resonant path $1 \rightarrow 2 \rightarrow 3 \rightarrow 4 \rightarrow 5$ above the critical frequency. For this example, summing the paths $1 \rightarrow 5$, $1 \rightarrow 3 \rightarrow 5$, and $1 \rightarrow 2 \rightarrow 3 \rightarrow 4 \rightarrow 5$ gives approximately the same result as the matrix solution.

This model can be adapted to other situations in buildings. For example, it could be used to model sound transmission between two adjacent rooms with flanking transmission via a short external corridor instead of a ceiling void. Plates 2 and 4 would then form the flanking walls into the corridor and because the sound reduction indices for the doorsets in these walls can be awkward to predict it would be convenient to incorporate laboratory measurements.

5.3.1.2 Coupling loss factors

Some junctions between plates, or between plates and beams are too complex to model and it may be necessary to measure the coupling loss factor as described in Section 3.12.3.3. This approach effectively treats the junction as a black box of unknowns and there are three points that need to be considered.

The first point concerns the situation where an SEA model of an existing structure is being created to gain an insight into the sound transmission paths. This usually means that the measurement will be taken *in situ* rather than in the laboratory where there is some control over the flanking paths. Depending on the measurement technique, the measured coupling loss factor may quantify more than one transmission path. These other paths may be purely structural but there may also be paths that involve sound radiation into spaces such as cavities; some examples are shown in Fig. 5.26. Care is then needed to make sure that transmission paths are not included more than once in the model. If the coupling loss factor is found to play a role in an important transmission path, more measurements may be necessary to isolate the individual paths. The second point concerns the omission of subsystems from the SEA model. A common example is a plate junction where only bending waves are excited and only bending wave motion is measured on the source and receiving plates. We recall from Section 5.2.2 that in-plane waves can also be generated at the junction. Using these measurements in an SEA model that considers only bending waves will not usually cause significant errors when predicting vibration transmission across one junction between adjacent rooms. (The assumption here is that the total loss factor associated with the in-plane wave subsystems does not vary significantly from one situation to another.) However, the measured coupling loss factor is unlikely to be suitable when modelling transmission between more distant rooms with transmission paths involving several junctions. The third point concerns plates with low mode counts and low modal overlap because a single physical measurement on one junction may be significantly different to similar junctions (refer back to the examples in Section 5.2.3).

5.3.1.3 Total loss factors

For some subsystems it is difficult to predict the total loss factor accurately. However, the total loss factor can be critical in accurately quantifying the overall sound insulation; this occurs with

(a) Lightweight construction: Junction of walls and floors

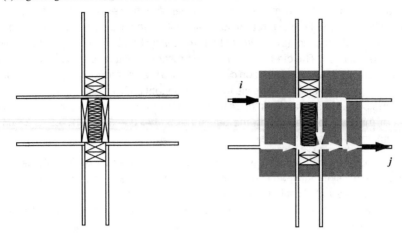

(b) Heavyweight construction: Junction of solid separating wall and external cavity wall with wall ties

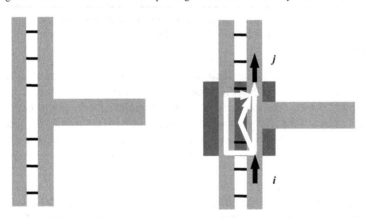

Figure 5.26

Examples of two complex junctions where measurement of the structural coupling loss factor, η_{ij}, could include more than one transmission path.

some cavities in walls and floors, and with some plates that form walls, floors, or ground floor slabs. Before including measurements of total loss factors in a model it is important to consider which part of the total loss factor is difficult to predict; the internal or the coupling loss factors. Internal loss factors are almost always determined from measurements, so this is rarely an issue. If it is a coupling loss factor then it is necessary to consider whether the coupling losses have been adequately dealt with in the model.

5.3.2 Models for direct and flanking transmission

Direct transmission across solid plates in the laboratory was modelled in Chapter 4 using a total loss factor for each plate that represented the losses when coupled to the rest of the laboratory structure. This assumed that each plate subsystem would lose energy into the laboratory structure but there would be negligible flow of energy from this structure into the plate. Now we want to assess flanking transmission it is necessary to create an SEA model of the building that will allow energy to flow in and out of each plate. A complete model will allow the total loss factor of each subsystem to be calculated from the internal loss factor and the sum of its coupling loss factors. However, most buildings contain some walls, floors and junctions that are awkward or not possible to include in the model. This section uses some examples to illustrate factors that need to be considered when restricting the model to only part of the flanking construction and when restricting calculations to a limited number of flanking paths.

5.3.2.1 Example: SEA model of adjacent rooms

To model direct and flanking transmission between adjacent rooms we start by looking at a T-junction. For this we can use the T-junction from Section 5.2.3.3 where the wave approach was used to calculate vibration transmission between rigidly connected masonry walls. We will shortly connect the same T-junction to other plates to form two rooms; hence in this example we will set the total loss factor of each plate subsystem to the values they will have when fully connected to these other plates. The SEA model consists of the source room, receiving room, and the three plates that form the T-junction. The predicted airborne sound insulation is shown in Fig. 5.27 from two SEA models; one with only bending wave motion, and the other

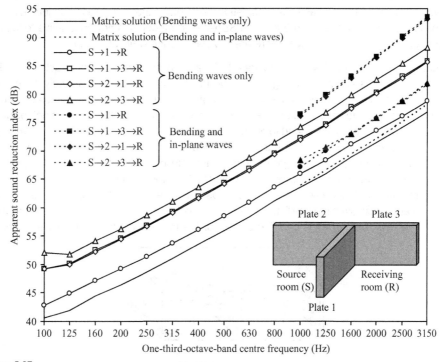

Figure 5.27

Predicted sound insulation between two adjacent rooms where all sound transmission occurs across a T-junction of masonry walls. The matrix solution and path analysis are shown for two SEA models (1) bending waves only and (2) bending and in-plane waves. In both models the plate subsystems used in the path analysis are the bending wave subsystems. Plate properties: $L_{x1} = 4.0\,\text{m}$, $L_{x2} = 3.5\,\text{m}$, $L_{x3} = 3.0\,\text{m}$, $L_{y1} = L_{y2} = L_{y3} = 2.4\,\text{m}$, $h_1 = 0.215\,\text{m}$, $h_2 = h_3 = 0.1\,\text{m}$, $\rho_{s1} = 430\,\text{kg/m}^2$, $\rho_{s2} = \rho_{s3} = 200\,\text{kg/m}^2$, $c_L = 3200\,\text{m/s}$, $\nu = 0.2$, total loss factors correspond to those in the fully connected system of nine plates in Fig. 5.28.

with bending and in-plane wave motion. Flanking transmission is evident from the fact that the sound insulation predicted with the matrix solution is lower than the direct path S→1→R. The bending and in-plane wave model is only shown in the high-frequency range where there are both quasi-longitudinal and transverse shear modes. Between 1000 and 2000 Hz the matrix solutions for the two models are approximately equal; it is only above 2500 Hz that they differ by >1 dB. However, the three flanking paths involving transmission between bending wave sub-systems across one junction (S→1→3→R, S→2→1→R, and S→2→3→R) indicate that whilst the resulting sound insulation is the same, the strengths of the individual transmission paths are quite different for the two models. Flanking path S→2→3→R becomes more important when both bending and in-plane waves are included. For adjacent rooms it is often the low- and mid-frequency ranges that are of most interest so it is simplest to consider only bending wave transmission. For masonry walls in the high-frequency range there are other factors that affect the accuracy such as thick plate theory, bonded surface finishes starting to vibrate independently of the base wall, and a decrease in vibration with distance. Although the coupling between plates and rooms can usually be estimated with thin plate theory up to $4f_{B(thin)}$, these other factors mean that there is no significant increase in accuracy with the bending and in-plane wave model for transmission between adjacent rooms. For both models, the matrix solution is within 0.5 dB of the sum of the direct path and the three flanking paths. This is noted because it indicates that with some constructions it may be possible to use path analysis to estimate the overall sound insulation using paths that only cross one junction; this approach is discussed in Section 5.4 for an SEA-based model. It is now instructive to look at a more realistic construction where there are several flanking walls and floors.

We now incorporate the T-junction (plates 1, 2, and 3) as part of a construction forming two adjacent rooms as shown in Fig. 5.28. The full construction has nine rigidly connected masonry/concrete plates for which the wave approach (bending waves only, thin plate theory) is used to calculate the transmission coefficients. The space outside the rooms is not included in the model so we only allow sound to be transmitted between the two rooms via the walls and floors. The concrete plates that form the ground floor and ceiling are identical except for the high internal damping given to the ground floor to simulate losses into the ground. Using the full construction we can now create smaller SEA models by removing certain plates, but ensuring that the remaining plates have the same total loss factor as when they are fully connected. Using this approach gives the predicted sound insulation for the separating wall, the separating wall with various flanking walls and floors, and for the full construction. These are shown in

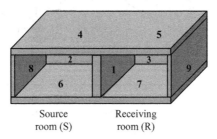

Source   Receiving
room (S)   room (R)

Figure 5.28

Two adjacent rooms formed from nine masonry/concrete plate subsystems. Source room: $4 \times 3.5 \times 2.4$ m. Receiving room: $4 \times 3 \times 2.4$ m. Plate properties: $h_1 = 0.215$ m, $h_2 = h_3 = h_8 = h_9 = 0.1$ m, $h_4 = h_5 = h_6 = h_7 = 0.15$ m. Plates 1, 2, 3, 8, and 9: $\rho = 2000$ kg/m³, $c_L = 3200$ m/s, $\nu = 0.2$, $\eta_{int} = 0.01$. Plates 4, 5, 6, and 7: $\rho = 2200$ kg/m³, $c_L = 3800$ m/s, $\nu = 0.2$, $\eta_{int} = 0.005$ for plates 4 and 5, $\eta_{int} = 3f^{-0.5}$ for plates 6 and 7 to simulate a highly damped ground floor. The total loss factor for each plate (apart from the highly damped ground floor plates) is within 2 dB of $\eta_{int} + f^{-0.5}$.

Fig. 5.29. The sound reduction index for the separating wall alone is higher than would usually be measured in the laboratory because it is rigidly connected to six other walls and therefore has a relatively high total loss factor. The sound insulation decreases as flanking walls and floors are added to the separating wall and more transmission paths are formed in models (b) to (f). Models (b), (c), (d), and (e) indicate the presence of important flanking transmission but do not give an adequate estimate of the sound insulation with model (f). This is particularly noticeable with this construction because flanking paths dominate over direct transmission via the separating wall. In this example, the difference between models (e) and (f) shows that the distribution of energy between the subsystems is significantly altered by including the back walls. This indicates that making *a priori* decisions on which flanking walls and floors can be excluded from a model is not always straightforward; in most cases it is better to include more, rather than less, of the construction.

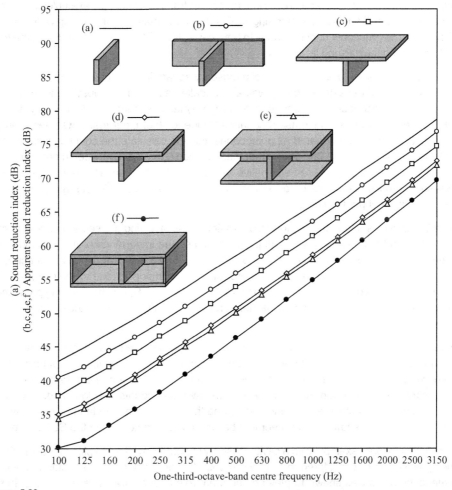

Figure 5.29

Sound insulation between two adjacent rooms predicted with the SEA matrix solution (bending waves only) when the separating wall is modelled as an isolated element in (a) and when connected to different combinations of flanking walls and floors in (b), (c), (d), (e), and (f).

For the isolated T-junctions in models (b) or (c), the matrix solution is within 1 dB of the sum of the direct path and the flanking paths that only cross one junction. With models (d), (e), and (f) there are many more flanking paths that cross more than one junction. Without incurring errors of at least a few decibels it is no longer possible to determine the same values as the matrix solution by summing the direct path and flanking paths that only cross one junction. There are usually a great number of paths that determine the sound insulation between adjacent rooms, as well as rooms that are far apart in a building; hence analysis of individual paths tends to become of limited use (Craik, 1996).

5.3.2.2 *Example: Comparison of SEA with measurements*

Validating models of direct and flanking transmission with measurements is usually easier in a flanking laboratory where there is more control over the quality of construction as well as the effect of the laboratory walls and floors on the flanking transmission. Two examples are used in this section: a solid masonry separating wall and a cavity masonry separating wall, both of which have cavity masonry flanking walls. In both cases the SEA model included both the test construction and the flanking laboratory.

Figure 5.30a shows an example with a solid masonry separating wall construction. The inner and outer leaves of the external cavity wall are modelled as separate subsystems. However the wall ties that connect these leaves have a high dynamic stiffness; so this is only appropriate above the mass–spring–mass resonance frequency (i.e. inner leaf – wall ties – outer leaf). The ties can then be treated as point connections to calculate the coupling loss factors (Section 4.3.5.4.1). Figure 5.30a shows that there is good agreement between measurements and the SEA models. Note that an alternative SEA model could potentially be used below the mass–spring–mass resonance by assuming that the inner and outer leaves act as a single plate, but this is not pursued here.

Figure 5.30b shows some of the short transmission paths involving the separating and flanking walls using the bending wave model. This indicates the importance of the flanking path via the wall ties in the external cavity wall, $S \to 2 \to 4 \to 3 \to R$ compared with the direct path $S \to 1 \to R$. Paths via the external cavities are relatively unimportant due to the cavity closer at the junction. Summing the direct path, path $S \to 2 \to 4 \to 3 \to R$, and the flanking paths that only cross one junction overestimates the standardized level difference from the matrix solution by $\approx 3\,dB$ in each frequency band.

Figure 5.30c shows the difference between measured and predicted velocity level differences with structure-borne excitation of walls 1, 2, and 4 using a shaker. Prediction of velocity level differences using SEA (wave approach) gives smooth curves that are in contrast to the fluctuating curves expected from masonry walls with low mode counts and low modal overlap. In the context of the measurement uncertainty and the assumptions made in using the wave approach, Fig. 5.30c shows good agreement between measurements and the SEA model.

Figure 5.31 shows an example with a cavity masonry separating wall on a split foundation. Only bending waves were considered in the SEA model. In Section 4.3.5.2.2, structural coupling via the foundations was shown to play an important role in direct transmission across these walls. The coupling loss factor for the foundation coupling, and reverberation times for the separating and flanking wall cavities were measured and incorporated in the model because they were difficult to predict accurately (Hopkins, 1997). The wall ties in the external cavity wall have a low dynamic stiffness giving a resonance frequency below 50 Hz. Comparing the path

(a) Comparison of an SEA model with airborne sound insulation measurements in a flanking laboratory

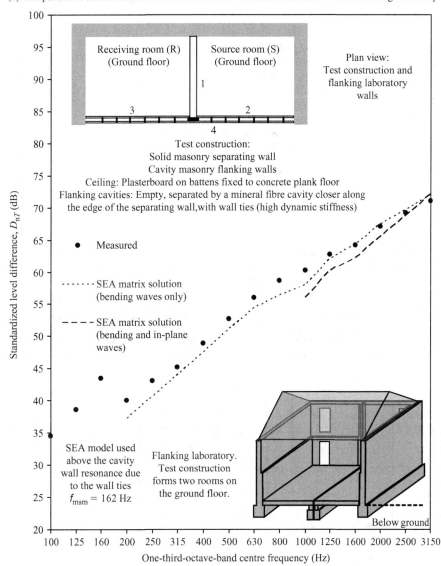

Figure 5.30

Direct and flanking transmission across a solid masonry wall construction. Separating wall: One leaf of 215 mm, 430 kg/m² solid masonry (plaster finish). Flanking cavity wall: Inner leaves of 100 mm, 150 kg/m² solid masonry (plaster finish). Outer leaf of 100 mm, 170 kg/m² solid brick. 100 mm cavity (2.5 wall ties/m², $s_{100mm} = 43.4$ MN/m). Ceiling and separating floor: 9.5 mm plasterboard fixed to 45 × 45 mm timber battens fixed to 150 mm, 300 kg/m² concrete slabs. Flanking laboratory: 560 mm, 900 kg/m² solid brick walls, 125 mm cast concrete ground floor slabs on hardcore. Measured data from Hopkins are reproduced with permission from ODPM and BRE.

S → 1 → 2 → R across the split foundations with the matrix solution in Fig. 5.31 indicates that this path is relatively unimportant. There are too many paths to make path analysis a practical option to determine the overall sound insulation. Hence it is the matrix solution that is needed and this gives a good estimate of the measured sound insulation.

(b) SEA path analysis between the source and receiving rooms (bending waves only)

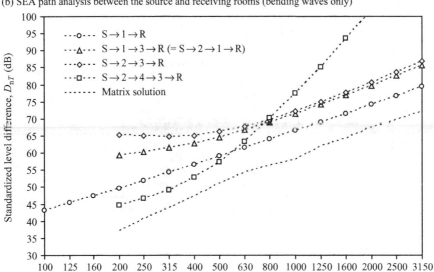

(c) Comparison of SEA model with measurements of velocity level differences (bending wave vibration) between walls. This is shown in terms of the difference between measured and SEA velocity level differences, $D_{v,ij}$, where i is the source subsystem. SEA model: 100–800 Hz (bending waves only), 1000–3150 Hz (bending and in-plane waves). The 95% confidence intervals correspond to the measurements.

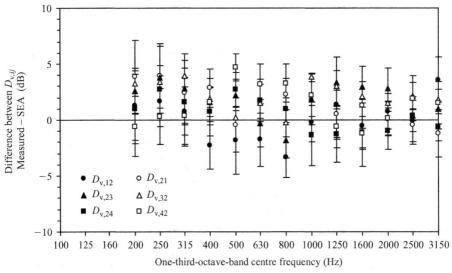

Figure 5.30

(*Continued*)

As with transmission suite measurements, the structure of the flanking laboratory also plays a role in determining the measured sound insulation. For the cavity separating wall this is obvious from the way that the foundations form part of the laboratory, but even the heavy walls and floors of a laboratory can participate in flanking transmission. Two options are usually

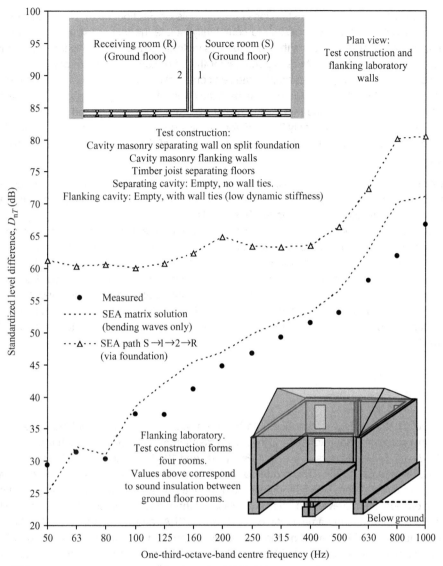

Figure 5.31

Direct and flanking transmission across a cavity masonry wall construction. Comparison of SEA model with airborne sound insulation measurements in a flanking laboratory. (*Note*: The frequency range is restricted to between 50 and 1000 Hz where measurements of the coupling loss factor were available for structural coupling between the separating cavity wall leaves via the foundations.) Separating cavity wall: Two leaves of 100 mm, 166 kg/m² solid masonry (plaster finish). 60 mm empty cavity (no wall ties). Flanking cavity wall: Inner leaves of 100 mm, 70 kg/m² solid masonry (plaster finish). Outer leaf of 100 mm, 154 kg/m² solid brick. 75 mm cavity (2.5 wall ties/m², $s_{75mm} = 1.7$ MN/m). Measured data from Hopkins are reproduced with permission from ODPM and BRE.

considered when building a test construction into a flanking laboratory: the first is to isolate the test construction from the laboratory using resilient layers, the second is to rigidly connect the test construction to the laboratory. The first option reduces the coupling losses from each wall or floor; if this causes low modal overlap then the coupling loss factors can be expected to vary significantly between similar constructions (see examples in Section 5.2.3). Note that for structural reasons it is not always possible to completely isolate the test construction on resilient layers. The second option means that the total loss factors and modal overlap factors are more representative of *in situ*; but there is potential for the laboratory structure to participate in the flanking transmission. In these examples the test construction was rigidly connected to the laboratory. The SEA models were used to estimate the effect of including the laboratory walls and floors on the measured sound insulation. The predicted effect on the sound insulation varied with frequency; the range was only 0.1–1 dB for the cavity wall, but 1–4 dB for the solid wall. This reinforces the point that it is better to include as much of the construction in the model as possible.

The examples in this section indicate that variation in the sound insulation performance of nominally identical separating walls or floors will not only be due to workmanship, but also due to different flanking constructions. This needs to be placed in the context of guidance documents that describe building constructions with the potential to achieve certain values of sound insulation. To avoid completely dictating the building design it is common to focus on the separating wall or floor, and the flanking walls or floors that are connected to it (e.g. see DIN 4109 1989; Homb *et al.*, 1983). Such guidance is usually based on field sound insulation measurements, so it is possible to make some statistical inference about the likelihood of achieving a certain value of sound insulation. Some account is therefore taken of workmanship and of walls and floors that are not directly connected to the separating element.

5.4 SEA-based model

For practical purposes it is very useful to have a model for direct and flanking transmission that can incorporate laboratory sound insulation measurements. This approach is taken in the model developed by Gerretsen (1979, 1986, 1994, and 1996) and subsequently implemented in the Standard EN 12354. The model for airborne and impact sound insulation can either be derived directly from classical diffuse field theories of sound and vibration, or by using SEA to provide the framework for the same classical theories (Gerretsen, 1979). The resulting model is affected by the same limitations that affect SEA; and this is made explicit here by referring to it as an SEA-based method. By deriving the model from an SEA perspective it is easier to see the links between SEA loss factors and laboratory measurements of the sound reduction index, structural reverberation times, and the vibration reduction index. The SEA framework also simplifies discussion of the roles of resonant and non-resonant transmission. The model is essentially the same as SEA path analysis between adjacent rooms, but the flanking transmission paths are restricted in their length. Each flanking path is restricted to vibration transmission across no more than one plate junction as shown in Fig. 5.32 (Gerretsen, 1979). This approach is quite commonly taken in documents giving construction guidance for sound insulation in buildings.

5.4.1 Airborne sound insulation

The aim is to determine the apparent sound reduction index; this is easily converted to other *in situ* sound insulation descriptors (e.g. D_n, D_{nT}) at the end of the calculation. Using SEA

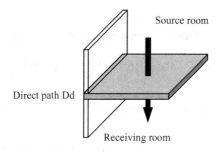

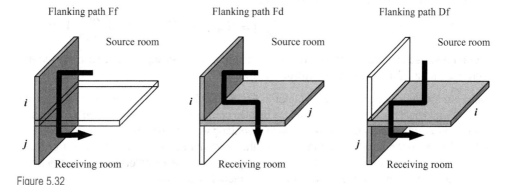

Figure 5.32

Direct and flanking transmission paths between two adjacent rooms (only one junction is shown). The length of each flanking path is restricted to vibration transmission across no more than one junction. Direct transmission via the separating element is indicated by D in the source room, and d in the receiving room. Flanking surfaces are indicated by F in the source room and f in the receiving room. Flanking paths Ff, Fd, and Df apply to airborne sound insulation. Flanking path Df applies to impact sound insulation.

path analysis, the apparent sound reduction index for P transmission paths is determined from Eq. 4.18 as

$$R' = 10\lg\left(\frac{E_S}{E_R}\right)_{\substack{\text{Due to}\\P\text{ paths}}} + 10\lg\left(\frac{V_R}{V_S}\right) + 10\lg\left(\frac{S}{A}\right) \tag{5.101}$$

where E_S and E_R are the energies for the source and receiving rooms respectively.

The transmission paths need to be combined outside the framework of SEA so it is useful to define a flanking sound reduction index, R_{ij} (Gerretsen, 1979). This is based on the ratio of the sound power, W_1, incident on a reference area, S_S, in the source room to the sound power, W_{ij}, radiated by flanking plate j due to sound power incident on plate i in the source room. This is given in decibels by

$$R_{ij} = 10\lg\left(\frac{W_1}{W_{ij}}\right) \tag{5.102}$$

Note that R_{ij} has not been written in terms of a transmission coefficient, τ_{ij}. This is because τ_{ij} has already been defined in Eq. 5.1 to describe bending wave transmission between plates i and j and it will shortly be needed again.

We will temporarily assume that the total loss factor of each plate connected *in situ* is the same as when its sound reduction index is measured in the laboratory; this assumption will shortly be removed when generalizing the model. By setting the reference area, S_S, to equal the area of the separating plate the apparent sound reduction index is determined using

$$R' = -10\lg \left(10^{-R_{Dd}/10} + \sum_{p=1}^{P} 10^{-R_{ij}/10} \right)$$ (5.103)

where R_{Dd} is the sound reduction index of the separating plate; a combination of non-resonant and resonant transmission.

It is assumed that the separating and flanking plates support a reverberant bending wave field with no significant decrease in vibration with distance across their surface. Using SEA path analysis, the flanking sound reduction index is

$$R_{ij} = 10\lg\left(\frac{E_S}{E_R}\right) + 10\lg\left(\frac{V_R}{V_S}\right) + 10\lg\left(\frac{S_S}{A}\right) = 10\lg\left(\frac{\eta_i\eta_j\eta_R}{\eta_{Si}\eta_{ij}\eta_{jR}}\right) + 10\lg\left(\frac{V_R S_S}{V_S A}\right)$$ (5.104)

where the subscripts are: S for the source room, R for the receiving room, i for plate i in the source room, and j for plate j in the receiving room.

There are two types of coupling loss factor in Eq. 5.104. Both of these involve a reverberant bending wave field on the plates; η_{Si} and η_{jR} involve coupling between a plate and a room, and η_{ij} involves vibration transmission resulting in a bending wave field on each of the two coupled plates. The former can be written in terms of the resonant sound reduction indices and the latter can be written in terms of a direction-averaged velocity level difference. Hence we can start to remove any direct reference to the loss factors in Eq. 5.104.

The resonant sound reduction index is given in terms of loss factors by Eq. 4.22. The plates are defined so that one side of plate i will face into the source room and one side of plate j will face into the receiving room. Hence regardless of whether the other side of plate i faces into the receiving room, or the other side of plate j faces into the source room it is useful to determine the resonant sound reduction indices using the same source and receiving room volumes and the same receiving room absorption as the adjacent rooms. This gives the resonant sound reduction indices for plates i and j as

$$R_{Resonant,i} = 10\lg\left(\frac{\eta_i\eta_R}{\eta_{Si}\eta_{iR}}\frac{V_R S_i}{V_S A}\right) \quad \text{and} \quad R_{Resonant,j} = 10\lg\left(\frac{\eta_j\eta_R}{\eta_{Sj}\eta_{jR}}\frac{V_R S_j}{V_S A}\right)$$ (5.105)

Both of the resonant sound reduction indices can now be rewritten purely in terms of η_{Si} and η_{jR} by using the consistency relationship (Eq. 4.2) and the fact that $\eta_{iS} = \eta_{iR}$ and $\eta_{jS} = \eta_{jR}$; this gives

$$R_{Resonant,i} = 10\lg\left(\frac{\eta_i\eta_R n_i}{\eta_{Si}^2 n_S}\frac{V_R S_i}{V_S A}\right) \quad \text{and} \quad R_{Resonant,j} = 10\lg\left(\frac{\eta_j\eta_R n_S}{\eta_{jR}^2 n_j}\frac{V_R S_j}{V_S A}\right)$$ (5.106)

We now need to combine the two resonant sound reduction indices in such a way that Eq. 5.104 can be rewritten in terms of the resonant sound reduction indices. This is done by combining them as follows,

$$\frac{R_{Resonant,i}}{2} + \frac{R_{Resonant,j}}{2} = 10\lg\left(\frac{\eta_R}{\eta_{Si}\eta_{jR}}\right) + 10\lg\left(\frac{V_R}{V_S A}\right) + 5\lg\left(\frac{\eta_i\eta_j S_i S_j n_i}{n_j}\right)$$ (5.107)

This allows Eq. 5.104 to be given in terms of the resonant sound reduction indices whilst leaving vibration transmission between the two plates in terms of a coupling loss factor,

$$R_{ij} = \frac{R_{Resonant,i}}{2} + \frac{R_{Resonant,j}}{2} + 10\lg\left(\frac{1}{\eta_{ij}}\right) + 5\lg\left(\frac{\eta_i\eta_j n_j}{n_i}\right) + 10\lg\left(\frac{S_S}{\sqrt{S_i S_j}}\right)$$ (5.108)

The next step is to look at vibration transmission between plates i and j for which we wish to remove η_{ij} and use a velocity level difference instead. To do this we take a blinkered view of the entire construction and focus on the two coupled plates to create a two-subsystem SEA model for plate subsystems i and j. Under the assumption that with excitation of subsystem i there is negligible power flow back from subsystem j to i, the coupling loss factor can be written in terms of the velocity level difference, $D_{v,ij}$ (Section 3.12.3.3)

$$D_{v,ij} = 10 \lg \left(\frac{\eta_j}{\eta_{ij}} \right) - 10 \lg \left(\frac{m_i}{m_j} \right) \tag{5.109}$$

Equations 5.108 and 5.109 indicate that if we also calculate the velocity level difference, $D_{v,ji}$, and write it in terms of η_{ij} using the consistency relationship we will have all the relevant terms that we need to rewrite Eq. 5.108 in terms of velocity level differences. For excitation of subsystem j, $D_{v,ji}$, is calculated in a similar way to $D_{v,ij}$ giving

$$D_{v,ji} = 10 \lg \left(\frac{\eta_i}{\eta_{ji}} \right) + 10 \lg \left(\frac{m_i}{m_j} \right) = 10 \lg \left(\frac{\eta_i \eta_j}{\eta_{ij} \eta_i} \right) + 10 \lg \left(\frac{m_i}{m_j} \right) \tag{5.110}$$

$D_{v,ij}$ and $D_{v,ji}$ can now be combined to give the direction-averaged velocity level difference, $\overline{D_{v,ij}}$, as

$$\overline{D_{v,ij}} = \frac{D_{v,ij} + D_{v,ji}}{2} = 10 \lg \left(\frac{1}{\eta_{ij}} \right) + 5 \lg \left(\frac{\eta_i \eta_j \eta_j}{\eta_i} \right) \tag{5.111}$$

Substituting Eq. 5.111 in Eq. 5.108, we can now rewrite Eq. 5.104 purely in terms of the resonant sound reduction indices and the direction-averaged velocity level difference,

$$R_{ij} = \frac{R_{\text{Resonant},i}}{2} + \frac{R_{\text{Resonant},j}}{2} + \overline{D_{v,ij}} + 10 \lg \left(\frac{S_S}{\sqrt{S_i S_j}} \right) \tag{5.112}$$

5.4.1.1 Generalizing the model for in situ

Now it is useful to remove the assumption that the total loss factor of each plate *in situ* is the same as when the sound reduction index is measured in the laboratory. This requires the structural reverberation time of each plate to be measured whilst it is installed in the laboratory for the sound reduction index measurement. It is then necessary to predict the structural reverberation time for each plate when installed *in situ*. The resonant sound reduction index of each plate measured in the laboratory is converted to the *in situ* value using

$$R_{\text{situ}} = R_{\text{Resonant}} - 10 \lg \left(\frac{T_{s,\text{situ}}}{T_{s,\text{lab}}} \right) \tag{5.113}$$

where the structural reverberation times *in situ* and in the laboratory are $T_{s,\text{situ}}$ and $T_{s,\text{lab}}$ respectively. If the resonant sound reduction index has been predicted then $T_{s,\text{lab}}$ corresponds to the value used in the prediction model.

For the direct path, $R_{\text{Dd,situ}}$ needs to include both non-resonant and resonant transmission as measured in the laboratory; however the conversion in Eq. 5.113 only applies to the resonant component. This may require estimating the non-resonant component of the sound reduction index so that it can be removed. The conversion is then carried out on the resonant component; and after the conversion the non-resonant component is re-introduced.

Laboratory measurements of vibration transmission between plates also need to be converted to a direction-averaged velocity level difference that corresponds to the *in situ* situation. This is achieved by defining a vibration reduction index, K_{ij} (Gerretsen, 1996) given by Eq. 3.252. This can either be measured in the laboratory or predicted. From Eq. 3.252, the direction-averaged velocity level difference for the *in situ* situation is given by

$$\overline{D_{v,ij,\text{situ}}} = K_{ij} - 10\lg\left(\frac{L_{ij}}{\sqrt{a_{i,\text{situ}}a_{j,\text{situ}}}}\right) \tag{5.114}$$

where $a_{i,\text{situ}}$ and $a_{j,\text{situ}}$ are the absorption lengths for plates i and j *in situ*, and L_{ij} is the junction length between elements i and j.

The *in situ* structural reverberation time for Eq. 5.113 and the *in situ* absorption lengths for Eq. 5.114 are related to the *in situ* total loss factor. For masonry/concrete plates that are rigidly connected on all sides, estimates for the *in situ* total loss factor are discussed in Section 2.6.5.

Various aspects relating to the measurement of K_{ij} are discussed in Section 3.12.3. To calculate K_{ij} it is useful to note the relationship between the vibration reduction index and the coupling loss factor. This can be determined from Eq. 5.111. By assuming that i and j are solid homogenous plates, K_{ij} can be written in terms of the critical frequencies as

$$K_{ij} = 10\lg\left(\frac{1}{\eta_{ij}}\right) + 5\lg\left(\frac{c_0^2 L_{ij}^2 f_{c,j}}{\pi^4 S_i^2 f_{c,i} f_{\text{ref}} f}\right) \tag{5.115}$$

from which the relationship to the bending wave transmission coefficient (Eq. 5.1) is

$$K_{ij} = 10\lg\left(\frac{1}{\tau_{ij}}\right) + 5\lg\left(\frac{f_{c,j}}{f_{\text{ref}}}\right) \tag{5.116}$$

where f_{ref} is a reference frequency of 1000 Hz.

Hence K_{ij} can be calculated using the wave approach (bending waves only) or by using a numerical method as discussed in Section 5.2.

As most walls and floors have linings on one or both sides it is useful to incorporate laboratory measurements of the sound reduction improvement index. We recall that the resonant sound reduction index is used to replace the coupling loss factor between a plate and a room. To use laboratory measurements of the sound reduction improvement index for a lining; we need to use the resonant sound reduction improvement index (Sections 3.5.1.2.2 and 4.3.8.2). With reference to the coupling loss factor (Eq. 4.21) the lining can be viewed as modifying the radiation efficiency of the plate. In practice there may be other effects due to interaction between the lining and the base wall or floor (Section 4.3.8.2). One of these concerns the mass–spring–mass resonance frequency which means that measurements of $\Delta R_{\text{Resonant}}$ can have highly negative values near the resonance. A practical solution is to use ΔR instead of $\Delta R_{\text{Resonant}}$ and accept that it sometimes gives an overestimate; hence the *in situ* value is given by

$$\Delta R_{\text{situ}} = \Delta R_{\text{Resonant}} \approx \Delta R \tag{5.117}$$

The *in situ* sound reduction index of the separating plate is now given by

$$R_{Dd} = R_{Dd,\text{situ}} + \Delta R_{D,\text{situ}} + \Delta R_{d,\text{situ}} \tag{5.118}$$

and the *in situ* flanking sound reduction index for each path is

$$R_{ij} = \frac{R_{i,\text{situ}}}{2} + \Delta R_{i,\text{situ}} + \frac{R_{j,\text{situ}}}{2} + \Delta R_{j,\text{situ}} + \overline{D_{v,ij,\text{situ}}} + 10\lg\left(\frac{S_S}{\sqrt{S_i S_j}}\right) \tag{5.119}$$

The apparent sound reduction index is calculated using Eqs 5.118 and 5.119 in Eq. 5.103.

5.4.2 Impact sound insulation

The model for impact sound insulation can be derived from SEA path analysis in a very similar way to airborne sound insulation. It is assumed that the plates support a reverberant bending wave field with no significant decrease in vibration with distance across their surface. As with airborne sound insulation, the resonant sound reduction indices and the direction-averaged velocity level difference are used to replace the coupling loss factors. The main difference is the inclusion of the power input from the ISO tapping machine; this was previously discussed in Sections 3.6.3 and 4.4.1.

The aim is to determine the normalized impact sound pressure level. In order to describe the different transmission paths a flanking normalized impact sound pressure level, $L_{n,ij}$, is defined as the normalized impact sound pressure level due to sound radiated by flanking plate j (wall/floor) in the receiving room with excitation of floor plate i by the ISO tapping machine. The normalized impact sound pressure level *in situ* can then be given by

$$L'_n = 10 \lg \left(10^{L_{n,d}/10} + \sum_j 10^{L_{n,ij}/10} \right) \tag{5.120}$$

where $L_{n,d}$ is the *in situ* normalized impact sound pressure level for direct transmission via the separating floor. Note that $L_{n,d}$ is not relevant to horizontally adjacent rooms.

The normalized impact sound pressure level for the floor measured in the laboratory is converted to the *in situ* value using

$$L_{n,\text{situ}} = L_n + 10 \lg \left(\frac{T_{s,\text{situ}}}{T_{s,\text{lab}}} \right) \tag{5.121}$$

where the structural reverberation times *in situ* and in the laboratory are $T_{s,\text{situ}}$ and $T_{s,\text{lab}}$ respectively. Note that if L_n has been predicted then $T_{s,\text{lab}}$ corresponds to the value used in the prediction model.

The *in situ* normalized impact sound pressure level for direct transmission via the separating floor is

$$L_{n,d} = L_{n,\text{situ}} - \Delta L_{\text{situ}} - \Delta R_{\text{Resonant}} \tag{5.122}$$

and the *in situ* flanking normalized impact sound pressure level for each path is

$$L_{n,ij} = L_{n,\text{situ}} - \Delta L_{\text{situ}} + \frac{R_{\text{Resonant},i}}{2} - \frac{R_{\text{Resonant},j}}{2} - \Delta R_{j,\text{situ}} - \overline{D_{v,ij,\text{situ}}} - 5 \lg \left(\frac{S_i}{S_j} \right) \tag{5.123}$$

where ΔL_{situ} applies to a floor covering ($\Delta L_{\text{situ}} = \Delta L$ for concrete floors), $\Delta R_{\text{Resonant}}$ applies to a ceiling and $\Delta R_{j,\text{situ}}$ (Eq. 5.117) applies to a lining facing into the receiving room on plate j.

5.4.3 Application

The aim is to use the SEA-based model to make a link between sound insulation measured in the laboratory and *in situ*. This link is made via the resonant sound reduction index. The sound reduction index that is measured in the laboratory represents a combination of non-resonant and resonant transmission below the critical frequency; it only represents resonant transmission at and above the critical frequency. For this reason it is useful to write the equations in terms of $R_{Resonant}$ to avoid giving the impression that it applies to all measured R values at all frequencies. When using measured R values in the model, the valid frequency range depends on the critical frequencies of the plates. To allow predictions over a wide frequency range the model is well-suited to plates with critical frequencies in the low-frequency range; this applies to many masonry/concrete plates for which the measured R can often be used as a reasonable estimate for $R_{Resonant}$ below the critical frequency (see examples in Section 4.3.1.3 for solid masonry walls). This is in marked contrast to lightweight plates with high critical frequencies where non-resonant transmission dominates over the majority of the building acoustics frequency range (see examples in Section 4.3.1.3 for glass and plasterboard). For some lightweight walls and floors it may be possible to predict the non-resonant component and extract it from the measured sound reduction index; this assumes that there is no significant decrease in the bending wave vibration level across their surface. This complication does not apply to solid homogenous plates (lightweight or heavyweight) for which the resonant sound reduction index can be predicted (Section 4.3.1.1).

We now consider the link with the resonant sound reduction index in terms of the assumption that each plate has a reverberant bending wave field. For thick plates (Section 4.3.1.4), plates formed from certain types of hollow bricks/blocks (Section 4.3.2.3) and sandwich panels (Section 4.3.6), there are thickness resonances and dilatational waves that give pronounced dips in the measured sound reduction index. The SEA-based model uses the resonant sound reduction index to determine bending wave vibration on plate i that is transmitted across a junction to give a reverberant bending wave field on plate j. Hence at frequencies where plate motion other than bending wave motion determines the measured sound reduction index, the measured R no longer represents $R_{Resonant}$. Therefore the valid frequency range for plates which are not solid and homogeneous is limited to frequencies where non-resonant transmission is negligible and where sound radiation is only due to bending wave motion. Some external flanking walls are formed from a number of rigid and resilient layers for thermal purposes (Lang, 1993). These are not usually referred to as sandwich plates or identified as a base plate with a lining. However, they exhibit mass–spring–mass resonances which also give rise to dips in the measured sound reduction index where the measured R does not represent $R_{Resonant}$.

The limitations discussed above stem from the fact that different sound transmission mechanisms are 'contained within' the measured sound reduction index. A plate–cavity–plate system not only 'contains' different transmission mechanisms but also different transmission paths (Section 4.3.5). Any extension of the SEA-based model to lightweight or heavyweight cavity walls or floors requires knowledge of the transmission paths both in the laboratory and *in situ* because the structural coupling between the plates may be different in each case. In addition, there may be important flanking paths that cross more than one junction, or involve cavities in the flanking construction. If an SEA model is suitable for a plate–cavity–plate system then it is simpler to work within the SEA framework (see example in Section 5.3.2.2). However, SEA models are not always available for direct and flanking transmission across lightweight walls and floors; particularly those that are built from layers of boards and resilient components. Lightweight walls and floors sometimes show a significant decrease in vibration across the element when they are excited along a junction line or by the ISO tapping machine; in such cases it may be reasonable to only consider short flanking paths because the long paths are less important. Ongoing research shows that there is the potential to modify the SEA-based model for lightweight constructions (e.g. see Guigou-Carter *et al.*, 2006; Nightingale, 1995; Schumacher and Sass, 1999; Villot and Guigou-Carter, 2006).

In discussing limitations of SEA and the SEA-based model it is important to note that there are many flanking paths and the ones that cannot be easily modelled are sometimes unimportant. As an example we can consider a construction in which a plate–cavity–plate system is conveniently incorporated into the SEA-based model. Consider airborne sound insulation across a lightweight separating floor, such as a timber joist floor, where masonry/concrete walls i and j form the flanking path Ff (shown on Fig. 5.32). Direct transmission across the timber joist separating floor may be rather complex to predict, so we use the laboratory measurement of the sound reduction index to give R_{Dd} for the model. Depending on joist orientation and the way they are fixed to the walls it may be reasonable to assume that the joists will have negligible effect on flanking path Ff; hence plates i and j can be treated as an in-line junction. In some cases it can also be assumed that vibration transmission between the joists and the flanking walls is negligible so that flanking paths Fd and Df can be ignored.

In the above example the number of paths was reduced in order to make a partial assessment of the flanking transmission. With the SEA-based model it is assumed that transmission paths involving more than one junction will have negligible effect on the overall sound insulation. With SEA this assumption can be checked by comparing path analysis with the matrix solution; this was previously done in Section 5.3.2. Flanking paths that cross more than one junction may be relatively weak, but as there are so many of them between adjacent rooms in a building they can be significant when they are combined together. Hence without the matrix solution there is uncertainty over the importance of the longer flanking paths. For buildings comprising only rigidly connected solid masonry/concrete walls and floors, SEA models indicate that if the airborne sound insulation for the direct path is X dB, and for all transmission paths it is Y dB (SEA matrix solution), then the sound insulation predicted using just the direct path and the flanking paths involving one junction is approximately $(X + Y)/2$ dB in any frequency band over the building acoustics frequency range (Craik, 2001). So if $X - Y$ is 10 dB, the SEA-based model will overestimate the airborne sound insulation by 5 dB. Whilst this indicates the potential importance of paths that cross more than one junction, any bias error will differ between different types of construction.

It is important to note that bias errors have not been found in some comparisons of measured and predicted single-number quantities for airborne and impact sound insulation (Gerretsen, 1979; Pedersen, 1999); other comparisons do indicate a bias error but this could be attributed to the input data (Metzen, 1999). Good agreement between the SEA-based model and measurements has also been shown for the sound insulation spectrum (Gerretsen, 1994). There are some confounding factors that complicate the comparison of SEA and the SEA-based model in real buildings, particularly dwellings. Firstly, these do not usually have large flanking walls that are devoid of openings such as doors and windows; for some types of construction, including or excluding these walls in a model will make a significant difference. Secondly, masonry/concrete walls and floors can have low mode counts and low modal overlap in the low and mid-frequency ranges. This can result in large differences in vibration transmission between similar junctions of different size walls and floors; in addition there is a tendency for the wave approach to overestimate the strength of vibration transmission (Section 5.2.3.3). These two issues are not avoided by measuring vibration reduction indices or coupling loss factors on junctions in the laboratory. This is because the inclusion of coupling parameters from isolated junctions into larger SEA models does not always improve their accuracy when the plates that form each junction have low mode counts and low modal overlap (see example in Section 5.2.3.6). An additional complication occurs when these measurements are made on a junction *in situ* because there may be unwanted vibration transmission via paths involving more than one junction (refer back to Fig. 3.96).

Both SEA and the SEA-based models have their limitations; it is a case of working with their strengths and trying to work around their weaknesses. There are many walls, floors, and linings that require laboratory measurements because they cannot be modelled easily or accurately with a complete SEA model. Therefore any insight gained into flanking transmission with the SEA-based model is useful; an allowance can often be made for neglecting the longer flanking paths. A partial assessment of the flanking transmission that identifies one important flanking path can still be used to make design decisions and to help find solutions to existing sound insulation problems. An example of this is given in Section 5.4.4.

The SEA-based model is not restricted to one type of construction, but just as with SEA, it is often necessary to restrict its use to certain frequency ranges. Use of the SEA-based model can be considered in two separate categories: the first is to provide insight that helps solve a sound insulation problem, the second is to predict single-number quantities for airborne or impact sound insulation. The former is the easiest because there is plenty of flexibility; the SEA-based model can be combined with SEA path analysis, other prediction models, and other measurements. Sound insulation problems are often solved by looking at specific parts of the frequency range. It is therefore of less concern that we do not have a model which covers the entire building acoustics frequency range; it is possible to use a mixture of measurements, predictions and empiricism. This is not the case when a single model is required to calculate single-number quantities for regulatory purposes. In most countries there are preferred or 'traditional' types of wall and floor construction for which there are a limited number of practical or feasible combinations. By using a database of field and laboratory measurements it is often possible to make simplifications or empirical corrections to the SEA-based model that will give a reasonable estimate of the *in situ* sound insulation in terms of the average single-number quantity. As an example of this we consider the fact that the measured K_{ij} can be frequency-dependent with large fluctuations in the low-frequency range that depend on the dimensions and how the plates are connected. However, to simplify calculations a frequency-average K_{ij} value is often used in the SEA-based model. Using such an approach is a pragmatic solution to an awkward problem; but care needs to be taken to ensure that it does not give a reasonable estimate of the average single-number quantity for the wrong reason. The desire to have one model for all types of construction means that the SEA-based model tends to evolve into a calculation procedure. This procedure can be a step removed from the original model and so the links to the original assumptions may be changed or severed. This often suits the intended purpose once the procedure is validated and kept within well-defined constraints. It also provides a pragmatic, simple solution for predicting sound insulation in buildings which can be quite complex for even the simplest constructions.

5.4.4 Example: Flanking transmission past non-homogeneous separating walls or floors

This example concerns non-homogenous separating walls or floors that have a significant decrease in vibration across their surface; these are awkward to include in SEA or SEA-based models. However, a partial assessment of the combined direct and flanking transmission may be possible when the flanking elements are homogenous and support a reverberant bending wave field. An example of this occurs with airborne sound insulation of a beam and block

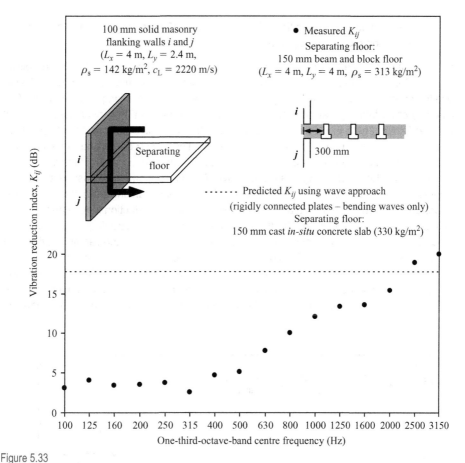

Figure 5.33

Vibration reduction index (path Ff) for a beam and block floor with masonry flanking walls (measured) compared with a solid homogenous isotropic concrete plate (predicted). Measured data from Hopkins are reproduced with permission from ODPM and BRE.

separating floor with masonry flanking walls (Hopkins, 2004). The decrease in vibration with distance across this beam and block floor was previously shown in Fig. 2.42.

In this example there are two aspects that are difficult to model, direct transmission across the separating floor and vibration transmission across the junction. This makes it more conveni- ent to use the SEA-based model and incorporate laboratory measurements. The measured vibration reduction index is shown in Fig. 5.33 for path Ff across a T-junction; note that we are ignoring paths Fd and Df which involve the non-homogenous floor. A prediction is also shown for the same masonry flanking walls but with a homogeneous concrete separating floor; this provides an example of a similar floor that can support a reverberant bending wave field without a significant decrease in vibration. The measured K_{ij} is frequency-dependent and shows no indication that it can be modelled as a junction of rigidly connected plates using either of the wave approaches in Section 5.2. In the low-frequency range this measurement may not be representative of an ensemble of similar junctions because $M_{av} < 1$ and $N_s < 5$. However, in this case there are large differences between the measured K_{ij} with the beam and block floor and the predicted K_{ij} with a concrete floor. This acts as a useful reminder of the errors that can be incurred by making gross assumptions about walls, floors and junctions.

We can now calculate the apparent sound reduction index for the combination of the direct path Dd with flanking paths Ff. This is only a partial assessment of the flanking transmission because the overall sound insulation will potentially be reduced by flanking paths Fd and Df as well as many other flanking paths involving more than one junction. For this example we will assume that flanking path Ff is identical on all four sides of the floor; in practice K_{ij} will differ slightly on two sides because of the beam orientation. The *in situ* beam and block floor will usually have a lining on each side so we can use a laboratory measurement of the floor with a lightweight floating floor and ceiling treatment to give R_{Dd}. It is reasonable to assume that these lightweight linings will not change the measured K_{ij}. The masonry flanking walls are assumed to have a bonded surface finish (e.g. plaster). The resonant sound reduction index can therefore be predicted because the walls can be treated as homogeneous isotropic plates; the *in situ* total loss factor for each wall is estimated to be $0.01 + 0.5f^{-0.5}$, and the radiation efficiency is calculated using method no. 3 (Section 2.9.4.3). The various sound reduction indices are shown in Fig. 5.34. R' is significantly lower than R_{Dd} due to the four flanking transmission paths (Ff). This provides an adequate basis on which to investigate various changes to the design that could increase the sound insulation, such as using wall linings on the flanking

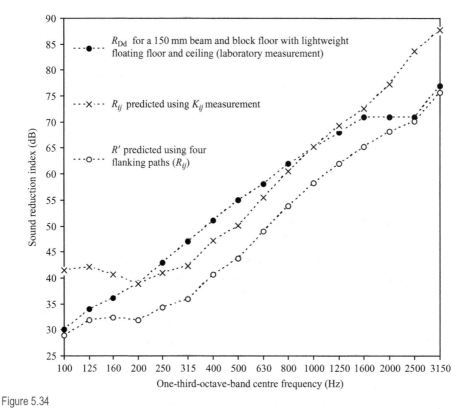

Figure 5.34

Sound reduction indices for a beam and block floor: direct airborne sound insulation (R_{Dd}) and predicted sound insulation *in situ* (R') due to flanking path Ff with masonry flanking walls on all four sides of the floor.

walls. The fact that we have been able to identify significant flanking transmission indicates the benefit in using SEA or an SEA-based model even when all the paths cannot be included.

There are many different types of beam and block floor; K_{ij} measurements tend to indicate significant differences compared to junctions of rigidly connected, solid, homogeneous plates (e.g. see Cortés et al., 2002).

References

Anon. (1989). DIN 4109 Supplement 1:1986 Sound control in buildings – Design examples and calculation procedure, *Deutsches Institut für Normung*.

Bosmans, I. (1998). Analytic modelling of structure-borne sound transmission and modal interaction at complex plate junctions, PhD thesis, Katholieke Universiteit, Leuven, Belgium. ISBN: 9056821318.

Briaud, J.L. and Cepert, P. (1990). WAK test to find spread footing stiffness, *Journal of Geotechnical Engineering*, **116**, 415–431.

Clarkson, B.L. and Ranky, M.F. (1984). On the measurement of the coupling loss factor of structural connections, *Journal of Sound and Vibration*, **94** (2), 249–261.

Cortés, A., Craik, R.J.M. and Esteban, A. (2002). Impact sound insulation: measurement and prediction in hollow constructions of the Basque country, *Proceedings of Forum Acusticum 2002*, Seville, Spain.

Craik, R.J.M. (1981). Damping of building structures, *Applied Acoustics*, **14**, 347–359.

Craik, R.J.M. (1996). *Sound transmission through buildings using statistical energy analysis*, Gower, Aldershot. ISBN: 0566075725.

Craik, R.J.M. (2001). The contribution of long flanking paths to sound transmission in buildings, *Applied Acoustics*, **62** (1), 29–46.

Craik, R.J.M. and Evans, D.I. (1989). The effect of workmanship on sound transmission through buildings: Part 2 – Structure-borne sound, *Applied Acoustics*, **27**, 137–145.

Craik, R.J.M. and Galbrun, L. (2005). Vibration transmission through a frame typical of timber-framed buildings, *Journal of Sound and Vibration*, **281**, 763–782.

Craik, R.J.M. and Osipov, A.G. (1995). Structural isolation of walls using elastic interlayers, *Applied Acoustics*, **46**, 233–249.

Craik, R.J.M. and Thancanamootoo, A. (1992). The importance of in-plane waves in sound transmission through buildings, *Applied Acoustics*, **37**, 85–109.

Craik, R.J.M., Steel, J.A. and Evans, D.I. (1991). Statistical energy analysis of structure-borne sound transmission at low frequencies, *Journal of Sound and Vibration*, **144** (1), 95–107.

Craven, P.G. and Gibbs, B.M. (1981). Sound transmission and mode coupling at junctions of thin plates, Part I: Representation of the problem, *Journal of Sound and Vibration*, **77** (3), 417–427.

Cremer, L., Heckl, M. and Ungar, E.E. (1973). *Structure-borne sound*, Springer-Verlag. ISBN: 0387182411.

Cuschieri, J.M. (1990). Structural power-flow analysis using a mobility approach of an L-shaped plate, *Journal of the Acoustical Society of America*, **87** (3), 1159–1165.

de Vries, D., van Bakel, J.G. and Berkhout, A.J. (1981). Application of SEA in building acoustics: a critical note, *Proceedings of Internoise 81*, Amsterdam, The Netherlands, 465–469.

Fahy, F.J. and Mohammed, A.D. (1992). A study of uncertainty in applications of SEA to coupled beam and plate systems. Part 1: Computational experiments, *Journal of Sound and Vibration*, **158** (1), 45–67.

Fothergill, L.C. and Hargreaves, N. (1992). The effect on sound insulation between dwellings when windows are close to a separating wall, *Applied Acoustics*, **35**, 253–261.

Gerretsen, E. (1979). Calculation of the sound transmission between dwellings by partitions and flanking structures, *Applied Acoustics*, **12**, 413–433.

Gerretsen, E. (1986). Calculation of airborne and impact sound insulation between dwellings, *Applied Acoustics*, **19**, 245–264.

Gerretsen, E. (1994). European developments in prediction models for building acoustics, *Acta Acustica*, **2**, 205–214.

Gerretsen, E. (1996). Vibration reduction index, K_{ij}, a new quantity for sound transmission at junctions of building elements, *Proceedings of Internoise 96*, Liverpool, UK.

Gibbs, B.M. and Gilford, C.L.S. (1976). The use of power flow methods for the assessment of sound transmission in building structures, *Journal of Sound and Vibration*, **49** (2), 267–286.

Guigou-Carter, C., Villot, M. and Wetta, R. (2006). Prediction method adapted to wood frame lightweight constructions, *Building Acoustics*, **13** (3), 173–188.

Guyader, J.L., Boisson, C. and Lesueur, C. (1982). Energy transmission in finite coupled plates, Part 1: Theory, *Journal of Sound and Vibration*, **81** (1), 81–92.

Hodges, C.H. and Woodhouse, J. (1989). Confinement of vibration by one-dimensional disorder, I: theory of ensemble averaging, II: A numerical experiment on different ensemble averages, *Journal of Sound and Vibration*, **130** (2), 237–268.

Hodges, C.H., Nash, P. and Woodhouse, J. (1987). Measurement of coupling loss factors by matrix fitting: an investigation of numerical procedures, *Applied Acoustics*, **22**, 47–69.

Homb, A., Hveem, S. and Strøm, S. (1983). Lydisolerende konstruksjoner: Datasamling og beregningsmetode, *Norges byggforskningsinstitutt*. ISBN: 8253601875.

Hopkins, C. (1997). Sound transmission across a separating and flanking cavity wall construction, *Applied Acoustics*, **52** (3/4), 259–272.

Hopkins, C. (1999). Measurement of the vibration reduction index, K_{ij}, on free-standing masonry wall constructions, *Building Acoustics*, **6** (3&4), 235–257.

Hopkins, C. (2002). Statistical energy analysis of coupled plate systems with low modal density and low modal overlap, *Journal of Sound and Vibration*, **251** (2), 193–214.

Hopkins, C. (2003a). Vibration transmission between coupled plates using finite element methods and statistical energy analysis. Part 1: Comparison of measured and predicted data for masonry walls with and without apertures, *Applied Acoustics*, **64**, 955–973.

Hopkins, C. (2003b). Vibration transmission between coupled plates using finite element methods and statistical energy analysis. Part 2: The effect of window apertures in masonry flanking walls, *Applied Acoustics*, **64**, 975–997.

Hopkins, C. (2004). Airborne sound insulation of beam and block floors: Direct and flanking transmission, *Building Acoustics*, **11** (1), 1–25.

Kawai, Y., Fukuyama, T. and Tsuchiya, Y. (2004). A chart for estimating the distance attenuation of flanking sound passing through open windows in the exterior wall of adjoining rooms and its experimental verification, *Applied Acoustics*, **65**, 985–996.

Keane, A.J. and Manohar, C.S. (1993). Energy flow variability in a pair of coupled stochastic rods, *Journal of Sound and Vibration*, **168** (2), 253–284.

Kihlman, T. (1967). Transmission of structure-borne sound in buildings, *Report 9/67, National Swedish Institute for Building Research*.

Lalor, N. (1990). Practical considerations for the measurement of internal and coupling loss factors on complex structures, *ISVR Technical Report No.182, Institute of Sound and Vibration Research*, University of Southampton, UK.

Lang, J. (1993). Measurement of flanking transmission in outer walls in test facilities, *Applied Acoustics*, **40**, 239–254.

Langley, R.S. and Heron, K.H. (1990). Elastic wave transmission through plate/beam junctions, *Journal of Sound and Vibration*, **143** (2), 241–253.

Ljunggren, S. (1985). Transmission of structure-borne sound from a beam into an infinite plate, *Journal of Sound and Vibration*, **100** (32), 309–320.

Lyon, R.H. (1975). *Statistical energy analysis of dynamic systems: Theory and applications*, Cambridge: MIT Press.

Lyon, R.H. and Eichler, E. (1964). Random vibration of connected structures, *Journal of the Acoustical Society of America*, **36** (7), 1344–1354.

Manohar, C.S. and Keane, A.J. (1994). Statistics of energy flows in spring-coupled one-dimensional subsystems, *Philosophical transactions of the Royal Society, London*, **A346**, 525–542.

McCollum, M.D. and Cuschieri, J.M. (1990). Bending and in-plane wave transmission in thick connected plates using statistical energy analysis, *Journal of the Acoustical Society of America*, **88** (3), 1480–1485.

Mees, P. and Vermeir, G. (1993). Structure-borne sound transmission at elastically connected plates, *Journal of Sound and Vibration*, **166** (1), 55–76.

Metzen, H. (1999). Accuracy of CEN-prediction models applied to German building situations, *Building Acoustics*, **6** (3/4), 325–340.

Nightingale, T.R.T. (1995). Application of the CEN draft building acoustics prediction model to a lightweight double leaf construction, *Applied Acoustics*, **46**, 265–284.

Osipov, A. and Vermeir, G. (1996). Sound transmission in buildings with elastic layers at joints, *Applied Acoustics*, **49** (2), 141–162.

Pedersen, D.B. (1995). Estimation of vibration attenuation through junctions of building structures, *Applied Acoustics*, **46**, 285–305.

Pedersen, D.B. (1999). Evaluation of EN12354 part 1 and 2 for Nordic dwelling houses, *Building Acoustics*, **6** (3/4), 259–268.

Petyt, M. (1998). *Introduction to Finite Element Vibration Analysis*, Cambridge University Press, Cambridge. ISBN: 0521266076.

Rébillard, E. and Guyader, J.L. (1995). Vibrational behaviour of a population of coupled plates: hypersensitivity to the connexion angle, *Journal of Sound and Vibration*, **188** (3), 435–454.

Roland, J. (1988). New investigation methods in building acoustics, *Proceedings of Internoise 88*, Avignon, France, 395–400.

Schumacher, R. and Sass, B. (1999). Flanking sound transmission by timber-framed glass facades, *Building Acoustics*, **6** (3 & 4), 309–323.

Sewell, E.C. and Savage, J.E. (1987). Effect of associated walls on the sound insulation of concrete party floors, *Applied Acoustics*, **20**, 297–315.

Simmons, C. (1991). Structure-borne sound transmission through plate junctions and estimates of SEA coupling loss factors using the finite element method, *Journal of Sound and Vibration*, **144** (2), 215–227.

Steel, J.A. (1994). Sound transmission between plates in framed structures, *Journal of Sound and Vibration*, **178** (3), 379–394.

Steel, J.A. and Craik, R.J.M. (1994). Statistical energy analysis of structure-borne sound transmission by finite element methods, *Journal of Sound and Vibration*, **178** (4), 553–561.

Timoshenko, S.P. and Woinowsky-Krieger, S. (1959). *Theory of plates and shells*, McGraw-Hill. ISBN: 0070858209.

Villot, M. and Guigou-Carter, C. (2006). Measurement methods adapted to wood frame lightweight constructions, *Building Acoustics*, **13** (3), 189–198.

Wester, E.C.N. and Mace, B.R. (1996). Statistical energy analysis of two edge-coupled rectangular plates: Ensemble averages, *Journal of Sound and Vibration*, **193** (4), 793–822.

Wester, E.C.N. and Mace, B.R. (1999). Ensemble statistics of energy flow in very irregular systems using a wave approach, *Proceedings of Internoise 99*, Florida, USA.

Wilson, R. and Craik, R.J.M. (1995). Sound transmission via the foundation of a cavity wall, *Building Acoustics*, **2** (4), 569–583.

Wöhle, W., Beckmann, Th. and Schreckenbach, H. (1981). Coupling loss factors for statistical energy analysis of sound transmission at rectangular structural slab joints. Parts I and II, *Journal of Sound and Vibration*, **77** (3), 323–344.

Woodhouse, J. (1981). An introduction to statistical energy analysis of structural vibration, *Applied Acoustics*, **14**, 455–469.

Zienkiewicz, O.C. (1977). *The finite element method*, McGraw-Hill Company, London.

Appendix: Material properties

This appendix contains indicative material properties for use with the equations in the main text; manufacturers should be consulted for the material properties of any specific product. Note that the properties in Tables A2, A3, and A4 will not always allow a definitive comparison of similar products because of the variation that exists between different manufacturers in different countries.

Table A1 Properties of gases

Gas	Ratio of specific heats, γ	Molar mass, M(kg/mol)	Phase velocity at NTP* (m/s)	Density, ρ at NTP* (kg/m^3)
Air (dry)	1.41	0.02895	345	1.205
Argon	1.67	0.040	319	1.662
Carbon dioxide	1.33	0.044	271	1.842
Nitrogen	1.41	0.028	350	1.165
Oxygen	1.41	0.032	328	1.331
Sulphur hexafluoride	1.33	0.146	149	6.2

*Normal temperature and pressure (20°C, 1.013×10^5 Pa).

Table A2 Material properties

Material name	Density, ρ (kg/m³)	Quasi-longitudinal phase velocity[a], c_L (m/s)	Poisson's ratio, ν	Internal loss factor (bending waves), η_{int}	$h.f_c$(m.Hz) (assuming $c_0 = 343$ m/s)
Aircrete/Autoclaved Aerated Concrete (AAC) blocks (solid) connected with mortar or thin joint compound (Hopkins)	400–800	1900 (Typical range: 1600–2300)	0.2[b]	0.0125	34.1
Aluminium (Heckl, 1981)	2700	5100	0.34	≤ 0.001	12.7
Bricks (solid) connected with mortar (Hopkins)	1500–2000	2700	0.2[b]	0.01[b]	24.0
Calcium-silicate blocks (solid) connected with thin joint compound (Schmitz et al., 1999)	1800	2500	0.2[b]	0.01	25.9
Chipboard (Hopkins)	760	2200	0.3[b]	0.01[b]	29.5
Clinker concrete blocks (solid) connected with mortar (Rindel, 1994)	1030	1850	0.2[b]	0.01[b]	35.1
	1720	2200	0.2[b]	0.01[b]	29.5
Clinker concrete slabs (Rindel, 1994)	1725	1910	0.2[b]	0.01[b]	34.0
Concrete – cast in situ (Hopkins)	2200	3800	0.2[b]	0.005[b]	17.1
Dense aggregate blocks (solid) connected with mortar (Hopkins)	2000	3200	0.2[b]	0.01[b]	20.3
Expanded clay blocks (solid) connected with mortar (Hopkins)	800	2300	0.2[b]	0.007[b]	28.2
Glass (Hopkins)	2500	5200	0.24	0.003–0.006	12.5
Lightweight aggregate blocks (solid) connected with mortar (Hopkins)	1400	2200	0.2[b]	0.01[b]	29.5
Medium Density Fibreboard (MDF) (Hopkins)	760	2560	0.3[b]	0.01[b]	25.3

Material	Density (kg/m³)	c_L (m/s)	ν	η	
Mortar (Maysenhölder and Horvatic, 1998)	1600	2450	0.2	0.013	26.5
Oriented Strand Board (OSB) (Hopkins)	590	2570[c]	0.3[b]	0.01[b]	25.2
Perspex, plexiglass (Hopkins)	1250	2350	0.3[b]	–	27.6
Plaster – gypsum based (Hopkins)	650	1610	0.2[b]	0.012[b]	40.3
Plasterboard – natural gypsum (Hopkins, 1999)	860	1490	0.3[b]	0.0141	43.5
Plasterboard – combination of flue gas gypsum and natural gypsum (Hopkins, 1999)	680	1810	0.3[b]	0.0125	35.8
Plasterboard – gypsum with glass fibre and other additives (Hopkins)	800	2010	0.3[b]	–	32.3
Plywood (Birch) (Hopkins)	710	3850	0.3[b]	0.016	16.8
Sand-cement screed (Hopkins)	2000	3250	0.2[b]	0.01	20.0
Steel	7800	5270 (Fahy, 1985)	0.28 (Fahy, 1985)	≤0.0001 (Heckl, 1981)	12.3
Timber (soft wood) used for joists, studs or battens (Hopkins)	440	5000	0.3[b]	–	13.0

[a] Values can be used as estimates for beams or plates.

[b] Estimate.

[c] This material is usually orthotropic with values between 2200 and 3500 m/s depending on the direction. The value quoted here is $c_{L,eff.}$.

Table A3 Dynamic stiffness per unit area of resilient materials measured according to ISO 9052-1

Material name	Density (kg/m^3)	Nominal uncompressed thickness (mm)	Dynamic stiffness per unit area, s'(MN/m^3)
Closed-cell polyethylene foam (Hopkins)	45	5	115
Expanded polystyrene (Hopkins)	14	50	78
Expanded polystyrene – pre-compressed (Hopkins)	10	50	68
Mineral wool – rock (Hopkins)	60	30	10
	80	30	11
	100	30	14
	140	30	19
Mineral wool – glass (Hopkins)	36	13	28
		25	11
	75	25	12
		40	7
Rebond foam (reconstituted open cell foam) (Hopkins and Hall, 2006)		15	12
	64	20	9
		25	7
	96	15	16

Table A4 Dynamic stiffness of wall ties

Wall tie	Cavity width, X(mm)	Dynamic stiffness, $s_{X\mathrm{mm}}$(MN/m)
Butterfly tie (Hopkins et al., 1999) (described in BS 1243:1978)	50	1.7
Double-triangle tie (Hopkins et al., 1999) (described in BS 1243:1978)	50	16.1
Vertical-twist tie (Hopkins et al., 1999) (described in BS 1243:1978)	50	94.0
Vertical-twist tie (proprietary) (Hall et al., 2001)	100	43.4

References

Fahy, F.J. (1985). *Sound and structural vibration. Radiation, transmission and response*, London: Academic Press ISBN: 0122476700.

Hall, R. and Hopkins, C. (2001). Dynamic stiffness of wall ties used in masonry cavity walls: measurement procedure, *BRE Information Paper IP3/01, BRE,* Watford, England ISBN: 1860814611.

Hall, R., Hopkins, C. and Turner, P. (2001). The effect of wall ties in external cavity walls on the airborne sound insulation of solid separating walls, *Proceedings of ICA 2001*, Rome, Italy.

Heckl, M. (1981). The tenth Sir Richard Fairey Memorial lecture: sound transmission in buildings, *Journal of Sound and Vibration*, **77** (2), 165–189.

Hopkins, C. (1999). Building acoustics measurements, *Proceedings of the Institute of Acoustics*, **21** (3), 9–16.

Hopkins, C. and Hall, R. (2006). Impact sound insulation using timber platform floating floors on a concrete floor base, *Building Acoustics*, **13** (4), 273–284.

Hopkins, C., Wilson, R. and Craik, R.J.M. (1999). Dynamic stiffness as an acoustic specification parameter for wall ties used in masonry cavity walls, *Applied Acoustics*, **58**, 51–68.

Hopkins, C. Courtesy of ODPM and BRE.

Maysenhölder, W. and Horvatic, B. (1998). Determination of elastodynamic properties of building materials, *Proceedings of Euronoise 98*, Munich, Germany, 421–424.

Rindel, J.H. (1994). Dispersion and absorption of structure-borne sound in acoustically thick plates, *Applied Acoustics*, **41**, 97–111.

Schmitz, A., Meier, A. and Raabe, G. (1999). Inter-laboratory test of sound insulation measurements on heavy walls. Part 1 – Preliminary test, *Building Acoustics*, **6** (3/4), 159–169.

Standards

The following list of acoustic Standards relate to the measurement and prediction of sound and vibration; these were current at the time of writing and are referenced in the text. The reader is advised that other Standards exist and that the Standards listed below will be updated over time.

ISO 140-1:1997 Acoustics – Measurement of sound insulation in buildings and of building elements – Part 1: Requirements for laboratory test facilities with suppressed flanking transmission, *International Organization for Standardization.*

ISO 140-2:1991 Acoustics – Measurement of sound insulation in buildings and of building elements – Part 2: Determination, verification and application of precision data, *International Organization for Standardization.*

ISO 140-3:1995 Acoustics – Measurement of sound insulation in buildings and of building elements – Part 3: Laboratory measurements of airborne sound insulation of building elements, *International Organization for Standardization.*

ISO 140-4:1998 Acoustics – Measurement of sound insulation in buildings and of building elements – Part 4: Field measurements of airborne sound insulation between rooms, *International Organization for Standardization.*

ISO 140-5:1998 Acoustics – Measurement of sound insulation in buildings and of building elements – Part 5: Field measurements of airborne sound insulation of facade elements and facades, *International Organization for Standardization.*

ISO 140-6:1998 Acoustics – Measurement of sound insulation in buildings and of building elements – Part 6: Laboratory measurements of impact sound insulation of floors, *International Organization for Standardization.*

ISO 140-7:1998 Acoustics – Measurement of sound insulation in buildings and of building elements – Part 7: Field measurements of impact sound insulation of floors, *International Organization for Standardization.*

ISO 140-8:1997 Acoustics – Measurement of sound insulation in buildings and of building elements – Part 8: Laboratory measurements of the reduction of transmitted impact noise by floor coverings on a heavyweight standard floor, *International Organization for Standardization.*

ISO 140-9:1985 Acoustics – Measurement of sound insulation in buildings and of building elements – Part 9: Laboratory measurement of room-to-room airborne sound insulation of a suspended ceiling with a plenum above it, *International Organization for Standardization.*

ISO 140-10:1991 Acoustics – Measurement of sound insulation in buildings and of building elements – Part 10: Laboratory measurement of airborne sound insulation of small building elements, *International Organization for Standardization.*

ISO 140-11:2005 Acoustics – Measurement of sound insulation in buildings and of building elements – Part 11: Laboratory measurements of the reduction of transmitted impact sound by floor coverings on lightweight reference floors, *International Organization for Standardization.*

ISO 140-12:2000 Acoustics – Measurement of sound insulation in buildings and of building elements – Part 12: Laboratory measurement of room-to-room airborne and impact sound insulation of an access floor, *International Organization for Standardization.*

ISO 140-14:2004 Acoustics – Measurement of sound insulation in buildings and of building elements – Part 14: Guidelines for special situations in the field, *International Organization for Standardization.*

ISO 140-16:2006 Acoustics – Measurement of sound insulation in buildings and of building elements – Part 16: Laboratory measurement of the sound reduction index improvement by additional lining, *International Organization for Standardization.*

ISO 140-11:2005 Acoustics – Measurement of sound insulation in buildings and of building elements – Part 11: Laboratory measurements of the reduction of transmitted impact sound by floor coverings on lightweight reference floors, *International Organization for Standardization.*

ISO 140-18:2006 Acoustics – Measurement of sound insulation in buildings and of building elements – Part 18: Laboratory measurement of sound generated by rainfall on building elements, *International Organization for Standardization.*

ISO 717-1:1996 Acoustics – Rating of sound insulation in buildings and of building elements – Part 1: Airborne sound insulation, *International Organization for Standardization.*

ISO 717-2:1996 Acoustics – Rating of sound insulation in buildings and of building elements – Part 2: Impact sound insulation, *International Organization for Standardization.*

ISO 15186-1:2000 Acoustics – Measurement of sound insulation in buildings and of building elements using sound intensity – Part 1: Laboratory measurements, *International Organization for Standardization.*

ISO 15186-2:2003 Acoustics – Measurement of sound insulation in buildings and of building elements using sound intensity – Part 2: Field measurements, *International Organization for Standardization.*

ISO 15186-3:2002 Acoustics – Measurement of sound insulation in buildings and of building elements using sound intensity – Part 3: Laboratory measurements at low frequencies, *International Organization for Standardization.*

ISO 9614-1:1993 Acoustics – Determination of sound power levels of noise sources using sound intensity – Part 1: Measurement at discrete points, *International Organization for Standardization.*

ISO 10848-1:2006 Acoustics – Laboratory measurement of the flanking transmission of airborne and impact sound between adjoining rooms – Part 1: Frame document, *International Organization for Standardization.*

ISO 10848-2:2006 Acoustics – Laboratory measurement of the flanking transmission of airborne and impact sound between adjoining rooms – Part 2: Application to light elements when the junction has a small influence, *International Organization for Standardization.*

ISO 10848-3:2006 Acoustics – Laboratory measurement of the flanking transmission of airborne and impact sound between adjoining rooms – Part 3: Application to light elements when the junction has a substantial influence, *International Organization for Standardization.*

ISO 18233:2006 Acoustics – Application of new measurement methods in building and room acoustics, *International Organization for Standardization.*

Technical specifications for measurement equipment

IEC 1043:1993 Electroacoustics – Instruments for the measurement of sound intensity – Measurement with pairs of pressure sensing microphones, *International Electrotechnical Commission.*

IEC 61260:1995 Electroacoustics – Octave-band and fractional-octave-band filters, *International Electrotechnical Commission.*

IEC 61672-1:2002 Electroacoustics – Sound level meters. Part 1: Specifications, *International Electrotechnical Commission.*

ISO 5348:1998 Mechanical vibration and shock – Mechanical mounting of accelerometers, *International Organization for Standardization.*

Reverberation time and absorption

ISO 3382:1997 Acoustics – Measurement of the reverberation time of rooms with reference to other acoustical parameters, *International Organization for Standardization.*

ISO 354:2003 Acoustics – Measurement of sound absorption in a reverberation room, *International Organization for Standardization.*

ISO 10534-1:1996 Acoustics – Determination of sound absorption coefficient and impedance in impedance tubes. Part 1: Method using standing wave ratio, *International Organization for Standardization.*

ISO 10534-2:1998 Acoustics – Determination of sound absorption coefficient and impedance in impedance tubes. Part 2: Transfer-function method, *International Organization for Standardization.*

Prediction

EN 12354-1:2000 Building acoustics – Estimation of acoustic performance of buildings from the performance of elements – Part 1: Airborne sound insulation between rooms, *European Committee for Standardization.*

EN 12354-2:2000 Building acoustics – Estimation of acoustic performance of buildings from the performance of elements – Part 2: Impact sound insulation between rooms, *European Committee for Standardization.*

EN 12354-3:2000 Building acoustics – Estimation of acoustic performance of buildings from the performance of elements – Part 3: Airborne sound insulation against outdoor sound, *European Committee for Standardization.*

EN 12354-4:2000 Building acoustics – Estimation of acoustic performance of buildings from the performance of elements – Part 4: Transmission of indoor sound to the outside, *European Committee for Standardization.*

Material properties

ISO 9052-1:1989 Acoustics – Method for the determination of dynamic stiffness – Part 1: Materials used under floating floors in dwellings, *International Organization for Standardization.*

ISO 9053:1991 Acoustics – Materials for acoustical applications – Determination of airflow resistance, *International Organization for Standardization.*

ISO/PAS 16940:2004 Glass in building – Glazing and airborne sound insulation – Measurement of the mechanical impedance of laminated glass, *International Organization for Standardization.*

Other Standards

ISO 1683:1983 Acoustics – Preferred reference quantities for acoustic levels, *International Organization for Standardization.*

ISO 266:1997 Acoustics – Preferred frequencies, *International Organization for Standardization.*

ISO 1996-1:1982 Description and measurement of environmental noise. Part 1: Guide to quantities and procedures, *International Organization for Standardization.*

ISO 9613-1:1993 Acoustics – Attenuation of sound during propagation outdoors. Part 1: Calculation of the absorption of sound by the atmosphere, *International Organization for Standardization.*

ISO 7626-2:1990 Method for experimental determination of mechanical mobility. Part 2: Measurements using single-point translation excitation with an attached vibration exciter, *International Organization for Standardization.*

ISO 7626-5:1994 Method for experimental determination of mechanical mobility. Part 5: Measurement using impact excitation with an exciter which is not attached to the structure, *International Organization for Standardization.*

EN 12758:2002 Glass in building – Glazing and airborne sound insulation – Product descriptions and determination of properties, *European Committee for Standardization.*

IEC 60721-2-2:1988 Classification of environmental conditions – Part 2: Environmental conditions appearing in nature – Precipitation and wind, *International Electrotechnical Commission.*

British Standards

British Standards can be obtained from BSI Customer Services, 389 Chiswick High Road, London W4 4AL. Tel: +44 (0)20 8996 9001. Email: cservices@bsi-global.com

Index

Index